World Soils Book Series

Series Editor

Alfred E. Hartemink
Department of Soil Science, FD Hole Soils Laboratory
University of Wisconsin–Madison
Madison, WI
USA

The World Soils Book Series publishes peer-reviewed books on the soils of a particular country. They include sections on soil research history, climate, geology, geomorphology, major soil types, soil maps, soil properties, soil classification, soil fertility, land use and vegetation, soil management, soils and humans, soils and industry, future soil issues. The books summarize what is known about the soils in a particular country in a concise and highly reader-friendly way. The series contains both single and multi-authored books as well as edited volumes. There is additional scope for regional studies within the series, particularly when covering large land masses (for example, The Soils of Texas, The Soils of California), however, these will be assessed on an individual basis.

Sheleme Beyene • Alemayehu Regassa •
Bipin B. Mishra • Mitiku Haile
Editors

The Soils of Ethiopia

Springer

Editors
Sheleme Beyene
School of Plant and Horticultural Sciences
Hawassa University
Hawassa, Ethiopia

Alemayehu Regassa
Department of Natural Resource Management
Jimma University
Jimma, Ethiopia

Bipin B. Mishra
Department of Soil Science
Bihar Agricultural University
Sabour, Bhagalpur, Bihar, India

Mitiku Haile
Department of Natural Resource Management and
Environmental Protection
Mekelle University
Mekelle, Ethiopia

ISSN 2211-1255 ISSN 2211-1263 (electronic)
World Soils Book Series
ISBN 978-3-031-17011-9 ISBN 978-3-031-17012-6 (eBook)
https://doi.org/10.1007/978-3-031-17012-6

This Springer imprint is published by the registered company Springer Nature Switzerland AG
The registered company address is: Gewerbestrasse 11, 6330 Cham, Switzerland

Foreword

Ethiopia is endowed with a diverse range of land resources in terms of topography, climatic conditions, geologic features, vegetation, and land use patterns. Consequently, there is a great ecological diversity, which in turn gave rise to the development of 32 major agroecological zones. Each of the land use patterns and ecosystem types is in part determined by the soils that nourish and shape the aboveground vegetation. Owing to the diversities in the pedogenetic processes and factors, there are about 70% of the WRB reference soil groups recognized globally. For this reason, Ethiopia is sometimes referred to as "the soils museum". Developed from the more recent volcanics, i.e., the trap series volcanics of the quaternary period, Ethiopian soils are relatively resilient and perform ecosystem and productive functions almost indefinitely with proper management. However, the long-term economic and ecological sustainability of the soils in different AEZs is dependent on knowledge and management of the soil resources. In addition, soil resource information is crucially essential for sustainably managing agroecosystems, forests, and grasslands, as well as soils exposed to pollution due to urbanization, industrialization, and agrochemical use in the country.

The research institutes, soil testing laboratories, higher learning institutions, and various development partners have conducted activities that generated valuable soil data across Ethiopia. However, the data are scattered and inaccessible calling for harmonization and exchange among the various actors. Thus, having a book on "Ethiopian Soils" is a long overdue. I, being professionally a Soil Scientist and an author of the book *Soils of the Ethiopian Highlands: Geomorphology and Properties,* fully realize the daunting task of collecting and collating existing data in a volume. In this regard, I confirm this new book on *The Soils of Ethiopia* as the first complete reference material on the highland and lowland soils of the country. What is special about this volume is its interdisciplinary nature exploring a broad range of topics including climate, geology, indigenous soil classification systems, fertility status, land evaluation and land use planning, future soil management use issues, and covering all the 32 agroecological zones of the country.

The broad and complex fields of the 21 soil types (earlier identified by the soil map of Ethiopia at 1:2 m scale) in the country are brought under systematically arranged 13 chapters. It is my sincere hope that this book will be a treasured document for the academia, researchers, and experts in the development sectors, and will help document the importance of soil for the well-being and prosperity of the nation. It is my firm belief that the information presented in the volume will contribute to our efforts in sustainably managing the nation's soil resources to achieve national food self-sufficiency, economic development, and poverty alleviation.

On behalf of the Government of the Federal Democratic Republic of Ethiopia and that of the Ministry of Agriculture, I would like to extend my gratitude to the authors and financers that made this work possible. Our thanks are due to the reviewers of the draft manuscripts that ensured the quality of the volume in its current form.

Eyasu Elias, Ph.D.
Professor of Soil Science and State Minister
Ministry of Agriculture
Addis Ababa, Ethiopia

Preface

Soils have always attracted, and continued to attract the attention of people, due to their essential functions in food production. The traditional construction of terraces by the Konso people which is registered as "World's Heritage" by UNESCO is the indication of how the ancient Ethiopians try to conserve their soils. A single individual does not have the knowledge and experience to adequately explain the diverse soil types in Ethiopia. Thus, soil scientists from academia, research, and development institutions were teamed to assemble information in the different chapters. The included information and level of detail in each chapter, therefore, vary reflecting the authors' interest and experience. The list of team members is presented hereunder.

The book contains 13 chapters which deal in-depth with various aspects of soils. The first two chapters (1 and 2) delve into the introductory aspect and the history of soil science education and research in Ethiopia. Chapter 3 examines the interaction of climate with soils with a focus on Ethiopia. Climate is one of the five soil-forming factors responsible for soil formation and development. Understanding of climate and climate dynamics is used to characterize soils and know soil resources for proper utilization and management. On the other hand, soils play a key role in regulating climate and are used as the main resource in climate change adaptation and mitigation practices. This chapter documents the interaction of climate and soil resources in Ethiopia, major impacts of climate on soil, and the importance of soil for climate change adaptation and mitigation practices.

Chapter 4 explores how geology and geomorphology are intimately linked to soil formation and its characteristics, the influence of geological factors on soils through the lithologic varieties which serve as in situ or transported parent materials and through the tectonic phenomena. Tectonic processes play an important role in sculpturing geomorphic features which control soil formation through relief, weathering, and sedimentation. Therefore, geologic and geomorphologic processes affect soil development, both spatially and temporally.

Soil classification is very important for countries like Ethiopia where agriculture is the backbone of the economy. An overview is given of the concepts of scientific and indigenous soil classification systems is presented in Chap. 5. Farmers in different regions and localities of Ethiopia have their own indigenous soil classification systems. Different names are given for the local soil types mainly on the basis of easily field identifiable morphological properties. Chapter 6 dwells on discussing the major soil types of Ethiopia. It unravels the mystery of the diversity of soils in Ethiopia. The chapter elucidates the fact that the wide ranges in the soil-forming factors in the country have given rise to the occurrence of different Soil Reference Groups with varying extent and spatial distribution. It further presents the evolution of soil mapping activities in the country by different organizations and institutions. Chapter 7 deals with the morphological, physicochemical, biological, and mineralogical properties of major soils of Ethiopia. There is quite a significant variation in the Ethiopian soils' chemical and mineralogical properties. The variability in soil chemical properties of Ethiopian soils suggests the need for varied management, reclamation, and utilization practices.

Soil fertility is fundamental to the enhancement of agricultural productivity and food security. Chapter 8 documents the soil fertility status under different agroecologies, soil quality and soil health, and their status in Ethiopia, as well as soil sustainability and soil resilience.

Soil sustainability and soil resilience are important concepts that are introduced to the field of Soil Science more recently. Soil sustainability is used to assess the impacts of soil management on functional integrity of a soil, while soil resilience is used to evaluate soil's ability to bounce back from a shock. As such, the two concepts are interrelated in that soil resilience can be used to measure soil sustainability. Soils differ in their types and degrees of degradation. This calls for the need to develop and implement specific soil management practices that are relevant to the specific problems of the soils. Chapter 9 explores the different management practices for the challenges of major soil types. For instance, the specific managements required for acid, salt-affected soils, and vertisols. On top of these, integrated soil fertility management (ISFM) is also beneficial for improving soil health, crop yields, and the livelihood of farmers.

Chapter 10 examines the nature and current state of land evaluation and land use planning practices in Ethiopia. Agriculture, which is largely dependent on the efficient use of rural land uses, is one of the most important economic sectors in Ethiopia maintaining human survival, despite population pressures and diminishing land productivity. Land evaluation and land use planning are strategic tools for assessing and introducing the best land management alternatives to prevent land use dynamics and land resource degradation. In particular, the need for, history, basics, principles, approaches, practices, challenges, and opportunities of land evaluation and land use planning in Ethiopia are presented in this chapter.

Chapter 11 explores soil resource and human interactions; the links between civilization and soil fertility and wealth; and soil in relation to human nutrition and health. Soil is indubitably one of the most exploited natural resources since the start of agriculture and, as such, there exists an intricate relationship between the two. It has been at the center of society's civilization. However, over the years, inappropriate human actions have resulted in its degradation. Soil influences human nutrition through the quantity and quality of food produced on it. On the other hand, direct contact with contaminated soil can cause many health concerns of various natures. Large number of medicines, most of them antibiotics, are also extracted from organisms living in the soil. Soil resource degradation from industrial by-products has become a serious concern in Ethiopia mainly due to industrialization and rapid urbanization. Chapter 12 presents the contribution of soils to industry and the impact of industry on soils. Due to the disposal of untreated and/or partially treated effluents from various industries, urban wastes, and the use of agrochemicals, the level of soil pollution in Ethiopia has reached an alarming stage.

Anthropogenic effects including the depletion of soil organic matter, erosion, acidity, salinity and/or sodicity, and pollution, cause major challenges to future advances in soils. Chapter 13 explores global and national advances in soil science data collection, analysis, and interpretation; direction in soil science education and research; and challenges and opportunities for future advances in soil science in Ethiopia.

The aim of this book is to provide an explanatory review of the soils under different agroecological zones of the country and to summarize the existing knowledge concerning soils' formation and management. Furthermore, some of the topics in the book provide important insights related to future sustainable soil use. Since Soil Science is young and under continuous evolution, it is believed that the information provided in this book will be periodically updated for practical uses.

Hawassa, Ethiopia Sheleme Beyene
Jimma, Ethiopia Alemayehu Regassa
Sabour, India Bipin B. Mishra
Mekelle, Ethiopia Mitiku Haile

Acknowledgements

Professor Bipin B. Mishra deserves due acknowledgement for initiating the preparation of the book. Heartfelt gratitude goes to Prof. Dr. Alfred E. Hartemink for suggesting the book would be an addition to the Series which motivated the start of the preparation. The German Cooperation's Integrated Soil Fertility Management project implemented by the Deutsche Gesellschaft für Internationale Zusammenarbeit (GIZ) GmbH; BENEFIT REALISE project; the Ministry of Agriculture; and Bahir Dar, Haramaya, Hawassa, and Jima universities are highly acknowledged for providing financial support for the write-shop meetings during the preparation of the book.

The Ethiopian Agricultural Transformation Agency (ATA), GIZ, CASCAPE and BENEFIT REALISE projects are thanked for allowing the use of their data from survey studies carried out in different parts of the country. Special thanks go to Mr. Ashenafi Ali who had tirelessly worked in the acquisition of the data from ATA, CASCAPE and BENEFIT REALISE projects and made the interpretations and prepared the figures from the data. Mr. Kiflu Gudeta also deserves thanks for sharing documents pertaining to land administration, land use planning, and land-related policies and laws. We also owe a debt of gratitude to all those involved in collection of the data.

All corresponding and contributing authors deserve acknowledgements for their time and professional preparation of the different chapters. Mr. Ashenafi Worku is thanked for his tremendous help in formatting the draft before submission. We are grateful to Prof. Eyasu Elias, the State Minster of the Ministry of Agriculture, for his encouragement, providing financial support for editing, and willingness to write the foreword note amidst his tight schedules.

Contents

Contributors

Bekele Abebe School of Earth Sciences, Addis Ababa University, Addis Ababa, Ethiopia

Girma Abera School of Plant and Horticultural Sciences, Hawassa University, Hawassa, Ethiopia

Ashenafi Ali Department of Geography and Environmental Studies, Addis Ababa University, Addis Ababa, Ethiopia

Mohammed Assen Department of Geography and Environmental Studies, Addis Ababa University, Addis Ababa, Ethiopia

Dereje Ayalew School of Earth Sciences, Addis Ababa University, Addis Ababa, Ethiopia

Bobe Bedadi School of Natural Resource Management and Environmental Sciences, Haramaya University, Dire Dawa, Ethiopia

Sheleme Beyene School of Plant and Horticultural Sciences, Hawassa University, Hawassa, Ethiopia

Teklu Erkossa Deutsche Gesellschaft Für Internationale Zusammenarbeit (GIZ), Addis Ababa, Ethiopia

Endalkachew Fekadu College of Agriculture, Woldia University, Woldia, Ethiopia

Berhan Gessesse Entoto Observatory and Research Center (EORC), Ethiopian Space Science and Technology Institute, Addis Ababa, Ethiopia

Kibebew Kibret School of Natural Resources Management & Environmental Sciences, Haramaya University, Dire Dawa, Ethiopia

Alemayehu Kiflu School of Plant and Horticultural Sciences, Hawassa University, Hawassa, Ethiopia

Girma Mamo Ethiopian Institute of Agricultural Research, Addis Ababa, Ethiopia

Bipin B. Mishra Soil Science, Bihar Agricultural University, Sabour, Bhagalpur, India

Alemayehu Regassa Department of Natural Resources Management, Jimma University, Jimma, Ethiopia

Yihenew G. Selassie Department of Natural Resources Management, Bahir Dar University, Bahir Dar, Ethiopia

Gebeyehu Taye Department Land Resources Management and Environmental Protection, Mekelle University, Mekelle, Ethiopia

Wondwosen Tena Department of Plant Biology and Biodiversity Management, Addis Ababa University, Addis Ababa, Ethiopia

Degefie Tibebe Hydroclimate and Geospatial Modelling, Water and Land Resource Center, Addis Ababa University, Addis Ababa, Ethiopia

Amanuel Zenebe Department of Natural Resource Management and Environmental Protection, Mekelle University, Mekelle, Ethiopia

Abbreviations

AAS	Atomic absorption spectrophotometer
AAU	Addis Ababa University
ADLI	Agricultural Development Led Industrialization
ADSWE	Amhara Design and Supervision Works Enterprise
AEZ	Agroecological Zone
AGRA	Alliance for Green Revolution in Africa
AI	Artificial intelligence
ALES	Automated Land Evaluation System
ALUP	Assistance to Land Use Planning Project
ARTP	Agricultural Research and Training Program
ARU	Agronomic Response Unit
ATA	Agricultural Transformation Agency
ATVET	Agriculture-oriented Technical and Vocational Education Training
BENEFIT	Bilateral Ethiopia Netherlands Effort for Food Income and Trade
BENEFIT-REALISE	Bilateral Ethiopia Netherlands Effort for Food Income and Trade-Realising Sustainable Agricultural Livelihood Security in Ethiopia
CA	Conservation agriculture
CASCAPE	Capacity building for Scaling up of evidence-based best Practices in Agricultural Production in Ethiopia
CDE	Center for Environment and Development
CEC	Cation Exchange Capacity
CHIRPS	Climate Hazards Group InfraRed Precipitation with Station data
CIAT	International Center for Tropical Agriculture
CLIFOOD	Climate Effects on Food Security
CoW	Coalition of the Willing
CRV	Central Rift Valley
CSA	Central Statistic Authority
CV	Coefficient of Variance
DAP	Diammonium phosphate
DAs	Development Agents
DEM	Digital Elevation Model
DSM	Digital Soil Mapping
DST	Decision Support Tool
DTPA	Diethylenetriamine Pent Acetate
EARO	Ethiopian Agricultural Research Organization
EC	Electrical Conductivity
ECDSWCo	Ethiopian Construction Design and Supervision Works Corporation
ECe	Electrical conductivity of extract from a saturated paste
EIAR	Ethiopian Institute for Agricultural Research
EIP	Environmental Impact Assessment Proclamation
EMA	Ethiopian Mapping Authority

ENSO	El Niño–Southern Oscillation
EPA	Environmental Protection Authority
ESIF	Ethiopian Strategic Investment Framework
ESP	Exchangeable Sodium Percentage
Ethio-GIS	Ethiopian Geographic Information System
EthioSIS	Ethiopian Soil Information System
FAIR	Finable, Accessible, Interoperable and Reusable
FAO	Food and Agriculture Organization of the United Nations
FC	Field Capacity
FCC	Fertility Capability Classification
FDRE	Federal Democratic Republic of Ethiopia
FFW	Food For Work
FR	Friable
GAC	Global Affairs, Canada
GDP	Gross Domestic Product
GERD	Grand Ethiopian Renaissance Dam
GHG	Green House Gases
GIS	Geographic Information System
GIZ	Deutsche Gesellschaft für Internationale Zusammenarbeit
GPS	Global Positioning Systems
GTP	Growth and Transformation Plan
GW	Giga Watt
HARC	Holetta Agricultural Research Center
HCH	Hexachlorocyclohexane
HWSD	Harmonized World Soil Database
IAHR	Animal Health Research
IAR	Institute of Agricultural Research
IBC	Institute of Biodiversity Conservation
ICBA	International Center for Biosaline Agriculture
IDRC	International Development Research Center
IECAMA	Imperial Ethiopian College of Agriculture and Mechanical Arts
ILRI	International Livestock Research Institute
ILUP	Integrated Land Use Planning
IRENA	International Renewable Energy Agency
ISFM	Integrated Soil Fertility Management
ISLE	Intelligent System for Land Evaluation
ISRIC	International Soil Reference and Information Centre
ISSS-ISRIC-FAO	International Society of Soil Science-International Soil Reference and Information Centre-Food and Agricultural Organization
ITCZ	Inter-tropical Convergence Zones
IUSS	International Union of Soil Science
IUSS WRB	International Union of Soil Sciences-World Reference Base for Soil Resources
JATS	Jimma Agricultural and Technical School
LCC	Land Capability Classification
LCs	Land Characteristics
LEFSA	Land Evaluation and Farming Systems Analysis
LLPPA	Local Level Participatory Planning Approach
LMU	Land Mapping Units
LQs	Land Qualities
LULC	Land Use and Land Cover
LUPRD	Land Use Planning and Regulatory Department

LURs	Land Use Requirements
LUTs	Land Utilization Types
MAM	March-April-May
masl	Meter above sea level
MBI	Menagesha Biotech Industry
ME	Medium
MERET	Managing Environmental Resources to Enable Transition
MFI	Modified Fournier Index
Mha	Million Hectares
ML	Machine Learning
MoA	Ministry of Agriculture
MoANR	Ministry of Agriculture and Natural Resources
MoARD	Ministry of Agriculture and Rural Development
MoCT	Ministry of Culture and Tourism
MODIS	Moderate Resolution Imaging Spectroradiometer
MoFED	Ministry of Finance and Economic Development
MoI	Ministry of Information
MoWR	Ministry of Water Resources
NDVI	Normalized Difference Vegetation Index
NEDECO	Netherlands Engineering Corporation
NFIU	National Fertilizer Input Unit
NGO	Non-Government Organization
NORAD	Norwegian Agency for Development Cooperation
NRM	Natural Resource Management
NSRC	National Soil Research Center
NSTC	National Soil Testing Center
OC	Organic carbon
OFRA	Optimizing Fertilizer Recommendations for Africa
OM	Organic Matter
ONRS	Oromia National Regional State
ORDA	Organization for Rehabilitation and Development in Amhara
OWWDSE	Oromia Water Works Design Supervision Enterprise
PASDEP	Plan for Accelerated and Sustained Development to End Poverty
PBS	Percent Base Saturation
PGPR	Plant Growth Promoting Rhizobacteria
pH	Power of hydrogen
PL	Plastic
POPs	Persistent Organic Pollutants
ppm	part per million
PSNP	Productive Safety Net Programs
PWP	Permanent Wilting Point
RARCs	Regional Agricultural Research Centers
RARIs	Regional Agricultural Research Institutes
RAW	Readily Available Water
REALISE	Realising Sustainable Agricultural Livelihood Security in Ethiopia
RSTL	Regional Soil Testing Laboratories
SAERAR	Sustainable Agriculture and Environmental Rehabilitation of Amhara Region
SAR	Synthetic Aperture Radar
SART	Sustainable Agricultural Rehabilitation in Tigray
SCRP	Soil Conservation Research Project
SDPRP	Sustainable Development and Poverty Reduction Program

SIDA	Swedish International Development Agency
SLM	Sustainable Land Management
SLMP	Sustainable Land Management Practices
SMAF	Soil Management Assessment Framework
SNNPR	South Nations, Nationalities, and People's Region
SOM	Soil Organic Matter
SON	September–October–November
SOTER	SOIL and TERRAIN
SPALNE	Soil and Plant Analytical Laboratories Network of Ethiopia
SPOT	Satellite Pour l'Observation de la Terre
SPSS	Statistical Package for Social sciences
SQ	Soil Quality
SQI	Soil Quality Index
SRTM	Shuttle Radar Topography Mission
SSA	Sub-Saharan Africa
SSDS	Soil Science Division Staff
SSPM	Soil Spatial Prediction Models
SSSA	Soil Science Society of America
SST	Sea Surface Temperature
SSTAs	Sea Surface Temperature Anomalies
ST	Sticky
SWC	Soil and Water Conservation
SWCD	Soil and Water Conservation Department
TAMS/ULG	Technical Assistance and Management Services/University Leadership Group
TN	Total Nitrogen
TNRS	Tigray National Regional State
UMB	Norwegian University of Life Science
UN	United Nations
UNCCD	United Nations Convention to Combat Desertification
UNDP	United Nations Development Project
UNEP	United Nations Environment Programme
UNESCO	United Nations Education and Science Cooperation
US	United States
USBR	United States Bureau of Reclamation
USDA	United States Department of Agriculture
USSLS	United States Salinity Laboratory Staff
VAM	Vesicula Arbuscular Mycorrhiza
Vf	Very fine
VPL	Very plastic
VST	Very sticky
WE	Weak
WFP	World Food Program
WLRC	Water and Land Resource Center
WoSIS	World Soil Information Service
WRB	World Reference Base for Soil Resources
XRD	X-ray diffraction

1 Introduction

Sheleme Beyene

Abstract

Ethiopia has a great ecological diversity, ranging from tropical to temperate climatic conditions. The altitude in the country ranges from −126 to 4620 masl, and there are 32 major agro-ecological zones. Agriculture is one of the most promising sectors but has been slowed down by periodic drought, overgrazing, deforestation, limited use of and access to external inputs, and high population density. About 35.7% of the total land mass is allocated to agricultural use. The existence of diverse environments resulted in 21 soil types. The country has several major rivers and lakes, and significant groundwater resources. There are also extensive mineral resources including precious metals, industrial minerals, construction materials, and agricultural input types. The national energy balance of Ethiopia has been predominated by two energy resources—hydropower and biomass use. The Grand Ethiopian Renaissance Dam (GERD) which is expected to generate about 15,720 GW hr per year will increase the electricity supply to the population and power export potential of the country. Indigenous soil classification in the country is as old as agriculture. Currently, the World Reference Base for Soil Resources (WRB) is widely practiced. Significant amounts of fertile topsoil are lost due to water erosion. Vast areas of the highlands are affected by soil acidity that is either strongly or moderately acidic and about 10% of the total land area is also affected by salinity and/or sodicity. The current rate of urbanization, industrialization, and the extent of agro-chemical use are raising soil pollution that can be a major concern in future soil issues. Agricultural development in the highlands is crucial both for ensuring the incomes of populations in these regions as well as the availability of food in food deficit regions. The development priority in these regions is to introduce improved technologies, soil and water conservation practices, and management of acid soils. Land is a common property of the Nations and shall not be subjected to sale or transfer to other means of exchange. Although comprehensive and integrated land use policy at federal or regional levels does not exist, various government ministries issued policies that directly or indirectly relate to or affect land use. Huge efforts and investments were made to prepare digital maps of soil fertility parameters from data collected through intensive soil sampling campaigns in cultivated and arable lands. Countrywide and regional land suitability evaluations are being conducted for the preparation of land use plans in various geographic regions.

Keywords

AEZ • Acidity • Erosion • Land resources • Land policy • Land use planning • WRB

1.1 Background

Ethiopia is endowed with abundant agricultural resources and has diverse agro-ecological zones. Agriculture is the mainstay of the Ethiopian economy, accounting for 35.45% of the country's gross domestic product (GDP) (https://www.statista.com). The country has about 51.3 million hectares of arable land but just over 20% is currently cultivated, mainly by the smallholders. Although agriculture is one of Ethiopia's most promising sectors, growth of the sector has been hampered by periodic drought, overgrazing, deforestation, and high population density that led to massive soil degradation. Decades of rapid population growth contributed to soil degradation through over-farming and deforestation. The degraded area on the highlands is about 27 million hectares, of which 14 million hectares are very seriously eroded, with 2 million hectares having reached "a

S. Beyene (✉)
School of Plant and Horticultural Sciences, Hawassa University, P.O. Box 5 Hawassa, Ethiopia
e-mail: shelemeb@gmail.com

S. Beyene et al. (eds.), *The Soils of Ethiopia*, World Soils Book Series,
https://doi.org/10.1007/978-3-031-17012-6_1

point of no return" (FAO 1984a). The degradation results in a vicious cycle of depletion of soil organic matter—decline in environmental quality and nutrient depletion—decline in crop yields and productivity—food insecurity, malnutrition, and hunger (Lal 2004).

Severe organic matter depletion is driven by the complete removal of crop residues from fields as livestock feed or burning for household energy (Fig. 1.1). Most cultivated soils are poor in organic matter content due to low amounts of organic materials applied to the soils and complete removal of the biomass from the fields (Yihenew 2002). The loss of soil organic matter resulted in nutrient exhaustion and lower crop yields (Gelaw et al. 2014). Continuous tillage coupled with the removal of plant residues depletes OC, total N, available P, K, S, and other nutrients and also limits the soil microbial diversity, abundance, and activity. The nutrient balances in the farming systems are generally negative (Haileslassie et al. 2005). Accordingly, the balances for nitrogen, phosphorus, and potassium are –122, –13, and –82 kg ha^{-1} per year, respectively, signifying the large rates of macronutrient depletion.

Ethiopia has a great ecological diversity, ranging from tropical to temperate climatic conditions. The altitude in the country ranges from −126 to 4620 masl and there are 32 major agro-ecological zones. The agro-ecological zones (AEZ) were determined with a crop suitability approach considering temperature, moisture regime, soil characteristics, and topography differences. About 51% of the total land area is categorized under arid, semi-arid, and sub-moist zones and the remaining half is moist humid zones. The major crop growing areas are sub-humid, humid, moist, sub-moist, and semi-arid climatic zones (Sect. 8.2.2).

1.2 Land Resources

Ethiopia has diverse land resources, varied topography, and complex geology that accentuates the unevenness of the surface; a highland complex of mountains and bisected plateau characterizes the landscape. The altitude of the country ranges from −126 masl at the Dalol depression to 4,620 masl at the highest Ras Dejen Mountain. The Great East African Rift Valley bisects the plateau in the northern, and the east and central highlands of the country. There are 21 soil types including Leptosols, Nitisols, Vertisols, Cambisols, Luvisols, Calcisols, Fluvisols, Gypsisols, Alisols, Solonchacks, Acrisols, Regosols, Andosols, Arenosols, Phaeozems, Lixisols, Chernozems, Gleysols, Solonetz, Ferralsols, and Planosols (Appendix Table A.6).

From a total of 1,013,798.4 km^2, Alehegne (2014) reported that nine land use and cover types including cultivated land (28.9%), forest (5.1%), woodland (15.6%), bushland (6%), shrub land (32.9%), bamboo (0.5%), dry and wet grazing land (10.4%), and water body (0.6%) were identified. However, the land use and land cover are subjected to a continuous change, and there has been a steady rate of change from forestland and rangeland to arable farmland (Biazin and Sterk 2013; Tsegaye et al. 2010; Garedew et al. 2009). The World Bank (2013) estimated that about 35.7% of the total land mass is allocated for agricultural use (i.e., arable, under permanent crops, and permanent pastures). The closed natural high forest is around 3.5% of the land area and is unevenly distributed in the country. Oromia, SNNPR, and Gambella regions account for 95% of the total high forest area (Bekele 2011; FAO 2010). Savannas, grasslands, and shrublands dominate the

Fig. 1.1 Removal of maize stover after harvest from farm in Toga area on the main road from Hawassa to Shashemane (Photo by Sheleme Beyene)

Table 1.1 Estimated mineral resources of Ethiopia (presence and deposit)

Type Metal	Quantity (tons)	Type industrial	Quantity (tons)	Type industrial	Quantity (tons)
Gold	401.15	Kaolin	48,740,773	Gypsum	56,420,000
Silver	201.47	Limestone	70,521,350,000	Talc	120,000
Platinum	12.5	Coal	267,923,954	Kyanite	10,000,000
Tantalum	5000	Graphite	460,000	Bromine	420,000
Tantalum concentrate	2976.5	Phosphate	181,000,000 (presence)	Salt (NaCl)	290,168,000
Niobium concentrate	1373.8	Bentonite (m^3)	6,473,143 45,105,000 (to be verified)	**Agricultural input**	
Iron	300,145,000	Sulfur	2,800	Potash	>290,168,000
Copper	> 50,000,000	Diatomite	45,105,000	Zeolite	100,000
Zinc	30,000,000	Dolomite	129,870,000	**Construction**	
Lithium	419,116	Quartz	500,000	Marble	92,400,000
Nickel	17,000,000	Feldspar	530,000	Granite	6,250,000
Lead	4,648	Silica-sand	811,400,000	Scoria	In Billion
Chromite	149,406	Pumice	In Billion	Ignimbrite, basalt	Many Billion

Source Ministry of Mines, Mineral Resource Directorate ten-year plan (2020)

lowlands, providing important resources for extensive livestock production (pastoralism). However, the loss of grasslands due to conversion to farmland enhances the dominance of less productive and non-palatable plant species. In Afar region, for example, over 1.3 million hectares of land have been taken over by the invasive species *Prosopis julilora* (Flintan 2011).

The country has several major rivers and lakes, and significant groundwater resources including 12 major river basins, 22 lakes, and large areas of wetlands (Alehegne 2014). There are also extensive mineral resources including precious metals, industrial minerals, construction materials, and agricultural input types (Table 1.1). Lege Dembi is the largest gold mine in the country, and there is also secondary enriched (placer) gold which has been mined traditionally for years back to biblical times. The cement factories in the country are using high quality limestone, clay, gypsum, and pumice as raw materials for cement production. Large input of construction minerals such as sand, gravel, scoria, crushed stones, aggregates, and pumice are used by the construction industry. Other mineral products including platinum from laterite, industrial minerals, gemstones (opal, peridot, and other precious stones), and decorative and construction materials also exist in the southern, western, central, and northern regions of the country (Ministry of Mines and Energy 2009). Additionally, jewelry mines including emerald, aquamarine, quartz, and others (tourmaline, opal, olivine, apatite, amazonite, augite, jasper, etc.,) exist in different parts of the country. Energy mines such as oil shale, geo-thermal resource, and natural gas are available.

The national energy balance of Ethiopia has been predominated by two energy resources—hydropower and biomass use. The latter had been the source of 80% of the energy needs and is a major factor in the depletion of the country's biomass resources. Alternative potential biomass sources include agricultural residues such as coffee husk (214,299 tonnes/year), enset, cotton stalk residue (89,000 tonnes), chat, municipal solid wastes, sawmill residues, energy plant, e.g., jatropha, and agro-industrial by-products such as biogas and ethanol. Exploration of these alternatives requires investment (Guta 2012).

The introduction of hydroelectric power and the large surge in capacity since 2005 raised electricity generating capacity by many folds on a per capita basis (Chamberlin and Schmidt 2011). Joy (2013) reported that electricity is said to be accessible to around 52% of the population, with two (2) million households connected. About 94% of electricity is produced by hydropower, 4% by wind and geothermal, and 2% by diesel stand-by. The Grand Ethiopian Renaissance Dam (GERD) which is expected to generate about 15,720 GW hr per year is expected to increase the electricity supply to the population and the power export potential of the country.

1.3 Soil Forming Factors

The main factors, which influence the development of soils, are climate, topography, parent material, time and biological parameters such as vegetation, land use, and the activity of organisms. Varied combinations of the soil forming factors lead to specific sets of soil forming processes that bring about change in soil properties. The great variations in the age and character of the parent materials coupled with the diverse physiographic and climatic conditions encourage the formation of different soil types (Hurni et al. 2007; Mesfin 1998).

Parent material: The parent materials have a great influence on the characteristics and types of soils. The distributions of soils are partly related to their proximate parent materials. The major geological materials of the highlands are tertiary and quaternary basalt and deposits and granite rocks, whereas gypsum, sandstone, and limestone form major geologic features in the lowlands (Fig. 6.10; Tefera et al. 1996). Soils of the central, eastern, and northern highlands are developed on hard basalt rocks or on materials derived from them (MoA 1987).

In the southwestern parts, most soils are formed from crystalline basement rocks. Andosols are developed from quaternary pyroclastic and volcanic ash deposits and pumice, found in the rift valley (Fig. 1.2) and on the high mountains such as the Simien mountains. Fluviatile deposits located along the course of main rivers and in foothills and depressions lead to the formation of Fluvisols. The lowland soils are developed on sandstone, limestone, and alluvial deposits. Moreover, on lacustrine deposit and depression sites with seasonal water cover Gleysols or their variants such as Gleyic Vertisols/Cambsiols/Rgosols are found.

Topography: Topography is the major differentiating factor of genesis and distributions of soils in the country (Mitiku 1987; FAO/UNDP 1984a, b). A rise in altitude is associated with a decrease in soil depth (Mesfin 1998) and an increase in soil organic carbon content (Hurni 1983; Murphy 1968). Gently sloping terrain units are occupied by well-drained Nitisols, Acrisols and Luvisols (FAO/UNDP 1984a, b), and steep and colluvial foot slopes by Leptosols, Cambisols, and Regosols (Virgo and Munro 1978). The thin soil depth occurring in steeper and convex slopes is an indication of rapid removal of weathered material (Virgo and Holmes 1977).

On the plateaus (<2500 m.a.s.l.), poorly drained Vertisols dominate (Mitiku 1987; Tegene 1982, 1996, 1997, 1998; Berhanu 1982). At intermediate high altitudes (2500–3000 m.a.s.l) Phaeozems occur on gently to steeply sloping terrain (Kefeni 1995; Veneema and Paris 1986; Paris 1985) and Andosols are found in the high altitudes >3000 m.a.s.l (Mohammed and Belay 2008; Weigel 1988; Bono and Sieler 1984; Frei 1978) and in the rift valley lowlands, conditioned by local geologic features (Chap. 5). The poorly drained flatter and depression sites are covered with Gleysols and found mainly around major lake basins.

Climate: The increase in rainfall with a rise in altitude increases leaching and leached soils are common in the high mountains (Murphy 1968). Highly leached soils dominate the high rainfall parts of the southern and western highlands (FAO/UNDP 1984a, b). Areas with high rainfall and high temperatures throughout the year have deep soils as a result of rapid weathering rates. In the northern, central, eastern, and southeastern highlands, the influence of climate is subtle. In these parts of the country, less leached and base-rich

Fig. 1.2 Andosol developed from volcanic ash in south central rift valley, Toga area on the main road from Hawassa to Shashemane (Photo by Sheleme Beyene)

soils dominate due to slow rate of soil formation (Hurni 1983). In the intermediate altitudes, semi-temperate and temperate forests such as *Juniperous procera* and *Podocarpus gracilor* are growing on Phaeozems and Leptosols with or without mollic A and/or umbric horizons, but generally with a high organic carbon content (EMA 1988; Murphy 1968).

On the other hand, drier semi-arid and arid lowlands of the country are occupied by soils with high salt and alkali materials that make them saline–sodic soils. These soils could be described as Solonchaks, Gypsisols, Calcisols and Solontez and occur mainly in the Ogaden lowlands, the Afar depression, and in large drier parts of the rift valley.

Organisms: The activities of organisms cause lateral mixing of soil organic matter, soil material, and possibly plant nutrients. In addition, burrowing animals dig out soil from subsoil and bring it to the surface, which makes deposit as termite mounds on the surface of the soils. These features are typically found in the south (Fig. 1.3) and central (Fig. 6.11) rift valleys and other parts of the country. These areas, depending on local factors, are covered with Andosols, Phaeozems, Calciosls, Cambisols, or Regosols. In such parts, where there is a good natural drainage, there is a partial removal of carbonates and residual carbonates accumulate in subsoil horizons, which are responsible for the formation of Calcisols and/or soils with some degrees of calcic properties.

The climatic-climax tropical woodlands and forests growing in high rainfall and high-temperature soils of southern and southwestern highlands (Daniel 1977, 1988) are associated with deep leached soils. In the uncultivated parts of the high plateau experiencing high rainfall, *Festuca* (tussock) and *Lobelia* grow on Andosols and/or Phaeozems. In the low rainfall parts, *Acacia spp*. and *Croton macrostachys* form the major vegetation types (Mitiku 1987).

1.4 Soil Classification

Indigenous soil classification in the country can be as old as agriculture. Indigenous knowledge and practices of soil classification and management have been transferred from generation to generation with little or no modifications. Farmers in different parts of the country classify soils based on easily recognizable properties such as soil color, particle size, thickness (depth), water holding capacity, and fertility status (Esayas 2005). Farmers' indigenous classification serves for crop suitability evaluation, proper allocation of land use, and application of the required inputs for soil fertility management. Thus, indigenous soil classification is not a pedogenetic classification in nature. However, the correlation of indigenous and scientific classification systems can serve to facilitate technology transfer between different localities, though the soils are named differently in various regions of the country (Chap. 5).

The history of soil classifications in Ethiopia has been related to the chronology of soil survey and mapping missions. In the 1920s, the soils were classified into four classes

Fig. 1.3 Termite mounds on either side of the main road from Mega to Moyale **a)** Calisol formed on depression, and **b)** Cambisol on gentle slope (Photo by Alemayehu Kiflu)

(Chernozems, Tropical Prairie soils, Laterites, and other Red Tropical soils) following Shantz and Marbut (1923). In the 1960s, the soils were classified into ten major soil groups/associations based on the D'Hoore classification system (D'Hoore 1964). Both Soil Taxonomy and the FAO-UNESCO classification systems have been used separately or in combination in the country. However, nowadays, the World Reference Base for Soil Resources (WRB) is widely practiced. The preliminary country-wide soil studies in the 1980s and 1990s employed the FAO-UNESCO (1974) system, its revised version (FAO 1988), and the World Reference Base for Soil Resources (FAO 1998). Soils of most major river basins covering the entire country (in the 1980s and 1990s), sub-basins, and development corridors in the 2000s and 2010s were classified following the FAO-UNESCO systems and IUSS-WRB recent editions (Ali et al. 2020; Leenaars 2020a, b; Elias 2016; Mesfin 1998; FAO 1984a, b). Nowadays, very limited area-wide soil resource studies employ the Soil Taxonomy system, except postgraduate-based soil studies that commonly use both Soil Taxonomy and IUSS-WRB systems (Zewdie 2013; Esayas 2005; Fikre 2003).

1.5 Soil Diversity

Ethiopia has diverse environments in terms of relief, climatic conditions, geologic features, vegetation, and land use patterns. The existence of such diverse environments results in the formation and presence of different soil types. Generally, 21 Soil Reference Groups exist in the country, of which 9 dominant soil types cover 90% of the land surface. These include Leptosols (30%), Nitisols (12%), Vertisols (11%), Cambisols (9%), Luvisols (8%), Calcisols (9%), Fluvisols (4%), Gypsisols (8%), and Alisols (3%) (Appendix Table A.6).

The distribution of the soils is largely governed by the interactions and variations in topography, climatic and biotic features (Assen 2003; Eylachew 1987; Mitiku 1987). Consequently, soils with dark organic matter-rich A horizons predominate in the cooler high altitudes above 2000 masl. In areas where removal of materials induced by steeper topography is accelerating water flow and gravity, shallow soils predominate, whereas deeper soils cover the deposition lands. Along the course of several rivers and streams, soils with fluviatile materials are common while soils with saline–alkaline characteristics cover the drier lowlands.

The dominant soil types are widely distributed in different parts of the country occupying varying physiographic positions. Leptosols exist on steeper topographic positions and are used as hilly grasslands and forest and natural vegetation cover (Appendix Table A.7). On the other hand, Gypsisols, Solonchaks, and Calcisols largely cover moisture deficient parts of the country. These soil types are mainly used for extensive pastoral farming and in the recent period, are being used for irrigation agriculture. Alisols are found in high rainfall areas where intense leaching processes typically characterize their formation, and are chemically poor to be used for crop farming (Appendix Table A.30–A.34).

Nitisols occur mainly in the humid southern, western, eastern, and northwest central highlands where intense to intermediate weathering processes associated with high-rainfall and high-temperature conditions prevail (Fig. 6.12). Vertisols are among the most widely distributed soil types and are found in both highland and lowland parts of the country (Appendix Table A.13). Cambisols are found in different agro-climatic zones and occupy gently and strongly sloping landforms (Appendix Table A.15). In parts where slopes are relatively steep, Cambisols give way to the occurrence of Leptosols and/or Regosols. Thus, they form a geographical continuum with Vertisols and/or Luvisols towards lower slope positions and Leptosols/Regosols towards upper slope positions.

Luvisols are found in different environments and are one of the major agricultural soils of the country. They are suitable for growing different crops depending on the agro-ecological conditions of their occurrence (Appendix Table A.19). Fluvisols occupy gently sloping landforms, between 1 and 5%, e.g., the Middle Awash Valley of the Amibara irrigation lands. They are also one of the productive soils mainly with the help of irrigation. However, flood might affect and damages crops cultivated in these soils associated with their positions in the landscape. Fluvisols are considered to be fertile, as they regularly receive fresh sediments containing high contents of soil organic matter (SOM) and exchangeable cations (Appendix Table A.24).

1.6 Threats to Soils

High population growth, erosion, soil acidity, grazing pressure, expansion of salt-affected soil, and soil pollution are some of the major threats to soils in the country. The country is struggling with challenges such as improving production and productivity for the ever-increasing population, under smallholder and fragmented landholdings (Fig. 1.4), with low input and traditional farming practices. When suitable land for cultivation runs out, people move into marginal lands of low productivity, for example, steep slopes (Muluneh 2010). The high population pressure, continuous and steep slope cultivation (Fig. 1.5), low vegetation cover, deforestation, and inadequate soil conservation practices cause annual soil loss of about 1.5 billion metric tons in the highlands of Ethiopia (Girma 2001). The ever-increasing land use change (Tsegaye et al. 2010) and the rugged topography in different parts of the country are aggravating the rates of soil erosion, soil fertility reduction, crop yield decline, and food insecurity.

Soil erosion is a major part of land degradation that affects the physical, chemical, and biological properties of soils and results in on-site nutrient loss and off-site sedimentation of water resources (Hurni 1993). Erosion is the major cause of on-site topsoil loss (Figs. 1.6 and 1.7) and significant amounts of fertile topsoil are lost, mainly due to water erosion, leading to severe depletion of nutrients and organic matter in upstream areas (Sheleme 2017) and over-deposition in downstream. The soil loss due to erosion in the highlands is high, varying between 42 and 175.5 t/ha/yr (Adimassu et al. 2017). Cultivation of steep hillsides in the highland parts of the country is marginally productive but caused erosion of the soil leading to loss of fertility in the topsoil and reduction in soil depth, which can have an adverse effect on crop yields. Also, reduced water storage capacity in the soil makes crops less able to withstand drought.

Vast areas of land in the western, southern, southwestern, northwestern, and even the central highlands of the country are affected by soil acidity (Mesfin 2007). The soils in these areas are either strongly or moderately acidic. Currently, about 43% of Ethiopian soils are classified as acidic soils (EthioSIS 2014), of these about 28.1% of the soils are strongly acidic soils (pH 4.1–5.5). It has become a serious threat to crop production in the highlands of the country with high rainfall (Sheleme 2016; Abdena et al. 2007). Crop yield loss due to acidity ranges from 20 to 80% depending on the degree of acidity, agronomic practices, and agro-ecologies (Tilahun 2019). Soil acidity also affects emergence and survival, legume nodulation and root growth, as well as microbial growth, especially if soil pH ($CaCl_2$) is less than 4.5. In strongly acidic soil (pH < 5), aluminum and manganese become more soluble and toxic in most soils, and deficiencies of essential plant nutrients such as P, Ca, K, Mg, and Mo arise (Wang et al. 2006). Consequently, some barley and wheat-growing farmers in southern highland areas have shifted to producing oats, a crop more tolerant to soil acidity (Wassie and Shiferaw 2009).

Liming and use of vermicompost improve the physical, chemical, and biological properties of soils and increase crop production on acid soils (Mathewos 2020; Fig. 1.8) by raising the pH, Ca, and Mg concentrations, and P availability and thereby improving nutrient uptake by plants. There have been four lime crushers engaged in the production of lime but the capacity of the crushers to meet the required amount is by far below the required level. The lime distribution system is also inefficient resulting in price build-up from the production site to the farmers' fields, mainly due to the price of transportation making the farm gate price high for the farmers (Gete et al. 2010).

High grazing pressure leads to loss of desirable or palatable forage species, encroachment of non-palatable species, and topsoil erosion due to overall ground cover loss. Denuding of vegetation by overgrazing exposes the soil to increased solar radiation and erosion by wind and water. Loss of surface horizon and organic matter results in crusting of the soil surface, which further prevents water infiltration and seed germination. The absence of land management

Fig. 1.4 Fragmented land holdings by smallholder farmers on the road from Tulobolo to Kela town in central Ethiopia (Photo by Tesfaye Abebe)

Fig. 1.5 Cultivation on steep slopes between Shone and Adilo, Southern Ethiopia (Photo by Tesfaye Abebe)

Fig. 1.6 Soil degradation by erosion around Adilo, on the main road from Shashemane to Wolayita Sodo (Photo by Sheleme Beyene)

practice in such areas leads to further land degradation leaving the landscape increasingly vulnerable to drought and rainfall-related erosion. When grasslands have been converted to farmland and then abandoned, less productive and non-palatable plant species take over. In Afar region, for example, over 1.3 million hectares of land have been taken over by the invasive species *Prosopis julilora* (Flintan 2011).

There is a widespread occurrence of salt-affected (saline, saline–sodic, and sodic) soils and soda waters and soda lakes in the irrigated arid and semi-arid areas of Ethiopia (Qureshi et al. 2019; Sileshi 2016; Abebe et al. 2015; Kidane et al. 2006; Heluf 1985). The total land area affected by salinity and/or sodicity is estimated at 11,033,000 ha, that is, nearly 10% of total land. From the total land area of salt-affected soils (44 Mha), approximately 33 Mha is dominantly affected by salinity, 8 Mha by salinity–sodicity, whereas 3 Mha is dominantly affected by sodicity problems (Habtamu 2013). This may be due to the massive expansion of irrigated agriculture by saline water. Following the poor practice of

Fig. 1.7 Loss of soil by erosion in Boricha district, Sidama region (Photo by Tesfaye Abebe)

Fig. 1.8 Growth of barley (*Hordeum vulgare* L.) as influenced by applications of lime and vermicompost in Bule district, southern Ethiopia (Photo by Mathewos Mesfin)

irrigation management, salinity has emerged as a major problem responsible for the reduction of land productivity and natural resources degradation.

The current rate of urbanization, industrialization, and the extent of agro-chemical use could result in soil pollution which will be a major concern in the future soil issues. High concentrations of heavy metals, such as zinc (Zn), chromium (Cr), nickel (Ni), cobalt (Co), and lead (Pb), which were higher than the internationally acceptable limit for soil, were found in the soil samples of the dumpsite of Addis Ababa and nearby open land (AGP 2011). This problem is of increasing concern with increasing industrial activity, agricultural chemicals, and/or municipal waste disposals (EIP 2002). For instance, the Awash River is highly polluted from chemicals used for spraying crops (Abule et al. 2005) and the discharge of untreated residential, commercial, and industrial wastewater in Addis Ababa. The river has levels of nitrogen, biochemical oxygen demand (BOD), and Pb, which are up to three times higher than the World Health Organization's maximum acceptable level (Redda 2003).

1.7 Soils and Climate

Ethiopia is a country of natural contrast; characterized by the largest proportion of (≈45%) elevated land masses in Africa. The complex topography of the country strongly defines both rainfall and temperature patterns, by modifying the influence of the large-scale ocean–land–atmosphere pattern, thus creating diverse localized climates. The country is distinctive and the most rugged country in Africa, and its climate is substantially governed by local factors in which the topography is powerful. Climate has a significant influence on soil development and properties and is therefore responsible for the existence of widely varying soils in the country.

Leptosols are the most abundant soil type found in most of the agroclimatic zones. It is mainly found in warm–moist, warm–sub-moist, and tepid–moist agroclimatic zones that are characterized by steep slope topography. Vertisols exist in warm–moist, tepid–moist, warm–sub-humid, and cool–sub-humid agroclimatic zones with valley bottom topography. Cambisols are found in warm–sub-moist and warm–semi-arid agroclimatic zones though they could develop under any climatic conditions. Luvisols and Nitisols are found in southwestern and eastern highlands with warm–moist, warm–sub-humid, tepid–moist, and tepid–sub-humid agroclimatic zones. Moisture availability in the range from moist to sub-humid moisture regimes is clearly associated with the development of Nitisols.

1.8 Roadmap of Agriculture in Highland Soils

Constraints on agricultural production and food security issues in highlands are not as severe as in the drought-prone regions. This, however, does not mean the people living in these regions are not exposed to poverty or totally free from vulnerability to food shortages. These regions play a crucial role in national food crop production, where agricultural development is viewed more broadly from the standpoint of national food security.

Accelerating agricultural development in the highlands is therefore important both for ensuring the incomes of populations in these regions as well as the availability of food in food-deficit regions. Year-round cropping and improvement of water utilization will be key development tasks. The priority in these regions is to introduce improved technologies, soil and water conservation practices, and management of acid soils (MoFED 2003).

Improved Technologies

The crop production in the highlands can be increased through the application of improved technologies, i.e., improved seeds, fertilizer, and tools and by harnessing rainwater. Production of high-value products may also be necessary. The practical experience of the people living in such areas gravitates towards the alternative where products such as coffee, spices, and false banana (*enset*) are grown. The government does not dictate the farmers to produce any specified crops but put forward proposals and make available technology packages for products that are most advantageous for the regions (MoFED 2003).

Different agricultural development strategies and/or activities may receive priorities in these regions but these do not mean that nothing else will be produced. Hence, many types of agricultural activities are undertaken and auxiliary products are produced alongside certain key priority products that command much of the attention of farmers and development workers (MoFED 2003).

Watershed Management

The most common types of physical measures of land management implemented in the highlands are hillside terracing, soil and water conservation (SWC) structures on cultivated lands (i.e., soil bunds, *Fanyajuu*, stone bund, soil faced stone bands, and grass strips), bench terraces, micro basins, semi-circle terracing, and trenches. Since the 1980s, various nationwide SWC initiatives supported by multiple donors have contributed to the current coverage of terraces (WLRC 2018). These initiatives include Food-for-Work (1973–2002), Managing Environmental Resources to Enable Transition to more sustainable livelihoods (MERET 2003–2015), Productive Safety Net Programs (PSNP 2005–present), Community Mobilization through free-labor days (1998–present), and the National Sustainable Land Management Project (SLMP 2008–2018).

The coverage of existing terraced landscape is about 7.7 Mha, which is about 23% of the 34 Mha that need treatment by physical soil and water conservation measures. Moreover, the existing physical conservation structures are mainly concentrated in the northern, northeastern, and eastern highlands of the country (WLRC 2018). Additionally, the "Green Legacy" that started in 2019 and covers the different parts of the country by tree planting, and agroforestry as one of the indigenous land management practices being extensively applied in many parts of the highlands.

Liming

The highlands experience high amount of rainfall and the soils in these areas are either strongly or moderately acidic. The government is putting emphasis on liming of these soils. There have been four lime crushers engaged in the production of lime, though the capacity of the crushers to meet the required amount is by far below the required level. For instance, the annual demand of 191,962 tons, the capacity of these crushers is limited to only 9,828 tons annually, and lime use is restricted to about 5,100 ha (Gete et al. 2010). In the GTP II, the government rehabilitated about 226,000 ha of agricultural land by expanding the production, distribution, and promotion of lime by smallholder farmers (MoANR 2015).

1.9 Land Ownership and Land Use Policy

1.9.1 Land Ownership

The Federal Democratic Republic of Ethiopia (FDRE) Constitution underlines that it is the duty of government to hold, on behalf of the People, land, and other natural resources and to deploy them for their common benefit and development (FDRE 2005). Likewise, the Constitution, under Article 92, further provides that the design and implementation of development programs and projects shall not damage or destroy the environment. It further stipulates that the right to the ownership of rural and urban land, as well as of all natural resources, is to be exclusively vested in the state and in the people. Thus, land is a common property of the Nations, Nationalities, and Peoples of Ethiopia and shall not be subjected to sale or to other means of exchange. Furthermore, the regions were entrusted the power to administer land and natural resources. The administration by regional governments is on behalf of the people and the role of the federal government in this respect was specified.

Peasant farmers/pastoralists engaged in agriculture for a living shall be given rural land free of charge. Also, any citizen of the country who is 18 years of age or above and who wants to engage in agriculture for a living shall have the right to use rural land, while children who lost their mothers and fathers due to death or other situations shall have the right to use rural land through legal guardians until they attain 18 years of age. Likewise, any person who is a member of a peasant farmer, semi-pastoralist, and pastoralist family having the right to use rural land may get rural land from his family by donation, inheritance, or from the competent authority. The rural land use right of peasant farmers, semi-pastoralists, and pastoralists shall have no time limit and the holder shall have the right to transfer his rural land use right through inheritance to members of his family. However, the duration of rural land use rights of other holders shall be determined by the rural land administration laws of the regions.

Additionally, the Constitution also stipulates that any holder of rural land shall be given a holding certificate that indicates size of the land, land use type and cover, level of fertility, and boarders, as well as the obligation and right of the holder. The holder of rural land is obliged to use and protect his land and to notify the competent authority when he abandons his land use right. When the land gets damaged, the user of the land shall lose his use right. Peasant farmers, semi-pastoralists, and pastoralists who are given holding certificates can lease to other farmers or investors the land from their holdings a size sufficient for the intended development in a manner that shall not displace them, for a period of time to be determined by rural land administration laws of the regions.

The government may redistribute land or use it for a public purpose in which case compensation is paid for capital invested and any improvement made on the land. The communal rural land holdings can be changed to private holdings as may be necessary by the government. Private investors that are engaged in agricultural development activities shall have the right to use rural land in accordance with the investment policies and laws at the federal and regional levels. Governmental and non-governmental organizations and social and economic institutions shall also have the right to use rural land in line with their development objectives (FDRE 2005).

1.9.2 Land Use Policy

There is no comprehensive and integrated land use policy at federal or regional levels, though various government ministries issued policies that directly or indirectly relate to or affect land use. The human–environment interaction and associated land use/land cover changes for the last millennia reflect that the human–agriculture interface has left their marks on the landscape (Gete et al. 2014). Fragmentation of land and intensification of land use is occurring, with subsequent reduction of fallowing practice or abandonment of fallow periods (Shiferaw 2011; Garedew et al. 2009; Abebe 2005). Currently, significant unplanned land use change is taking place in all regional states. Towns are mushrooming without a plan for agricultural lands. Farmers are building houses on their farmlands situated near main roads and selling them. Farmers are changing croplands to "eucalyptus" and "chat" growing fields. A lack of appropriate land use policies has been highlighted as a key root cause of these trends and the resulting negative impacts.

Land use issues have been one of the concerns of the governments since the Emperor Haile Selassie period. Different attempts were made in establishing institutions to deal with land use issues and to develop master land use plans. The attempts of developing land use-related activities, like the river basin studies, continued in a scattered and uncoordinated manner. Land use issues become even more complicated after the adoption of a federal form of government in 1994 because regional states are given the power to administer lands within their jurisdiction and land use planning activities have been taken up at federal and regional levels, and in different government sectors at federal and regional levels. This fragmented approach resulted in undesired and unplanned land use changes and damage to natural resources.

There are currently various land use-related policies and laws developed by different sectors. The government issued different development strategic plans and programs including the Agriculture Led Industrialization (ADLI 1992) and the Rural and Agricultural Development Policies, Strategies and Tactics (MoI 2002); and the medium- and long-term strategic plans that expound ADLI, i.e., the Growth and Transformation Program (MoFED 2010) and the previous strategic plans, i.e., PASDEP (MoFED 2006), SDPRP (MoFED 2002), and the Ethiopian Strategic Investment Framework (Sustainable Land Management) (MoARD 2010).

Sector ministries and regional bureaus also issued various strategic papers and policies that are related to one or more aspects of land use. These include the National Biodiversity Strategic and Action Plan (IBC 2014), the Environment (EPA 1997), the Water (MoWR 2001), the Forest development and Conservation (MoARD 2007), the Wildlife Development and Conservation (MoARD 2005), and the Tourism Development (MoCT 2009). There are also laws that impact land use issues under the Rural Land Administration and Use (FDRE 2005), like the expropriation of lands for public purposes and payment of compensation laws, mining proclamation, the biosafety laws, the investment proclamation, and the Urban Planning Proclamation.

These policies and laws that deal with or relate to land use can be classified into three broad categories. The first types lay basic frameworks and guide all other land use-related policies. The second types are land use laws that provide some guidance on land use planning, while the third types are sector policies that are issued to advance sector interests and programs, like the forest development and conservation policy, the national culture policy, the energy policy, and sector strategic plans.

The Agriculture Development Led Industrialization (ADLI) policy is the overarching development policy of the FDRE that aspires to achieve poverty reduction and industrialization through the development of smallholder agriculture. The ADLI was modified, without abandoning its basic principle of development through smallholder agriculture and incorporated into the strategic development plans of the government via the Sustainable Development for Poverty Reduction Program (SDPRP), a Plan for Accelerated and Sustained Development to End Poverty (PASDEP), and the Growth and Transformation Plan (GTP).

The SDPRP promoted agricultural development and poverty reduction in rural areas by strengthening agricultural extension services, training farmers in Farmers Training Centers, water harvesting and irrigation, improved marketing opportunities, restructuring peasant cooperatives, and supporting micro-finance institutions. The PASDEP advanced ADLI with the objective of attaining accelerated industrial growth and urban development through the commercialization of agriculture as well as rural development and ensuring food security. The GTP reaffirming the smallholder agriculture as a source of agricultural development intends to develop smallholder commercialized farmers who are supposed to engage, with private agricultural investors, in producing marketable farm products for domestic and export markets. Under the GTP, a shift to production of high-value crops with a special focus on high productivity areas, intensified commercialization, and support for the development of private investment in large-scale commercial agriculture "in the lowland areas where land for large-scale commercial farming is in demand" will be pursued. Expanding irrigation development and improving natural resources conservation are also part of the strategic direction under the GTP. In pastoral areas, the GTP has given priority to water resources development for local community and livestock; and resettlement of pastoralists on a voluntary basis in areas suitable for irrigation.

The Environmental Protection Agency (EPA) issued an Environmental Policy in 1997 that has the overall policy goal of "improving and enhancing the health and quality of life of all Ethiopians and to promote sustainable social and economic development through the sound management and use of natural, human-made, and cultural resources and the environment as a whole so as to meet the needs of the present generation without compromising the ability of future generations to meet their own needs (EPA 1997)."

1.10 Land Evaluation

A primitive attempt was made to evaluate and measure the quality of land in conjunction with the transfer of land to soldiers during the reign of Emperor Tewodros (1855–1868) (FAO 2014; FAO 1984a), and the quality of land was divided into four categories ranging from "Highly productive" to "Least productive" land. The Ministry of Agriculture was established in 1890, during the reign of Emperor Menelik II, to issue land use and forest protection regulations. In 1974, international consulting firms conducted numerous catchment-based studies comprising diverse land suitability studies. Additionally, various institutions made fragmented efforts between 1950 and 1970 to generate biophysical baseline information and land suitability studies (Esayas and Debele 2006).

Land evaluation exercises in the country serve as a connection between basic resource surveys and agricultural and other land development plans. However, most of the land evaluation practices in Ethiopia were not possible to completely combine economic and social analysis with physical land evaluation (Gessesse 2013; Kassa et al. 2010; Assen 2003; Yizengaw and Verheye 1995). The first division of soil survey and land evaluation was formed in 1971 under the then Institute of Agricultural Research (IAR), and land suitability studies garnered substantial institutional legitimacy. In 1973, the unit was transferred to the Ministry of Agriculture's Land Use Planning and Regulatory Department (LUPRD) (Esayas 2001; Debele 1980). The LUPRD conducted countrywide and regional land suitability evaluations for the preparation of land use plans in various geographic regions. Additionally, the river basin-based land resource assessments and land evaluation began in the 1980s under the Ministry of Water Resources.

The government established the project "The Ethiopian Agricultural Transformation Agency (ATA)/Soil Information System (EthioSIS)" under the overall Soil Fertility Road Map of the Ministry of Agriculture in 2010 mainly to conduct soil fertilitymapping to reformulate fertilizer recommendations (Ali et al. 2020). Accordingly, huge efforts and investments were made to prepare digital maps of soil fertility parameters from data collected through intensive soil sampling campaigns in cultivated and arable lands. The approach was, however, limited to auger-based composite sampling mainly in the top 20 cm for annual crops and in some cases at 20–50 cm soil depth for perennial crops. The digital soil fertility maps were predicted and soil property maps were developed using Digital Soil Mapping (DSM)-Soil Spatial Prediction Models (SSPM). The ATA/EthioSIS project finalized and released the digital soil fertility and fertilizer formulation maps/atlas for almost all regions and collected approximately 100,000 soil datasets for about 13 soil nutrients (N, P, S, K, Ca, Mg, Na, Fe, Cu, Zn, B, Mn, Si) and 5 soil properties (pH, EC, CEC, OC, $CaCO_3$,) in the top 20-cm soil depth.

1.11 Land Use Planning

Land in Ethiopia is being exploited without respect for its environmental, social, and economic appropriateness as well as natural resource protection (Gebeyehu et al. 2017). It is also used without adequate planning and management across numerous economic sectors, such as in agriculture, forest, livestock, water development, wildlife, and tourism, leading to severe deterioration of natural resources and the environment, as well as creating a low level of economic development, environmental degradation, and poverty trap. A mix of land-related geo-ecological and biophysical limitations are the major drivers of the low level of productivity and the food shortage in the country.

Starting in 1973, the Land Use Planning and Regulatory Department (LUPRD) of the Ministry of Agriculture conducted countrywide and regional land suitability evaluations for the preparation of land use plans in various geographic regions. The Ministry of Water Resources also conducted several reconnaissance land suitability studies for major river basins of the country to prepare development master plans including Abay in 1998 (MoWR 1998), Omo-Gibe in 1996 (MoWR 1996), Wabeshebele in 2007 (MoWR 2007), Genale Dawa in 2006 (MoWR 2006), and Rift valley lakes in 2007 and 2010 (MoWR 2010). The river basin master plans are quite extensive and deal with the utilization of various natural resources as well as socioeconomic elements of land based on severe site suitability assessments, even though the primary goal of these studies was to develop water resources.

Since 2000, various land evaluation and use planning studies have been completed and/or are currently being conducted in some regional states, with land evaluation studies being one of the key components. The Construction Design and Supervision Works Corporation (ECDSWCo), as well as the Oromia Water Works Design Supervision Enterprise (OWWDSE), and the Amhara Design and Supervision Works Enterprise (ADSWE), conducted the majority of these studies (Ali et al. 2020). The planning efforts are aimed at "Integrated Land Use Planning" or "Land Suitability Evaluation for Various Irrigated Cropping", which includes all sectoral activities in the study region and results in a multi-sectoral and multidisciplinary plan. The plans are based on bio-physical, economic, social, and environmental suitability assessments and have been undertaken in Borena, Harerge, East Amhara, Oromiya Special Zone, Kuraz Fentale, and Tibila at various scales. Furthermore, land evaluation studies were also carried out in various locations for the partial fulfillment of graduate-level university degrees, and many of them are remarkable for large-scale level land evaluation exercises in the country (Gessesse 2013; Kassa et al. 2010; Assen 2003).

1.12 Conclusions

Agriculture is one of the most promising Ethiopian resources but has been slowed down by periodic drought, overgrazing, deforestation, and high population density. It is estimated that about 35.7% of the total land mass is allocated for agricultural use. Severe organic matter depletion is driven by the complete removal of crop residues from fields as livestock feed or burning for household energy. Thus, most cultivated soils are poor in organic matter content due to low amounts of organic materials applied to the soils and complete removal of the biomass from the fields. There should be alternative mechanisms for energy sources to reduce residue removal for feed and household fuel. The existence of diverse environments resulted in 21 soil types, of which 9 dominant soil types cover 90% of the land surface. Indigenous knowledge and practices of soil classification and management transferred from generation to generation with little or no modifications. Currently, the World Reference

Base for Soil Resources (WRB) is widely practiced in the country. The untapped indigenous knowledge and experiences should be documented and related with scientific classification systems.

The country has several major rivers and lakes, and significant groundwater resources including 12 major river basins and 22 lakes and large areas of wetlands. These should be designed in using for irrigating crops to increase production and partially fulfill the food demand of the increasing population. Additionally, these resources could also be used to increase the energy supply and decrease dependence on the use of biomass for fuel, which results in soil organic matter depletion and land degradation. The national energy balance of Ethiopia has been predominated by two energy resources—hydropower and biomass use. The latter had been the source of 80% of its energy needs and is a major factor in the depletion of the country's biomass resources. The Grand Ethiopian Renaissance Dam (GERD) which is expected to generate about 15,720 GW hr per year is expected to increase the electricity supply to the population and power export potential of the country.

The high population pressure, continuous and steep slope cultivation, low vegetation cover, deforestation, and inadequate soil conservation practices cause annual soil loss of about 1.5 billion metric tons in the highlands. The ever-increasing land use change and the rugged topography in different parts of the country are aggravating the rates of soil erosion, soil fertility reduction, crop yield decline, and food insecurity. Significant amounts of fertile topsoil are lost due to erosion leading to severe depletion of nutrients and organic matter in upstream areas and over-deposition downstream. Agricultural development in the highlands is important both for ensuring the incomes of populations in these regions as well as the availability of food in food-deficit regions. Priority should be given to the introduction of improved technologies, soil and water conservation practices, and management of the highlands. The soil conservation practices through mass mobilization and the Green Legacy should be accelerated to save the agricultural lands in these regions. The soil acidity problem is very critical in the highlands, where the soils are either strongly or moderately acidic. The use of agricultural lime, vermicompost, biochar, and other organic inputs could improve the physical, chemical, and biological properties of soils and increase crop production on acid soils by raising the pH, Ca and Mg concentrations, and P availability and thereby improving nutrient uptake by plants. Hence, integrated soil fertility management is one of the options to combat the negative impacts of acidity and increase crop production on these soils.

About 10% of the total land area is affected by salinity and/or sodicity which is partly due to the massive expansion of irrigated agriculture by saline water. Following the poor practice of irrigation management, salinity has emerged as a major problem responsible for the reduction of land productivity and natural resources degradation. Management practices that reduce the level of salts including the use of gypsum and phosphogypsum together with leaching with quality irrigation water should be practiced. The current rate of urbanization, industrialization, and the extent of agro-chemical use are raising soil pollution which will be a major concern in the future soil issues. This calls for awareness creation on waste management to save our soils.

Land is a common property of the nations and is not subjected to sale or transfer to other means of exchange. Huge efforts and investments made to prepare digital maps of soil fertility parameters from data collected through intensive soil sampling campaigns in cultivated and arable lands were commendable. Countrywide and regional land suitability evaluations were conducted for the preparation of land use plans in various geographic regions. These should be implemented for proper use of the land based on its suitability.

References

Abdena D, Negassa W, Tilahun G (2007) Inventory of soil acidity status in crop lands of central and western Ethiopia. Utilization of diversity in land use systems: sustainable and organic approaches to meet human needs. Tropentag, October 9–11, 2007, Wizenhausen, Germany

Abebe F, Alamirew T, Abegaz F (2015) Appraisal and mapping of soil salinity problem in Amibara irrigation Farms, middle awash basin, Ethiopia. Int J Innov Sci Res 13:298–314

Abebe S (2005) Land use and land cover change in the headstream of Abay Watershed, Blue Nile Basin, Ethiopia. Masters Dissertation, AAU

Abule E, Snyman H, Smit G (2005) Comparisons of pastoralists perceptions about utilization in the Middle Awash Valley of Ethiopia. J Environ Manage 75

Adimassu Z, Langan S, Johnston R, Mekuria W, Amede T (2017) Impacts of soil and water conservation practices on crop yield, run-off, soil loss and nutrient loss in Ethiopia: review and synthesis. Environ Manage 59(1):87–101. https://doi.org/10.1007/s00267-016-0776-1

ADLI, Agriculture Development Led Industrialization (1992) Transitional Government of Ethiopia, Addis Ababa

AGP (2011) Agricultural growth program (AGP) of Ethiopia: Baseline report, Addis Ababa

Alehegne D (2014) Land administration to nurture development (Land): importance of land use policy in Ethiopia. Tetra Tech for the United States Agency for International Development, USAID Contract No. AID-OAA-I-12–00032/AID-663-TO-13–00005, under the Strengthening Tenure and Resource Rights (STARR) Indefinite Quantity Contract (IQC), Addis Ababa, p 29

Ali A, Tamene L, Erkossa T (2020) Identifying, cataloguing, and mapping soil and agronomic data in Ethiopia. CIAT Publication No. 506. International Center for Tropical Agriculture (CIAT). Addis Ababa, Ethiopia, p 42

Assen M (2003) Land suitability evaluation in the Jelo catchment, Chercher highlands, Ethiopia. Thesis (Ph.D.). The University of the Free State
Bekele M (2011) Forest plantations and woodlots in Ethiopia. African Forest Forum, Nairobi
Tegene B (1982) A study of a soil toposequence in northwestern Alem Gena woreda. Thesis, Addis Ababa University, Addis Ababa, MSc
Tegene B (1996) Characteristics and landscape relationships of vertisols and VerticLuvisols of Melbe, Tigray, Ethiopia. SINET Ethiop J Sci 19(1):93–115
Tegene B (1997) Variabilities of soil catena on degraded hill slopes of wtiya catchment, Welo, Ethiopia. SINET J Sc 20(2):151–175
Tegene B (1998) Pedogenesis and soil-geomorphic relationships on the Piedmont slopes of Wurgo Valley, South Welo, Ethiopia. SINET Ethiop J Sci 21(1):91–111
Berhanu D (1982) Detailed soil survey and land suitability Evaluation of Holeta Agricultural Research Station. Soil survey and evaluation Report No.2, Institute of Agricultural Research, Addis Ababa
Biazin B, Sterk G (2013) Drought vulnerability drives land-use and land cover changes in the Rift Valley dry lands of Ethiopia. Agr Ecosyst Environ 164:100–113
Bono R, Seiler W (1984) The soils of AnditTid research unit (Ethiopia): classification, morphology and ecology. RR 2, Soil Conservation Research Project (SCRP), Addis Ababa
Chamberlin J, Schmidt E (2011) Ethiopian agriculture: a dynamic geographic perspective. ESSP II Working Paper No. 017. IFPRI, USA
Daniel G (1977) Aspects of climate and water budget in Ethiopia. Addis Ababa University Press, Addis Ababa
Daniel G (1988) Environment and development in Ethiopia. In: Penrose A (ed) An examination of the issues behind the famine in Ethiopia. International Institute for Relief and Development, food for the hunger international, Geneva
Debele B (1980) Survey of studies conducted about soil resources appraisal and evaluation for rural development in Ethiopia. Institute of Agricultural Research, Addis Ababa
D'Hoore JL (1964) Soil map of Africa. Sheet 3. Pub. No. 93, Joint Project No. 11, Commission for Technical Cooperation in Africa. Lagos, Nigeria
EIP (2002) Environmental Impact Assessment Proclamation No. 299/2002, Federal Negarit Gazette, 9thYear No. 11, Addis Ababa
Elias E (2016) Soils of the ethiopian highlands: geomorphology and properties. CASCAPE Project, ALTERRA, Wageningen UR, the Netherlands. (library.wur.nl/WebQuery/isric/2259099)
EMA, Ethiopian Mapping Authority (1988) National atlas of Ethiopia. Berhanenaselam printing press, Addis Ababa
EPA, Environment Protection Authority (1997) Environmental policy of Ethiopia. The Federal Democratic Republic of Ethiopia, Addis Ababa
Esayas A (2001) Application of Geographic Information System (GIS) for soil resource study in Ethiopia. In: Proceedings of the national sensitization workshop on agro metrology and GIS, 17–18 December 2001, Addis Ababa, Ethiopia, pp 162–169. Esayas (2001) Application of Geographic Information System (GIS) for soil resource study in Ethiopia
Esayas A (2005) Characteristics, genesis and classification of ressish soils from sidamo region of Ethiopia. PhD Thesis, University of Putra Malaysia
Esayas A, Debele B (2006) Soil survey in Ethiopia: the past, present and future. In: Proceedings of the seventh conference of the Ethiopian Society of Soil Science on soils for sustainable development, 27–28, April 2006, Addis Ababa, Ethiopia, pp 61–79
EthioSIS, Ethiopian Soil Information System (2014) Soil fertility status and fertilizer recommendation Atlas for Tigray Regional State. Agricultural Transformation Agency (ATA). Addis Ababa
Eylachew Z (1987) Study on phosphorus status of different soil types of Chercher highlands, Southeastern Ethiopia. PhD. Thesis, Justus-Liebig Universitat, Giesen
FAO-UNESCO (1974) FAO-Unesco Soil Map of the World: vol 1, Legend. Unesco, Paris
FAO (1988) Soil map of the world. Revised legend, by FAO–UNESCO–ISRIC
FAO (2014) Land use planning guidelines-Ethiopia. FAO/MoA Technical Cooperation Project TCP/ETH/3402. Technical Document 1. Rural Land Administration and Utilization Directorate of the Ministry of Agriculture. Addis Ababa
FAO (2010) State of the World's Forests 2010. Rome
FAO (1998) The Digital Soil and Terrain (SOTER) Database and Maps of North-East Africa. Land and water digital media series 2. Food and Agriculture organization of the United Nations. FAO, Rome
FAO (1984a) Assistance to land use-planning, Ethiopia: geomorphology and soils. Field Document AG DP/ETH/78/003. The United Nations Development Programme and Food and Agriculture Organization. FAO, Rome
FAO (1984b) Assistance to land use-planning, Ethiopia: provisional soil association map of Ethiopia. Field document No 6. The United Nations Development Program and Food and Agriculture Organization. FAO, Rome
FDRE, Federal Democratic Republic of Ethiopia Constitution (2005) Rural land administration and land use Proclamation No. 456/2005, Federal Negarit Gazeta of FDRE, 11th year No. 44, 15 July 2005
Fikre M (2003) Pedogenesis of major volcanic soils of the southcentral rift valley region, Ethiopia. MSc Thesis, University of Saskatchewan. Saskatoon, Canada, pp 128
Flintan F (2011) Broken lands: broken lives? Causes, processes and impacts of land fragmentation in the rangelands of Ethiopia, Kenya and Uganda. A report for REGLAP, Kenya. Internet. http://landportal.info/resource/kenya/broken-lands-broken-lives-causes-processes-and-impacts-land-fragementation-rangelands
Frei E (1978) Andepts in some high mountains of East Africa. Geoderma 21:119–131
Garedew E, Sandewall M, Soderberg U, Campbell BM (2009) Land-use and land-cover dynamics in the Central Rift Valley of Ethiopia. Environ Manage 44:683–694
Gebeyehu Z, Woldegiorgis SB, Belete AD, Abza TG, Desta BT (2017) Ethiopia's move to a national integrated land use policy and land use plan. Paper prepared for presentation at the 2017 World Bank Conference on Land and Poverty. The World Bank, Washington DC, March 20–24
Gelaw AM, Singh BR, Lal R (2014) Soil organic carbon and total nitrogen stocks under different land uses in a semi-arid watershed in Tigray, Northern Ethiopia. Agr Ecosyst Environ 188:256–263. https://doi.org/10.1016/j.agee.2014.02.035
Gessesse B (2013) Characterization and modelling of landscape transformation for optimizing agricultural land-use in the Ethiopian highlands: a case study of Modjo Watershed. PhD Dissertation, Addis Ababa
Gete Z, Tibebu K, Woldie A, Venusia G, Heinimann A, Hurni K, Hurni H, Berhanu D (2014) Land administration to nurture development (Land), Land use and land cover change trends in Ethiopia and policy imperatives. Tetra Tech for the United States Agency for International Development, USAID Contract No. AID-OAA-I-12–00032/AID-663-TO-13–00005, under the Strengthening Tenure and Resource Rights (STARR) Indefinite Quantity Contract (IQC), Addis Ababa, p 20

Gete Z, Getachew A, Dejene A, Shahidur R (2010) Fertilizer and soil fertility potential in Ethiopia: constraints and opportunities for enhancing the system. IFPRI Working paper. 42

Girma T (2001) Land degradation: a challenge to Ethiopia. International Livestock Research Institute, Addis Ababa, Ethiopia, pp 815–823

Guta D (2012) Assessment of biomass fuel resource potential and utilization in Ethiopia: sourcing strategies for renewable energies. Int J Renew Energy Res 2(1)

Habtamu AA (2013) Impact of salinity on tolerance, vigor, and seedling relative water content of haricot bean (Phaseolus *vulgaris* L.) cultivars. J Plant Scie 1(3):22–27

Haileslassie A, Priess J, Veldkamp E, Teketay D, Lesschen JP (2005) Assessment of soil nutrient depletion and its spatial variability on smallholders' mixed farming systems in Ethiopia using partial versus full nutrient balances. Agr Ecosyst Environ 108(1):1–16. https://doi.org/10.1016/j.agee.2004.12.010

Heluf GK (1985) Investigation on salt affected soils and irrigation water quality in Melka Sedi-Amibara Plain, Rift Valley Zone of Ethiopia. MSc Thesis, School of Graduate Studies, Addis Ababa University. Addis Ababa, Ethiopia, p 131. https://www.statista.com. Accessed 25 February 22

Hurni H, Amare B, Herweg K, Portner B, Veit H (2007) Landscape Transformation and Sustainable Development in Ethiopia. Background Information for the Study Tour through Ethiopia, 04–20, September 2006Hurni H, Amare B, Herweg K, Portner B, Veit H (2007) Landscape transformation and sustainable development in Ethiopia. Background information for the study tour through Ethiopia, 04–20 September 2006. Center for Development and Environment, University of Bern, Institute of Geography, Bern, Switzerland

Hurni H (1993) Land degradation, famine, and land resource scenarios in Ethiopia. In: Pimental D (ed) World soil erosion and conservation. Cambridge, pp 27–61

Hurni H (1983) Soil erosion and soil formation in agricultural ecosystems: Ethiopia and Northern Thailand. Mt Res Dev 3 (2):131–142

IBC, Institute of Biodiversity Conservation (2014) Ecosystem approach. Internet. http://et.chm-cbd.net/biodiversity/ecosystem-approach

Joy O (2013) Earth, wind and water: Ethiopia bids to be Africa's powerhouse. Internet. http://edition.cnn.com/2013/11/08/business/earth-wind-water-ethiopia/

Kassa T, Van Rompaey A, Poesen J (2010) Land suitability assessment for different irrigation methods in Korir Watershed, Northern Ethiopia. J Drylands 3(2):214–219

Kefeni K (1995) The soils of the Anjeni Area-Gojam Research Unit, Ethiopia. RR 27, Soil conservation Research Project (SCRP), Addis Ababa

Kidane G, Abebe F, Heluf GK, Fentaw A, Wondimadegne C, Hibstu A, Asegid A, Messele F, Mohammed B (2006) Report of the national task force on assessment of salt affected soils and recommendations on management options for sustainable utilization; Ethiopia, AA, ed. Minister of Agriculture and Rural Development, Addis Ababa, Ethiopia, p 120

Lal R (2004) Soil carbon sequestration impacts on global climate change and food security. Soils-The final Frontier. www.sciencemag.org. Science 304:1623–1627

Leenaars JGB, Elias E, Wösten JHM, Ruiperez González M, Kempen B (2020a) Mapping the major soillandscape resources of the Ethiopian Highlands using random forest. Geoderma 361. https://doi.org/10.1016/j.geoderma.2019.114067

Leenaars JGB, Ruiperez González M, Kempen B, Mantel S (2020b) Semi-detailed soil resource survey and mapping of REALISE woredas in Ethiopia. Project report to the BENEFIT-REALISE program. December 2020b, ISRIC-World Soil Information, Wageningen

Mathewos M (2020) Effects of lime and vermicompost on selected physicochemical properties of acid soils and growth and yield of barley (*hordeum vulgare* l.) in Gedeo zone, southern Ethiopia. MSc Thesis, School of Graduate Studies, Hawassa University, p 127

Mesfin A (2007) Nature and management of acid soils in Ethiopia. Haramaya University, Haramaya, Ethiopia, p 82p

Mesfin A (1998) Nature and management of Ethiopian soils. International Livestock Research Institute (ILRI), p 272

Ministry of Mines and Energy (2009) National report on mining. A report to the UN Commission on Sustainable Development (UNCSD), New York. Ministry of Mines and Energy, FDRE, Addis Ababa

Mitiku H (1987) Genesis, characteristic and classification of the Central Highland soils of Ethiopia. PhD Thesis, State University of Ghent, Belgium

MoA, Ministry of Agriculture (1987) A simple system for defining agroecological zones in Ethiopia, Final Report. MOA, Addis Ababa

MoANR, Ministry of Agriculture and Natural Resource (2015) Assessment of fertilizer consumption and use by crop in Ethiopia. IFRC, Addis Ababa

MoARD, Ministry of Agriculture and Rural Development (2010) Ethiopian strategic investment framework for sustainable land management. The Federal Democratic Republic of Ethiopia, Addis Ababa

MoARD, Ministry of Agriculture and Rural Development (2007) Forest development. Federal Democratic Republic of Ethiopia, Addis Ababa, Conservation and Utilization Policy and Strategy

MoARD, Ministry of Agriculture and Rural Development (2005) Wildlife development. The Federal Democratic Republic of Ethiopia, Addis Ababa, Protection and Utilization Policy and Strategy

MoCT, Ministry of Culture and Tourism (2009) Tourism development policy. The Federal Democratic Republic of Ethiopia, Addis Ababa

MoFED, Ministry of Finance and Economic Development (2010) Growth and transformation plan (GTP) 2010/11–2014/15 Government of Federal Democratic Republic Ethiopia, Addis Ababa

MoFED, Ministry of Finance and Economic Development (2006) Plan for accelerated and sustained development to end poverty (PASDEP) 2005/06–2009/10. Government of Federal Democratic Republic Ethiopia, Addis Ababa

MoFED, Ministry of Finance and Economic Development (2003) Rural development policy and strategies. MoFED Policy and Planning Department, Addis Ababa

MoFED, Ministry of Finance and Economic Development (2002) Sustainable development and poverty reduction program (SDPRP), 2002/03–2004/05. Government of the Federal Democratic Republic Ethiopia, Addis Ababa

Mohammed A, Belay T (2008) Characteristics and classification of the soils of the Plateau of Simen Mountains National Park (SMNP), Ethiopia. SINET Ethiopian J Sci 31(2):89–102

MoI, Ministry of Information (2002) Agriculture development policies, strategies, and tactics. Government of the Federal Democratic Republic Ethiopia, Addis Ababa

MoWR, Ministry of Water Resources (2010) Rift Valley Lakes basin integrated resources development master plan study. Phase 3. Final Report. Western Lake Ziway Sub-basin Integrated Watershed Management Feasibility Study. Volume 1: Main Report

MoWR, Ministry of Water Resources (2007) Wabeshebele Gennale-Laga Wata/ Sure, Rivers basin Integrated development master plan study. Addis Ababa

MoWR, Ministry of Water Resources (2006) Genale Dawa River Basin integrated resource development master plan study project. Master Plan Report: Land Evaluation/Land Suitability, Addis Ababa

MoWR, Ministry of Water Resources (2001) Ethiopian water sector policy & strategy. The Federal Democratic Republic of Ethiopia, Addis Ababa

MoWR, Ministry of Water Resources (1998) Abay River Basin Integrated Development Master Plan Project Phase 2: Land Resources Development, Addis Ababa

MoWR, Ministry of Water Resources (1996) Integrated development of Omo-Gibe River Basin Master Plan Study, vol XI F1, F2, F3, Addis Ababa, Ethiopia

Muluneh A (2010) Synthesis of research on land use and land cover dynamics in the Ethiopian highlands. UNU-Land Restoration Training Programme, Reykjavik, Iceland

Murphy HF (1968) A report on the fertility status and other data on some soils of Ethiopia, Experiment station bulletin No. 44. College of Agriculture, Haile Selassie I University press, Dire-Dawa

Paris S (1985) Soil survey of the Borkena area (Wello). AG/DP/ETH/82/010, Field Document 7, MOA/LUPRD, FAO/UNDP, Addis Ababa

Qureshi AS, Tesfaye, EM, Melese MG (2019) Extent characterization and Management strategies for the salt affected soils in Ethiopia project report 2 submitted to International Fund for Agricultural Development (IFAD)

Redda T (2003) A literature-based study into the expected pollution of the Awash River, Ethiopia (TOX-80316), preliminary toxic risk evaluation of the Awash River (TX-80328) (Unpublished Master's thesis). Wageningen University, The Netherlands

Shantz HL, Marbut CF (1923) The vegetation and soil of Africa. Amer Geog Soc Res. Series 13 Newyork

Sheleme B (2017) Topographic positions and land use impacted soil properties along Humbo Larena-Ofa Sere toposequence, Southern Ethiopia. J Soil Sci Environ Manage 8(8). https://doi.org/10.5897/JSSEM2017.0643

Sheleme B (2016) Effects of lime and phosphorus applications on soil chemical composition, growth and nutrient uptake of maize (*Zea mays* L.) in acid soils of Gununo, southern Ethiopia. J Sci Dev 4 (1):1–9

Shiferaw A (2011) (2011) Evaluating the land use and land cover dynamics in Borena Woreda of South Wollo Highlands, Ethiopia. J Sustain Dev Afr 13:87–107

Sileshi A (2016) Status of salt affected soils, irrigation water quality and land suitability of Dubti/Tendaho Area, North Eastern Ethiopia. PhD Dissertation, Haramaya University. Haramaya, Ethiopia

Tefera T, Tadiwos C, Workineh H (1996) Geology of Ethiopia. (Explanatory note of geological map of Ethiopia 1:2,000,000). Second Edition, Federal Democratic Republic of Ethiopia, Ministry of Mines

Tilahun A (2019) Managing acid soils for reclaiming livelihoods in Ethiopia. ICRISAT. http://oar.icrisat.org/11409/1/Managing%20acid%20soils_Ethiopia_02%20May%202019_low.pdf

Tsegaye D, Moe S, Vedeld P, Aynekulu E (2010) Land use/cover dynamics in Northern Afar rangelands, Ethiopia. Agr Ecosyst Environ 139:174–180

Veneema JH, Paris S (1986) Soils of Bichena area (Gojam): Assistance to land use planning Ethiopia. AG: DP/ETH/82/010, field Document 10, LUPRD/MOA, FAO/UNDP, Addis Ababa

Virgo KJ, Holmes DA (1977) Soil and landform features of mountainous terrain in South Thailand. Geoderma 18:207–225

Virgo KJ, Munro RN (1978) Soil and erosion features of the Central Plateau region of Tigray, Ethiopia. Geoderma 20(2):131–157

Wang J, Raman H, Zhang G, Mendham N, Zou M (2006) Aluminium tolerance in barely (*Hordeum vulgare* L.): physiological mechanisms, genetics and screening methods. J Zhejiang Univ Sci 7:769–787

Wassie H, Shiferaw B (2009) Mitigation of soil acidity and fertility decline: challenges for sustainable livelihood improvement: evidence from the southern region of Ethiopia. In: Sustainable land management and poverty alleviation, pp 131–143. Ethiopian Development Research Institute, Addis Ababa

Weigel G (1988) Some soils of the lower area of the Simen Mountains National Park. RR10, Soil Conservation Research Project (SCRP), Addis Ababa

WLRC, Water and Land Resource Centre (2018) Mapping Soil and Water Conservation (SWC) Coverage of 2016, Modelling SWC Requirements and Composite Land Management Practices of Ethiopia and SLMP sites. Addis Ababa

World Bank (2013) Data agricultural land (percent of land area). http://data.worldbank.org/indicator/AG.LND.AGRI.ZS. World Soil Resources Report No. 60. Rome

Yihenew GS (2002) Selected chemical and physical charateristics of soils of adet research center and its testing sites in northwestern Ethiopia. Ethiopian J Nat Resour 4:199–215

Yizengaw T, Verheye W (1995) Application of computer captured knowledge in land evaluation, using ALES in Central Ethiopia. Geoderma 66(3–4):297–311

Zewdie E (2013) Properties of major agricultural soils of Ethiopia. Lambert Academic Publishing, Germany

2 History of Soil Education and Research

Sheleme Beyene and Alemayehu Regassa

Abstract

Agricultural education and research in Ethiopia began at Ambo and Jimma Institutes of Agriculture; and the Jimma Institute was later extended to College of Agriculture at Alemaya in 1957. The first university-level agricultural training program, with a 4-year curriculum leading to a Bachelor of Science degree in General Agriculture, was started in September 1953. The MSc program in Soil Science started at Haramaya University in 1998 and, currently, 16 universities in the country offer training postgraduate at MSc and PhD levels in soil science. The first small-scale agricultural experiments were initiated in 1952 by the government with the help of foreign experts. In 1956, an agricultural experiment station was established at Debrezeit. Well-organized agricultural research began with the establishment of the Institute of Agricultural Research (IAR) in 1966 as a semi-autonomous institute with financial support from UNDP and FAO. In 1997, the IAR was restructured by establishing EARO, which was later rechristened to the present-day Ethiopian Institute of Agricultural Research (EIAR). The National Soil Laboratory was established in 1967 with the assistance of UNDP/FAO. The laboratory was originally mandated to assist farmers with fertilizer application. Soil conservation research in Ethiopia began with the initiation of the Soil Conservation Research Project (SCRP) in 1981 within the framework of the then Soil and Water Conservation Department (SWCD) of the Ministry of Agriculture. The first national digital soil mapping was undertaken in 2012 by the Ethiopian Soil Information System (EthioSIS).

S. Beyene (✉)
School of Plant and Horticultural Sciences, Hawassa University, P.O. Box 5 Hawassa, Ethiopia
e-mail: shelemeb@gmail.com

A. Regassa
Department of Natural Resource Management, Jimma University, P.O. Box 307 Jimma, Ethiopia

Keywords

Agriculture education • Institute of agricultural research • Experimental station • Graduate programs • Soil conservation • Soil laboratory

2.1 Introduction

Agricultural education and extension work in Ethiopia began in 1931 with the establishment of the Ambo Agricultural School which is one of the oldest institutions and the first agricultural high school offering general education with major emphasis on agriculture. The school offered training to students and demonstrated the potential effects of improved varieties and agricultural practices to the surrounding farmers. Limited extension activities started after the creation of the Ministry of Agriculture in 1943 (Belay 2008). The services rendered were more regulatory in nature and included providing advice in soil conservation through the grow-more-trees campaign, better variety of seeds and seedlings, cleaning and seed selection, protection of game fish, preservation of hides and skins, etc. (Haileselassie 1959).

The real agricultural extension and research work began in the early 1950s following the establishment of the Imperial Ethiopian College of Agriculture and Mechanical Arts (IECAMA) at Alemaya (now Hramaya University) with the assistance of the United States under the Point Four Program (Belay 2000). The academic program of the Alemaya College of Agriculture was instituted on the Land Grant College system with three fundamental but related responsibilities which are training of high-level manpower, promotion of agricultural research, and dissemination of appropriate technologies.

The government established 25 college-level agriculture-oriented Technical and Vocational Education Training (ATVET) diploma programs with the capacity to

S. Beyene et al. (eds.), *The Soils of Ethiopia*, World Soils Book Series,
https://doi.org/10.1007/978-3-031-17012-6_2

train roughly 30,000 students each year for the agricultural extension program to overcome the shortage of extension personnel (Byerlee et al. 2007). However, the ATVET training programs are different from those offered by universities, and they are of recent origin with very limited research and outreach programs (Belay 2008).

2.2 Soil Science Education

The training programs in agriculture and related fields, after successful completion of secondary school include Diploma, BSc/DVM, MSc/MVSc, and PhD programs. Diploma program (2 years of training) is destined to produce essentially middle-level technicians in a variety of subjects, the BSc/DVM undergraduate degree program leads to a first degree after 3–5 years of study, and MSc/MVSc and PhD programs are graduate programs leading to Master's and Doctor's degree, respectively. The higher level training in agriculture including soil science started at Jimma Agricultural and Technical School and transferred to the Imperial College of Agriculture and Mechanical Arts (IECAMA) which was established at Alemaya (Belay 2008). This university-level agricultural education began in the early 1950s, following the "Agreement for a Co-operative Agricultural Education Program between The Imperial Ethiopian Government and the Government of the United States of America" which was signed in Addis Ababa in 1952. The agreement laid down the foundations for the establishment of Jimma Agricultural and Technical School (JATS) and the Imperial Ethiopian College of Agriculture and Mechanical Arts (IECAMA) known as "Alemaya College of Agriculture", now Haramaya University.

Another agreement was also signed between the Technical Co-operation Administration of the United States Department of State and the Oklahoma Agricultural and Mechanical College (now Oklahoma State University) in 1952 which gave the latter the mandate to establish and operate the college, establish and operate a nationwide system of agricultural extension, set up agricultural research and experiment stations, and furnish technicians and administrative staff to start the college. The college was established at Alemaya (525 km east of Addis Ababa) and its academic program was modeled on the land-grant college system with three fundamental but related responsibilities: training of highly skilled workers, promotion of agricultural research, and dissemination of appropriate technologies (Belay 2000).

The first university-level agricultural training program, with a 4-year curriculum leading to a Bachelor of Science degree in General Agriculture, was started in 1953 at the Jimma Agricultural and Technical School (JATS). The JATS was to serve as an interim site where students complete their first- and second-year course requirements. The IECAMA started enrollment of the first batch of students in 1956. The original curriculum of the college was to produce graduates with a BSc degree in General Agriculture. However, following the felt needs of the country, additional programs of study at bachelor's, master's, and doctoral degree levels were introduced. Since the second half of the 1970s, the following junior colleges of agriculture and other agriculture-related institutions were set up.

2.2.1 The Start of Agricultural Education and Its Expansion

Ambo and Jimma Colleges of Agriculture: The Ambo Agricultural Institute was established in 1931. Ambo and Jimma Institutes of Agriculture were primarily intended for the training of Agricultural Technicians. The Institutes were secondary schools enrolling students who had completed grade 8 and providing them 4 years of general education with major emphasis on agriculture until 1966. In 1967, these schools became institutes of agriculture giving 2 years' diploma training in General Agriculture and were put under the Ministry of Agriculture. These institutes were upgraded to the College of Agriculture level in 1978. The Jimma College of Agriculture was renamed by the university as the Jimma College of Agriculture and Veterinary Medicine in 2002, while Ambo College of Agriculture was upgraded to university level in 2002.

Debre Zeit Junior College of Agriculture: This college was founded at the Debre Zeit (now Bishoftu) Agricultural Experiment Station, which was established in 1953 to serve as the first experimental station of the IECAMA. The station was later developed as an autonomous agricultural experiment station under the auspices of Addis Ababa University (AAU). In September 1977, a 2-year diploma-granting institution was attached to the station and the whole unit was renamed Debre Zeit Junior College of Agriculture and Agricultural Research Centre. The junior college program of the center was discontinued in 1984 and transferred temporarily to Alemaya. The program was phased out at the end of the 1987/1988 academic year (Belay 2008).

Awassa College of Agriculture: The college was founded in July 1976 in Awassa (now Hawassa) 275 km south of Addis Ababa. It was formerly under the administration of AAU and was reorganized as an independent institution in 1994 under the administration of the Ministry of Education. The college started BSc programs in 1988 and following the establishment of Debub University in 2000 (now Hawassa University), the college became the nucleus and one of the constituent parts of the university. The college now runs different programs in agriculture and related fields at BSc, MSc, and PhD degree levels.

Wondo Genet College of Forestry: The college is located 264 km south of Addis Ababa and was opened in early 1978. The college used to offer only a 2-year diploma program in Forestry till the second half of the 1990s. At the end of the 1997/1998 academic year, the 4-year BSc training program in Forestry was transferred from the then Alemaya University (now Haramaya University) to the Wondo Genet College of Forestry. The college was initially administered by the Ministry of Agriculture and then by the Higher Education Main Department of the Ministry of Education. Following the establishment of Debub University (now Hawassa University) in 2000, the college became one of the constituent parts of the university. The college now runs different programs in Forestry and Natural Resource Management at BSc, MSc, and PhD degree levels.

Faculty of Dry Land Agriculture and Natural Resources: The Faculty of Dry Land Agriculture and Natural Resources started its training program in 1993 as part of a newly opened college, namely the Mekele University College, which is located in Tigray (783 km north of Addis Ababa). This faculty trains students in Dry Land Agriculture and was under the Higher Education Main Department of the Ministry of Education till 2000 when it became a constituent part of Mekele University. The college started MSc and PhD training programs in the 2003 and 2011 academic years, respectively.

2.2.2 The Launch of Graduate Programs in Soil Science

Until the middle of the 1970s, university-level education in agriculture and related fields was offered in Alemaya College of Agriculture, and Ambo and Jimma Institutes of Agriculture. The graduate programs in areas of Agriculture were started in Alemaya College of Agriculture under Addis Ababa University in 1979. The Alemaya College was raised to an autonomous university status and named the Alemaya University of Agriculture (now Haramaya University) in the late 1980s and the soil science program was first launched in 1998. Following this, Hawassa University's College of Agriculture and Wondo Genet College of Forestry and Natural Resources; Mekele University's Faculty of Dryland Agriculture and Natural Resources; Bahir Dar University's College of Agriculture and Environmental Sciences; and Jimma University's College of Agriculture and Veterinary Medicine started to offer trainings in the areas of soil science and related fields at graduate level. Currently, about 16 universities have graduate programs in the area of soil science and related fields (Table 2.1).

2.3 Establishment of Agricultural Research Systems

The agricultural research system in Ethiopia has been evolved over time. Rudimentary form of agricultural research activities is traced back to the early 1930s, though agricultural research took roots with the establishment of the Ambo and Jimma Colleges of Agriculture in 1947 and the Imperial College of Agriculture and Mechanical Arts (now Haramaya University) in 1953 with triple functions of education, research, and extension by the USA Land-Grant University model. In 1956, an agricultural experiment station was established at Debre Zeit (now Bishoftu), 50 km southeast of Addis Ababa. This was later merged with the then College of Agriculture at Alemaya a year later (Tsedeke et al. 2004) as a semi-autonomous institute with financial support from UNDP and FAO (Beintema and Menelik 2003). Other agricultural research centers that were established at various times include the Plant Protection Research Center at Ambo —established in 1972 and merged with the Institute of Agricultural Research in 1995; Plant Genetic Resources Center—founded in 1974 and later became the Biodiversity Institute; Forestry Research Center (1975); Wood Utilization Research Center (1979); National Soils Laboratory (1989); and Institute of Animal Health Research (IAHR 1992).

2.3.1 Institute of Agricultural Research and Research Centers

The Institute of Agricultural Research (IAR) was established in 1966 as a public semi-autonomous organization with financial support from UNDP and FAO (Beintema and Menelik 2003). It was established with a national mandate of developing a national policy for agricultural research, conducting research activities on all disciplines of agriculture at research centers established in all important agricultural zones of the country, and coordinating agricultural research at the national level. In addition to IAR, there were also a number of other national research centers established during the 1970s such as the Plant Protection Research Center, the Plant Genetic Resources Center, the Forestry Research Center, the Wood Utilization Research Center, the National Soils Laboratory, and the Institute of Animal Health Research (Cohen 1987).

Table 2.1 The graduate programs at different universities in Ethiopia

University	MSc program and starting year	PhD program and starting year
Haramaya	Soil Science, 1998 Climate Smart Agriculture and Biodiversity Management (Soil and Water), 2017	Soil Science, 2002
Hawassa	Soil Science, 2004 Soil Management & Agroforestry, 2008 Watershed Management, 2008	Soil Science, 2010
Bahir Dar	Land Resource management, 2009 Soil Science, 2017 Soil and Water Conservation, 2015	Soil Science, 2015
Mekelle	Soil Science, 2003	Soil Science 2011
Jimma	Soil Science, 2012	
Gondar	Soil Science, 2013	
Debre Berhan	Soil and Water Conservation, 2016	
Ambo	Soil Science, 2017	
Wollega	Soil Science & Agricultural Chemistry, 2017	Soil Science, 2020
Metu	Soil Science, 2018	
Wolkitie	Soil Science, 2018	
Arba Minch	Soil and Water Management, 2019	
Oda Bultum	Agroforestry and Soil Management, 2019	
Woldiya	Soil Science, 2019	
Bule Hora	Soil Science, 2020	
Wachemo	Soil Science, 2021	

Source Personal communications

The IAR had two types of research approaches: commodity and zonal research approaches. The commodity research approach was designed to address selected strategic crops at a national level, while the zonal approach was intended to address production constraints specific to the particular agricultural development zones. In general, the national agricultural research system largely followed a team approach composed of breeders, agronomists, plant pathologists, entomologists, agricultural economists, and soil scientists.

In 1993, some IAR centers were decentralized to create independent research centers run by the respective regional governments and became the Regional Agricultural Research Centers (RARCs) under their respective regional bureaus of agriculture. Accordingly, seven of the nine regional states of the country, namely the Afar, the Amhara, the Gambella, the Oromia, the Somali, the Southern Nations Nationalities and People, and the Tigray regions have established their respective Regional Agricultural Research Institutes (RARIs). The RARCs conduct research that addresses the specific needs of the particular region promoting multidisciplinary researches at the regional level.

In 1997, the IAR was restructured by establishing the Ethiopian Agricultural Research Organization (EARO), which was later rechristened to the present-day Ethiopian Institute of Agricultural Research (EIAR) and merged all the existing agricultural research institutions including the original IAR research centers at Holetta, Nazreth, Jima, Bako, Melka Werer, Ambo, Kulumsa, Pawe, and Debre Zeit Agricultural Research Center (previously under Alemaya University), the Biodiversity Institute, the Forestry Research Center, the Wood Utilization Research Center, the Institute of Animal Health Research, and the National Soils Laboratory (all previously under the Ministry of Agriculture). The newly established research centers through the World Bank Agricultural Research and Training Program (ARTP) include—Jijiga (Somali Region), Shiket (Afar Region), Jinka (Southern Region), Humera (Tigrai Region), Sekota (Amhara Region), and Yabello (Oromia Region) were also included. The mandate of the EIAR as stated in the Federal Negarit Gazeta (1997) underlines its responsibilities for generating, improving, and adapting technologies and coordinating, encouraging, and assisting research activities in order to fulfill the current and long-term agricultural requirements of the country. In 2014, the Ethiopian Agricultural Research Council was established to provide a national coordination role to the country's National Agricultural Research System which can be regarded as a truly Agricultural Research Council model.

2.3.2 Establishment of Field Trials

Field trials provide a meeting ground where research and extension personnel can come together to attack/combat the problems limiting crop production. The first small-scale agricultural experiments were initiated in 1952 by the Ethiopian government with the help of foreign experts. In 1956, an agricultural experiment station was established at Debrezeit (now Bishoftu). A year later, the station was merged with the college of agriculture at Alemaya (Berhanu and Ochtman 1974). Since the Freedom from Hunger Campaign program began in 1967, a large number of trials (fertilizer and others) were carried out in Ethiopia. Sixteen field trial sites were established covering the northwestern, central, northeastern, and southeastern administrative zones during the 1986 growing season. The trials were not replicated and were very simple so as to attain their demonstrative value of introducing the peasant farmers to fertilizers, improved varieties, weed control, and aspects of planting times and seeding rates (Deckers 1986). Widespread on-station and on-farm trials were conducted in the 1970s in different parts of the country with the objective of determining the rate, sources, and methods of fertilizer applications that established the use of Urea (46-0-0) for nitrogen (N) and DAP (di-ammonium phosphate) (18-46-0) for phosphorus (P) as the only artificial fertilizers in the country (NFIU 1993). Various organizations including IAR and the National Fertilizer Trials Program conducted fertilizer trials from 1971 to 1990 at research stations and on-farm. Higher learning institutes also conducted trials on responses of improved crop varieties to nutrient levels (FAO 1997; ADD/NFIU 1991). Based on this fact, few trials were conducted on selected stations from 1975 to 1990, IAR recommended fertilizers based on specific soil and crop types. However, minimal efforts were made with regard to the extrapolation of the results to a wider range of environments.

2.3.3 Soil Conservation Research

Soil conservation research in Ethiopia is an experimental and quantitative approach that began with the initiation of the Soil Conservation Research Project (SCRP) in 1981 within the framework of the then Soil and Water Conservation Department (SWCD) of the Ministry of Agriculture. This project was launched upon the request of the government to guide the massive soil and water conservation programs supported by the World Food Program (WFP) since the 1974 famine. Seven research stations were set up in a wide range of agroecological zones since year 1981 and additional basic data were generated from test plots on erosion processes (Grunder 1986). The Food-For-Work (FFW) project, which took place over a period of 15 years (1980–1994), is the most widely known SWC intervention. The FFW project was modified and become the Local Level Participatory Planning Approach (LLPPA) and later the Managing Environmental Resources to Enable Transitions (MERET) project. Under the MERET project alone, approximately 1 million hectares (Mha) of farmland and 0.3 Mha of hillside were covered with different types of SWC structures, such as farmland and hillside terraces (MERET 2013; Nedassa et al. 2011). Another substantial program is the Productive Safety Net Program (PSNP) which has been implemented continuously since 2005 with an annual budget of USD 500 million (Gilligan et al. 2009). The Sustainable Land Management Program (SLMP) under the MoA has also made large investments in SWC interventions. In addition to projects led by the government with support from international funding agencies, NGOs, such as Sustainable Agricultural Rehabilitation in Tigray (SART), Sustainable Agriculture and Environmental Rehabilitation of Amhara Region (SAERAR), and the Organization for Rehabilitation and Development in Amhara (ORDA) have also made significant investments in SWC.

The SCRP was aimed at supporting soil and water conservation activities in the country. Its specific objectives were to monitor soil erosion damage, soil loss and runoff, catchment sediment loss and runoff; develop viable models of soil loss, runoff, catchment runoff, and sediment and productivity losses for the research areas, and to test their applicability for larger areas; develop ecologically sound, economically viable, and socially acceptable conservation measures and approaches in different research regions; and train project personnel and to have research fellows to improve the country's research and implementation capability in soil conservation. The SCRP catchments and surrounding areas were conserved through mass mobilization with food-for-work and other incentives. The SCRP project was decentralized in 1996 and finally phased out in 1998. The research sites are now managed by regional research systems.

2.3.4 Establishment of National Soil Laboratory

The National Soil Service Laboratory was established in 1967 with the assistance of UNDP/FAO under the department of crop production and protection within the Ministry of agriculture. In 1979, the National Soil Laboratory was structurally organized under the Land Use Planning and Regulatory Department (LUPRD) and was mainly meant to support soil mapping activities of the department. The National Soil Laboratory is also mandated to give soil laboratory services to assist farmers with fertilizer application. Furthermore, it is also responsible to assist Regional Soil Testing Laboratories (RSTL) with regard to the provision of trainings to technicians on soil, water, and plant analysis; soil sampling; interpretation of results and calibration; acid soil management; soil mapping; preparation of purchase specification on instruments and reagents; and general laboratory management.

In 1997, it was renamed "The National Soil Research Center (NSRC)" and became one of the Federal Research Centers of EARO under the Soil and Water Research Directorate where it was mainly engaged in soil and water research activities and to a lesser extent in development-oriented analytical services and soil resource assessment studies. The NSRC also served as the coordinator of the Soil and Plant Analytical Laboratories Network of Ethiopia (SPALNE) which was established in 1990, with 22 member laboratories belonging to different research, higher learning, and development organizations in the country.

The NSRC had both technical and administrative and finance sections. The technical sections include soil survey, characterization and land evaluation, soil fertility and plant nutrient management, soil biology and microbiology, soil chemistry, soil physics, soil mineralogy, and soil database management and training. The laboratory is well equipped with essential analytical instruments for soil chemical, physical, mineralogical, and biological characterization, including plant and water analysis. In addition to the ordinary analytical instruments like spectrophotometers, flame photometers, and atomic spectrophotometers, X-ray diffractometer and differential thermal analyzer for soil mineralogical analysis are also available. Moreover, the soil survey and data management sections have GIS and other field surveying, cartographic, and data management facilities. It has got also a large greenhouse for conducting basic plant nutrition and soil amendment studies.

2.3.5 Recent Advances—Digital Soil Mapping

In digital soil mapping, many initiatives have been undertaken globally to map soil resources. The Ethiopian Soil Information System (EthioSIS), a project launched by the Ethiopian Government's Agricultural Transformation Agency (ATA) in 2012, is the first national institution to undertake a digital soil mapping in Ethiopia. EthioSIS was mandated with conducting soil fertility mapping to reformulate fertilizer recommendations. EthioSIS, in collaboration with different stakeholders, has been pursuing a rapid development program on the assessment of the soil resources of the country to establish a national soil resources database, and assess the nutrient status of agricultural lands to produce soil fertility map of many districts and come up with solid, evidence-based and targeted recommendations for fertilizer applications and other management interventions (ATA 2014). The project has finalized and released the digital soil fertility and fertilizer formulation maps/atlas for almost all regions and has collected approximately 100,000 soil datasets. The digital soil fertility maps are predicted soil property maps developed using Digital Soil Mapping (DSM)-Soil Spatial Prediction Models (SSPM). The EthioSIS predicted digital soil fertility and fertilizer recommendation maps/atlas, in general, do not present the prediction accuracy for each soil fertility parameter and geographic region. Moreover, soil data acquisition/sampling year versus map publication year was not properly indicated in the EthioSIS maps.

2.4 Conclusions

The history of agricultural education and extension work in Ethiopia dates back to the 1930s with the establishment of the Ambo Agricultural School. The higher level training in agriculture including soil science started at Jimma Agricultural and Technical School and later transferred to the Imperial College of Agriculture and Mechanical Arts (IECAMA) which was established at Alemaya. On the other hand, the real agricultural research took roots with the establishment of the Ambo and Jimma Colleges of Agriculture in 1947 and the Imperial College of Agriculture and Mechanical Arts (now Haramaya University) in 1953. Graduate program in soil science started at Alemaya University of Agriculture (now Haramaya University) in 1998. Currently, 16 universities in the country offer postgraduate trainings in Soil Science and related fields.

Well-organized agricultural research began with the establishment of the Institute of Agricultural Research (IAR) in 1966. Widespread on-station and on-farm trials were conducted in the 1970s in different parts of the country to determine the rates, sources, and methods of fertilizer applications. The Institute of Agricultural Research (IAR) has undergone structural evolutions. Accordingly, EARO was established in 1997 and was later rechristened to the present-day Ethiopian Institute of Agricultural Research (EIAR). The initiation of the Soil Conservation Research Project (SCRP) in 1981 within the framework of the then Soil and Water Conservation Department (SWCD) of the Ministry of Agriculture marked the beginning of experimental and quantitative research in soil and water conservation. The first ever digital soil mapping in Ethiopian was undertaken by the Ethiopian Soil Information System (EthioSIS), a project launched by the Ethiopian Government's Agricultural Transformation Agency (ATA) in 2012. The project released the digital soil fertility and fertilizer formulation maps/atlas for almost all regions and reformulate fertilizer recommendations.

Higher learning institutions, research organizations, NGOs, other government sectors, and development partners are making tremendous efforts in researches in different fields of soil sciences for years. However, the results of such efforts are sometimes shelved in different offices, serve only for the partial fulfillment of graduation requirements of MSc and PhD students, and/or reporting to donors. Thus, academia, research, and other organizations should team up in conducting multidisciplinary researches on soils to come up with long-lasting and sustainable solutions to the problems existing in the soils of the country.

References

ADD/NFIU, Agricultural Development Department/National Fertilizer Inputs Unit (1991) Results of fertilizer trials conducted on major cereal crops (1986–1989). In Joint Working Paper No. 34. Ministry of Agriculture (MoA), Addis Ababa, pp 1–88

ATA, Agricultural Transformation Agency (2014) Soil fertility mapping and fertilizer blending. ATA, Addis Ababa

Beintema NM, Menelik S (2003) Agricultural science and technology indicators: Ethiopia. ASTI Country Brief No. 9. IFPRI, ISNAR and EARO. http://www.asti.cgiar.org/pubs-africa.htm (PDF). Accessed 10 February 2004

Belay K (2008) Linkage of higher education with agricultural research, extension and development in Ethiopia. High Educ Pol 21:275–299. https://doi.org/10.1057/palgrave.hep.8300139

Belay K (2000) Empowering agricultural labor in Ethiopia: the challenges to training and development. Africa Dev 25(1 and 2):161–190

Berhanu D, Ochtman LHJ (1974) Soil resource appraisal and evaluation studies for rural development in Ethiopia, A country report presented at the east African soil correlation committee Nairobi, pp 13–16

Byerlee D, Spielman DJ, Alemu D (2007) Policies to promote cereal intensification in Ethiopia: a review of evidence and experience, International Food Policy Research Institute Discussion Paper 00707. IFPRI, Washington, DC

Cohen JM (1987) Integrated rural development. The Ethiopian experience and debate. The Scandinavian Institute of African Studies. Motala Grafiska, Motala

Deckers JA (1986) New trials program for agricultural development. pp 33–39. In: Beyene D (ed) Proceedings of soil science research in Ethiopia, 11–14 Feb. 1986, Addis Ababa, Ethiopia. IAR, Addis Ababa

FAO, Food and Agriculture Organization of the United Nations (1997) National fertilizer and inputs unit, project findings and recommendations. Report Prepared for the Government of Ethiopia. FAO, Rome

Gilligan DO, Hoddinott J, Taffesse AS (2009) The impact of Ethiopia's productive safety net program and its linkages. J Dev Stud 45 (10):1684–1706

Grunder M (1986) Soil conservation research in Ethiopia. A paper presented at the First International Workshop on African Mountains and Highlands, Addis Ababa, Ethiopia, 18–27 October 1986, p 14. (Published in Mountain Research and Development, vol 8, nos 2/3, 1988)

Haileselassie B (1959) An analysis of administrative organization and extension services with suggestions for application in Ethiopia. Unpublished MSc, Thesis, Cornell University

MERET, Managing Environmental Resources to Enable Transition to More Sustainable Livelihoods (2013) MERET News, No. 13. Ministry of Agriculture (MoA), Addis Ababa

Nedassa B, Seyoum L, Chadhokar PA (2011) Proceedings of government-donor consultative meeting, MERET project, May, 2011. Ministry of Agriculture (MoA), Addis Ababa

NFIU, National Fertilizer Inputs Unit (1993) Agronomic feasibility of the proposed recommendations and comparison with the previous recommendations. National Fertilizer Inputs Unit (NFIU) of the Ministry of Agriculture. General Paper No.17, Addis Ababa

Proclamation No. 79 (1997) A proclamation for the establishment of the Ethiopian agricultural research organization. 3rd Year No.42. 5th June 1997. Addis Ababa

Tsedeke A, Abera D, Amare M (2004) The Ethiopian NARS: evolution, challenges and opportunities. ASARECA Strategic planning paper, 26–28 Feb., 2004, Addis Ababa

3 Climate

Degefie Tibebe, Girma Mamo, Amanuel Zenebe, and Ashenafi Ali

Abstract

Climate is one of the five soil-forming factors responsible for soil formation and development. Ethiopia has complex topography that defines various distinct climate patterns, by modifying the influence of the large-scale ocean-land–atmosphere pattern, thus creating diverse localized climates. The chapter highlights the global, regional, and local factors that make the climate distinctly dynamic. Further, it describes the climate with respect to agroclimatic zones. It is found that Ethiopia has 33 distinct agroclimatic zones and responsible for having 17 soil types across the country. The chapter further highlights how climate influences on soil properties in the country particularly focus on soil organic carbon and soil pH. In addition, impact of current and anticipated future climate on soil erosion, acidity, salinity, and sodicity are assessed. With respect to climate change, soil plays a key role in regulating climate, and used as the main resource in climate change adaptation and mitigation practices. In this regard, potential and proven land management practices for restoring degraded soil and maintaining soil health are reviewed in the Ethiopian context.

D. Tibebe (✉)
Hydroclimate and Geospatial Modelling, Water and Land Resource Center, Addis Ababa University, P.O. Box 1176 Addis Ababa, Ethiopia
e-mail: degefie.t@wlrc-eth.org

G. Mamo
Ethiopian Institute of Agricultural Research, P.O. Box 2003 Addis Ababa, Ethiopia

A. Zenebe
Department of Natural Resource Management and Environmental Protection, Mekelle University, P.O. Box 231 Mekelle, Ethiopia
e-mail: amanuel.zenebe@mu.edu.et

A. Ali
Department of Geography and Environmental Studies, Addis Ababa University, P.O. Box 1176 Addis Ababa, Ethiopia

Keywords

Climate as soil-forming factor • Climate patterns • Agroclimatic zones • Climate change

3.1 Introduction

3.1.1 Climate of Ethiopia

Completely lying within the tropical latitudes and relatively compact with similar north–south and east–west direction, Ethiopia forms the tropical part of the Greater Horn of Africa. Ethiopia is a country of natural contrast, characterized by the largest proportion of (≈45%) elevated land masses in Africa. The complex topography of the country strongly defines both rainfall and temperature patterns, by modifying the influence of the large-scale ocean-land–atmosphere pattern, thus creating diverse localized climates. Three traditional mega climatic zones have been well known. These are Kolla (warm semi-arid), less than 1500 m above sea level; Woinadega (cool sub-humid temperate zone), 1500–2400 m above sea level; and Dega (cool and humid zone), greater than 2400 m above sea level. Later, with the population growth and expanded, agricultural activities two more zones were added at the extreme ends of the agroclimatic spectrum. These are Bereha (hot arid) and Wurch (cold and moist).

Spatially, rainfall in Ethiopia is characterized by a decreasing trend in the direction from west to east, south-north, west-northeast, and west–east. The lowlands in the southeast and northeast most covering approximately 55% of the country's land area experience arid and semi-arid climate. Annual rainfall ranges from less than 300 mm in the southeastern and northwestern lowlands to over 2,000 mm in the southwestern (southern portion of the western highlands). The eastern lowlands get rain twice a year, in April–May and October–November, with two dry periods in between. The total annual precipitation in this regime varies

S. Beyene et al. (eds.), *The Soils of Ethiopia*, World Soils Book Series,
https://doi.org/10.1007/978-3-031-17012-6_3

from 500 to 1,000 mm. The driest of all regions is the Denakil Plain, which receives less than 500 mm and sometimes none at all.

Rainfall variability in Ethiopia is governed by global, regional, and local factors. Globally, the large-scale sea surface temperature (SST) anomaly over central and equatorial Pacific, where warming or cooling events are associated with deficit or excess rainfall, is a determinant factor. The striking correspondence of the positive sea surface temperature anomalies (SSTAs) with the drought incidences in the 1960s, 1970s, 1980s, and 1990s in the Sahel region is evident. Accordingly, there have been several studies (Kassahun 1987) examining connections between GCMs outputs and Ethiopian rainfall variability, notably through SSTAs of the global oceans. This involves either warming or cooling of the ocean surface, mainly the eastern Pacific, Atlantic, and Indian seas that have a substantial influence on the tropical climate system.

Segele and Lamb (2005), Segele et al. (2009a, b), and Diro et al. (2011) also reported the presence of correlation between SSTs over the southern Atlantic Ocean and Gulf of Guinea and Ethiopian June–September rainfall (long rains), More expressively, the increasing central and equatorial Pacific SST (ENSO warm phase) is associated with drought condition during the long rains. In contrast, the relative cooling of the central and equatorial Pacific "La Niña" or the cold ENSO phase is associated with "below-normal" rainfall during the March–April–May (MAM) season, but in "above-normal" rainfall during the long rains. El Nino is a warming of the surface water of the eastern and central Pacific Ocean, occurring every 4 to 12 years and causing unusual global weather patterns. An El Niño is said to occur when the trade winds that usually push warm surface water westward weaken, allowing the warm water to pool as far eastward as the western coast of South America. "ENSO", referring to the El Niño/Southern Oscillation, the interaction between the atmosphere and ocean in the tropical Pacific that results in a somewhat periodic variation between below-normal and above-normal sea surface temperatures and dry and wet conditions over the course of a few years is very important in Ethiopian rainfall pattern (Fig. 3.1).

Regionally, the tropical climate systems that cause rainfall seasonality over Ethiopia at large are the Inter-Tropical Convergence Zones (ITCZ). The ITCZ migration between an extreme northward location of 15°N in July and an extreme southward location of 15°S in January determines the rainfall seasonality in coupling with the topographic interaction (Segele and Lamb 2005), More expressively, the northward advance of the ITCZ produces orographic rains in March–April–May over southwestern, south-central, and southeast Ethiopia. Low pressure over South Sudan draws in a moist flow from the Indian Ocean and Gulf of Aden (Segele et al. 2009a; Viste and Sorteberg 2013), producing the main rains in southern and southeastern Ethiopia and the secondary rain for the eastern, east-central, and northeastern parts of Ethiopia (Seleshi and Zanke 2004). Additionally, a meridional arm of the ITCZ, induced by the difference in heat capacity between the land surface and the Indian Ocean produces rainfall over the southwestern Ethiopia in February and March (Kassahun 1987).

During the long rains in particular (June–September), the Arabian and the Sudan low-pressure ITCZ moves to the extreme northern Ethiopia along 15°N due to the mountainous topography over central part of Ethiopia. This causes the moisture-laden flows from the Atlantic and Indian Oceans through westerly wind systems (Viste and Sorteberg 2013), producing rains over most parts of Ethiopia, except for the drier condition over the southern and southeastern lowlands. Likewise, during September–October–November (SON), the rainfall over Ethiopia retreats toward the south, following the southward migration of ITCZ, thus providing small rains for the southern part of the country. During December–February short rains, the ITCZ is located well south of Ethiopia and the country predominantly falls under the influence of dry warm and cool northeasterly winds. These dry air masses originate either from the Saharan anticyclone and/or from the ridge of high pressure extending into Arabiam from a large high over Central Asia (Siberia) (Kassahun 1987; Gissila et al. 2004).

Ethiopia is also broadly divided into four rainfall regimes (Fig. 3.2 left). Regime A experiences a quasi-bimodal (type-1 or two maxima), i.e., the first peak in April (March–May) and the second peak in August (June–September). Region B has one wet season during June to September, of which the corresponding length of growing period decreases upward, i.e., b1, b2, and b3, February/March to October/November, April/May to October/November, and June/July to August/September, with the upward declining trend. Region C experiences two wet seasons (i.e., the relatively long rains during Feb-May and short rains during September–October to November. Region D is characterized by irregular rainfall pattern (diffused) with sporadic rain prevailing from August/September to January/February; therefore, region D has no well-defined seasons (Haile 1988).

Locally, given that Ethiopia is distinctive and the most rugged country in Africa, its climate is substantially governed by local factors in which the topography is powerful. Mountain ranges create barriers that alter wind and precipitation patterns. Accordingly, topographic barriers such as mountains and hills force the prevailing winds up and over their slopes. While rising, air also cools, and cooler air is capable of holding less water vapor than warmer air. As air cools, the water vapor is forced to condense, thus depositing rain on windward slopes. This creates an effect known as a rain shadow on their leeward (protected) sides, where the air contains very little moisture (Gemechu 1988). For instance, the Belg rainfall (March–May) is restricted to the east,

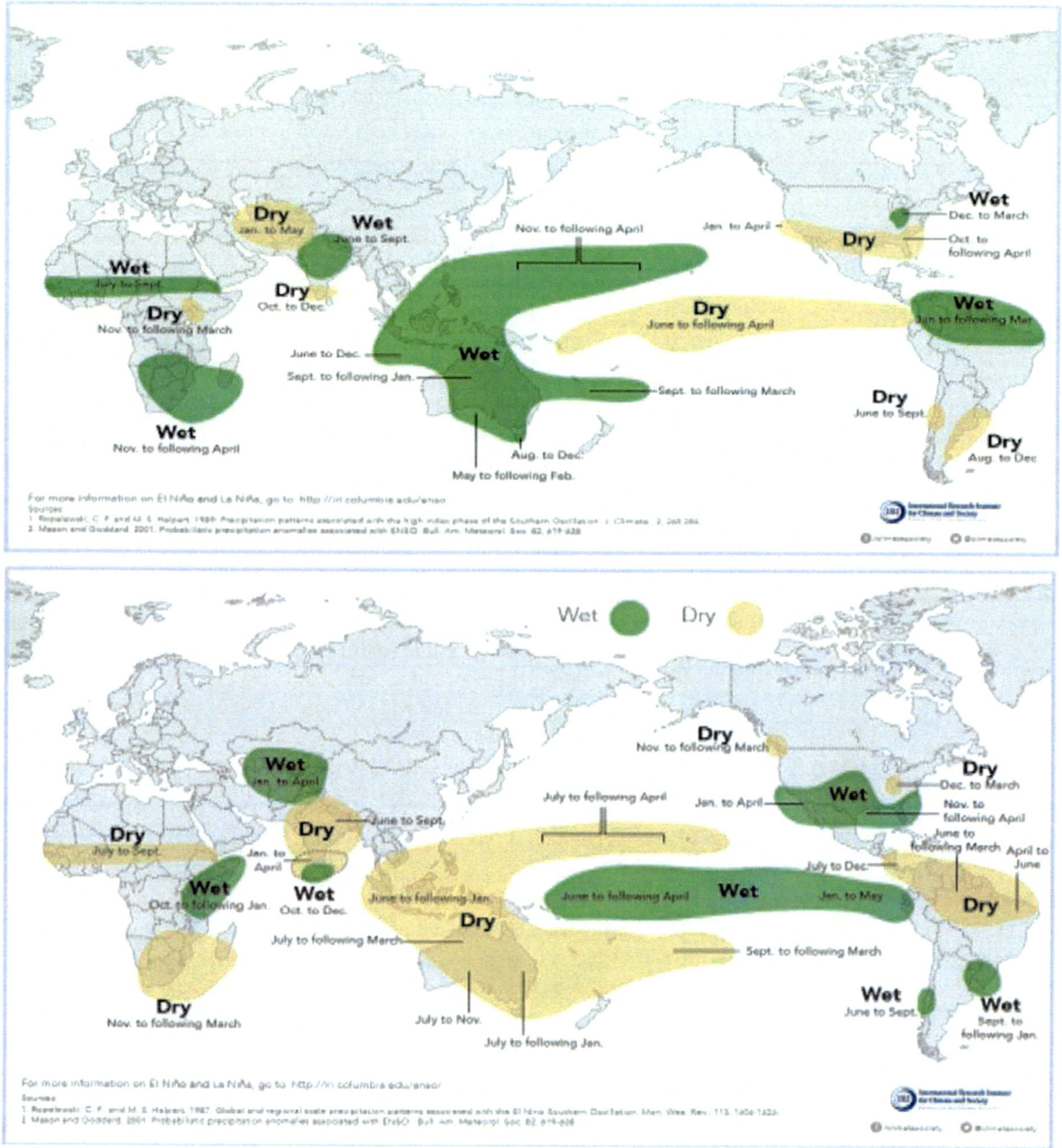

Fig. 3.1 IRI tele-connection of ENSO with regional rainfall performance during El Nino (upper panel) and La Nina (lower panel) (*Source* www.IRI.colombia.edu)

southeast, and southern parts of Ethiopia due to orography, while the northward advance of the ITCZ produces orographic rains in March–May (Belg) over southwestern, south, and southeastern Ethiopia (Segele et al. 2009a). By July, most of the Highlands experience the main rainy season known locally as "keremt", which generally lasts to around mid-September.

Temperatures are also greatly influenced by the rapidly changing altitude in Ethiopia. This varies from about 35 °C, in the northeast lowland (Denakil, 125-m mean to less than 7.5 °C over the north and central highland (Ras Dashen, 4620 m.a.s.l). Many studies confirm that there is a significantly increasing trend in minimum and maximum temperature over Ethiopia, with mean temperature increases by

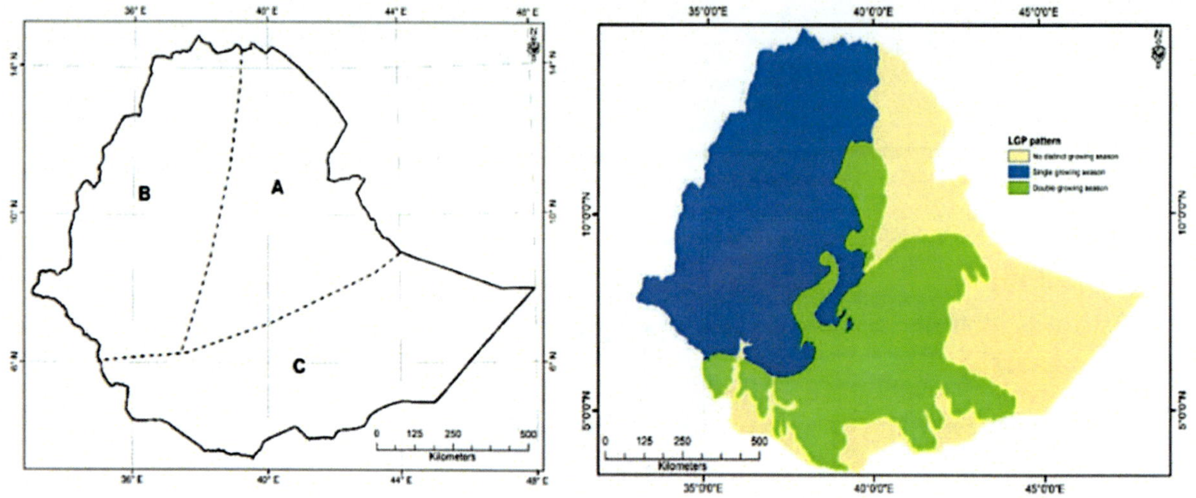

Fig. 3.2 Rainfall regimes of Ethiopia (left), *Source* NMA 1996, LGP pattern (right) (*Source* Own analysis)

0.45 °C per decade. Relatively higher increasing trend is observed in the eastern and northern part of the country (MEFCC 2015).

3.1.2 Climate, Soil Formation, and Development

The transformation of parent materials into soil is deeply dictated by climate. Climate is one of the five soil-forming factors responsible for soil formation and development. It has significant influence on the properties of soil. Generally, soils in warmer or wetter climates are more developed than soils in cooler or drier climates. The development of soil can be determined from looking at the profile. Soils that are more developed have more horizons and deeper horizons than soils that are less developed. Wet conditions favor leaching, or moving deeper with water, of clay and other minerals so that E and B horizons develop. Warm conditions promote the chemical and biological reactions that develop parent material into soil. In a dry climate, the A horizon would be very thin because there are few plants to become organic matter, and the C horizon would still be present, with nutrients still locked into minerals, because there is not enough water to promote weathering and leaching of minerals, or development of a B horizon. In a tropical environment, the soil can become so leached that there are very few nutrients available from soil minerals.

In Ethiopia, climate has a significant influence on soil development and properties and therefore responsible for having widely varying soils in the country. Because Ethiopia is located in the tropical latitudes, its areas of lower elevation experience climatic conditions typical of tropical savanna or desert, while higher elevations experiencing weather typical of temperate zones. Generally, the wet and warm condition in the west, hot and dry condition in the east, and strong rainfall seasonality in the central Ethiopia are some of the main climatic factors responsible for having different soil development and properties. For instance, the very wet and warmer climate in the West, South West, and the West and East escarpment of central rift valley of the country favors for the development of deep soil having distinct soil horizons. Soils like Nitisols and Luvisols are good example and abundantly found in these areas. The reddish color and acidic property of these soils are associated with the high rainfall condition. In hot dry areas of Ethiopia, particularly in the North East, Afar region, soils like Regosols are developed. The hot and dry climatic conditions are responsible for very weakly developed mineral soils with very shallow depth, sandy texture, or with fluvic properties. In the desert areas of the Eastern Lowland and Denakil Plain, soils like Calcisols, Arenosols and other saline soils are developed which are favored by the hot and dry climatic conditions. The bottom part of the central rift valley of Ethiopia is characterized by strong climatic seasonality as distinct dry and wet periods, and this climate seasonality responsible for the development of soils like Vertisols which is abundantly found in this part of the country. Vertisols is also found in the northwestern lowlands of Ethiopia, the very west of Gambela and some parts in Somali regions.

3.1.3 Soil as a Climate Regulator

Next to ocean, soil is the largest pool of carbon, with the amount of carbon stored in soils being over three times that is found in the atmosphere globally (Batjes 1996; Jobbagy and Jackson 2000; Lal 2004). Soil is therefore an important component of the carbon cycle, with changes in soil management practices reducing emissions of carbon-containing gases into the atmosphere.

The regulation of the global climate is an important ecosystem function that soils perform through carbon storage and a reduction of greenhouse gas emissions, referred to as the climate regulation soil function (Schulte et al. 2014). Soils that are wetter or denser hold heat and stabilize the surroundings from temperature changes more so than drier and looser soils. Broadly, soils play a regulating role of the global climate, serving both as source and sink for greenhouse gases (GHG) such as carbon dioxide (CO_2), and nitrous oxide (N_2O) (Ciais et al. 2013; Le Quéré et al. 2018); thus reducing the negative impact of climate change, as well as improve mitigate soil degradation by altering the soil–water balance (runoff, infiltration, percolation, and drainage) across the landscape, vegetation regeneration, and abundant above-ground canopy.

In the context of Ethiopia, soil-based emission reduction is one of the strategies the country is following (FDRE 2011). Currently, Ethiopia emits close to 150 Mt CO_2e from different sectors, and out of it 7% is from soil-based crops production (FDRE 2011). On the other hand, the country has great potential to sequester carbon. Accordingly, Ethiopia has strategized to limit its GHGs emission in 2030 at 145 Mt CO_2e or less, compared to the projected 400 Mt CO_2e to be emitted to the atmosphere under the business-as-usual national development plan (CRGE 2011 or FDRE 2011). CO_2 removal by sequestering carbon in soils and trees is likely to be an essential element of any climate change mitigation scenario that achieves safe climate stabilization (Griscom et al. 2017; Smith 2016). Therefore, improving the soil organic matter contents helps in improving the soil structure and aggregate formation that are of high water-holding capacity and nutrient rich. Currently, about 2 million hectares of land are being under soil and water conservation interventions in Ethiopia (MoA 2021) that modify the climate, under the theme "climate smart agriculture practices".

3.2 Climatic-Soil Variability

3.2.1 Spatial Rainfall Variability and Soil Moisture Regime

Soil moisture is one of the most important soil climate parameters, and it facilitates the four basic soil-forming processes: translocations, transformations, additions, and losses of soil constituents in a soil profile. These processes determine the chemical, morphological, and physical properties of soil such as the variation of texture with depth. Moisture in soil contributes to weathering processes, and indicators of these processes are preserved by the soil profile in the form of observable and measurable soil characteristics (O'Geen et al. 2010). Other soil morphologic processes including abrupt accumulation of clay in the subsoil, development of soil structure, and presence of cemented layers are also associated with soil moisture condition (Fritsch and Fitzpatrick 1994).

Understanding soil moisture characteristics and its spatial variability is very essential in terms of knowing the soil-forming processes across geographical areas. Soil moisture is majorly influenced by rainfall condition, and thus understanding the spatial rainfall variability is very necessary. Accordingly, Ethiopia is characterized by spatial rainfall variability with higher rainfall amount receiving areas in the west and southwest, whereas low rainfall receiving areas in the northeast and southeast. The annual rainfall ranges from less than 300 mm in the southeastern and northwestern lowlands to over 2,000 mm in the southwestern highlands. This rainfall pattern is responsible for having different moisture regimes in the country. In this regard, Aridic, Ustic, and Udic are the type of moisture regimes found in the country (Fig. 3.3).

Aridic soil moisture regime is defined as a soil moisture regime that has no water available for plants for more than half the cumulative time that the soil temperature at 50 cm below the surface is >5 °C, and has no period as long as 90 consecutive days when there is water for plants while the soil temperature at 50 cm is continuously >8 °C. Soils that have an aridic (torric) moisture regime normally occur in areas of arid climates. A few are in areas of semi-arid climates and either have physical properties that keep them dry, such as a crusty surface that virtually precludes the infiltration of water, or are on steep slopes where runoff is high. There is little or no leaching in this moisture regime, and soluble salts accumulate in the soils if there is a source. In Ethiopian case, most of the south eastern lowland and Afar triangle areas characterized by Aridic soil moisture regimes and characterized as arid and semi-arid and hot climate.

Udic moisture regime is one in which the soil moisture control section is not dry in any part for as long as 90 cumulative days in normal years. If the mean annual soil temperature is lower than 22 °C and if the mean winter and mean summer soil temperatures at a depth of 50 cm from the soil surface differ by 6 °C or more, the soil moisture control section, in normal years, is dry in all parts for less than 45 consecutive days in the 4 months following the summer solstice. In addition, the udic moisture regime requires, except for short periods, a three-phase system, solid–liquid-gas, in part or all of the soil moisture control section when the soil temperature is above 5 °C. The udic moisture regime is common to the soils of humid climates that have well-distributed rainfall; have enough rain in summer so that the amount of stored moisture plus rainfall is approximately equal to, or exceeds, the amount of evapotranspiration; or have adequate winter rains to recharge the soils. Water moves downward through the soils at some time in normal years. In Ethiopia, southwestern highland parts, the Arsi-Bale-Sidama Highland areas, the Shewa Plateau, the North Central Massifs, and the Gojjam massive are areas

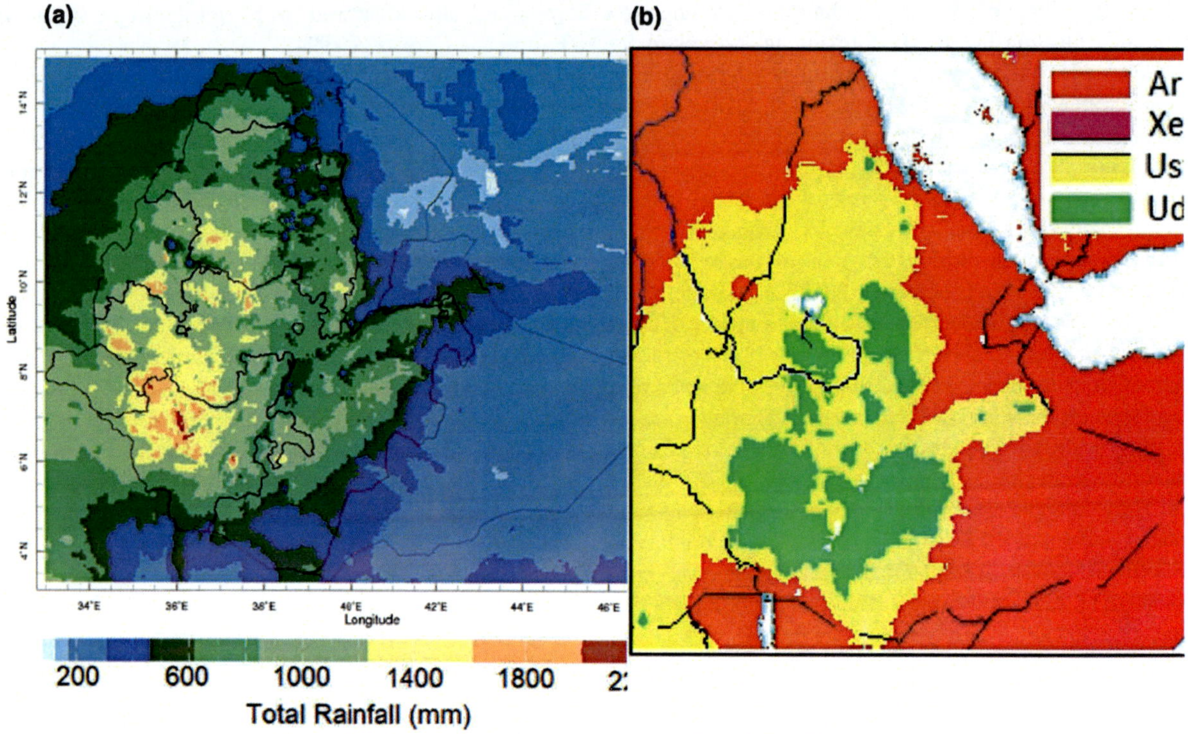

Fig. 3.3 Ethiopian rainfall (**a**), *source* NMA maproom, soil moisture regime (**b**) (*Source* USDA-NRCS)

characterized by Udic soil moisture regime. These areas are characterized as high-rainfall-receiving regions.

The ustic moisture regime is intermediate between the aridic regime and the udic regime. Its concept is one of moisture that is limited but is present at a time when conditions are suitable for plant growth. If the mean annual soil temperature is 22 °C or higher or if the mean summer and winter soil temperatures differ by less than 6 °C at a depth of 50 cm below the soil surface, the soil moisture control section in areas of the ustic moisture regime is dry in some or all parts for 90 or more cumulative days in normal years. It is moist, however, in some part either for more than 180 cumulative days per year or for 90 or more consecutive days. In Ethiopia, most of the western lowland part including Gambela, Benishangul, Western Amhara and Western Tigray areas, the Tigray Plateau in the north, the Hararghe Plateau, and the Borena Plains are characterized by this soil moisture regime.

3.2.2 Spatial Air Temperature Variability and Soil Thermal Regime

Soil temperature is one of the soil climate characteristics which influence different process in soil. It knows that soil temperature alters the rate of organic matter decomposition and mineralization of different organic materials. It also affects soil–water content, its conductivity, and availability to plants.

Soil temperature is majorly influenced by air temperature of the atmosphere. To understand the soil temperature distribution, it is essential to know first the spatial distribution of air temperature in Ethiopia. In this regard, the air temperature of Ethiopia is characterized as the average temperature ranges from 25° to 30 °C in the southeast and northeast lowlands, and 15 °C to 20 °C in the central highlands. There are some pocket areas where the mean temperature is less than 10 °C. For instance, the Arsi-Bale-Sidama Highland and Ras Dejen Mountains and its surrounding in the north are some of the pocket areas where the mean annual temperature is less 4 °C. This temperature pattern is responsible for having different soil temperature regimes including isomesic, isothermic, isohyperthermic, and isomegathermic.

Isomesicis defined as the normal mean soil temperature is greater than or equal to 8 °C but less than 15 °C and the difference between the normal mean summer soil temperature and the normal mean winter temperature is less than or equal to 5 °C. This soil temperature regime is found in some pocket areas of Ethiopia particularly in Semien and Bale mountain areas. Isothermic soil temperature regime is characterized as the normal mean soil temperature which is greater than or equal to 15 °C but less than 22 °C and the difference between the normal mean summer soil temperature and the normal mean winter temperature is less than or equal to 5 °C. Most of the southeastern and western

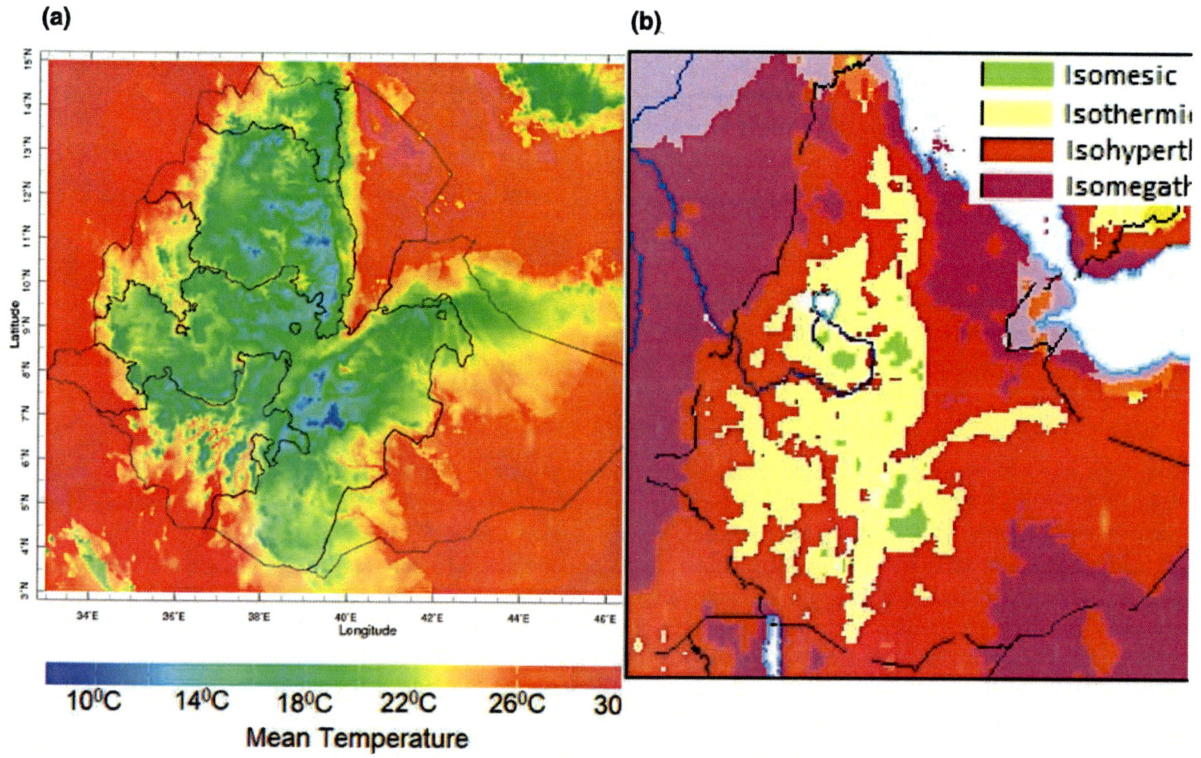

Fig. 3.4 Mean annual temperature (**a**), *source* NMA maproom, soil temperature (**b**) (*Source* USDA-NRCS)

highlands in Ethiopia are characterized by isothermic soil temperature regimes. Isohypertermic is soil temperature regime having the normal mean soil temperature which is greater than or equal to 22 °C but less than 28 °C and the difference between the normal mean summer soil temperature and the normal mean winter temperature is less than or equal to 5 °C, whereas isomegathermic soil temperature regime is characterized by the normal mean soil temperature which is greater than 28 °C and the difference between the normal mean summer soil temperature and the normal mean winter temperature is less than or equal to 5 °C. Most of the dry and hot areas located in the peripheral parts of Ethiopia are characterized by isohypertermic temperature regimes with some areas in the southeastern lowland and Afar triangle having isomegathermic soil temperature regimes (Fig. 3.4).

3.3 Agroclimatic Zones and Soil

3.3.1 Agroclimatic Zones of Ethiopia

Climate plays an important role in soil formation; soils differ widely from one major climatic zone to another. Climates with their high temperature and rainfall yield deep strongly weathered and leached soils with low nutrient contents. Climates with low precipitation and high evaporation give rise to soils that contain variable amounts of easily soluble components such as calcium carbonate or gypsum that is left behind after evaporation of water from the soil. The importance of climate for soil formation was recognized by early soil scientists, especially in Russia, where the concept of "zonality" was developed. This concept distinguished "zonal" soils (soils corresponding to the major climatic belts of the Earth), "azonal" soils (soils not yet in equilibrium with the present-day climatical conditions), and "intra-zonal" soils (soils that are strongly subject to local conditions other than climate).

Climate classification basically organizes single or multiple climatic parameters to identify the similarities of a region (Griffiths 1976). The formula of classification sometimes becomes very complex. Different climatologists have used various techniques according to their needs for climate classification all over the world. Some renowned classifications are Hutchinson et al. (2005), White (1998), Le Houérou (1996), Zuo (1996), Zuo et al. (1996a, b), Fisher et al. (1995), FAO (1981), Koppen (1936). Accordingly, an agroclimatic zone is developed for Ethiopia by using 7 moisture regimes and 6 thermal regimes that yield 33 distinct agroclimatic zones (Table 3.1 and Fig. 3.5a). This agroclimatic zone is used to characterize soils of Ethiopia in terms of soil type and soil properties. The agroclimatic zones

Table 3.1 Agroclimatic zones

		Hot	Warm	Tepid	Cool	Cold	Very cold
		(>27.5)	(21–27.5)	(16–20)	(11–15)	(7.5–10)	(<7.5)
		1	2	3	4	5	6
A	Arid (<45)	A1	A2	A3	A4	A5	A6
SA	Semi-arid (45–60)	SA1	SA2	SA3	SA4	SA5	SA6
SM	Sub-moist (60–120)	SM1	SM2	SM3	SM4	SM5	SM6
M	Moist (120–180)	M1	M2	M3	M4	M5	M6
SH	Sub-humid (180–240)	SH1	SH2	SH3	SH4	SH5	SH6
H	Humid (240–300)	H1	H2	H3	H4	H5	H6
PH	Per-humid (>300)	PH1	PH2	PH3	PH4	PH5	PH6

range from hot arid which is found in the Afar triangle to very cold humid in Bale Mountain. The major agroclimatic zones in terms of area coverage are warm moist, warm sub-moist, and warm semi-arid. These agroclimatic zones are found mainly in the lowland peripheral areas of the country in all directions. Further, the hot arid and warm arid agroclimatic zones are largely found in Afar triangle and southeastern lowland part of Ethiopia. The western highlands of the country mainly classified as warm sub-humid, tepid humid, and hot sub-humid agroclimatic zones. Most of the central part of the country characterized by tepid moist and tepid sub-humid agroclimatic zones while the western lowland is characterized by warm moist and hot moist agroclimatic zones.

3.3.2 Agroclimatic Zone Versus Soil Type

According to ISRIC (2020) soil grid data v2.0, Ethiopia has 17 soil types which are found across the country, of which five soil types, namely, Leptosols, Vertisols, Cambisols, Luvisols, and Nitosols make up about 95% of the soils in Ethiopia (Fig. 3.5b). Leptosols is the first most abundantly available soil types in the country. This soil type is found in most of the agroclimatic zones so that the development is less dependent on climate compared with other soil-forming factors. However, this soil is majorly found in warm moist, warm sub-moist, and tepid moist agroclimatic zones that are characterized by steep slope topography. Next to Leptosols, Vertisols is the second most abundantly available soils in Ethiopia particularly in warm moist, tepid moist, warm sub-humid, and cool sub-humid agroclimatic zones with valley bottom topography. These moist-to-sub-humid moisture regimes indicate that Vertisols soil development mainly depends on availability of moisture in the soil and having strong rainfall seasonality, which are distinct dry and wet seasons. The third soil type which is majorly found in Ethiopia is Cambisols. This soil is particularly found in warm sub-moist and warm semi-arid agroclimatic zones though Cambisols could be developed under any climatic conditions. Luvisols and nitosols are the fourth and fifth abundantly available soils in the country and majorly found in southwestern and eastern highland of Ethiopia with warm moist, warm sub-humid, tepid moist, and tepid sub-humid agroclimatic zones. Moisture availability in the range from moist-to-sub-humid moisture regimes is clearly associated with the development Nitisols.

3.3.3 Agroclimatic Zone Versus Soil Properties

Soil Organic Carbon

Climate is one of the determinant factors that influences the distribution of soil organic carbon on the earth surface. This is particularly described through a relationship between climate and vegetation, and vegetation and organic carbon. In general, moist-to-humid rainfall and warm-to-tepid temperature favors for different vegetation to grow and this vegetation contributes to the soil organic matter pool and further into soil organic carbon. Accordingly, the spatial distribution of soil organic carbon in Ethiopia follows vegetation density pattern of the country which itself is influenced by the spatial rainfall distribution. Higher vegetation density is found in the west and southwest of the country which is governed by high rainfall and warmer temperature whereas the vegetation is decreased toward southeast and northeast as the rainfall amount also decreases. In terms of agroclimatic zones, warm moist, warm sub-humid, tepid sub-humid, tepid humid, and warm humid zones are characterized by very high-to-high soil organic carbon. Warm sub-moist, warm moist, and tepid moist are dominated by moderate soil organic carbon. Those agroclimatic zones characterized by dry and hot climate have low soil organic carbon.

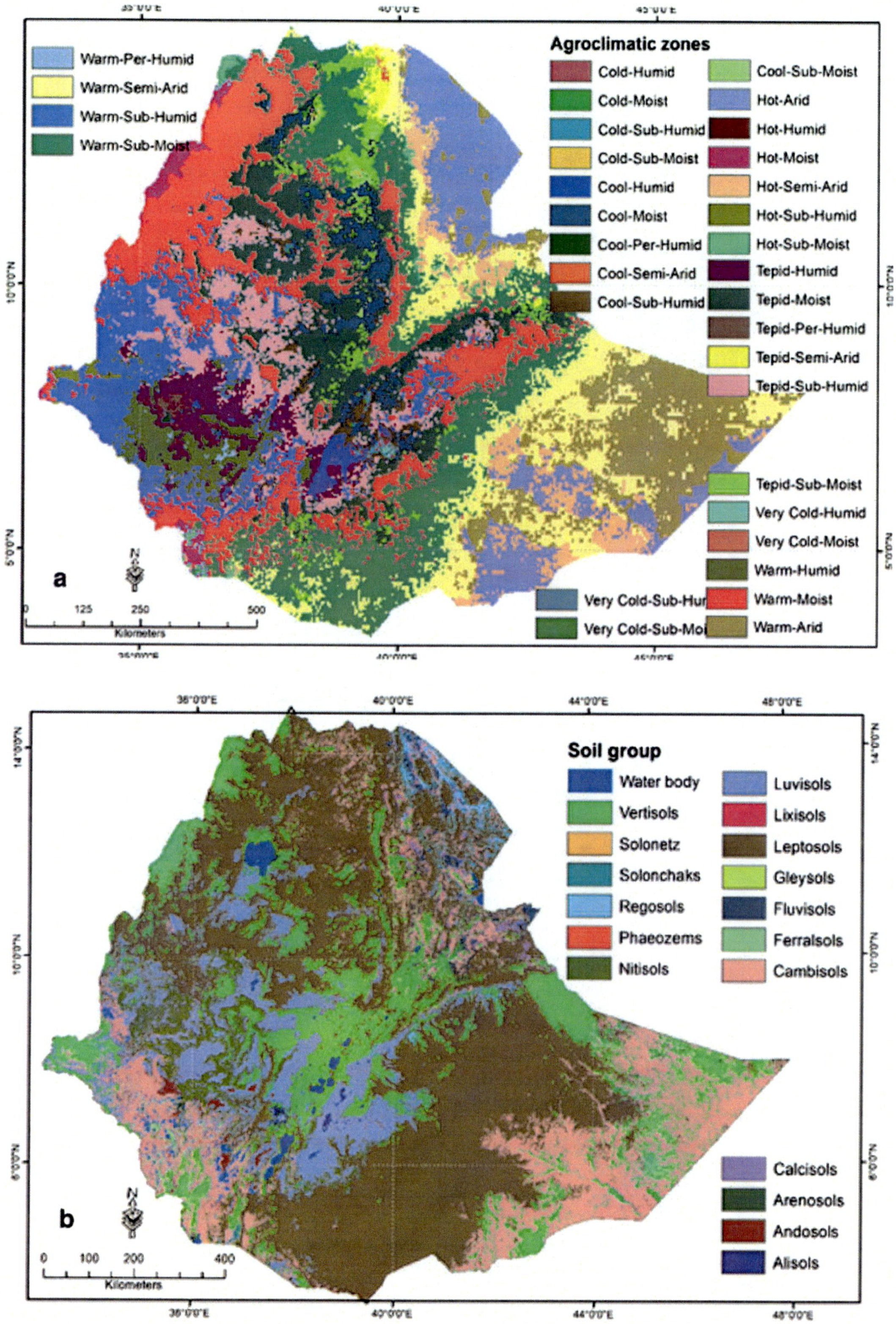

Fig. 3.5 Agroclimatic zones of Ethiopia (**a**), *Source* own analysis, Major soil type (**b**), (*Source* Map extracted and compiled by the author using a digital map from SoilGrids250m version 2.0: https://soilgrids.org//)

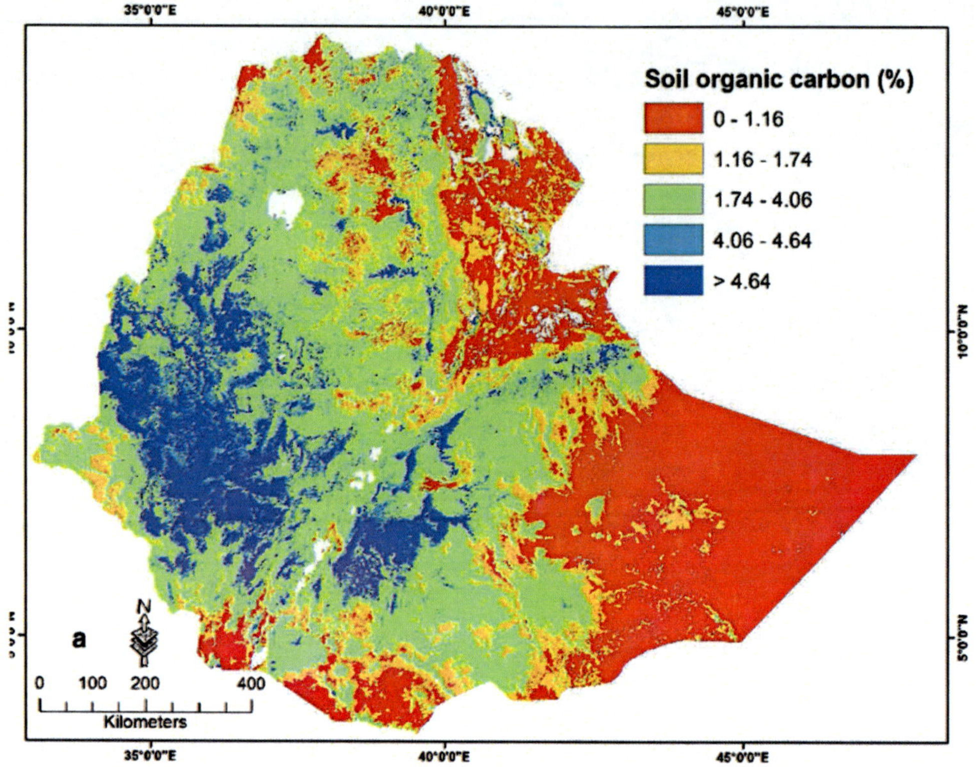

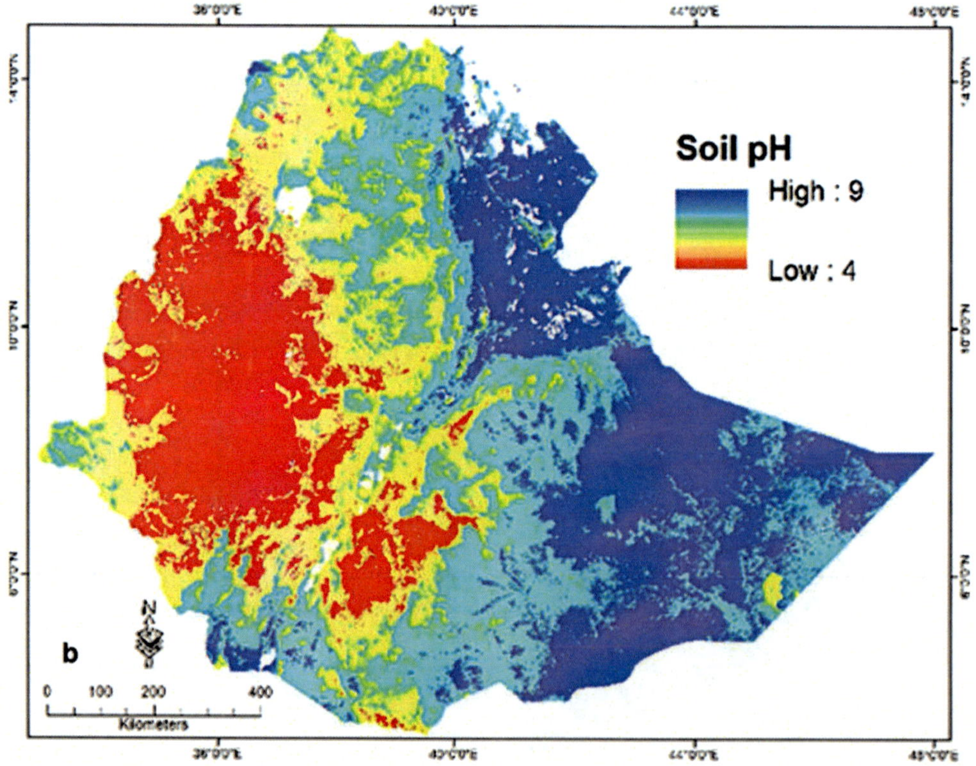

Fig. 3.6 Soil organic carbon (0–20 cm depth) (**a**) and soil pH (**b**) (*Source* Map extracted and own compilation for top 20cm depth by author from SoilGrids 2017:https://files.isric.org/soilgrids/former/2017-03-10/data/)

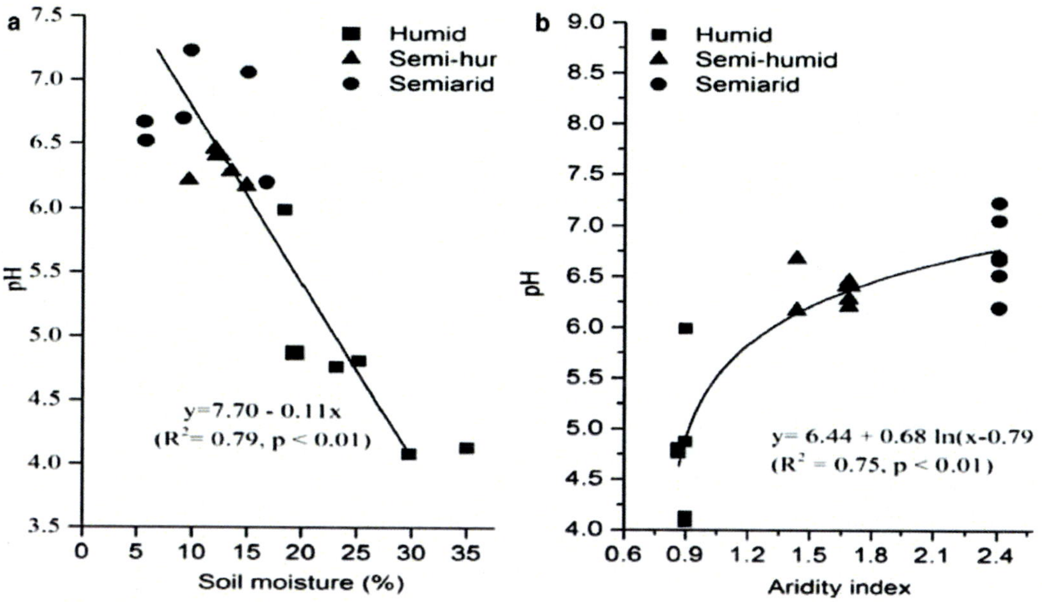

Fig. 3.7 Soil moisture versus soil pH (**a**), aridity index versus soil pH (**b**) (*Source* Jinbo et al. 2015)

Soil pH

Inherent factors that affect soil pH include climate, mineral content, and soil texture. Natural soil pH reflects the combined effects of the soil-forming factors (parent material, time, relief or topography, climate, and organisms). The pH of newly formed soils is determined by the minerals in the parent material. Temperature and rainfall affect the intensity of leaching and the weathering of soil minerals. In warm, humid environments, soil pH decreases over time through acidification due to leaching from high amounts of rainfall. In dry environments where weathering and leaching are less intense, soil pH may be neutral or alkaline (Fig. 3.7).

In Ethiopia, estimated soil pH ranges from 4 to 9. Lower soil pH values are mostly found in the west and southwestern parts of the country. Medium pH values are found in the central and higher pH values in eastern part of the country, particularly northeast and southeast (Fig. 3.6b). Generally, pH value increases from west to east direction. This spatial pattern is strongly associated with the rainfall pattern of the country as rainfall amount decreases from west to east. In terms of agroclimatic zones, low pH soils are majorly found in moist-to-humid moisture regimes and warm-to-tepid thermal regimes. High pH soils are found in semi-arid-to-arid moisture regimes and warm-to-hot thermal regimes.

3.4 Climate Impact on Soil

3.4.1 Physical Impact

Rainfall Erosivity and Soil erosion

Soil erosion is the detachment, transportation, and deposition of soil by water or wind. The rate of erosion depends on the erosive power of the rainfall "erosivity" at large. More intense rainfall events generally mean quicker ponding at the soil surface and increased runoff, which in turn cause more severe soil erosion, an issue that has become a critical challenge in food security of Ethiopia (Hurni 1983).

Annually, Ethiopia losses more than 1.5 × 109 tons of fertile soil by heavy rain and flood with an associated loss of 1.5 × 106 ton crop production (Hurni et al. 2015; Girma 2001). The soil loss due to erosion was estimated to cost the country's economy by 1–1.5 million tons of grain production per year. According to some estimates, from the 60 million hectares of agriculturally productive land, about 27-million hectares are significantly eroded and 2 million hectares of land are irreversibly lost in Ethiopia.

Rainfall erosivity is one of the major factors for soil erosion. Rainfall erosivity depends on the rainfall, and the spatial rainfall pattern tells how rainfall erosivity would be.

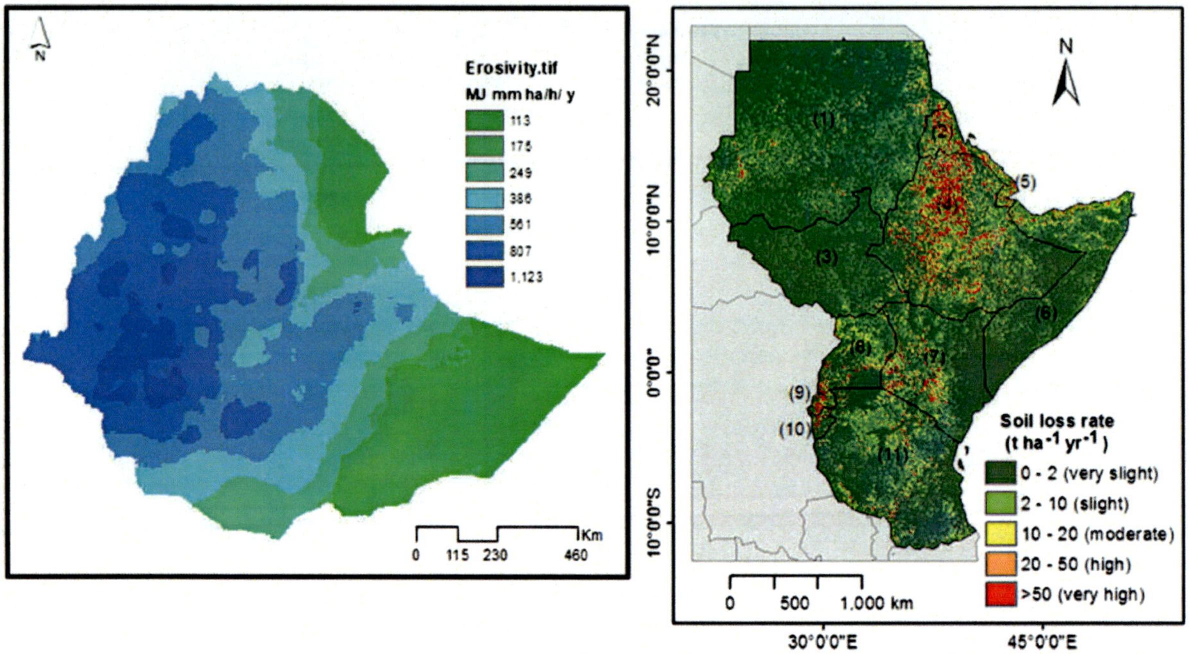

Fig. 3.8 Rainfall erosivity (left panel), soil erosion (right panel) (*Source* Alemaw et al. (2017)

Table 3.2 Mean annual rainfall and erosivity of Ethiopia based on CHIRPS data (1983–2014)

Rainfall classes	Mean annual rainfall (mm)	Average of mean annual rainfall (mm)	Annual rainfall erosivity (MJ mm ha/h/y)	Area (Km^2)	Percent
1	129–302 mm	216	113	254,537	22
2	302–350 mm	326	175	27,830	2
3	350–566 mm	458	249	128,209	11
4	566–835 mm	701	386	174,388	15
5	835–1189 mm	1012	561	274,486	24
6	1189–1711 mm	1450	807	243,137	21
7	1711–2313 mm	2012	1123	60,727	5

Rainfall erosivity is computed for Ethiopia using CHIRPS data shows that the value ranges from 113 to 1123 MJ, mm ha^{-1} $year^{-1}$. Higher erosivity is estimated in the west and southwest of the country whereas the low erosivity is computed in the North east and south east of the country. The rainfall erosivity follows the spatial pattern of the spatial rainfall pattern of the country that rainfall erosivity increases from east to west, from northeast to west, and from southeast to west (Fig. 3.8). Overall, very high erosivity risks are found along the highland part of the country as eastern and western mountain massive, while the lowland part is less risky, because of the low rainfall. In general, rainfall erosivity increases as rainfall amount increases, and proportional areas of the country fall under low, medium, and high rainfall erosivity regimes (Table 3.2).

Alemaw et al. (2017) computed rainfall erosivity for eastern Africa including Ethiopia using the Modified Fournier Index (MFI). They used a 5 × 5 km resolution multi-source rainfall product (Climate Hazards Group Infra-Red Precipitation with Stations, CHIRPS). They found out that rainfall erosivity revealing high spatial variability both declining and increasing trends. Observed decline in rainfall erosivity should not be translated into a decrease in the risk of soil erosion, as factors other than rainfall erosivity such as vegetation cover and soil type may also play an important role (Panagos et al. 2015; Vrieling et al. 2010). Rainfall erosivity increases positively with the amount of rainfall, which may also reduce vegetation cover and hinders surface soil protection. In a recent study, however, Meshesha et al. (2014) reported a general decreasing trend of erosivity opposite to the trend in soil erosion in the Central Rift Valley of Ethiopia and attributed this to factors such as land use/land cover changes coupled with steep topography.

Studies in the northern Ethiopian highlands (Nyssen et al. 2005) and the Central Rift Valley of Ethiopia (Meshesha et al. 2014) revealed larger erosive power of rainfall as compared with elsewhere in the world. Niang et al. (2014) indicated a likely increase in precipitation and extreme precipitation events throughout the twenty-first century, particularly in the highlands.

Soil erosion is also affected by climate change as the rainfall erosivity is changed in response to change in rainfall intensity and rainfall amount. According to EPCC (2015), it is anticipated that increase in precipitation by 4% to 12% by 2100 compared to the 1975–2005 baseline. Further McSweeney et al. (2008) reported that larger share of total precipitation will fall during heavy precipitation events especially from July to December. This is expected to lead to increased incidence of extreme events with severe droughts in 1 year, and heavy flooding with soil erosion and landslides in the next (Aragie 2013). These changes in rainfall amount and intensity potentially change to higher erosive power and lead to higher soil erosion.

3.4.2 Chemical Impact

Soil acidity and alkalinity

Worldwide, many soils have been changed into the class of problematic category due to different natural hazards and poor agricultural practices (Shoghi et al. 2019). A significant increase in rainfall leads to increase in leaching, loss of nutrients, with the simultaneously increasing acidification, depending on the buffering pools existing in soils. Soil acidity is one form of land degradation, adversely affecting sustainable crop production and resulting in lower crop yields. Therefore, soils in humid tropics become acidic naturally due to the leaching of basic cations and mainly replaced by Al (Agegnehu et al. 2019; Zhang et al. 2019; Abate et al. 2017) due to high rainfall and accumulation of bicarbonate and sodium ions in an area soil.

Management wise, the use of nitrogen (N) fertilizers in the form of ammonia is a source of acidification (Fageria and Nascente 2014; Guo et al. 2010). When ammonium fertilizers are applied to the soil, acidity is produced, but the form of N removed by the crop is similar to that found in fertilizer. Hydrogen is added in the form of ammonia-based fertilizers (NH_4), urea-based fertilizers [(CO $(NH_2)_2$], and as proteins (amino acid) in organic fertilizers. Transformation of such sources of N fertilizers into nitrate (NO_3) releases hydrogen ions (H^+) to create soil acidity. In reality, N fertilizer increases soil acidity by increasing crop yields, thereby increasing the amount of basic elements being removed by crop harvest without incorporation. Hence, application of fertilizers containing NH^4 or even adding large quantities of organic matter to a soil can ultimately increase soil acidity and lower pH (Guo et al. 2010).

In Ethiopia, soil acidity is widespread in the western part of Ethiopia, where the lush type of rainfall occurs, Therefore, soil acidity declines from west to east, west to northeast, and west to south eastern (Fig. 3.9) with acid-affected soils covering southern, northwestern, southwestern, and central parts (Mosissa 2018) associated with inherently acidic soils such as Nitisols, Alisols, and Fluvisols (Agegnehu et al. 2019; Kidanemariam et al. 2013). In Ethiopia about 43% of the agricultural land is acidic (Ethio SIS 2014).

In Ethiopia, the productivity of acid soils is low and declines rapidly due to their poor nutrient availability, poor soil microbial activities, coupled with Al toxicity (Agegnehu et al. 2021). For example, in barley, wheat, and faba bean growing areas of western, central, and southern highlands, farmers have shifted to producing acid-tolerant crops like oats (Haile and Boke 2011). Specifically, in areas with strong soil acidity (pH < 5.5), the growth of crops is constrained due to high concentration of aluminum (Al) and manganese (Mn), and the resulting deficiency of phosphorous (P), magnesium (Mg), calcium (Ca), and potassium (K) (Agegnehu et al. 2019; Mosissa 2018; Abate et al. 2017; Kidanemariam et al. 2013).

Similarly, the modeling research to predict the extent and severity of soil acidity based on 109,704 soil pH samples collected from soil laboratories and compiled from various studies in which rainfall, altitude, slope gradient, soil, and land cover were considered to generate multivariate interpolated soil pH surfaces (Gizaw et al. 2021). The model outputs showed that 47% of the country's total area and 45% of the rain-fed areas are acidic (pH < pH < pH). Therefore, integrated acid soil management efforts should be given a priority to soil acidity areas of western, central, northwestern, and southwestern parts of the country. Thus, appropriate and integrated land management technique that enable ameliorating the impact of soil acidity is a critical challenge.

Soil salinity and Sodicity

Soil salinity and sodicity are two core problems most constraining the productivity of the high potential land/soil resources (Wudu and Mahider, 2020). In terms of genesis, soil salinization occurs in areas where evaporation exceeds rainfall, while increased subsoil drying increases concentration of salts in the soil solution. These include soil salinity and sodicity in the dry land areas, particularly the irrigated low land areas.

Soil salinization is one of the major threats to the sustainable food security worldwide. Saline soils are the

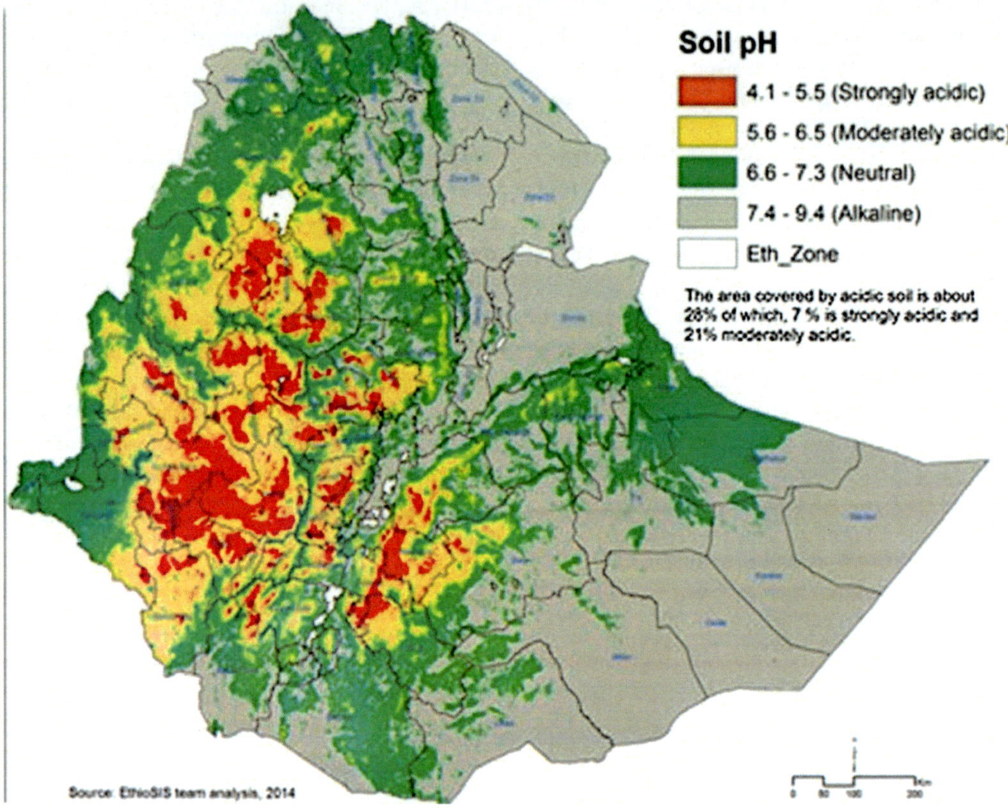

Fig. 3.9 Extent and distribution of soil acidity in Ethiopia (2014) (*Source* Ethio SIS)

non-sodic one containing soluble salts in quantities great enough to interfere with the growth and productivity of most crop plants (Kidane et al. 2006). The electrical conductivity of the saturation extracts (ECe) at 25 °C of saline soils is greater than 4 dS/m, ESP is less than 15, and their pH value is less than 8.5. Chemically, saline soils are composed of the ions Cl^-, SO_4^{2-}, Na + , Ca^{2+}, Mg^{2+}, and small amounts of NO_3^-, HCO_3^-, and K^+, with soluble Na^+ seldom exceeding the sum of the other cations, and thus not adsorbed to any significant extent (Kidane et al. 2006; Tanji 1990). In context, accumulation of soluble salts (minerals and salts more soluble than gypsum: $CaSO_4.2H_2O$) is the single most important soil-forming process operative in arid and semi-arid regions, where ET exceeds rainfall (Cooke et al. 1993).

Ethiopia ranks first from among the top ten salt-affected countries in Africa. It is estimated that salt-affected lands (salinity and sodicity) cover a total area of 11 million ha, being the highest in any African country (Fantaw 2007). Most of these soils are concentrated in the plain lands of arid, semi-arid, and desert regions of the Rift Valley system, including Afar, the Somali Lowlands, the Denakil Plain, and valley bottoms throughout the country (Daba and Bedadi 2016; Fantaw 2007).

Figure 3.10 shows the map of the electrical conductivity of saturation extract (ECe) of the saline soil spatial distribution in Ethiopia. From the ECe map, it can be noted that very high and high of saline soil are found in some pocket areas in Somali regions, while the hot arid bottom, warm arid flat, and warm arid bottom are dominantly characterized by high and very high saline soils. Also, substantial amount of lands is characterized by moderate salinity in hot arid flat, hot arid slope, and warm semi-arid slope agroecologies. In contrary, most of the western, northwestern, and southern parts of the country are free from salinity. On the other hand, the development of large-scale irrigation projects in Rift Valley in the absence of proper drainage systems for salinity control has resulted in increasing severity and rapid expansion of soil salinity and sodicity problems leading to complete loss of land for crop cultivation in these areas. Restoration of salt-affected lands into productive lands and

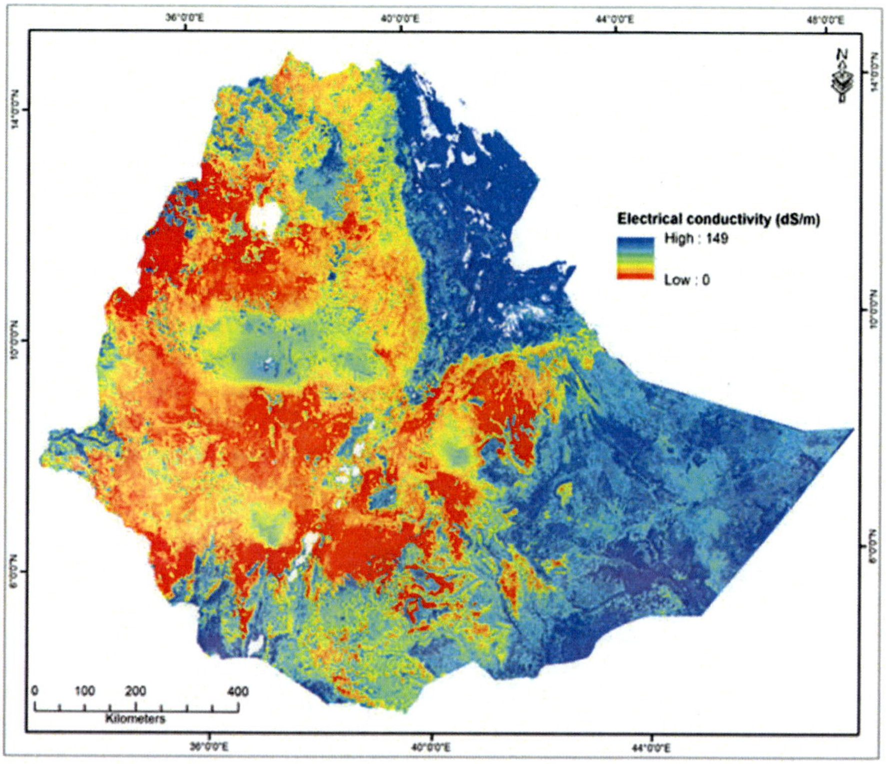

Fig. 3.10 Spatial distribution of electrical conductivity (0–20 cm depth) (*Source* Map extracted and own analysis for the top 20cm depth by the author, from soil property mapsof Africa at 250 m resolution: https://www.isric.org/projects/soil-property-maps-africa-250-m-resolution)

Table 3.3 Summary of classification of salt-affected soils based on their chemical properties

Salt-affected soil type	Electrical conductivity of saturation extracts (ECe) at 25 °C (dS/m)	Saturation (%) of cation exchange capacity with Na (ESP)	Reaction (pH value)
Saline soil	>4	<15	<8.5
Saline sodic soil	>4	<15	<8.5
Sodic (alkalin) soil	>4	<15	8.5–10
Non-saline non-sodic	>4	<15	Neutral

protection of newly developed areas from the spread of salinity is therefore of paramount importance (Table 3.3).

Most of the export crops such as cotton, sugarcane, citrus fruits, banana, and vegetables are being produced in the Rift Valley. To cope with salinity, plants have developed several adaptive mechanisms including altered growth pattern, osmotic adjustment, and ion homeostasis (Flowers and Colmer 2008). These complex traits are extensively reported in both salt-sensitive (glycophytes) and salt-resistant (halophytes) plants (Zhu 2001; Tester and Davenport 2003).

3.4.3 Biological Impact

Soil microorganisms

Climate is one of the important factors that determines the soil microbial community composition and their functions. The impact of climate affects death and disturbance, metabolic activity, reduction (stimulation) of biomass, diversity, and composition leads to extinct/shift, having negative or positive results on its physiology and greenhouse gases emission. As the temperature increases, microbial community structures are altered and processes like respiration, fermentation, and methanogenesis are also accelerated. Soil moisture is also responsible for having different enzymatic activities (dehydrogenase and phenol oxidase) which is considered an important factor controlling soil processes. Enzymes involve in C cycle; thus, it is important in understanding the C dynamics. Significant positive correlation between enzymatic activities (dehydrogenase and phenol oxidase) and soil respiration suggested enzymes have significant role in soil respiration. On the other way, microorganisms are affecting climate. They accelerate global warming through organic matter decomposition and finally increase the flux of CO_2 into atmosphere (Weiman 2015; Swati et al. 2014; Castro et al. 2010; Bardgett et al. 2008; Fierer and Schimel 2003). Microbial decomposition of soil carbon is producing positive feedback to rising global temperatures. Microbial biomass and enzymes are powerful tool to stimulate warming because they decompose carbon-based organic matter efficiently and release toxic compounds to the environment. At the same time, preventing climate change. Temperature directly affects enzyme activity and microbial physiological property. Efficiency of soil microorganisms in using carbon determines the soil carbon response to climate change (Allison et al. 2010; Bardford et al. 2008; Steinweg et al. 2008; Friedlingstein et al. 2006). Microbial community composition, abundance, and function are altered when microbes are exposed to new extremes in environmental condition, that is, environmental change or global warming/climatic disturbance has an effect on microbial ecology, ecosystem structure, and function. Moreover, significant changes also happen through over time in their functional genes and traits. This kind of effect/influence occurs on each biogeochemical cycle (Sayer et al. 2017; Yergeau et al. 2012).

In Ethiopia, much has not been investigated about the relationship between climate and the distribution of terrestrial microbial community composition and their functions. However, the wide ranges of topographic and climatic factors, parent material, and land use in Ethiopia have resulted in extreme variability of soils (FAO 1984e). The different altitudinal ranges, climate and vegetation as well as crop types favor a great diversity of microbial community in Ethiopia.

3.5 Soil for Climate Change Mitigation and Adaptation

3.5.1 Increasing Soil Health as Climate Change Adaptation

Soils are formed over long periods of time (FAO 2013a). They are made up of differing proportions of weathered rock, decayed plant and animal matter, and a diversity of living plants and animals. Due to differences in local geology, topography, climate, vegetation, and human management often over thousands of years, soils are highly variable, both across landscapes and in depth. The diversity and abundance of life that exists within the soil is greater than in any other ecosystem. A handful of soil can contain billions of different organisms that play a critical role in maintaining soil structure, health, and ensuring functioning, including incorporation and breakdown of organic matter, which enhances soil and makes available nutrients for plant uptake.

Nowadays, soils are degraded due to repetitive tillage and cultivation over decades, with reduced or no fallow periods and removal of most organic matter after each harvest for fodder, forage, or fuel, leading to a progressive diminution of soil organic matter (SOM) and essential nutrients for plant growth. Thus, after each harvest most plant materials (that are largely made up of carbon and nitrogen) are removed from the land. As these have not been replaced over recent decades through appropriate organic matter management (addition of compost and/or manure), soils have become degraded, leading to declining productivity (due inter alia to soil acidification, loss of soil organisms and their biological functions that breakdown SOM, poor nutrient recycling, damage to soil structure and moisture infiltration/holding capacity) as plant growth is compromised. These degraded soils are prone to increased rates of runoff of valuable rainwater and as a result erosion of topsoil.

With climate change, rainfall levels are expected to decline in many places or to occur in more intense events, while both evaporation and transpiration rates are projected to increase (FAO 2013a). These changes will reduce the availability of soil moisture for plant growth. The higher temperatures will also increase the rate of SOM decomposition, especially near the soil surface, which will affect the soil's potential capacity to sequester carbon, retain water (ibid.), and supply vital nutrients to plants. Figure 3.3 shows a farmer hand tilling his soil, which has been the traditional method of preparing land for planting by small-holders but is now recognized not only as a very onerous task but also as opening up the soil to these degradation pressures.

There are already adapted proven land management practices which enhance the ability of soils to store nutrients and water (FAO 2013a; TerrAfrica 2009), improving soil health and thereby reducing the impacts of changes in weather and climate. Practices such as reduced tillage, improved soil cover, and rotations that replenish soil organic matter and nutrients are associated with conservation agriculture systems. These practices can build resilience in farming systems and reduce vulnerability to climate or other shocks, and their widespread adoption has the potential to make major contributions to the achievement of food security local and national levels and to development goals.

Conservation agriculture

In Ethiopia, conservation agriculture (CA) has been conducted with respect to improving soil–water balance. For instance, formation of semi-permanent raised beds were found to reduce water runoff (Araya et al. 2015) and opening rip-lines in CA system increased water infiltration compared with conventional practices and has led to higher maize and wheat yields (Liben et al. 2017; Araya et al. 2015). In context, the established plots of CA on sandy loam and loam soils improved soil–water content in the top 0–30 cm soil layer, compared to the conventional tillage during the main growing season (Burayu et al. 2006). Recent CA study in the semi-arid Central Rift Valley (CRV) also showed that stored soil water at 0 to 100 cm depth at physiological maturity of maize was 21% more with CA as compared to conventional ploughing (Liben et al. 2017). Similarly, conservation tillage study on vertisols in the dry lands showed constantly higher soil–water storage (0–80 cm soil depth) during the growing season.

Watershed management

Since the beginning of the 1990s, watershed management approaches that integrate SWC, intensified natural resource use have been implemented in several micro-watersheds of Ethiopia (SLMP 2013a; Haregeweyn et al. 2012; MoARD 2006). The concept of participatory watershed development and management emphasizes a multi-disciplinary and multi-institutional approach for multiple interventions (German 2007). A holistic watershed or catchment development integrates various interventions within a catchment while utilizing synergies and reducing competitiveness among interventions. The result from three experimental watersheds located in the Ethiopian highland, namely, Anjeni, AnditTid, and Maybar watersheds (graph from BSNP) showed that out of the 13 years' time series (i.e., in the period of 1987–1993, 1996–1998, 2000, and 2008), 6 years (1987–1992), and 20 years (i.e., in the period of 1988–1989, 1991–1993, 1995, and 2000– 2013) rainfall and runoff datasets at the plot level revealed overall water yield improvement.

Bunding improves soil–water conservation in those areas where yields are constrained by moisture deficiency. Sorghum yield data for Kobo (Welo), semi-arid region indicate moisture conservation practices more than double the yields during drought years but no significant difference when there was enough rainfall. In Hararghe Highlands even during the years with above-average rainfall, moisture conservation practice has given significantly higher sorghum yields as well as crop response to fertilizer application is much higher at lower dosage of fertilizers when coupled with availability of moisture (Hawando 1986). Hawando found in Hararghe that on the average yields of plots with tied ridges were 57% higher for maize and 25% higher for sorghum than of those without tied ridges. The increased yields were attributed both to increased water conservation and to increasing soil depths in areas where yields may already be constrained by shallow soil depth (Fig. 3.11).

Ciampalinia et al. (2012) reported that in the Aksum area, soil–water conservation measures have been applied for centuries, most likely first implemented during the Aksumite Kingdom (400 BC to 800 AD). The traditional terraces in Konso Area constitute a spectacular example of a living cultural tradition stretching back 21 generations (more than 400 years) (Beshah 2003).

In Tigray, SWC interventions have been the commonest types of adaptation responses, with about 522, 600 ha of land already covered by some form of SWC interventions, mainly stone bunds from 1991 to 2002 (Nyssen et al. 2007). The most commonly reported impact is reduction in runoff and increase in sediment deposition (461 cases) followed by improved crop yield (61 cases) and enhanced vegetation (49 cases). Other impacts reported to a lesser extent also included impacts on income, livestock productivity, soil fertility, carbon stocks, and others.

The following table summarizes the types of soil and water conservation in the face of reducing the impacts of rainfall erosivity, mainly in the drylands of Tigray. The marked diversity of SWC interventions is because of the diverse agro-ecological setting in northern Ethiopia that requires different solutions (Table 3.4).

On the contrary, Hurni et al. (2005) indicated the potential benefits of soil and water conservation efforts in the Ethiopian humid-highlands work to a limited extent, compared with the semi-arid areas. Similarly, Taye et al. (2015) studied the evolution of the effectiveness of stone bunds and trenches to adapt runoff and soil loss in the semi-arid Ethiopian highlands is limited and they concluded that these measures are only fully effective in the first few years of their construction.

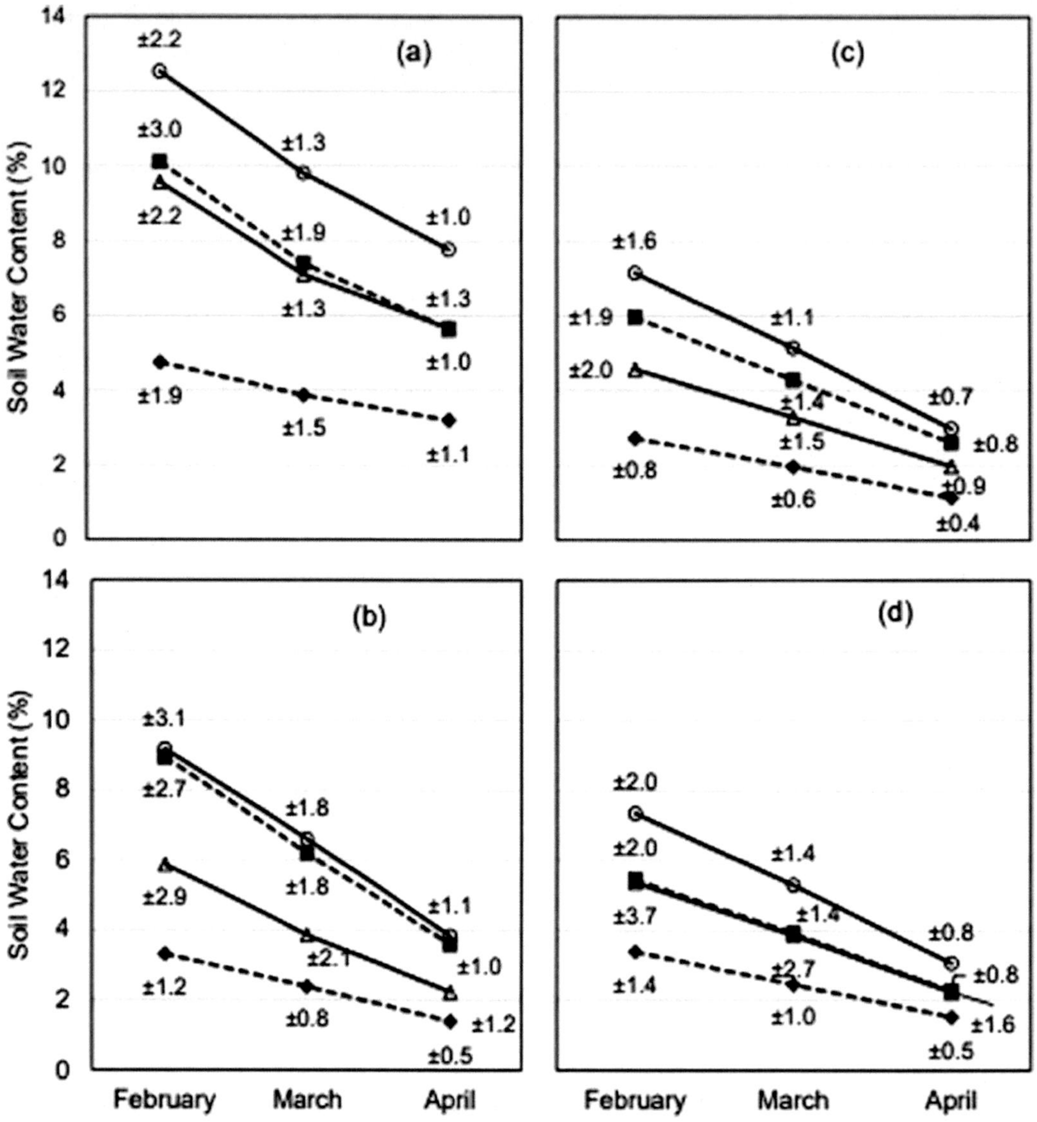

Fig. 3.11 Soil water content on the four sites **a** Teshi; **b** RubaFeleg; **c** Michael Emba; and **d** Enda Chena. The numbers close to the points indicate the standard deviations

3.5.2 Soil Carbon Sequestration Potential as Climate Change Mitigation

Over the last two decades, the sequestration of carbon in soils has often been advocated as a solution to mitigate the steady increase in the concentration of CO_2 in the atmosphere, one of the most commonly mentioned causes of climate change. However, designing a strategy to manage soils to mitigate climate change is not straightforward, as a thorough understanding of relevant processes at play is necessary. It has been a widely established knowledge that soil organic matter is capable of acting both as a source and sink of carbon in the biosphere. Evidence on soil carbon sequestration (be it empirically or modeling) is scanty in Ethiopia. However, according to a study conducted in the Ethiopian highlands, the mean C stock in the 0–20 cm soil depth showed significant variation with agroclimatic zone ranging from 47.1 (±0.08) t ha^{-1} in warm semi-arid-arid to 137.5 (±0.25) t ha^{-1} in cold sub-moist-humid (Abegaz et al. 2022).

In Ethiopia, limited local studies confirmed the SOC improvement with of Conservation Agriculture (CA) practices. On a conservational tillage experiment established on sandy loam and loam soils in the dry land areas of the Central Rift Valley of Ethiopia (CRV), soil organic matter on weight basis was higher under conservation tillage (1.6%)

Table 3.4 Soil and water conservation activities implemented in Ethiopia

Soil and water conservation type	Description of the intervention	Objectives	Target sites or areas applied	References
Bench terracing	Holding soil moisture is still not well established in arid areas prone to desertification and soil erosion	Reducing runoff and drought risks		
Tied ridges	Small earthen ridges, 15–20 cm high, with an upslope furrow which accommodates runoff from a catchment strip between the ridges	Conserve water on farm, reduce amount of moisture lost due to excessive plowing	Farmlands	Grum et al. (2017a)
Terrewah	Traditional plowing followed by making every 1.5–2 m contour furrows. Furrows are made at 2–4 m intervals along the contour the same day after planting, especially practiced for teff (Eragrostis teff)	To conserver moisture on farm and prevent runoff	Farmlands	Araya et al. (2016a)
Derdero	Bed furrows prepared along contour using the traditional marasha (Plowshare) at the last tillage operation or after farmers broadcast seeds over the farmland	To conserving water	Farmlands	Nyssen et al. (2010)
Shilshalo	A traditional plowing where contour furrows are created within the standing crop (mostly sorghum and maize) during second weeding operation	Conserve moisture on farm and prevent runoff, reduce weed infestation Sorghum and maize	Fields	Nyssen et al. (2010)
Crop residue mulching	Leaving crop residue on the farm, after harvest so that they serve as mulch against evaporation of moisture	Conserve residual moisture	Drought-affected farmlands	Araya and Stroosnijder (2010)
Earthen trenches/trenches	Ditches dug on the path of an erosion or a catchment	Trap sediment, reduce runoff or erosion On high-erosion areas	Grazing lands, hillsides)	Taye et al. (2014)
Stone bunds/terraces	Stone structures constructed along a contour in a slopping land (usually hillsides and slopping farmlands)	Reduce erosion or runoff On hillsides	Sloppy farmlands	Nyssen et al. (2008b)
Soil bunds	Soil-based structures constructed along a contour in a slopping land (usually only on slightly sloped farmlands)	Reduce erosion and runoff	Slightly slopping farmland	Teshome et al. (2013)
Gabion check dams	Check dams constructed using rocks and reinforced by gabion or wire mesh for increased strength or resistance	Control gulley expansion, trap sediment, and conserve water	Gullies, streams, and eroded areas	Mekonnen et al. (2015)
Micro-dam reservoirs	Small dams built in catchments and valleys to reserve water during the rainy season for using it in the dry season	Harvesting and reserving water for irrigation and livestock use	On catchments and valley bottoms	Berhane et al. (2016a, b)
Percolation pond	Pits and ditches of various sizes and shapes, dug or constructed for the sole purpose of allowing runoff to percolate	Increase groundwater availability in downstream areas	Wastelands, degraded areas upstream	Grum et al. (2017b)
Deep trench	Deep trenches at lower part of a catchment or across the slope of a catchment dug to harvest water and silt during farmland sides	Control runoff, trap silt, and enhance groundwater recharge processes	Road sides, degraded areas	Woldearegay et al. (2015)
Hand-dug wells	Wells dug deep into the ground to access and collect groundwater	Provide water for domestic use and irrigation	Backyards, near farmlands in the lower section of catchments	Woldearegay and Van Steenbergen (2015)
Spate irrigation	The diversion of seasonal floods to irrigate farmlands	Utilize excessive flood water for growing crops		Hiben and Tesfa-alem (2014)
Low-cost family drip irrigation	Low-cost family-based drip irrigation systems that use small plastic tubes and water-conserving structures	Save water saving water	Backyard and highly fertile soils	Waktola (2007)

Source Balehegn et al. (2019)

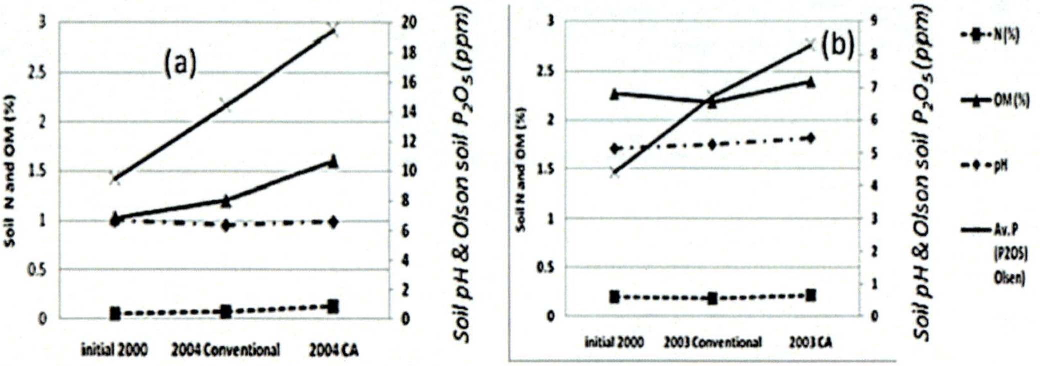

Fig. 3.12 Chemical properties of soils in conventional and CA systems at **a** MARC and **b** Jimma. The secondary vertical axis shows the soil pH and Olson soil P_2O_2 (ppm). Computed from Worku et al. (2006) and Bogale et al. (2003)

at a depth of 0–15 cm, compared to conventional tillage (1.2%) (Worku et al. 2006). A meta-analysis by Abera et al. (2021) reported that implementation of conservation agriculture practices showed significant increases of SOC (24%) and agricultural productivity (18%) and a significant decrease of soil erosion (45%). Another experiment conducted on smectite-rich clay mineral Vertisol at Chefe Donsa in the central highlands showed an increasing trend in SOM content in reduced tillage (Erkossa 2011). Similarly, a higher level of SOC 16 g kg^{-1} was reported in CA fields, compared to 12 g kg^{-1} in conventional plowed fields at 0–0.05 m surface soil depth at Melkassa (Liben et al. 2018). Thirty-year crop growth simulation study in seven different agroecologies revealed 33% maize grain yield advantage by combined use of N fertilizer, crop rotation and conservation tillage (Liben et al. 2020). The modeling result also showed a slowdown in the rate of SOC and N decline over time.

Retention of crop residues and diversification of crop species grown in sequence or associations under no-tillage affect nutrient cycling and availability. It has been found that no-tillage helped to conserve more nitrogen (Bessam and Mrabet 2003) and resulted in increased extractable phosphorus and exchangeable potassium concentrations in the upper root-zone similar to the finding by Bogale et al. (2003) from Ethiopia (Fig. 3.12). Total nitrogen content increased from low-to-medium level (0.13%) in conservation tillage while it remained under low category (0.07%) in conventional tillage (Worku et al. 2006).

A study conducted at Bako in Ethiopia on clay loam soils reported appreciable improvement of soil organic C and total N content as well as extractable P and exchangeable K for zero-tillage with 5-year residue retention (Debele et al. 2007). The larger total N values under no-tillage than conventional tillage imply N immobilization in microbial biomass near the soil surface, leaving less N available for mineralization or leaching that is slow-release overtime.

3.5.3 Climate Change and Soil Resilience

Healthy soils support food, energy, and water security for people and it is a key medium to reducing greenhouse gas (GHG) emissions from agriculture. These facets are threatened by increasing rates of soil degradation underpinned by a complex interaction between natural vulnerability and anthropogenic activities. Particularly, the rise in global surface temperature is linked to increased occurrence of extreme weather events, which is predicted to intensify with continued warming. This exacerbates soil degradation resulting in a destabilized soil system leading to fracturing of the provisioning, regulatory, and supporting ecosystem services provided by healthy systems. Restoring degraded soils is the way for sustainable provision of all soil-related ecosystem services including soil carbon sequestration. Fortunately, there are as many potential soil solutions as there are sources of soil degradation that are promoting healthy soils so that increasing soil resilience.

There are many ways to develop health and resiliency to soils. The best way is managing soils as nature itself manages soils. Undisturbed soils or less disturbed soils tend to be more resilient to changes in climate than soils that have been cultivated. Nature manages soils by keeping soil to be covered, letting many species function together, reliance on soil organisms to till the soil and closed loop of organic input, breakdown, and re-uptake of nutrients. These strategies could be used to devise mechanisms to build health and resiliency back into degraded soils or to maintain the integrity of soils that are already healthy. Some of these strategies which can be employed to manage and build soil's resilience are keeping soils be covered using multi-species cover crops, diversifying species, minimize tilling practices, and building organic matter of soils. These soil health strategies can be practiced at any scale. Focusing on the health and resiliency of soils will not only create a healthier system but will prepare landscapes for changing patterns in the future.

3.6 Conclusions

Ethiopia has a complex climate system which is the result of interaction of global, regional, and local climate-forming factors. The complex climate system together with wide ranges of topographic factor, parent material, and land use has in Ethiopia resulted in extreme variability of soils. Climate is also impacting soils in Ethiopia majorly through soil erosion, acidification, and salinization. Anticipated climate change in the future may have considerable impacts on the soil functions to perform, and more importantly the future use of soils often needs significant adaptations in order to meet the challenges of changing climate. Changes in temperature and rainfall patterns can have a great impact on the organic matter and processes that take place in our soils, as well as the plants and crops that grow from them. Soil is also a source and sink of carbon, thus restoring degraded soils through proven land management practices contributing to climate change mitigation. This also leads to increase in fertility of the soil through carbon sequestration, and thus enhances the ability of soils to store nutrients and water improving soil health and thereby reducing the impacts of changes in weather and climate.

References

Abate E, Hussein S, Laing M, Mengistu F (2017) Soil acidity under multiple land-uses: assessment of perceived causes and indicators, and nutrient dynamics in small-holders' mixed-farming system of northwest Ethiopia. Acta Agric Scand Sect B Soil Plant Sci 67 (2):134–147. https://doi.org/10.1080/09064710.2016.1230227

Abegaz A, Ashenafi A, Tamene L, Abera W (2022) Smith JU (2022) Modeling long-term attainable soil organic carbon sequestration across the highlands of Ethiopia. Environ Dev Sustain 24:131–5162. https://doi.org/10.1007/s10668-021-01653-0)

Abera W, Tamene L, Abegaz A, Hailu H, Piikki K, Söderström M, Sommer R (2021) Estimating spatially distributed SOC sequestration potentials of sustainable land management practices in Ethiopia. J Environ Manage 286:112191

Agegnehu G, Amede T, Erkossa T, Yirga C, Henry C, Tyler R, Nosworthy MG, Beyene S, Sileshi GW (2021) Extent and management of acid soils for sustainable crop production system in the tropical agroecosystems: a review. Acta Agric Scand Sect B Soil Plant Sci. https://doi.org/10.1080/09064710.2021.1954239

Agegnehu G, Yirga C, Erkossa T (2019) Soil Acidity management. Ethiopian institute of agricultural research (EIAR). Addis Ababa

Alemaw FA, Yasuda H (2017) Shimizu K (2017) Spatial distribution and temporal trends of rainfall and erosivity in the Eastern Africa region. Hydrol Process 31:4555–4567. https://doi.org/10.1002/hyp.11378

Allison SD, Wallenstein MD, Bradford MA (2010) Soil-carbon response to warming dependent on microbial physiology. Nat Geosci 3:336–340. https://doi.org/10.1038/ngeo846

Aragie EA (2013) Climate change, growth, and poverty in Ethiopia. The Robert S. Strauss center for international security and law working paper No (3), Texas university Austin

Araya T, Nyssen J, Govaerts B, Deckers J, Cornelis WM (2015) Impacts of conservation agriculture-based farming systems on optimizing seasonal rainfall partitioning and productivity on Vertisols in the Ethiopian drylands. Soil Tillage Res 148:1–13

Araya T, Nyssen J, Govaerts B, Baudron F, Carpentier L, Bauer H, Lanckriet S, Deckers JA, Cornelis W (2016a) Restoring cropland productivity and profitability in northern Ethiopian drylands after nine years of resourceconserving agriculture. Exp Agric 52:165–187

Araya A, Stroosnijder L (2010) Effects of tied ridges and mulch on barley (Hordeum vulgare) rainwater useefficiency and production in northern Ethiopia. Agric Water Manag 97:841–847

Balehegn M, Haile M, Fu C, Wu L (2019) Ecosystem-based adaptation in Tigray, northern Ethiopia: a systematic review of interventions, impacts, and challenges. https://doi.org/10.1007/978-3-319-71025-9_117-1

Bardgett RD, Freeman C, Ostle, NJ (2008) Microbial contributions to climate change through carbon cycle feedbacks. ISMEJ 2: 805–814. https://goo.gl/KGPJEk

Batijes N (1996) Total carbon and nitrogen in the soils of the world. Eur J Soil Sci 47:151–163. https://doi.org/10.1111/j.1365-2389.1996.tb01386.x

Berhane G, Gebreyohannes T, Martens K, Walraevens K (2016a) Overview of micro-dam reservoirs (MDR) inTigray (northern Ethiopia): challenges and benefits. J Afr Earth Sci 123:210–222

Berhane G, Kebede S, Gebreyohannes T, Martens K, van Camp M, Walraevens K (2016b) An integratedapproach for detection and delineation of leakage path from Micro-Dam Reservoir (MDR): a case study from AratoMDR, northern Ethiopia. Bull Eng Geol Environ 75:193–210

Beshah T (2003) Understanding Farmers: Explaining soil and water conservation. In Konso W, Wello E. PhD Thesis. Wageningen University and Research Centre ISBN 90–5808–795–6

Bessam F, Mrabet R (2003) Long-term effect of tillage and cropping systems on particulate organic matter of a Calcixeroll soil of semi-arid Morocco. Proceedings 16th Triennial conference of international soil tillage research Organisation, pp 144–149

Bogale T, Eshetu T, Tulu T, Abebe M (2003) Comparative study of tillage and crop residue management in continuous maize-legume mixed cropping systems. Paper presented at conservation tillage workshop. June 19, 2003, Nazareth, Ethiopia

Bradford MA, Davies CA, Frey SD, Maddox TR, Melillo JM (2008) Thermal adaptation of soil microbial respiration to elevated temperature. Ecol Lett 11:1316–1327

Burayu W, Chinawong S, Suwanketnikom R, Mala T, Juntakool S (2006) Conservation tillage and crop rotation: win-win option for sustainable maize production in the dryland, central rift valley of Ethiopia. Kamphaengsaen Acad. J.4

Castro HF, Classen AT, Austin EE, Norby RJ, Schadt CW (2010) Soil microbial community responses to multiple experimental climate change drivers. Appl Env Microbiol 76: 999–1007. Link: https://goo.gl/UQ1d9N

Ciais P, Sabine C, Bala L, Bopp L, Brovkin V, Canadell J (2013) Carbon and other biogeochemical cycles. In: Stocker TF, Qin D, Plattner G-K, Tigmor M, Allen SK, Boschung J et al (eds) climate change 2013: the physical science basis. Contribution of working group I to the fifth assessment report of the intergovernmental panel on climate change. Cambridge University Press, Carmbridge; New York, NY, pp 465–570

Ciampalinia R, Billi P, Ferrari G (2012) Soil erosion induced by land use changes as determined by plough marks and field evidence in the Aksum area (Ethiopia). Agr Ecosyst Environ 146:197–208

Cooke R, Warren A, Goudie A (1993) Desert geomorphology. Press Limited with University of California, London, p 526

Daba A, Bedadi B (2016) Studies on soil physical properties of salt affected soil in Amibara Area. Central Rift Valley of Ethiopia. 3:8–17

Debele T, Du Preez CC, Ceronio GM (2007) Effect of tillage system and nitrogen fertilization on yield and yield components of maize in western Ethiopia. South African J Plant Soil 24(2):63–69

Diro GT, Grimes DIF, Black E (2011) Teleconnection between Ethiopian summer rainfall and sea surface temperature: part I-observation and modeling. ClimDyn 37:103–119. https://doi.org/10.1007/s00382-010-0837-8

EPCC (2015) First assessment report. Published by the Ethiopian Academy of Sciences, Working Group I. Physical Science Basis

Erkossa T (2011) Tillage effects on physical qualities of a Vertisol in the central highlands of Ethiopia. Afr J Environ Sci Technol 5 (12):1008–1016

EthioSIS, Ethiopian Soil Information System (2014) Soil fertility mapping and fertilizer blending. Agricultural transformation agency (ATA) report. Ethiopia soil information system (Ethio SIS). Ministry of Agriculture, Addis Ababa

Fageria N, Nascente A (2014). Management of soil acidity of south American soils for sustainable crop production. https://doi.org/10.1016/B978-0-12-802139-2.00006-8

Fantaw A (2007) An overview of salt-affected soils and their management status in Ethiopia. A paper presented in the 3rd international workshop on water management project, Haramaya University, Ethiopia

FAO (1981) Report on the agro-ecological zones project. Vol 1. Methodology and results for Africa; Vol 2, Results for Southwest Asia; Vol 3, Methodology and results for South and Central America; Vol 4, Results for Southeast Asia. [FAO] World Soil Resources Report No. 48/1–4

FAO (1984e) Farming system development information note. Rome

FAO (2013a) Climate smart agriculture sourcebook. food and agriculture organization. Rome, Italy. Available from: http://www.fao.org/docrep/018/i3325e/i3325e.pdf

FDRE, Federal Democratic Republic of Ethiopia (2011) Ethiopia's Climate Resilient Green Economy Strategy. Addis Ababa

Fierer N, Schimel JP (2003) A Proposed mechanism for the pulse in carbon dioxide production commonly observed following the rapid rewetting of a dry soil. Soil Sci Soc Am J 67:798–805. Link: https://goo.gl/8PQDoJ

Fisher G, de Pauw E, van Veldhuizen HT, Nachtergaele FO, Antoine J (1995) A provisional world climate resource inventory based on the length of growing period concept. In: Batjes NH, Kauffman JH, Spaargaren OC (eds) Proceedings of a workshop on national soil reference collections and databases (NASREC), Vol 3. Papers and Country Reports. ISRIC: Wageningen, The Netherlands, pp 30–43

Flowers TJ, Colmer TD (2008) Salinity tolerance in halophytes new Phytologist 179:945–963. https://doi.org/10.1111/j.1469-8137.2008.02531.x

Friedlingstein P, Cox P, Betts R, Bopp L, Bloh WV (2006) Climate–carbon cycle feedback analysis: Results from the C4MIP model intercomparison. J Clim 19:3337–3353. Link: https://goo.gl/YyadFk

Fritsch E, Fitzpatrick RW (1994) Interpretation of soil features produced by ancient and modern processes in degraded landscapes. A new method for constructing conceptual soil-water-landscape models. Aust J Soil Res 32:889–907

Gemechu D (1988) Environment and development in Ethiopia. In: Penrose A (ed) Beyond the famine: an examination of the issues behind famine in Ethiopia. Geneva

German L (2007) Participatory integrated watershed management: evolution of concepts and methods in an ecoregional program of the eastern African highlands. Agric Syst 94:189–204

Girma T (2001) Land degradation: a challenge to Ethiopia. International Livestock Research Institute, Addis Ababa, Ethiopia. Pp 815–823

Gissila T, Black E, Grimes DIF, Slingo MJ (2004) Seasonal forecasting of the Ethiopian summer rains. Int J Climatol 24:1345–1358. https://doi.org/10.1002/joc.1078

Gizaw DG, Kassawmar TT, M, Zeleke G (2021) Extent and distribution of surface soil acidity in the rainfed areas of Ethiopia. Land Degrad Dev. https://doi.org/10.1002/ldr.4113

Griffiths CL (1976) Some new and notable Amphipoda from southern Africa. Ann South African Museum 72:11–35

Grum B, Assefa D, Hessel R, Woldearegay K, Kessler A, Ritsema C, Geissen V (2017a) Effect of in situ water harvesting techniques on soil and nutrient losses in semi-arid northern Ethiopia. Land Degrad Dev 28:1016–1027

Grum B, Woldearegay K, Hessel R, Baartman JE, Abdulkadir M, Yazew E, Kessler A, Ritsema CJ, Geissen V (2017b) Assessing the effect of water harvesting techniques on event-based hydrological responses and sediment yield at a catchment scale in northern Ethiopia using the Limburg Soil Erosion Model (LISEM). Catena 159:20–34

Griscom B, Adams J, Ellis P, Houghton R, Lomax G, Miteva D, Schlesinger W, Shoch D, Siikamäki J, Smith P, Woodbury P, Zganjar C, Blackman A, Campari J, Conant R, Delgado C, Elias P, Gopalakrishna T, Hamsik M, Fargione J (2017) Natural climate solutions. Proc Natl Acad Sci.https://doi.org/10.1073/pnas.1710465114

Guo JH, Liu XJ, Zhang Y, Shen JL, Han WX, Zhang WF, Christie P, Goulding KW, Vitousek PM, Zhang FS (2010) Significant acidification in major Chinese croplands. Science 327:1008–1010

Haile T (1988) Causes and characters of drought in Ethiopia. Ethiop. J Agri Sci 10(1–2):85–97

Haile W, Boke S (2011) Response of Irish Potato (Solanum tuberosum) to the application of Potassium at Acidic Soils of Chencha, Southern Ethiopia. Int J Agric Biol 13:595–598

Haregeweyn N, Berhe A, Tsunekawa A (2012) Integrated watershed management an effective approach to curb land degradation: A case study of the Enabered watershed, northern Ethiopia. J Environ Manage 50:1219–1233

Hawando T (1986) Agricultural research in regions with erratic rainfall distribution in Ethiopia. Part I. Rainfall characteristics and its implications on soil and water management practices and crop production patterns in eastern Ethiopia. FAO/TCP/ETH/4403 Projec,. 70 p

Hiben MG, Tesfa-Alem G (2014) Spate irrigation in Tigray: the challenges and suggested ways to overcomethem. In: Flood-based Farming for food security and adaption to climate change in Ethiopia: potential andchallenges, International Water Management Institute (IWMI). Colombo, Sri Lanka, pp 137–148

Hurni H (1983) Soil erosion and soil formation in agricultural ecosystems: Ethiopia and Northern Thailand. Mt Res Dev 3 (2):131–142

Hurni H, Kebede T, Zeleke G (2005) Implications of changes in population, land use and land management for surface runoff in the upper basin area of Ethiopia. Mt Res Dev 25(2):147–154. https://doi.org/10.1659/0276-4741(2005)025[0147:TIOCIP]2.0.CO;2

Hurni K, Zeleke G, Kassie M, Tegegne B, Kassawmar T, Teferi E, Moges A, Tadesse D, Ahmed M, Degu Y, Kebebew Z (2015) Soil degradation and sustainable land management in the rainfed agricultural areas of Ethiopia: an assessment of the economic implications. Report for the economics of land degradation initiative

Hutchinson MF, McIntyre S, Hobbs RJ, Stein JL, Garnett S, Kinloch J (2005) Integrating a global agro-climatic classification with bioregional boundaries in Australia. Glob Ecol Biogeogr 14:197–212

ISRIC, International Soil Reference Center (2020) Soil map of Ethiopia at 250m Resolution. World Soil Information, Ministry of Agriculture, Ethiopia

Jinbo Z, Yu X, Lei M, Christoph M, and Zucong C (2015) Variations of soil N transformation and N_2O emissions in tropical secondary forests along an aridity gradient. J Soils Sediments. https://doi.org/10.1007/s11368-015-1121-7

Jobbágy E, Jackson R (2000) The Vertical distribution of soil organic carbon and its relation to climate and vegetation. Ecol Appl 10:423–436. https://doi.org/10.2307/2641104

Kassahun B (1987) Weather systems over Ethiopia. Proceedings first technology conference on meteorological research in eastern and Southern Africa, Nairobi, Kenya, UCAR, pp 53–57

Kidane G, Abebe F, Heluf GK, Fentaw A, Wondimadegne C, Hibstu A, Asegid A, Messele F, Mohammed B (2006) Report of the national task force on assessment of salt affected soils and recommendations on management options for sustainable utilization; Ethiopia, A.A., Ed.; Minister of Agriculture and Rural Development: Addis Ababa, Ethiopia, 120p

Kidanemariam A, Gebrekidan H, Mamo T, Fantaye KT (2013) Wheat crop response to liming materials and N and P fertilizers in acidic soils of Tsegede highlands, northern Ethiopia. Agric Forest Fish 2:126–135

Koppen W (1936) Das geographisca system der Klimate. In Handbuch der Klimatologie, Vol. 1, W Koppen, G Geiger (eds). C. Gebr, Borntraeger: Berlin, 1– 44.

Lal R (2004) Soil carbon sequestration to mitigate climate change. Geoderma 123:1–22. https://doi.org/10.1016/j.geoderma.2004.01.032

Le Houérou HN (1996) Climate change, drought and desertification. J Arid Environ 34:133–185

Le Quéré C, Andrew RM, Friedlingstein P, …Walker AP, Wiltshire A J, Wright R, Zaehle S, Zheng B (2018) Global carbon budget. Earth Syst Sci Data 10:2141–2194https://doi.org/10.5194/essd-10-2141

Liben F, Birhanu T, Tola Y, Wortmann C, Kim H, Mupangwa W (2018) Conservation agriculture effects on crop productivity and soil properties in Ethiopia. Agron J.https://doi.org/10.2134/agronj2017.07.0384

Liben F, Hassen S, Weyesa B, Wortmann CS, Kim H, Sime M, Yeda GG, Beshir B (2017) Conservation agriculture for maize and bean production in the central rift valley of Ethiopia. Agron J 109:2988–2997. https://doi.org/10.2134/agronj2017.04.0072

Liben F, Wortmann CS, Tirfessa A (2020) Geospatial modeling of conservation tillage and nitrogen timing effects on yield and soil properties. Agric Syst 177:1–10. https://doi.org/10.1016/j.agsy.2019.102720

McSweeney C, New M, Lizcano G (2008) Cambodia climate change profile. Oxford University, Oxford, UK. Available at http://country-profiles.geog.ox.ac.uk

MEFCC (2015) The first transformational and growth plan (GTPI) (2011–2015). Implementation concluding report, Addis Ababa

Mekonnen M, Keesstra S, Baartman J, Ritsema C, Melesse A (2015) Evaluating sediment storage dams:structural off-site sediment trapping measures in northwest Ethiopia. Cuadernos de Investigación Geográfica 41:7–22

Meshesha D, Tsunekawa A, Tsubo M, Haregeweyn N, Adgo E (2014) Evaluating spatial and temporal variations of rainfall erosivity, case of Central Rift Valley of Ethiopia. Theoret Appl Climatol 119. https://doi.org/10.1007/s00704-014-1130-2

Ministry of Agriculture (MOA) (2021) Ethiopia's Rural Productive Safety Net Programme's contribution to sustainable agriculture and to NDC in climate change adaptation and mitigation (unpublished)

MoARD (2006) Watershed management guidelines. Addis Ababa: agriculture sector support project (ASSP), ministry of agriculture and rural development

Mosissa F (2018) Progress of soil acidity management research in Ethiopia. Adv Crop Sci Technol 6.https://doi.org/10.4172/2329-8863.1000377

Niang I, Ruppel OC, Abdrabo MA et al (2014) Africa. In: Barros VR, Field CB, Dokken DJ et al (eds) Climate change 2014: impacts, adaptation, and vulnerability. Part B: regional aspects. Contribution of Working Group, pp 1199–1265

Nyssen J, Poesen J, Gebremichael D (2007) On-site evaluation of stone bunds to control soil erosion on cropland in northern Ethiopia. Soil Tillage Res 94:151–163

Nyssen J, Vandenreyken H, Poesen J, Moeyersons J, Deckers J, Haile M, Salles C, Govers G (2005) Rainfall erosivity and variability in the Northern Ethiopia Highlands. J Hydrol 311:172–187. https://doi.org/10.1016/j.jhydrol.2004.12.016

Nyssen J, Poesen J, Moeyersons J, Haile M, Deckers J (2008b) Dynamics of soil erosion rates and controllingfactors in the northern Ethiopian highlands–towards a sediment budget. Earth Surf Process Landf 33:695–711

Nyssen J, Govaerts B, Araya T, Cornelis WM, Bauer H, Haile M, Sayre K, Deckers J (2010) The use of themarasha ard plough for conservation agriculture in Northern Ethiopia. Agron Sustain Dev 31:287–297

O'Geen AT (2010) Research connects soil hydrology and stream water chemistry in California oak woodlands. Calif Agric 64:78–84

Panagos P, Borrelli P, Meusburgerb K, van der Zanden E, Poesen J, Alewell C (2015) Modeling the effect of support practices (P-factor) on the reduction of soil erosion by water at European scale. Environ Sci Policy 51:23–34

Sayer EJ, Oliver AE, Fridley JD, Askew AP, Mills RTE (2017) Links between soil microbial communities and plant traits in a species-rich grassland under long-term climate change. Ecol Evolut 7:855–862. Link: https://goo.gl/dmZLPm

Schulte RPO, Creamer RE, Donnellan T, Farrelly N, Fealy R, O'Donoghue C (2014) Functional land management: a framework for managing soil-based ecosystem services for the sustainable intensification of agriculture. Environ Sci Policy 38:45–58. https://doi.org/10.1016/j.envsci.2013.10.002

Segele Z, Lamb PJ, Leslie LM (2009a) Seasonal-to-interannual variability of Ethiopia/Horn of Africa monsoon. Part I: associations of Wavelet-Filtered large-scale atmospheric circulation and global sea surface temperature. J Climate 22:3396–3421. https://doi.org/10.1175/2008JCLI2859.1

Segele ZT, Lamb PJ (2005) Characterization and variability of Kiremt rainy season over Ethiopia. Meteorol Atmos Phys 89:153–180. https://doi.org/10.1007/s00703-005-0127-x

Segele ZT, Lamb PJ, Leslie LM (2009b) Large-scale atmospheric circulation and global sea surface temperature associations with Horn of Africa June-September rainfall. Int J Climatol 29:1075–1100. https://doi.org/10.1002/joc.1751

Seleshi Y, Zanke U (2004) Recent changes in rainfall and rainy days in Ethiopia. Int J Climatol 24:973–983. https://doi.org/10.1002/joc.1052

Shoghi KS, Pannell DJ, Thamo T, White B, Polyakov M (2019) Soil acidity, lime application, nitrogen fertility, and greenhouse gas emissions: optimizing their joint economic management. Agric Syst 176,:102684, ISSN 0308-521X. https://doi.org/10.1016/j.agsy.2019.102684

SLMP (2013a) Sustainable land management project (SLMP) II revised final draft document on environmental and social management framework (ESMF) September 2013. Addis Ababa, Ethiopia: SLMP

Smith P (2016) Soil carbon sequestration and biochar as negative emission technologies. Glob Change Biol 22(3):1315–1324

Steinweg JM, Plante AF, Conant RT, Paul EA, Tanaka DL (2008) Patterns of substrate utilization during long-term incubations at different temperatures. Soil Biol Biochem 40:2722–2728. https://doi.org/10.1016/j.soilbio.2008.07.002

Swati T, Ramesh S, Shaily J (2014) Effect of climate change on plant-microbe interaction: an overview. Europ J Molec Biotechnol 5:149–156. Link : https://goo.gl/tvo1AC

Tanji KK (1990) The nature and extent of agricultural salinity problems. In: Tanji KK (ed), Agricultural salinity assessment and management. ASCE manual reports on engineering practices, 71, pp 1–41

Taye WG, Poesen J, Vanmaercke M, Bas W, Martens L, Teka D, Nyssen J, Deckers J, Vanacker V, Haregeweyn N, Hallet V (2015) Evolution of the effectiveness of stone bunds and trenches in reducing runoff and soil loss in the semi-arid Ethiopian highlands. Z Geomorphol 59:477–493. https://doi.org/10.1127/zfg/2015/0166

Taye G, Poesen J, Vanwesemael B, Vanmaercke M, Teka D, Deckers J, Goosse T, Maetens W, Nyssen J,Hallet V (2014) Effectiveness of soil and water conservation structures in reducing runoff and soil loss for differentland use and slope gradients: case study from northern Ethiopia. EGU General Assembly conference abstracts

TerrAfrica (2009) Advancing sustainable land management in Sub-Saharan Africa our land, our wealth, our future, our hands TerrAfrica annual report. New Partnership for Africa's Development (NEPAD)

Tester M, Davenport R (2003) Na+ Tolerance and Na+ transport in higher plants. Ann Bot 91:503–527. https://doi.org/10.1093/aob/mcg058

Teshome A, Rolker D, de Graaff J (2013) Financial viability of soil and water conservation technologies innorthwestern Ethiopian highlands. Appl Geogr 37:139–149

Viste E, Sorteberg A (2013) Moisture transport into the Ethiopian highlands. Int J Climatol 333:249–263. https://doi.org/10.1002/joc.3409

Vrieling A, Sterk G, De Jong S (2010) Satellite-based estimation of rainfall erosivity for Africa. J Hydrol 395:235. https://doi.org/10.1016/j.jhydrol.2010.10.035

Waktola DK (2007) Rainwater harvesting in Ethiopia: capturing realities and harnessing opportunities.Rainwater Urban Design 2007:484

Weiman S (2015) Microbes help to drive global carbon cycling and climate change. Microb Mag 10(6):233–238

White DH (1998) A global agro-climatic analysis of the distribution and production of livestock commodities. Economic Evaluation Unit Working Paper No. 30, Australian Centre for international agricultural research, p 74

Woldearegay K, van Steenbergen F (2015) Shallow groundwater irrigation in Tigray, northern Ethiopia:practices and issues. In: Lollino G, Arattano M, Rinaldi M, Giustolisi O, Marechal J-C, Grant GE (eds) Engineeringgeology for society and territory - volume 3: river basins, reservoir sedimentation and water resources. SpringerInternational Publishing, Cham

Woldearegay K, van Steenbergen F, Perez MA, Grum B, van Beusekom M (2015) Water harvesting fromroads: climate resilience in Tigray, Ethiopia. IRF Examiner. 12 (Winter 2017): 1–7

Worku B, Sombat C, Rungsit S, Thongchai M, Sunanta J (2006) Conservation tillage and crop rotation: win-win option for sustainable maize production in the dryland, central rift valley of Ethiopia Kamphaengsaen Acad. Journal 4(1):48–60

Wudu A, Mahider W (2020) Soil salinity and its management options in Ethiopia. Edition 2 |Article 1042 ScienceForecast Publications LLC

Yergeau E, Sanschagrin S, Maynard C et al (2012) Microbial expression profiles in the rhizosphere of willows depend on soil contamination. ISME J 8:344–358 (2014). https://doi.org/10.1038/ismej.2013.163

Zhang M, Riaz M, Zhang L, El-desouki Z, Jiang C (2019) Biochar induces changes to basic soil properties and bacterial communities of different soils to varying degrees at 25 mm rainfall: more effective on acidic soils. Front Microbiol 10. https://doi.org/10.3389/fmicb.2019.01321.ISSN=1664-302X

Zhu JK (2001) Plant salt tolerance. Trends Plant Sci 6:66–71. https://doi.org/10.1016/S1360-1385(00)01838-0

Zuo H (1996) Agroclimatic analysis for mainland East Asia by a GIS Approach. PhD thesis, Australian National University, Canberra, Australia, 264 p

Zuo H, Hutchinson MF, McMahon JP, Nix HA (1996a) Developing a mean monthly climatic database for China and Southeast Asia. In: Proceedings of an international workshop on matching trees and sites, Bangkok, Thailand, 27–30 March 1995. Canberra: Australian Centre for International Agricultural Research, 1995

Zuo H, McMahon JP, Nix HA, Hutchinson MF (1996b) A GIS–based agro–climatic classification for mainland East Asia. In: Proceedings of the second Australian conference on agricultural meteorology, The University of Queensland, 1–4

4 Geology and Geomorphology

Bekele Abebe, Dereje Ayalew, Alemayehu Regassa, and Gebeyehu Taye

Abstract

Ethiopia is characterized by a wide variety of geological formations and geomorphological features. The major rock types on the basis of chemical and mineralogical composition of Ethiopia include the Precambrian metamorphic rocks with associated plutonic igneous rocks which form the basement, the Paleozoic-to-Mesozoic sedimentary rocks, Cenozoic basic and felsic volcanic rocks with intercalated sedimentary rocks. Internal earth processes including metamorphism, folding, faulting, Paleozoic–Mesozoic sedimentation, and Cenozoic volcanism set the ground for surface processes. Erosion and deposition modified the physiography and landforms, which, in turn, influenced soil formation and distribution in Ethiopia. The major geomorphologic elements are summits and crests straddling the nearly flat surfaces of the plateau volcanic pile; plateaus (the western and eastern highlands without the overlying volcanic massifs); flat-gently sloping Rift Valley (main Ethiopia Rift and the Afar depression); the Ogaden Plains; and low-lying areas in the west, southwest, and southern peripheral parts of the country; bottom areas which include the lake basins and areas lying below 500 m a.s.l. The major soil types vary in response to parent materials and landforms among other factors of soil formation. Rock types affect soil texture and mineralogy while landform affects spatial distribution of soils due to complex erosion and depositional processes. Nitisols and Vertisols dominate on basalt formation whereas leptosols exclusively dominate hill slope irrespective of the geological formation. Understanding parent materials and land form is crucial toward understanding, mapping, and managing soil resources.

Keywords

Geology • Geomorphology • Landform • Rock types • Soils

4.1 Introduction

Ethiopia has diverse geological structures and physiography that exhibit remarkable variety and mosaic of landscapes (Schluter 2000; Henricksen et al. 1984a, b). The landscape and landforms of Ethiopia resulted from the Precambrian-Mesozoic events of deformation, erosion, and sedimentation overprinted in most parts of the country by the Cenozoic uplift, volcanism, and faulting associated with the opening of the East African Rift. The Ethiopian volcanic plateau was formed when extensive eruptions deposited ~350,000 km^3 of basalt (flood basalt; marking the appearance of the Ethiopian-Afar plume super swell) over a <2 Ma period (Mohr and Zanettin 1988; Mohr 1983). Several large shield volcanoes subsequently developed on the surface of the flood basalt (Mohr and Zanettin 1988), which in turn produced ~1/5 of the total volume of basaltic lava distributed in discrete outcrops across the flood basalt. Deposits from these (younger) shield volcanoes are generally more felsic, thinner, and less continuous than the extensive flood basalts that they cover.

The rocks exposed in Ethiopia are Pre-Cambrian, Jurassic, Cretaceous, Tertiary, and Quaternary in age. Tertiary and Quaternary volcanism has been extensive in Ethiopia.

B. Abebe (✉)
School of Earth Sciences, Addis Ababa University, P.O. Box 1176 Addis Ababa, Ethiopia
e-mail: bekele.abebe@aau.edu.et

D. Ayalew
School of Earth Sciences, Addis Ababa University, P.O. Box 1176 Addis Ababa, Ethiopia

A. Regassa
Department of Natural Resource Management, Jimma University, P.O. Box 307 Jimma, Ethiopia

G. Taye
Land Resource Management and Environmental Protection, P.O. Box 231 Mekelle, Ethiopia
e-mail: gebeyehu.taye@mu.edu.et

S. Beyene et al. (eds.), *The Soils of Ethiopia*, World Soils Book Series,
https://doi.org/10.1007/978-3-031-17012-6_4

The Ethiopian Plateau, which occupies the western half of the country, is generally above 1000 m high, except for the major river valleys, and much of the area is over 2000 m high. The Simien Mountains in the northern part of the Ethiopian Plateau include peaks higher than 4000 m. The Eastern Plateau in the eastern part of the country generally exceeds 1000 m in elevation. The Ethiopian Rift Valley lies between the two plateaus. The southern section of the rift system consists of a series of internal drainage basins and lakes. The northern part, known as the Afar Region, includes the Danakil Desert with extensive salt plain and a number of closed, saline lake basins. Geomorphologic features bear the imprint of complex tectonic, volcanic, erosional, and slope processes. The location and form of mountain ranges, escarpments, intermontane and rift basins, alluvial planes, and peneplanation surfaces are all controlled by the combined actions of external and internal earth processes.

Geology can be considered as the background factor for soil formation. Soil parent materials determine the rate of soil formation, the composition, texture, structure, and nutrients available to plants. All kinds of parent material are subjected, in varying degrees, to the other soil-forming factors acting simultaneously with different intensities and efficiencies. Several studies have examined the differences in various soil properties under specific parent materials over particular regions (e.g., Gruba and Socha 2016; Cathcart et al. 2008; Chaplot et al. 2003), but results did not show clear universal trends (Gray et al. 2016). This is because a given soil is the product of interactive effects of the five soil-forming factors and several pedogenic processes operating over geological time. For instance, in Western Ethiopia, soils are exclusively developed on Trap series volcanics, Precambrian basement, colluvial and alluvial parent materials. However, irrespective of the diversity of parent materials, soils tend to be similar over large area due to similarity in climate and vegetation condition (FAO 1984). Several studies have shown that parent material has a strong influence on chemical weathering rates, profile depth, clay content, and cation exchange capacity (Driese et al. 2003; Dekayir and El-Maataoui 2001; Palumbo et al. 2000; Vaselli et al. 1997; Venturelli et al. 1997). Direct relationship exists between the texture and mineralogy of parent material and the texture of sand fraction, mineralogy, and the base status of the soils. The clay mineralogy of the soils is controlled by parent material composition, drainage, and the pH of the chemical system (Keller 1964). In the crystalline rocks, such as granite, base-poor soils including Lixisols and Acrisols are formed (FAO 1984). The volcanic basalt rocks give way to the development of base-rich soils including Vertisols, Leptosols and Phaeozems (Mesfin 1998). A detailed geological study thus appears to be important in a study designed to determine soil variations in the landscape and the reasons for these variations (Teramoto et al. 2001).

Geomorphology influences soil formation through its effects on climate and vegetation and differential transport and deposition of eroded materials. Therefore, the role of major and local geomorphological characteristics of a region is, especially, of paramount importance in determining the nature of transported parent materials of soils. This, in turn, controls the ultimate products—soils. Landforms control much of the distribution of soils in the landscape to such an extent that soils of markedly different properties may develop from the same parent material and soils with similar characteristic can form on different parent materials. In the flat landscape areas of the rift valley systems, the Omo Basin, Gambella, northwestern Ethiopia, Ogaden, and the lower reaches of wide river valleys, the soils range from poorly drained to moderately drained. These areas are mainly depositional surfaces where erosion is not a threat to the soils and the environment. Erosion from surrounding slopes also provides additional sediment material to the valley bottoms. Except the intermontane basins and marginal grabens, the mountainous terrain of the western and eastern plateaus and the rift margins are characterized by different landforms with differing runoff, erosion, sunlight, precipitation, and vegetation. These differences create their own influences on soil physical and chemical properties. In general, good understanding of the relationship between soil, geology, and geomorphology is important for understanding the occurrence of soil in the landscape, thus allowing the prediction of soil distribution across landscape.

4.2 Geological Distribution

The main rock types of Ethiopia are (1) the Precambrian metamorphic rocks with associated plutonic igneous rocks which form the basement; (2) the Paleozoic-to-Mesozoic sedimentary rocks; (3) the Cenozoic basic and felsic volcanic rocks; and (4) the sedimentary rocks, associated with the Cenozoic volcanics (Fig. 4.1).

4.2.1 Precambrian Rocks

The Precambrian basement rocks of Ethiopia are exposed in the peripheral parts (Fig. 4.2a) of all corners of the country (e.g., Asrat et al. 2001; Kazmin et al. 1978). This part of the Ethiopian geologic history consists of intensely folded, faulted, and tilted rocks of sedimentary and igneous origin which were metamorphosed to varying degrees as well as acidic-to-ultramafic intrusive rocks (Abbate et al. 2015 and references therein). The metamorphic rocks comprise in low-grade rocks including slates and phyllites, intermediate-grade schists, and high-grade gneisses of varied compositions (Fig. 4.2b). The low-grade metamorphic rocks

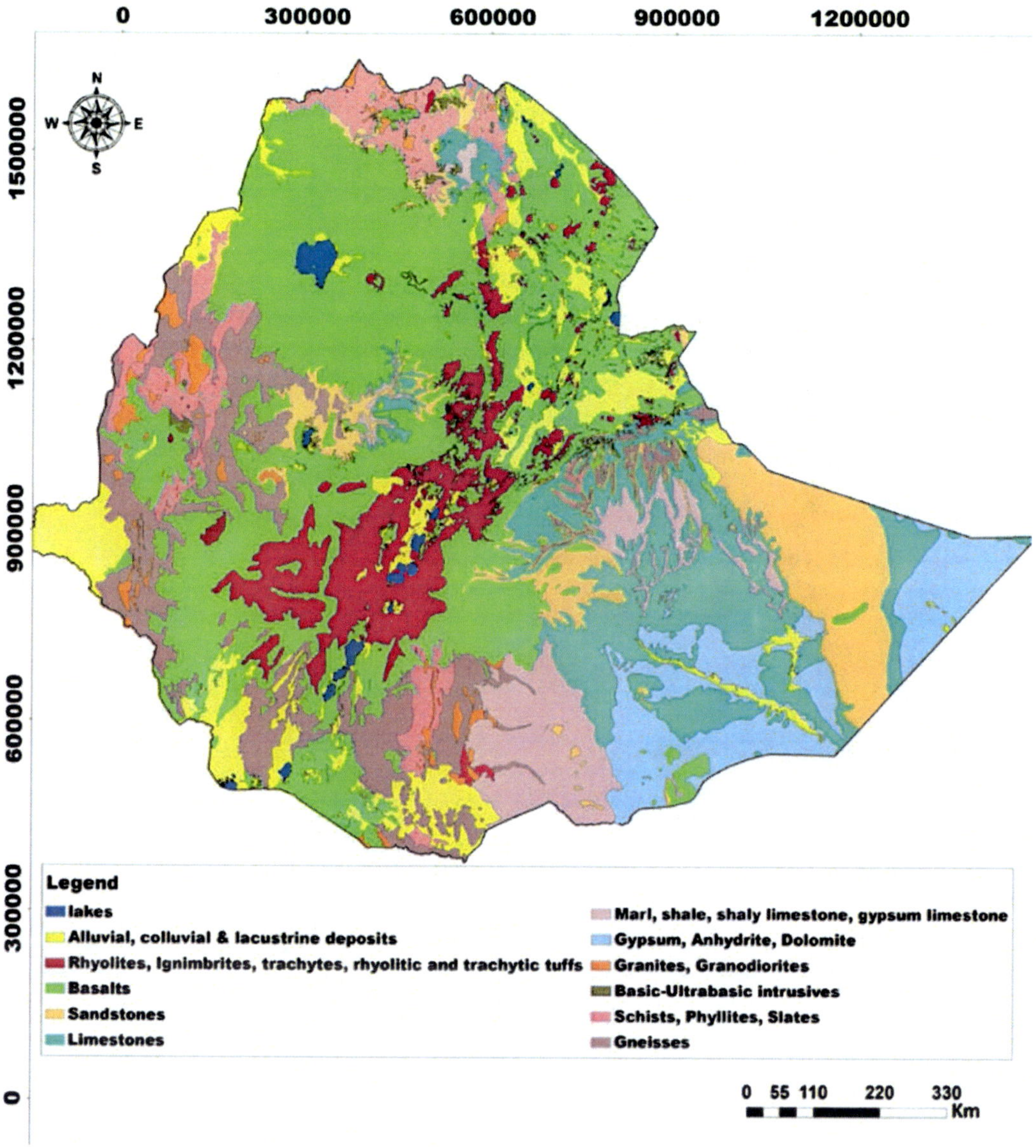

Fig. 4.1 Geologic map of Ethiopia (modified from Tefera et al. 1996a, b. *Geological Survey of Ethiopia. Reproduced with permission. All rights reserved*)

consist mainly of graphitic and tuffaceous slates, phyllites; graphite, sericite, chlorite, carbonate, biotite, graphite, talc schists, metaconglomerates, and meta sandstones (Fig. 4.2c).

These are associated with the high-grade gneisses of granitic, granodioritic, amphibolite-mica, quartzo-feldspathic, pyroxene-garnet, and other compositions (Tefera et al. 1996a, b; Kazmin 1972).

The crystalline basement rocks are intruded by pre-syn- and post-tectonic granite (Fig. 4.3), granodiorite, diorite, tonalite and gabbro as well as dolerites and pegmatite veins (Fig. 4.3a).

4.2.2 Paleozoic and Mesozoic Sedimentary Rocks

A large part of the Paleozoic-to-Mesozoic sedimentary rocks are exposed on the eastern Ogaden, the dissected plateau areas in the Blue Nile River Basin and in northern Tigray around Mekele (Fig. 4.4) described as Mekelle outlier. The typical succession of the basin includes from bottom to top three formations: (i) lower sandstone or Adigrat sandstone; (2) Antalo limestone; and upper sandstone (Amba Aradam Sandstone) covered with flood basalt.

The Paleozoic Era in Ethiopia is represented mainly by peneplanation of the Precambrian basement morphology and

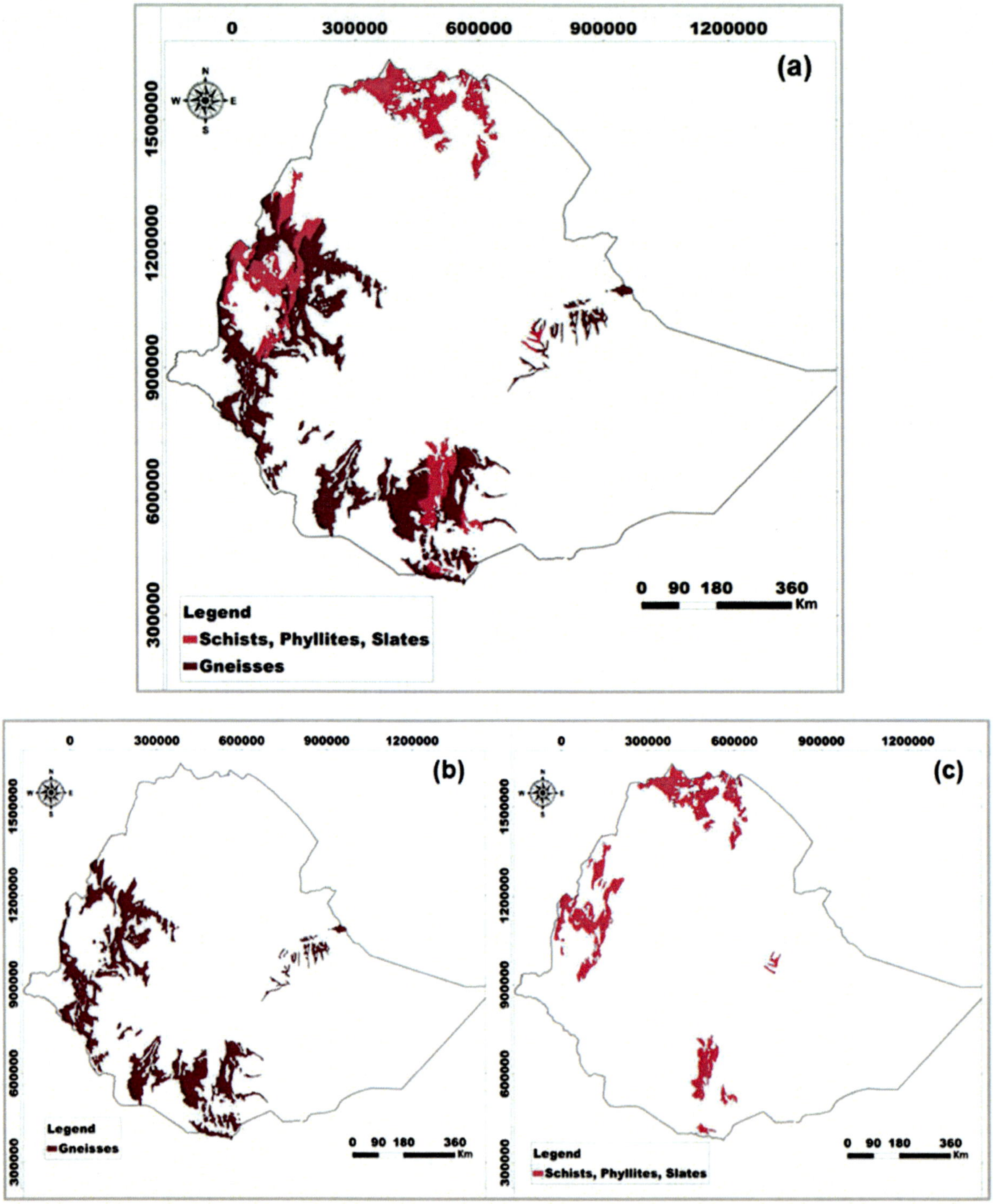

Fig. 4.2 Distribution of Precambrian rocks in Ethiopia, **a** low- and high-grade metamorphic rocks with acidic-to-ultramafic intrusions; **b** high-grade rocks including gneisses and migmatites; **c** schists, phyllites, and slates (*modified from* Tefera et al. 1996a, b. *Geological Survey of Ethiopia. Reproduced with permission. All rights reserved*)

Fig. 4.3 Examples of Precambrian basement rocks in Ethiopia: **a** Granitic gneiss cut by basaltic dikes (Dengego-Dire Dawa road); **b** Granites of Babile areas (*photo courtesy of Bekele Abebe*)

a few patchy sedimentary outcrops resting unconformably on the tilted and eroded basement (Bussert and Schrank 2007). These sediments are mainly fluviatile lower Enticho sandstone overlain by glacial fluvio-lacustrine Upper Enticho sandstones and glacials (Bussert and Schrank 2007; Sacchi et al. 2007; Russo et al. 1994; Dow et al. 1971).

Following the Paleozoic peneplanation and localized sedimentation, regional marine transgression and regression led to the invasion of the Horn of Africa from the northeast and east by the sea (e.g., Bosellini 1989). Continental clastic sediments were the first to be deposited and these formed the Adigrat Sandstone lying over the Permo-Triassic sediments. They were mainly deposited in fluvial or piedmont zones, but also in fluvio-lacustrine and deltaic environments (e.g., Wolela 2008; Bosellini et al. 1997, 2001; Beauchamp 1977). The Adigrat Sandstone has a widespread occurrence in Ethiopia with correlative units in the whole of East Africa and Arabia (Abbate et al. 2015). It attains a maximum thickness of 700 m (Fig. 4.5a).

Conformably overlying the Adigrat sandstone, up to 1,000-m-thick shallow marine carbonates, namely, the Antalo limestone and Hamanlei Formation were deposited in the Ogaden, Blue Nile, and around Mekele areas (Hunegnaw et al. 1998; Blandford 1870). Marls, shales, gypsum, anhydrite, gypsiferous, and shaly units also occur in the Mesozoic sequence (see Fig. 4.6). The carbonate deposits are overlain by another clastic unit, called the Amba Aradam sandstone, having a maximum thickness of 200 m (Bosellini et al. 2001). It is associated with lenses of quartz conglomerates and red shales.

4.2.3 Cenozoic Volcanic Rocks

The development of the East African Rift System is accompanied by concomitant volcanic episodes which produced acidic, intermediate, and basic lavas and pyroclastics (Fig. 4.7a, b).

The Ethiopian Continental flood volcanism represents massive outpourings of basaltic lava flows with some rhyolitic lavas and pyroclastic rocks at 31–29 Ma (Ukstins et al. 2002; Baker et al. 1996). However, older basalts (35–45 Ma) occur in southern Ethiopia and these are related to the Kenyan plume (Rogers et al. 2000). The flood basalt eruption produced extensive horizontal lava flows intercalated with ignimbrites, forming the western and eastern highland plateaus (Fig. 4.8). These flows were covered by Mid-Upper Miocene shield volcanoes and plugs which, together with the rift faults, resulted in the rugged topography (Fig. 4.9).

Both volcanism and faulting propagated north–south along the Ethiopian Rift, hence forming border faults of differing ages (Wolfenden et al. 2004). After the establishment of the boundary between the rift floor and the plateaus, volcanic and tectonic activities became localized within the rift floor. Cenozoic volcanic activity in Ethiopia continued to occur in pulses producing mainly acidic and basic lavas as well as pyroclastics with minor intermediate varieties. During the Quaternary, volcanism and faulting became restricted mainly to the rift axis which is marked by regularly spaced axial acidic volcanoes, widespread pyroclastics, basaltic cones, and lava flows (Abebe et al. 2007 and references

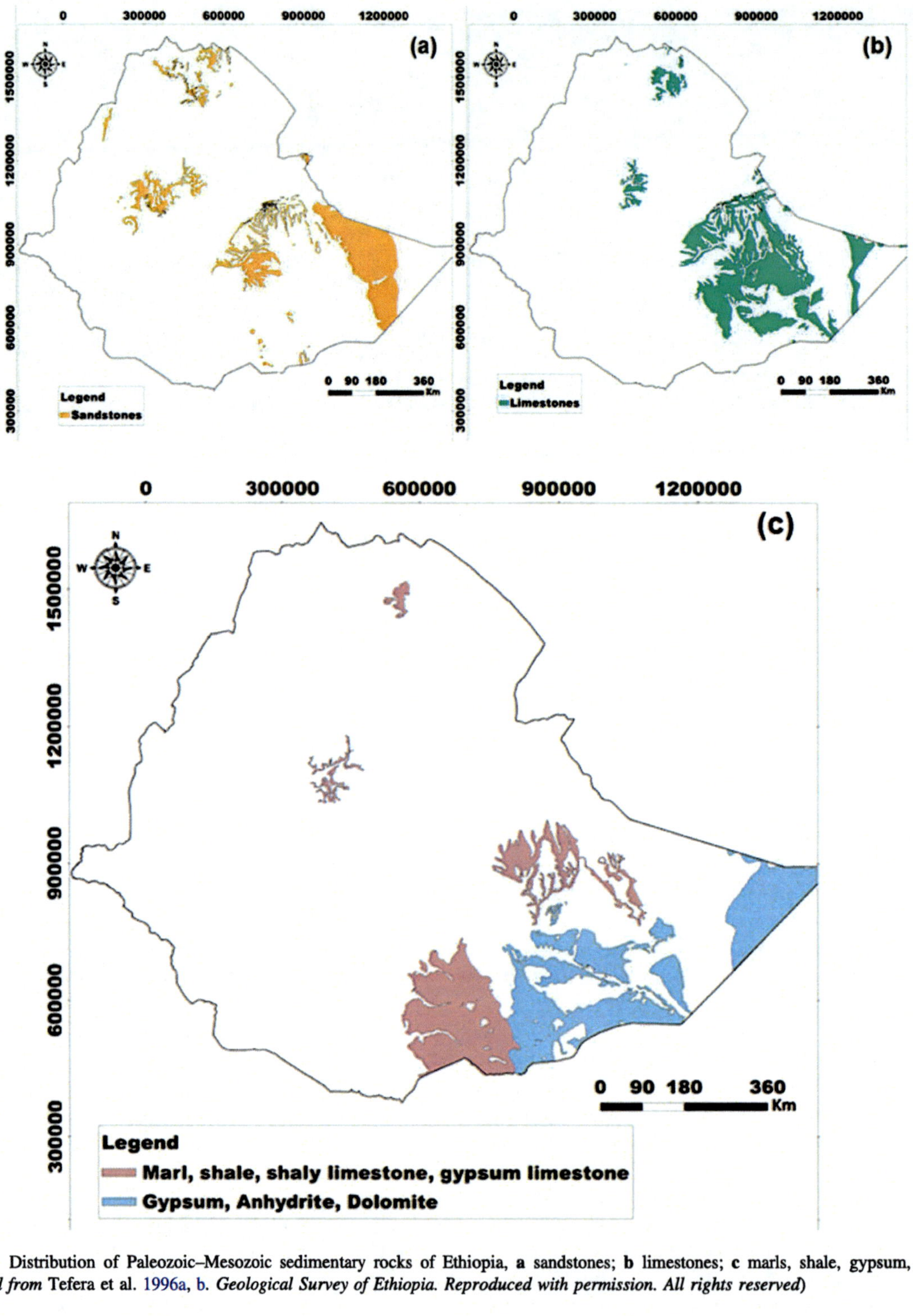

Fig. 4.4 Distribution of Paleozoic–Mesozoic sedimentary rocks of Ethiopia, **a** sandstones; **b** limestones; **c** marls, shale, gypsum, dolomite (*modified from* Tefera et al. 1996a, b. *Geological Survey of Ethiopia. Reproduced with permission. All rights reserved*)

Fig. 4.5 Lower (Adigrat) sandstone (**a**) and Limestone (**b**) in the Abay Gorge (*photo courtesy of Bekele Abebe*)

therein). The structurally controlled grabens and half grabens served as sites of sediment accumulation which became parent materials for subsequent soil formation (Fig. 4.10).

Associated with the continental flood basalt sequence of Ethiopia, fluvio-lacustrine sediments, consisting of mudstone, siltstone, lignites, and sandstones, occur in different parts of the country (Sembroni et al. 2019; Abbate et al. 2014). These intravolcanic sediments range in thickness from a few meters to ca. 200 m.

4.2.4 Quaternary Colluvial, Alluvial, and Lacustrine Deposits

This widespread unit comprises colluvial, alluvial, and lacustrine and slope deposits. It has an extensive distribution, mainly along the Ethiopian Rift from the Omo-Chew Bahir area to northern Afar (Fig. 4.11). Other occurrences are in the low lands on the western and south-western borders of the country. Moreover, this class of soil parent material is mapped along the alluvial flood plains of major rivers of the country.

Fig. 4.6 Limestone and shale intercalation in the Abay Gorge (*photo courtesy of Bekele Abebe*)

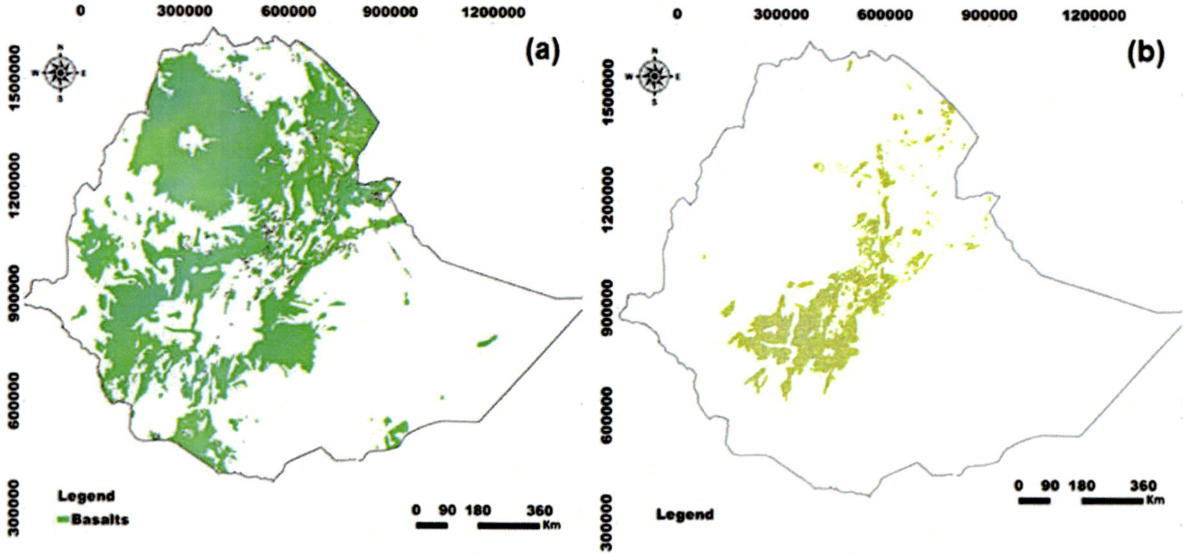

Fig. 4.7 Distribution of basic and felsic rocks in Ethiopia: **a** flood basalts, shield volcanoes, basaltic cones, and fissural flows of the rift valley; **b** acidic and intermediate rocks including rhyolite lava flows and domes, trachyte flows, domes and plugs, pumice, ignimbrite, and tuff (*photo courtesy of Bekele Abebe*)

4.3 Major Geologic Structures

The geologic history of Ethiopia is marked by various episodes of ductile and brittle deformation. During the Precambrian, there were multiple phases of folds, thrusts, and strike slip deformations. Major structures have strike directions of NE-SW, NNE-SSW, and NW–SE. Paleozoic–Mesozoic sedimentary basins in Ethiopia were controlled by NW–SE striking faults. Both the Precambrian weaknesses and Paleozoic–Mesozoic faults partly determined the configuration of the Cenozoic fault system of the Ethiopian Rift.

In Ethiopia, there are different sets of faults and lineaments which define basins and guide major rivers

Fig. 4.8 Horizontal basaltic lava flows of the plateau with minor ignimbrite levels (Afar window, edge of western plateau) (*photo courtesy of Bekele Abebe*)

Fig. 4.9 Rugged topography created by sedimentary and volcanic rocks deeply incised by the Abay River (*photo courtesy of Bekele Abebe*)

Fig. 4.10 Some typical landforms of the Ethiopian Rift **a** step faults, grabens, and half grabens along the eastern rift margin, Asela; **b** aligned basaltic cones and Holocene basaltic flows near Kone (Main Ethiopian Rift); **c** flat lying rift floor and Fantale Volcano on the background; **d** Dobi salt plain, Afar (*photo courtesy of Bekele Abebe*)

(Fig. 4.12). Such structures also determine the flow paths of rivers which, in turn, accumulate extensive colluvial and alluvial flood plain deposits. The faults also determined the geometry of grabens and half grabens which became one of the major landforms in Ethiopia. The main structural trends in the Main Ethiopian Rift and Afar are NE-SW, NNE-SSW, and NW–SE with a few E-W lineaments and faults on the plateaus, rift margins, and the rift floor.

4.4 Geology and Soil

Soils are regarded as products of geology and the weathering processes to which rock types are subjected. Though it is difficult to identify the effect of single soil-forming factor on soil properties, as the five factors of soil formation act in an integrated manner to shape soil, the effect of parent material is reflected on soil texture, color, fertility (content of basic cations), and soil pH. Several studies have shown that parent material has a strong influence on chemical weathering rates, profile depth, clay content, and cation exchange capacity (Driese et al. 2003; Dekayir and El-Maataoui 2001; Palumbo et al. 2000; Vaselli et al. 1997; Venturelli et al. 1997). The parent material includes all kinds of bedrocks and any type of unconsolidated sediments, such as alluvial, colluvial, lacustrine, eolian, and slope deposits. The properties and formation of soils are determined to a great extent by the rate and nature of chemical weathering at the parent material-regolith interface (Alemayehu et al. 2014). Differences in various soil properties under specific parent materials over particular regions have been reported by several authors (e.g., Gruba and Socha 2016; Cathcart et al. 2008; Chaplot et al. 2003). The effect of parent materials on soil properties is through the rates and products of weathering (Weil and Brady 2017; Eghbal et al. 2018). In particular, a remarkable influence of parent material on the resulting soil properties can easily be seen at the early stage of soil development and on the soils of dry regions. Since geology influences texture of the resulting soil type, this in turn influences several of the physical and chemical properties of soils.

In the crystalline rocks, such as granite, base-poor soils including Lixisols and Acrisols are formed (FAO 1984). The volcanic basalt rocks give way to the development of base-rich soils including Vertisols, Leptosols, and Phaeozems (Mesfin 1998). Basalt is one of the most common types of rocks in Ethiopia and so are the black, uniform texture,

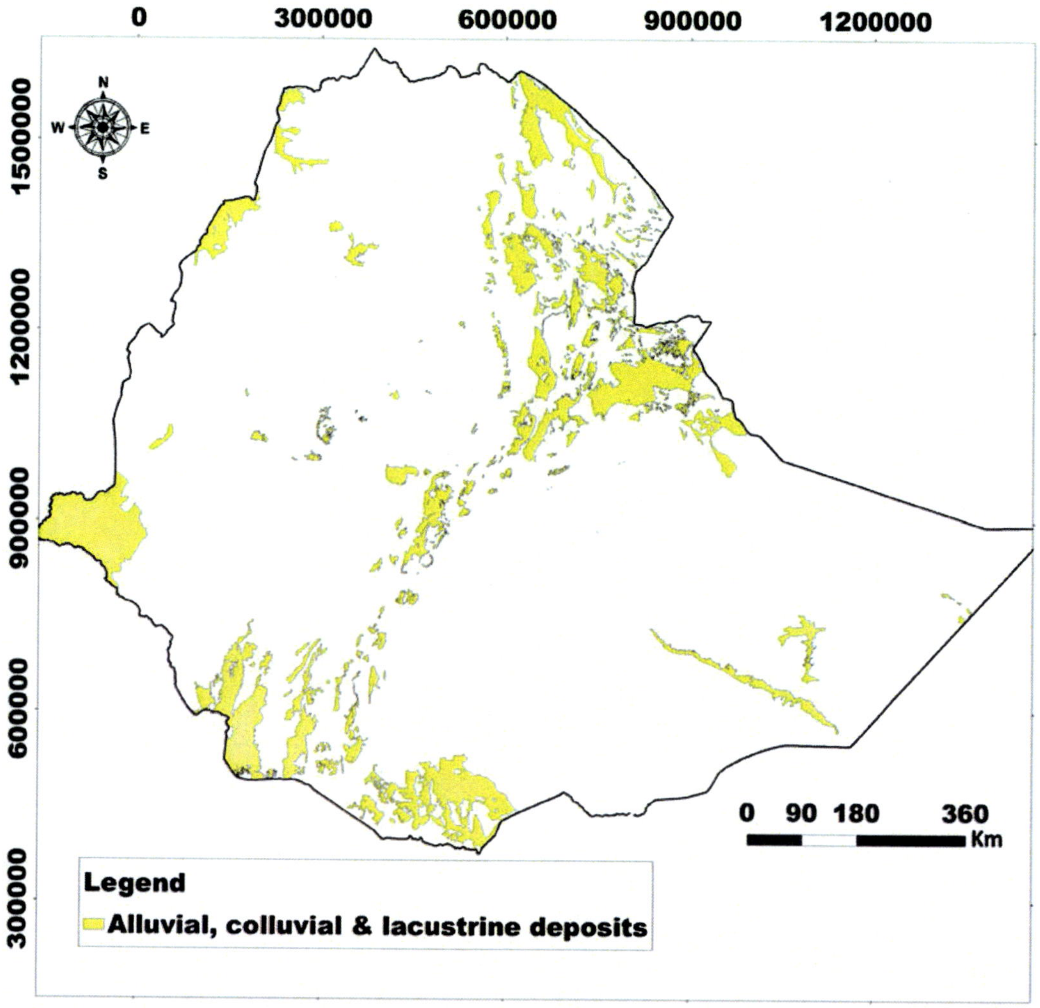

Fig. 4.11 Distribution of alluvial, colluvial, and lacustrine deposits in Ethiopia (*courtesy of Bekele Abebe*)

coarse structure, and dark gray clay soils which develop on the basalt, or the alluvium from it. Basalt releases much calcium and will produce only clay and silt when it weathers to soil. It tends to generate very fertile soils because it also provides phosphorus, along with significant amounts of iron, magnesium, and calcium. The physical properties of basalt reflect its relatively low silica content and typically high iron and magnesium content (USGS Volcano Hazards Program-Glossary 2015).

To understand the distribution of soils along the different types of geological formation, an overlay analysis has been conducted. Accordingly, soil types such as Vertisols, Nitisols, Luvisols, Acrisols, Regosols, Ferralsols, Andosols, and Leptosols on slope are derived from basalt lithology. Alluvial, colluvial, and lacustrine deposits give rise to the development of Fluvisols, Luvisols, Calcisols, Vertisols, Gleysols, and Regosols. Leptosols, Fluvisols, Arenosols, Alisols, Cambisols, and Andosols develop from rhyolites, trachytes, rhyolitic, and trachytic tuff formations. Cambisols, Leptosols, and Arenosols correspond to sandstone formation while Calcisols, Regosols, Cambisols, and Verisols correspond to limestone formation. Weathering of Marl and shaly limestone produces Vertisols, Leptosols, Gypsisols, and Calcisols. The distribution of Gypsum, Cambisols, Calcisols, Solonchaks, and Solonetz coincides with Gypsum, Anhydrite, and Dolomite parent materials under dry climatic conditions in the country. Coarse textured Granite and Granodiorites weather slowly into less developed sandy Arenosols, and well-developed Acrisols and also Ferralsols, Lixisols, and Nitisols on stable landscape with minimum rate of soil erosion and deposition. Basic-ultrabasic intrusive rocks produce Nitisols among others. Gneisses, Schists, Phyllites, and slates weather and produce Nitisols, Lixisols, Vertisols, Solonchaks, Solonetz, Phaeozems, and Regosols

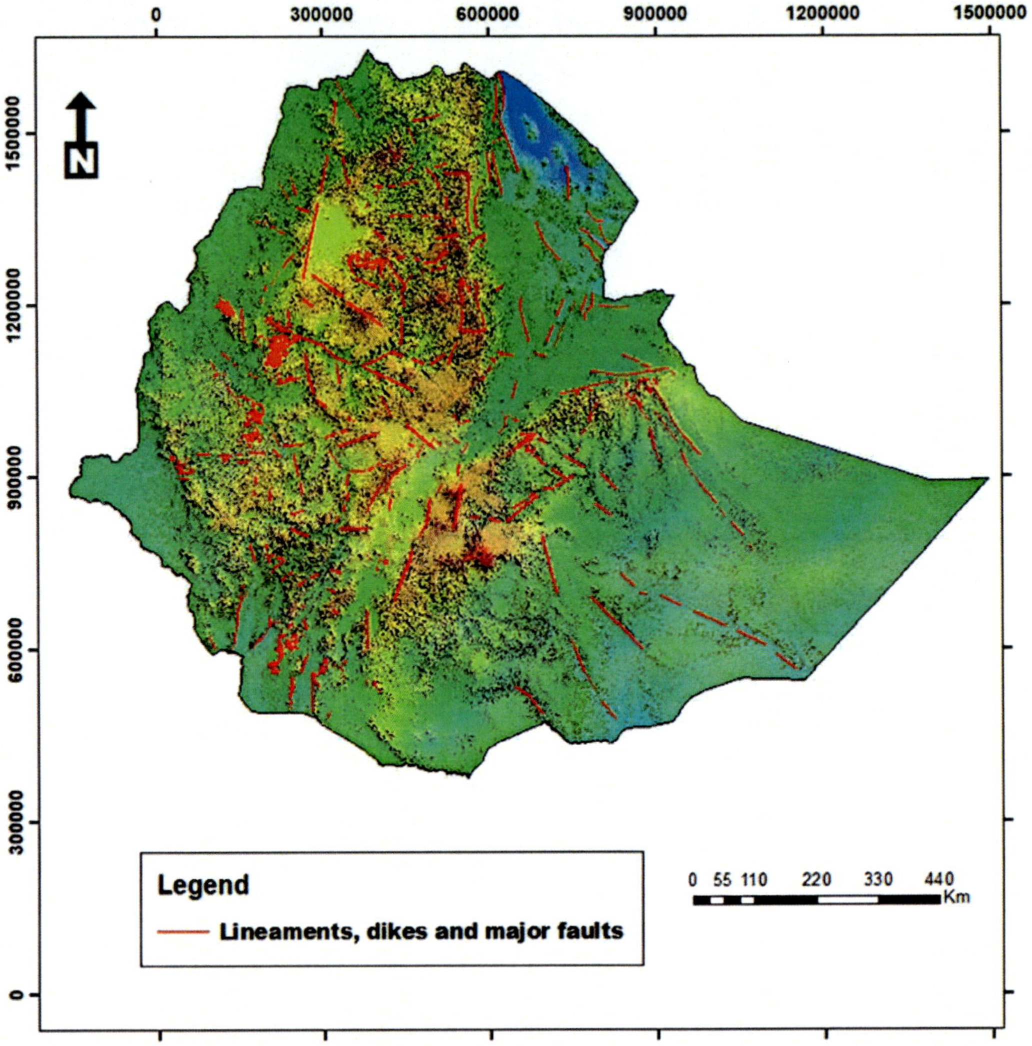

Fig. 4.12 Major geologic structures in Ethiopia showing that large structures on the plateaus are mainly NW–SE and the rift structures are NE-SW, NNE-SSW, and NW–SE (*courtesy of Bekele Abebe*)

among other soil types depending on the effects of local climate, vegetation, and land use (see Figs. 4.11, 4.12 and 4.13).

4.5 Geomorphology

The geomorphology of Ethiopia is intimately related to its underlying geology. The landscape and landforms have resulted from the Precambrian-Mesozoic events of deformation, erosion, and sedimentation overprinted in most parts of the country by the Cenozoic uplift, volcanism, and faulting associated with the opening of the East African Rift. Except where deeply dissected by subsequent denudation and apart from the volcanic piles of the Trap Series, the Ethiopian plateau shows a flat horizontal surface which expresses the presence of a peneplained Precambrian basement rocks. The Rift System divides the uplifted Ethiopian Massif into two units, the Western Plateau and the Eastern Plateau. The former includes all the highlands between the Rift scarps in the east and the Sudan border in the west. On both sides, this plateau has been subject to erosional recession starting from the original tectonic scarps. Except for the major river valleys, the whole of this region lies above 1000 m. and about half of it reaches altitudes above 2000 m.

According to Leenaars (2019), Ethiopia is subdivided into four physiographic features consisting of (Fig. 4.14).

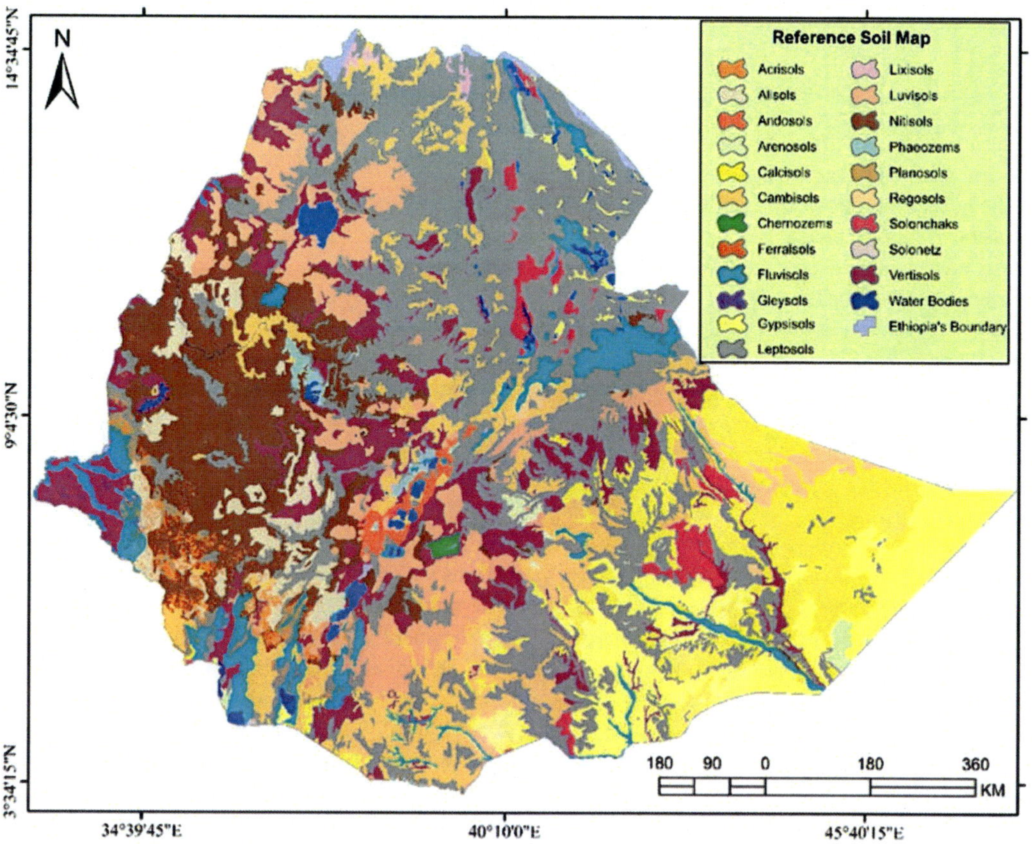

Fig. 4.13 Soil map of Ethiopia (based on FAO/ISRIC/SOTER 1997)

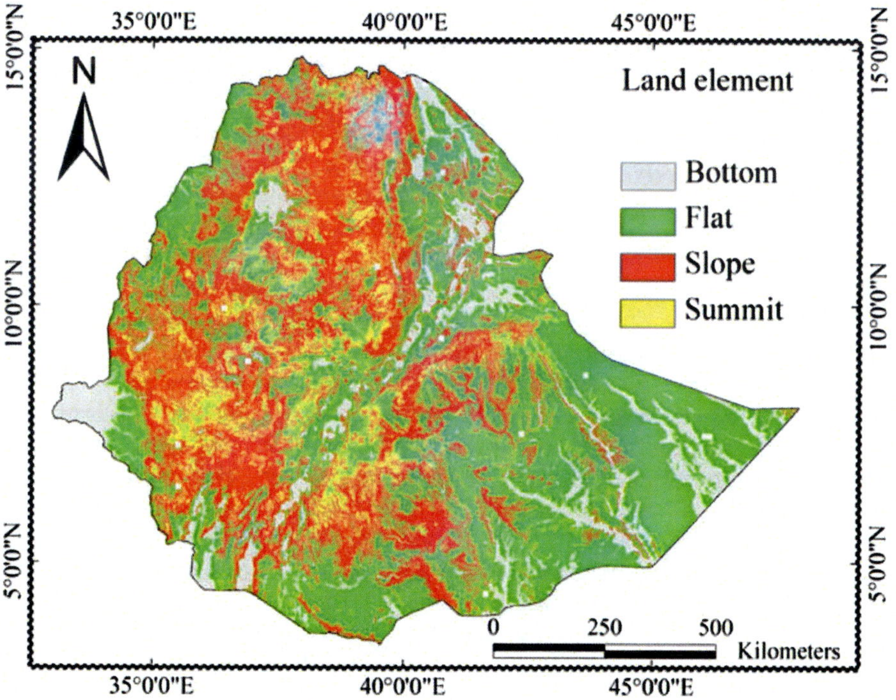

Fig. 4.14 Major landforms of Ethiopa (After Lenaars 2019)

(i) Summits and crests straddling the nearly flat surfaces of the plateau volcanic pile (Mesa landscape).
(ii) Plateaus (the western and eastern highlands without the overlying volcanic massifs).
(iii) Flat-gently sloping Rift Valley (Main Ethiopia Rift and the Afar depression); the Ogaden Plains; and low-lying areas in the west, southwest, and southern peripheral parts of the country.
(iv) Bottom areas which include the lake basins and areas lying below 500 m a.s.l.

The summits and crests include volcanic constructs of varying sizes emplaced over the sub-horizontal flood basalt flows. Shield volcanoes, elongate ridges, small-to-intermediate size volcanic centers, and volcanic plugs are included in this category (Fig. 4.14). What is defined as plateau generally lies above 2,000 m a.s.l. and is formed by successive eruption of horizontal or sub-horizontal basalts and subordinate pyroclastic rocks accumulated over great areas. The class named as flat occupies a large area of the rift floor, the gently inclined surfaces following the drainage systems of the major rivers outward from the central highlands and most of the rift floor and graben systems. The bottom category occupies the lakes and other level areas generally lying below 500 m a.s.l. in the rift floor, some river valleys in the Ogaden Plains, Omo and Gambella Basins. Superimposed on these major geomorphological classes are diverse landforms of differing sizes and geometry, including volcanic centers, rift and intermontane basins, horsts, pediments, erosional and fault scarps, flood plains, simple and coalescent alluvial fans, river valleys, dunes fields, and river terraces. Differences in each of these landform classes, such as, for example, the facing of slopes (aspects) and height which affect local climates and vegetation, add to the diversity of the landforms in Ethiopia. These differences, in turn, determine soil formation and profile development at the local scale.

4.6 Geomorphology and Soil

Geomorphology influences soil formation through its effects on climate and vegetation and differential transport and deposition of eroded materials. Landforms control much of the distribution of soils in the landscape, to such an extent that soils of markedly different properties may develop from the same parent material and soils with similar characteristic can form on different parent materials.

The elevation in Ethiopia ranges from −125 m at Dallol to 4,550 m Ras Dejen (Fig. 4.15). Within this range of elevation, there are diverse lithologic units, large- and small-scale landforms with their specific altitude, steepness,

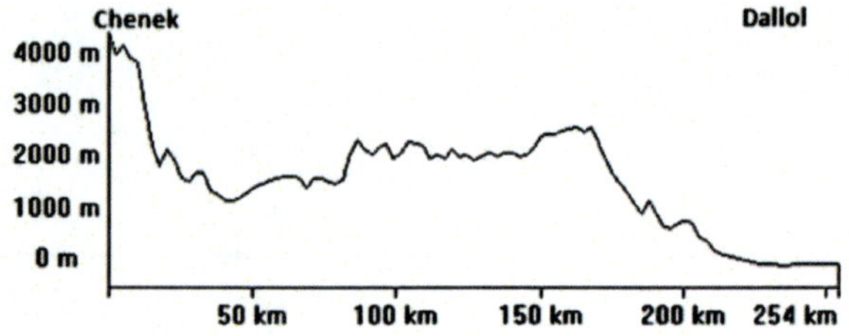

Fig. 4.15 West–east topographic profile from Chenek (Semien Mts.) to Dallol (*Courtesy of Bekele Abebbe*)

and facing of slopes (Fig. 4.15) all together affecting soil development.

The physiographic region of the plateau designated as slope (highland) developed mainly on the sub-horizontal flood basalts capped by volcanic massifs forming summits and ridge crests. These regional landforms are characterized by deep river valleys, nearly level ground and mountain slopes with differing altitude, orientation, and steepness. The rift margins has a width of up to 40 km with step-like arrangement of fault blocks where vertical displacement brought down the elevation to less than 1,800 m. The continuous decrease in elevation from the border faults to the rift floor is locally interrupted by antithetic faults which produce elongated and broad marginal grabens. The rift floor is low lying and nearly flat surface interrupted by dominantly felsic volcanoes, scoria cones, and faults. The latter create nearly level surfaces called grabens and half grabens (Fig. 4.16b, c).

These landforms influence runoff, drainage, erosion, deposition, vegetation, precipitation as well as the collection of solar energy which all together influence types and properties of the resulting soils. In the flat landscape areas of the rift system, Omo Basin, Gambella, northwestern Ethiopia, Ogaden, and the lower reaches of wide river valleys, the soils range from poorly drained to moderately drained. These areas are mainly depositional surfaces where erosion is not a threat to the soils and the environments. Erosion from surrounding slopes also provides additional sediment material to the valley bottoms.

Except the intermontane basins and marginal grabens, the mountainous terrain of the western and eastern plateaus and the rift margins are characterized by different landforms with differing runoff, erosion, sunlight, precipitation, and vegetation. These differences create their own influences on soil physical and chemical properties.

As with parent material, it is only in the highly unlikely situation where all soil-forming factors are constant, that one can make direct correlation between landform (topography) and soil type and this applies to all other soil-forming factors. However, there are certain conditions where soils under the same topography and parent materials coevolve

Fig. 4.16 Some examples of landforms in the rift valley and on the plateaus: **a** Trachytic plugs forming crests on the sub-horizontal plateau (Kabe, Wollo); **b** the main western escarpment of the Afar rift, Dessie; **c** faulted, flat lying terrain in the Main Ethiopian Rift, Metehara; **d** Flat landscape near Shinile area, northeast of Dire Dawa (*photo courtesy of Bekele Abebe*)

producing soils of the same type or series and described by soil topo sequence. The effect of landform (topography) on soil formation is through its effect on distribution of water and sediment on the landscape, erosion, and deposition processes and mass movement. Steep sloping areas serve as sources of water and sediment enriching valleys and floodplains. In the context of Ethiopia, where large part of the country's landmass can be categorized as steep slope, the effect of topography on soil formation and profile development is considerable. On steep slopes where soil erosion by water is rampant and vegetation growth is sparse, soils remain shallow and more ston, like Leposols and Cambisols (Fig. 4.17). On the contrary, valley bottoms and flat areas receive clay- and organic-matter-rich sediment from upland and favor more vegetation growth leading to the development of deep fertile Fluvisol, Luvisol, and Vertisol profiles. There is a clear relation among topography (elevation)-climate-vegetation and soil types in Ethiopia (Birhanu et al. 2021). On hill slopes with accelerated soil erosion, Van de Wauw et al. (2008) found Skeletic Cambisols and Pellic Vertisols for basalt-dominated landscapes while Calcic Regosols, Clluvic Calcaric Cambisols, and Calcic Vertisols on limestone-dominated landscapes. On flat plateau surfaces in the south-western Ethiopia, the relationship between topography and soil type is quite complex owing to erosion, deposition, and localized landslide processes.

About 21 dominant soil types were mapped in Ethiopia (Fig. 4.13 and Table 4.1). The geomorphological relationship of these soils, their area coverage, and their coverage proportion for the different landforms are indicated (Table 4.1). The results illustrate that water bodies (areas of land covered by water), Fluvisols, Gleysols, have large (more than 50%) area coverage and also Solonetz has 43% coverage on the bottom landform. Most of the soils do have their large area coverage of more than 50% on flat landform (Table 4.1). From this, it can be deduced that flat landform is less subjected to erosion and favors sediment deposition providing a more stable landform for the development of most soils. Soils including Acrisols, Alisols, Leptosols, Luvisols, and Nitisols have large area coverage on slopy

Fig. 4.17 Hill slope erosion and sediment deposition in the valley bottoms leading to the formation of shallow soils on hill slope on one hand and a deep and fertile soil development in the valleys on the other hand **a** May-Muk watershed Tigray **b** Alluvial deposit along Tirare River Waghimra zone Amhara region (*photo courtesy of Bekele Abebe*)

Table 4.1 Soil-landform relationships: soil distribution and occurrences on different landforms, their area coverage, and proportion on different landforms (NA means not available)

Soil type	Bottom		Flat		Slope		Summit	
	Area (ha)	Area (%)	Area (ha)	Area (%)	Area (ha)	Area (%)	Area (ha)	Area (%)
Acrisols	55812.50	3.83	312475.00	21.45	1008418.75	69.24	79737.50	5.47
Alisols	76175.00	2.55	656537.50	21.95	1923443.75	64.31	334681.25	11.19
Andosols	111237.50	15.89	472387.50	67.47	113262.50	16.18	3237.50	0.46
Arenosols	100450.00	15.65	413387.50	64.41	116518.75	18.16	11406.25	1.78
Calcisols	1135668.75	10.86	8432406.25	80.63	841831.25	8.05	48381.25	0.46
Cambisols	586162.50	5.54	6013206.25	56.86	3202031.25	30.28	773925.00	7.32
Chernozems	13568.75	9.73	111806.25	80.18	575.00	0.41	13500.00	9.68
Ferralsols	868.75	11.63	6593.75	88.28	6.25	0.08	NA	0.00
Fluvisols	2560856.25	54.58	1829531.25	38.99	267943.75	5.71	33437.50	0.71
Gleysols	60368.75	67.17	28643.75	31.87	850.00	0.95	6.25	0.01
Gypsisols	2665893.75	31.11	5533318.75	64.57	370831.25	4.33	NA	0.00
Leptosols	1649337.50	4.97	10292156.25	31.02	19528475.00	58.86	1710206.25	5.15
Lixisols	12643.75	6.43	101806.25	51.75	52487.50	26.68	29781.25	15.14
Luvisols	273787.50	3.15	1941450.00	22.34	5038150.00	57.98	1436706.25	16.53
Nitisols	530856.25	3.91	3773081.25	27.79	6861343.75	50.54	2411406.25	17.76
Phaeozems	20912.50	4.52	175231.25	37.84	153481.25	33.14	113443.75	24.50
Planosols	18.75	13.04	106.25	73.91	6.25	4.35	12.50	8.70
Regosols	157712.50	12.42	1037518.75	81.70	71093.75	5.60	3618.75	0.28
Solonchaks	603818.75	31.24	1224981.25	63.37	103312.50	5.34	825.00	0.04
Solonetz	7556.25	43.57	9381.25	54.09	406.25	2.34	NA	0.00
Vertisols	2477050.00	20.82	6659993.75	55.97	1666875.00	14.01	1095000.00	9.20
Water Bodies	336000.00	79.97	47131.25	11.22	25881.25	6.16	11143.75	2.65

landform (Table 4.1). Acrisols, Alisols, and Luvisols are commonly distributed in forested areas of south-western Ethiopia whereas shallow Leptosols are the most predominant soil in the country covering large part of slope landform. Nitisols derived from basalt parent materials are also common on sloping areas of the western and central highlands of Ethiopia where the effects of parent rocks and geomorphology on soils properties are significantly affected by climate particularly rainfall and vegetation (Coltorti et al. 2019; FAO 1984).

Soil types such as Ferralsols, Gypsisols, and Solonetz do not develop on summit landform and this is attributed to conditions of soil formation in addition to the nature of parent materials. Summit landforms are covered with Trap series formation. Soil development started with rapid rate of weathering, and intensive leaching conditions on this landscape are less favorable for the formation of Gypsisols. Gypsisols are derived from gypsum containing parent materials under dry climatic condition and Solonetz developed from sodium-rich parent material and limited leaching of the salts allowing for progressive accumulation of salts in the soil. Though most soils have area coverage on summit landform, no soil has its predominating area coverage on the summit landform.

4.7 Geopedology

Pedological studies worldwide proved that pedogenic processes are integral parts of landscape evolution and geomorphic processes play a major role in the genesis and distribution of soils (Huggett 1975; Ruhe 1975; Daniels et al. 1971). Topography affects climate, vegetation, movement of materials, and energy (Jenny 1941). Formation and distribution of soils in Ethiopia are largely influenced by agro-ecological zones underlain by geology and physiography (Bruggeman 1984; Henricksen et al. 1984a, b). Soil surveys conducted in different parts of the country have shown a distinctive and typically recurring soil, slope, and geomorphic relationships (Belay 1997; Paris 1985; Bruggeman 1984; Henricksen et al. 1984a, b).

Various studies have also reasserted that a great number of geomorphic processes involved in landform genesis and parent materials of varying origin and chronology (Tadesse 2014; Solomon and Mulugeta 2000; Tadesse and Tsegaye 2000; Ayalew and Moore 1989; Henricksen and Wijntje-Bruggeman 1984). Soils developed on Trap series basalts in less humid regions in southern Ethiopia, northwestern margin of Rift Valley, and Central Ethiopia with higher proportion of clay fraction in both surface and subsurface horizons (Abera et al. 2016; Fritzsche et al. 2007; Essayas et al. 2006; Negassa 2001). On the contrary, development of dissimilar soils such as Vertisols, Luvisols, and Acrisols on varieties of parent materials including basalts, sandstones, granites, dolerite intrusions, and granites in southern eastern, northeastern, northwestern, and central Ethiopia with higher proportion of clay both in surface and subsurface horizons were reported by other authors (Henricksen and Wijntje-Bruggeman 1984). In basalt-dominated highland of Tigray of Northern Ethiopia, geology, mass movement, and erosion predominantly influenced soil development and variability (Van de Wauw et al. 2008).

A transect study in Didessa watershed, Western Ethiopia indicated that the effect of parent materials on soil development varied with elevation. In the low elevation position, Vertisols were developed on alluvium and colluvium parent materials. At the higher elevation, Alisols were developed on Cenozoic basalts whereas Luvisols were developed on Cenozoic basalts and granitic gneiss. In the midland and highland, the effect of local topography seems to have a more dominating effect on the soil development than parent materials (Abdenna et al. 2018). Therefore, a good understanding of the relationship between soil, geology, and geomorphology is important for understanding the occurrence of soil in the landscape, thus allowing the prediction of soil distribution across landscape.

4.8 Conclusions

Ethiopia is characterized by complex geological formations and landforms due to both internal and external geological and geomorphological processes acting over geological time. Soils are products of soil-forming factors acting on the parent material over time. Parent rock and landform are bases for soil formation. The influence of parent material on the resulting soil types is through its effect on texture and mineralogical composition. Landforms also play a critical role in soil formation as it influences erosion and deposition as well as spatial distribution of water which also determine the nature of vegetation all together affecting nature and properties of soil. Basalt-dominated lithology gives rise to the development of Verisols and Nitisols while steep topography covering significant landmass of the country is predominantly covered with shallow Leptosols and cambisols. Understanding the relation between soils, on one hand, and geology and geomorphology, on the other, is important to predict spatial distribution of soils and their management.

References

Abbate E, Bruni P, Ferretti M-P, Delmer C, Laurenzi M-A, Hagos M, Bedri OL, Sagri M, Libsekal Y (2014) The East Africa Oligocene intertrappean beds: regional distribution, depositional environments and Afro/Arabian mammal dispersals. J Afr Earth Sci9 9:463–489

Abbate E, Bruni P, Sagri M (2015) Geology of Ethiopia: a review and geomorphological perspectives. In: Billi P (ed), Landscapes and landforms of Ethiopia. World Geomorphological Landscapes, pp 33–64

Abdenna D, Yli-Halla M, Muktar M, Lemma W (2018) Soil classification of humid Western Ethiopia: a transect study along a toposequence in Didessa watershed. Catena 163:184–195

Abebe B, Acocella V, Korme T, Ayalew D (2007) Quaternary faulting and volcanism in the Main Ethiopian Rift. J Afr Earth Sci 48:115–124

Abera T, Dagne W, Ernest S, Msanya BM, Tolessa D, Haekoo K (2016) Pedological characterization, fertility status and classification of the soils under maize production of Bako Tibe and Toke Kutaye Districts of Western Showa, Ethiopia. Ethiopian J Appl Sci Technol 7(1):1–17

Alemayehu R, Van Daele K, De Paepe P, Dumon M, Deckers J, Asrat A, Van Ranst E (2014) Characterizing weathering intensity and trends of geological materials in the Gilgel Gibe catchment, southwestern Ethiopia. J Afr Earth Sci 99:568–580

Asrat A, Barbey P, Gleizes G (2001) The Precambrian geology of Ethiopia: a review. Afr Geosci Rev 8:271–288

Ayalew T, Moore JM (1989) The Gore Gambella Geotraverse, Western Ethiopia. A report to the international development research centre on a cooperative project between Addis Abeba Universily. The Ethiopian institute of geological surveys and the Ottawa-Carleton geosciences centre. Canada, Addis Ababa

Baker J, Snee LM, Menzies MA (1996) brief Oligocene period of flood volcanism in Yemen, Earth Planet. Sci Lett 138:39–55

Beauchamp J (1977) La série sédimentaire en Ethiopie central and oriental. PhD thesis, University Claude Bernard Lyon, 419 p

Belay T (1997) Variability of soil catena on degraded hillslopes of Watiya Catchment, Welo, Ethiopia. SINET: Ethiop. J Sci 20:151–175

Birhanu L, Bekele T, Tesfaw B, Sebsebe D (2021) Relationships between topographic factors, soil and plant communities in a dry Afromontane forest patches of Northwestern Ethiopia. PLoS ONE 16(3). https://doi.org/10.1371/journal.pone.0247966

Blandford WT (1870) Observations on the geology and zoology of Abyssinia, made during the progress of the British expedition to that country in 1867–1868. Macmillan, London

Bosellini A (1989) The continental margins of Somalia: their structural evolution and sequence stratigraphy. Mem Sci Geol Padova 41:373–458

Bosellini A, Russo A, Assefa G (2001) The Mesozoic succession of Dire Dawa, Harar Province, Ethiopia. J Afr Earth Sci 32:403–417

Bosellini A, Russo A, Fantozzi PL, Assefa G, Solomon T (1997) The Mesozoic succession in the Mekele outlier (Tigrai province, Ethiopia). Mem Sci Geol Padova 49:95–116

Bruggeman W (1984) Assistance to Land Use Planning Ethiopia. Provisional soil association map of Ethiopia (1:2,000,000). The provisional military government of socialist Ethiopia

Bussert R, Schrank E (2007) Palynological evidence for a latest Carboniferous-Early Permian glaciation in Northern Ethiopia. J Afr Earth Sc 49(4–5):201–210

Cathcart J, Cannon K, Heinz J (2008) Selection and establishment of Alberta agricultural soil quality benchmark sites. Can J Soil Sci 88:399–408

Chaplot V, Van Vliet-Lanoe B, Walter G, Curmi P, Cooper M (2003) Soil spatial distribution in the Armorican Massif, western France: effect of soil-forming factors. Soil Sci 12:856–868

Coltorti M, Pieruccini P, Arthur K, Arthur J, Curtis M (2019) Geomorphology, soils and palaeosols of the Chencha area (Gamo Gofa, south western Ethiopian Highlands). J Afr Earth Sc 151:225–240

Daniels RB, Gamble EE, Cady JG (1971) The relation between geomorphology and soil morphology and genesis. Adv Agron 23:51–88

Dekayir A, El-Maataoui M (2001) Mineralogie et bilan de matiere de l'alteration supergene d'un basalte alcali du Moyen Atlas Marocain. J Afr Earth Sc 32:619–633

Dow DB, Beyth M Hailu T (1971) Paleozoic glacial rocks recently discovered in northern Ethiopia. Geol Mag 108:53–60

Driese SG, Jacobs JR, Nordt LC (2003) Comparison of modern and ancient Vertisols developed on limestone in terms of their geochemistry and parent material. Sed Geol 157:49–69

Eghbal MK, Hamzehpour N, Farpoor MH (2018) Geology and geomorphology. In-Soils of Iran, Springer International Publishing, pp 35–56

Essayas A, Zauyah S, Hanafi M, Rosenani A (2006) Genesis and classification of sesquioxide soils from volcanic rocks in sub-humid tropical highlands of Ethiopia. Geoderma 136:682–695

FAO (1984) Assistant to land use planning, Ethiopia. geomorphology and soils. Addis Ababa, Ethiopia pp.409

FAO/ISRIC/SOTER (1997) Soils of East Africa, 1:1 million scale. User's Guide Version 1.0. FAO, Rome

Fritzsche F, Zech W, Guggenberger G (2007) Soils of the main Ethiopian rift valley escarpment: a transect study. Catena 70:209–219

Graya J, Bishopb TFA, Wilford JR (2016) Lithology and soil relationships for soil modelling and mapping. Catena 147:429–440

Gruba P, Socha J (2016) Effect of parent material on soil acidity and carbon content in soils under silver fir (Abies alba Mill.) stands in Poland. Catena 140:90–95

Henricksen BL, Ross S, Shltan T, Wijntje-Bruggeman H (1984a) Assistance to land use planning Ethiopia: Soils and geomorphology. Addis Ababa, Ethiopia, United Nation Development Program and Food and Agricultural Organization of the United Nations, Rome

Henricksen BL, Ross S, Tilimo S, Wijntje-Bruggeman H (1984b) Geomorphology and soils: report prepared for the government of Ethiopia. Addis Ababa, Ethiopia: The Food and Agricultural Organization of the United Nations

Hugget RJ (1975) Soil landscape systems: a model of soil genesis. Geoderma 13:1–22

Hunegnaw A, Sage L, Gonnard R (1998) Hydrocarbon potential of the intracratonic ogaden basin, se Ethiopia. J Petrol Geol 21(4):401–425

ISRIC, International Soil Reference Center (2020) Soil map of Ethiopia at 250m Resolution. World Soil Information, Ministry of Agriculture, Ethiopia

ISRIC-world soil information, ministry of agriculture, Ethiopia

Jenny H (1941) Factors of soil formation: a system of quantitative Pedology. McGraw-Hill

Kazmin V (1972) Geology of Ethiopia; explanatory note to the geological map of Ethiopia, 1:2,000,000. Ministry of Mines, Addis Ababa, Ethiopia, 14 p

Kazmin V, Shifferaw A, Balcha T (1978) The Ethiopian Basement: stratigraphy and possible manner of evolution. Geol Rundsch 67 (2):531–546. https://doi.org/10.1007/bf01802803

Keller WD (1964) Processes of origin and alteration of clay minerals. In: Rich CI, Kunze GW (eds) Soil Clay Mineralogy, A Symposium.. Durham, N. C., SeemanPrintery, 3–76.p

Leenaars JGB (2019) Geomorphic map of Ethiopia at 50 m resolution
Mesfin A (1998) Nature and management of Ethiopian soils. Alemaya University of Agriculture, Ethiopia, p 272p
Ministry of agriculture land use planning and regulatory department. Addis Ababa: United Nations Development Program, Food and Agricultural Organization of the United Nations
Mohr P (1983) Ethiopian flood basalt province. Nature 303:577–584
Mohr P, Zanettin B (1988) The Ethiopian flood basalt province. In: Macdougall JD (ed) Continental Flood Basalts. Kluwer Academic, Dordrecht, pp 63–110
Negassa W (2001) Assessment of important physicochemical properties of Nitosols under different management systems in Bako Area, Western Ethiopia. M.Sc. Thesis, Alemaya University, Alemaya, p 109
Palumbo B, Angelone M, Bellanca A, Dazzo C, Hauser S, Neri R, Wilson J (2000) Influence of inheritance and pedogenesis on heavy metal distribution in soils of Sicily, Italy. Geoderma 95:247–266
Paris S (1985) Soil survey of the Borkena Area (Welo) AG: DP/ETH/82010, Field doc.7 LUPRD/UNPD/FAO, Land Use Planing and Reg Dept Min Agr Addis Ababa, 168 p
Rogers N, Macdonald RJ, Godfrey F, Rhiannon G, Martin S, Barbara B (2000) Two mantle plumes beneath the East African rift system: Sr, Nd and Pb isotope evidence from Kenya Rift basalts. EPSL 176:387–400
Ruhe RV (1975) Geomorpholoy: Heomorphic processes and surficial geology. Houghton Mifflin Co., Boston
Russo A, Getaneh A, Atnafu B (1994) Sedimentary evolution of the Abbai River (Blue Nile) Basin, Ethiopia. N J B Geol Palaeo Mh 5:291–308
Sacchi R, Alene M, Barbieri M, Conti A (2007) On the Palaeozoic Tillite of the Adigrat Group (Tigrai, Ethiopia). Periodico Di Mineralogia 76(2–3):241–251
Schluter T (2000) Geological atlas of Africa. Springer, Nairobi, Kenya
Sembroni A, Molin P, Dramis F (2019) Regional geologyof the Dogu'a Tembien massif. In: Geo-trekking in Ethiopia's Tropical Mountains —The Dogu'a Tembien District. Springer Nature. ISBN 978-3-030-04954-6
Solomon G, Mulugeta H (2000) Geology of Nekemt Area. Memoris No. 14. GeologicalSurvey of Ethiopia, Addis Ababa
Tadesse A (2014) Geology of the Arjo Area. Basic geoscience mapping directorate. Geological Survey of Ethiopia, Addis Ababa.
Tadesse A, Tsegaye A (2000) Geology of Gimbi Area. Memoris 15, Geological Survey of Ethiopia, Addis Ababa
Tefera M, Chernet T, Haro W (1996a) Geological map of Ethiopia, 2nd Edition, Ethiopian Mapping Authority
Tefera T, Tadiwos C, Workineh H (1996b) Geology of Ethiopia. (Explanatory note of geological map of Ethiopia 1:2,000,000). Second Edition, Federal Democratic Republic of Ethiopia, Ministry of Mines
Teramoto ER, Lepsch IF, Vidal Torrado P (2001) Relations of soil, geomorphic surfaces and geological substrate in Ribeirão Marins Waterbasin. (Piracicaba-SP). Scientia Agricola 58(2):361–371
Ukstins P, Renne E, Wolfenden J, Baker M, Menzies (2002) Matching conjugate volcanic rifted margins: 40Ar/39Ar chrono-stratigraphy of pre- and syn-rift bimodal flood volcanism in Ethiopia and Yemen, Earth Planet. Sci Lett 198:289–306
USGS Volcano Hazards program "Basalt"- Glossary. USGS (2015)
Van de Wauw J, Baert G, Moeyersons J, Nyssen J, De Geyndt K, Taha N, Zenebe A, Poesen J, Deckers J (2008) Soil–landscape relationships in the basalt-dominated Highlands of Tigray. Ethiopia Catena 75:117–127
Vaselli O, Buccianti A, De Siena C, Bini C, Coradossi N, Angelone M (1997) Geochemical characterization of ophiolitic soils in a temperate climate: a multivariate statistical approach. Geoderma 75:117–133
Venturelli G, Contini S, Bonazzi A, Mangia A (1997) Weathering of ultramafic rocks and elements mobility at Mt. Prinzera, Northern Apennines. Italy. Mineral Mag 61:765–778
Weil RR, Brady NC (2017) The Nature and properties of soils, 15th edn, Pearson Education
Wolela A (2008) Sedimentation of the Triassic-Jurassic Adigrat sandstone formation, Blue Nile (Abbai) basin, Ethiopia. J Afr Earth Sci 52:30–42
Wolfenden E, Ebinger C, Yirgu G, Deino A, Ayale D (2004) Evolution of the Northern Main Ethiopian Rift: Birth of a Triple junction. Earth Planet Sci Lett 224:213–228. https://doi.org/10.1016/j.epsl.2004.04.022

Soil Classification

5

Alemayehu Regassa, Ashenafi Ali, and Gebeyehu Taye

Abstract

Soil classification is a systematic categorization of soils into groups at varying levels of generalization according to their morphological, physical, mineralogical, and chemical properties. Soil morpho-physicochemical properties which are interaction results of soil-forming factors and processes serve as marks or evidence of definitions used for soil groupings. The FAO/UNESCO Soil Map of the World and Soil Taxonomy of the United States are by far the most common classification systems used worldwide. Ethiopia has no national soil classification system developed and adapted in the country. Both soil taxonomy and the FAO-UNESCO classification systems are used in combination or in isolation in the country. However, nowadays the World Reference Base for Soil Resources (WRB) is widely adopted. Farmers in different localities and regions of Ethiopia have different names, as is the case elsewhere in the world, for the local soil types on the basis of color, particle size, thickness, and fertility.

Keywords

Indigenous soil classification • Soil-forming factors • Soil-forming processes • Soil taxonomy • World reference base for soil resources

A. Regassa (✉)
Department of Natural Resources Management, Jimma University, P.O. Box 307 Jimma, Ethiopia
e-mail: alemayehu.regassa@ju.edu.et

A. Ali
Department of Geography and Environmental Studies, Addis Ababa University, P.O. Box 1176 Addis Ababa, Ethiopia
e-mail: ashenafi2010ali@gmail.com

G. Taye
Department Land Resources Management and Environmental Protection, Mekele University, P.O. Box 231 Mekelle, Ethiopia
e-mail: gebeyehu.taye@mu.edu.et

5.1 Introduction

Soil classification is a century-old science that deals with the systematic categorization of soils into groups at varying levels of generalization according to their morphological, physical, mineralogical, and chemical properties (Buol et al. 1997). Modern classification systems classify soils based on quantitative characteristics defined as diagnostic horizons, properties, and materials. Soil morpho-physicochemical properties which are interaction results of soil-forming factors and processes serve as marks or evidence of definitions used for soil groupings (Silva et al. 2017; Bockheim 2014; Esu 2010; Brady and Weil 2015). The FAO/UNESCO Soil Map of the World and Soil Taxonomy of the United States are by far the most common classification systems used worldwide (Buol et al. 1997; Landon 1991).

Soil classification is very important for countries like Ethiopia where agriculture is the backbone of the economy. When viewed from the perspective of local authorities in charge of the soil surveys, in most of the reports, dissertations, and publications released from Research and Higher Learning Institutes, the soil taxonomy system of classification is followed. However, reports and maps issued by line Government Ministries are mainly according to the FAO-UNESCO system of classification.

Farmers locally have a broad knowledge to describe and classify soils. Local land use and management decisions are largely dictated by local soil classification. Farmers differentiate soils by naming them with respect to observed and experienced unique properties. However, indigenous classification could not differentiate the soils on the bases of pedogenetic difference as occur in the subsurface soil unless translated from the soil surface soil. Linking indigenous and scientific systems can then serve to facilitate technology transfer from similar soils outside the local area, which have been named differently.

S. Beyene et al. (eds.), *The Soils of Ethiopia*, World Soils Book Series,
https://doi.org/10.1007/978-3-031-17012-6_5

The history of soil classifications in Ethiopia has been related to the chronology of soil survey and mapping missions. In the 1920s the soils of Ethiopia have been classified into four classes (Chernozems, Tropical Prairie soils, Laterites, and other Red Tropical soils) by Shantz and Marbut (1923). In the 1960s Ethiopian soils were classified into 10 major soil groups/associations based on D'Hoore classification system (D'Hoore, 1964). However, D'Hoore made a serious attempt to relate to the United States system of soil taxonomy. In 1972, Donahue regrouped D'Hoore soils group into predominant soil orders, using the United States system of soil taxonomy, namely, Aridisols, Entisols, Ultisols, and Vertisols. Further, in 1975, Wesphal, attempted to classify soil regions in Ethiopia in terms of the U.S.D.A. 7th approximation Soil Classification (Wesphal 1975). After these notable nationwide efforts, nowadays very limited soil resource studies employ the US soil taxonomy system except postgraduate-based soil studies that commonly use both FAO-UNESCO/IUSS-WRB and soil taxonomy systems. However, the FAO-UNESCO and its successor- the IUSS-WRB system of soil classification is the most widely used in Ethiopia as there are no validated national soil correlation and classification systems.

5.2 History of Soil Classification in Ethiopia

As a natural body soil is not uniform along the landscape and significantly varies from place to place. These variations are the results of the different soil-forming factors and pedogenic processes acting on parent materials over a period of time and being reflected on the resulting soil properties and types. In response to such variability, the classification of soil resources is aimed at organizing knowledge in an ordered and structured manner so that relationships among individuals and classes of soils can be understood; properties of specific soil type can be known. Classification can also be used to establish classes for practical and applied purposes in predicting behavior of specific soils, identifying their best uses, estimating their productivity and research and technology transfer (Weil and Brady 2017). Soil classification based on uses/suitability such as forest soil and grassland soils and based on parent material such as basal soils and limestone soils do not reflect details of soil properties for scientific-based communication. Different national and international based soil classifications systems were proposed to facilitate such communication among scientists and soil managers since the 1870s.

Among the different national and international soil classification and correlation systems, two enjoy very wide international recognition: US Soil Taxonomy and FAO system of soil classification (David 2001). Both systems have been revised a number of times based on the latest findings. The US Soil Taxonomy system, completely new in design and nomenclature, was created and continues to evolve in the United States (Soil Survey Staff 1999). The system was developed to serve the national cooperative soil survey of the United States. However, in its history, the Taxonomy system envisioned the continuing need for a system that would include all soils in the world and would accommodate new knowledge. Similarly, the FAO-UNESCO system of classification has been revised a number of times since its first publication in 1975 and now reached IUSS-WRB 2014 update of 2015 version.

In Ethiopia, there were several attempts to classify and map soils of the country dating back to the 1920s (Berhanu 1994). These early classification and mapping attempts were based on relations among soils on the one hand and geology, vegetation, climate, and physiography on the other hand (Elias 2016a, b; Esayas 2005). However, the relationship among soils and factors of soil formation remains highly uncertain in the context of the country. Effective soil survey, classification and mapping in Ethiopia started in the 1960s to support agricultural development interventions. These surveys were mainly to map soils of a specific area, and conduct soil surveys at river basins and national levels for different purposes (Elias 2016a, b; Esayas 2005). However, as there is no nationally developed soil classification system, most of the soil survey, classification, and mapping were carried out using a combination of both soil taxonomy and the FAO-UNESCO systems (Esayas 2005). However, the FAO-UNESCO-ISRIC (including the recently revised IUSS-WRB 2014 (update of 2015) version enjoying wide use in Ethiopia. In fact, most government agencies entrusted with the responsibility for mapping and inventorying the soil resources of the country invariably adopt the system. This might be attributed because of its universality and simplicity (Elias 2016a, b; Esayas 2005; Debele 1980).

5.3 Soil Classification Systems

5.3.1 World Reference Base (WRB) for Soil Resources

The IUSS-WRB system of soil classification is the most widely used soil classification system in Ethiopia. This is because the classification system is simple and has less data requirement and is suited to the classification of complex soil resources like the case in Ethiopia showing an extreme spatial variation of soil resources in response to complex physiography and geological formation (Van de Wauw et al. 2008).

In, Ethiopia several area-wide soil studies employing various soil classification systems. However, the preliminary country-wide soil studies in 1980s and 1990s employ the

FAO-UNESCO (1974), its revised version (FAO 1988), and the World Reference Base for Soil Resources (ISSS-ISRIC-FAO 1998) soil correlation and classification systems. Soils of most of the major river basins covering the entire country (in the 1980s and 1990s), sub-basins, and development corridors in 2000s and 2010s were classified based on the FAO-UNESCO systems and IUSS-WRB recent editions (Ali et al. 2020; Leenars 2020a, b; Elias 2016a, b; Taye et al. 2013; Esayas and Debele 2006; Mesfin 1998; FAO 1984a, b; Debele 1980).

5.3.2 Soil Taxonomy

A soil classification in Ethiopia has been related to the history of soil survey and mapping initiatives. In 1968 the U.S. Soil Conservation Service prepared a soil map of the world, using the new taxa of soil orders and suborders. The map was presented for Africa including Ethiopia (Donahue 1972). The soils were combined into the 7 soil orders out of the potentially existing 10 soil orders (Oxisols, Aridisols, Alfisols, Entisols, Inceptisols, Ultisols, and Vertisols) because Histosols, Mollisols, and Spodosols, did not occur in large enough to be shown on this map. In 1972, Donahue regrouped soils into predominant soil orders, using the United States system of soil taxonomy, namely, Aridisols, Entisols, Ultisols, and Vertisols. Further, in 1975, Wesphal, attempted to classify soil regions in Ethiopia in terms of the U.S.D.A. 7th approximation Soil Classification (Westphal 1975).

In Ethiopia, after these notable nationwide efforts, nowadays very limited area-wide soil resource studies employ the US soil taxonomy system except for postgraduate-based soil studies that commonly use both soil taxonomy and IUSS-WRB systems (Zewdie 2013; Esayas 2005; Fikre 2003; Mesfin 1998).

5.3.3 Indigenous Soil Classification Systems

Indigenous soil classification in Ethiopia can be as old as agriculture as it is in the other parts of the world. Indigenous knowledge and practices of soil classification and management are transferred from generation to generation with little or no modifications. Farmers from different parts of the country classify soils based on similar properties such as soil color, particle size, thickness (depth), water holding capacity, and fertility status but soil names remain different (Esayas 2005). In the Tigray region of the Northern Ethiopia, indigenous soil classification indicated different soil classes as "*Walka*" Vertisolwhich is hard to work with when dry "*Hutsa*" Cambisol easy to work, "*Rekir*" infertile and shallow soils mostly derived from limestone parent material or soils on a slope "*Reguid*" deep and fertile soils derived from alluvial sediment along river terraces or on hillslope derived from basal (Corbeels et al. 2000). In addition to fertility status of soils, farmers also use physical soil properties such as color, texture, depth, water holding capacity, and topographic locations to classify soils. A study by Amare (2017) in Burie and JabiTehnan districts of Amhara region indicated indigenous soil classification as "*Tikur Afer*", "*DebayAfer*" and "*Borebore Afer*" corresponding to Vertisol, Leptosol and Nitisol, respectively.

In the southern part of Tigray, Northern Ethiopia indigenous soil classification indicated several soil types. Soils having vertical properties are generally named as "*Walka*", soils in the valley bottom that are affected by water with Gleyic and fluvic properties are named as "*Danshel*", fertile soils on slope derived from basalt parent material "Hamed Qoriya" while fertile cambisol on mid-slope is named as "Boda" (Zenebe et al. 2015). On the other hand vertisols dominated by cambic, calcic, or skeletic properties are named as "*Hawas Walka*" means mixed vertisols. Farmers of the area are also known for water and sediment harvesting, such harvested sediment is named as "*Hamediliqua*" means borrowed soil (Amanuel et al. 2015; Taye 2008).

Based on their study on farmers' perception and knowledge of soils in the three districts of Amhara region, Kebede (2016) reported that farmers classify their soils into four major groups based on soil texture, color, and coarse fragment content as "*Walka Afer*", "*Bunama Afer*", "*Key Afer*", and "*Chincha Afer*". "*Walka Afer*" corresponds to black and high clay-containing soils with surface cracks observed in the field, while "*Bunama Afer*" is a brownish soil having good drainage, medium fertility, and workability. On the other hand, "Key Afer" is reddish soil that is workable, deep, and medium in terms of fertility status and "*Chincha Afer*" is shallow soils on a slope containing considerable proportion of coarse fragment content. Fekadu (2018), reported that farmers soil classification is used for allocation of crop types and required inputs for soil fertility management in Jimma zone of South Western Ethiopia. Accordingly, soil types of the three districts of Omo Nada, Limu Seka, and Gera are classified as "*Dimaa*" red soil, Guracha/Megala soil having dark to reddish-brown color, Dalacha/Supee soils with grayish color and "*Koticha*" soils having black color mainly based on soil color.

5.4 Correlation of Soil Classification Systems

Farmers' indigenous knowledge of soil classification correlates with the scientific study of soil physicochemical properties. Ethnopedological study in southwestern highlands of Ethiopia indicated that red soils locally named as

Biyyee diimaa were correlated with Luvisols and Nitisols reference soil groups according to WRB and Alfisols of USDA soil taxonomy. On the other hand, fertile soils which are rich in organic matter are named as Biyyee *Gurraacha* which literally means black soils. These soils are correlated with Phaeozems according to the WRB system of classification and Mollisols according to soil taxonomy. Moderately fertile soils were classified as *Gurraacha Diimaa* meaning dark red. These soils are correlated with Nitisols and Cambisols according to WRB and Alfisols and Inceptisols according to soil taxonomy (Fekadu 2018). In the Amhara region, Burie and Tehnan districts of northwestern high lands of Ethiopia, the black cotton soils, Vertisols are named as *Tikur Afer*. Shallow soils in the hills are classified as *Debay soil* and the fertile reddish-brown agriculturally highly suitable soils are classified *Borebore soils*. These soils correlate with Entisols and Leptosols and Alfisols and Nitisols of soil taxonomy and WRB, respectively. In all the localities the nomenclature is structured in terms that infer the potential of the soil classes for land use.

5.5 Conclusion

There is no national soil classification system developed and adapted in Ethiopia. Soil classification system followed in the different soil survey works varies considerably. As most of the soil survey works were done by expatriate consultants from different countries such as the former Russia, United States, France, United Kingdom, and FAO the soil classifications depend on the local authority in charge of the studies. soil surveys carried out by the US consultants are classified mainly according to soil taxonomy, whereas the rest of the studies, followed the FAO-UNESCO system of classification. Thus, it can be stated that both the soil taxonomy and the FAO-UNESCO classification systems are followed almost in isolation in the country. Farmers in different regions and localities of Ethiopia have different names, as is the case elsewhere in the world, for the local soil types on the basis of color, particle size, thickness, and fertility. There is however quite a good correlation of farmers' indigenous knowledge of soil classification with the scientific study of soil physicochemical properties.

References

Ali A, Tamene L, Erkossa T (2020) Identifying, cataloguing, and mapping soil and agronomic data in Ethiopia. CIAT Publication No. 506, International Center for Tropical Agriculture (CIAT), Addis Ababa, Ethiopia. https://hdl.handle.net/10568/110868

Amanuel A, Gebresamuel G, Girma A (2015) Characterization of agricultural soils in cascape intervention Woredas in Southern Tigray, Ethiopia, p 284

Amare S (2017) Farmers' knowledge on soil classification and fertility management and its relation with scientific methods the case of Burie and Jabi Tehnan Districts, North Western Ethiopia. Master Thesis Bahir Dar University, Ethiopia

Berhanu D (1994) The soils of Ethiopia: annotated bibliography. Technical handbook no. 9. Regional Soil Conservation Unit (RSCU), Swedish International Development Authority (SIDA)

Bockheim JG (2014) Cryopedology. Progress in soil science. Springer, New York

Brady NC, Weil RR (2015) The nature and properties of soils, 15th edn. Pearson Education Inc., Upper Saddle River, New Jersey, USA

BuolSW HFD, McCracken RJ, Southard RJ (1997) Soil genesis and classification, 4th edn. Iowa State University Press, Ames

Corbeels M, Shiferaw A, Haile M (2000) Farmers' knowledge of soil fertility and local management strategies in Tigray, Ethiopia. Managing Africa's Soils No. 10

David GR (2001) Lecture notes principles of soil classification. International Institute for Aerospace Survey and Earth Sciences (ITC), Enschede, The Netherlands

Debele B (1980) The physical criteria and their rating proposed for land evaluation in the highland region of Ethiopia. Land Use Planning and Regulatory Department, Ministry of Agriculture, Addis Ababa, Ethiopia

D'Hoore JL (1964) Soil map of Africa. Sheet 3. Pub. No. 93, Joint Project No. 11, Commission for Technical Cooperation in Africa. Lagos, Nigeria, 1964

DonahueR (1972) Ethiopia: taxonomy, cartography and ecology of soils. African Studies Center

Elias E (2016a) Soils of the Ethiopian highlands: geomorphology and properties. CASCAPE Project, ALTERRA, Wageningen UR, The Netherlands. library.wur.nl/WebQuery/isric/2259099

Elias E (2016b) Soils of the Ethiopian highlands: geomorphology and properties. ALTERA Wageningen University Research Centre, The Netherlands. ISBN:978-99944-952-6-9

Esayas A (2005) Characteristics, genesis and classification of reddish soils from Sidamo Region of Ethiopia. PhD Thesis, Universiti Putra Malaysia

EsuIE (2010) Soil characterization, classification and survey. Heinneman Educational Books Publishers, Nigeria, pp 232

FAO–UNESCO (1974) FAO–UNESCO Soil map of the world, vol 1. Legend. UNESCO, Paris

FAO (1984a) Assistance to land use-planning, Ethiopia: geomorphology and soils. Field Document AG DP/ETH/78/003. The United Nations Development Programme and Food and Agriculture Organization. FAO, Rome

FAO (1984b) Assistance to land use-planning, Ethiopia: provisional soil association map of Ethiopia. Field document No 6. The United Nations Development Program and Food and Agriculture Organization. FAO, Rome

FAO (1988) FAO-UNESCO Soil map of the world: revised legend with corrections and updates. Food and Agriculture Organization and International Soil Resource Information Centre. Wageningen, The Netherlands, pp 146

Fekadu D (2018) Farmers' perceptions of soil fertility and management in Omo Nada, LimuSeka and Gera Districts of Jimma Zone, South Western Ethiopia. Msc thesis

Fikre M (2003) Pedogenesis of major volcanic soils of the south central rift valley Region, Ethiopia. MSc Thesis. University of Saskatchewan. Saskatoon, Canada, pp 128

ISSS-ISRIC-FAO (1998) World reference base for soil resources. ISSS, ISRIC, FAO, Rome

Landon JR (1991) Booker tropical soil manual. Longman Scientific and Technical Essex, UK, pp 474

Leenaars JGB, Elias E, Wösten JHM, Ruiperez González M, Kempen B (2020a) Mapping the major soillandscape resources of the Ethiopian highlands using random forest. Geoderma 361. https://doi.org/10.1016/j.geoderma.2019.114067

Leenaars JGB, Ruiperez González M, Kempen B, Mantel S (2020b) Semi-detailed soil resource survey and mapping of REALISE Woredas in Ethiopia. Project report to the BENEFIT-REALISE programme. Dec 2020b, ISRIC-World Soil Information, Wageningen

Mesfin A (1998) Nature and management of Ethiopia soils. Alemaya University of Agriculture, Ethiopia

Shantz HL, Marbut CF (1923) The vegetation and soils of Africa. National Research Council and 793 American Geographical Society, New York

Silva YJAB, do Nascimento CWA, Biondi CM, van Straaten P, de Souza Júnior VS, da Silva YJAB, dos Santos CA, do Carmo Trezena de Araújo J (2017) Influence of metaluminous granite mineralogy on the rare earth element geochemistry of rocks and soils along a climosequence in Brazil. Geoderma 306:28–39

Soil Survey Staff (1999) Soil taxonomy. A basic system of soil classification for making and interpreting soil surveys, 2nd edn. U. S. Department of Agriculture, Natural Resources Conservation

Taye G (2008) Effects of agricultural technologies on farm productivity and income in the highlands of Tigray, Ethiopia: dryland eco-farm approach. Master Thesis Norwegian University of Life Science (UMB), pp 81

Taye G, Poesen J, Van Wesemael B, Vanmaercke M, Teka D, Deckers J, Goosse T, Maetens W, Nyssen J, Hallet V, Haregeweyn N (2013) Effects of land use, slope gradient, and soil and water conservation structures on runoff and soilloss in semi-arid northern Ethiopia. Phys Geogr 34:236–259

Van de Wauw J, Baert G, Moeyersons J, Nyssen J, De Geyndt K, Taha N, Zenebe A, Poesen J, Deckers J (2008) Soil–landscape relationships in the basalt-dominated Highlands of Tigray, Ethiopia Ethiopia, Catena, vol 75, pp 117–127

Weil R, Brady N (2017) The nature and properties of soils, 5th edn. Pearson Education

Westphal E (1975) Agricultural systems in Ethiopia. Agriculture Research Report (Versl. landbouwk. Onderz.) 826, (x) + 278, p 16, figs, 103 photographs, 10 maps in separate cover, 9 tables, ZZZ refs, 2 appendices. ISBN 90 220 0556 9

Zewdie E (2013) Properties of major agricultural soils of Ethiopia. Lambert Academic Publishing

6 Major Soil Types

Alemayehu Regassa, Mohammed Assen, Ashenafi Ali, and Berhan Gessesse

Abstract

Ethiopia has diverse soil resources. Natural conditions, such as parent material, climate, topography, biotic and land use/land cover changes are largely responsible for creating regional and local differences in soil types and characteristics. In Ethiopia, the history of soil studies can go back as early as the 1920s although a well-organized and systematic soil survey was started in the 1960s. The soil survey works since 60 s can be broadly categorized into three: area specific, river basins, and general country-wide surveys. The area-specific studies concentrated on potential agricultural areas for either rainfed or irrigated agricultural development. Systematic area-specific and basin-wide soil surveys in the country were commenced in the late 1950s. So far, there is only one country-wide soil study carried out in Ethiopia, soil association map of Ethiopian (1:2 million scale). This was carried out in the early 1970s as a component of the Land Resources Inventory of Ethiopia for the preparation of the Master Land Use Plan for Ethiopia. Despite the presence of several types of soils, only nine soil types cover 90% of the land surface of Ethiopia. These major types of soils are Leptosols (30%), Nitisols (12%), Vertisols (11%), Cambisols (9%), Luvisols (8%), Calcisols (9%), Fluvisols (4%), Gypsisols (8%) and Alisols (3%).

Keywords

Soil survey • Soil map • Soil forming factors • Soil forming processes • Pedogenesis • Geogenesis

6.1 Introduction

Soils are the result of the interaction of the environmental factors. The main factors, which influence the development of soils, are climate, topography, parent material, time, and biological parameters such as vegetation, land use, and the activity of organisms. Varied combinations of the soil forming factors lead to specific sets of soil forming processes that bring about change in soil properties. The spatial variations of environmental conditions and their interactions result in a spatial differentiation of soil characteristics (Buol et al. 1989). The five soil forming factors in combination with various forms of anthropogenic activities act on and affect soil properties in a given locality. Regional patterns of climate, vegetation, and parent material can be used to predict the kinds of soil in large areas. On the other hand, the local patterns of topography or relief, parent material and time and their relationships to vegetation and microclimate can be used to predict the kinds of soil in small areas.

Soil surveys provide a scientific inventory of the soils occurring within a specified land area and involve the systematic examination, description, classification, and mapping of such soils. Sustainable use of soils depends on the inherent characteristics of such a land. Therefore, the characterization and classification of soils in a manner that will ease communication and transfer of knowledge about such soils to farmers and other stakeholders is essential.

A. Regassa (✉) · M. Assen · A. Ali · B. Gessesse
Department of Natural Resources Management, Jimma University, P.O. Box 307 Jimma, Ethiopia
e-mail: alemayehu.regassa@ju.edu.et

M. Assen
e-mail: moh_assen@yahoo.com

A. Ali
e-mail: ashenafi2010ali@gmail.com

B. Gessesse
e-mail: berhanavu@gmail.com

A. Ali
Department of Geography and Environmental Studies, Addis Ababa University, P.O. Box 1176, Addis Ababa, Ethiopia

B. Gessesse
Entoto Observatory and Research Center (EORC), Ethiopian Space Science and Technology Institute, P.O. Box 1176 Addis Ababa, Ethiopia

S. Beyene et al. (eds.), *The Soils of Ethiopia*, World Soils Book Series,
https://doi.org/10.1007/978-3-031-17012-6_6

In Ethiopia, coupled with great variations in the age and character of the parent materials, the diverse physiographic and climatic conditions encourage the formation of different soil types as these factors trigger the abundance and intensity of soil building processes so that due to the heterogeneity of soil forming factors and processes, different types of soils in different stages of development exist (Mesfin 1998).

6.2 History and Extent of Soil Survey

6.2.1 Pioneer Soil Survey Initiatives

In Ethiopia, systematic area-specific and basin-wide soil survey missions were commenced in the late 1950s. Among early efforts, Murphy (1968a, 1963, 1968b) studied the general fertility status of soils in Ethiopia by collecting some 2,600 samples along the main roads across the country. However, he gives no classification attempt and soil types rather than the characterization of soils based on soil color and chemical properties.

The first ever-useful information on Ethiopian soils had been supplied by Shantz and Marbut in 1923 followed by D'Hoore in 1964 (Wesphal 1975; Donahue 1972). D'Hoore published a soil map of Africa at 1:5 million scale in which 10 major soil groups/associations were identified in Ethiopia (D'Hoore 1964). In parallel, between the late 1950s and 1974, there had been many fragmented (not country-wide) efforts by various institutions, commissioning either foreign technical aid agencies or private foreign consultants, to generate soil information about Ethiopia. The aim of most studies was for assessing the agricultural development potential of each soil. Accordingly, about 11 major soil studies were conducted including studies of the Blue Nile River Basin (studied between 1959 and 1962), Awash River Basin (between 1961 and 1964), Wabi Shebelle River Basin (between 1969 and 1970), Rift Valley Lakes Basin (between 1973 and 1974), Setite Humera area (between 1973 and 1974); Gidabo sub-basin morpho-pedologial survey in 1976; Gamabela Agricultural Development Project in 1974.

These studies varied widely in scope, scale, and approach of the survey (Abayneh and Berhanu 2006; Debele 1980; Donahue 1972). Regarding the soil survey approach and methods, in the late 1950s and 1970s, aerial photo interpretation was extensively employed to facilitate soil survey missions, delineate different landforms, and locate soil observation points. For instance, a soil survey across the Setit Humera area covering about 1 million hectares of land was conducted on 1:20,000 scale aerial photographs from which 77 mosaics were compiled and reduced to 1:100,000 scale (Debele 1980). Most aerial photo interpretations were conducted manually with the help of a bench and/or pocket stereoscope followed by manual delineation of homogeneous land units or land facets that guided actual field soil investigations and soil unit's delineation.

In Ethiopia, soil survey has gained institutional recognition after the establishment of the Institute of Agricultural Research (IAR) in 1966 with UNDP/FAO assistance to conduct agricultural research at a national level. The first soil survey and land evaluation section were established in 1971 under the then IAR (Debele 1980). The section was mandated to carry out soil surveys and aid in providing soil and land evaluation data to agricultural research, development planning, and conservation of land and water resources. The section conducted detail soil survey in 1972 in Gode and Holeta, in 1974 in Jima, and in 1977 in Bako agricultural research stations. Further, the section was giving advisory and coordination roles to the government of Ethiopia in guiding soil survey and land evaluation studies by expatriate consultants and/or consulting firms and ensuring a coordinated approach in matters of soil and land classifications.

The soil survey unit of IAR was later transferred to the Land Use Planning and Regulatory Department (LUPRD) of the Ministry of Agriculture (MoA) in 1973 (Elias 2016; Esayas and Debele 2006; Ashenafi 2001). The unit under LUPRD, funded by UNDP with the technical assistance of FAO, conducted a number of soil surveys at different scales (Abayneh 2001). Accordingly, after several early period fragmented soil survey missions, country-wide geomorphology and soils map of Ethiopia at a scale of 1:1 million was produced in 1984 by the then LUPRD, through funding by UNDP and with the technical assistance of FAO (Ali et al. 2020; Abayneh and Berhanu 2006; FAO 1984a, b). In parallel, the then LUPRD of the MoA, conducted serious of soil survey missions and produced other area-specific soil maps and reports at 1:50,000 scale (e.g., soil of Borkena, Bichena, and Hossaina) and 1:250,000 scale (Soils of Menagesha, Hayikoch and Butajira, and Yerer and Kereyu administrative areas) (FAO 2014).

Since the decentralization of the federal government into regional states in 1993 the soil survey unit of the then LUPRD abandoned its role and apparently no institution was there at a national level for the soil resource studies of the country until 1999 (Elias 2016; Ashenafi et al. 2001). However, in 1999, the former IAR or Ethiopian Agricultural Research Organization (EARO) now renamed as Ethiopian Institute of Agricultural Research (EIAR) had reinitiated soil survey of agricultural research centers by establishing a soil survey and land evaluation research section under the then National Soil Research Center of EARO (Abayneh 2001). Accordingly, between 1999 and 2008, the section had conducted a series of soil survey missions and generated soil resource information for about 30 main federal and regional agricultural research centers, sub-centers, and testing sites (Ali et al. 2020).

6.2.2 Soil Survey in Major River Basins, Sub-Basins, and Development Corridors

Land resource studies of river basins started in 1980s under the Ministry of Water Resources for Integrated Development Master Plan preparation. The ministry commissioned various soil survey missions to foreign and national consultants and conducted a number of reconnaissance soil surveys (i.e., at the scale of 1:250,000) for major river basins of the country (Appendix Table A.2) including the Rift Valley lakes basin in 2007 and 2010; Genale Dawa basin in 2006; BaroAkobo basin in 1998; Wabeshebele basin in 2003; Mereb basin in 2001; Tekeze basin in 1999; Abay basin in 1998; Omo-Gibe basin in 1996 (MoWR—Halcrow 2010). These basin studies cover around 85% of the country at a scale of 1:250,000 (Abayneh 2001). Following the integrated development master plan studies, a number of feasible irrigation, groundwater development, and hydropower projects were identified and various soil surveys were conducted at semi-detail and detail levels. These circumstances have also enabled to the generation of immense soil data for each project site which is currently archived at the Ministry of Water and Energy (Ali et al. 2020).

Recently, as of 2000s, federal and regional waterworks design and supervision enterprises have conducted quite a large number of soil survey studies to meet soil data demands in the water and energy sector and growth corridor development projects implementation. The enterprises are public consulting firms and have been commissioned by various clients to conduct studies on many irrigations, hydropower, basin/sub-basin, and growth corridor development plan preparation projects. Accordingly, huge soil information has been collected by these firms. For instance, several site-specific soil survey investigations (at scales varying from 1: 2,500 to 1: 25,000) have been carried out for more than 45 irrigation and multi-purpose projects by the Ethiopian construction design, and supervision works corporation (Ali et al. 2020). Similarly, regional waterworks design and supervision enterprises, such as Oromia, Tigray, and Southern regions water works design and supervision enterprises, and Amhara design and supervision works enterprises have conducted various basin and sub-basin level soil surveys. These surveys were conducted in semi-detail (e.g., Borena, Hararge, East Amhara development corridors) and detail/ intensive soil surveys (e.g., Oromia Special Zone soil studies; Kuraz irrigated sugarcane development project soil studies; Soil studies for a local development plan in Fentale and Tibila irrigation development sites) (Ali et al. 2020). Accordingly, these soil resource inventory missions have generated voluminous soil data from various geographic areas of the country.

6.2.3 The Present Soil Survey Missions: Ethiopian Soil Information System (EthioSIS) and Digital Soil Mapping (DSM) Efforts

In order to to address many of the soil fertility management and fertilizer use-related problems, the Government of Ethiopia-MoA/Agricultural Transformation Agency (ATA) in 2010 established the project "the Ethiopian Soils Information System (EthioSIS)".Accordingly, huge efforts and investments were made to conduct soil survey missions and prepare digital soil property maps, at 250 m resolution, from about 100,000 soil datasets collected through intense topsoil sampling campaigns and digital soil mapping techniques.

The design was to make only soil fertility evaluation and fertilizer formulation without soil resource mapping. The approach followed was limited to auger-based composite sampling mainly in the top 20 cm for annual crops and in some cases 20–50 cm. Currently, the ATA/EthioSIS project has finalized and released the digital soil fertility and fertilizer formulation maps/atlas (Appendix Table A.4) of Ethiopia for 13 soil nutrients (N, P, S, K, Ca, Mg, Na, Fe, Cu, Zn, B, Mn, and Si) and 5 soil properties (pH, EC, CEC, OC, and $CaCO_3$) (EthioSIS 2014).

Similar soil survey efforts by various development partners have been ongoing to complement the existing digital soil information system of Ethiopia. In this regard, the Bilateral Ethiopia Netherlands Effort for Food Income and Trade (BENEFIT) Partnership is a notable one. BENEFIT-Cascape project (2013–2015) as a collaborator of MoA/ATA has produced a map-database of the major soil types/class maps (250 m) classified according to the World Reference Base (WRB 2015) -reference soil groups, prefix and suffix qualifiers of the landscapes of the 30 CASCAPE intervention-districts/weredas using DSM technique-Random Forest (Leenaars et al. 2020a, b; Leenaars et al. 2016). Similarly, in 2019, BENEFIT-REALISE along with the MoA initiated wereda-wide soil resource characterization and mapping task at 1:50,000 scale in 15 BENEFIT-REALISE intervention / districts/weredas.

The objective of the study was to prepare a semi-detailed soil-class map for 15 food-insecure weredas. Accordingly, Reference-Soil-Group (RSG) with varying numbers of principal and supplementary qualifiers was classified according to the World Reference Base (WRB 2015) and was modeled and mapped at 50 m resolution by random-forest (Leenaars et al. 2020a, b).

In Ethiopia, several DSM efforts have been implemented by different initiatives to generate spatially explicit soil information. Most of the national DSM initiatives have been undertaken with the collaboration of international partners such as ISRIC/Wageningen University and Research, and AFSIS. Further, the current DSM application efforts vary in scope and approach. For instance, early DSM studies conducted for modeling and mapping soil properties followed geo-statistical methods such as ordinary kriging, as implemented in ArcGIS environment to develop Woreda-based soil fertility and fertilizer recommendation atlas (Laekemariam et al. 2016; Lelago et al. 2016; Mekonnen 2014; Abate 2015; Belete 2014). However, the national/region-wide digital soil property modeling and mapping by the ATA-EthioSIS project employed MODIS selected spectral bands and derived vegetation indices, 90-m SRTM-DEM derived topographic indices along with satellite-based climatic variables, and statistical/machine learning techniques.

In a nutshell, in most national DSM initiatives, spatially explicit covariates derived from remote sensing products including MODIS products, SRTM-DEM, CHRIPS satellite images, were intensively used to extract proxies of soil forming factors and correlate with point observations (Lennaars et al. 2021; Leenaars et al. 2016). In addition, remote sensing imageries have been widely used to estimate the different soil properties, such as soil functional properties using Landsat and SRTM-based approaches (Vågen et al. 2013), soil salinity mapping using Landsat products (Zewdu et al. 2015), soil moisture estimation using MODIS products (Dagnet 2009). However, very limited studies apply radar-based remote sensing products in predicting soil properties.

In this regard the most notable study by Getachew et al. (2019) incorporates the combinations of dual-polarized Sentinel-1 SAR data, normalized difference vegetation index (NDVI), and digital elevation model as input parameters for soil moisture monitoring in the Upper Blue Nile basin. Similarly, a study by Getachew et al. (2020) implement a combined approach using Sentinel-1 SAR and Landsat sensors products for residual soil moisture retrieval over agricultural fields of the Upper Blue Nile basin.

6.3 The Soil Maps at Different Times

6.3.1 Pre-1975 Soil Maps

National soil maps provide an important archive depicting soil science theory and ideas behind the application of soils information at the time the maps were created (Brevik and Hartmink 2012). In Ethiopia, soil maps have been generated since as early as 1923 in a very general manner by Shantz and Marbut at a scale of 1:25 million as part of the publication "The vegetation and soils of Africa" (Wesphal 1975). There were four map units including Chernozems, Tropical Prairie soils, Laterites, and other Red Tropical soils.

The second soil map of Africa was published by D'Hoore in 1964 at 1:5 million scale and identified 10 major soil groups/associations in Ethiopia (D'Hoore 1964). The map was based upon multicounty soil surveys made with several intensities of detail and with varying criteria for classification. This publication consists of a colored soil map of Africa with 63 map units (Donahue 1972). The map depicts D'Hoore major soil groups/associations including (Wesphal 1975; D'Hoore 1964): Raw mineral soils-formed from rock an rock debris rich in ferromagnesian minerals (Aa), raw mineral soils-not differentiated (Ac), raw mineral soils-formed from desert detritus—sands (Ar), Lithosols (skeletal soils) and lithic soils-not differentiated (Bd), sub desert soils-not differentiated (Bf), Juvenile soils on recent deposits—On riverine and lacustrine alluvium (Bo), Verstisols and similar soils-derived from rocks rich in ferromagnesian minerals (Da), Verstisols and similar soils—Of topographic depressions, not differentiated (Dj), Brown soils or arid and semiarid tropical regions-not differentiated (Gb), Ferruginous tropical soils (fersiallitic soils) not differentiated (Jd), Ferrisols-Humic (Ka), Ferrisols—On rocks rich in ferromagnesian minerals (Kb), Halomorphic soils—Saline soils, alkali soils, and sahne alkali soils (Mb), Hydromorphic soils-mineral hydromrphic soils (Na), an Hydromorphic soils-organic hydromrphic soils (Nb).

D'Hoore (1964) made a serious attempt to relate the map units of his map to the new United States system of soil taxonomy, but much African information was lacking. This is attributed to the United States taxonomy system requirements for more laboratory and field data on soil characterizations than were available for most of Africa.

The third soil map of Africa, on a scale of 1 to 88 million, appeared as a part of "Soils of the World: Probable Occurrence of Orders and Suborders," Soil Conservation Service, United States Department of Agriculture, May 1968. A revised edition, on a scale of 1 to 47 million, was published in January 1971. In 1972, Donahue constructed a soil map of Ethiopia (Fig. 6.1) from the field and laboratory data on 158 soil samples collected from 29 pedons in 4 provinces across diverse ecologies in Ethiopia, and from field observations over an 18-month period of travel and study. This map represents a consolidation of the 18 map units by D'Hoore into 4 units based upon probable predominant soil orders, using the United States system of soil taxonomy, namely, Aridisols, Entisols, Ultisols, and Vertisols. According to Donahue map units generalization, Aridisols comprise 50% of Ethiopia and occur mostly in the eastern, and to a less extent in the southern, parts of the country. Aridislos are classified by D'Hoore as desert detritus—sands, sub desert soils-not differentiated, and brown soils of arid and semiarid tropical soils-not differentiated. Entisols are predominantly rocky, occur mostly in the mountainous west of Ethiopia,

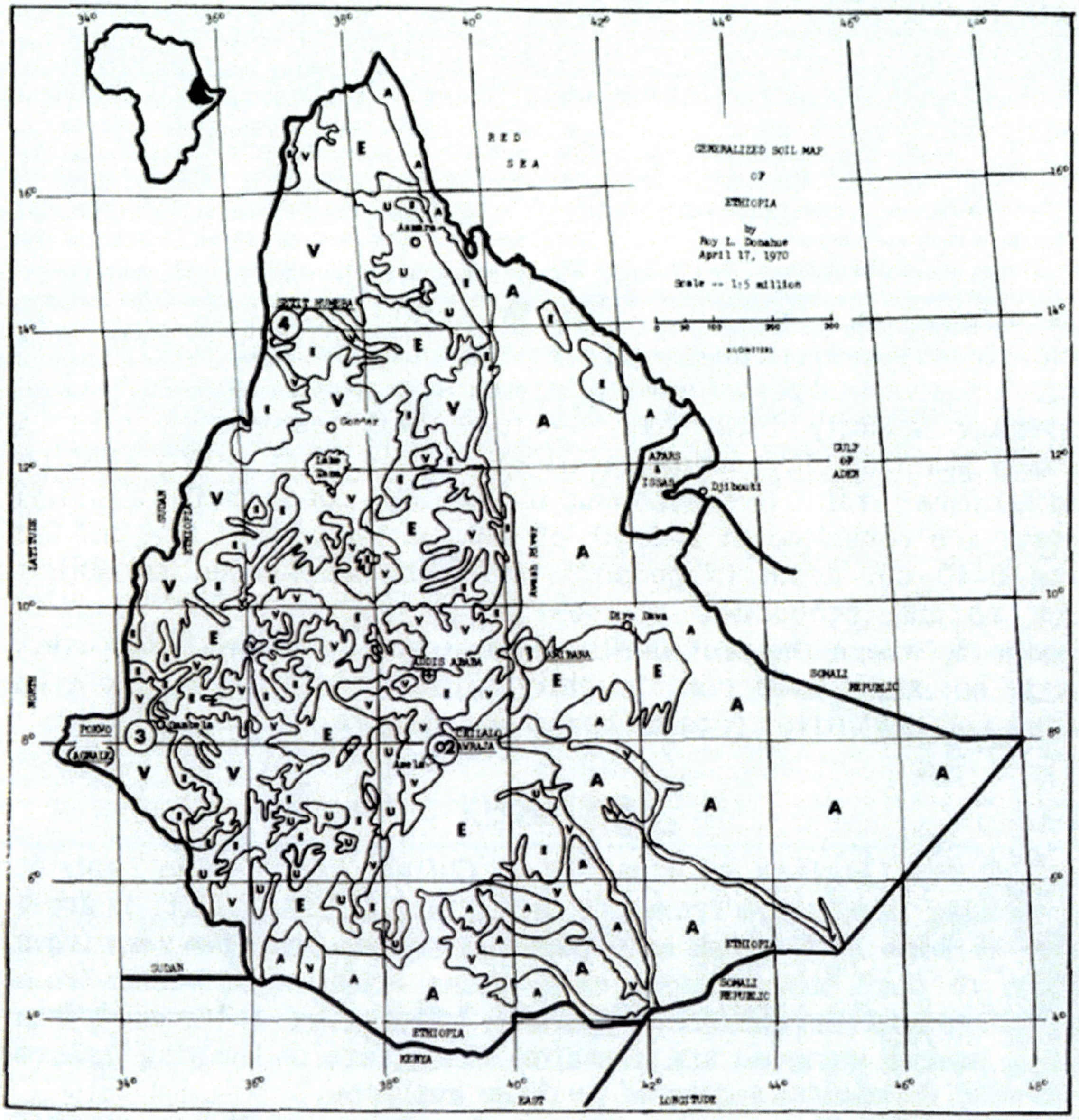

Fig. 6.1 Donahue's constructed soil map of Ethiopia. A: predominantly Aridisols, E: predominantly Entisols, U: predominantly Ultisols, V: predominantly vertisols. (*Source* Donahue 1972)

and total 25% of the area. Ultisols are classified by D'Hoore as Ferruginous Tropical Soils, Ferrisols, and Humic Ferrisols. They occur in the southwest and total 6% of Ethiopia. Vertisols on the map comprise 19% of the country, occur at intermediate elevations in the western half of Ethiopia, and include D'Hoore's Vertisols, various soils from alluvium, and a small area classified by D'Hoore as Organic Hydromorphic Soils.

In 1975, Wesphal, attempted to delimit the following broadly outlined soil regions in Ethiopia in terms of the U.S. D.A. Soil Classification (7th approximation, 1960) (Note: Although soil cartographic representations were not produced, the province/geographic region mapping and naming follows the then administrative boundaries):

(i) Oxisols, Ultisols and Vertisols. In the Eastern Highlands shallow Inceptisols are found.
(ii) **Soils of the Abbay Trough**. In alluvial and colluvial material Vertisols and Inceptisols are found with, more to the west, Entisols in association with Vertisols (in topographic depressions).
(iii) **Soils of the Danakil plains and the Rift Valley**. The Danakil Plains have Aridisols, with salinity occurring in the Kobar Sink. In the Rift Valley itself, west of Awash, Aridisols are found in association with Vertisols (in topographic depressions), whereas to the south also Inceptisols ("Brown forest soils") and Mollisols ("Chernozem") are present. The Awash River valley is an exceptional area in the Rift Valley with a large plain of alluvial soils (with Entisols) near Nazret where sugarcane is grown.
(iv) **Soils of the Somali plateau.** This region has mainly Aridisols with soils rich in gypsum in the extreme south-east of the Ogaden.
(v) **Soils of the "Crystalline Highlands"**. In the area of west Tigre and north Begemdir soils are stony and shallow. They belong to the Orders Ultisols and Alfisols, which are in the extreme north even more rocky and shallow, whereas to the west dry Vertisols (in topographic depressions) and Inceptisols. In western Wellega soils are deep and belong to the Orders Oxisols and Ultisols; in south Sidamo mainly Aridisols occur.

In parallel, in the early 1970s, the UNESCO/FAO produced the "Soil Map of Africa" at a scale of 1:5 million (FAO 1974). The map contains a considerable number of details and the recent soil classification system at that time. Accordingly, the then soil survey and land evaluation section of the Institute of Agricultural Research (IAR) decided to use it as a base map on which to produce a generalized soil map of Ethiopia. The section enlarged the UNESCO/FAO map to a scale of 1:2 million and superimposed on a topographic base for internal use until further validation work was conducted. Although the map was not extensively validated, it was one of the early efforts by national experts to provide broad country-wide soil information to wider soil data users. This preliminary map (Fig. 6.2) identified 13 major groups of soil associations including salt flats (Debele 1980). The major soil groups in terms of area coverage include Regosols (22.88%), Cambisols (20.57%), Nitosols (13.87%), Yermosols (11.72%), Xerosols (10.37%), Lithosols (7.09%), Acrisols (3.67%), Vertisols (3.91%), Arenosols (2.33%), Fluvisols (1.21%), Andosols (0.74%), Ferralsols (0.58%), Solonchaks (0.22%), and Salt flat (0.27%).

The most dominant mapping units identified by the IAR generalized soil map of Ethiopia include the western plateau, with altitudes above 1000 m, the vast area of soil associations in which Cambisols are dominant. In the western section of this plateau, they are largely Humic Cambisols associated with Acrisols, whereas northwards they are mostly Eutric Cambisols associated with Luvisols. The Xeresols and Yermosols are common on and along the semiarid to arid borders of both the Eastern and Western Plateau, as well as in vast stretches of the Rift Valley. Often they occur side by side with the Vertisols. The vast areas on the Ethiopian highlands, in particular Northwest, West, Southwest, and South of Addis Ababa, have been mapped as Eutric Nitosols in various associations with other soils. Finally, the vast areas of Regosols and Lithosols patterns are exhibited on the steep mountain, hill, and valley slopes. In addition, there are other vast areas in southeastern Ethiopia where Calcaric Regosols are particularly common on the limestone formations, where arid conditions prevail at that time.

6.3.2 The Geomorphology Period

After several early periods of fragmented soil survey missions and adaptation of coares global soil maps to the national spatial domain, country-wide geomorphology, and soils map of Ethiopia (Fig. 6.3) at a scale of 1:1 million was produced in 1984 by the then LUPRD/MoA. The map was created as part of FAO/UNDP assistance to the preparation of a Master Land Use Plan, which included an inventory of the land resource as a vital component (Ali et al. 2020; Abayneh and Berhanu 2006; FAO 1984a, b). The map was produced through geomorphic interpretation of 71 scenes of Land sat imagery and overlay with existing soil, climate and vegetation information, and topographic and geologic maps followed by field traverses and checking inaccessible areas. Accordingly, a total of 380 landscape units presumed to be homogeneous in terms of geomorphology, soils, and vegetation were identified (FAO 1984a, b). The primary premise was that soil mapping is founded on the definition and characterization of relatively homogeneous landscape units. The landscape was further divided into geomorphic units (GMUs), which were made up of three main components. These are (1) residual, structural, volcanic, or alluvial/colluvial landform genesis; (2) topographic features or relief characteristics such as high relief hills, undulating to rolling plateaus, dissected side slopes, etc.; and (3) geology/parent

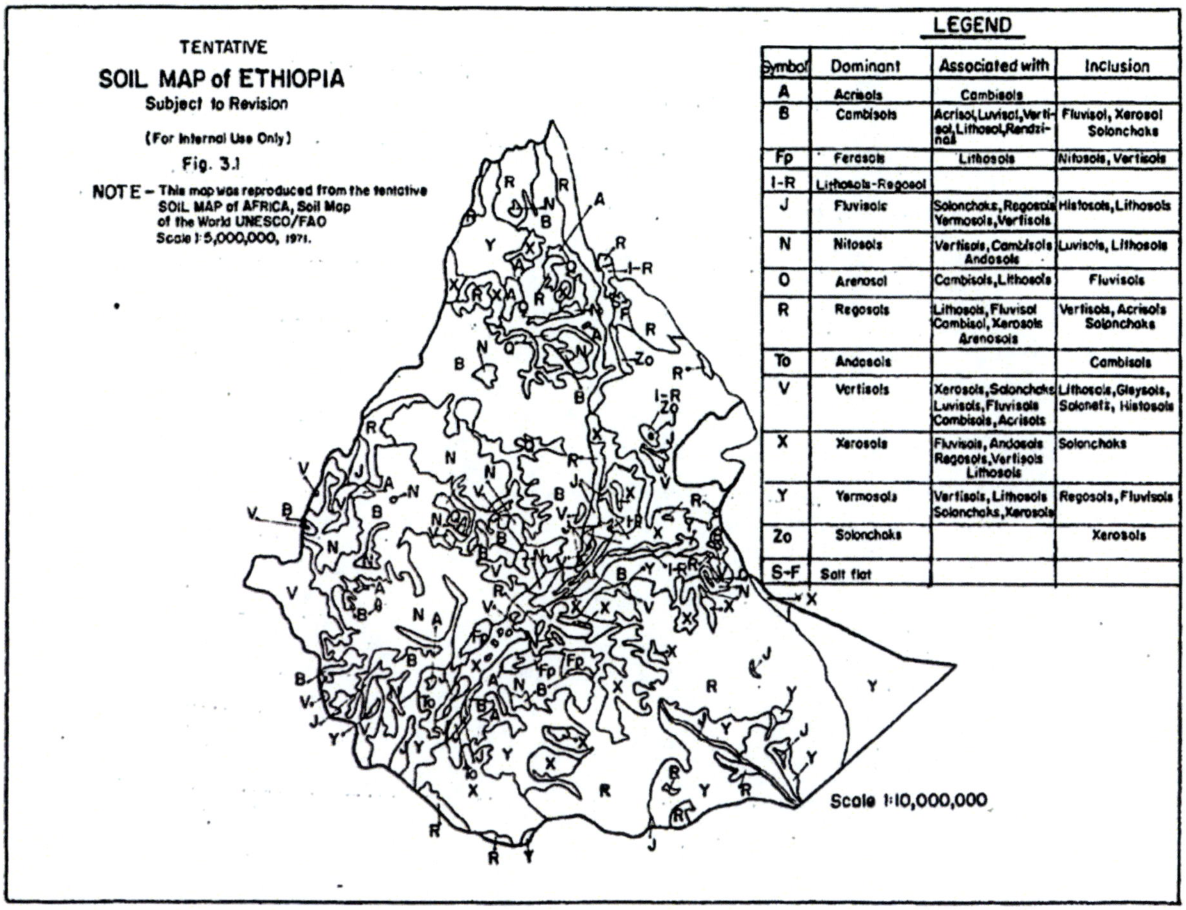

Fig. 6.2 Soil map of Ethiopia by Institute of Agriculture Research (Debele 1980)

materials such as volcanic, limestone, sandstone, felsic Precambrian basement, or alluvium/colluvium. The FAO-UNESCO legend of 1974 was used to classify soil types. The map recognizes the prevalence of 18 primary soil groups across the country, each with different soil types connected with different landscape units.

In 1983, the Soil Association Map of Ethiopia at a scale of 1: 2000 000 was developed in an attempt to update the 1974 FAO/Unesco soil map of the world, which was published at 1:5,000,000 scale. The map is based on the geomorphology and soils map, which was developed at 1:1,000,000 (Esayas and Debele 2006; FAO 1984a, b). However, it included some new information generated after the completion of the geomorphology and soils map. The map and legend are formatted in the same way as the 1974 FAO/UNESCO World Soil Map for soil nomenclature. However, a further subdivision of the soil units is not attempted. The legend has around 160 different map units, which are made up of individual soil units or associations of soil units.

When a map unit is not homogenous, it is made up of dominant soil and associated soils, with the latter covering at least 20% of the map unit's surface area. Important soils that cover less than 20% of the map unit's surface area are added as inclusions. For each association, the dominating soils' textural class and slope class are listed. When hard rock or indurate layers are found at shallow depths, phases are utilized to signify stoniness, salinity, alkalinity, or flooding risk. Nineteen (19) significant soil types were discovered on the provisional soil association map, including Lithosols (17%), Cambisols (16.5%), Nitosols (12%), Vertisols (11%), Xerosols (8%), Luvisols (7%), and Solonchaks (5%) (Fig. 6.4).

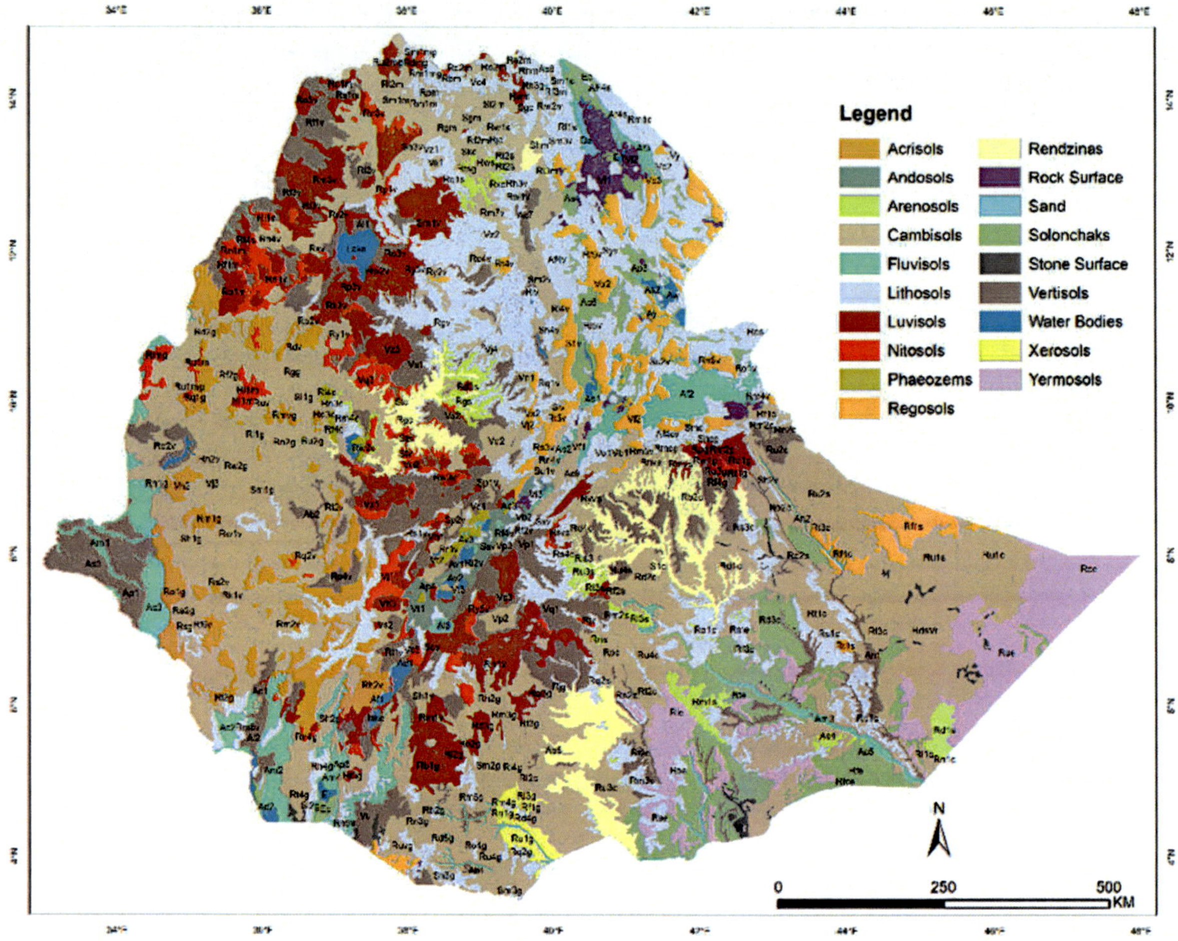

Fig. 6.3 Soil map of Ethiopia (FAO 1984a). (Source: Map prepared by author using digital map obtained from CDE-EASAPP, 2006)

6.3.3 Recent National Soil Mapping Initiatives

The MoA/ATA-Ethiopian Soils Information System (EthioSIS) project has put forth a lot of effort to create national digital soil property maps using about 100,000 topsoil datasets obtained through intensive topsoil sampling campaigns and digital soil mapping methodologies. The EthioSIS project (2012–2019) produced national soil property/fertility maps (Fig. 8.2) with a resolution of 250 m for 13 soil nutrients (N, P, S, K, Ca, Mg, Na, Fe, Cu, Zn, B, Mn, and Si) and 5 soil properties (pH, EC, CEC, OC, and CaCo3) in the top 20 cm (EthioSIS 2019; EthioSIS 2014).

In 2012, the Center for Environment and Development (CDE) attempted, through a thesis research study titled "A National Soil Model of Ethiopia: A geo-statistical technique to develop a national soil map of Ethiopia," to update Ethiopia's national soil information.

The major goal of the research was to design and document a process for more precisely locating existing soil data using a 90 m SRTM-DEM topographic base positioning and terrain unit delineations (Brunner 2012). The result is a set of 1:500,000 scale synthetic soil layers. The maps are not meant to be used in the field (since they are synthetic maps created by modifying several modern sets of procedures), but the model's tools and methods provide a variety of starting points for further development and adaptation in Ethiopia (Brunner 2012).

Development partners have also run a number of projects to assist the MoA efforts to improve the existing digital soil information system. In line with this BENEFIT-Cascape project (2013–2015) has produced a map-database (Fig. 8.2) of the major soil types/class maps (250 m) classified according to the World Reference Base (WRB 2015)—reference soil groups, prefix and suffix qualifiers of the landscapes of the 30 Cascape intervention—Weredas using DSM technique-Random Forest (Leenaars et al. 2020a; Leenaars et al. 2016). About 97% of the agricultural soils in CASCAPE woredas are classified, based on IUSS WRB 2006,

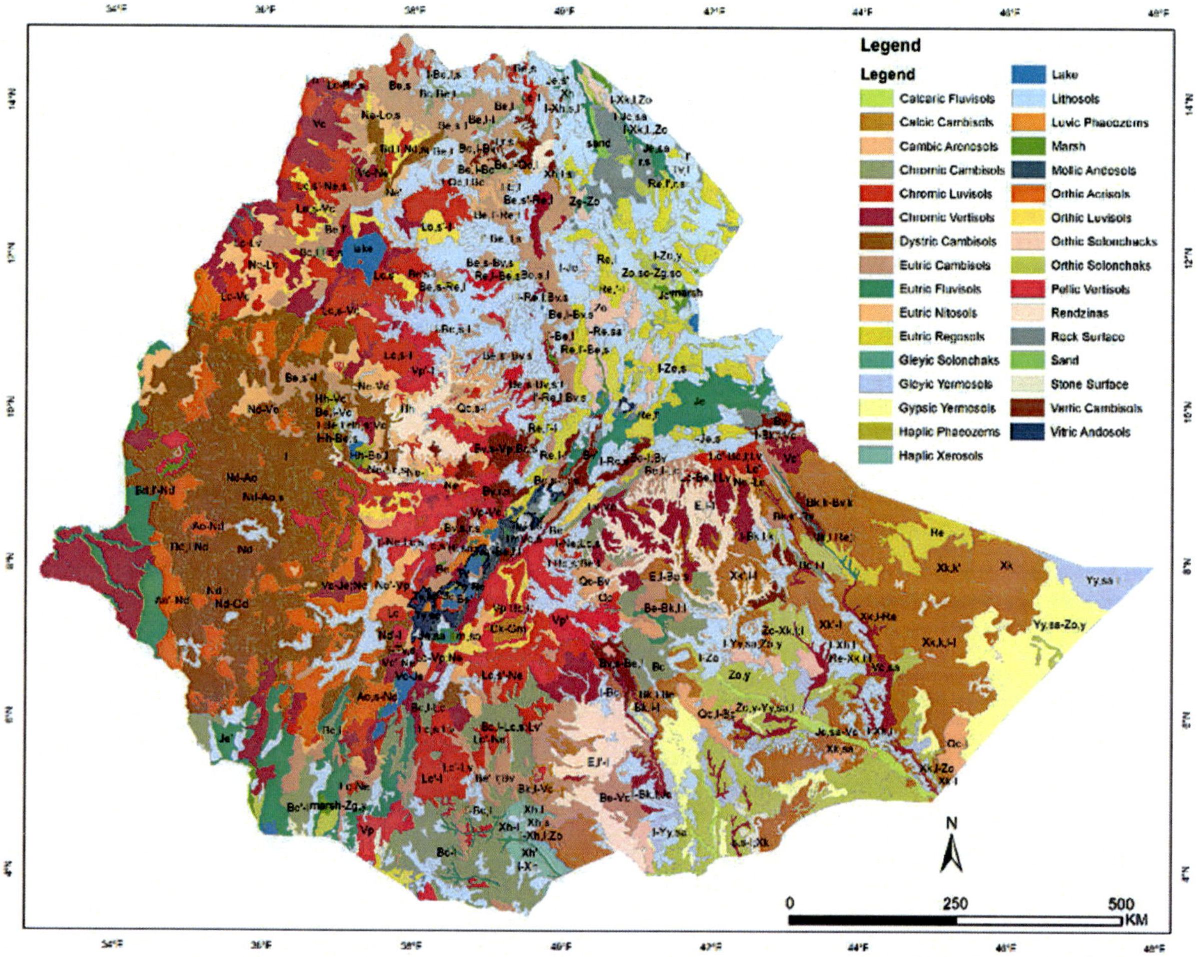

Fig. 6.4 Provisional soil association map of Ethiopia (FAO 1984b). (Source: Map prepared by author using digital map obtained from CDE-EASAPP, 2006)

into five soil types: Nitisols (31%), Vertisols (27%); Leptosols (26%), Luvisols (11%) and Planosols (2%). Other soil groups that have local agricultural importance but with limited spatial coverage include Regosols, Alisols, Andosols, Arenosols, Fluvisols, Phaeozems, Acrisols, Calcisols, and Gleysols (Elias 2016).

Similarly, in 2019, BENEFIT-REALISE along with the MoA has generated digital soil resource maps (Figs. 6.5, 6.6, 6.7) at 1:50,000 scale in 15 Weredas. Spatial soil-class distribution was modeled at 50 m resolution by random forest. Accordingly, the semi-detail soil maps composed of the Reference-Soil-Group (RSG), and varying numbers of principal and supplementary qualifiers (PQ & SQ) are classified according to the World Reference Base (WRB 2015; Leenars et al. 2020b).

Digital soil mapping, in Ethiopia, has been implemented in a variety of ways, each with its own scope and methodology. Early DSM investigations used a geo-statistical method called ordinary kriging, which was applied in the ArcGIS environment to model soil attributes (Laekemariam et al. 2016; Lelago et al. 2016; Mekonnen 2014; Abate 2015; Getachew 2015; Belete 2014). The ATA-EthioSIS project, on the other hand, used MODIS spectral bands and derived vegetation indices, 90-m SRTM-DEM generated topography indices, satellite-based meteorological variables, and statistical/machine learning techniques to predict regional digital soil fertility. Most national DSM programs relied heavily on spatially explicit remote sensing products such as MODIS, SRTM-DEM, and CHRIPS satellite images to derive proxies of soil forming processes and connect with point measurements (Leenaars et al. 2016).

Remote sensing imageries have been widely used to estimate various soil properties in Ethiopian landscapes, such as soil functional properties using Landsat and SRTM-based approaches (Vagen et al. 2013), soil salinity mapping using Landsat products (Zewdu et al. 2015), and

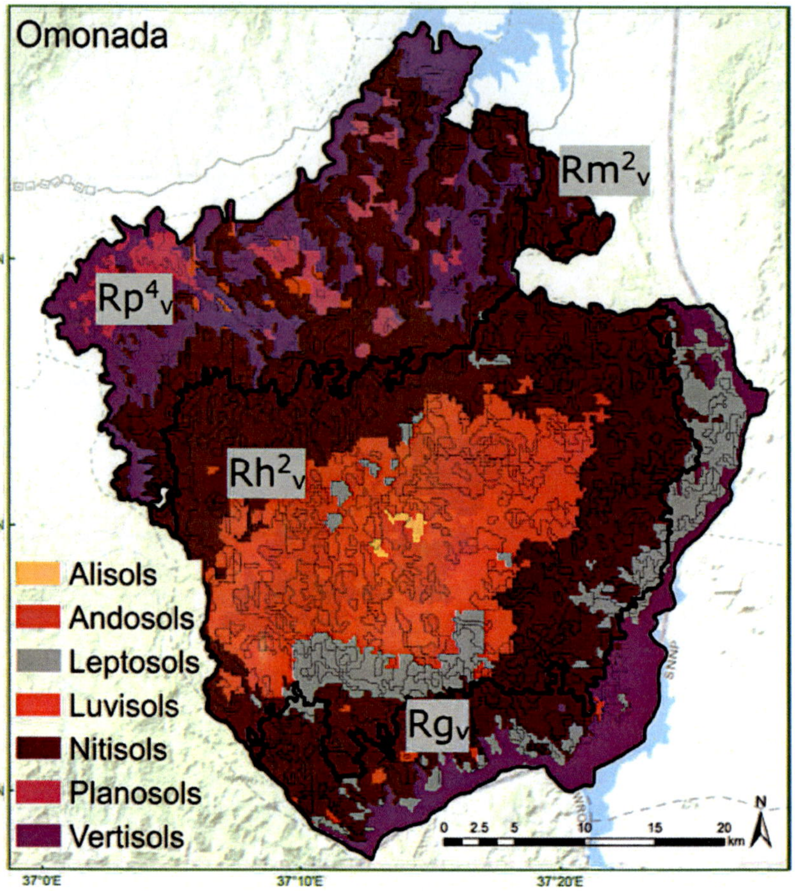

Fig. 6.5 Polygon map of the major soil-landscape resources in Omonada woreda at the level of RSGs (RSGs indicated by color, major landscapes by bold black polygons, and landscape facets by thin black polygons). Rgv: Major river gorges, canyons, and escarpments including very steep side slopes and plateau terraces, Rh2v:high to mountainous relief hills, Rm2v: moderate to high relief hills, Rp4v: undulating to rolling plateaus. (*Source* Leenaars et al. 2020a)

soil moisture estimation using MODIS products (Dagnet 2009). Only a few studies, however, have used radar-based remote sensing products to forecast soil attributes. Getachew et al. (2019) use dual-polarized Sentinel-1 SAR data, normalized difference vegetation index (NDVI), and digital elevation model as input parameters for soil moisture monitoring in the Upper Blue Nile basin. Similarly, Getachew et al. (2020) use a combined technique for residual soil moisture retrieval over agricultural fields in the Upper Blue Nile basin, using Sentinel-1 SAR and Landsat sensor outputs.

6.3.4 Regional and Global Soil Maps Layering

Besides various national soil resource investigation studies, several endeavors are being made globally and regionally to coordinate soil information generation, sharing, and improving access to soil information by various institutions. Among the existing efforts, the most commonly applied products in the Ethiopia context include Harmonized World Soil Database (HWSD), Soil and Terrain (SOTER) database, World Soil Information Service (WoSIS), and Africa Soil Information Service (AFSIS) (Ali et al. 2020). These global and regional initiatives are designed to avail soil information (soil maps and point profile data) at global and regional scales though most initiatives are based on the dataset collected by national efforts.

At a regional level in 1997, the FAO-ISRIC SOTER initiative has produced polygon-based digital soil and terrain map of Ethiopia at 1:1mln scale (FAO/ISRIC/SOTER 1997). Compared to the geomorphology and soils, and soil association map of Ethiopia, the SOTER map is considered as the latest nationwide soil and terrain information of Ethiopia (Elias 2016). The SOTER technique entails the definition of areas with a uniform set of soil and topography features, as well as the creation of an attribute database for mapping units based on well-defined distinguishing criteria (e.g., landform, lithology, local hydrological conditions). The identification of sections of land with a characteristic, often the repetitive pattern of landform; lithology, surface form, slope, parent material, and soil is at the heart of the SOTER technique. SOTER units are land parcels that have been identified in this way. As a result, the SOTER mapping

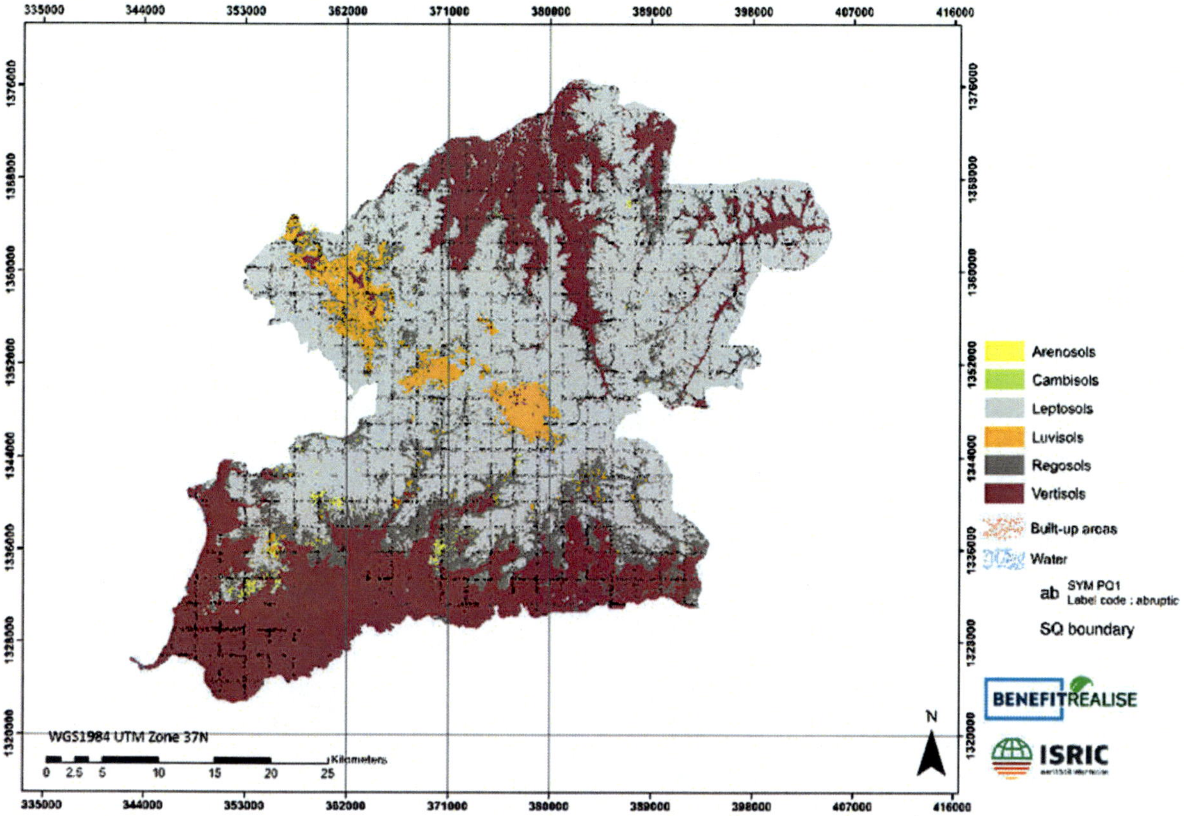

Fig. 6.6 Polygon map of the major soil-landscape resources in Libokemkeme woreda. (Source: Leenars et al. 2020b)

approach is similar to physiographic soil mappings in many ways; each SOTER unit reflects a unique combination of terrain and soil characteristics. Ethiopia's SOTER map is made up of 18 Reference Soil Groups (RSGs) and 41 soil units (Fig. 6.7) classified based on the updated 1988 FAO-UNESCO legend (FAO-ISRIC 1997).

The Africa Soil Information Service (AFSIS) is another latest regional initiative (2009–2016) that aims to fill a major gap in soil spatial information in Africa. To this end, new soil data were collected and combined with collated and harmonized soil legacy data from over 18,000 locations in Africa of which about 1,820 legacy profiles were sourced from Ethiopia. The AfSIS project employed a digital soil mapping technique (3D regression-kriging based on random forests) and prepared 3D maps of a great variety of soil properties layering Ethiopia at spatial resolutions up to 250 m (Leenaars et al. 2014; Leenaars 2012). The 3D predictions of soil properties include - organic carbon, pH, sand, silt and clay fractions, coarse fragments, bulk density, cation exchange capacity, total nitrogen, exchangeable acidity, exchangeable bases (Ca, Mg, K, Na), and extractable aluminum-at 250 m spatial resolution at either two or six standard soil depths (Hengl et al. 2015, 2017).

Among the global soil mapping initiatives layering in Ethiopia, SoilGrids facilitate consistent predictions and provide various soil property and class maps (Poggio et al. 2022). It is a system for global digital soil mapping that makes use of global soil profile information and covariate data to model the spatial distribution of soil properties and classes (currently at 250 m resolution) across the globe. Soil properties predictions are made at six standard depths. Maps of the following soil classes and properties are freely available: World Reference Base (IUSS 2006) soil groups (Fig. 6.8), pH, soil organic carbon content, bulk density, coarse fragments content, sand content, silt content, clay content, cation exchange capacity (CEC), total nitrogen as well as soil organic carbon

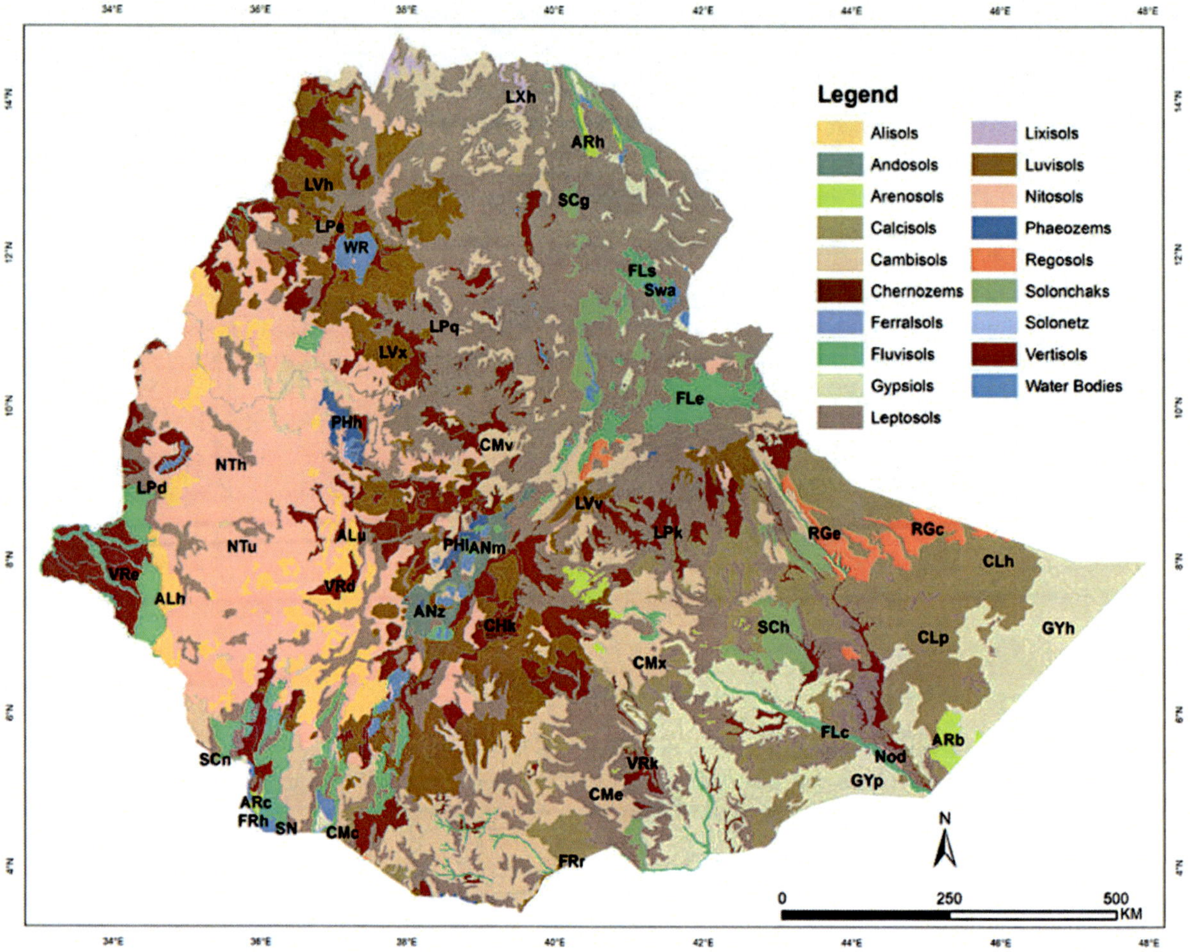

Fig. 6.7 Soil and terrain map of Ethiopia. Source: Map prepared by author using digital map from FAO/ISRIC/SOTER (1997)

density and soil organic carbon stock. Soil Grids have been updated since the previous release in 2017, currently, the new updated version 2020 is freely available for users (Poggio et al. 2022; Hengl et al. 2017).

6.4 State Factors

Soils are the result of interaction of the environmental factors. The spatial variations of environmental conditions and their interactions result in a spatial differentiation of soil characteristics (Buol et al. 1989). In fact, little attention has been paid to soil genesis and classification studies in Ethiopia, but research into this aspect would create an opportunity to understand and solve problems in land resources management (Mitiku 1987). Major factors affecting the state of Ethiopian soils are briefly discussed as hereunder.

Parent material

In Ethiopia, distributions of soils are partly related to their proximate parent materials. The major geological materials of the Ethiopian highlands are tertiary and Quaternary basalt and deposits and granite rocks, whereas gypsum, sandstone and limestone form major geologic features in the lowlands (Fig. 6.9; Tefera et al. 1996).

Most soils of the central, eastern, and northern highlands are developed on hard basalt rocks or on materials derived from them (Ministry of Agriculture 1987). In the southwestern parts, most soils are formed from crystalline basement rocks. Andosols are developed from Quaternary pyroclastic and volcanic ash deposits and pumice, found in the Ethiopian Rift Valley (Fig. 6.10) and on the high mountains such as the Simein mountains and elsewhere. Fluviatile deposits located along the course of main rivers and in foothills and depressions lead to the formation of

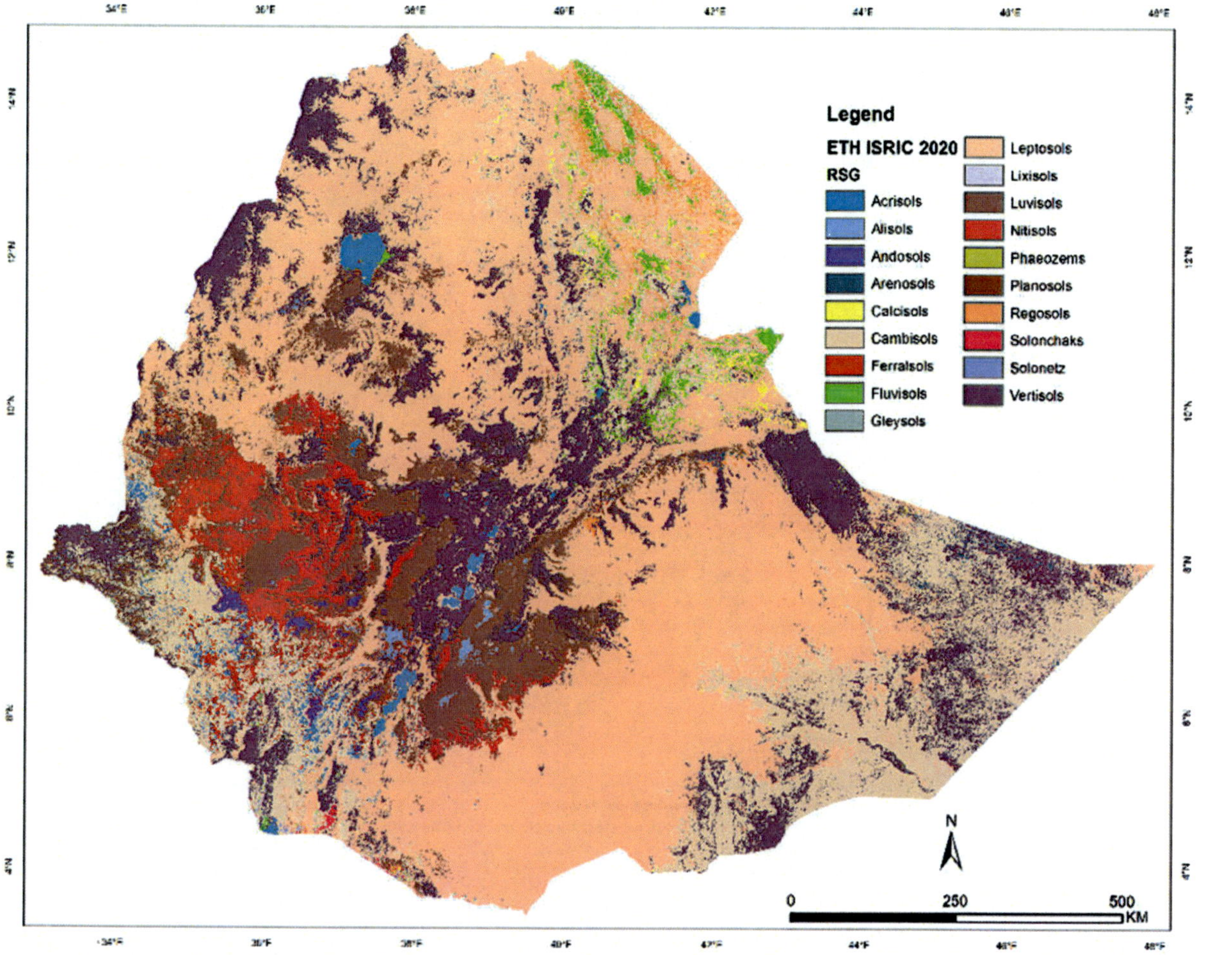

Fig. 6.8 WRB-Reference soil group map of Ethiopia (SoilGrids250m version 2.0: https://soilgrids.org//). Source: Map extracted and compiled by author using digital map from SoilGrids250m version 2.0: https://soilgrids.org//

Fluvisols. Ethiopian lowland soils are developed on sandstone, limestone, and alluvial deposits.

Moreover, on lacustrine deposit and depression sites with seasonal water cover Gleysols or their variants such as Gleyic Vertisols, Cambsiols and Rgosols are found. Therefore, under the Ethiopian conditions, parent materials have a great influence on the characteristics and types of soils. The general relationships between soils and their proximate parent materials are summarized in Appendix Table A.5.

Topography

Throughout the Ethiopia highlands, topography is the major differentiating factor of genesis and distributions of soils (Fikru 1990; Mitiku 1987; FAO/UNDP 1984a, b; Belay 1982). Gently slopping terrain units are occupied by well-drained Nitisols, Acrisols and Luvisols (FAO/UNDP 1984a, b) and steep and colluvial foot slopes by Leptosols, Cambisols and Regosols (Virgo and Munro 1978). In the latter parts, active erosion and deposition leave little time for soil development (Venema and Paris 1986, 1987; Virgo and Munro 1978).

On the plateaus (<2500 m.a.s.l.), poorly drained Vertisols dominate (Mitiku 1987;Belay 1982; 1996; 1997a; 1998; Berhanu 1982). At intermediate high altitudes (2500 to 3000 m.a.s.l) Phaeozems occur on gently to steeply sloping terrain (Kefeni 1995; Venema and Paris 1986; Paris 1985) and Andosols are found in the high altitudes (>3000 m.a.s.l) (Mohammed and Belay 2008; Weigel 1988; Belay 1995; Bono and Sieler 1984; Frei 1978) and in the Rift Valley lowlands, conditioned by local geologic features. The poorly drained flatter and depressions sites are covered with Gleysols and found mainly around major lake basins such as Tana and Hawasa lakes.

Topographically induced soil variability is the result of the processes of the surface and subsurface water flow and erosion and deposition distributions in the landscape. Hillside processes remove finer particles and soluble salts of Ca,

Fig. 6.9 Geological map of Ethiopia). (Source: Map re-drawn by author using digital map from Tefera et al. 1996)

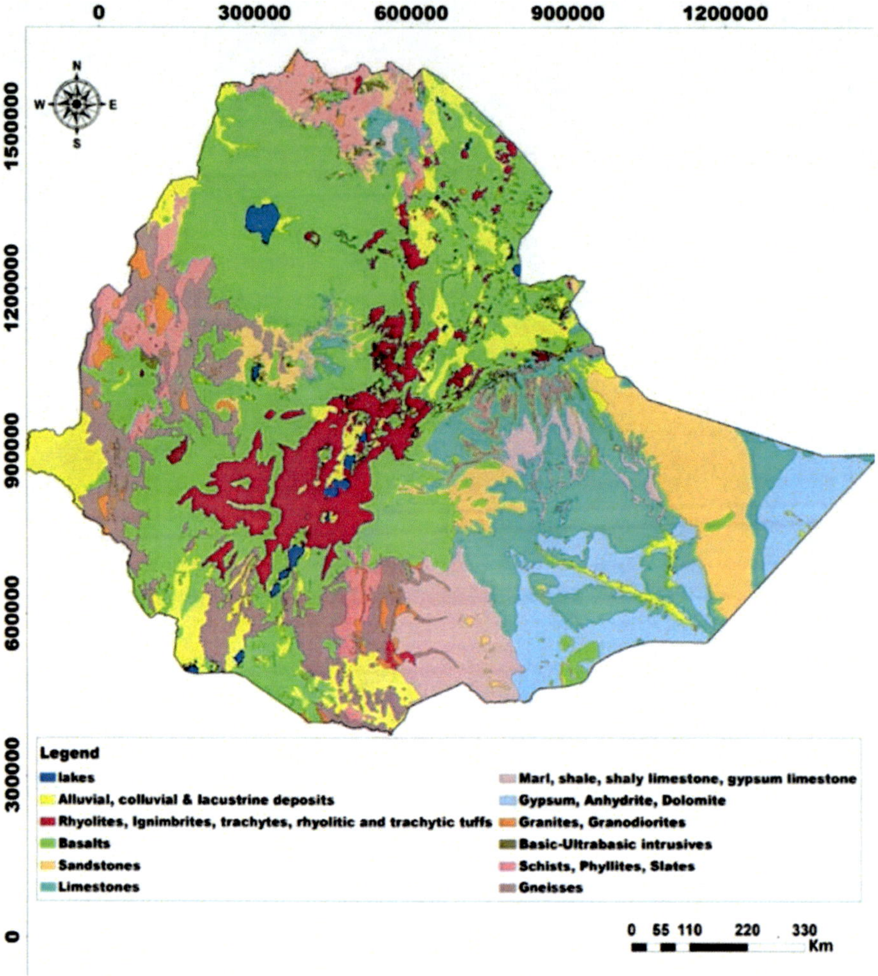

Fig. 6.10 Volcanic ash forming Andosols' parent material, Upper Ethiopian Rift Valley (Shashemene area) (Photo by Mohammed Assen)

Mg, and Na from upper steeper slopes (Agbu et al. 1989; Parsons 1978). Consequently, upper slope soils are freely drained as indicated by redder hues (7.5 YR or 5YR, 2.5YR) (Belay 1982; Mitiku 1987). Rapid runoff and shallow soils limit water infiltration and retard soil development. These soils are unstable (Parsons 1978). The thin soil depth occurring in steeper and convex slopes of Ethiopia is an indication of rapid removal of weathered material (Virgo and Holmes 1977).

On the other hand, materials eroded from steep upper slopes are deposited in concave flatter slopes (Parsons 1978). Concentrations of fine particles, soluble minerals, and organic carbon characterize these parts (Nizyeimana and Bicki 1992). These lower slopes are areas of water accumulation and are either poorly or imperfectly drained. The soils have yellowish or yellowish red hues (10YR) (Nizyeimana and Bicki, 1992; Mitiku 1987), and high pH reflecting the concentrations of salts (El-Hassanin 1985) and leading to calcification (Birkeland 1999; Buol et al. 1989).

Climate

Climate influences the intensity of weathering and leaching (Birkeland 1999). Since climate varies over short distances in the highlands of Ethiopia, its impact on soils and soil properties also varies. In the high rainfall parts of the southern and western highlands of Ethiopia, highly leached soils dominate (FAO/UNDP 1984a, b; FAO-Unesco 1977). Areas of Ethiopia with high rainfall and high temperatures throughout the year have deep soils as a result of rapid weathering rates. In these parts of the country, soil formation rates are estimated at 22 t ha-1 yr-1 (Hurni 1983). In the northern, central, eastern, and southeastern highlands of Ethiopia, the influence of climate is subtle. In these parts of the country, less leached and base-rich soils dominate and the rate of soil formation here is slow (Hurni 1983).

On the other hand, drier semiarid and arid lowlands of the country are occupied by soils with high salt and alkali materials that make them to be saline-sodic soils. These soils include Solonchaks, Gypsisols, Calcisols, and Solonte. They occur mainly in the Ogaden lowlands, the Afar depression, and in large drier parts of the Ethiopian Rift Valley.

Vertical variations in climate affect soils in a similar way to lateral variations. The increase in rainfall with a rise in altitude increases leaching and leached soils are common in the high mountainous areas of Ethiopia (Murphy 1968a). Low temperatures slow down organic matter decomposition and soil formation (Fanning and Fanning 1989). In Ethiopia, rates of soil formation are estimated at less than 2 t ha-1 yr-1 for altitudes >3000 m.a.s.l (Hurni 1983). Consequently, a rise in altitude is associated with a decrease in soil depth (Mesfin 1998) and an increase in soil organic carbon content (Hurni 1983; Murphy 1968a).

Organism

Organisms including both plants and animals as well as man-made activities such as land use types play an important role in the formation and distribution of soils (Fanning and Fanning 1989). Vegetation is also a result of the environment. The climatic-climax tropical woodlands and forests growing in high rainfall and high-temperature soils of southern and south western of Ethiopian highlands (Daniel 1977, 1988) are associated with deep leached soils.

In the intermediate altitudes, semi-temperate and temperate forests such as Juniperous procera and Podocarpus gracilor are growing on Phaeozems and Leptosols with or without mollic A and / or umbric horizons, but generally with a high organic carbon content (EMA 1988; Murphy 1968a).

In the uncultivated parts of the Ethiopian high plateau experiencing high rainfall, Festuca (tussock) and Lobelia grow on Andosols and/or Phaeozems. In the low rainfall parts, Acacia spp. and Croton macrostachys form the major vegetation types (Mitiku 1987). In the very high altitudes of the country, Erica arborea interspersed with Hypericum revoltum grow on Andosols (Mohammed and Belay 2008; Frei 1978). On the other hand, when natural forests are turned to man-made uses such as farming, soils are degraded and lost their natural quality. As a result, in the long term, naturally existed mollic and/or umbric horizons would be transformed to ochric A horizons. This causes a change even in the soil types, as the characteristics of mollic/ umbric horizons are modified by the impacts of land use activities.

Animals move from one soil to another by carrying soil materials and depositing their fecal and carried material, largely responsible for mixing materials. This activity causes lateral mixing of soil organic matter, soil material, and possibly plant nutrients. In addition, burrowing animals dig out soils from the subsoil and bring them to the surface, which makes deposit as termite mounds in the surface of the soils. These features are typically found in the savanna lands of Central Rift Valley, on either side of Mojdo-Hawasa main highway. These areas, depending on local factors, are covered with Andosols, Phaeozems, Calciosls, Cambisols, or Regosols. In such parts, where there is a good natural drainage, there is a partial removal of carbonates and residual carbonates accumulate in subsoil horizons responsible for formations of calcisols and/or soils with some degrees of calcic properties.

6.5 Soil Forming Processes

Understanding of the soil forming processes contributes to a better characterization of soils (IUSS working Group WRB 2014). Two different, non-exclusive processes have been suggested that could be at the base of the formation of any soil: the geogenetic process and the pedogenetic process, the latter can be distinguished into physical pedogenetic process and chemical pedogentic processes (Driessen et al. 2001). The geogenic versus pedogenic arguments concern the relative importance of geological controls and processes such as sedimentary layering and erosion/deposition as opposed to pedological processes such as translocation and bioturbation. While geogenic processes are driven by geomorphic or geologic factors external to the soil, the pedogenetic factors are internal to the soil. For example, texture contrasts arising due to translocation of clays or bioturbation are primarily pedogenic. In the succeeding sub-sections, a brief account of the pedogenetic and geogenetic processes underlying the major soils of Ethiopia is given.

6.5.1 Pedogenetic Processes

In a pedogenetic study conducted in the Kulumsa watershed of the Arsi highlands, Southeast Ethiopia the main soil forming processes that has been operating by the soil development at the lower altitude areas were calcification, OM accumulation, humification (melanization), braunification, eluviation and illuviation that led to the formation of argic (argillic) and cambic subsurface horizons and mollic surface horizons (epipedons) which are diagnostic horizons of Kastanozems, Chernozems and Luvisols. On the other hand, at the higher altitude, the dominant pedogenetic processes appeared to be intensive leaching (soluvation) accompanied by decalcification, lessivage, braunification, melanization, faunal pedoturbation and mass-wasting (solifluction), that led to the formation of soils such as Retisols and Alisols (Alemayehu 2018). In the didessa watershed, western Ethiopia, a transect-based pedological study indicated that Melanization, cheluviation (chelation) of organo-mineral substances, oxidation of iron, braunification, rubification, ferrugination, leaching of iron compounds and cations, clay translocation, de-alkalization, and acidification were the major pedological processes in the upland soils. Erosion was also observed at shoulder slope positions. Melanization, calcification, and pedoturbation were the conspicuous pedological processes in the low-lying topographic position at the footslope in the landscape. Alisols, Luvisols and Vertisols were the major reference soil groups formed by these pedogentic processes.

6.5.2 Geogenetic Processes

Geogenetic processes are those processes that result in a coarser fraction covering a finer fraction, like the selective sedimentation or colluvial deposition of a sandy layer on a more clayey layer, selective erosion of the finer soil fractions in the upper part or sheet wash of a lighter textured soil over a layer with more clayey soil. In the study of the genesis of Planosols in the highlands of southwestern Ethiopia, Van Ranst et al. (2011) indicated that the abrupt textural change between the silty loam textutred albic horizon and the underlying heavy clayey horin is this soil is formed by the geogenetic process. Counter-indicative to the pedogenetic process, ferrolysis previously ascribed to be responsible for the formation of these soils by the several authors, the authors found out that at the point of abrupt textural change there are relatively high pH, presence of a sizeable reserve of weatherable minerals, feldspars, and volcanic glass. It has been concluded therefore that these soils were formed by the deposition of volcanic ash on top vertisols. In soil characterization study within the Geba catchment, Northen Ethiopia, Nyssen et al. (2019) indicated that Planosols close to the Adgirat cliff was formed by the geonetic process where the coarse surface layer is a colluvial deposit from the cliff while the clayey subsurface horizon is from the Edaga Arbi tillites that outcrop under the sandstone cliff.

6.6 Soil Types

Previous sections discussed that Ethiopia has diverse environments in terms of its relief, climatic types, geologic features, vegetation, and land use patterns. The existence of such diverse environments results in the formation and presence of different soil types and soil characteristics, conditioned by the given available time for the formation and existence of particular soil characteristics. The local and regional variability of these factors has resulted in the formation of a variety of soil types. Consequently, this variation in soil characteristics and thereof in soil types is recognized in relation to biotic and land use patterns, geologic, climatic, and relief characteristics of the country. Whereas other factors are also important, as reported in many studies, distribution of the soils of Ethiopia is largely governed by the interactions and variations in its topography, climatic and biotic features (Mohammed 2003; Eylachew 1987; Mitiku 1987). Consequently, in the high cooler altitudes, mainly at above 2000 masl, soils with dark organic matter rich A horizons predominate. Those parts, where removal of materials, induced by steeper topography, is accelerating water flow and gravity, are largely occupied by shallow soils. Deeper soils cover the different portions of deposition lands resulting in diverse textural characteristics, ranging from coarse textured to clay dominated types of soils. Along the course of several rivers and streams, soils with fluvatile materials are common. On the other hand, drier lowlands contain soils with saline-alkaline, calcic and gypsic characteristics.

Despite the presence of several types of soils, only nine soil types cover 90% of the land surface of Ethiopia. These major soil types are: Leptosols (30%), Nitisols (12%), Vertisols (11%), Cambisols (9%), Luvisols (8%), Calcisols (9%), Fluvisols (4%), Gypsisols (8%), and Alisols (3%) (Appendix Table A.6).

These dominant soil types are widely distributed in different parts of the country mainly governed by their local and regional factors of soil formation. In terms of their agricultural significance, Nitisols, Vertisols, Cambisols, Luvisols, and Fluvisols are major cultivated soil types. Leptosols are shallow soils, limited by hard rock and steeper topographic feature, and are less devoted to crop farming. Where they are found, these soils are largely used as hilly grasslands and forest and natural vegetation cover. On the other hand, Gypsisols, Solonchaks and to some extent Calcisols and other dry land soils largely cover moisture deficient parts of the country, and are less used for crop cultivation. These types of soils are occupied by extensive pastoral farming and in the recent period, irrigation agriculture is expanding toward these areas. Alisols are found in high rainfall areas where intense leaching processes typically characterize locally, and are chemically poor to be used for crop farming.

The major soil types of Ethiopia can be generally categorized as given in Appendix Table A.6. These soil types are discussed in relation to their major characteristics, distribution, and agricultural properties.

Fig. 6.11 Typical Nitisol profile from Didessa district, Buno Bedele, southwest Ethiopia showing characteristic shiny ped faces (Photo by Alemayehu Regassa)

Leptosols

These soils occur in the steep and convex slope forms and eroded landscapes of Ethiopia. However, they also occur in lower slope positions where hard rocks limit the formation and presence of deep soil. Soil depth in Leptosols varies depending on the depth of underlying lithology and topographic characteristics. Leptosols of the drier, calcic, and gypsic geologic composition have high pH values, but the wetter and cultivated Leptosls have low pH values and even become acidic, whereas virgin soils are moderately acidic (Appendix Tables A.7 and A.8). Contents of soil organic matter and total nitrogen of Leptosols vary in relation to their land use scenarios. As can be seen from Appendix Table A.7, cultivation has also lowered CEC, pH and available P of Leptosols.

Nitisols

Nitisols cover about 12% (approximately about 14,100,000 ha) of land area of the country. These soils occur mainly in the humid southern, western, eastern, and northwest central highlands. Nitisols are found in parts where there is intense to intermediate weathering processes associated with high rainfall and high-temperature conditions of Ethiopia, e.g., in its southern and western highlands. This gives these soils to have deep well weathered horizons and mainly occur on sloppy to gently sloppy topographic positions (Fig. 6.11).

The dark brown surface horizon in these soils is due to high organic matter content whereas their well-drained properties largely give their subsoil to reddish or yellowish-reddish color patterns. The subsoil horizons of Ethiopian Nitisols are characterized by prominent shiny pedsurfaces, making it one of the distinctive features of these soils. The clay distribution in the subsoil horizons is more or less uniform and has low silt:clay ratio (Appendix Table A.10).

As many other soils of Ethiopia, Nitisols are dominated by bivalent cations, and the magnitude of exchangeable cations is common in the order of Ca > Mg > K > Na. These soils are commonly poor in available phosphorus and nitrogen contents. In tropical soils, the content of soil organic matter is very low to low which was also observed in many Nitisol zones of Ethiopia (Appendix Tables A.10,

A.11, and A.12). These soils have base saturation of commonly below 50%, which reflects their occurrence in high leaching-prone environments. The value of pH in these soils also ranges from slightly acidic to moderately acid. However, Nitisols tend to have high levels of nutrient storage capacity and this makes them to be one of the fertile soils of Ethiopia.

The good subsoil structures make Nitisols to be one of the favorable soils of thigh rainfall areas of Ethiopia for farming. Annual crops such as maize and perennial coffee are grown on these soils in the eastern, southern, and western highlands. Also, another common plant locally called Enset (false banana) grows in southern, central, and south western highlands of Ethiopia. This plant is considered as one of the major food security crops in these parts of the country. The north western highland Nitisols are largely occupied in annual crops like maize, wheat, and *teff*.

Vertisols

Vertisols occupy about 11% (12,110,000 ha.) land area commonly on the lower positions of the landscaped positions and formations of finer materials are favored. Owing to their occurrence in this particular landscape, Vertisols have thick depth, which in some cases reach as deep as 200 cm (Appendix Table A.13). A typical feature of Vertisols of Ethiopia is a uniform strong to very strong and coarse to very coarse soil structure with prominent characteristic slickensides (Fig. 6.12). They are clayey in texture with clay content exceeding 40% throughout the depth of the soil profiles of most studied profiles.

The clay mineralogy is dominated by smectitic as discussed in some studies (e.g., Mitiku 1987; Eylachew 1987). The dry bulk density values of the surface horizons of Ethiopian Vertisols commonly range from 0.95 to 1.5 g cm^{-3}, which can be accepted as the value for cultivated soils (Appendix Table A.13).

Fig. 6.12 Typical Vertisol profile with characteristic slickensides on the flat terrain of Ener and Enamore district in the Guragie high lands, southwest Ethiopia (Photo by Alemayehu Regassa)

Vertisols form major cultivated soils of Ethiopia depending on their local environmental features. In high rainfall areas, they form major shallow-rooted *teff* land in both major and small rainy Ethiopian Vertisols are cultivated with irrigated maize, cotton, and sugarcane crops as seasons. The lowland well as many other fruit crops. The irrigated cultivated Vertisols are often subjected to impacts of salinity due to over application of flood irrigation.

Depending on the rainfall patterns and geological characteristics, soil reaction of Ethiopian Vertisols ranges from slightly acid to alkaline characteristics (Appendix Tables A.13 and A.14). This soil tends to be slightly acidic in high rainfall and moderately alkaline in low rainfall and probably carbonates rich area parts of the country. Thus, soil-pH variation of Ethiopian Vertisols is directly related to the leaching impact of available rainfall. As a result, the presence of some levels of none leached $CaCO_3$ content in some horizons of Vertisol profiles, e.g., northern Ethiopian highland Vertisols partly indicates the low level of leaching efficacy. Also, in the absence of calcareous parent materials, source of carbonates in the soils could be weathering of basalt rocks, which have Ca-rich primary minerals. As a consequence, Ethiopian lowland Vertisols tend to have calcic properties as they receive weathered and eroded soluble ca-materials from their counter highland regions of the country. These low land Ethiopian Vertisols tend to be slightly alkaline to alkaline in pH conditions.

The Vertisols of Ethiopia have very low organic matter and total nitrogen contents. However, this varies depending on the local land use patterns, where uncultivated and/or followed Vertisols have better levels of organic matter and nitrogen contents.

Calcium and magnesium form predominant exchangeable bases throughout Vertisols of Ethiopia. In Vertisol profiles and horizons, the amount of exchangeable calcium (Ca) ranges from 22.48 to 69.75 $cmol_c$ kg^{-1}soil. This accounts for 52% to over 100% of the total cation exchange capacity (CEC). In subsoil horizons, it varies from 31.42 to 48.00 $cmol_c$ kg^{-1}soil, covering 77 and 83% of the total CEC in that order. In the lower and deeper horizons (A3/AC), it ranges between 32.85 and 101.50 $cmol_c$ kg^{-1}soil.

The content of exchangeable magnesium (Mg) covers between 6 and 36% of the total surface of CEC (Appendix Tables A.13 and A.14) and shows important variations with the depth of Vertisols, e.g., in the A1 horizons it varies from 2.53 $cmol_c$ kg^{-1}soil to 11 and 15.79 $cmol_c$ kg^{-1} soil in deeper horizons. Under the Ethiopian Vertisols, it appears that in horizons where exchangeable Ca is relatively less predominant, the proportion of exchangeable Mg increases. These two exchangeable cations together cover 69% to

100% of the total CEC in the horizons of the Vertisols. Dudal (1965) and Mohr (1971) conclude that this feature typically characterizes profiles of Vertisols.

Cation exchange capacity (CEC) is high throughout Ethiopian Vertisols although it shows a slight variation both within and between the many studied Ethiopian Vertisols. Thus, values of CEC tend to show slight variation within and between profiles, and reveal a slight increase with depth in many of the Vertisols profiles of Ethiopia (Appendixs Tables A.13 and A.14). The values of CEC kg^{-1} clay in many of the profiles are commonly over 80 reaching up to 103. A CEC value of 80 $cmol_c$ kg^{-1}clay to 150 $cmol_c$ kg^{-1} clay is a typical property of smectite/ montmorillonite clay mineralogy (Birkeland1999). The prevalence of values below 80 CEC $cmol_c$ kg^{-1} clay in some Ethiopian Vertisol profiles implies intermingling of other forms of clay mineralogy although montmorillonite could still be dominant. However, Vertisols could have CEC values as low as 50 $cmol_c$ kg^{-1} clay (Young 1976). Eylachew (1987) shows that Vertisols as well as other soil orders of the Ethiopia highland soils contain predominantly and/or fully smectite/ montmorillonite clay mineralogy. In many of the Ethiopian Vertisol profiles, the base saturation is over 70% and increases with depth, whereas in many of lower horizons reach to nearly 100%. However, some high rainfall Vertisols, e.g., western and central highlands of Ethiopia, tend to have < 50% base saturation, making them to be classified as Dytric Vertisols.

Vertisols mainly developing on alluvial materials tend to have low levels of phosphorus (Ahmad 1983). As a consequence, Ethiopian Vertisols commonly have low levels of phosphorus (P) in general. Great variations in the level of P exist between regional and local Vertisols. However, in most of them, values of P show a declining trend with depth. This reflects the association of availability and its variations largely with the contents of soil organic matter. Under Ethiopian soils, phosphorus content is also related to soil texture, pH, and contents of exchangeable Ca and $CaCO_3$. There may be a complex chemical reaction of P with Ca to form Ca-phosphate in Vertisols (Dudal1965) as clay particles of Vertisols may contain significant contents of calcium, mainly in the dryland lowland Vertisols. In the Vertisols of Ethiopia, the levels of P also decrease with an increase and decrease in pH values. Thus, the contents of available P could be expected to be low at high and low pH values since pH is directly associated with high levels of exchangeable Ca and total aluminum and iron contents, respectively.

Cambisols

Cambisols cover about 9% (10, 100,000 ha.) of Ethiopian landmass. These soils are formed under diverse processes and microenvironments. These soils occur in different agro-climatic zones of Ethiopia and occupy gently and strongly sloping landforms. In parts where slopes are relatively steep, Cambisols give way to the occurrence of Leptosols and/ or Regosols. Thus, they form a geographical continuation of Vertisols and/or Luvisols toward lower slope positions and Leptosols/ Regosols toward upper slope positions, reflecting a presence of some properties of the adjoining soils in respective transitional zones.

Thus, they possess diverse morphological, physical, and chemical characteristics both within and between environments of the country. In spite of the existence of certain differences, all variants of Cambisols possess distinct characteristics that clearly separate them from other soil groups, e.g., the presence of sequential horizon arrangements of A/Bw/BC/C or A/Bw/BCk/C (Fig. 6.13). Commonly, a decrease and/or irregular variation in clay content with depth of profiles characterizes all types of Ethiopian Cambisols and allows recognition of these soils. As can be seen from Fig. 6.13, the subsoil horizon of the profile has maintained rock structures and slightly weathered materials. These indicate that Cambisols show evidence of alteration in color, structure, and consistence in relation to the underlying or parent materials resulting from soil forming processes.

Depending on their environmental occurrences, Cambisols of Ethiopia are cultivated in different crops and are considered as one of the major Ethiopian cultivated soils. However, their low water holding capacity limits their use for crop farming, and in some drier environments, these soils demand supplementary irrigation for successful farming activities. In a typical drier and semi-dry environment, Cambisols are devoted to pastoral and/or fully or partial irrigation crop framing.

Influenced by their environmental and assumed parent materials, color characteristics of Cambisols show variations within a profile, as well as between different types of Cambiols of Ethiopia (Appendix Tables A.15–A.17). Whereas their surface horizons can vary from dark brown or black in organic matter rich areas, to grayish or brown in degraded and cultivated soils, the lower horizons mainly forming Cambic horizon largely have brownish color patterns. As with their environments, total and /or effective depth of Cambisols varies from environment to environment depending on slope gradient, positions, and depth and/or thickness of underlying materials. The total depth of Cambisols ranges from 50 to 200 cm in the very deeper Cambisols (Appendix Table A.15).

The soil structure and consistence characteristics of most Ethiopian Cambisols are related to their texture and level of organic matter content. Topsoil soil horizons with high clay content reveal moderate to strong medium and coarse angular or subangular blocky structures. Horizons with high contents of silt and sand are associated with weak and fine subangular blocky structures. Topsoil horizons with a relatively high organic matter level exhibit strong and medium subangular blocky structures in their subsoil horizons. Hard

Fig. 6.13 Typical Cambisols at the foot slope of Mygudo hill, Omo Nada district, southwest Ethiopia (Photo by Alemayehu Regassa)

and very hard dry consistence characterizes soils with high contents of clay and Cambisols with high contents of sand and coarse fragments show slightly hard dry consistence.

Depending on their local and regional locations, drainage characteristics of Cambisols vary from well to moderately well drained. The presence of impermeable hard rock or layer causes the presence of poorly drained Cambisols. Thus, the occurrence of mottles is observed in parts related to the underlying impermeable hard rock and gentler slope, which may cause a slow removal of excess water at high rainfall times.

The textures of Cambisols vary from clayey to silt loam (Appendix Tables A.15–A.17). The texture of these soils becomes coarser with depth, and invariably clay content is lower in subsoil cambic horizons. These textural characteristics of Cambisols vary in relation to their positions in the landscape and parent materials as well as degree of their weathering stages. Cambisols of Ethiopia occupy higher slope positions, eroded and formed on weathering resistant materials have commonly coarse texture features. Under the Ethiopian situations, values of bulk densities commonly increase with depth of Cambisol profiles as with the sand content of Cambisols.

Both field capacity and permanent wilting point of Cambsiols, show variations between the different environments and sites. The content of field capability of Cambisols generally decreases with depth, conditioned by their levels of organic matter and clay content as well as subsoil structure characteristics. Hence, subsoil horizons with low clay contents and high coarse fragments, and low contents of soil organic matter have low contents of field capacity, and as a result, these types of soils have low available water contents.

The location and environment and land use history of Cambisols determine many of their chemical characteristics. Depending on their locality, soil reaction ranges from slightly acid to alkaline conditions (Appendix Tables A.15–A.18). As their tropical locations, both total nitrogen and soil organic matter (SOM) content of Cambisols is generally low to very low. Cambisols regularly receiving deposition materials commonly have a relatively high level of contents of surface soil organic matter and total nitrogen, and as a result, both of these contents regularly decrease with the depth of Cambisols.

As the other Ethiopian soils, Cambisols of Ethiopia are uniformly predominated by divalent cations. Consequently, throughout the profiles exchangeable Ca and Mg together contain over 50% of CEC of the soil. Exchangeable Ca and Mg form, respectively, the first and second dominant cations in the cation exchange complex of Cambisols of Ethiopia.

The proportions of monovalent cations in the cation exchange complex of Cambisols are unimportant to be mentioned, as in many other Ethiopian soils. They together contain a very small proportion of cation exchange sites varying from 1 to 4%. However, in some Cambisols of northern Ethiopian highlands e.g., Tigray highlands, exchangeable Na may contain as high as 20% of subsoil exchange complex. This later condition shows a presence of drier environments causing little leaching of the soils. Generally, CEC is very high throughout Cambisols. Most Ap horizons have values of CEC of over 35.0 Cmol(c +) kg^{-1} soil by ammonium acetate pH 7 method.

Although it remains high, CEC shows a slight decreasing pattern in the cambic horizons. In most of Cambisols, levels of P were low, which is typical characteristics of tropical soils (Young 1976) including Ethiopian soils. It also reveals a decreasing pattern with the depth of profiles. Great variations in the level of P exist both within and between different environments of Cambisols. The existence of a better P level in the Ap horizons is related to the presence of a relatively high content of soil organic matter. Moreover, Cambisols of Ethiopia have high levels of percentage base saturation, making them to be fertile soils of Ethiopia.

Luvisols

Luvisols are soils with an argic subsurface horizon, which has a distinctly higher clay content than the overlying horizon (WRB 2006). Therefore, clay content differentiation forms one of the distinct features in Luvisol profiles (Fig. 6.14).

Fig. 6.14 Luvisols showing characteristic clay enriched subsurface (argic) horizon with diffuse horizonation in Ener and Enamore district, southwest Ethiopia (photo by Alemayehu Regassa)

As studied in Ethiopia, Luvisols have deep to moderately deep profiles and occupy gentler to flatter slopes where clay illuviation is favored. They occupy about 8,700,000 ha (8%) of the total land area of Ethiopia, forming one of the major agricultural soils of the country. These soils are found under different Environments and are suitable for different crops depending on their agro-ecological patterns. The soils are cultivated in major annual crops widely grown in different agro-climatic patterns of Ethiopia.

Clay content of subsoil horizons of Luvisols reaches as high as over 60%, whereas some underlying horizons of Luvisols might have as low as 20% clay content (Appendix Tables A.19–A.21). Consequently, these soils have coarser surface horizons than their counter subsoil horizons. Therefore, silt: clay ratio shows variations mainly between surface and subsurface horizons, where it is likely to be lower in the subsoil horizons. The high clay content and textural differentiation may be caused by the accumulation of clay content in their subsurface horizon. In many Ethiopian Luvisols, this distinct higher clay of the B-horizon is attributed to various factors, such as illuvial accumulation of clay and selective surface erosion of clay or destruction of clay in the surface horizon.

The profiles of Luvisols may contain the presence of different proportions of weatherable materials in their surface and subsurface horizons. Moreover, variation in soil depth, along the genetic horizons, is also not uncommon. Their deep profiles are probably related to their positions in the landscape, and have common features to soils found both in their upper positions, e.g., Cambisols *vs* Vertisols.

As shown in Appendix Tables A.19–A.21, pH values of Ethiopian Luvisols varied from moderately acidic to moderately alkaline. Luvisols have high cation exchange capacity (> 24 cmol(c +) kg^{-1} 1 M NH_4OAc at pH 7.0). Thus, Luvisols have high activity of clay (Tables A.19). They have high cation exchange capacity, with exceptionally higher CEC values in the subsoil horizons. This is related to the high clay content of the subsoil horizons, and as a result, clay content determines many of the characteristics of Luvisols. The high clay content retards the percolation of water, which in the end may lead to surface water accumulation, and thereof erosion. These soils require special soil and water management methods to bring them under agricultural use.

Exchangeable Ca and Mg, respectively, are the dominant cations in the exchange site. These are followed by changeable K and Na, in that order, and the presence of changeable Na in Luvisols is little to be mentioned. In many of the Luvisols studies made in Ethiopia, these soils have high base saturation (by1M NH_4OAc) between 20 and 100 cm from the surface, and regarded as one of the Ethiopian fertile soils. However, high rainfall Luvisols as described in central highlands (Appendix Tables A.19–A.21). Low base saturation levels in Luvisols occur in different environments of the country.

Available P forms one of the agricultural limiting factors of Luvisols. As observed in appendix Table A.19, these soils commonly have lower available P content, which requires applications of P fertilizers to farm Luvisols.

Regosols

Regosols are soils that do not meet the definitive criteria of any other soil groups (WRB, 1998, 2006). They are identified as having only medium textured and moderately deep to deep ochric A horizon. Regosols cover about 1.13% (1,300,000 ha) of land area of Ethiopia. The quality and characteristics of these soils in Ethiopia vary depending on their location in the landscape and assumed parent material (Fig. 6.8)as well as agro-ecology. Ethiopian Regosols located on fluviatile materials tend to contain better levels of soil organic matter, total nitrogen and available phosphorus. On the other hand, those Regosols found on eroded topographic features have poor levels of total nitrogen, organic matter with low water holding capacity, and these make these soils to be less suitable for crop production. In parts where they occur in reliable rainfall patterns, Regosols are used for crop production. However, the drier and shallower Regosols of Ethiopia are turned to extensive pastoral farming, e.g., central Rift Valley of Ethiopia and northern semiarid highlands. Crop farming in parts of these soils is largely limited by their shallow depth and low water holding capacities.

In parts where the amount of organic matter is better, topsoil color varies between very dark grayish and black, whereas Regosols developed on eroded landscapes have a brownish or light brown color. Subsoil horizons of Regosols containing buried material have dark brownish color patterns. This is an indication that the soil color pattern of Ethiopian soils is largely determined by the contents of soil organic matter (Appendix Table A.23). The color pattern of Regosols, therefore, provides information about the chemical and physical characteristics of the soil horizons, as is true to other Ethiopian soils too.

The depth of the Regosols varies from shallow on eroded landscapes to deep and very deep on deposited and concave slope positions (Appendix Table A.23). The structure type of Regosols is related to the contents of soil organic matter and textural classes. It changes from strong coarse subangular in the clayey surface horizons to strong coarse angular blocky in their organic matter poor subsoil horizons. As a result, consistence tends to be extremely hard dry, firm moist, and very sticky, very plastic wet when clay content is high. The dry consistence of these soils would vary from extremely hard to slightly hard and the wet consistence from slightly sticky and slightly plastic to very sticky and very plastic, again related to the amount of clay particle size classes. An appreciable amount of gravel could occur in concave slopes and eroded landscape positions, which are basically related to the presence of deposited gravels.

Regosols found on deposited materials have irregular particle size class variations (Appendix Table A.23), whereas those found on eroded and steep landscapes would be dominated by coarse soil particle size classes. As observed in the eastern and north eastern highlands of the country, the textural class of Regosols varies from clayey to silty clay loam. Bulk density, total porosity, and water contents vary in relation to variations in textural characteristics and show irregular distribution with the depth of Regosols. As a result, surface horizons of Regosols have lower bulk density values than their subsoil horizons. In Regosols, where soil texture class shows irregular variation, soil bulk density values also vary likewise. Closely and inversely related to values of bulk density, total porosity shows unsystematic patterns of distributions. Soil moisture contents in Regosols are largely governed by available contents of soil organic matter and clay and their depths. The shallower and coarse textured Regosols, in most cases, have low water holding capacities, which is a major limiting factor of crop farming. The soil reaction characteristics of Ethiopian Regosols vary between slightly acid and alkaline, which shows a slight increase with depth of some Regosols.

Regosols have low to moderate levels of organic matter and total nitrogen. Those located on depression and deposition sites have relatively better amounts of soil organic matter than other Regosols located on eroded and steeper topographic positions. As with the other soils of Ethiopia, deposition site Regosols have irregular contents of organic matter and nitrogen with their depths. Exchangeable Ca and Mg form predominant constituents of Regosols in the cation exchange sites followed by changeable K and Na. As with contents of organic matter and total nitrogen, exchangeable cations reveal an irregular variation with the depth of these soils.

Whereas exchangeable K and Na together contain 1 to 3% of the CEC, exchangeable Ca and Mg together contain 60–70% of the exchange site of Regosols.The value of CEC is high in all Ethiopian Regosols and reaches as high as 60 $cmol_c\ kg^{-1}$ soil, and also in many of Regosols, levels of base saturation are high. Similar to other Ethiopian soils, Regosols have variable levels of available phosphorus, and this content decreases with depth of the Regosol profiles, and varies depending on land use history and agroecology.

Fluvisols

Fluvisols largely occupy flatter areas along river and stream courses and depositional sites. These soils cover about 4% (4,691,062 ha.) of the Ethiopian land area. The slope gradient in these soils varies between 1 and 5%. As many of these soils develop on deposition landform, they have deep solum reaching as deep as 200 cm. As a result of this, they largely contain deposited gravels indicating deposition of fluviatile material, clearly separating the horizons of a profile from each other. Where they are cultivated, Fluvisols are one of the productive soils mainly with the help of irrigation. However, the flood might affect and damage crops cultivated in these soils associated with their positions in the landscape, e.g., the Middle Awash Valley of the Amibara irrigation lands, Ethiopia. Fluvisols are considered to be Ethiopian fertile soils as they regularly receive fresh sediments containing high contents of soil organic matter (SOM) and exchangeable cations (Appendix Table A.24).

Many of the characteristics of Ethiopian Fluvisols are related to the contents of available soil organic matter and textural class. The moist color of Fluvisol profile varies from black to dark grayish brown in their surface horizons to very dark gray in the subsoil horizons (Appendix Tables A.24–A.26). Governed by their formations, mainly their environments and site, Fluvisols have variable textural classes throughout their depths (Appendix Tables A.24 and A.25). In recent deposition sites, silt forms the dominant particle size classes and sand particle size classes become dominant in upper slope positioned Fluvisols. The recent sediments are typically characterized by stratified textural classes and organic matter, whereas Fluvisols formed on historical sediments have obliterated textural classes, but show slightly stratified contents of exchangeable bases mainly the divalent cations of Ca and Mg, cation exchange capacity (CEC) and pH values. Therefore, stratification features of Ethiopian Fluvisols can be seen from depth-wise distribution of both biochemical and physical characteristics of each horizon.

As with other characteristics, water contents of Fluvisols show irregular variations with the depth of profile. Horizons with high contents of clay and organic matter typically have better water holding capacity than other horizons dominated by high contents of sand and low levels of organic matter.

The soil reaction of the Fluvisol profile varies from slightly acid to moderately alkaline and alkaline. This makes them to be suitable for the cultivation of many crops (Appendix Tables A.24 and A.25). A remarkable feature of many Ethiopian Fluvisols profiles is the unsystematic variation of the exchangeable cation with the depth of the Fluvisols (Appendix Tables A.24 and A.25). Throughout the horizons of Fluvisols, CEC is high reaching as high as 50–60 cmolc kg-1 soil and shows unsystematic distributions with the depth of the profile. The calculated CEC values against clay content are also high (Appendix Tables A.24 and A.25). This is an indication that Ethiopian Fluvisols are characterized by the dominance of expansible clay mineralogy. Depending on their sites and agro-ecology patterns, there is a great variation of percentage base saturation (PBS) between Ethiopian Fluvisols. In those found in high rainfall areas, PBS tends to be low (<50%) and others found in semiarid and low rainfall agro-ecologies have high levels of PBS.

As a consequence, PBS of Fluvsiols varies from < 50% to over 80% reaching as high as 100 in some of the Fluvisol horizons and revealing unsystematic variation with depth of the profile (Appendix Tables A.24 and A.25). Contents of available P in the profile of the Fluvisols vary from low to high levels throughout the horizons and show unsystematic variations with depth (Appendix Tables A.24 and A.25). Unlike other Ethiopian soils, these soils have better available P levels as they receive deposition materials including P from upper positions of the landscape.

Phaeozems

Phaeozems are estimated to cover 0.4% (463,000 ha.) of its total land area in Ethiopia. Where cultivated, they form the most fertile soils of the country for their high SOM and high base status levels. However, Phaeozems located on steeper slopes are susceptible to erosion, which degrades the natural quality of these soils if disturbed by human activities.

A large proportion of Phaeozems occur at high altitudes, where the environment favors the accumulation of soil organic matter. As a result, a large proportion of Phaeozems are found in temperate and subhumid climate zones of Ethiopia. These soils may occur on a steeper slope gradient reaching as steep as 45% and occurs in both gentle and very steep slope forms. In most of the cases, their occurrence in the very steep slopes is associated with the presence of natural vegetation cover, e.g., *juneprus procera*. In the lower altitudes, the occurrence of Phaeozems is associated with virgin savanna grasslands. In cultivated areas, the depth of these soils may be truncated and organic matter degraded, which may no more qualify for a mollic A horizon (Appendix Table A.27).

Owing to the high level of organic matter, Phaeozmes have black surface horizons which may change to very dark grayish brown subsurface horizons in their subsoil horizons. The soil depth of Phaeozmes varies in relation to slope characteristics and land use history. It generally varies between 10 and 150 cm. The structure characteristics are closely related to SOM content. In the surface horizons, it consists of moderate fine granular and even crumb and where SOM content decreases, structure consists of moderate medium subangular blocky. Again in relation to available contents of SOM, consistence is slightly hard and slightly sticky in most surface horizons and hard to slightly sticky, slightly plastic to plastic in subsurface horizons.

The texture characteristics of Phaeozems vary depending on land use history, and positions in the landscape (Appendix Table A.27). Many studies confirm that cultivated degraded soils have clay loam textural class that changes to clayey soil texture in its subsoil horizon. The uncultivated Phaeozems have loam textural classes, and in some cases, they are dominated by silt size particle size classes. As a result, silt/clay ratios of Phaeozems reach up to 2.0 mainly in their surface horizons.

As the surface soils of Phaeozmes have high levels of SOM, they have low values of bulk densities, which commonly recorded between 1 and 1.2 cm^{-3} in their surface horizons, but increase to slightly over 1.2 cm^{-3} in their subsurface horizons. Phaeozmes have good water holding capacity, which may reach up to 50% on the surface but decreases in their subsoil horizons. The decreasing pattern of water holding capacities with depth reflects the decreasing pattern of SOM. Therefore, variations of many of the Phaeozems characteristics are related to variation and level of soil organic matter.

Soil reaction of surface horizons of Phaeozems is moderately acidic, but slightly increases in their subsoil horizons. The relatively low pH values in these soils are related to relatively high values of SOM contents, which release organic acids through the decomposition of organic substances.

As many of Ethiopian Phaeozems are found in diverse climatic zones of both low and high rainfall temperate / savanna climatic regions, the soils contain intermediate levels of soluble elements. Exchangeable Ca and Mg form predominant cations, which together contain up to over 90% of the exchange site, with less contents of exchangeable K and Na, together covering less than 2% of the exchange site of the soils. The levels of P are moderate to high and vary in relation to their land use history. The uncultivated Phaeozmes have higher values of available P than the cultivated ones. The value of available P in Phaeozems decrease with depth, implying that in Phaeozems, high amounts of available P are related to high amounts of SOM as well as the presence of good soil fertility quality in surface horizons of Phaeozems.

Calcisols

Calcisols are common in calcareous parent materials, which are widespread in arid and semiarid environments of Ethiopia. They are also found in local depression sites where lime is deposited mainly eroded from upper landscape positions. Calcisols cover about 9% (11000000 ha) of the land surface of Ethiopia. A large proportion of Calcisols are found in the drier lowlands of Ethiopia, which makes them be less suitable for crop farming in their dry moisture environments. As mentioned above, these drier lowland Calcisols of Ethiopia are recently turned to cultivation with the help of irrigation, such as in the Awash River basin. Calcisols also occur in parts of the Ethiopian Rift Valley, where the formation and accumulation of lime are favored, such as in and around Ziway, Alage area of central Rift Valley of Ethiopia (Fig. 6.8). The parent material in the lower parts of the landscape is mostly alluvial and colluvial consisting of mainly base-rich weathered deposits mainly of highly calcareous sands and gravel (Ahmed et al. 2013). On the other hand, Calcisols are also found in the semi-drier highland areas of the northern highlands of Ethiopia as in the Tigray highlands. Rhyolitic volcanics are the parent material of these soils in the northern Ethiopian highlands. Moreover, Calcisols also occur in the depression sites of south eastern highlands, e.g., in the irrigation sites, where calcium-rich minerals are deposited from upper slopes and deposited in depression areas. Therefore, Calcisols are widely distributed both in the highlands and lowlands although a large portion of these soils occur in the semi-drier lowlands, such as the central and middle Awash valley and foot slopes of eastern mountains, e.g., around Jigjiga and similar environments.

Moist colors of the soils are different within and between pits of the same and different locations (Appendix Tables A.27, A.28, and A.29). In large cases, the lower horizons of Calcisols have whitish colors coated by white lime. Thus, soil color of Calcisols ranged from very dark gray (2.5Y3/1) to light reddish brown (2.5Y6/3). In other cases, it varies from dusky red (10R 3/2), very dark gray (7.5YR 3/1), and to pinkish white (10R 8/2). Depending on the organic matter level, surface horizons varied from dark to dark brown or brownish color.

Calciols commonly have well-developed granular and blocky structures throughout their profiles (Appendix Tables A.28–A.30). This is caused by the available lime content which enhances the aggregate formation of the soil particles. The moist consistence of these soils is dominated by friable, whereas the wet consistence of most horizons was slightly sticky and slightly plastic. The friable and non/slightly sticky and non/slightly plastic consistencies indicate the low clay contents of the soils and their ease to till. In contrast, the sticky, very sticky, plastic a very plastic consistencies show the presence of high clay and low organic carbon contents and difficulty to till, clearing showing wide variability in characteristics of Calcisols. The presence of very sticky and very plastic consistency could be indicative of the presence of smectitic clay minerals in the soils.

As studied, the texture of these soils is dominantly loamy, with clay content increasing with depth in some soils. However, their subsoil is largely dominated by powdery or concrete lime that gives the soil to have calcic horizon. Owing to their lime content, pH level of Ethiopian Calcisols is high and increases with depth following the accumulation pattern of the content of lime. Therefore, these soils are alkaline with pH values ranging between 8.0 and 8.5.

Calcisols in the lowland have low contents of organic matter and total nitrogen. Available P is low for all Calcisols except in parts where the surface horizon is probably related to the application of P fertilizer. The result of the analysis of soil samples of Calciols (Appendix Table A.30) affirmed that the soils in this group have high to very high $CaCO_3$ values. This has led to the development of Calcaric diagnostic material and/or a calcic diagnostic horizon. The CEC and exchangeable K values are in the medium to high range. The EC values indicate that the soils are slightly saline, while ESP values are low and very low and not sufficient to cause sodicity hazard (Appendix Tables A.28–A.30).

Alisols

Alisols cover about 3% (3,000,000 ha) of the total land area of Ethopia. However, Alsiols tend to be major local soils, where they are found. Under the Ethiopian situations, these soils are formed in high rainfall areas, where there is high leaching intensity. As a result, Alisols have low fertility status and are dominated with exchangeable aluminum. In Ethiopia, they are mainly devoted to natural forest cover, and where cultivated tea plant, such as Wush wush tea plantation areas of western Ethiopia, forms the major cultivated plant. On the other hand, Alisols are cultivated with slash and burn methods consisting of shifting cultivation, traditional subsistence farming systems, e.g., western Ethiopian and Gambela areas.

Alisols have deeper solum, which may reach as deep as 200 cm (Appendix Tables A.31–A.35). The surface texture of the Alisols is clay loam, loam, and sandy loam. The surface drainage of Alisols is moderately drained in the flat area and rapidly and well drained in the elevated positions of the landscape (Fig. 6.15). The uncultivated Alisols have high levels of organic matter, reaching as high as 21%. On the other hand, cultivated Alisols exhibit low contents of organic matter. This low content of organic matter in cultivated Alsiols is related to cutting-burning process of natural vegetation cover for subsistence farming, as experienced in some Alsiols of highlands of western Ethiopia, where these soils are largely found. The contents of total nitrogen also vary in relation to land use history of Alsiols. In the uncultivated Alsiols, relatively better levels of total nitrogen is found than in cultivated Alisols (Appendix Tables A.31–A.35). Therefore, uncultivated surface horizons of Alsiols have total N contents reaching as high 1.1%, but this characteristic is lowered in cultivated surface horizons of Alsiols. Therefore, contents of organic matter and total nitrogen can be considered as a measure of status and quality of Alisols of Ethiopia.

Fig. 6.15 Deeply weathered and well drained Alisols in the Kersa district, southwest Ethiopia (Photo by Alemayehu Regassa)

Alisols are considered one of the leached soils of Ethiopia. Up on cultivation, it is severely affected by erosion mainly rill and gully formations. Consequently, these soils contain poor levels of exchangeable base cations, thus dominated with acid forming cations, such as aluminum, as indicated by its name.

Therefore, soil-pH values of these soils are below 4.8. This level of pH is a measure of a presence of a significant proportion of acid forming base minerals, such as aluminum and H + ions. They have low contents of available phosphorus as well as low base saturation levels, making them to be one of the infertile Ethiopian soils (Appendix Tables A.31–A.35). Alisols have high-activity clay throughout their argic horizons and a low base saturation at certain depths.

Arenosols

Arenosols consists of sandy soils and are found both in dry and wet climatic lands of Ethiopia. These soils cover about 0.6% (64,000 ha) of land surface. In dry areas Arenosols show minimal profile development because soil forming processes are at a standstill during long periods of drought and/or because the parent material is of young age and/or less weathered. In the wet climatic regions, Arenosols have better profile differentiation. Arenosols occur on the steeper slopes and on coarse textured colluvium in the highlands where sandstones have been exposed largely in the gorges of southwestern Wello, at the base of Mt. Ras Dejen, Tigray highlands (Appendix Table A.35) and in northeastern Bale mountains. In the drier landscapes, Arenosols are largely found in northern Rift Valley and large parts of Ogaden lowlands. As these soils have low water accumulation capacities, mainly due to their sandy textures (leading to rapid permeability), they are less devoted to crop farming, but largely used for extensive grazing. Crops like *teff* and linseed are cultivated in Arenosols mainly in good rainfall periods and areas.

Arenosols in the highland wet zone of Ethiopia, e.g., northern highlands. Tigray has better developed but still ochric surface horizons over a substratum that may have thin iron coatings throughout, or contain lamellae of illuviated humus, clay or iron compounds that are too thin, too few, or contain too little humus to qualify as a diagnostic horizon. In these northern highlands, Arenosols occur on the upper slopes and convex ridge crests of undulating sandstone plateaux. Locally, the soil occurs on middle slopes. The slopes range from one to five percent. The soils are formed on colluvial and in situ weathered sandstone material, e.g., Tigray highlands. The soils have variable base saturation with calcium and magnesium as the dominant cations accompanied with a very low amount of cation exchange capacity, total nitrogen and organic matter.

Gleysols

As discussed earlier, Gleysols are formed under waterlogged conditions produced by a rising groundwater. Under the Ethiopian situations, they are commonly found in wetland environments and depression sites. As a consequence, Gleysols are found locally throughout Ethiopia, irrespective of climate or parent material. A large proportion of Gleysols can be found in the central highlands where Vertisols have not developed, for example on the coarse textured pyroclastics in and around Mt. Batu site. On the other hand, Gleysols occur along major rivers in the drier parts of the country (Table A.35 and A.36), and commonly have sodic and/or saline phases (FAO 1984a, b). The relatively lower altitudes and lower slopes position are attributed to the poor drainage of soil. As a result, most part occupied by Gleysols, is at least seasonally flooded and some parts are permanently waterlogged and form marshy and high rainfall area. These soils provide key ecosystem services such as wildlife habitat, recreation, and filtration of pollutants in runoff, e.g., around Lakes Tana and Hawasa basins. Apart from these uses, Gleysols are used as water reserve sites and/or dryland grazing lands. As a general fact, Gleysols have hydromorphic properties.

In Ethiopia, these soils cover about 0.08% (90,000 ha) land surface area. Beyond their hydromorphic properties, which may include the presence of a histic H horizon, dominant neutral hues, saturation by groundwater and/or mottling or other evidence of reduction, Gleysols may be extremely variable with regard to chemical and physical properties (Appendix Tables A.37 and A.38). Generally, they are commonly relatively high in clay content and high in organic matter content and have good inherent fertility (FAO 1984a, b). The soil matrix color varies from very dark grayish brown to very dark gray with the presence of prominent mottles. This suggests the stagnic and gleyic conditions of the soil and indicates prolonged period of anaerobism. The surface texture of the soil is loam and there is a gradual increase in clay content with depth of profiles.

The good chemical composition of the soil was manifested in the available phosphorous, high cation exchangeable capacity, and base saturation. Exchangeable calcium is the dominant cation composing the exchangeable bases (Appendix Tables A.37 and A.38). The dryland Gelysols develop on river basin courses, where it is regularly flooded by rivers outside its bank. These Gleysols contain a significant amount of calcium carbonate (Appendix Tables A.37 and A.38), such as in lower courses of Wabeshebele river basins in the Ogaden lowlands.

Ferralsols

Ferralsols are deeply weathered red or yellow soils of the humid south and southwestern high rainfall (with deep intense weathering zone) parts of Ethiopia. Ferralsols are old soils or are soils that are developed in strongly weathered parent materials of mainly southwestern, western, south central and southern Ethiopian highlands (Figs. 6.15 and 6.16). In Ethiopia, Ferralsols do not form major soils, which

Fig. 6.16 Acrisol profile in Yaballa, Buno Bedele district, southwest Ethiopia (Photo courtesy of Alemayehu Regassa)

cover only 0.01% (7,000 ha) land surface of the country. They are intensively cultivated for wild coffee. Apart from the coffee plant, natural forest trees cover Ferralsols and elsewhere, they are used for slash-burn traditional or shifting farming systems. As Ferralsols are located in high rainfall and intense weathering sites, they are dominated with clay particle size classes with little weatherable minerals (Appendix Tables A.39–A.41). These soils largely contain residual metal oxides and are highly characterized by the leaching of mineral nutrients (Appendix Tables A.39–A.41), they have low fertility and require additions of lime and fertilizer if they are to be used for agriculture. Thus, the soils are poor in exchangeable base minerals, caused by intense leaching, which makes the soils to have lower pH values. With the dominance of metallic oxides coupled with lower pH conditions, Ferralsols are poor in available phosphorus. Moreover, the soils are dominated by kaolinite and/or sesquioxides clay minerals (Appendix Tables A.39–A.41). Oxide composition (%) of Ferralsols.

Acrisols

Acrisols are characterized by a presence of Argic B-horizon, dominance of stable low-activity clay and generally with little base forming minerals (Appendix Tables A.42–A.47). Acrisols cover about 1.3% (1,100,000 ha) of land area in Ethiopia. In terms of their environments and formations, these soils are very similar to Ferralsols. Ethiopian Acrisols largely consist of Kaolinitic and gypsite minerals. As a result, Acrisols are found in high rainfall parts of south and southwestern highlands of Ethiopia (Appendix Tables A.39–A.41). These soils are located in pockets of local sites and do not form continuous soil-landscape cover in the country.

Acrisols are highly leached and infertile. Thus, these soils have less agricultural value and largely covered with natural forest trees of tropical Ethiopian humid climates. Where they are cultivated, they are used for slash-burn shifting cultivation and rotation subsistence farming systems. Morphologically, most Acrisols have a thin, brown, ochric surface horizon, particularly in areas with pronounced dry seasons; darker colors are found where periodic waterlogging retards mineralization of soil organic matter. The existence of little weatherable minerals and the comparable contents of Fe-, Al-, and Ti-oxides with Ferralsols are the important mineralogical characteristics of Acrisols.

In the high rainfall western highlands of Ethiopia, many Acrisols in low landscape positions show signs of periodic water saturation, and their surface color is darkened by soil organic matter. However, deeper horizons of Acrisols are typically characterized by reddish color mainly due to the presence of Fe-Al oxides. Moreover, Acrisols have poor chemical properties. They typically have low pH values, low contents of exchangeable bases, and are poor in available phosphorus. These make Acrisols of Ethiopia to be unsuitable for farming. Levels of plant nutrients are low and aluminum toxicity and P-sorption capacity are strong limitations. As biological activity is low in Acrisols, natural regeneration capacity of surface soil that was degraded by mechanical operations is very slow.

Lixisols

Lixisols are soils dominated with reddish or yellow colors. In Ethiopia, Lixisols are found in very high rainfall areas of its western highlands, such as Dizi area of Illubabor zone. As a result, Lixisols are found in very similar environments to Alisols, Acrisols and Ferralsols. They, thus, could be described as one of the highly leached soils of Ethiopia. Areas with Lixisols that are still under natural savanna or open woodland vegetation are widely used for low volume grazing. In Ethiopia, Lixisols cover about 0.18% (200,000 ha) of land surface of Ethiopia.

The morphological characteristics of Lixisols includes a thin, brown, ochric surface horizon over a brown or reddish brown argic (Bt) horizon that often lacks clear evidence of clay illuviation other than a sharp increase in clay content over a short vertical distance (Appendix Tables A.48–A.49). Lixisols comprise soils that have higher clay content in the subsoil than in the top soil as a result of pedogenetic processes (especially clay migration) leading to an *argic* subsoil horizon (Appendix Tables A.48–A.49).

Lixisols of Ethiopia have higher base saturation and accordingly somewhat stronger structure than normally found in Acrisols (Appendix Tables A.48–A.49). They have low-activity clay at a certain depth of the Lixisols profile. The moisture holding properties of Lixisols are slightly better than those of Ferralsols or Acrisols with the same contents of clay and organic matter. In addition, Lixisols are strongly weathered soils with low levels of available nutrients and low nutrient reserves. However, the chemical properties of Lixisols are generally better than of Ferralsols and Acrisols, because of their higher soil-pH and the absence of serious Al-toxicity. The absolute amount of exchangeable bases is generally not more than 2 cmol(+) kg^{-1} fine earth on account of the low cation exchange capacity of Lixisols (Appendix Tables A.48–A.49). The low absolute level of plant nutrients and the low cation retention by Lixisols makes recurrent inputs of fertilizers and/or lime, which would be a precondition for continuous cultivation.

Planosols

Planosols are largely found and concentrated in south central and southern parts the Omo-Ghibe River basin. Planosols make up atleast 130 ha land, making less than 0.01% of the total land area, and do not form major soils of the country.

Fig. 6.17 Vertic planosol profile with characteristic bleached silty loam texture horizon overlying heavy clayey vertical subsurface horizon from Gilgel Gibe catchment, southwest Ethiopia (photo by Alemayehu Regassa)

But these soils are environmentally important, i.e., if they are disturbed, the environment is very sensitive for environmental degradation, e.g., sensitive to erosion resulting in the formation of severe gullies. As a result, these soils are less suitable to crop farming and other forms of man-made disturbance, which leads them to the initiation and development of deep wide gullies. A large proportion of Plansols is used for pastoral farming, mainly in the southern parts of the country, and this is a favorable land use for Planosols.

The most consistent characteristic morphological features of all Ethiopian Planosols are their clay-poor "albic" E horizons that abruptly overlie a clayey subsurface horizon (Fig. 6.17). Thus, most Planosols exhibit an AEB horizon sequence (Appendix Tables A.50–A.51). A process of the reduction occurring in wet periods causes the formation of iron and manganese nodules in the E horizon. As a result, Fe and Mn concretions are commonly observed in the albic (eluvial) E horizons in south central Ethiopian highlands, e.g., Wulbareg–Siltie and Omo Nada districts. The thickness of the eluvial horizons, as well as the depth of Planosol profiles vary depending on several factors, e.g., thickness of pre-weathered or deposited weathered materials as well as their positions in the landscape. On the other hand, poorly developed Planosol profiles are also studied in the northern highlands of Ethiopia, e.g., developing on sand stones in Tigray highlands. Plansols of northern Ethiopia have clayey layer (below a depth of 40 cm) and a higher effective cation exchange capacity - ECEC (15.9 cmolc/kg), but lower organic carbon and N contents.

On the other hand, poorly developed Planosol profiles are also found in the northern highlands, e.g., those developing on sand stones in Tigray highlands. Plansols of northern Ethiopia have clayey layer (below a depth of 40 cm) and a

Fig. 6.18 Andosol profile with darkened organic matter rich (melanic) surface horizon overlying weathered volcanic material in Omo Nada district, southwest Ethiopia (Photo by Alemayehu Regassa)

higher effective cation exchange capacity—ECEC (15.9 $cmol_c$/kg), but lower organic carbon and N contents.

Andosols

Andosols develop on recent deposited pyroclastic and volcanic ash materials. They are found both in the Rift Valley and on the high mountains of the country e.g., top of northwest and northeast, and south-east Ethiopian highlands (Appendix Table A.53). These high mountain Andosols are dominantly darkened by organic matter, largely leading to the formation of umbric A horizon, with chroma of less than 2 (Fig. 6.18).

However, the Rift Valley Andosols have a mollic A horizon having chroma of up to 3 (Appendix Tables A.54–A.55). Generally, mountain Andosols are inherently low in pH and high P adsorption capacities, mainly due to the dominance of Fe and Al complexes, originating from their parent materials. Andosols cover about 0.62% (700,000 ha) of the surface land area. As determined by diverse environments, the mountain Andosols (mainly found at over 3000 masl) develop on silty deposited material underlain by weathered soil material (Appendix Tables A.52 and A.53).

Where they are found, these soils are locally important agricultural soils. The mountains Andosols are cultivated in potato and barley, mainly conditioned by the local climatic characteristics. The cultivation of mountain Andosols are further limited by their topographic conditions such as their occurrence on steeper slopes.

The Rift Valley Andosols are cultivated in maize and sorghum. The uncultivated Andosols, both in the mountains and Rift Valley areas, are used as grazing and pasture lands. The mountain Andosols have dark topsoil horizons whose morphology was marked by isotropic black color (10YR 2/1, moist); silt loam to loam texture, moderate medium crumb structure, slightly to non-plastic and slightly sticky wet, firm moist, and soft dry consistency. Andosols of the high mountains have low clay content dominated with silt size particle size classes. The dark (topsoil) horizon varied from 50 to 100 cm, whereas the total soil depth reaches up to 200 cm (Appendix Tables A.52 and A.53).

The dark A horizon in some cases graded to a dark brown (7.5YR 3/2, moist) subsoil (B-horizon) and in others to hard rock, a lithic or paralithic contact depending on degree of weathering of underlying material as well as local slope characteristics. The underlying dark brown B-horizon, where it existed, had a sandy loam to clayey texture, fine to coarse rounded weakly to strongly weathered gravel, strong coarse to medium subangular blocky structure. The Rift Valley Andosols are dominated with sand particle size classes, making them to have sandy loam texture (Appendix Tables A.54 and A.55).

The Rift Valley Andosols largely develop on volcanic ash and pumice materials. As a general fact, Andosols have a better level of soil organic matter and total nitrogen than other soils of their surroundings. However, both soil organic matter and total nitrogen are considered to be higher for the high mountains Andosols than their counter Rift Valley Andosols. In the uncultivated Mountain Andosols, surface soil organic matter reaches as high as 35%, and nitrogen levels are recorded up to over 1%. On the other hand, in the cultivated high mountain Andosols of Ethiopia, both organic matter and total nitrogen contents are considerably lowered, as indicated in (Appendix Tables A.52 and A.53). In the cultivated Ethiopian Rift Valley Andosols (Appendix Tables A.52 and A.53), soil organic matter attains a maximum of 6%, and total nitrogen attains up to 0.23% reaching as low as 0.06%. As a consequence, any disturbance form of Andosols makes them to loose their inherent soil quality, requiring an important care to use these soils for cultivation purposes. In any environments of Andosols, the contents of soil organic matter and total nitrogen considerably decreased with their depths.

Soil pH levels in the mountain Andosols are commonly between 5 and 6.2, and this level is low for surface soils (Appendix Tables A.54 and A.55). Thus, there is a slight increase in pH levels with a depth of highland (mountain) Andosols. The relatively low surface pH levels of the Mountain Andosols are related to the presence of Al and Fe oxides and high amounts of rainfall, resulting in leaching of exchangeable bases. Therefore, the Mountain Andosols of Ethiopia commonly have low percentage base saturation, low contents of exchangeable Ca and Mg. However, the Rift Valley Andosols have pH levels of over 6.5 in their surface horizons, increasing with their depth reaching as high as 9. These Rift Valley Andosols have high percentage base saturation levels (Appendix Tables A.54 and A.55). On the other hand, in any environments, Andosols have high CEC.

Saline-sodic soils

Salt affected soils are generally referred to as arid and semiarid soils, mostly considered as saline-sodic soils. Thus, their formation, occurrence, and characteristics are largely conditioned by the local climatic, topographic, and geological features of the arid and semiarid areas. Apart from these, the formation of saline-sodic soils is intensified by faulty irrigation systems, e.g., middle and lower Awash Valley. In general, saline-sodic soils accumulate salt, sulfate, carbonates, and chlorides at or close to the surface horizons. In some areas, deeper subsoil horizons contain many salt materials from their overlying horizons. As arid and semiarid soils, these soils, in Ethiopia, are found extensively in eastern (Ogaden lowland-Wabeshebele basin) and Rift Valley zones (Dankil depression, lower, Middle, and Upper Awash valley) and Chamo basin parts of the country. Improper irrigation systems that took place for decades in parts of the Middle and lower Awash Valley, recent expansion of floriculture around Zeway area have intensified the occurrences of saline-sodic soils in the country. However, there is no significant study made on the classification and mapping of saline-sodic soils of Ethiopia. Where studies are available, these focused only on the determination of contents of saline-sodic properties forming minerals and elements, and lack mapping of any systematic morphological, physical and chemical characteristics. The saline-sodic soils of Ethiopia, together, cover about 11,033,000 ha surface land, and widely distributed in different parts of the country. The world map recognizes these semiarid and arid soils as Calcisols (discussed above), Durisols (not reported to occur in Ethiopia), Gypsiols (expected to occur in pockets of Ethiopian semiarid and arid lands), Solonchaks (discussed below) and Solontez (less mapped soils of Ethiopia). The following discussions on saline-sodic soils are accounted based on some available studies made in selected arid and semiarid parts.

Solonchaks

Solonchaks are soils with a high concentration of soluble salts (commonly > 4 dS/M ECe), as a result of which, they

Fig. 6.19 Flood irrigation on the soils of the Middle Awash River Basin (Photo courtesy of Almaz Deche)

are largely confined to the arid and semiarid climatic zones of Ethiopia. In Ethiopia, Solonchaks are found in parts where evaporation exceeds precipitation, such as in extreme dry lands of the north, eastern, and some parts of the Rift Valley of the country. In the Rift Valley, Solonchaks occur in sites of inadequately managed irrigation systems, e.g., the Middle Awash Valley of Ethiopia. Also, the existence of irrigation-induced Solonchaks have been reported from recent irrigation farms of Ogaden lowlands, as in the Wabesheble-Ghenale irrigation project sites. Therefore, in Ethiopia, Solonchaks are caused by both natural causes and man-made activities. In Ethiopia, Solonchaks cover less than 2% (2,000,000 ha) of land surface of the country. Solonchaks accessed to irrigation and other forms of watercourse or on the main flood plains of river systems are used for the production of monoculture banana, mango, or maize, whereas the majority of those located distant from the watercourse are predominantly used for agroforestry, e.g., gum production and extensive pastoralism components.

In most of the cases, Solonchaks have AC or ABC-profiles, often with gleyic properties at some depth. Ethiopian Solonchaks with a deep-water table, e.g., Solonchaks of Ogaden lowlands have the greatest accumulation of salts at some depth below the surface, which could be referred to as internal Solonchak. The surface horizons for most Pedons have Munsell moist color of chroma ≤ 2 darker, value ≤ 2 and chroma ≤ 3, and in some cases value of ≤ 6 (dry); thus, Solonchaks have very dark brown mollic A horizons in some of their profiles.

Solonchaks have a stable soil structure that is brought by the high salt content of the soil. The surface and subsurface horizons of Solonchaks have salty crystals, but in most cases, salt powder covers its surface horizons, caused by faulty irrigation systems (Fig. 6.19). The textural classes of the surface layers ranged from silt loam, silt clay, and to clay and the bulk density of the soils is in the range of 1.09 g cm^{-3} in the A horizon to 1.65 g cm^{-3} in the B horizon.

Chemically, saline soils are composed of the ions Cl^-, $SO4^{2-}$, Na^+, Ca^{2+}, Mg^{2+} and small amounts of NO_3^-, HCO_3^-, and K^+, with soluble Na^+ seldom exceeding the sum of the other cations and thus not adsorbed to any significant extent (Appendix Tables A.56–A.58). As solonchaks of Ethiopia are found in the drier landscape, they tend to have low contents of organic matter and total nitrogen (Appendix Tables A.56–A.58). Thus, the dark color of some of its horizons is largely caused by the dispersal of an available small amount of organic matter in the soil matrix. However, the mollic A horizons have total nitrogen contents of 0.18% and organic matter of 5% (Appendix Tables A.59–A.60), and both decrease with depth. The pH-H2O values of Solonchaks vary from 5.88 to 8.48 (Appendix Tables A.56–A.58). The exchange complex of the soils is dominated by Ca followed by Mg, K, and Na (Appendix Tables A.57and A.58). Reports indicate that cation exchange capacity (CEC) of Solonchaks ranged from 28.94to 56.35 cmol(+) kg^{-1}, of which can be generally described as high to very high CEC values for surface and subsurface horizons, indicate these soils to be good agricultural soil. As a result, the major limitations of these soils are their location in moisture limitation and salt feature formation areas.

Other Saline-Sodic soils

Similar to Solonchaks, other saline-sodic soils are also found in dryland environments, and mainly where salts and alkali materials are brought to the surface through evaporation and/or deposited deep in subsoil horizons. Based on the chemical composition of these soils, they can be identified as saline-sodic and/or sodic soils in such types of soils, sulfides and chlorides are the predominant salts (Appendix Tables 6.61, 6.62, and 6.63). The presence of sulfide materials in huge quantities causes for the existence of high sodicity/alkalinity levels (Appendix Table 6.61) and in places where drainage properties of these regions are poor, these salt and alkali substances result in productive land abandoning for agricultural purposes. For instance, in the Middle Awash Valley of the Ethiopian Rift Valley, a significant portion of land cultivated for commercial crops, e.g., banana, sugarcane, and cotton is out of crop cultivation. These soils, to the extreme cases, can be identified as Solontez, or sodic soils. In Ethiopia, Solontez covers about 0.02% (17,000 ha) surface land area. This small coverage report of these soils in Ethiopia is due to the little study of them, and by no means may indicate their insignificant coverage of the land area of the country.

The saline-sodic soils have the properties of both saline and sodic characteristics. As depicted in Appendix Table 6.61, they are characterized by ECe of more than 4 dS/m, an ESP of over 15 and pH value usually below 8.5. Saline-sodic soils have the problems of both salinity and sodicity. Thus, saline-sodic soils have high contents of salts and soluble ions. In Ethiopia, these soils are found in the irrigation lands of the Middle and lower Awash valley, such as Melka-sedi-Worer and Dubti irrigation lands.

The Sodic soils are non-saline soils but with high sodium contents in their surface and/or subsoil horizons (Appendix Tables 6.61, 6.62, and 6.63). Sodic soils contain excessive quantities of exchangeable Na in their exchange complex. They have an ESP of greater than 15 (or SAR > 13), an ECe less 4 dS/m and a pH reading often > 8.5 (Appendix Table s6.59 and 6.60). Sodic soils consist mostly of the anions Cl -, $SO4_2^-$, HCO_3^- and small amounts of CO_3^{2-}. The presence of free CO_3^{2-} and HCO_3^- and lack of calcium carbonate to buffer pH causes strong alkaline reaction (pH value between 9 and 11), which tends to reduce the solubility and availability of Ca and Mg and increase the solubility of Na.

Under the Ethiopian situations, they are found within the Rift Valley irrigation sites such as Methara, Wonji, Chamo, and Zeway sites. Apart from their manufactured causes, sodic soils are also naturally occurred around the Rift Valley lakes and drier lands, e.g., eastern and northeastern parts of the country.

Gypsisols

Gypsisols of Ethiopia occur in the very arid lands of the country, where gypsum (most soluble mineral) and other highly soluble elements can exist under semi-dry and extreme drier soil and environmental conditions. These soils tend to occur in the Ogaden lowlands bordering Somalia and some pockets of drier Rift Valley areas such as northern Rift Valley of Ethiopia. As these occur in very drylands, they are less studied and less reported under the Ethiopian conditions. Consequently, Gypsisols are mainly mapped by associating expected soil characteristics to their soil forming factors mainly climatic, geologic, and topographic features of their expected occurrence sites.

Gypsisols cover about 7.7% (8,600,000 ha) of the country's total land area. As these soils largely covered the water deficit parts of the country, they are less suitable for any form of crop and even animal farming. However, they are used as corridor zones for pastoralists and in parts where rivers cross Gypsisols, grasses growing along the rivers are used for grazing animals.

Gysisols have a minimal soil profile development profile. These soil, where existed, have a surface and/or underlain gypsic horizon. Their total profile is dominated by high contents of gypsum, which could be recognized as primary and/or secondary types, depending on the local climatic and topographic features.

6.7 Conclusions

Soil development, expression of pedogenic processes, and properties varied with topography, climate, parent materials, vegetation, land use, and local site characteristics across varying physiographic regions of the country. The wide ranges in the soil forming factors in Ethiopia have had the result that soils are variable throughout the country. Whereas other factors are also important, as reported in many studies, distributions of the soils of Ethiopia are largely governed by the interactions and variations in its topography, climatic and biotic features. In Ethiopia, soil maps have been generated since as early as 1923 in a very general manner by Shantz and Marbut at a scale of 1:25 million as part of the publication "The vegetation and soils of Africa". After several early periods of fragmented soil survey missions and adaptation of coares global soil maps to the national spatial domain, country-wide geomorphology, and soils map of Ethiopa at a scale of 1:1 million was produced in 1984 by the then LUPRD/MoA.

Throughout the Ethiopia highlands, topography is the major differentiating factor of genesis and distributions of soils. Well drained Nitisols, Acrisols, Luvisols, and steep occupy gently slopping terrain units and colluvial foot slopes by Leptosols, Cambisols and Regosols. In the latter parts, active erosion and deposition leaves little time for soil development. On the plateaus (<2500 m.a.s.l.), poorly drained Vertisols dominate. At intermediate high altitudes. Phaeozems occur on gently to steeply sloping terrain and

Andosols are found in the high altitudes. The poorly drained flatter and depressions sites are covered with Gleysols. Despite the presence of several types of soils, only nine soil types cover 90% of the land surface of Ethiopia. These major types of soils are Leptosols (30%), Nitisols (12%), Vertisols (11%), Cambisols (9%), Luvisols (8%), Calcisols (9%), Fluvisols (4%), Gypsisols (8%), and Alisols (3%). In terms of their agricultural significance, Nitisols, Vertisols, Cambisols, Luvisols, and Fluvisols are major cultivated soil types.

References

Abate N, Kibret K, Gebrekidan H, Esayas A (2015) Characterization and classification of soils along the toposequence at the Wadla Delanta Massif, north central highlands of Ethiopia. J Ecol Nat Environ 6(9):304–320

Abayneh E, Berhanu D (2006) Soil survey in Ethiopia: past, present and the future. In: Teklu E, Michael M (eds) Proceedings of the 8th conference of the Ethiopian society of soil science, pp 1–10

Abayneh E (2001) Application of geographic information system (GIS) for soil resource studin Ethiopia. In: Proceedings of the national sensitization workshop on agro metrology and GIS, 17–18 December 2001. Addis Ababa, Ethiopia, pp 162–169

Agbu PA, Ojangu AG, Olson KR (1989) Soil-landscape relationships in the Sokoto Rima Basin Nigeria. Soil Sci 148(2):132–139

Ahmad N (1983) Vertisols. In: Wilding LP, Smeck NE, Hall GF (eds.) Pedogenesis and soil taxonomy II. The soil orders. Elsevier Science publishers B.V., Amsterdam

Ahmed R, Rasha A, AwekeMG SB, Hernán F, Tran L, Amado L, Sabil DM, Erica M, Marta R, Hishe WS, AlfredKT MSY, Maria GG, MuktarMY VA (2013) Soil mapping and classification: a case study in the Tigray Region, Ethiopia. J Agric Environ Int Dev 107(1):73–99

Alemayehu R (2018) Characterization of planosols in the highlands of Southern and Southwestern Ethiopia, Addis Ababa

Ali A, Tamene L, Erkossa T (2020) Identifying, cataloguing, and mapping soil and agronomic data in Ethiopia. CIAT Publication No. 506. International Center for Tropical Agriculture (CIAT), Addis Ababa, Ethiopia, p 42

Belay T (1996) Characteristics and Landscape relationships of Vertisols and VerticLuvisols of Melbe, Tigray Ethiopia. SINET Ethiop J Sci 19(1):93–115

Belay T (1997a) Variabilities of soil catena on degraded hillslopes of Watiya catchment, Welo Ethiopia. SINET Ethiop J Sci 20(2):151–175

Belay T (1998) Pedogenesis and soil-geomorphic relationships on the Piedmont slopes of Wurgo Valley South Welo, Ethiopia. SINET Ethiop J Sci 21(1):91–111

Belay T (1982) A study of a soil toposequence in northwestern Alem Gena woreda. MSc Thesis, Addis Ababa University, Addis Ababa

Belay T (1995) Morphological, physical, and chemical characteristics of Mollic Andosols of Tib Mountains, Central Ethiopia highland. SINET Ethiopia J Sci 18(2):143–169

Belete T (2014) Fertility mapping of soils of Abay-Chomen District, Western Oromia, Ethiopia. MSc Thesis, Haramaya University, Dire Dawa

Berhanu D (1982) Detailed soil survey and land suitability evaluation of holeta agricultural research station. Soil survey and evaluation Report No.2, Institute of Agricultural Research, Addis Ababa

Birkeland PW (1999) Soils and geomorphology, 3rd edn. Oxford University Press, New York and Oxford

Bono R, Seiler W (1984) The soils of AnditTid research unit (Ethiopia): classification, morphology and ecology. RR 2, Soil Conservation Research Project (SCRP), Addis Ababa

Brevik EC, Hartemink AE (2010) Early soil knowledge and the birth and development of soil science. Catena 83(2010):23–33

Brunner M (2012) A national soil model of ethiopia: a geostatistical approach to create a national soil map of Ethiopia on the basis of an SRTM 90 DEM and SOTWIS soil data. A Master's Thesis Submitted to the University of Bern, Switzerland

Buol SW, Hole FD, Mccracken DR (1989) Soil genesis and classification, 3rd edn. Iowa State University Press, Ames

CDE-ESAPP (2006) Support to the rehabilitation of a national land use planning database for Ethiopia. Centre for development and environment Eastern and Southern Africa partnership programme (ESAPP), centre for development and environment, institute of geography, University of Berne, Switzerland

DagnenetF (2009) Sattlite remote sensing for soil moisture estimation: Gumara catchment, Ethiopia. MSc thesis, ITC, The Nethrelands

Daniel G (1977) Aspects of climate and water budget in Ethiopia. Addis Ababa University Press, Addis Ababa

Daniel G (1988) Environment and development in Ethiopia. In: Penrose A (ed) An examination of the issues behind the famine in Ethiopia. International Institute for Relief and Development, food for the hunger international, Geneva

Debele B (1980) Survey of studies conducted about soil resources appraisal and evaluation for rural development in Ethiopia. Institute of Agricultural Research, Addis Ababa

D'Hoore JL (1964) Soil map of Africa. Sheet 3. Pub. No. 93, Joint Project No. 11, Commission for Technical Cooperation in Africa. Lagos, Nigeria

Donahue RL (1962) Ethiopia. Taxonomy, cartography and ecology of soils. Michigan State Univ., AfricanStud. Center & Inst. Int. Agric., Comm. Ethiopian Stud., Occasional Papers Series, Monograph 1

Driessen P, Deckers J, Spaargaren O, Nachtergaele F (2001) Lecture Notes on the major soils of the world (Rome, Italy)

Dudal R (1965) Dark clay soils of tropical and subtropical regions. FAO Agricultural Development paper No. 83, FAO, Rome

El-Hassanin AS (1985) Soil erosion in the Awash River Basin, Ethiopia. In: International symposium on the African Horn, Institute of African Research and Studies, University of Cairo, Cairo

Elias E (2016) Soils of the Ethiopian highlands: geomorphology and properties. CASCAPE Project, ALTERRA, Wageningen UR, the Netherlands. Retrieved from, library.wur.nl/WebQuery/isric/2259099

EthioSIS (2014) Soil fertility and fertilizer recommendation atlas of Tigray region. Ministry of agriculture (MOA) and agricultural transformation agency (ATA)

Ethiopian Mapping Authority EMA, 1988 Ethiopian Mapping Authority EMA (1988) National Atlas of Ethiopia. Berhanenaselam printing press, Addis Ababa

Eylachew Z (1987) Study on phosphorus status of different soil types of Chercher highlands, Southeastern Ethiopia. Ph.D. Thesis, Justus-Liebig Universitat, Giesen

Fanning DS, Fanning MCB (1989) Soil: morphology, genesis and classification. Wiley and sons Inc., New York

FAO (1984b) Assistance to land use-planning, ethiopia: provisional soil association map of Ethiopia. Field document No 6. The United Nations Development Programme and Food and Agriculture Organization. FAO, Rome

FAO (1984a) Assistance to land use-planning, ethiopia: geomorphology and soils. Field document AG DP/ETH/78/003. The United Nations Development Programme and Food and Agriculture Organization. FAO, Rome

FAO (2006) Guideline for soil description, 4th Edn. Food and Agriculture Organization of the United Nations. Rome, Italy, p 110

FAO/ISRIC/SOTER (1997) Soils of East Africa, 1: 1 million scale. User's guide version 1.0. FAO, Rome. (library.wur.nl/WebQuery/isric/27879

FAO/UNDP (1984a) Ethiopia: Provisional soil association map of Ethiopia: Assistance to land use planning (1: 2,000,000). FAO/UNDP, ETH/78/003, LUPRD/MOA, Addis Ababa

FAO/UNDP (1984b) Geomorphology and soils of Ethiopia: Assistance to land-use planning (1:1,000,000). FAO/UNDP, ETH/78/003, LUPRD/MOA, Addis Ababa

FAO-UNESCO (1974) Soil Map of the World (1:5000, 000 scale). Volume I: Legend. United Nations Educational, Scientific and Cultural Organization, Paris, France, p 65

FAO-UNESCO (1977) Soil map of the world (1:5,000,000), Volume VI. Africa, UNESCO, Paris

Fikru A (1990) Soil resources of Ethiopia. In: IAR (ed.)Natural resources degradation: a challenge to Ethiopia. First natural Resources Conservation conference, IAR, AA

Frei E (1978) Andepts in some high mountains of East Africa. Geoderma 21:119–131

Getachew A, TadesseT GB, Yigrem Y (2019) Soil moisture monitoring using remote sensing data and a stepwise-cluster prediction model: the case of Upper Blue Nile Basin, Ethiopia. Remote Sens 11 (2):125. https://doi.org/10.3390/rs11020125

Getachew A, Tadesse T, Gessesse B, Yigrem Y, Melesse A (2020) Combined use of Sentinel-1 SAR and landsat sensors products for residual soil moisture retrieval over agricultural fields in the Upper Blue Nile Basin, Ethiopia. Sensors 20(11):3282. https://doi.org/10.3390/s20113282

Hengl T, Leenaars JGB, Shepherd KD, Walsh HMG, MamoT, (2017) Soilnutrient maps of Sub-Saharan Africa: assessment of soil nutrient content at 250 m spatialresolution using machine learning. Nutr Cycl Agroecosyst 109:77–102. https://doi.org/10.1007/s10705-017-9870-x

Hengl T, Heuvelink GBM, Kempen B, Leenaars JGB, Walsh MG, Shepherd KD (2015) Mapping soil properties of Africa at 250 m resolution: random forests significantly improve current predictions. PLoS One 10(6):e0125814. https://doi.org/10.1371/journal.pone.0125814 Retrieved fro http://afsoilgrids250m.isric.org/

Hurni H (1983) Soil erosion and soil formation in agricultural ecosystems: Ethiopia and Northern Thailand. Mt Res Dev 3 (2):131–142

IUSS Working Group WRB (2006) World reference base for soil resources: a framework for international classification, correlation and communication. World Soil Resources Report 103. International Union of Soil Science, Rome

IUSS Working Group WRB (2014) World reference base for soil resources: a framework for international classification, correlation and communication. International Union of Soil Science, Rome (Update of 2015). Retrieved from http://www.fao.org/3/i3794en/I3794en.pdf

Kefeni K (1995) The soils of the Anjeni Area-Gojam Research Unit, Ethiopia. RR 27, Soil conservation Research Project (SCRP), Addis Ababa

Laekemariam F, KibebewK TM, Hailu S (2016) Accounting spatial variability of soil properties and mapping fertilizer types using geostatistics in Southern Ethiopia. Commun Soil Sci Plant Anal 49 (1):124–137. https://doi.org/10.1080/00103624.2017.1421656

Leenaars JGB, Oostrum AJM, Ruiperez GM (2014) Africa soil profiles database, version 1.2. A compilation of georeferenced and standardised legacy soil profile data for Sub-Saharan Africa (with dataset). ISRIC Report 2014/01. Africa Soil Information Service (AfSIS) project and ISRIC–World Soil Information, Wageningen, The Netherlands. Retrieved from www.isric.org/data/africa-soil-profiles-database-version-01-2

Leenaars JGB, Eyasu E, Wösten H, Ruiperez GM, Kempen B, Ashenafi A, Brouwer F (2016) Major soil-landscape resources of the cascape intervention woredas,Ethiopia: Soil information in support to scaling up of evidence-based best practices in agricultural production (with dataset). CASCAPE working paper series No. OT_CP_2016_1. Cascape. Retrieved from https://edepot.wur.nl/428596

Leenaars JGB, Ruiperez GM, Kempen B, Mantel S (2020b) Semi-detailed soil resource survey and mapping of REALISE woredas in Ethiopia. Project report to the BENEFIT-REALISE programme. December 2020a, ISRIC–world soil information, Wageningen

Leenaars JGB, Elias E, Wösten JHM, Ruiperez GM, Kempen B (2020a) Mapping the major soil landscape resources of the Ethiopian Highlands using random forest. Geoderma 361. https://doi.org/10.1016/j.geoderma.2019.114067

Leenaars JGB (2012) Africa soil profiles database, version 1.0. A compilation of georeferenced and standardized legacy soil profile data for Sub-Saharan Africa (with dataset). ISRIC Report 2013/03. Africa Soil Information Service (AfSIS) project and ISRIC–World Soil Information, Wageningen, The Netherlands. Retrieved from www.isric.org/data/africa-soil-profiles-database-version-01-0

Lelago A, Mamo T, Haile W (2016) Assessment and mapping of status and spatial distribution of soil macronutrients in kambata tembaro zone, southern Ethiopia. Adv Plants Agric Res 4(4):305–317. https://doi.org/10.15406/apar.2016.04.00144

Mekonnen M (2014) Fertility mapping of soils in ChehaWoreda, Gurage Zone, southern Ethiopia. MSc Thesis, Haramaya University, Ethiopia

Mesfin A (1998) Nature and Management of Ethiopian soils. International Live Stock Research Institute (ILRI), Addis Ababa

Ministry of Agriculture (MoA) (1987) A simple system for defining agroecological zones in Ethiopia, Final Report. MOA, Addis Ababa

Ministry of Water Resource (MoWR), BECOM (1998) Abay River basin integrated development master plan project phase 2: land resources development. Addis Ababa

Ministry of Water Resource (MoWR), Halcrow (2007) WabeshebeleGennale-Laga Wata/sure, rivers basin integrated development master plan study. Addis Ababa

Ministry of Water Resource (MoWR), Halcrow (2010) Rift valley lakes basin integrated resources development master plan study. Phase 3. Final Report. Western Lake Ziway Sub-basin Integrated Watershed Management Feasibility Study, vol 1, Main Report

Ministry of Water Resource (MoWR), Lahmeyer (2006) Genale Dawa river basin integrated resource development master plan study project. Master plan report: land evaluation/land suitability. Addis Ababa

Ministry of Water Resource (MoWR), NEDECO (1999) Tekeze river basin integrated development master plan study

Mitiku H (1987) Genesis, characteristic and classification of the Central Highland soils of Ethiopia. Ph.D. Thesis, State University of Ghent, Belgium

Mohammed A, Belay T (2008) Characteristics and classification of the soils of the Plateau of Simen Mountains National Park (SMNP), Ethiopia. SINET Ethiop J Sci 31(2):89–102

Mohammed A, Solomon T (2012) Characteristics and fertility quality of the irrigated soils of Sheneka, Ethiopia. Ethiop J Nat Resourc (EJNR) 12(1 & 2):1–22

Mohammed A (2003) Land suitability evaluation in the Jelo Catchment, Chercher highlands, Ethiopia, PhD Thesis, University of the Free state, Bloemfontein

Mohr P (1971) The geology of Ethiopia. Haile Selassie I University Press, Addis Ababa

Murphy HF (1963) Fertility and Other Data on Some Ethiopian Soils. Exp. Sta. Bui. No. 4. Halle Sellassie I University, Alemaya, Ethiopia. 16

Murphy HF (1968a) A report on the fertility status and other data on some soils of Ethiopia. Exp. Sta. Bui. No. 44, Oklahoma State University, Stillwater, Oklahoma

Murphy HF (1968b) A report on the fertility status and other data on some soils of Ethiopia, Experiment station bulletin No. 44. College of Agriculture, Haile Selassie I University press, Dire-Dawa

NizeyimanaE BTJ (1992) Soil and soil-landscape relationship in the North Central region of Rwanda, East Central Africa. Soil Sci 153 (3):225–236

Nyssen J, Tielens S, Gebreyohannes T, Araya T, Teka K, Van de Wauw J (2019) Understanding spatial patterns of soils for sustainable agriculture in northern Ethiopia's tropical Mountains. PLoS One 14(10):e0224041. https://doi.org/10.1371/journal.pone.0224041

Paris S (1985) Soil survey of the Borkena area (Wello). AG/DP/ETH/82/010, Field Document 7, MOA/LUPRD, FAO/UNDP, Addis Ababa

Parsons RB (1978) Soil-geomorphology relationships in Mountains of Oregon, USA. Geoderma 21:25–39

Poggio L, de Sousa LM, Batjes NH, Heuvelink GBM, Kempen B, Ribeiro E, Rossiter D (2022) SoilGrids 2.0: producing soil information for the globe with quantified spatial uncertainty

Tefera M, Chernet T, Haro W (1996) Geological map of Ethiopia, 2nd Edition, Ethiopian Mapping Authority

Vågen G, Leigh AW, Assefa A, Kiros MH (2013) Landsat-based approaches for mapping of land degradation prevalence and soil functional properties in Ethiopia. Remote Sens Environ 134:266–275

Van Ranst E, Dumon M, Tolossa AR, Cornelis J-T, Stoops G, Vandenberghe RE, Deckers J (2011) Revisiting ferrolysis processes in the formation of Planosols for rationalizing the soils with stagnic properties in WRB. Geoderma 163:265–274

Veneema JH, Paris S (1986) Soils of Bichena area (Gojam): assistance to land use planning Ethiopia. AG: DP/ETH/82/010, field Document 10, LUPRD/MOA, FAO/UNDP, Addis Ababa

Venema JH, Paris S (1987) Soils of Hosaina study area: assistance to land use planning Ethiopia. AG: DP/ETH/82/010, Field Document 13, LUPRD/MOA, FAO/UNDP, Addis Ababa

Virgo KJ, Holmes DA (1977) Soil and landform features of mountainous terrain in South Thailand. Geoderma 18:207–225

Virgo KJ, Munro RN (1978) Soil and erosion features of the Central Plateau region of Tigray, Ethiopia. Geoderma 20(2):131–157

Weigel G (1988) Some soils of the lower area of the Simen Mountains National Park. RR10, Soil Conservation Research Project (SCRP), Addis Ababa

Westphal E (1975) Agricultural systems in Ethiopia. Agric.Res. rep.826,ISBN 90 220 0556 9,2718p

World Reference Base for Soil Resources WRB (1998) World soil resources reports 84. FAO/ISSS-AISSS-IBG/ISRIC, FAO, Rome

World Reference Base for Soil Resources WRB (2006)A framework for international classification, correlation and classification Reports 103. FAO/IUSS /ISRIC, FAO, Rome

Young A (1976) Vertisols. In: tropical soils and soil survey. Cambridge University Press, Cambridge and New York, pp 180–191

Zewdu S, Suryabhagavan KV, Balakrishnan M (2015) Geo-spatial approach for soil salinitymapping in Sego irrigation farm, South Ethiopia. J Saudi Soc Agric Sci 16:91–97. Cambridge University Press, Cambridge

Soil Properties

7

Alemayehu Regassa, Kibebew Kibret, Yihenew G. Selassie, Alemayehu Kiflu, and Wondwosen Tena

Abstract

Ethiopian soils greatly vary in their morphological, physico-chemical and mineralogical properties due to the wide variations in soil forming factors and the associated processes. The different soil types of the country show great variation in their morphological properties such as depth, color, structure and consistence. Across the country, the sand content varies from 1 to 75%; silt from 9 to 60%, and clay from 10 to 86%. The clay content of Ethiopian soils ranges between 10 and 86%. About 48.65% of the country's total area is covered by soils whose clay content is greater or equal to 35%, indicating that most. About 0.38% of the country's total area is covered by soils whose bulk density in the upper 0–20 cm is less than 1.0 g/cm^3. And about 5.36% of the country's total area is covered by soils with total porosity values that vary from 60 to 71.06%. When it comes to soil chemical properties, soil reactions in Ethiopian soils can be broadly grouped into three categories. The soils in the highlands of north-western, western and central parts of Ethiopia are predominantly acidic. The Rift Valley, the lowlands of eastern and southern regions and arid dry highlands of Tigray are dominated by alkaline soils. Ethiopia's highlands, covering approximately 40% of the nation, historically have extremely low soil carbon content. About 36.39%, 26.17%, 33.42%, and 2.37% of the Ethiopian landmass has soil organic carbon content of very low, low, medium, high and very high respectively. About 48.92% of the soils of the country have high, 25.45% have very high and 19.07% have medium CEC levels. Only 3.36% of the soils have low CEC levels. Soil salinity is one of the major land degradation problems in Ethiopia. The country stands first in Africa in the extent of salt affected soils. About 44 million ha (36% of the total land area) is potentially susceptible to salinity problems of which 11 million ha have already been affected by different levels of salinity. Soil sodicity is the problem of dry areas in Ethiopia mainly Afar and northern parts of the Somali Regions. About 4,250 km^2 of sodic soils exist in Ethiopia. Relative abundances of the different minerals identified in the different soil types of the country based on the intensity of X-ray diffraction peaks for the different minerals are quite variable among the different soil reference groups. Among others, Ethiopian Nitisols contain a large quantity of 1:1 minerals like kaolinite and a small quantity of 2:1 clay minerals such as smectite Illite and mixed-layer minerals. Many Ethiopian Vertisols contain illite as the dominant mineral while smectite dominates in some Vetisols. The Luvisols have a large amount of 1:1 clay minerals mainly kaolinite with high-intensity diffraction peaks.

Keywords

Morphological • Physical • Chemical • Mineralogical • Biological properties

A. Regassa (✉)
Department of Natural Resources Management, Jimma University, P.O. Box 307 Jimma, Ethiopia
e-mail: alemayehu.regassa@ju.edu.et

K. Kibret
School of Natural Resources Management & Environmental Sciences, Haramaya University, P.O. Box 138 Dire Dawa, Ethiopia
e-mail: kibebew.kibret@haramaya.edu.et

Y. G. Selassie
Department of Natural Resources Management, Bahir Dar University, P.O. Box 79 Bahir Dar, Ethiopia
e-mail: yihenewgs@gmail.com

A. Kiflu
School of Plant and Horticultural Sciences, Hawassa University, P.O. Box 5 Hawassa, Ethiopia
e-mail: alemanchy@gmail.com

W. Tena
Department of Plant Biology and Biodiversity Management, Addis Ababa University, P.O. Box 1176 Addis Ababa, Ethiopia
e-mail: wondtena@gmail.com

S. Beyene et al. (eds.), *The Soils of Ethiopia*, World Soils Book Series,
https://doi.org/10.1007/978-3-031-17012-6_7

7.1 Introduction

Soil properties result from soil formation processes which are governed by the factors such as climate, topography, parent material, vegetation and time (Jenny 1941). A well organized and documented data bank on soil properties and related site characteristics is inevitable for one to be able to advise both current and potential land users on how to use the land in the best possible way (Dinku et al. 2014).

Soil physical properties are characteristics, processes or soil reactions that are caused by physical forces and that can be described by, or expressed in, physical terms or equations. Examples of important physical properties include texture, structure, bulk and particle densities, porosity, and soil–water retention and transmission characteristics. Soil texture is one of the fundamental and intrinsic properties of soils, affecting almost all other soil properties. Specifically, it influences soil properties that are important to agriculture.

Given the wide variations in soil forming factors and the associated processes, there exists a high variation of particle size distribution in Ethiopian soils. Across the country, the sand content varies from 1 to 75%; silt from 9 to 60%, and clay from 10 to 86%. In Ethiopia, bulk density is one of the commonly measured physical soil parameters. About 0.38% of the country's total area is covered by soils whose bulk density in the upper 0–20 cm is less than 1.0 g/cm^3. Therefore, total porosity is estimated from bulk density and particle density assuming a particle density value of 2.65 g/cm^3. Values of total porosity estimated from this approach range between 71.06 and 36.79%. The low bulk density values of the upper 0–20 cm layer have resulted in generally high total porosity values across soils of the country. About 5.36% of the country's total area is covered by soils with total porosity values that vary from 60 to 71.06%. About 57.02% of the country's area has soils with porosity values that range between 50 and 60, with 37.63% of the area covered with soils whose total porosity varies from nearly 37 to 49%. Soil depth is a result of slope in a topographic position that influenced soil formation and development through its effects on water movement down the slope causing erosion and deposition as well as the infiltration and the percolation of water deep into the soil at the lower topographic positions (Mulugeta and Sheleme 2017; Sheleme 2017; Dinku et al. 2014; Alemayehu and Sheleme 2013).

Chemical Properties of soils are those characteristics, processes or reactions that are caused by mainly chemical activities and that can be described by or expressed in chemical terms or equations. Soil chemical properties are mainly described by soil parameters like soil reaction, cation exchange capacity (CEC), base saturation, cation/anion exchange processes, soil salinity and soil sodicity. Soil chemical properties affect various activities in the soil including nutrient retention, availability and uptake, nitrogen dynamics, soil biological activity and decomposition of organic matter.

Acidic conditions occur in soil with parent material high in elements such as silica (rhyolite and granite), high levels of sand with low buffering capacities (ability to resist pH change), and in regions with higher amounts of precipitation (Foster et al. 2016). In Ethiopia, the major soil forming factors and management practices giving rise to the increase in soil acidity in the country involve climatic factors such as high rainfall, temperature, topographic factors, soil parent materials, intensive mono-cropping and lack of technological inputs in the agricultural sector to mitigate the problem (Mesfin 1998).

Soil minerals play a significant role in dictating the suitability and behavior of soil for various land uses. It is one of the principal soil forming factors (Jenny 1941) and therefore a key determinant of basic soil functional properties. Therefore, understanding soil mineralogy is essential to understanding many facets of land use and is often a key to solving specific agricultural and environmental problems.

7.2 Morphological Characteristics

Soil depth

Due to the existence of diversified topography, the soils of Ethiopia have different depths (Fig. 7.1). The presence of rugged toposequence in a very short distance makes the soils to be shallow in the upper positions and deeper in the lower areas (Sheleme 2017). The variation in soil depth is a result slope of the topographic positions that influenced soil formation and development through its effects on water movement down the slope causing erosion and deposition as well as the infiltration and the percolation of water deep into the soil at the lower topographic positions (Alemayehu and Sheleme 2013; Dinku et al. 2014; Mulugeta and Sheleme 2017; Sheleme 2017).

Leptosols cover the largest areas in Ethiopia. They have usually shallow depths of less than 30 cm (Fig. 7.2) with A/AC/C or A/C horizon combinations (Zewdie 2013; Legass and Assen 2012). The surface horizon sometimes could have lithic contact (Zewdie 2013). As the sand and silt particles are dominant, they have a consistency of nonsticky/plastic or slightly sticky/plastic when wet, firm moist, common medium pores and clear to the abrupt smooth boundary (Legass and Assen 2012; Assen and Tegene 2008). These soils have different types of surface structures, which showed a close relationship with the parent material. Those from fluvic materials have massive structures whereas those developed

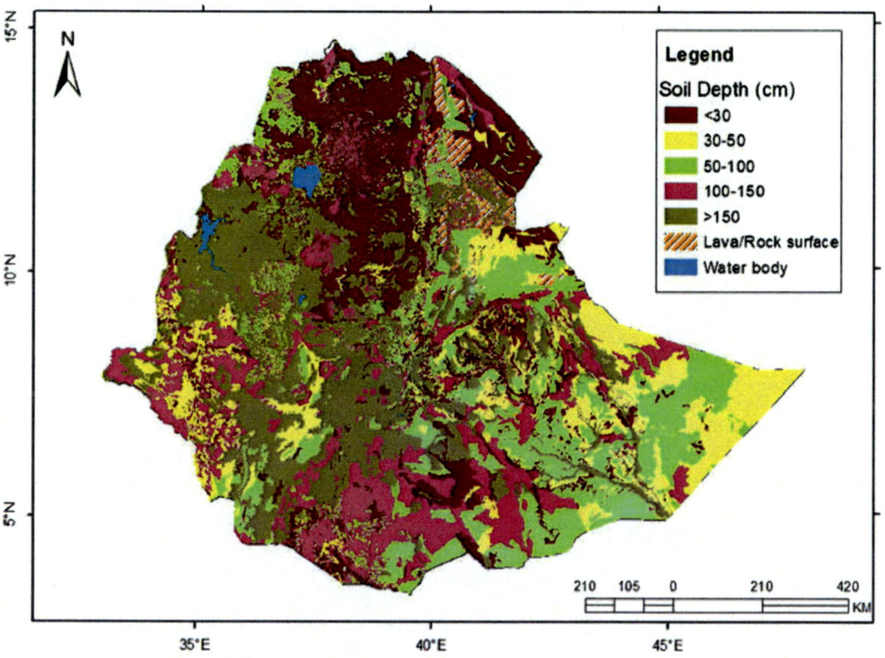

Fig. 7.1 Soil depth map of Ethiopia. (*Source* Map extracted and own analysis by author from FAO/ISRIC/SOTER (1997)

Fig. 7.2 Leptosols in Limu Saka district, south-western Ethiopia (Photograph courtesy of Alemayehu Regassa)

from in-situ weathering rocks gave a crumby or sub-angular blocky structure that breaks into granular (Zewdie 2013). They are marked by dark brown and brown dry color and very dark brown to dark brown when moist (Fig. 7.2).

Nitisols exist across western, southeastern and southern parts of the country (Zewdie 2013) and are formed on old land surfaces where the soils tend to be leached to various degrees. They have a diversified pore system and are characterized by a well-developed structure with common shiny ped faces usually called nitric property (Gizachew et al 2015; Zewdie 2013) (Fig. 7.3).

A soil color pattern in Nitisols is usually red. They have dark to very dark brown on the surface to red in subsoil horizons (Gizachew et al 2015; Zewdie 2013; Assen and Yilma 2010). Their structure also ranged from granular in the topsoil for most cultivated soils and sub-angular for uncultivated ones to sub-angular, angular blocky, or columnar in the subsoil that has a strong relationship with clay translocation (Zewdie 2013). Due to high clay content in their subsurface horizon, the dry soil consistency is almost hard on the surface and hard on Bt horizons while the moist and wet soil consistency is dominated by friable and sticky and plastic respectively (Assen and Yilma 2010; Zewdie 2013; Gizachew et al. 2015; Adhanom and Teshome 2016) (Table 7.1).

Ethiopian Vertisols vary in their color widely. Color differences between Vertisols are due to the differences in drainage status. A more reddish hue or stronger chromas of relatively better-drained Vertisols are found in eastern Ethiopia which is due to higher contents of free iron oxides (Zewdie 2013). On the other hand, a poorly drained central Ethiopian plateau Vertisols hue is less red and their chromas are weaker (Alemayehu 2018; Fig. 7.4).

Moisture variation also influences the size of structural aggregates, intensity or strength of slickenside, width and depth of tonging and cracks and frequency of gilgai formation and color. Vertisols from high precipitation areas showed the massive upward movement of parallelepipeds with dominant prominent slickenside during the rainy seasons and develop deep and wide cracks (Fig. 7.5). Moreover, a wide range of cracks between 5 and 10 cm with depth the

Fig. 7.3 Typical structures of a nitic horizon showing characteristic polyhedric nut-shaped peds with shiny faces (Photo courtesy of Alemayehu Regassa)

Table 7.1 Morphological features and classification of Ayehu area Nitisols (Fisseha and Gebrekidan 2007, Symbol FAO 2006)

Depth (cm)	Horizon	Boundary	Color		Structure	Consistency		
			Dry	Moist		Dry	Moist	Wet
0–25	Ap	AS	5YR 3/3	7.5YR 3/2	Weak, fine granular	s	vfr	sssp
25–50	Bt1	GS	5YR 3/4	5YR 3/4	Weak, fine and medium sub-angular blocky	h	fr	sp
50–100	Bt2	GS	5YR 4/6	5YR 4/3	Moderate, medium sub-angular blocky	h	fi	sp
100–200+	Bt3	–	5YR 4/3	2.5YR 4/6	Moderate, Fine angular blocky	vh	fi	sp

depth of cracks extending more than 50 cm downward could and gilgai formation could be also dominant (Fig. 7.6).

Cambisols are the only soil type whose development is controlled by age and types of parent materials as a result of slow weathering processes. The process of horizon differentiation is at a very low rate. In most cases after the development of a cambic subsurface horizon. Ethiopian Cambisols are formed due to water erosion or slow weathering parent material (Zewdie 2013) and therefore they could be easily differentiated by their color from the C-horizon. Cambisols are also identified by the amount of weatherable minerals and the absence of the signs of advanced pedogenesis evidence moreover they become structured and yellowish-brown to reddish in color (Zewdie 2013; Legass and Assen 2012). In most cases, the black/very dark gray subsoil color gradually changes into a brown/very dark brown color in the lower horizons due to a decrease in the contents of organic matter in some and the effect of iron oxidation in others. Some Cambisols are also classified as young soils and they receive fresh sediments every year during regular floods as a result of which they have stratified layers with or without organic matter.

Luvisols are generally formed on the rolling topography and occur at low or moderately high and higher altitudes

Fig. 7.4 In-situ appearance of the four Vertisols **a)** Wonji **b)** Ginchi **c)** Sheno and **d)** Alemaya (Photo courtesy of Eylachew Zewdie 2013)

Fig. 7.5 Slickenside formation of Vertisols Ener and Enamore, southern Ethiopia (Photo courtesy of Alemayehu Regassa)

Fig. 7.6 Crack formation of Vertisols at Ginchi (Photo courtesy of Eylachew Zewdie 2013)

where wet season precipitation exceeds evapotranspiration. Luvisols show a similar pattern of transformation, but a slight color difference was observed as a result of the difference in organic matter. Luvisols do not show significant textural differences downward. They have a slightly sticky and plastic consistency throughout the profile when wet. The boundaries between horizons and within the Bt are diffuse and smooth. The subsurface horizons were relatively darker in color due to the strong clay translocation process, which is evidenced by the clay skin in Bt horizons (Abay et al. 2015; Alem et al. 2015). Luvisols have a granular structure at the surface whereas the subsurface horizons have angular or sub-angular blocky structures (Abay et al 2015; Zewdie 2013). Luvisols are red in color, usually, Reddish-brown dominates the profile which could be due to the occurrence of various hydrated iron oxide forms.

Alisols have a very deep profile (>200 cm). They are characterized by clear smooth and gradual smooth boundaries at the surface and subsurface respectively. Moreover, these soils have shown great variability in relation to soil color patterns but in general, are red in color and darker on the surface and as compared to subsurface horizons owing to the relatively higher organic matter contents in the surface horizons (Sheleme 2017). The dominant red color in these soils suggests the presence of iron compounds in the different states of oxidation and hydration (Sheleme 2017; Alemayehu et al. 2016). The soil structure in Alisols varied from weak, fine to medium granular and sub-angular blocky at the surface to moderate, fine to coarse angular blocky in the subsurface layers.

7.3 Physical Properties

7.3.1 Principles

Soil physical properties determine the physical fertility of a soil and thus influence to a greater extent a soil's suitability for agricultural, environmental, and engineering purposes or uses. The soil's physical properties influence physical processes such as the supporting capability, movement, retention, and availability of water and nutrients to plants, ease in penetration of roots, and flow of heat and air. They also influence the chemical and biological properties of soils (Phogat et al. 2015) and, hence, maintenance of soil health. Soil's physical properties are crucial to agricultural sustainability. Soil's physical properties are important in soil's role as a vast reactor that transforms, deactivates, denatures, or detoxifies chemicals. In general, soil's physical properties are important to agricultural sustainability through enhancing efficient use of water and nutrients, maintaining environmental quality through purification of water and air, and helping in fighting climate change through reduction of greenhouse gases emission and enhancing sequestration. Therefore, looking at the current stress on the soil as a natural resource for food security and safety, due emphasis is needed for maintaining soil physical fertility by adding organic materials, introduction of legumes in rotation, adoption of conservation tillage, and other management practices.

7.3.2 Soil Texture

The term soil texture has both qualitative and quantitative meanings in it. Qualitatively, it describes the 'feel' of the soil material when it is worked with fingers (often when rubbed between thumb and forefinger). The sand particles are coarse and gritty; the silt particles are smooth and floury, while the clay particles feel sticky. In quantitative terms, it denotes the precisely measured distribution of particle sizes and the proportions of the various size ranges of particles composing a given soil (Phogat et al. 2015; Geering and So 2006; Lal and Shukla 2004; Hillel 2004). As such, soil texture is an intrinsic attribute of the soil and the one most often used to characterize its physical makeup (Hillel 2004). Although there are many systems of classifying the soil separates into three size classes (often varying with country and professional institutions- Hodgson 1974), all of them consider the upper limit of a soil material's size at less than 2 mm.

Soil texture is one of the fundamental and intrinsic properties of soils, affecting almost all other soil properties. Specifically, it influences soil properties that are important to agriculture (e.g., ease of cultivation, nutrient- and water holding capacity and their transmission), the environment (e.g., susceptibility to pollution), and engineering purposes (e.g., packing of particles, hardsetting) (Geering and So 2006). Because of its static nature, it is influenced little by tillage or other manipulations unless it is drastic (Chestworth 2008). The overall textural designation of a soil as determined from the relative proportion of its sand, silt, and clay contents is called *textural class*. According to the USDA system, there are twelve textural classes, falling under three broad primary textural groups: *sandy* (sand, loamy sand, and sandy loam), *loamy* (loam, silt loam, silt, sandy clay loam, clay loam, silty clay loam) and *clayey* (sandy clay, silty clay, and clay) (Phogat et al. 2015). The clay fraction influences soil behavior the most owing to its greater surface area per unit mass. The combination of the three fractions in various configurations is what constitutes the matrix of the soil (Hillel 2004).

Given the wide variations in soil forming factors and the associated processes, there exists a high variation of particle size distribution in Ethiopian soils. Across the country, the sand content varies from 1 to 75%; silt from 9 to 60%, and clay from 10 to 86%. However, only 0.023% of the country's total area is covered by soils with sand content greater or equal to 70%. Although areas with high sand content soils are distributed throughout the country, the dominant ones are located in the Afar and Somali Regional States. These places are areas with low rainfall and high evapotranspiration resulting in limited chemical weathering and dominantly physical weathering.

Like the sand content, the silt content also varies greatly across soils in different parts of the country. Available data indicates that this fraction ranges between 9 and 60%. About 0.05% of the country's total area is covered by soils whose silt content is between 40 and 60%, with the remaining large area of the country occupied by soils whose silt content is between 9 and 39%. Areas with silt content between 9 and 21% are also areas with high sand content in the lowlands of Somali Regional State. The high silt content soils are found along the Great Rift Valley system.

The clay content of Ethiopian soils ranges between 10 and 86%. About 48.65% of the country's total area is covered by soils whose clay content is greater or equal to 35%, indicating that most Ethiopian soils contain high amount of clay. Generally, soils in the western part of the country contain higher amount of clay, while those in the Somali and Afar Region are characterized by high sand content.

In terms of textural class, ten of the 12 USDA textural class names are present in Ethiopian soils, with sand and silt the missing textural class names. In terms of area coverage, clay loam (36.15%), followed by sandy clay loam (27.28%) and clay (20.82%), are the most dominant textural classes, confirming the high clay content of most soils in Ethiopia (Fig. 7.7).

7.3.3 Bulk Density

Bulk density is defined as the ratio of mass of an oven-dry soil to its bulk volume, which includes the volume of solids as well as pores. As a dynamic soil property, it is affected, among others, by soil texture, structure, organic matter content, soil depth, and management. Under natural conditions, coarse-textured soils tend to have higher bulk density values (ranging from 1.4 to 1.9 Mg/m^3) than medium and fine-textured soils (0.9–1.4 Mg/m^3), while volcanic soils that are made up of special clay minerals that are amorphous have lower bulk density values (often less than 1.0 Mg/m^3). For the same textural class, soils with higher organic matter content are expected to have lower bulk density since organic matter weighs less per unit volume. Similarly, aggregated soils tend to have lower bulk density than non-aggregated soils, for aggregation increases porosity. Generally, loose, well-aggregated, porous soils and those rich in organic matter have lower bulk density values. It typically increases with soil depth since subsurface layers are more compacted and have less organic matter, less aggregation, and less root penetration compared to surface layers.

The practical values of soil bulk density data include its use in the evaluation of the degree of compactness and hence

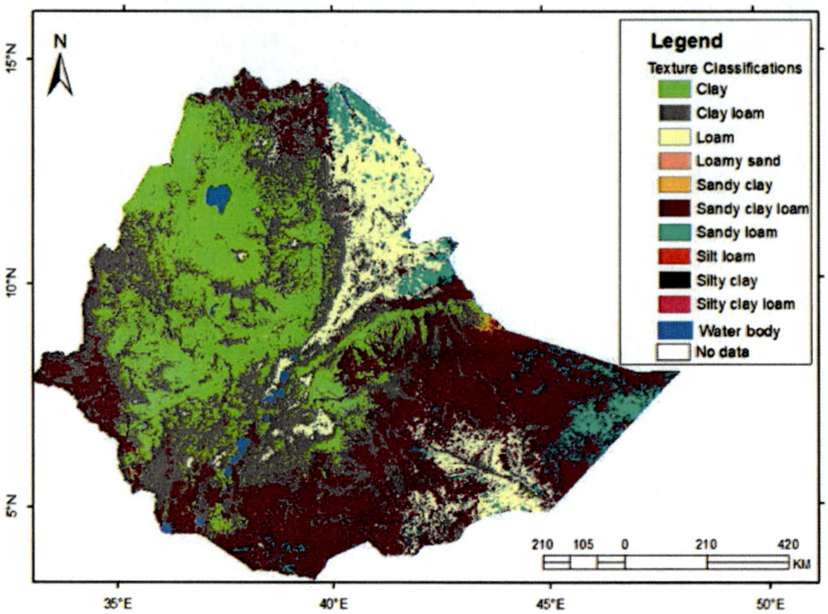

Fig. 7.7 Surface soil textural class map of Ethiopia (*Source* Map extracted and own analysis for top 20 cm from individual sand, silt and clay maps, by author from SoilGrids 2017: https://files.isric.org/soilgrids/former/2017-03-10/data/)

as an indirect measure of soil structure, an indicator of soil aeration status, calculation of the mass of soil per unit area and depth, and for converting soil water content and nutrient values from the mass to volumetric basis. More recently, it has been used, together with other soil properties, for developing pedotransfer functions that predict soil–water retention at a given potential, hydraulic conductivity, and penetration resistance (Barros and van Lier 2014). Furthermore, it is being used as an input parameter in models that deal with sediment and nutrient transport.

In Ethiopia, bulk density is one of the commonly measured physical soil parameters. In Ethiopia, the use of heavy machines such as tractors for tillage operations is not something very common. Most farmers in Ethiopia use a simple implement locally called *Maresha* for tillage or land preparation. This is a small implement and its weight is not likely to cause heavy compaction below the plow layer. Whenever compaction occurs due to tillage using *Maresha*, it is the result of repeated tillage at the same depth rather than the weight of this tillage implemented. This is justified by the relatively low bulk density values that range from 0.77 to 1.68 g/cm^3 (Fig. 7.8) recorded across soils in the country. About 0.38% of the country's total area is covered by soils whose bulk density in the upper 0–20 cm is less than 1.0 g/cm^3. On the other hand, about 8.13% of the country's total area contains soils whose bulk density values are greater than 1.47 g/cm^3 and thus can restrict root growth in clay soils with clay content greater than 45%. None of the bulk density values are in the range that poses restriction to root penetration in sands and loamy sands. The results suggest that compaction may not be a serious problem in the 0–20 cm depth of Ethiopian soils.

7.3.4 Particle Density

Particle density, variously called *mean particle density*, *true density* or *density of solid particles*, is computed from the ratio of the mass of soil particles to the volume occupied by the soil solids (Huang et al. 2012). The particle density of a soil is a weighted average over the distribution of soil minerals present (Skopp 2012). It is one of the static soil's physical properties. Particle density is affected by mineralogical composition and organic matter content alone; it is not affected by soil structure or texture. Typical particle density values in soils range from 2.6 to 2.8 g/cm^3 for mineral soils and around 1.0 g/cm^3 for organic soils. In terms of mineralogical composition, soils containing appreciable amount of heavy minerals, such as iron oxides and other heavy minerals, tend to have higher values of particle density. For soils dominated by quartz, an average value of 2.65 g/cm^3 is often used as a representative value for mineral soils. Although particle density gives few insights into soil physical processes, it is used in the calculation of total porosity and particle size distribution.

In Ethiopia, particle density is not a routinely measured soil parameter. Most studies that require the use of particle density in the calculation of total porosity or particle size

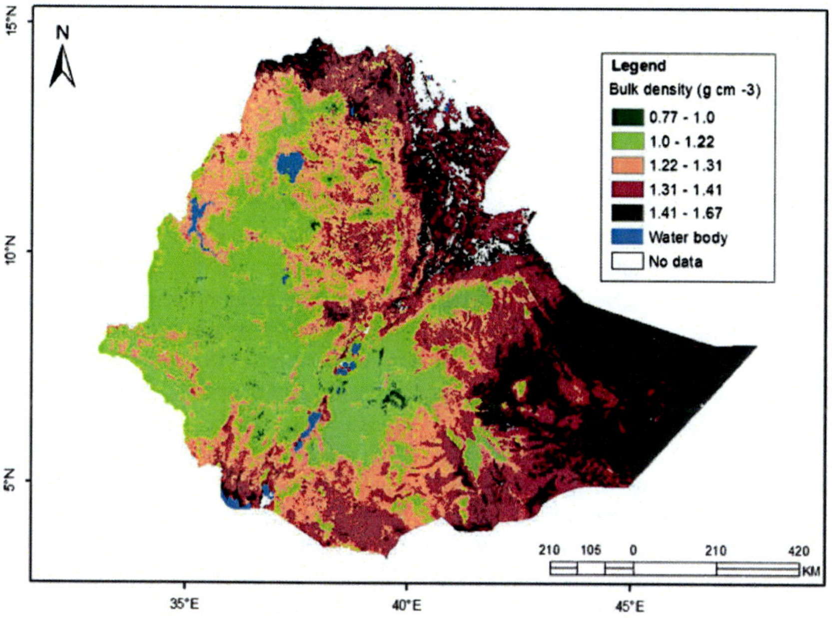

Fig. 7.8 Soil bulk density map of Ethiopia (*Source* Map extracted and own compilation for top 20 cm depth by author from ISRIC SoilGrids 2017: https://files.isric.org/soilgrids/former/2017-03-10/data/)

distribution often use the typical value of 2.65 g/cm^3 assuming that mineral soils in the country are formed from minerals dominated by silica (quartz). The few studies that determined particle density using the pycnometer method revealed that it varies between 2.57 and 2.91 g/cm^3 in the surface layers of profiles and 2.52–2.84 g/cm^3 in the subsurface layers.

7.3.5 Porosity

Total porosity is defined as the volume occupied by pores per unit volume of soil. Alternatively, is also estimated from bulk density and particle density. It is an index of relative pore volume in soil and is generally expressed as a percentage. It is a dynamic property and, is affected by salient soil properties such as texture and structure. Factors that affect bulk density also affect particle density, working in the opposite direction. Under natural soils, packing density, particle shape, and cementing agents affect porosity. Typical values of porosity may range from 0.30 to 0.70 (Nimmo 2004). However, observed porosities can be as great as 0.8–0.9 in a peat (extremely high organic matter) soils. Size distribution and connectivity of pores are important attributes that influence the movement of water, air, and other fluids; the transport and the reaction of chemicals; and the residence of roots and other biotas (Nimmo 2004). Porosities can be classified based on their mode of formation as textural and structural or based on their size as macro, meso, and micro porosity.

In Ethiopia, facilities for direct measurement of total porosity or any of its attributes are not available. As a result, it is calculated from values of dry bulk density and particle density. Furthermore, particle density is not a routinely measured parameter in Ethiopia. Very few studies have measured particle density. Therefore, total porosity is estimated from bulk density and particle density assuming a particle density value of 2.65 g/cm^3. Values of total porosity estimated from this approach range between 71.06 and 36.79%. The low bulk density values of the upper 0–20 cm layer have resulted in generally high total porosity values across soils of the country. About 5.36% of the country's total area is covered by soils with total porosity values that vary from 60 to 71.06%. About 57.02% of the country's area has soils with porosity values that range between 50 and 60, with 37.63% of the area covered with soils whose total porosity varies from nearly 37 to 49%. However, these values are only total porosity and they do not give any information about the pore size distribution and their connectivity, which are very important for fluid movement and root penetration. Similar to bulk density, porosity is very much affected by how soils are managed or used. Available evidences indicate that, under managed ecosystem, values are generally lower in areas where there is overgrazing and higher in forested areas with minimal disturbance.

7.3.6 Soil Hydraulic Properties

The most important soil hydraulic properties include water retention characteristics, available water capacity, hydraulic conductivity, and infiltration. These properties indicate the interaction between water and the solid matrix. Therefore, they are affected by various inherent soil properties such as soil texture, structure, bulk density, porosity, organic matter, density and viscosity of water, and interactions of these properties with each other. Furthermore, soil manipulations (e.g., tillage, agronomic practices, compaction, etc.) that influence the inherent soil properties such as structure have a remarkable impact on these properties. The capability of the soil to provide ecosystem services and maintain environmental quality is determined to a greater extent by soil hydraulic properties, for these properties control the rate of water entry and flow, water retention, the fate of solutes including nutrients and pollutants, availability of water to plants, and crop growth. As a consequence, these properties are used as inputs in models that predict or simulate different processes, such as predicting the plant–water relationship, water movement and storage, designing irrigation and drainage systems, leaching of nutrients, runoff and catchment management, crop breeding planning, and assessing the productive potential of soils.

7.3.6.1 Soil–Water Characteristic/Retention Curve

A soil–water characteristic curve describes the amount of water retained in a soil under equilibrium at a given matric potential. The soil matric potential is the energy with which the water is held in the soil. The soil–water retention curve, being one of the main hydraulic properties, expresses the relationship between the matric potential and the water content of a soil under equilibrium. It can be considered as the soil's fingerprint since the shape of the curve is related to various physical and chemical soil properties, which are unique for each soil. However, the soil–water retention curve for a given soil is not unique, which means that a soil can have at least two distinct curves depending on whether the curve is derived from an initially saturated soil (desorption curve) or an initially dry soil which is saturated by progressively reducing the suction (sorption curve).

This hydraulic property is affected by morphological properties (e.g., texture and structure), nature and amount of colloidal materials (organic as well as inorganic clay-sized colloids), clay mineralogy, and many chemical properties (e.g., cation exchange capacity, iron content, sodium adsorption ration, calcium carbonate, etc.) (Rawls et al. 1991; Reeve and Carter 1991; Williams et al. 1990; Pachepsky 1989; Lambooy 1984; Baumer and Brasher 1982; Kutilek 1973; Archer and Smith 1972; Sharma and Uehara 1968; Cagauan and Uehara 1965; Ekern 1963; El Ashkar et al. 1956; Croney and Coleman 1954; Childs 1940). The relative importance of these different factors varies across a range of suctions. The amount of water retained in the wet range or low suctions (0–100 kPa) of the curve depends primarily upon the capillary effect and, in non-shrinking soils, on pore size distribution, and hence is strongly affected by the structure of the soils. In sandy soils, which contain many macro pores, the majority of the water is released at low suctions. On the contrary, in clay soils, which contain a large proportion of micro and meso pores, a small amount of water is released at low suction and retains a large proportion of their water (Fig. 7.9). On the other hand, water retentions at high suctions are a function of adsorption and are thus influenced by the specific surface of the soil material, which is dependent on the texture of a soil. Processes like

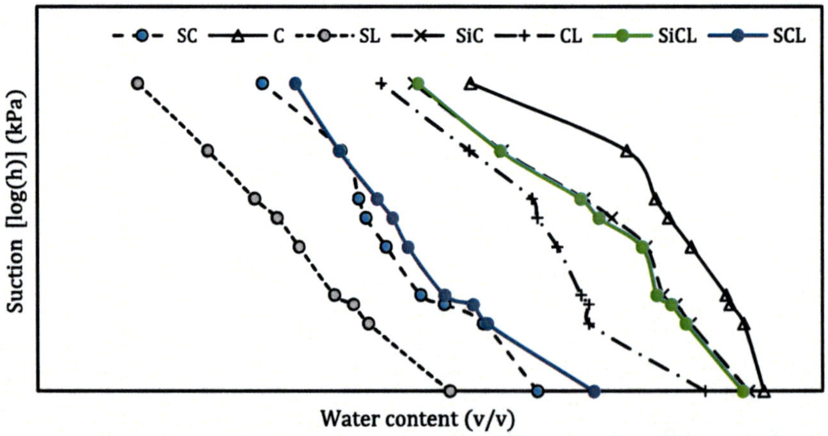

Fig. 7.9 An example of soil–water retention curves of soils with different textural classes (*Source* Own compilation)

compaction and bulk density also affect water retention at low suctions due to their effect on the total pore space of a soil mainly by reducing the amount of large pores, which are responsible for retaining water at low suctions. Soil–water retention is needed for the study of plant available water, infiltration, drainage, hydraulic conductivity, irrigation, water stress of plants, and solute movement (Kern 1995). In non-swell-shrink soils, it is also used to derive information on pore size distribution.

On this curve, at least three points are important. These are saturation, field capacity, and permanent wilting point. Knowledge on water retained at these water potentials is used for different purposes that include, but not limited to, agronomic, engineering, and environmental applications. Saturation refers to a condition where all the pores are filled with water and corresponds to a suction that is close to zero. Saturation results in impeded aeration and, thus, has to be dealt with through drainage in agricultural fields. Furthermore, it can be used to calculate the upper bound or limit of plant available water (see equations below). Field capacity refers to the amount of water that remains in the soil after the excess gravitational water contained in the macro pores has drained away until the drainage becomes negligible, which might take 24–48 h in the absence of evaporation. However, it is well to note that attainment of this condition is affected by many factors that include, but are not limited to, transmission characteristics of the profile, soil stratification, swelling and shrinking, presence of pans, and occurrence of the groundwater table. It is a concept that is acceptable for readily drained soils, but hard to apply for heavy textured soils whose drainage could persist for months. The soil–water suction values generally used for field capacity approximation vary with soil texture, ranging from 5 kPa for coarse-textured soils to 10 kPa and 33 kPa for medium and fine-textured soils (McIntyre 1974).

Permanent wilting point, on the other hand, is defined as the water content at which the leaves of a growing plant reach a stage of permanent wilting from which they do not recover their turgor even if they are put in a saturated atmosphere. Although a suction of 1500 kPa based on wilting studies with dwarf sunflower is generally taken to be an approximation of permanent wilting point (Reeve and Carter 1991), it depends on a number of factors that include climatic conditions (evaporative demand of the atmosphere), soil conditions (water retention and transmission characteristics), and plant species (e.g., rooting density, root depth, and rate of root extension).

For most plants, the water retained between field capacity and permanent wilting point is called plant available water or available water capacity (AWC) although the ease of its extraction becomes increasingly difficult towards the permanent wilting point. This leads to the definition of readily available water (RAW), which refers to the water content between field capacity and a certain critical moisture content where the water is easily available to plants but decreases below this point. The following equations are pertinent:

$$\mathrm{AWC}\left(\frac{mm}{m}\right) = 1000 \times (\theta_{FC} - \theta_{\mathrm{PWP}})$$

$$\mathrm{AWC}_0\left(\frac{mm}{m}\right) = 1000 \times (\theta_s - \theta_{\mathrm{PWP}})$$

$$\mathrm{RAW}\left(\frac{mm}{m}\right) = 1000 \times (\theta_{FC} - \theta_{cr})$$

where:

AWC = available water holding capacity

θFC = water retained at field capacity (volumetric)

θPWP = water retained at permanent wilting point (volumetric)

AWC_0 = available water holding capacity corresponding to saturated water content

θs = water content at saturation (volumetric)

RAW = readily available water

θcr = critical moisture content, which has to be determined experimentally for each soil type, crop, and climate

1000 = conversion factor into mm/m.

If the soil contains n number of layers with their respective water retention, the total values of the respective parameters are calculated as the sum of the individual layers' values as follows:

$$\mathrm{TAWC} = \sum_{i=1}^{n} \mathrm{AWC}_i$$

$$\mathrm{TAWC}_0 = \sum_{i=1}^{n} \mathrm{AWC}_{0i}$$

$$\mathrm{TRAW} = \sum_{i=1}^{n} \mathrm{RAW}_i$$

where T stands for total, n represents the number of layers, and i represents the ith layer.

The rationale behind calculating AWC_0 is that in soils with low soil permeability at the deeper layers such as the B horizon, available water above the field capacity can often be utilized by plants and this value indicates the upper bound of available water capacity. Among other factors, available water capacity of a soil is influenced by soil texture. Generally, medium-textured soils that contain appreciable silt have higher available water capacity than either the coarse- or fine-textured soils. This is due to both relatively higher field capacity and lower PWP water content of medium-textured soils than the coarse- or the fine-textured soils. In the coarse-textured soils, the water retained at both field capacity and permanent wilting point is low, while in the fine-textured soils water retained at both FC and PWP is high.

Determination of soil–water retention at representative suction points to generate a representative SWR curve is not a common laboratory activity in Ethiopia. As a result, only limited studies (most of them student studies) have determined the SWRC of limited soils. Those that have measured it generated only the desorption curve. Results of the few studies that have determined the SWRC indicate the variation in the shape of the curve and water retained at any given potential owing to differences among soils in their salient characteristics and management that affect the relationship. Contributing factors include variations in particle size distribution, clay type, soil structure which affects pore geometry and size distribution, soil organic matter content, bulk density, and in some instances the content of silt plus clay. Some soils with very high clay content were found to retain a high amount of water at the suction points considered, while others with low clay and high sand contents retained relatively low water content. For the fine-textured soils, the slopes of the curves were very gradual or gently, while they are very steep for coarse-textured soils dominated by macro pores. The clear effect of soil texture on SWRC is illustrated in Fig. 7.9.

As discussed above, available water capacity is affected by, among other factors, texture of a soil. Data obtained from surface layers of 85 profiles opened in different parts of the country is presented in Table 7.2. The variations in AWC within the same textural class indicate the contributions of other factors such as clay type (clay mineralogy), organic matter content, and soil structure (bulk density, pore size distribution).

Table 7.2 Range of AWC for soils of different textural classes

Textural class	Range of AWC (mm/m)
Clay	62.50–297.20
Clay loam	43.70–193.20
Loam	82.40–106.73
Sandy clay	75.00–130.40
Sandy clay loam	87.00–167.90
Silty clay	100.99–148.00
Silty clay loam	133.00–170.00
Sandy loam	82.80–131.00

7.4 Soil Chemical Properties

7.4.1 Principles

Chemical Properties of soils are those characteristics, processes or reactions that are caused by mainly chemical activities and that can be described by or expressed in chemical terms or equations. Soil chemical properties are mainly described by soil parameters like soil reaction, cation exchange capacity (CEC), base saturation, cation/anion exchange processes, soil salinity and soil sodicity. Soil chemical properties affect various activities in the soil including nutrient retention, availability and uptake, nitrogen dynamics, soil biological activity and decomposition of organic matter. Hence, the aforementioned parameters are addressed in this sub-chapter.

Soil reaction is an indication of soil solution's (soil water together with its dissolved substances) acidity or alkalinity. It is measured in pH units indicating the concentration of hydrogen ions (H^+). It is described by the negative logarithmic scale of the hydrogen ion concentration [H^+], i.e. pH = −log [H^+]. Soil pH is measured on a scale from 0 to 14 and acidic solutions have a pH less than 7, while basic or alkaline solutions have a pH greater than 7. Exchangeable acidity and the amount of acid cations (aluminium and hydrogen) contained in the CEC are also used to determine soil acidity for liming purposes. Exchangeable acidity is measured in centi-mol of charges per kg ($cmol_c\ kg^{-1}$).

Acidic conditions occur in soil with parent material high in elements such as silica (rhyolite and granite), high levels of sand with low buffering capacities (ability to resist pH change), and in regions with higher amounts of precipitation (Foster et al. 2016). For example, granitic soils are acidic, and limestone-based soils are alkaline. However, soil pH can change over time and soils become acidic through natural processes as well as human activities. Rainfall and climatic conditions influence the pH of most soils. In humid climates, heavy rainfall percolates through the soil and leaches basic ions, such as calcium and magnesium, and replaces them with acidic ions such as hydrogen and aluminium. Conversely, alkaline soils are developed in arid and semi-arid temperate-zone regions where precipitation is less than potential evapotranspiration with annual rainfall less than 500 mm. In this situation, salts accumulate at the surface during evaporation as rainfall is not heavy enough to leach basic ions from soils in these areas (Foster et al. 2016). Hydrolysis of the carbonates (CO_3^{2-}) and bicarbonates (HCO_3^-) of basic cations like Ca^{2+}, Mg^{2+}, K^+ and Na^+ is the cause of soil alkalinity.

Cation exchange capacity (CEC) is the maximum quantity of total cations that a soil is capable of holding, at a given pH value, available for exchange with the soil solution (FAO 2021). CEC is used as a measure of soil fertility, nutrient retention capacity, and the capacity to protect groundwater from cation contamination. In other words, it is the ability of the soil's negatively charged colloids to hold onto nutrients and prevent them from leaching beyond plant roots (Chapman 1965). It is expressed as centimol of charges per kg ($cmol_c\ kg^{-1}$). Clay soils retain more nutrients than coarser/sandier soils, just as clay soils hold more water, because of the greater negatively charged surface area (greater number of cation exchange sites) to which cationic nutrients can be adsorbed. Organic matter also has enormous negatively charged sites that attract and hold positively charged particles. Thus, sandy soils rely on organic matter content to increase cation exchange capacity (Foster et al. 2016). The fraction of the base-forming cations (Ca^{2+}, Mg^{2+}, K^+, Na^+) that occupies positions on the soil colloids is called the base saturation percentage (FAO 2021). Landon (1991a, b) classified CEC ($cmol_c\ kg^{-1}$) into five categories: <5 = very low, 5–15 = low, 15–25 = medium, 25–40 high, >40 = very high. Similarly, Landon (1991a, b) classified base saturation (%) into <20 = low, 20–60 = medium, and >60 = high.

Soil salinity is a measure of the concentration of all the soluble salts in soil water, measured as electrical conductivity (EC) and expressed in deci-Siemens per metre ($dS\ m^{-1}$). The major soluble mineral salts are made from the cations: sodium (Na^+), calcium (Ca^{2+}), magnesium (Mg^{2+}), potassium (K^+) and the anions: chloride (Cl^-), sulfate (SO_4^{2-}), bicarbonate (HCO_3^-), carbonate (CO_3^{2-}), and nitrate (NO_3^-) (Zaman et al. 2018; Foster et al. 2016). Soils with electrical conductivity of extract from a saturated paste (ECe) that equals or exceeds 4 $dS\ m^{-1}$ at 25 °C are said to be saline (USSLS 1954). FAO (1988) also classified soil salinity for the same extraction method as non-saline (0–2), slightly saline (2–4), moderately saline (4–8), strongly saline (8–16) and very strongly saline (>16).

The sources of salts include saline parent materials, particularly weathering of soluble primary minerals, fossil salts of former marine and lacustrine deposits, atmospheric deposition, collection of saline sediments in catchment areas, irrigation waters and/or fertilization (Kidane et al. 2006). Salt affected soils are most common in arid and semi-arid regions, where evaporation exceeds precipitation and dissolved salts are left behind to accumulate, or in areas where removal of vegetation cover or excessive surface irrigation caused salts to leach and accumulate in low-lying places (Foster et al. 2016). The high salt concentration negatively affects soil's physical and chemical properties as well as soil microbial activity, thus causing a decline in soil productivity.

For agricultural purposes, salt affected soils are regarded as a class of problem soils that require special remedial measures and management practices. This is because the salt constituents of such soils produce harmful effects to plants by increasing the salt contents of the soil solution and the degree of saturation of the soil exchange complex by sodium (Kidane et al. 2006). Soil salinity affects plants through the osmotic effect. Soil water flows from higher osmotic potential (low-salt concentration) to lower osmotic potential (high-salt concentration). A soil solution with low osmotic potential due to the higher concentration of soluble salts compared to plant cells, will not allow plant roots to extract water from the soil, causing drought-like symptoms in the plants (Seelig 2000). The other problem associated with soil salinity is the specific ion effect. Certain ions like Na^+, Cl^-, $H_4BO_4^-$ and HCO_3^- are quite toxic to many plants and high levels of Na^+ compete for the uptake of essential nutrients like K^+.

Sodicity refers to an excess of exchangeable sodium in the soil (FAO 2021). Sodic soils tend to occur generally within arid to semi-arid regions and are often unstable, exhibiting poor physical and chemical properties, which impede water infiltration, water availability, and ultimately plant growth. Sodicity is measured by the Na^+ amount as sodium adsorption ration (SAR) or exchangeable sodium percentage (ESP). Soils with SAR of greater than 13 and ESP of above 15% are considered sodic (Brady and Weil 2015). Similarly, FAO (1988) classified soil sodicity based on ESP (%) as none to slight (<15), light to moderate (15–30), moderate to high (30–50), high to very high (50–70) and extremely high (>70).

Soil sodicity causes degradation of soil structure through the process called soil dispersion. The forces that hold clay particles together are greatly weakened when excessive sodium (monovalent cation) is adsorbed at the negative charges of clay particles, forming sodium-clay particles (Seelig 2000). Whereas, wet, sodium-clay particles get easily disintegrated or dispersed from the larger soil aggregates. Once dry, sodium-clay particles clog the soil pores (especially macro pores) and settle down in dense layers. The resulting poor physical structure then results in soils difficult to till, poor seed germination and restricted plant root growth (Foster et al. 2016). The adverse effect of exchangeable sodium on the growth and yield of crops in various classes occurs according to the relative crop tolerance to excess sodicity. Whereas the growth and yield of only sensitive crops are affected at ESP levels below 15, only extremely tolerant native grasses grow at ESP above 70 (FAO 1988). Saline sodic soils have both high soluble salts and sodium contents. These soils are affected by problems linked with both salinity and sodicity.

7.4.2 Soil Reaction (Acidity/Alkalinity)

Soil acidity affects the growth of crops because acidic soil contains toxic levels of aluminium and manganese and a

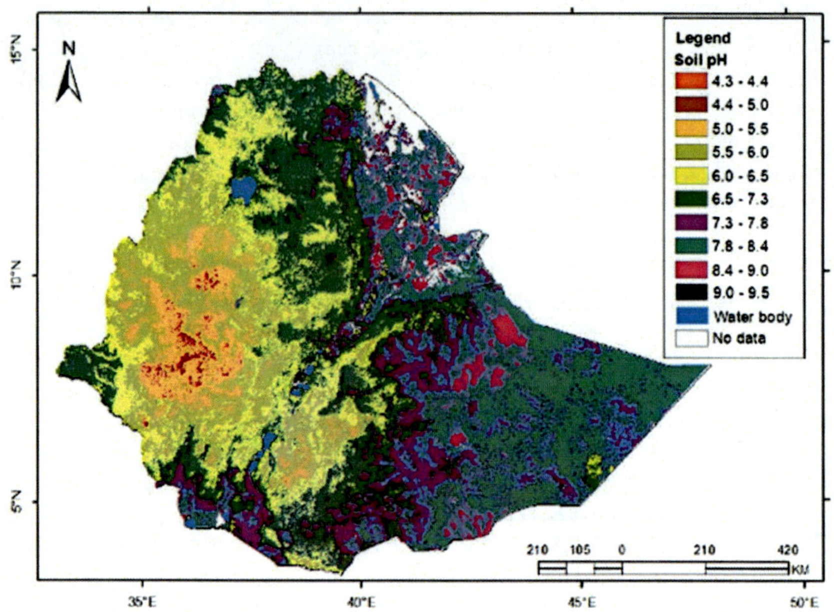

Fig. 7.10 Spatial distribution of pH in Ethiopian soils. (*Source* Map extracted and own compilation for top 20 cm depth by author from SoilGrids 2017: https://files.isric.org/soilgrids/former/2017-03-10/data/)

deficiency of essential plant nutrients such as P, N, K, Ca, Mg, and Mo (Wang et al. 2006). Soil reaction affects nutrient availability (Foster et al. 2016). Nitrogen, potassium, calcium, magnesium and sulfur are more available within soil pH of 6.5–8; while boron, copper, iron, manganese, nickel, and zinc are more available within soil pH of 5–7. At pH less than 5.5, high concentrations of H^+, aluminium and manganese in soil solution can reach toxic levels and limit crop production (Brown et al. 2008; Kidd and Proctor 2001). Phosphorus is most available within soil pH of 5.5–7.5. Hence, soil acidity severely affects the yields of many crops in high rainfall areas of Ethiopia (Angaw and Desta 1988) where the annual rainfall exceeds the potential evaporation (Wegene 2019). The major soil forming factors and management practices giving rise to the increase in soil acidity in the country involve climatic factors such as high rainfall, temperature, topographic factors, soil parent materials, intensive mono-cropping and lack of technological inputs in the agricultural sector to mitigate the problem (Mesfin 1998). About 40.9% of the Ethiopian total land is affected by soil acidity (Wegene 2019; Wassie and Shiferaw 2009; World Bank 1995). Of this area, about 27.7% is dominated by moderate to weak acid soils (pH in KCl of 4.5–5.5), and around 13.2% by strong acid soils (pH in KCl of <4.5) and one-third has aluminium toxicity problems (Mesfin 2007).

Soil reactions in Ethiopian soils can be broadly grouped into three categories. The soils in the highlands of north-western, western and central parts of Ethiopia (Gojjam, Wollega, Ilu Ababor, Wolayta, Sidama, Gedeo, Arsi and Bale) are predominantly acidic. The Rift Valley, the lowlands of eastern and southern Ethiopia, arid dry highlands of Tigray are dominated by alkaline soils (Fig. 7.10). The third category could be soils with neutral pH found in pocket areas linked to specific topographic positions and parent materials scattered all over the country. The share of different pH levels on Ethiopian soils has shown that about 24.38% of the total land area has neutral pH ranging from 6.6 to 7.3 followed by alkaline soils (23.58%) ranging from 7.9 to 8.4 and slightly acidic soils (14.98%) with pH ranging from 5.6 to 6.5. The two extreme soil pH values (<4.4 and >9) are represented the least (Table 7.3).

Distribution of pH in different soils is presented in Table 7.3. Acrisols are known to be one of the soils with highly acidic reactions. Selassie (2002) reported pH values as low as 4.81 in the Injibara area (Amhara Region). Nitisols are the other soils with acidic soil reactions that are abundantly available in Ethiopia. Eyasu (2016) reported that 80% of the Nitisols and Luvisols subgroups, found in the north-central and south-western highlands of Ethiopia, are very strongly to strongly acidic having pH of 4.5–5.5. Moreover, more than 80% of the landmasses originating from Nitisols could be acidic in nature, partly because of the leaching of basic cations (Eyasu 2016). Other reports indicated that Nitisols are moderately and strongly acidic in many places ranging from 4.82 in Merawi, Amhara Region (Selassie 2002) to 6.6 in Munesa eucalyptus plantation, SNNPRS (Betre et al. 2000). Wondwosen (2015) reported that dystric Nitisols of Bule District of Gedeo zone, SNNPRS and Nitisols of Burie District, West Gojjam Zone, ANRS (Selassie 2002) had low pH ranging from 4.6 to 4.75. High acidity with pH ranging from 4.49 to 5.2 and exchangeable acidity of 3.83 $cmol_c\ kg^{-1}$ were also reported by Isreal et al.

Table 7.3 Spatial distribution of pH in Ethiopian soils based on SoilGrids of ISRIC (2017)

Soil pH	Area	
	km^2	%
<=4.4	0.19	0.0000165
4.5–5.0	6,238.00	0.55
5.1–5.5	60,804.00	5.37
5.6–6.0	169,801.00	14.98
6.1–6.5	139,305.00	12.29
6.6–7.3	276,226.00	24.38
7.4–7.8	138,529.00	12.22
7.9–8.4	267,202.00	23.58
8.5–9.0	38,521.00	3.40
>9.0	411.00	0.04
No data	36,151.38	3.19
Total	1,133,190.00	100.00

(2018) in Nitisols of Abelo area Masha district, South-Western Ethiopia.

Betre et al. (2000), in their comprehensive study that focused on the effect of tree cover on soil pH, indicated that the type of forest cover affects the pH level of Nitisols. Soils covered by primary and indigenous broadleaf trees had significantly ($p < 0.05$) higher pH values (7.28) than the conifer species covered by *C. lusitanica* and *P. patula* (5.41–5.86) and eucalyptus plantation (6.02–6.60). Acidic soil under conifers trees and eucalyptus plantation was attributed to the significant leaching of basic cations and insufficient supply of cations from the plantation stands as leaf drops compared to the diverse species of the natural forest. Lower pH values under eucalyptus plantations have also been observed in other tropical countries like Tanzania (Lundgre 1978) and India (Balagopalan et al. 1991).

Luvisols in Ethiopia are known to have a moderately acidic reaction. These soils are characterized by having an argic horizon starting ≤ 100 cm from the soil surface (IUSS 2015). Selassie (2002) reported a pH value of 5.7 at Debretabor Luvisols (Amhara Region) and Shimelis et al. (2007) showed a pH value of 5.6 for similar soils at Tehocha-Wenchaser, South-west Shewa (Oromiya Region).

Cambisols, soils characterized mainly by their cambic horizon (IUSS 2015), are known to have varied pH values in different circumstances. The available literature has indicated pH values of Cambisols ranging from 4.14 in Tsekede area, northern Ethiopia (Abreha et al. 2012) to 8.38 at Sheneko irrigation farm of Bale Mountains (Mohammed and Solomon 2012). Endalkachew et al. (2018a, b) also mentioned that the pH of the Cambisols was within the range of very strongly acidic in the surface layer (4.5–5.0) to slightly acidic (6.1–6.5) in the bottom layer.

Vertisols have relatively higher pH in many locations because they occur on landscapes with lower slopes ranging from 0 to 8% (Jutsi 1988) and most frequently with slopes between 0 and 2% (Berhanu 1985). This situation is responsible for low surface erosion in Vertisols and the accumulation of eroded basic cations from the upper parts of landscapes. Soil pH values varying from 5.36 in Vertisols of Pawe, Ethiopia (Fekadu et al. 2012), 5.8 in Vertisols of Woreta and 6.2 in Vertisols of Bichena, Amhara Region (Selassie 2002), 8.7 in Bedessa, Chercher Highlands (Eylachew 1999) and 8.0–9.0 in Raya (Abayneh 2001) were reported. Similarly, slightly acidic pH values for Ginchi (6.31) and for Boche and Boroda (6.06–7.03) were reported for Vertisols of the Oromia Region (Getachew and Amare 2004; Amsal and Douglas 2001).

Andisols are soils mainly located in the Rift Valley and the highlands of Mount Ras Dashen and Guna. The Andisols of the Rift Valley around Ziway are characterized by high pH values ranging from 7.2 to 8.0 (Eylachew 2004). High pH values were the reflection of the sufficient amount of calcium that surpassed the formation of a large quantity of Al and Fe oxides. Moreover, the good buffering capacity of the minerals which has kept pace with the production of acidity, made the pH of the soils not to significantly drop.

Calcisols, with more than 15% calcium carbonate equivalent, were reported to have high pH (7.89) in the Sheneko irrigation farm of Bale Mountains with an increased trend with depth (Mohammed and Solomon 2012). The reason for the increase in pH with depth was due to leaching and accumulation of calcium carbonate down the soil depth. High pH values were recorded in Regosols (8.1) and Leptosols (6.5–7.7) of the Gerado catchment (Asmamaw and Mohammed 2012). Abayneh (2001) also reported alkaline soil reaction with pH values ranging from 7.4 to 9.0 in Raya Valley Fluvisols. Mohamed and Tessema (2013) reported a pH of 8.3 (strongly alkaline) in surface soils of Awash River Basin at Fursa small scale irrigation farm, Afar National Regional State, north-eastern Ethiopia. Based on the results of the study in Babile low lands, Eastern Ethiopia, moderately alkaline to alkaline soil reactions (7.89–8.54) were reported (Assefa et al. 2019). Reports of Kidane et al. (2006) showed that Melka Sedi-Amibara Plain sampled on an abandoned cotton field had a pH of 7.2

Regarding exchangeable acidity, the available literature is scanty. However, in soils studied, exchangeable acidity values are low (Table 7.4) except for Acrisols of Injibara (Selassie 2002), Nitisols of Masha (Iseael et al. 2018) and Nitisols of Illu Ababor area (Jafer and Gebresilassie 2017). Endalkachew et al. (2018a, b) also mentioned that

Table 7.4 Status of pH and exchangeable acidity in different Ethiopian soils

Soil group	pH (H_2O)	Exchangeable acidity (cmol$_c$ kg^{-1})	Location	Source
Acrisols	4.81	2.40	Injibara, ANRS	Selassie (2002)
Nitisols	5.38	0.24	Adet, ANRS	Selassie (2002)
	4.82	0.56	Merawi, ANRS	
	5.33	0.44	Finoteselam, ANRS	
	4.97	0.32	Mota, ANRS	
	5.5	Na	Adet, ANRS	Asgelil Dibabe (2000)
	5.38–5.48	Na	Adet, ANRS	Minale et al. (2001)
	5.8	Na	Tenocha-Wenchasher, South-West Shewa, ONRS	Shimelis et al. (2007)
	6.2–7.5	Na	Bedessa, Chercher Highlands	Eylachew (1999)
	5.93	Na	Bekoji, Arsi	Yesuf Assen and Duga Debele (2000)
	7.28	Na	Munessa primary forest, Southern Ethiopia	Betre et al. (2000)
	5.41–5.86	Na	Munessa conifer plantation, Southern Ethiopia	Betre et al. (2000)
	6.02–6.60	Na	Munessa eucalyptus plantation, Southern Ethiopia	Betre et al. (2000)
	5.4	Na	Munessa primary forest, Southern Ethiopia	Yeshanew Ashagrie and Wolfgang Zech (2013)
	5.3	Na	Munessa eucalyptus forest, Southern Ethiopia	Yeshanew Ashagrie and Wolfgang Zech (2013)
	4.5–6.0	Na	Negade Soro and Mengagesha, Central Ethiopia	Amsal and Douglas (2001)
	4.55	Na	Bako, Central Ethiopia	Girma and Ravishankar (2004)
	4.5–6.0	Na	Boche, Central Ethipia	Amsal and Douglas (2001)
	4.49–5.2	3.83	Masha District, South-western Ethiopia	Iseael et al. (2018)
	4.6	3.81	Illu Ababor, ONRS	Jafer and Gebresilassie (2017)

(continued)

Table 7.4 (continued)

Soil group	pH (H_2O)	Exchangeable acidity ($cmol_c$ kg^{-1})	Location	Source
Luvisols	5.17	0.28	Debretabor, ANRS	Selassie (2002)
	5.60	Na	Tenocha-Wenchasher, South-west Shewa, ONRS	Shimelis et al. (2007)
Cambisols	5.90	0.11	Adet, ANRS	Selassie (2002)
	5.20	Na	Tenocha-Wenchasher, South-west Shewa, ONRS	Shimelis et al. (2007)
	6.0	Na	Gerado, Wollo	Asmamaw and Mohammed 2012
	5.93	0.28	Dabat, ANRS	Selassie (2002)
	6.72–8.38	Na	Sheneko irrigation farm of Bale Mountains	Assen and Solomon Tadesse (2012)
	4.14–6.33	1.36–2.58	Tsekedea, Tigray	Abreha et al. (2012)
	4.5–6.5	1.6	Lay Gayint, ANRS	Endalkachew et al. (2018a)
Vertisols	5.8	0.16	Woreta, ANRS	Selassie (2002)
	6.03	0.28	Bichena, ANRS	
	8.0–9.0	Na	Raya valley, TNRS	Abayneh (2001)
	8.0–8.7	Na	Bedessa, Chercher Highlands	Eylachew (1999)
	5.36–5.17	Na	Pawe, Ethiopia	Fekadu et al. (2012)
	6.31	Na	Ginchi, ONRS	Getachew and Amare (2004)
	6.06–7.03	Na	Boche and Borodo, Central Ethipia	Amsal and Douglas (2001)
Chromic Vertisols	8.1	Na	Lower Wabi Shebelle River Valley	Kidane et al. (2006)
Calcaric Gleysols	7.9	Na	Lower Wabi Shebelle River Valley	Kidane et al. (2006)
Andosols	7.2–8.0	Na	Central Rift Valley	Eylachew (2004)
Calcisols	7.89	Na	Sheneko irrigation farm of Bale Mountains	Assen and Solomon Tadesse (2012)
Plinthosols	6.0	Na	Tenocha-Wenchasher, South-west Shewa, ONRS	Shimelis et al. (2007)
Leptosols	5.6	Na	Tenocha-Wenchasher, South-west Shewa, ONRS	Shimelis et al. (2007)
	6.5–7.7	Na	Gerado, Wollo	Asmamaw and Mohammed 2012
Fluvisols	7.4–8.4	Na	Raya valley, TNRS	Abayneh (2001)
Lithosols	7.3–8.2	Na	Kuni, Chercher Highlands	Eylachew (1999)
Regosols	8.1	Na	Gerado, Wollo	Asmamaw and Mohammed 2012
Solonchacks) (Saline Soil	7.7	Na	Mekelle	Kidane et al. (2006)

(continued)

Table 7.4 (continued)

Soil group	pH (H_2O)	Exchangeable acidity (cmol$_c$ kg^{-1})	Location	Source
Solonetz (Saline sodic)	7.2	Na	Melka Sedi-Amibara Plain	Kidane et al. (2006)
Solonetz (Sodic soil)	7.8	Na	Borkena plain	Kidane et al. (2006)
Solonetz (Sodic soil)	8.4	Na	Humera, NW Ethiopia	Kidane et al. (2006)

Soil pH categories: Ultra acid <3.5; Extremely acid 3.5–4.4; Very strongly acid 4.5–5.0; Strongly acid 5.1–5.5; Moderately acid 5.6–6.0; Slightly acid 6.1–6.5; Neutral 6.6–7.3; Slightly alkaline 7.4–7.8; Moderately alkaline 7.9–8.4; Strongly alkaline 8.5–9.0; Very strongly alkaline >9.0 (USDA-NRCS 1998). Exchangeable acidity categories for Ethiopia (Cmol$_c$ kg^{-1}) = 0–2 = low; 2–4 = medium; 4–6 = high; 6–8 = very high; >8 = extremely high; Na = data not available

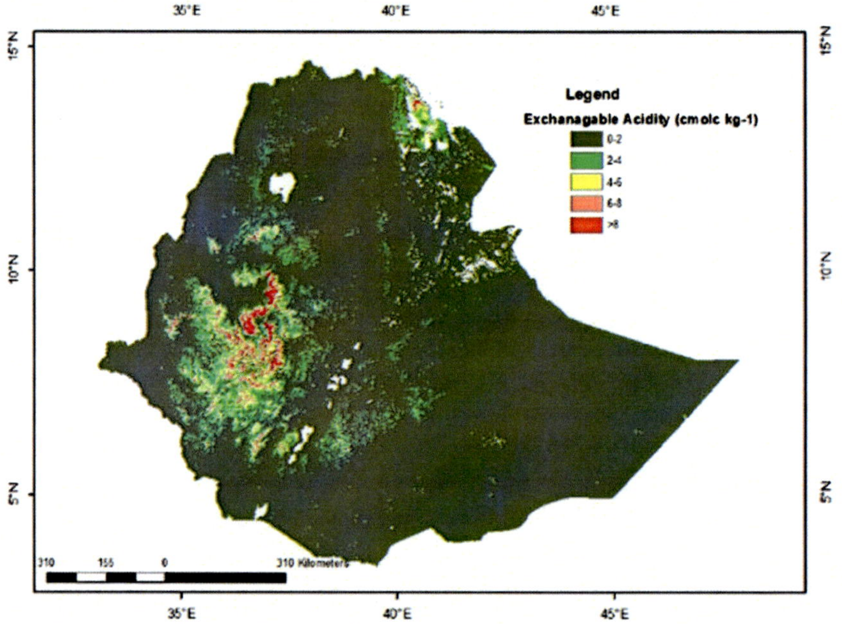

Fig. 7.11 Spatial distribution of exchangeable acidity in Ethiopian soils based on soil property maps of Africa at 250 m resolution (*Source* Map extracted and own analysis for the top 20 cm depth by the author, from soil property maps of Africa at 250 m resolution: https://www.isric.org/projects/soil-property-maps-africa-250-m-resolution)

Cambisols have exchangeable acidity of 1.6 cmol$_c$ kg^{-1}. Regarding the spatial coverage of exchangeable acidity in soils of Ethiopia (Fig. 7.11 and Table 7.4), the majority (90.20%) of Ethiopian soils have low exchangeable acidity levels (0–2 cmol$_c$ kg^{-1}). Low exchangeable acidity is common in soils of Tigray, South Gonder, Wollo, Gambella, the Rift Valley, and eastern, southern and western lowlands. Based on the categorization of exchangeable acidity, about 5.86% have medium exchangeable acidity levels (2–4 cmol$_c$ kg^{-1}) and 2.53% have high exchangeable acidity levels (4–6 cmol$_c$ kg^{-1}). Soils of south-western Ethiopia have very high (6–8 cmol$_c$ kg^{-1}) and extremely high (>8 cmol$_c$ kg^{-1}) exchangeable acidification levels; however, the coverage is low (1.42%) (see Table 7.5).

7.4.3 Cation Exchange Capacity and Base Saturation

Available data on CEC of Ethiopian soils is presented in Table 4. Higher CEC values were reported for Vertisols (Selassie 2002), Fluvisols (Abayneh 2001), Cambisols (Asmamaw and Mohammed 2012), Leptosols (Asmamaw and Mohammed 2012) and Regosols (Asmamaw and

Table 7.5 Spatial distribution exchangeable acidity in Ethiopian soils (0–20 cm depth) based on SoilGrids of ISRIC (2017)

Exchangeable acidity class* (Cmol$_c$ kg^{-1})	Area	
	km^2	%
0–2 (low)	994,987.60	90.20
2–4 (medium)	64,610.75	5.86
4–6 (high)	27,863.75	2.53
6–8 (very high)	11,049.44	1.00
>8 (Extremely high)	4,597.94	0.42
	1,103,110.00	100.00

Mohammed 2012). Conversely, CEC is low in Acrisols and Nitisols (Selassie 2002).

Vertisols, developed under flat land conditions, are known to have high CEC (Abayneh 2001). Shimelis et al. (2007) reported that Umbrisols, recognized by gleying properties with abundant course prominent diffuse and reddish brown mottles, have high CEC (48.6 cmol$_c$ kg^{-1}). High CEC (59.00 cmol$_c$ kg^{-1}) was recorded in Regosols of Gerado catchment due to the presence of expansible minerals as well as earlier stages of soil development (Asmamaw and Mohammed 2012). The same report indicated that Leptosols also had high CEC (52.00 cmol$_c$ kg^{-1}).

Soils covered by broadleaf natural forests were reported to have higher CEC than soils under coniferous and eucalyptus plantations in Nitisols of Munessa forest (Betre et al. 2000). Balagopalan et al. (1991) also reported that the litter from eucalyptus could be comparatively difficult to decompose due to its chemical constituents, which might also hinder activities of the decomposing microorganisms, survival and growth of underground herbaceous vegetation. However, as the age of the natural forest and plantations increases, CEC and base saturation also increase which suggests that as trees get older, they tend to drop more leaves that will be converted to organic matter and increase CEC and base saturation (Betre et al. 2000). CEC ranging from 36.0 to 63.0 cmol$_c$ kg^{-1} was reported by Endalkachew et al. (2018a) for Cambisols of Lay Gayint District of ANRS. According to Landon (1991a, b), the CEC of the soils in the watershed qualifies for the high and very high signifying that the soils have a better nutrient reserve. Pal and Selassie (2018) also reported CEC ranging from 40.6 to 61.7 cmol$_c$ kg^{-1} in Kesem Allaideghe plains, eastern Ethiopia.

Lower CEC is an attribute of acidic soils like Acrisols (Selassie 2002), Nitisols (Girma and Ravishankar 2004) and Plinthosols (Shimelis et al. 2007). Isreal et al. (2018) also reported low CEC (18.8–21.44 cmol$_c$ kg^{-1}) in Nitisols of Abelo area, Masha District, South-western Ethiopia that was explained by intensive weathering.

The available literature on base saturation indicated no clear association with the soil types of Ethiopia. Similar soils can have contrasting base saturation values in different locations. However, high levels of base saturation are linked to high levels of CEC (Table 7.6). Soils with high CEC have high base saturation in many instances (Asmamaw and Mohammed 2012; Eylachew 2004).

Table 7.6 Status of cation exchange capacity in Ethiopian soills

Soil group	CEC (cmol$_c$ kg^{-1})	BS (%)	Location	Source
Vertisols	45.80–60.20	Na	Raya valley, TNRS	Abayneh (2001)
	39.70	70.22	Woreta, ANRS	Selassie (2002)
	35.65	92.91	Bichena, ANRS	
	44–52	80.00	Gerado, Wollo	Asmamaw and Mohammed 2012
	26–25.2	Na	Pawe, Ethiopia	Fekadu et al. (2012)
	34.8–624	Na	Boche and Borodo, Central Ethipia	Amsal and Douglas (2001)
Cambisols	29.43	99.63	Adet, ANRS	Selassie (2002)
	29.91	99.98	Dabat, ANRS	
	27.60	25.00	Tenocha-Wenchasher, South-west Shewa, ONRS	Shimelis et al. (2007)
	42.00	Na	Gerado, Wollo	Asmamaw and Mohammed 2012
	27.43	Na	Tsekedea, Tigray	Abreha et al. (2012)
	36.02–53.44	19.0–63.0	Lay Gayint, ANRS	Endalkachew et al. (2018a)

(continued)

Table 7.6 (continued)

Soil group	CEC (cmol$_c$ kg^{-1})	BS (%)	Location	Source
Luvisols	28.91	45.19	Debretabor, ANRS	Selassie (2002)
	27.00	31.00	Tenocha-Wenchasher, South-west Shewa, ONRS	Shimelis et al. (2007)
Acrisols	29.16	35.69	Injibara, ANRS	Selassie (2002)
Nitisols	25.32	41.07	Adet, ANRS	Selassie (2002)
	28.47	34.07	Merawi, ANRS	
	28.35	53.74	Finoteselam, ANRS	
	28.59	39.85	Mota, ANRS	
	34.00	38.0	Tenocha-Wenchasher, South-west Shewa, ONRS	Shimelis et al. (2007)
	74.63	78.95	Munessa primary forest, Southern Ethiopia	Betre et al. (2000)
	42.50	42.0–73.75	Munessa conifer plantation, Southern Ethiopia	Betre et al. (2000)
	30.7–38.42	39.03–70.52	Munessa eucalyptus plantation, Southern Ethiopia	Betre et al. (2000)
	51.1	Na	Munessa primary forest, Southern Ethiopia	Yeshanew Ashagrie and Wolfagang Zech (2013)
	37.2	Na	Munessa Eucalyptus forest, Southern Ethiopia	Yeshanew Ashagrie and Wolfagang Zech (2013)
	22.60	35.0	Bako, Central Ethiopia	Girma and Ravishankar (2004)
	44.48	Na	Adet, ANRS	Asgelil Dibabe (2000)
	16.2–33.6	Na	Negado Sororo and Menagesha, Central Ethipia	Amsal and Douglas (2001)
	18.8–21.44	49.69	Masha District, South-western Ethiopia	Iseael et al. (2018)
Plinthosols	27.60	39.00	Tenocha-Wenchasher, South-West Shewa, ONRS	Shimelis et al. (2007)
Leptosols	23.60	30.00	Tenocha-Wenchasher, South-west Shewa, ONRS	Shimelis et al. (2007)
	52.00	100	Gerado, Wollo	Asmamaw and Mohammed 2012
Regosols	59.00	74.0	Gerado, Wollo	Asmamaw and Mohammed 2012
Fluvisols	35–60.3	Na	Raya valley, TNRS	Abayneh (2001)
Salonchaks	54.58	84.48	Shewarobit, ANRS	Okubay (2019)
Solonetz	33.71–49.97	Na	Abaya State farm, Southern Rift Valley, SNNRS	Haile et al. (2000)
Andisols	22.01–33.63	71–100	Central Rift Valley	Eylachew (2004)
Umbrisols	48.6	13	Tenocha-Wenchasher, South-west Shewa, ONRS	Shimelis et al. (2007)

CEC cmol$_c$ kg^{-1} categories: <5 = very low, 5–15 = low, 15–25 = medium, 25–40 high, >40 = very high (Landon 1991a, b). Base saturation (%) categories: <20 = low, 20–60 = medium, >60 = high (Landon 1991a, b). Na = data not available

Results from site specific studies showed that high base saturation (75.0) was recorded in Regosols of Gerado catchment (Asmamaw and Mohammed 2012); Cambisols of Dabat (99.98%), Cambisols of Adet (99.63%) and Vertisols of Bichena (92.91%) (Selassie 2002). Asmamaw and Mohammed (2012) indicated that Leptosols had high base saturation (100%) at Gerado area, Wollo. Andisols of the Central Rift Valley (Ziway area), developed from volcanic ash material, had high base saturation (71–100%) due to the nature of the parent material and high evapotransportation and poor quality of irrigation water that had left high amounts of carbonate and bicarbonate calcium (Eylachew 2004). Another report indicated that soils covered by broadleaf natural forests had high base saturation in Nitisols

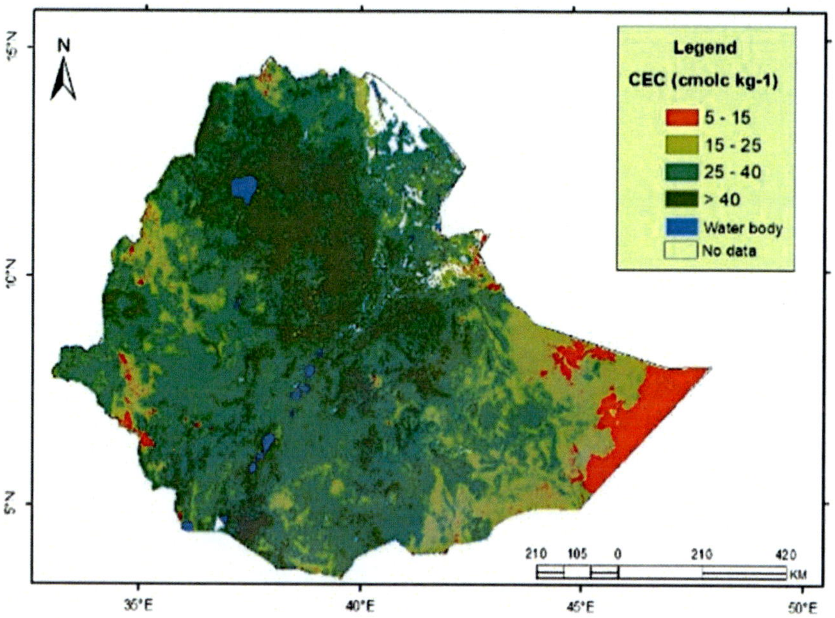

Fig. 7.12 Spatial distribution of CEC in Ethiopian soils based on ISRIC SoilGrids (2017) (*Source* Map extracted and own analysis for the top 20 cm depth by author from SoilGrids 2017: https://files.isric.org/soilgrids/former/2017-03-10/data/)

Table 7.7 Spatial distribution of CEC in Ethiopian soils (0–20 cm depth) based on SoilGrids of ISRIC (2017)

CEC ($cmol_c$ kg^{-1})	Area	
	Km^2	%
5–15	38,021.00	3.36
15–25	216,088.50	19.07
25–40	554,404.38	48.92
>40	288,438.13	25.45
No data	36,238.00	3.20
Total	1,133,190.00	100.00

of Munessa forest (Betre et al. 2000). Solonchaks of Shewarobit were also reported to have high base saturation (Okubay 2019).

Relatively lower base saturation levels were reported for Cambisols (25%) and Luviosls (31%) of Tenocha-Wenchasher, South-west Shewa, ONRS (Shimelis et al. 2007). Contradicting the report of Eylachew (2004), Mizota and Chapelle (1988); Ping et al. (1988) showed that Andisols generally have low base saturation. Umbrisols were reported to have low base saturation of 13% (Shimelis et al. 2007). Moreover, a base saturation value of 49.69% was reported in Nitisols of Abelo area, Masha District, South-western Ethiopia (Iseael et al. 2018). The principal cations occupying the exchange site were in the order of $Ca^{2+} > Mg^{2+} > K^+ > Na^+$ (Selassie 2002; Alem et al. 2015; Endalkachew et al. 2018a, b).

Generally, Landon (1991a, b) showed that CEC > 40 $cmol_c$ kg^{-1} and BS > 60% is considered very high and high, respectively. Similarly, FAO (1999) reported that soils with base saturation of >50% are regarded as fertile soils. Moreover, Asmamaw and Mohammed (2012) indicated that CEC above 40 $cmol_c$ kg^{-1} and base saturation above 80% are considered to be high in Ethiopia.

Looking at the geographic distribution of CEC, 48.92% of the soils of the country have high, while 25.45% have very high and 19.07% have medium CEC levels. Only 3.36% of the soils have low CEC levels (Fig. 7.12 and Table 7.7).

7.4.4 Soil Salinity and Sodicity

Soil salinity is one of the major land degradation problems in Ethiopia. The original source of soluble salts in the salt affected soils of Ethiopia is weathering of Na, Ca, Mg and K rich igneous rocks and their primary minerals occurring in the volcanic regions of the country (Heluf 1985, 1987, 1995). These include granites, feldspars and alumina-silicates of sodium and potassium, hyper alkaline silicic lavas, alkaline olivine-and dolerite-andesite basaltic magmas, carbonate, volcanic ash, tuff, pumice, and rhyolite parent materials.

The naturally salt affected areas are normally found in the arid and semi-arid lowlands and in the Rift Valley and other areas that are characterized by higher evapotranspiration rates in relation to precipitation (PGRC 1996; Paulos 2002; Geressu and Gezaghegne 2008; Okubay 2019). Hey are distributed in the lowlands of Ethiopia where the ratio of precipitation to evaporation is less than 0.75. This low rainfall amount cannot leach down the saline materials like chlorides, sulfates, carbonates and bicarbonates of Ca and Mg. Low levels of annual rainfall and high daily temperatures consequently contribute to high concentrations of soluble salts in lowland areas (Bekele 2005; Tenalem 2007; Sileshi 2015).

The soil salinity problems in Ethiopia also stem from the use of poor quality water coupled with the intensive use of soils for irrigation agriculture, poor on-farm water management practices, lack of adequate drainage facilities and shallow inherent saline ground water levels (Gebremeskel et al. 2018). Discharge to the groundwater by surplus irrigation water has caused a rise in the water table (0.5 m/year) in Middle Awash irrigated fields and caused secondary salinization in surface and subsurface soil horizons (Taddese et al. 2003). Another source of salinity for rivers and other sources of irrigation water is attributed to salts of marine origin. During the rainy season, water quality of River Wabi Shebelle for irrigation deteriorates as a result of very high flooding which dissolves soluble salts from parent materials of loose marine origin along its course (Merga and Ahmed 2019; Taddese 2001).

Nearly 50% of the country's land area is regarded as a marginal environment for crop production mainly due to soil and water salinity (Qureshi et al. 2018). In Ethiopia, about 44 million ha (36% of the total land area) is potentially susceptible to salinity problems of which 11 million ha have already been affected by different levels of salinity (Merga and Ahmed 2019). The salt affected soils are mainly concentrated in the Rift valley and rank 7th in percent of the total land of salt affected soil area among the various countries in the world (Tadelle 1993; Ruffeis et al. 2008; Qureshi et al. 2018; Merga and Ahmed 2019). Ethiopia stands first in Africa in the extent of salt affected soils in Africa with over 10,608,000 ha (ICBA 2018a). Okubay (2019) also reported similar results indicating that the total land area covered by salt affected soils in Ethiopia is estimated at about 11,033,000 hectares, and salinity occurs for the most part in the Rift Valley zone, where groundwater has been used as a source of irrigation water. FAO (1988) also reported the presence of 10,608,000 ha of salt affected soils in Ethiopia.

Results of specific studies (Table 7.8) indicated that salinity and sodicity problems in the Awash basin, which accounts for about one-third of the total irrigated area of the country, caused severe land degradation (Ruffeis et al. 2008; Paulos 2002). The soils of the Melka Sedi-Amibara Plain of the Middle Awash Valley are reported to be highly saline with ECe ranging from 16 to 18 dS m^{-1} (Qureshi et al. 2018). Sardo (2005) has also reported that about 80% of Dubti/Tendaho State farm is affected by soil salinity (that is, 27% saline, 29% saline sodic and 24% sodic soils). The main reason for this was the increase in groundwater levels due to excessive irrigation. Moderate salinity status (0.63–7.75 dS m^{-1}) was reported for surface soils of Awash River Basin at Fursa small scale irrigation farm, Afar National Regional State, North-eastern Ethiopia (Mohamed and Tessema 2013).

Salinization has been a major constraint related to irrigation agriculture in the country (Merga and Ahmed 2019), especially in the Awash valley (Tenalem 2007). For example, soil salinity has caused the abandonment of banana plantations in Amibara, cotton plantations in Melka Sedi and nearly 30 ha of farmland in Metahara sugar plantations due to a progressive rise of groundwater as a result of over irrigation (Abebe et al. 2015; Fentaw 1996; Tenalem 2007). Asfaw and Itanna (2009) also indicated that from the entire Abaya State Farm, 30% has already been salt affected.

Table 7.8 Spatial distribution of saline soils in four National Regional States of Ethiopia

EC (dS/m)	Afar NRS		Amhara NRS		Oromia NRS		Tigray NRS	
	km^2	%	km^2	%	km^2	%	km^2	%
Non-saline/water body/rock outcrop (<2)	40,787.09	41.96	137,421.72	88.29	287,768.25	88.70	48,066.77	97.29
Low saline (2–5)	26,915.98	27.70	4,902.92	3.15	17,292.05	5.33	0	0
Medium saline (5–10)	9,798.23	10.10	11,891.52	7.64	17,152.54	5.29	1339.20	2.71
High saline (10–15)	5,618.43	5.78	1,229.62	0.79	1,576.72	0.49	0	0
Extreme saline (>15)	14,084.96	14.50	202.34	0.13	713.74	0.22	0	0
Total	97,204.70	100.00	155,648.12	100.00	32,4428.69	100.00	100.00	100.00

Source ICBA (2018a, b, c, d)

Results of the study at Sego Irrigation Farm, South Ethiopia, indicated that 2.8% was strongly saline, 39.5% was moderately saline, 31.2% of the land area was slightly saline and the rest (26.6%) was non-saline (Shegena et al. 2017). From these, the areas most affected by salinity were on shallow water tables covered by Cambisols and Fluvisols. Pal and Selassie (2018) also reported ECe ranging from 0.9 to 8.0 dS m^{-1} in the Kesem Allaideghe plains irrigation project area, eastern Ethiopia.

In terms of area coverage, Afar National Regional State is the most salinity affected region in the country (ICBA 2018a). It is reported that around 30% of the soils of the region are medium to extreme saline followed by Amhara (8.56%) (ICBA 2018b) and Oromia (6%) National Regional States (ICBA 2018c) (Table 7.8, Fig. 7.13).

Soil sodicity is a problem of dry areas of Ethiopia mainly Afar and northern parts of the Somali Regions (Fig. 7.14). It is reported that 4,250 km^2 sodic soils exist in Ethiopia (ICBA 2018a; FAO 1988). However, the report of ISRIC SoilGrids (2017) indicated the coverage to be over 10,893.06 km^2 (Table 7.9). This area covers 0.96% of the total surface (0–15 cm) land area and of the country. Compared to the upper 0–15 cmsoil depth, the extent of sodicity was higher (1.5%) at the 15–60 cm depth which could be due to the leaching of sodium salts down the profile.

Sodicity problems were reported by Haile et al. (2000) in Abaya State Farm, Southern Rift Valley of Ethiopia which were caused by the presence of shallow saline/sodic groundwater and an insufficient management system. The report indicated that the soils contained a high concentration of carbonates and bicarbonates of sodium (5.42–19.62 $mmol_c\ L^{-1}$) with sodium adsorption ratio ranging from 7.02 in the upper soil layer (0–45 cm) to 35.62 in the lower soil depth (71–116 cm). Similarly, the same work reported ESP in the range of 11.92% at the upper soil depth (0–45 cm) to 86.63% at the depth of 116–162 cm. The study clearly indicated that as the soil depth increases and gets nearer to the groundwater, sodicity increases.

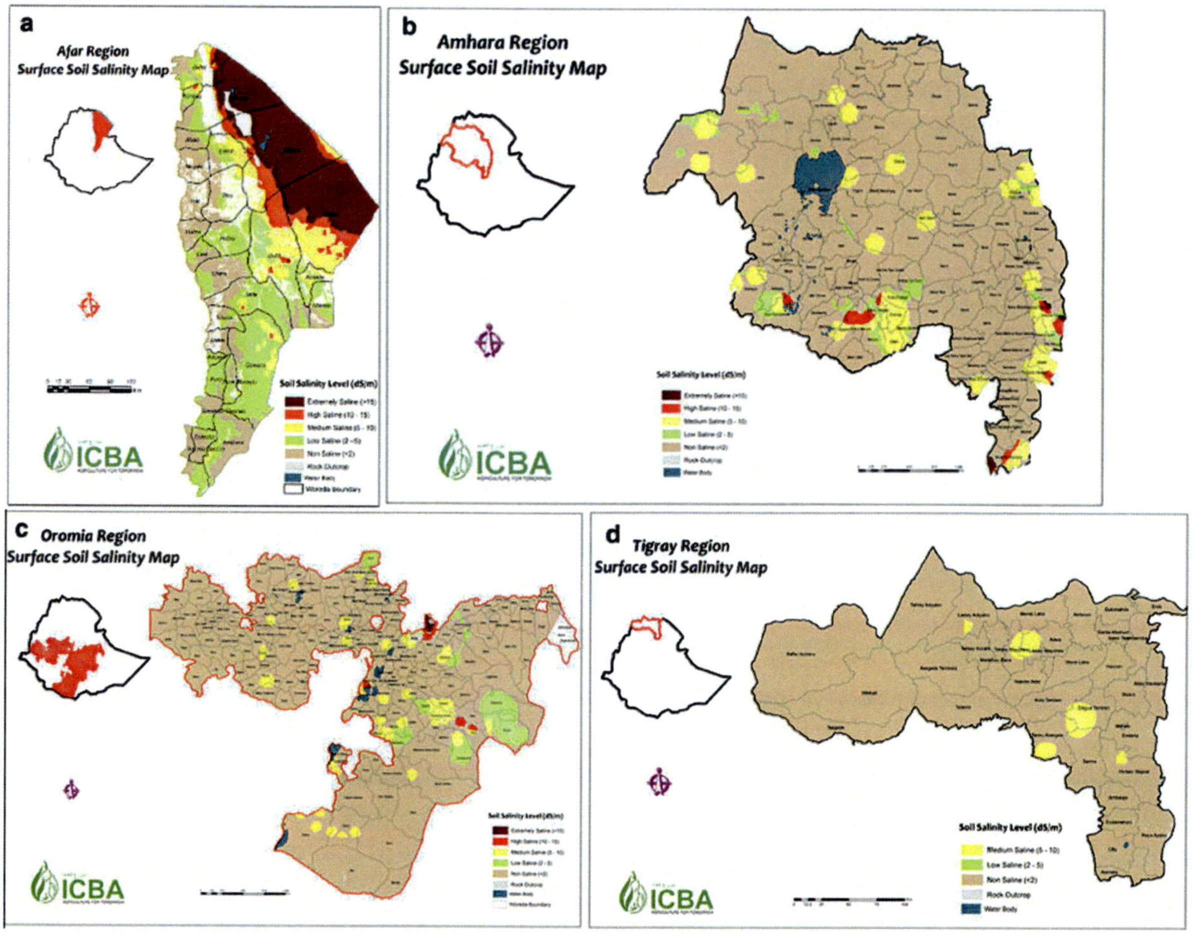

Fig. 7.13 Spatial distribution of salinity in soils of four National Regional States of Ethiopia (ICBA 2018a, b, c, d)

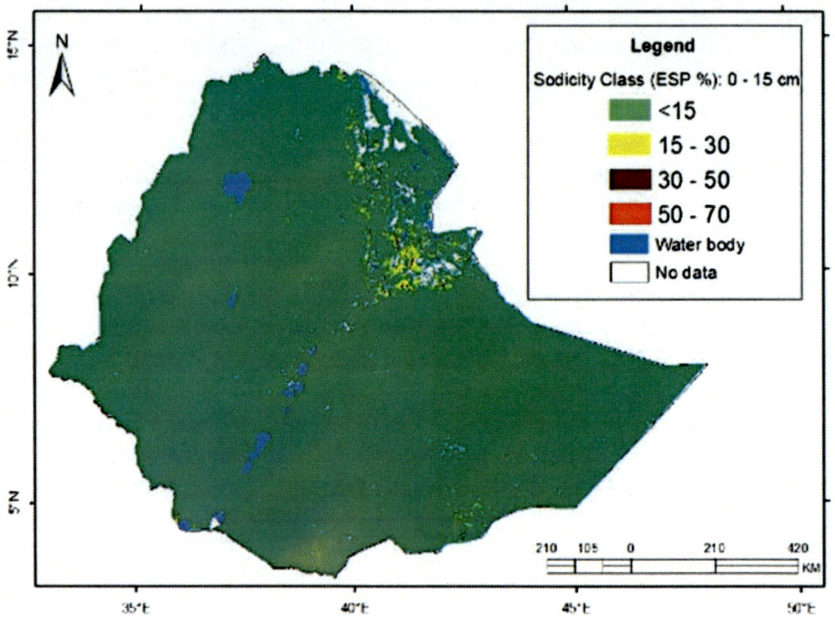

Fig. 7.14 Spatial distribution of sodicity (0–15 cm depth) in Ethiopia soils (ISRIC SoilGrids 2017, own analysis from Na and CEC) (*Source* Map extracted and compiled by author from ISRIC SoilGrids 2017, own analysis from Na and CEC: https://files.isric.org/soilgrids/former/2017-03-10/data/)

Table 7.9 Status of saline and sodic soils of Ethiopia

Soil group	ECe (dS/m)	ESP (%)	SAR	Location	Source
Solonchaks	5.11	Na	3.67	Shewa Robit, ANRS	Okubay (2019)
Solonchaks	16.0–18.0	Na	Na	Awash Valley	Qureshi et al. (2018)
Solonchaks	0.63–7.75	Na	Na	Fursa, Afar Region	Mohamed and Tessema (2013)
Solonchaks	30.3	Na	Na	Mekelle	Kidane et al. (2006)
Vertisols	0.15–0.69	2.8–29.2	Na	Raya valley, TNRS	Abayneh (2001)
Solonetz	0.7–1.7	11.92–86.63	7.0–35.6	Abaya State farm, Southern Rift Valley, SNNPRS	Haile et al. (2000)
Solonetz	Na	Na	37.0	Melka Sedi-Amibara, Afar Region	Qureshi et al. (2018)
Solonetz	2.79	Na	19.6	Meki-Zuwai Areas	Kidane et al. (2006)
Solonetz	0.66	Na	35.3	Yellen-Jeweha Areas, North Shewa Zone, ANRS	Kidane et al. (2006)
Solonetz	3.54	Na	18.0	Lower Omo River Valley	Kidane et al. (2006)
Saline sodic	4.0	Na	15.2	Babilie, Eastern Ethiopia	Assefa et al. 2019
Saline sodic	8.1–19.3	58.7–385.9	34.3–78.8	Lake Ziway area, ONRS	Meron (2007)
Saline sodic	16.6–18.6		37.1	Melka Sedi-Amibara Plain	Kidane et al. (2006)

ECe = electrical conductivity of saturated paste extract; ESP = exchangeable sodium percentage; SAR = sodium adsorption ratio. Salinity classification (Landon 1991a, b): ECe (dS/m) <4.0 = non-saline; 4–8 = slightly saline; 8–15 = saline; >15 = strongly saline. Sodicity classification (Brady and Weil 2015): ESP (%)/SAR/ECe (dS/m) = <15/13/- = non sodic; >15/13/<4 = sodic; >15/13/>4 = saline sodic. Na = data not available

Table 7.10 Spatial distribution of sodicity in Ethiopian soils based on SoilGrids of ISRIC (2017)

ESP (%)	0–15 cm soil depth		15–60 cm soil depth	
	Km^2	%	km^2	%
0 - 15	1,092,059.00	96.37	1,094,325.38	96.5[a]
15–30	9,204.81	0.81	12,037.79	1.06
30–50	1,613.00	0.14	3,312.79	0.29
50–70	75.25	0.01	1,661.72	0.15
No data	30,237.94	2.67	21,852.33	1.93
Total	1,133,190.00	100.00	1,133,190.00	100

Qureshi et al. (2018) reported an SAR of 37.0 in soils of Melka Sedi-Amibara Plain of the Middle Awash Valley. Abayneh (2001) showed that Vertisols can have high ESP reaching 29.2%. Pal and Selassie (2018) also reported ESP that ranged from 9.9 to 42.7% in the Kesem Allaideghe plains irrigation project area, eastern Ethiopia. Kidane et al. (2006) reported an SAR value of 19.56 for soils of the Meki-Zuwai area; 35.25 for Yellen-Jeweha Areas in the North Shewa Zone of ANRS and 18.0 for Lower Omo River Valley.

Availability of saline sodic soils was also reported by some studies. Electrical conductivity greater than 4 dS m^{-1} and SAR of 15.17 were reported in Babile low lands, eastern Ethiopia (Assefa et al. 2019). Meron (2007) also showed EC, SAR and ESP values ranging from 8.1 to 19.3 dS m^{-1}, 58.7 to 385.9 and 34.3 to 78.8%, respectively in saline sodic soils located at the south-western shore of Lake Ziway. Reports of Kidane et al. (2006) also showed that Melka Sedi-Amibara Plain sampled on abandoned cotton fields contains excessive soluble salts ranging between 16.6 and 18.6 dS m^{-1} and SAR of 37.06 (see Table 7.10).

[a] Non-sodic soils

7.5 Soil Organic Carbon and Biological Properties

7.5.1 Principles

Soil has a heterogeneous, diverse, and porous structure made up of liquid, gaseous, and solid phases. The liquid phase represents 25% of the total volume of soil and it carries out different kinds of relevant functions in the soil ecosystem. The gaseous phase also represents 25% of the total soil volume. The solid phase comprises 50% of the total soil volume. This phase is mainly represented by inorganic and organic solid components that make up 45% and 5%, respectively, of soil volume. The inorganic phase is made up of minerals classified as sand, silt, and clay particles. The organic component is made up of the nonliving phase commonly known as soil organic matter (SOM) and the living phase in which soil microorganisms and soil fauna and flora reside. Soil is an important habitat for thousands of associated organisms (Nieder and Benbi 2008).

Soils are some of the most species-rich habitats on earth, home to an abundance of species that enable soils to function and develop. Many of these species are essential for the functional diversity and resilience of the soil and the ecosystems that depend on the soil. Soil Biodiversity is an indicator of soil quality: a higher species diversity results in greater soil stability in terms of its capacity to perform key functions such as cycling of nutrients, assimilation of organic wastes, and maintenance of soil structure (Jónsson and Davíðsdóttir 2016; Jobbagy and Jackson 2001). Soil Biodiversity, soil organic carbon and soil organic matter are closely related but distinct. Soil Biodiversity reflects the mix of living organisms in the soil, including bacteria, fungi, protozoa, insects, worms, other invertebrates and vertebrates. These organisms interact with one another, as well as with plants and small animals, forming a web of biological activity (Orgiazzi et al. 2016). Most species live in the top 2–3 cm soil layer, where organic matter and root concentrations are highest. These soil dwellers provide food or nutrients that support organisms that live above and below ground. Soils also play critical roles in buffering and filtering freshwater ecosystems. Consequently, soils are extremely important to human societies (Aislabie and Deslippe 2013).

7.5.2 Soil Organic Carbon Status of Ethiopian Soil

Soil organic matter (SOM) encompasses all of the organic components of a soil, including living biomass (intact plant and animal tissues and microorganisms), dead roots and other plant residues and dead tissue, and soil humus (Laban et al. 2018). SOM and the debris of dead tissues left by plants and animals (detritus) are essential for biological activity in the soil and are the main source of energy, nutrients and habitat for the vast majority of soil organisms (Delgado-Baquerizo et al. 2017). Soil organic matter (SOM) is an important soil quality parameter. Relatively high SOM contents indicate relatively high buffer capacities for water and nutrients, good soil structure, increased water holding capacity and friability (Loveland and Webb 2003; Xiao 2015). Moreover, the decomposition of SOM releases nutrients that may be taken up by crops which in turn highly contribute to the natural soil fertility (Pekrun et al. 2003). Soil Organic Carbon makes up approximately 50–60% of SOM. An estimation of SOM is usually made by measurement of SOC, with the application of a factor of 1.72–2 for converting SOC to SOM (FAO 2017; Kibblewhite et al. 2016). Soils with high organic matter content are capable of supporting greater vegetation diversity, which in turn increases SOM and SOC while enhancing soil biodiversity. Although few studies have quantified these effects (Bernoux and Chevallier 2014), recent research has utilized SOC as a proxy for SOM and soil biological diversity, demonstrating that even marginal reductions in SOC content in the order of 1 percent can have a significant negative impact on soil natural capital and ecosystems services (Brady et al. 2015).

Low agricultural productivity caused by soil degradation is a serious challenge in Ethiopian highlands (Amare et al. 2006). Soil fertility depletion is the most pressing development challenge in Ethiopian agriculture for the sustainable crop, livestock and forest production. Land degradation and associated soil fertility depletion have been recognized as a major biophysical root cause for the declining per-capita food production in Ethiopia. The problem is aggravated by several factors which include among others, soil erosion, nutrient mining, soil acidity in the highlands and salinity in the lowlands, improper land use, low biomass production at the watershed level, low level of application of nitrogen and phosphorus and weak institutional support for managing soil fertility (Getachew et al. 2013).

Agricultural Transformation Agency (ATA) (2013) reported that most agroecosystems in Ethiopia had lost their soil organic carbon. Organic carbon has been highly depleted in Ethiopian soils across a wide coverage of the nation. Ethiopia's highlands, covering approximately 40% of the nation, historically have extremely low soil carbon content. In general, Fig. 8.3 shows that 36.39% (372,864.00 km^2), 26.17% (268,165.50 km^2), 33.42% (342,446.00 km^2), 1.66% (16,973.75km^2) and 2.37% (24,314.75km^2) of Ethiopian land is soil organic carbon content of very low, low, medium, high and very high. According to ATA (2016) categorized the organic carbon content of a soil to very low, low, medium, high and very high when the ranges fall between less than 1.16%, 1.16–1.74%, 1.74–4.05%, 4.05–4.64% and greater than 4.64%, respectively. Very high and high organic carbon content is mainly found at some central (Arsi and Balle), western (Wollega and Gambella) and some southern (Sidama and Gedio) parts of Ethiopia. Whereas very low and low organic carbon content is found at Eastern (Ethio Somali and Afar) and Northern (some parts of Amhara and Tigray) parts of Ethiopia. Most of the higher organic carbon content areas are situated at higher rainfall receiving areas and mostly the land uses are coffee and forest area. Low soil organic carbon content is particularly severe in Vertisols, Leptosols and Cambisols while Nitisols and Luvisols have moderately adequate levels of organic carbon (Eyasu 2016).

The amount of SOC in a terrestrial ecosystem is influenced by natural and anthropogenic factors (Sun et al. 2015). Human-induced land use change causes a particularly substantial loss of SOC (Fan et al. 2016; Gelaw et al. 2014). Land use change is associated with ecosystem carbon change (Getu et al. 2020) and drives negative impacts on climate and the environment. Numerous studies have shown that deforestation and land use change results in land degradation and poorer soil quality (Yimer et al. 2007; Gelaw et al. 2014; Meshesha et al. 2014; Sun et al. 2015). Likewise, several studies in the Ethiopian highlands reported higher SOC stocks in forest lands than in grazing lands and croplands (Abegaz et al. 2017; Miheretu and Abegaz 2017); Guteta and Abegaz 2017). In Ethiopia, the conversion of natural vegetation to croplands or plantations is increasing due to population pressure and socio-economic drivers. This has implications for biodiversity decline, land productivity, desertification, and SOC dynamics.

The study in 30 selected woredas in Ethiopia by Van Beek et al (2018) indicated that Soil carbon balance differences in climate (rainfall and temperature) and soil types were the main causes of variability. The highest depletion rates were found in Mekelle and Haramaya, because both were semi-arid areas with low biomass production that precipitates little options for organic matter circulation. Whereas in a great difference in the organic carbon content between Jimma and Mekele is explained by the high organic matter inputs in Jimma compared to Mekele. Farming systems in Jimma were dominated by coffee-based, mixed systems, whereas those in Hawassa were cereal based. Consequently, in Jimma more organic residues were available for composting that was applied to the fields. Similarly, the conversion of native forests and subsequent cultivation reduced the amount of SOC in the South-Eastern Highlands of Ethiopia because of lower supply and return of organic matter into the soil system (Fentaw et al.2007). On average, soil organic matter balances decreased by 37% per annum. This was about 4% to 7% of the total soil organic carbon stock, depending on the organic carbon content of the particular soil, a rate that was alarming by any standards.

Soil organic carbon balances were negative in the intensive but subsistence farming systems of the highlands of Ethiopia. On average, soil organic C balances were −3.7 tons $ha^{-1}y^{-1}$ that steadily decreased with time (Van Beek et al. 2018). Soil organic carbon is commonly regarded as a potential solution to climate change as carbon sequestration in soil is the largest potential sink for CO_2 (Lal 2010). Yet, in the agricultural soils of Ethiopia, with the common trends in soil organic C balance, C is not sequestrated. When balances are negative, organic C is released as CO_2 directly contributing to greenhouse gas emissions. Unbalanced organic C and nutrient balances in agricultural soils can be a precursor of land degradation and, eventually, land abandonment. The continuous soil mining results in lower soil organic matter content consequently reducing the nutrient and water holding capacity, and the biological activity of the soil. After prolonged depletion, soils fall prey to erosion and desertification or become unproductive and abandoned, which could no longer render the ecosystem service of carbon sequestration. This downward spiral is commonly known as the poverty trap. This is not an unthinkable scenario for Ethiopia where, especially in the central rift valley, the land is degraded and productivity has declined (Woldeamlak and Abebe 2013).

In Ethiopia, combined with sustained and improper agricultural management such as crop residues removal, steep slope cultivation, deforestation, use of animal manure (manure cake) as a source of energy for cooking and excessive livestock pressure, have led to a great extent to soil organic carbon depletion and eventually to the abandonment of land. Although farmers are aware of the problems of permanent and intensive cropping and of the need for organic inputs to restore soil fertility, they have a few alternatives at their disposal because of the persistent lack of inputs and alternative energy sources (EIAR and TARI 2011; Alemu and Kohlin 2008). The estimated annual nation-wide loss of nutrients through crop residue removals and dung burning is estimated to be equivalent to 200,000 MT of urea and 350,000 MT of DAP (IFDC 2012; MoARD 2010). The use of crop residue for livestock feeding and use of dung as fuel instead of fertilizer is estimated to reduce Ethiopia's agricultural GDP by 7% (Zenebe 2007) suggesting the lack of alternative fuel sources is a significant constraint for improving crop production.

Soil is a vital natural resource and its health is fundamental for sustainable agricultural production. Soil health is defined as "the capacity of soil to function as a living system", while soil quality is its "fitness for use" (FAO 2011). In an agricultural context, high soil quality means highly productive soil with low levels of degradation (Fuentes et al. 2009). Soil quality for sustainable crop production is related to soil health. Soils function to provide ecosystem services that include increased soil–water retention and availability, soil aggregation, nutrient cycling and storage, and microbial diversity and function (Doran and Zeiss 2000). As a living system, the soil consists of organisms whose activities include nutrient cycling, symbiotic relationships with plant roots, pest, weed and disease control, and soil aggregate formation and aeration which influence susceptibility to erosion and water infiltration (Turmel et al. 2015). A healthy soil is rich in organic matter which allows a high diversity of soil organisms to flourish and act as a reservoir of soil nutrients and moisture. The addition of regular inputs of organic amendment is necessary to increase or maintain soil organic matter content and thus contributes to soil health (FAO 2011).

7.5.3 Soil Biological Properties

Soil biodiversity plays a large role in agroecosystems by affecting crop quality, occurrence of soil-borne pests and diseases, nutrient cycling, and water transfer. It can also reflect disturbance and stress, as low soil biodiversity is often due to human-caused disturbance (Brussaard et al. 2007).

Ethiopia is diverse in terms of relief, climate, lithology, soils and agricultural systems. A combination of some of these has been used to stratify the region into agroecological zones, including the sub-humid zone, the humid zone, the highland zone and the arid and semi-arid zones. Because of the diversity of ecosystems within the country and its long history, Ethiopia is endowed with a wide diversity of fauna and flora. The country is also a center of origin and diversity for a number of crop and animal genetic resources, reflecting its long history of agriculture (IBC 2009). Ethiopia has a large soil biodiversity resource (Wolde-Meskel 2007), however, only a few of them have been explored extensively. Soil ecology, which deals with microflora, microfauna, mesofauna, and macrofauna is a young science in Ethiopia and some of these sectors have not been addressed in detail, especially the soil fauna. There are several publications, on various aspects of the microflora of the soils, mainly rhizobial bacteria and mycorrhizal fungus.

Soil fauna work as soil engineers, initiating the breakdown of dead plant and animal material, ingesting and processing large amounts of soil, burrowing 'biopores' for water and air movement, mixing soil layers, and increasing aggregation. Land use change may lead to changes in soil's physical, biological and chemical properties through their influence on various ecological processes. Different land use and management practices have an important impact on soil macrofauna biomass, abundance, and diversity which, in turn, affect their functional role in maintaining soil ecosystem processes resulting in differing soil chemical and physical properties.

Soil macrofauna such as earthworms; also plays an important role in the soil environment. Earthworms are described as ecosystem engineers because their effects on the soil ecosystem can last beyond their body size and lifetime.

They directly affect C and N cycles by consuming, storing, and cycling nutrients through their biomass, releasing in particular significant amounts of N through excretion and mortality (Whalen et al. 2000). Indirectly, they affect C and N cycles and soil aggregate stability by mixing organic matter in the soil through their gut and in their structures (casts, burrows and middens). These activities bring microorganisms in closer contact with organic matter, thus stimulating microbial processes such as decomposition and mineralization (Bossuyt et al. 2006).

In Ethiopia, no adequate information has been collected from areas with different land use practices so far. The results of a limited study by Berhanu (1991) do not make it possible to draw definite conclusions about the distribution and abundance of earthworms in Ethiopia. But it is believed that the worms are distributed throughout the country wherever there is moisture and plant litter, in the form of several genera. Cultivated land usually contains fewer earthworms than grassland. This could be due to mechanical damage during cultivation, the loss of the insulating layer of vegetation, or a decreased supply of food as the organic matter content gradually decreases. However, large differences in the earthworm population were noted between cultivated land to which dung was added annually and cultivated land which was left unmanured. This is due to the increase in the organic content of the soil that gives an advantage to the increment of the earthworm population in manured fields. Overgrazing and trampling by animals affected the normal activities and distribution of the worms on the hillside; but in the area which is protected from animal interference; large numbers of earthworms were observed. A greater number of earthworms were present during the rainy season, when the soil temperature is mild for worm multiplication and the soil is moist, promoting normal activities. On the other hand, worms migrate to deeper soil when the surface soil is too dry (bega) and lack of moisture can cause them to become quiescent or go into diapause.

Earthworms are domesticated and, when fed plant and animal wastes, they produce vermicompost, a process that has many advantages over conventional composting. This technology serves both social and environmental goals of sustainable agriculture and is widely employed mainly in developed countries, Canada, the United States, France, India, Australia, New Zealand, Cuba and Italy (Negash et al. 2018). For successful vermicomposting, the selection of suitable earthworm species is an important step and needs effort, especially its potential to feed upon different wastes which are suitable for worm growth should be known. In Ethiopia, vermicomposting was done using crop residue (sorghum and tef straw), industrial waste, fruit waste and khat waste as bedding materials (Yitagesu et al. 2018) using different earthworm species (essp. Esinia fetida) revealed variation between earthworms for their reproduction. However, Zerihun Getachew et al. (2018) reported that local (Debrezeit and Keshmando) and exotic (Eisenia fetida) earthworm species in different feed mixtures significantly affected the growth and production as well as the nutrient content of vermicompost produced.

Termites play a key role as decomposers of organic matter, nutrient cycling, and soil structure improvement in the savannah as well as in subtropical and tropical ecosystems (Ayuke 2010). Despite the potential beneficial role of termites, of the over 2800 described species, about 10% of these have been recorded as pests of crops, forestry, housing structures, and rangelands (Sileshi et al. 2008). Globally over 3,000 species of termites are known that are grouped under seven families and 281 genera (Grohmann et al. 2010). Africa is the richest continent in termite diversity, accounting for one-third of the species recorded worldwide (UNEP 2000).

The termite fauna of Ethiopia is not well known. At present 62 species belonging to 25 genera and four families have been recorded and 10 of the species are endemic (Demisachew et al. 2018). In Ethiopia, the genus *Macrotermes* is represented by two species *Macrotermes subhyalinus* (Rambur) and *M.herus* (Sjöstedt) (Abdurahman 1990; Abdurahman et al. 2010). Abdurahman (1990) stated that the mounds of *Macrotermes* termites found in western Ethiopia differ significantly in size and shape from those found in the Maki-Batu (Batu formerly Ziway) area of Central Ethiopia. The study by Daniel (2018)showed that the species builds low mounds with turrets and is absent in the Western part of the country (Western Wallaga Zone of Oromia National State), but found only in Adami Tullu Jido Kombolcha district and Arsi Negelle district of the Central Rift Valley of Ethiopia. *M. herus* builds low, closed and flattened dome-shaped mounds and it was the only *Macrotermes* species recorded from Western Ethiopia in the current study. The fauna includes Pan-African, East African, and West African or sub-Sahelian species, a single Palaearctic species and ten species endemic to Ethiopia and the neighbouring region (Cowie et al. 1990). The range of altitude in Ethiopia has a significant effect on distributions. For instance, Mucrotermes is not found in central Shoa, around Addis Ababa (2300 m), but is common in the Rift Valley around Meki, Zway and further south, and elsewhere below about 1800 m. Above 2000 m the fauna is very restricted, but Odontotermes is found in Addis Ababa and a single worker of a completely subterranean Odontotermes species was seen near Debre Birhan, northeast of Addis Ababa, at about 3200 m (Cowie et al. 1990).

Termite invasion is a new phenomenon which becomes a threat to rangeland management. The study conducted in Southern Oromia indicated that Borana rangeland has been under stress starting from the past few decades (Demisachew et al. 2018). Termite intensification in this area has been increasing from time-to-time. Even if termites are believed to be ecological engineers, the destruction of crops and fodder needs to be minimized at an optimum level. Ants comprise a single insect family (Formicidae), sister to the Apoidea (honeybees), within the insect order Hymenoptera. There are some 20 000 ant species, from which some 15 000 are described. The main characteristics uniting ants as a Family are eusociality, having a petiole and a metapleural gland, and female reproductive castes (queens) that can shed their wings after mating. Ants, along with termites and earthworms, have been named "ecosystem engineers". Due to their enormous abundance and large species diversity in natural environments (that is, functional diversity including a large spectrum of sizes, ecological traits and ecological strategies), ants heavily modify their surroundings and can influence soil functions in many direct and indirect ways.

Nematodes have received little research attention in the tropics particularly in sub-Saharan Africa as compared to temperate countries. Few studies in Ethiopia indicated that numerous factors could have contributed to the distribution of nematodes including soil type, cropping patterns, production systems, suitable climatic conditions and cultural practices employed. Soil type plays a significant role in nematode distribution. A sandy soil favors the abundance and greater damage of crops by many species of plant-parasitic nematodes in comparison with heavy clay soil (Afolami et al. 2014). A survey of the occurrence, distribution, and abundance of plant-parasitic nematodes associated with major khat-growing districts in the East Hararghe Zone identified eight plant-parasitic nematode genera: Criconema spp., Helicotylenchus spp., Hemicyclophora spp., Longidorus spp., Meloidogyne spp., Paratylenchus spp., Pratylenchus spp., and Rotylenchulus spp. associated with the khat crop (Seid et al. 2015). Among these, Pratylenchus, Meloidogyne, Helicotylenchus, and Longidorus were the most frequently encountered and abundant plant-parasitic nematode genera with 80, 60, 53.3, and 46.6% frequency of occurrence from soil, respectively. The distribution, population density and incidence of plant-parasitic nematodes of enset was determined and eleven plant-parasitic nematode taxa were identified, with Pratylenchus (lesion nematode) being the most prominent genus present with a prominence value of 1460 (Selamawit et al. 2020). Six nematode genera i.e., Pratylenchus, Rotylenchulus, Tylenchoryhnchus, Xiphinema, Ditylenchus and Tylenchus were identified from Faba bean growing area of Oromia and Amhara region of Ethiopia (Feyisa 2021). The most dominant nematode genera were xiphinema followed by Ditylenchus with 12% and 7% of occurrence respectively.

A survey of plant-parasitic nematodes in coffee was conducted on 132 sites of different agroecologies in Ethiopia. Plant-parasitic nematode genera recovered were: Helicotylenchus, Scutellonema, Rotylenchus, Xiphinema, Heterodera and Tylenchorhynchus. Helicotylenchus predominated throughout the area (65–74%), followed by Xiphinema (29–40%) (Mekete et al. 2008). Heterodera occurred at a frequency of 8–10% and reached the highest population density of any nematode taxa with 3310 juveniles per 100 g of soil. Nematode densities were generally lower in the dry season than in the wet season, however, relative abundance of nematode taxa was in the same order.

7.5.4 Soil Microbiology Studies

Even though, the use of rhizobium inoculants for improvement in N-fixation and productivity of grain legumes has been well established in developed countries; it is still in the developing stage in most parts of sub-Saharan Africa, including Ethiopia. Grain legumes are generally grown in low fertility soils in the country. The most frequently deficient nutrient is nitrogen, while nitrogen fertilizers are costly, inadequate, and not timely in supply, as well as associated with side effects on use. This makes rhizobia inoculants a cheaper, easier and safer option to improve the N_2-fixation and productivity of grain legumes. However, the use of inoculants in Ethiopia, though initiated as far back as the 1980s, still has a long way to go. The collection of root nodules, isolation of best rhizobial strains and evaluation and selection of superior types as inoculants is part-and-parcel of the research activities in the research institutes and universities in Ethiopia. To this end, soil microbiology courses were introduced in the universities starting from the early 1970s. Consequently, human and research capacity building in soil microbiology or legume-rhizobiology have been made at different higher learning and research institutes (Fassil et al. 2018).

It is nearly four decades since the beginning of research in *Rhizobium* inoculant technology in Ethiopia. Thus the pioneer post graduate work of Abebe (1982) at Addis Ababa University had been considered as the prelude to the introduction of research and development of effective strains of rhizobia and associated technologies. He extended the Rhizobium study at Melkassa Agricultural Research Center and collected root nodules of different legumes and characterized them (Abebe 1986). *Rhizobium* research shifted its center from Melkasa to the soil microbiology laboratory at Holetta Agricultural Research Center (HARC) in 1984, where much emphasis was given to inoculants associated with highland pulse crops. This practice continued till 1986 and, in 1990, in collaboration with the establishment of microbial laboratory at the National Soil Testing Center (NSTC) under the then Ministry of Agriculture (MoA), the rhizobial inoculant technology went up substantially and started to produce packages of biofertilizers at least for research purpose.

According to Fassil et al. (2018), most of the earlier researches focused on screening symbiotically effective *Rhizobia* from fewer highland pulse crops in Ethiopia. Thereafter, Addis Ababa, Haramaya, and Hawassa Universities played important roles in human capacity building and installing microbiology courses in their faculties. Many graduate students published their findings in peer-reviewed national and international journals, and tried to select isolates to recommend for technology incubations. From the year 2000 to 2015, over 1700 isolates were processed; more than 75 articles were published and made available to users by graduate students and faculty members of these universities. Universities also organized different workshops and conferences on rhizobial inoculant technology. Several research outputs, articles and posters were presented in these workshops and conferences.

Research by universities was extended to include the diversity and symbiotic association of other Plant Growth Promoting Microorganisms (PGPM), Phosphate solubilizers, bio-control agents and *Mycorrhiza* that are also important for plant health and productivity in relation to integrated soil fertility management. The research works were undertaken in collaboration with many local and international partners and financially supported (facilities, equipment, and chemicals) by the Ministry of Education, Ministry of Science and Technology and others.

7.5.4.1 Rhizobial Bacteria Symbiotic and Genetic Diversity

Symbiotic nitrogen fixation can be improved, among others, by being able to fix a selection of effective strains, host plant breeding for enhanced nitrogen fixation, and use of different agronomic methods that improve soil conditions for the crop, microbial symbionts and their favorable synergy (Montañe 2000). Several studies showed that Ethiopian soils harbor diversified and effective rhizobia associated with their legume hosts. The first research on Rhizobial bacteria diversity and *Symbiotic* association in Ethiopia dates back to the mid-1980s at the Nazret Research Station microbiology laboratory. Abebe (1986) made the first rhizobial collection and isolation work from different pulse growing locations. The different isolates were obtained from field pea, faba bean, lentil, haricot bean, soybean, chickpea and clover (*Trifolium spp.*) from diverse area. He isolates 110 strains of Rhizobium leguminosarum from *Pisum sativum, Vicea faba, & Lens esculenta;* 20 strains of *Rhizbium phaseoli* from *phaseolus vulgaris;* 328 strains of *R.* japonicum (cowpea type *Rhizobium* from *Glycine max,* and *Citer areietinum)* and 34 Strains of *R. trifoli* from Trifolium Sp. And then, in the 1980s, a systematic collection of thousands of nodules was made by the Holetta Agricultural Center to isolate more than 108 strains of which 23 faba bean strains were found to be superior, and a few isolates were promoted for field inoculation (Mamo and Dibabe 1994). Since then several workers and graduate students have been involved in the collection, isolation and characterization of native rhizobial inoculants harbored in Ethiopian soils.

Several previous studies showed that the nodule and rhizosphere of different legumes harbor different types of rhizobia and rhizobacteria with various plant growth promoting (PGP) properties in Ethiopia. The term "plant growth promoting rhizobacteria" (PGPR) is coined to refer to root colonizing bacteria that cause the increase in growth and yield and to differentiate them from other mechanisms found in rhizosphere that don't colonize roots or enhance plant growth (Gupta et al. 2000). Rhizobial isolation and their phenotypic and symbiotic characterization were conducted for various legumes of Ethiopia including faba bean (*Vicia faba* L.) (Abere et al. 2009, 2016; Alemayehu 2009; Ayneabeba et al. 2001; Asfaw and Angaw 2006; Dereje et al. 2015; Solomon and Fassil 2013; Zerihun and Fassil 2011; Anteneh 2012a), lentil (*Lens culinaris*) (Tena et al. 2016b; Adigo et al. 2015; Anteneh 2012b; Mulissa and Fassil 2011, field pea (*Pisum sativum* L.) (Amha and Fassil 2018; Kassa et al. 2015; Aregu et al. 2012), Fenugreek (*Trigonellafoenum-graecum* L.) (Mekasha et al. 2015), chickpea (*Cicer aeritinum* L.) (Mulissa and Fassil 2012; Mulissa et al. (2016); Demissie et al. 2018; Asfaw and Angaw 2006), Haricot bean (Mulugeta et al. 2013), Soybean (Diriba 2017; Abera and Assefa 2018), grass pea (Lathyrus sativus L.) (Amha and Fassil 2017; Mussa 2009) and common bean *(Phaseolus Vulgaris L.)* (Getaneh 2016), Cowpea (*Vigna unguiculata*) Kenasa et al. 2017), Pigeon pea (*Cajanus cajan*) (Tulu et al. 2018a). The rhizobial isolates showed diversity in nodulation, symbiotic nitrogen fixation, and nutrient utilization. The results of the study indicated that rhizobial isolates have shown wide diversity in their different C and N-sources utilization patterns and tolerance to salinity, high temperatures, acid and alkaline pH, heavy metals and antibiotics. Symbiotic and morphological characterization also showed a wide diversity among tested isolates. Over-all, these results confirmed the presence of a great diversity of rhizobia species in Ethiopia, inviting further exploration. Moreover, the differences in the symbiotic effectiveness of the test strains indicated the potential for selecting and using them as inoculants to improve the productivity of lentils in the country.

Rhizobacterial isolation and their PGP and molecular characterization were carried out for different legumes that included lentil (*Lens culinaris)* (Tena et al. 2016b; Mulissa et al. 2015); chickpea (*Cicer aeritinum* L.) (Mulissa et al. 2016; Tena et al. 2016a); common bean (*Phaseolus vulgaris* L.) (Getaneh 2016; Endalkachew et al. 2018a, b; Zerihun and Fassil 2011), soybean [*Glycine max* (L) (Diriba 2017), coffee (*Coffea arabica* L.) (Diriba et al. 2007, 2009, 2013), Grass pea (Lathyrus sativus L.) (Mussa 2009): Groundnut

(Arachis hypogaea L., Fabaceae), cowpea (Vigna unguiculata L. Walp.; Girmaye et al. 2017) and mung bean (Vigna radiata L. Wilczek; Tulu et al. 2018b). Woody and herbaceous legumes (Tulu et al. 2011). This study clearly shows that the characterization of symbionts of unexplored legumes growing in previously unexplored biogeographical areas will reveal additional diversity. Chickpea (Cicer arietinum L.) used to be considered a restrictive host that nodulated and fixed nitrogen only with Mesorhizobium ciceri and M. mediterraneum. Recent analysis in Ethiopia revealed that chickpea can also establish effective symbioses with strains of several other Mesorhizobium species such as M. ciceri, M. loti, M. haukuii, M. amorphae, M. muleiense, M. *abyssinicae, M. shonense* (Tena et al. 2016a; Gunnabo et al. 2020a). All Ethiopian chickpea nodulating strains had nearly identical symbiotic genes that grouped them in a single cluster with M. ciceri, M. mediterraneum and M. muleiense (Gunnabo et al. 2020a). The presence of two genetic groups (Kabuli and Desi genepools) did not affect interaction with Mesorhizobium strains (Gunnabo et al. 2020b). Thus our results suggest that efforts to find more effective strains may be more rewarding than aiming for the identification of superior combinations of strains and genotypes. However, Keneni et al. (2012), showed differences in chickpea genotypes in nodulation abilities.

Recent research findings showed that all grain legumes were responsive for rhizobial inoculation at different agroecologies of the country. The responsiveness of the grain legumes for rhizobial inoculation might be due to soil fertility depletion as a result of an intensive farming system or poor management of the farmlands to sustain the fertility of the soil and rhizobial population in the soil (Abere et al. 2016). For example, Aragaw (2014) reported the importance of inoculating soybean crops with an elite isolate of *Bradyrhizobium* sp. even if the crop grows in saline soils. A study made at Bako in 2006 confirmed that inoculation of soybean with rhizobia biofertilizer TAL-379 has shown a 53% grain yield increment over the uninoculated control (Solomon 2006).

Inoculation studies of Azospirillum isolates on pot-grown tef plants showed marked increases in height, grain yield, total shoot and root weight, root-shoot ratio and total grain nitrogen. An increase in grain yield up to 12% over uninoculated control was observed (Solomon et al. 2000). Since effectiveness in yield depends on cultivar and bacterial diversity, a thorough study on the interaction of various tef cultivars with Azospirillum isolates from different agroclimatic regions of the country under field conditions is necessary for better results. As inoculation of tef by VAM fungi (Tekalign 1987) and phosphate solubilizing fungi (Asfaw 1993) had shown to increase tef productivity, evaluation of mixed inoculation of these microorganisms and on cereal crops may further improve yield.

Biofertilizers may be defined as "substances which contain living microorganisms that colonize the rhizosphere or the interior of the plants and promote growth by increasing the supply or availability of primary nutrients to the target crops, when applied to soils, seeds or plant surfaces" (Mazid et al. 2011). Rhizobial inoculants have been used successfully in world agriculture for about 100 years. Rhizobial biofertilizers are also considered as one of the best and sustainable soil fertility management interventions in Ethiopia (Abere et al. 2016). Efforts have been devoted to identify and characterize the efficient forms of local and exotic rhizobial strains for major grain legumes. The currently available commercial strains were from local collections. In Ethiopia, the commercialization of strains is a recent phenomenon, only started since the year 2000 (Asfaw and Angaw 2006).

Currently, the National Soil Testing Center (NSTC), Menagesha Biotech Industry (MBI) plc and the Ethiopian Institute of Agricultural Research (EIAR) have capacities to produce rhizobia-based biofertilizer in Ethiopia. Research activities so far were focused on rhizobial inoculants collection, characterization, selection and evaluation. Strains were developed for biofertilizer packaging with baseline information on the symbiotic effectiveness of indigenous rhizobia of almost all important pulse crops and a few leguminous trees. Post graduate studies at Addia Ababa, Haramaya, and Hawassa Universities have played significant roles in these activities. As the result, more than 1000 symbiotically effective rhizobial and non-rhizobial isolates from different food legumes, forage legumes and forest tree legumes were examined (Fassil et al. 2018). There have been a number of inoculant strains under mass production for different legumes by NSTC, EIAR, Haromaya University, Hawassa University and MBI plc (Fassil et al. 2018).

7.5.4.2 Soil Fungi

In Ethiopia, mycorrhiza research in agronomy can go back to the mid-1980s (Tekalegn and Killham 1987). Several studies have been carried out in relation to the diversity and density of AMF on coffee and shade trees in montane forests (Muleta et al. 2007; Wubet et al. 2003), and in the dry deciduous woodlands of Northern Ethiopia (Birhane et al. 2010). The study (Zerihun et al. 2013) showed that the acacia species were characterized by relatively high AMF colonization and very high AMF diversity. A total of 41 AMF species in 14 genera and 7 families of the Glomeromycota were identified. The AMF species found in earlier studies of acacia trees belong mainly to the genera Glomus and Gigaspora (Yohannes and Assefa 2007; Michelsen 1993). AMF spore density and AM root colonization in acacia roots were influenced by soil factors such as available P and soil texture.

Morphological studies on AM association in different land uses and species had also been conducted in Ethiopia. Michelsen (1992) has investigated the mycorrhizal status of tree nursery seedlings in Ethiopia (Zebene 2003) and has also morphologically investigated AM associations on the traditional agroforestry system of southern Ethiopia. Diriba et al. (2008) investigated factors affecting the AMF spore abundance in the south-western Ethiopia Coffee arabica farming systems. Accordingly, they have found that sampling points, sites and depths, shade tree species and shade tree/coffee plant age significantly affected AMF spore abundance. Moreover, they have found that, compared with coffee monoculture system, coffee agroforestry system maintained a higher AMF spore abundance and more importantly, at a lower soil depth. Similarly, Tadesse and Fassil (2013) have morphologically investigated the AM fungi abundance and diversity in the rhizosphere of Coffea arabica shade trees of south-western Ethiopia. Mengsteab et al. (2013) morphologically investigated the role of agroforestry trees to transfer infective AMF to the associated annual crop. Mengsteab et al. (2014), using the trap plant maize, investigated the AMF infectivity of Faidherbia albida rhizospheric soil collected from different landuse types, viz., area enclosures, grazing, and cultivated lands. Zerihun et al. (2015) investigated the diversity and abundance of AM fungi across different land use types. They also compared the AMF abundance and diversity of soils from these land uses versus trap culture. Accordingly, they have found out that land uses with diversity in plants had better AMF abundance and diversity. They have also found out that AMF abundance and diversity varied between sites collected and trap culture soils.

In recent years, there have been a handful of AMF research activities conducted in Ethiopia (Tesfaye et al. 2006; Emiru et al. (2012, 2014); De Beenhouwer et al. 2015; Tesfaye et al. 2009; Beyene et al. 2016; Yoseph et al. 2017; Emiru et al. 2017, 2018; Fisseha et al. 2019). However, it can be concluded that AMF research and collection in Ethiopia is at its infant stage. Therefore, both AMF research and the collection of beneficial AMF species should get sufficient attention.

7.5.4.3 Soil Algae

Soil algaediversity and distribution are not well known and studied in Ethiopia. More research works on algaediversity and distribution in Ethiopia have focused on water bodies (Nega et al. 2020). Algal is associated with plant roots and produces hormones that stimulate root growth and promote the activity of useful microorganisms found in the root region (Adesalu and Olugbemi 2015). Cyanobacteria are the oldest photoautrophic component of Biological Soil Crusts communities which include bacteria, algae, fungi, lichens and mosses. These biocrusts occur on and within the top centimeters of the soil surface all over the world, including the most hostile environments like extremely arid and dryland areas where cyanobacteria are often the most important primary producers. Cyanobacteria play a key role in carbon and N-cycling, nutrient dynamics, and ecosystem productivity. They furnish nutrients to the soil as a good biofertilizer and stabilize soil structure by the production of the extracellular polysaccharides that aggregate soil particles, and sometimes increase their porosity and water holding capacity. Cyanobacteria have significant potential as biofertilisers. Application of dried cyanobacteria increases the growth and yield of maize, pepper, romaine lettuce and kale, and the fertility of soils in Ethiopia (Mulat et al. 2019; Muluneh and Zinabu 2013). Dried cyanobacterial biofertilizer also improved the nutritional qualities of the crops by increasing micronutrient (Zn and Fe) concentration, especially in the edible parts of kale and pepper.

7.6 Soil Mineralogy and Clay Minerals

Soil mineralogical composition is the most permanent and reliable record of the story of soil formation, including the nature of the parent material, weathering sequence and intensity. Mineralogy influences several properties of soil and its capability for different uses (Fikre 2003). Studies on the clay mineralogy of Ethiopian soils are very limited. Even the available ones are very fragmented and difficult to conclude at large.

A study on the clay mineral compositions of Ethiopian Vertisols indicates that there are differences between Vertisols due to the variation in drainage (Zewdie 2013). However, Vertisols retain most of the ions that have been liberated from the primary silicates. The x-ray diffraction patterns (Fig. 7.15) exhibited a higher illite mineral in many soils (Fikre 2003; Zewdie 2013) while smectite dominates in some Vetisols (Zewdie 2013; Mitiku 1987). The 2:1 mixed-layer minerals were identified as vermiculite/smectite and considered to represent weathering product of illite. In many Vertisols illite was the dominant mineral found in the saprolite and this may have a contribution to the stability of the physico-chemical characteristics of Vertisols of Ethiopia. However, chlorite and vermiculite are also found in the subsurface Horizons (Table 7.11).

In general, Vertisols in the highlands developed from basalt, limestone and granite, whereas in lowlands areas they developed from different colluvial/alluvial materials rich in

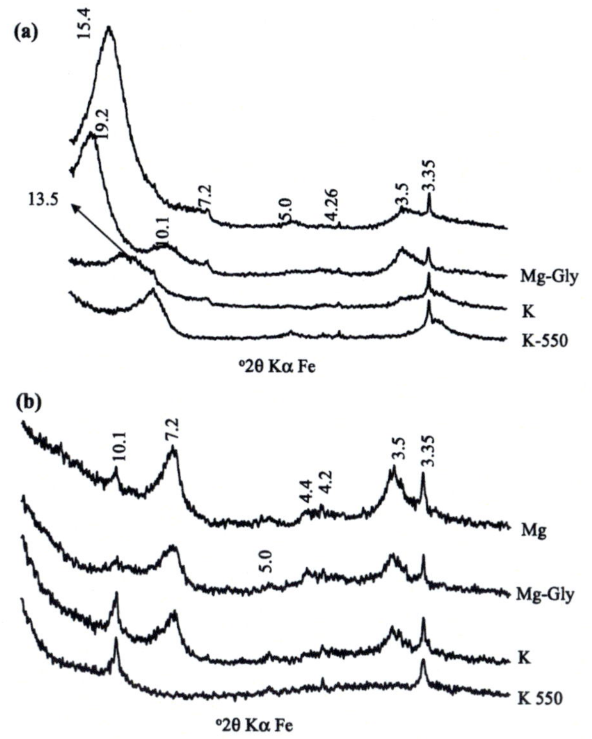

Fig. 7.15 X-ray diffractograms of Vertisols Ofaseri Wolayta **a** Ap, and **b** Cr horizons (Fikre 2003)

base cations (Zewdie2013; Mitiku 1987). This condition has helped the soils to maintain their mineralogical property that maintains the structural aggregates at stable condition over the years.

Andosols of Ethiopia are mainly found in the great Ethiopian rift valley areas. They contain quartz as predominant in their silt fraction (2–20 μm). Therefore, the tendency of 2:1 and 2:1:1 clay formation is high. The x-ray analysis indicated (Fig. 7.16) that there is a clear indication of the 1:1 type of clay mineral (halloysite) and weak expression of 2:1 type of clay mineral surface horizons. The diffuse XRD patterns and the field observations of pumice fragments support the idea that Andosols are dominated by slightly weathered volcanic glass, not crystalline phyllosilicates. The formation of crystalline clay minerals in volcanic ash soils of the Ethiopian Rift Valley can be attributed to the existence of marked dry and wet seasons that caused a weaker desilication process.

Ethiopian Nitisols contained a small quantity of 2:1 clay minerals such as smectite (Zewdie 2013; Fikre 2003) Illite and mixed-layer minerals. They rather contain a large quantity of 1:1 minerals like kaolinite (DeWispelaere et al. 2015; Fikre 2003) and sometimes even with poorer diffraction patterns than even Luvisols, suggesting a higher degree of interference from short-range order minerals (Fig. 7.17).

Some Nitisols also exhibit mixed-layer minerals that were identified as illite/chlorite and were considered to represent weathering products of mica, but at the low pH of these soils, it will be transformed to kaolinite/halloysite within a short period. The soils have a clayey texture, with a clay content increment down the subsurface horizon usually identified as Bt horizons which is evident from the Micromorphological features that the subsurface horizon has on illuvial clay (Fig. 7.18).

Luvisols are characterized by the relative accumulation of stable primary and secondary minerals. The quantity of 1:1 clay Luvisols are characterized by relative accumulation of stable primary and secondary minerals (Zewdie 2013). The clay assemblage is dominated by kaolinite, goethite, hematite and gibbsite in varying amounts. The Luvisols in Ethiopia exhibited a large amount of 1:1 clay minerals, mainly kaolinite with high-intensity diffraction peaks (Fig. 7.19).

Luvisols do not contain an expandable type of clay minerals. However, BA and Bt1 contained a small quantity of chlorite, vermiculite and a mixed-layer of chlorite/vermiculite (Zewdie 2013).

Cambisols are characterized by a high accumulation of easily weatherable minerals and the more resistant feldspars and micas. In most parts of Ethiopia, the development of the Cambisols is favored by slow and continuous water erosion in mountainous areas and occurs all over the country even in association with highly developed soils like Nitosols and Luvisols. The clay assemblage is dominated by smectite,

Table 7.11 Approximate mineral contents in the clay fraction (%) of Vertisols profile at Ginchi and C-horizons of Wonji, Sheno and Alemaya

Horizon	Smectite	Vermiculite	Chlorite	Mixed-Layer	Illite	Kaolinite
Bw1 (Ginchi)	14	23	41	8	12	2
Bw2 (Ginchi)	65	–	–	6	29	–
BCk (Ginchi)	–	–	–	–	100	–
CB (Wonji)	–	5	–	10	85	–
CBK (Sheni)	–	7	–	13	80	–
C (Alemaye)	–	–	–	5	95	–

Source (Zewdie 2013)

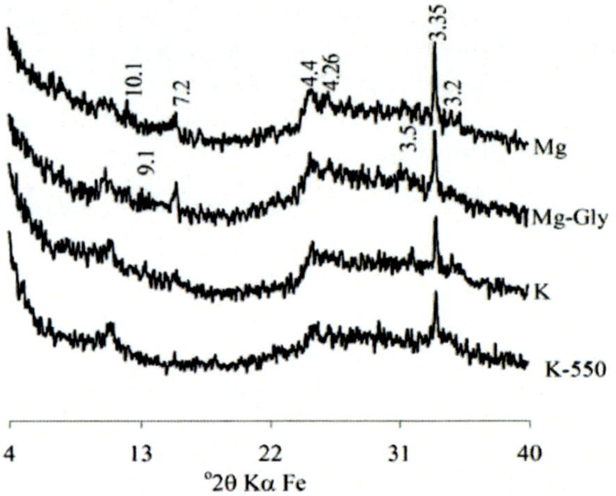

Fig. 7.16 X-ray diffractograms of the clay fractions (<2 μm) of a representative Andosols pedon of the Ethiopian rift Valley (Toga) (peaks values in Å) (Fikre 2003)

illite, and very few quantities of Kaolinite (Zewdie 2013). The Cambisols in Ethiopia exhibited a large amount of 2:1 clay minerals mainly smectite. Their XRD diffraction indicates that except for their clay minerals the primary minerals did not show the collapse of peaks (Fig. 7.20). This proves that there is quite a large amount of primary minerals.

The mineralogical analysis of Leptosols, Regosols and Fluvisols indicated that the soils contain relatively high amounts of weatherable minerals. The dominance of primary minerals in these soils makes them sensitive to erosion and drought (Zewdie 2013).

There is no big difference in mineralogical composition between Regosols and Leptosols. Both of them contain relatively high amounts of primary mineral and Kaolinite (Fig. 7.21). The existence of Kaolinite implied that these soils were developed from parent materials that are rich in hematite, limonite etc. However, the Regosols contain better amounts of 2:1 types of clay mineral compared with Leptosols.

The X-ray diffraction pattern for Planosols indicates that the clay minerals readily identified are dominated by kaolinite, illite and a small quantity of smectite (Fig. 7.22). Moreover, they also exhibit the presence of mixed minerals. Based on these observations alone, both kaolinite–smectite and illite–smectite mixed-layers might be present (Dumon et al. 2014).

Leptosols of Ethiopia mainly contain moderate as the predominant fraction in their whole soil fraction. They also contain pyroxenes, anorthoclase, plagioclase, moderate, quartz, and magnetite, and various types of clay-size minerals (Belay 2000). The most predominant clay mineral in latosols is chlorite but some mica, moderate, and sepiolite were also identified in the fraction (Table 7.12). Much of the chlorite appears to be pedogenic, although some of it might have been inherited from the mafic igneous rocks.

Much of the chlorite appears to be pedogenic, although some of it might have been inherited from the mafic igneous rocks. The high background of the diffractograms, observed in the patterns of the samples, provides additional evidence for the presence, of a considerable amount of amorphous minerals (Fig. 7.23).

Fig. 7.17 X-ray diffraction patterns for the fine clay fraction of Nitisols, in air-dry conditions (AD) and after glycolation (EG) d-values are in nm. Abbreviations are Gth: goethite, Ilt: illite, Kln: kaolinite, 2:1: open 2:1 mineral, hk0: diffraction band related to phyllosilicates (*Source* DeWispelaere et al. (2015))

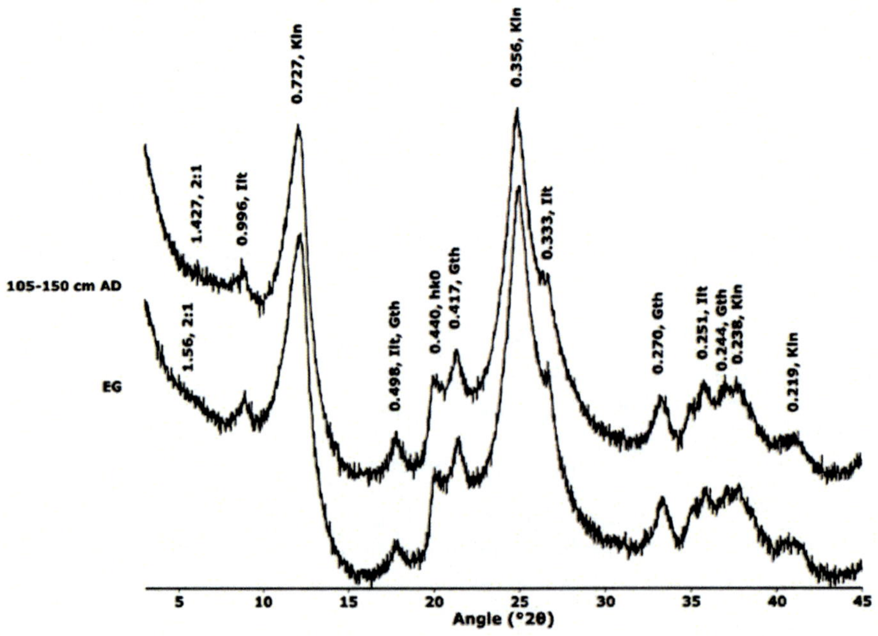

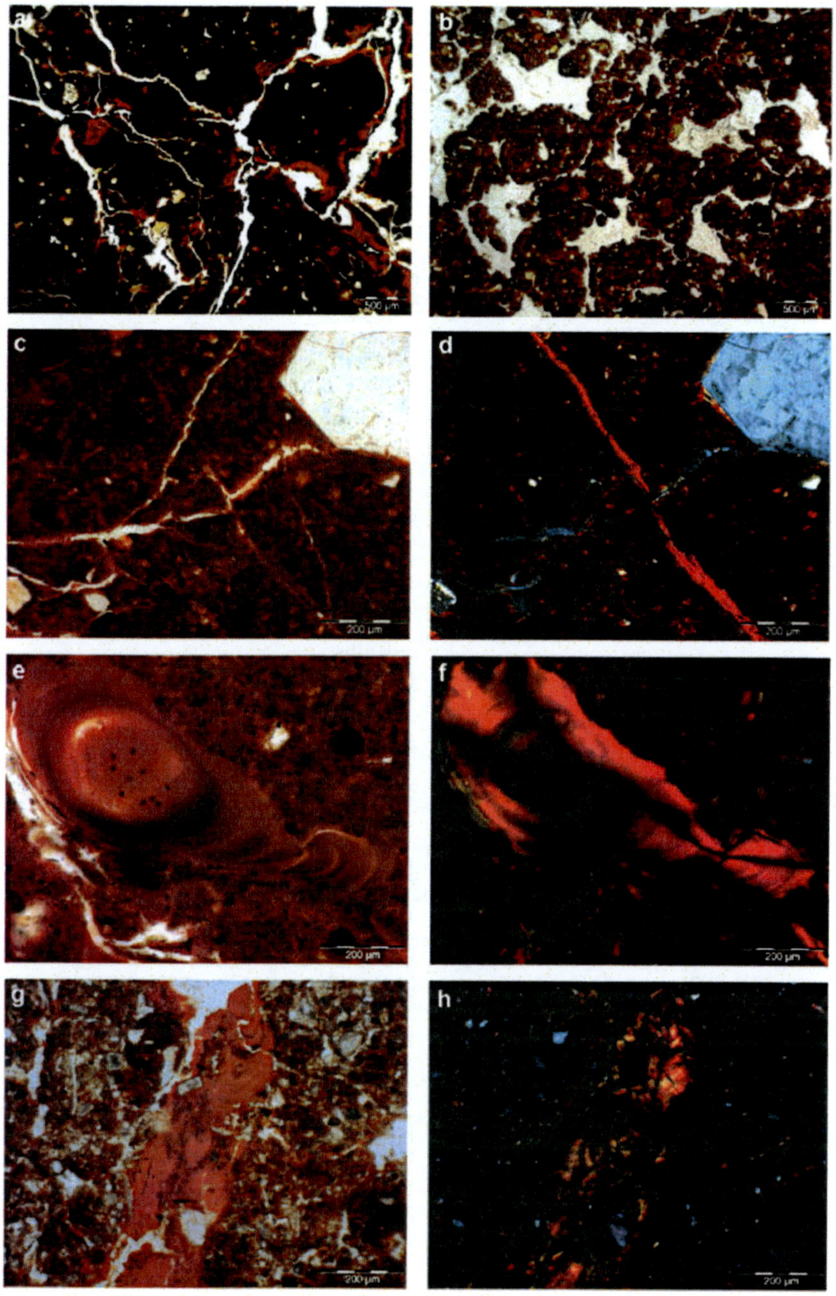

Fig. 7.18 Micromorphological features: of Nitosols **a)** sub-angular blocky microstructure, with abundant illuvial clay surrounding some of the structural units; **b)** infilling with granular, partly coalescent, aggregates; **c)** sub-angular blocky microstructure, showing several generations of planar voids, with clay infillings occupying the older pores; **d)** same as previous, in XPL; **e)** micro-laminated coating/infilling of clay with strong parallel alignment; **f)** same as previous, in XPL; **g)** coating/infilling with well-oriented clay and admixture of groundmass material including a phytolith; **h)** same as previous, in XPL (DeWispelaere et al. 2015)

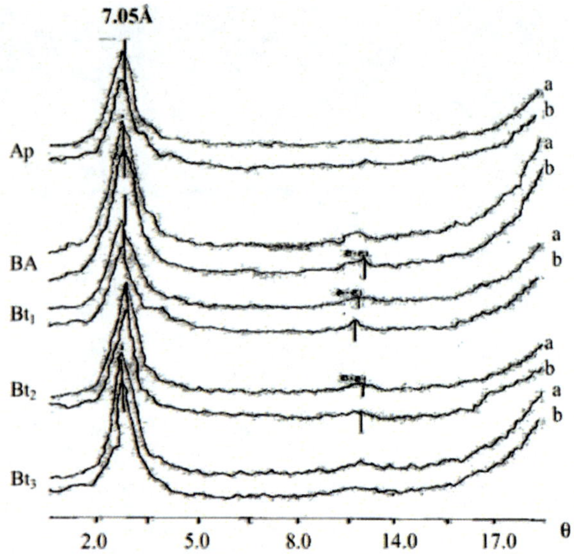

Fig. 7.19 X-ray diffractogram (Å) of the clay fractions (<2 μm) of representative Luvisols pedon of Ethiopia. (Zewdie 2013)

7.7 Conclusions

Ethiopian soils greatly vary in their morphological, physicochemical and mineralogical properties. Soil depth varies depending on slope gradient, curvature of landscape and types of proximate parent material. The steep gradient and convex curvature soils are shallow in depth while the gentle and concave slope positions are deep. Soilcolor is greatly variable among different horizons and references soil groups based on the differences in organic matter, land use and topographic attributes. Soil particle size distributions due to the wide variations in soil forming factors and the associated processes. Across the country, the sand content varies from 1 to 75%; silt from 9 to 60%, and clay from 10 to 86%. However, only 0.023% of the country's total area is covered by soils with sand content greater or equal to 70%. About 0.05% of the country's total area is covered by soils whose silt content is between 40 and 60%, with the remaining large area of the country occupied by soils whose silt content is between 9 and 39%. The clay content of Ethiopian soils

Fig. 7.20 X-ray diffractogram (Å) of the clay fractions (<2 μm) of representative Cambisols pedon of Ethiopia (0–30 cm). (Zewdie 2013)

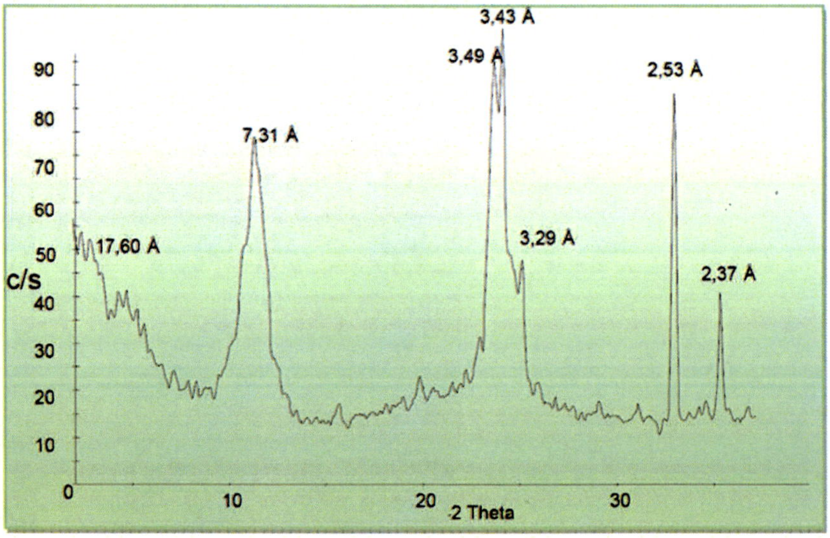

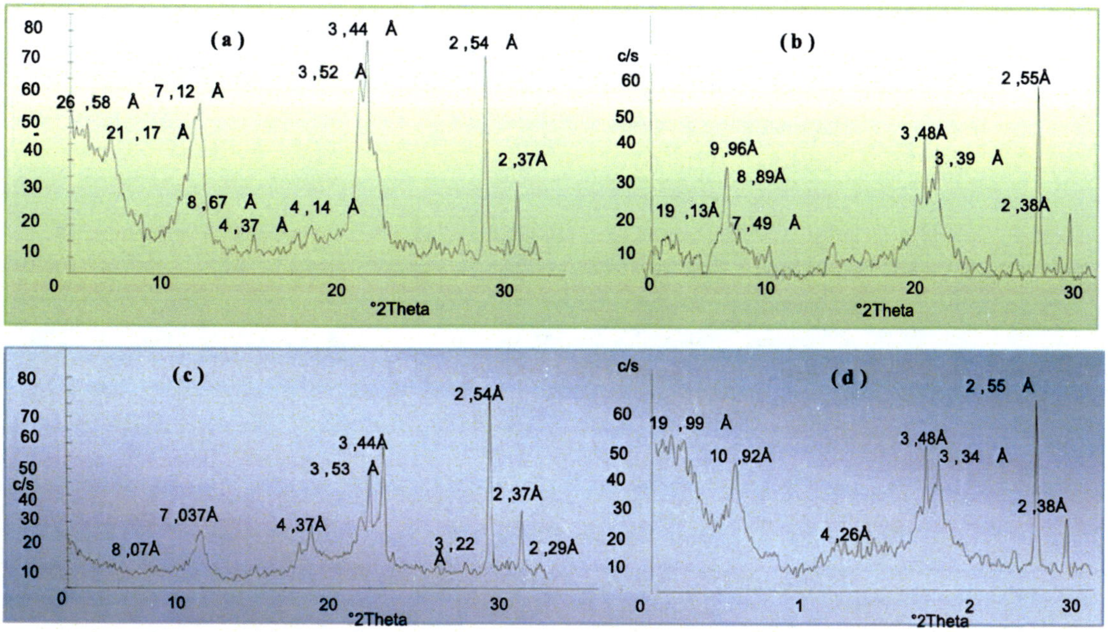

Fig. 7.21 X-ray differactogram (Å) of the clay fractions (<2 µm) of representative **a** Regosols, **c** Leptosols and **b** & **d** Fluvisols pedons of Ethiopia (0–30 cm) (Zewdie 2013)

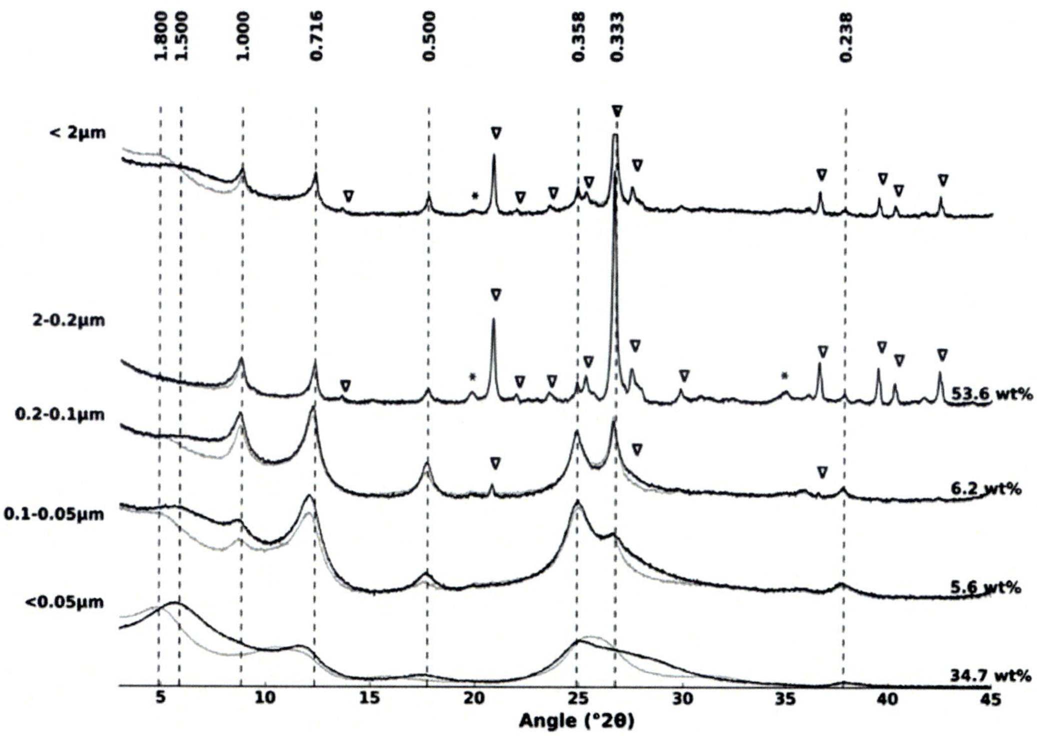

Fig. 7.22 Observed X-ray diffraction patterns for the bulk, pretreated b 2 fraction and sub-fractions of the 20–25 cm interval of the bleached horizon. Black and gray traces are patterns for air-dry conditions and after glycolation respectively. Quartz peak at 0.334 nm is clipped for clarity. An * indicates hk0 bands, open triangles are accessory quartz and feldspar reflections and a diamond indicates goethite reflections. Relative mass (wt.%) is indicated at the right of each sub-fraction pattern. Spacings of important 00 l reflections are in nm (Dumon et al. 2014)

Table 7.12 Relative abundance of minerals identified Leptosols of Gora Daget forest using X-ray diffractometer (Belay 2000)

A-whole soil <2 mm							
Soil Type	Depth (cm)	Pyrox	Plag	Anort	Mord	Quart	Mag
Phaeozem	21–53	**	**	**	*	**	*
Leptosols	22–27/33	*	*	*	****	*	*
Phaeozem	16–43	****	*	***	*	*	*
B silt fraction <16 micron							
Soil Type	Depth (cm)	Chlorite	Mica	Anorthoclase	Morderite		
Phaeozem	21–53	**	*	***	**	–	–
Leptosols	22–27/33	**	*	**	***	–	–
Phaeozem	16–43	**	*	****	*	–	–
C clay fraction <2 micron							
Soil type	Depth (cm)	Chlorite	Mica	Sepiolite	Morderite		
Phaeozem	21–53	*	–	*	–	–	–
Leptosols	22–27/33	*	*	–	*	–	–
Phaeozem	16–43	*	–	–	–	–	–

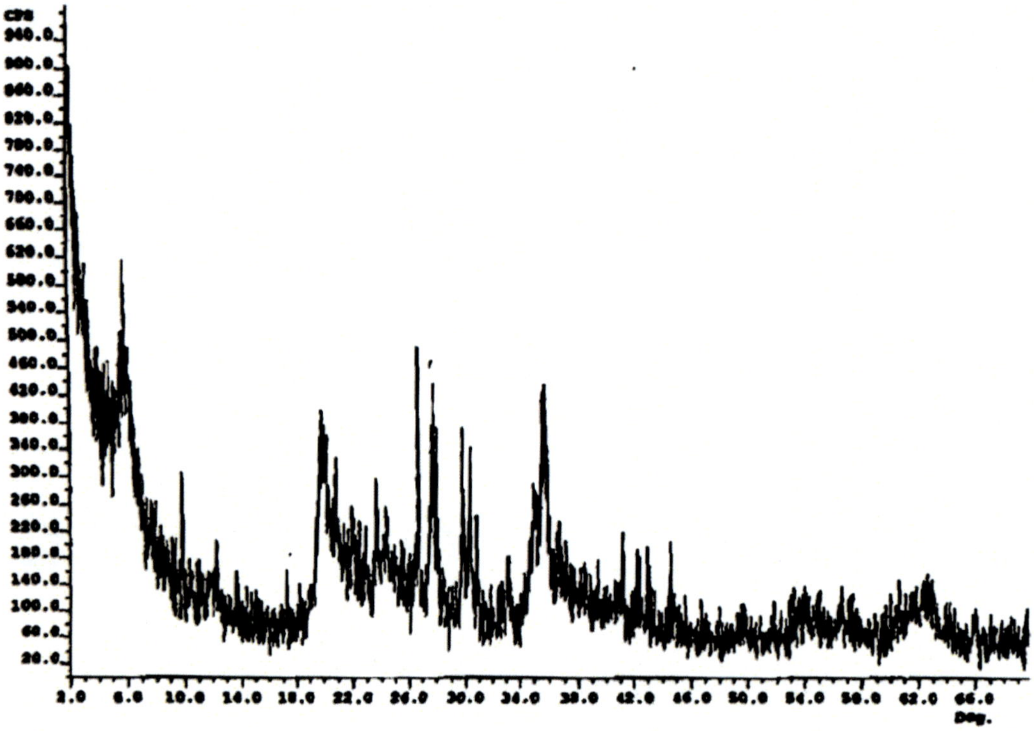

Fig. 7.23 X-ray diffractogram of clay fraction sampled from Leptosols (Belay 2000)

ranges between 10 and 86%. About 48.65% of the country's total area is covered by soils whose clay content is greater or equal to 35%, indicating that most Ethiopian soils contain high amount of clay. Generally, soils in the western part of the country contain a higher amount of clay, while those in the Somali and Afar regions are characterized by high sand content.

There is also quite a significant variation in the Ethiopian soil's chemical and mineralogical properties. Soil reaction ranges from strongly acidic in the highlands and high rainfall areas of central, western and north-western parts of the country to strongly alkaline in the lowlands of the Rift Valley and highlands of northern Ethiopia receiving low rainfall. Rainfall and topography are the major determining factors for soil reaction. CEC is low in acidic, highland, high rainfall area soils and it is high in young soils of the dry and lowland areas. Base saturation in many cases follows the natural pattern of CEC. Salinity and sodicity are problems of the lowlands having low precipitation but with high evaporation. These problems are common in the Rift Valley and eastern parts of the country where soil acidity is not prevalent. The variability in soil chemical properties of Ethiopia soils suggests the need for varied management, reclamation and utilization practices. Relative abundances of the different minerals identified in the different soil types of the country based on the intensity of X-ray diffraction peaks for the different minerals are quite variable among the different soil reference groups.

References

ATA, Agricultural Transformation Agency (2016) Soil fertility status and fertilizer recommendation atlas of amhara national regional state, Addis Ababa, Ethiopia

Abay A, Sheleme B, Fran W (2015) Characterization and classification of soils of selected areas in southern Ethiopia. J Environ Earth Sci 5:116–137

Abayneh E (2001) Soil physico-chemical characteristics of the raya valley soils. Ethiopian J Natl Resour 3:179–193

Abdurahman A, Abraham T, Mohammed D (2010) Importance and management of termites in Ethiopia. Pest Manage J Ethiop 14:1–20

Abdurahman A (1990) Foraging activity and control of termites in western Ethiopia, PhD thesis, University of London. 277pp. https://spiral.imperial.ac.uk/bitstream/10044/1/47734/2/Abdulahi-A-1990-PhD-Thesis.pdf

Abebe TF, Alamirew T, Abegaz F (2015) Appraisal and mapping of soil salinity problem in Amibara irrigation farms, Middle Awash basin, Ethiopia. Int J Innov Sci Res 13(1):298–314

Abebe A (1982) Screening for rhizobia from different important pulse crops in Ethiopia. MSc thesis, Department of Biology, Addis Ababa University, Addis Ababa, Ethiopia

Abegaz A, Tamene L, Abera W, Yaekob T, Hailu H, Nyawira SS, Sommer R (2020) Soil organic carbon dynamics along chrono-sequence land-use systems in the highlands of Ethiopia. Agric Ecosyst Environ 300:106997

Abebe A (1986) Culture collection of Rhizobium strains of important pulses in Ethiopia. Paper presented at IFS Workshop on Biological Improvement of Soil Fertility, 19–25 March 1986, Dakar, Senegal. http://horizon.documentation.ird.fr/exl-doc/pleins_textes/pleins_textes_4/colloques/24379.pdf

Abera T, Semu E, Debele T, Wegary D, Kim H (2015) Determination soil rhizobium populations, intrinsic antibiotic resistance, nodulation and seed yield of faba bean and soybean in Western Ethiopia. World J Agri Sci 11:311–324

Abere M, Heluf G, Fassil A (2009). Symbiotic effectiveness and characterization of rhizobium strains of faba bean (Vicia faba L.) collected from Eastern and Western Hararghe highlands of Ethiopia. EJNR 11(2):223–244

Abere M, Yifru A, Getahun M (2016) Response of grain legumes to inorganic and biological fertilizers applications in Ethiopia: a review. Ethiop J Natl Resour 16(1):43–67

Abreha KM, Heluf GK, Tekalign M, Kindie T (2012) Characterization of acidic soils in the highlands of Tsegedie, Tigray, Ethiopia. Ethiop J Natl Resour 12:143–168

Adesalu TA, Olugbemi OM (2015) Soil algae: a case study of two vegetable farmlands in Lagos and Ogun states, south west Nigeria. Ife J Sci 17(3):765–772

Adigo S, Sefinew T, Samuel A, Tadesse D, Solomon K (2015) Isolation and phenotypic characterization of phosphate solubilizing bacteria from lentil (Lens culnaris.) Rhizosphere Soils from Southern Parts of Tigray, Ethiopia. Intl J Microbiol Res 6(3):188–194

Adhanom D, Toshome T (2016) Characterization and classification of soils of aba-midan sub watershed in Bambasi Wereda, west Ethiopia. Int J Sci Res Publ 7:390–399

Afolami S, Solomon S, Daramola F (2014) Biodiversity of plant-parasitic nematodes of Sugarcane in Bacita, Nigeria. J Entomol Nematol 6(6):71–79

Agricultural Transformation Agency (ATA) (2013) Status of soil resources in Ethiopia and priorities for sustainable management. In: Proceedings of the global soil partnership (GSP) for eastern and southern Africa launching workshop Nairobi, pp 25–27

Aislabie J, Deslippe JR (2013) Soil microbes and their contribution to soil services. In: Dymond JR (ed) Ecosystem services in New Zealand–conditions and trends. Manaaki Whenua Press, Lincoln, New Zealand

Alem H, Kibebew K, Heluf GK (2015) Characterization and classification of soils of Kabe Sub-watershed in south Wollo Zone, northeastern Ethiopia. Afr J Soil Sci 3:134–146

Alemayehu W (2009) The effect of indigenous root-nodulating bacteria on nodulation and growth of faba bean (Vicia Faba) in the low-input agricultural systems of tigray highlands, Northern Ethiopia. MEJS 1 (2):30–43

Alemayehu K, Sheleme B (2013) Effects of different land-use systems on selected soil properties in southern Ethiopia. J Soil Sci Environ Manage 4:100–107

Alemayehu K, Sheleme B, Schoenau J (2016) Characterization of problem soils in and around the south central Ethiopian Rift Valley. J Soil Sci Environ Manag 7(11):191–203. https://doi.org/10.5897/JSSEM2016.059

Alemayehu K (2018) Soil classification, nature of soil phosphorus, and influence of reclamation on phosphorus forms in soils of the south-central Rift Valley region of Ethiopia. PhD dissertation, Hawassa University, College of Agriculture

Alemu M, Kohlin G (2008) Determinants of household fuel choice in major cities in Ethiopia. Working paper in economics No 399. School of Business, Economics and Law at University of Gothenburg, Sweden, p 19

Amare HS, Priess JA, Veldkamp E, Lesschen JP (2006) Smallholders' soil fertility management in the Central Highlands of Ethiopia: implications for nutrient stocks, balances and sustainability of agroecosystems. Nutr Cycl Agroecosyst 75:135–146

Amha G, Fassil A (2017) The effect of inter cross-inoculation host group rhizobia on the growth and nitrogen fixation of faba bean (Vicia faba L.) varieties in North Showa, Amhara Regional State, Ethiopia. J Agric Biotech Sustain Dev 10(2):25–33

Amha G, Fassil A (2018) The effect of inter cross-inoculation host group rhizobia on the growth and nitrogen fixation of faba bean (Vicia faba L.) varieties in North Showa, Amhara Regional State, Ethiopia. J Agric Biotechnol Sustain Dev 10(2):25–33

Amsal T, Douglas T (2001) Effects of fertilizer application on N and P uptake, recovery and use efficiency of bread wheat grown on two soil types in central Ethiopia. Ethiop J Natl Resour 3:219–244

Anteneh A (2012a) Characterization of symbiotic effectiveness of rhizobia nodulating faba bean (Vicia faba L.) Isolated from Central Ethiopia. Res J Microbiol 7(6):280–296

Anteneh A (2012b) Evaluation of symbiotic effectiveness and size of resident Rhizobium leguminosarum var. viciae nodulating lentil (Lens culinaris medic) in some Ethiopian soils. Int J Agr Agri R 2 (4):18–31

Angaw T, Desta B (1988) Summary of lime trials on different yield of crops. In: Beyene D (ed) Proceedings of soil science research in Ethiopia, Feb 1996. Addis Ababa, Ethiopia

Aragaw A (2014) Response of soybean to inoculation with Bradyrhizobium spp. in saline soils of Shinille plains, eastern Ethiopia. East Afr J Sci 8:79–90

Archer JR, Smith PD (1972) The relation between bulk density, available water capacity, and air capacity of soils. J Soil Sci 23:475–480

Aregu A, Fassil A, Asfaw HM (2012) Symbiotic and phenotypic characterization of rhizobium isolates of field pea (Pisum Sativum L.) fabaceae, From central and southern Ethiopia. Ethiop J Biol Sci 11(2):163–179. The Biological Society of Ethiopia, 2012 ISSN 1819-8678

Asgelil D (2000) Effect of fertilizer on the yield and nodulation pattern of faba bean on a Nitisols of Adet, Northwestern Ethiopia. Ethiop J Nat Resour 2(2):37–44

Asfaw HM (1993) The effect of phosphate solubilizing fungi on the yield of tef in P fixing soils. In: ESTC Research report 4th proceeding. Ethiopian Science and Technology Commination, Addis Ababa, pp. 25–27

Asfaw HM, Angaw T (2006) Food and forage legumes of Ethiopia: progress and prospects. In: Ali K, Kenneni G, Ahmed S, Malhotra R, Beniwal S, Makkouk K (eds) Proceedings of the workshop on food and forage legume, ICARDA/EARO, 22–26 September 2003, Addis Ababa, Ethiopia, pp 172–176

Asmamaw L, Mohammed A (2012) Characterization and classification of the soils of Gerado Catchment, north-eastern Ethiopia. Ethiop J Natl Resour 12:1–22

Assefa A, Heluf GK Kibebew K (2019) Assessment of the physico-chemical status soils and irrigation water qualities in eastern Ethiopia. For Res Eng Int J 3(3):91–104

Assen M, Tegene B (2008) Characteristics and classification of the soils of the plateau of simen mountains national park (Smnp), Ethiopia. Ethiop J Sci 31(2):89–102

Assen M, Yilma S (2010) Characteristics and classification of the soils of gonde micro-catchment, Arsi highlands, Ethiopia. Ethiop J Sci 33:101–116

Ayneabeba A, Asfaw H, Fassil A, Endeshaw B (2001) Studies of rhizobium inoculation and fertilizer treatment on growth and production of faba bean (Vicia faba L.) in some yield-depleted and yield sustained regions of Semen Shoa. SINET: Ethiop J Sci 24:197–211

Ayuke FO (2010) Soil macrofauna functional groups and their effects on soil structure, as related to agricultural management practices across agroecological zones of Sub-Saharan Africa. PhD thesis, Wageningen University, Wageningen, p 211

Balagopalan M, Thomas TP, Mary MV, Sankar S, Alexander TG (1991) Soil properties in teak, bomax and ecualyptus plantation of trchur forest division, Kerla. J Trop For Sci 5:26–34

Barros AHC, van Lier QJ (2014) Pedotransfer functions for Brazilian soils. In: Teixeira WG et al (eds) Application of soil physics in environmental analyses: measuring, modelling and data integration, progress in soil science. https://doi.org/10.1007/978-3-319-06013-2_6. ©Springer International Publishing Switzerland

Baumer OW, Brasher BR (1982) Prediction of water content at selected suctions. Am Soc Agri Eng Pap 82–2590

De Beenhouwer M, Van Geel M, Ceulemans T, Diriba M, Lievens B, Onnay O (2015) Changing soil characteristics alter the arbuscular mycorrhizal fungi communities of Arabica coffee (Coffea arabica) in Ethiopia across a management intensity gradient. Soil Biol Biochem 91:133–139

Bekele G (2005) The Nature and properties of salt affected soils in middle Awash Valley of Ethiopia. J Soil Sci 8:23–30

Belay T (2000) Characteristics and classification of soils of Gora Daget forest, south Welo Highlands, Ethiopia. Ethiop J Sci 23:35–51

Berhanu D (1985) The vertisols of Ethiopia: their properties, classification and management. Fifth meeting of the east African soil correlation and land evaluation sub-committee, 4–10 December 1983. Wad Medani, Sudan. World Soil Resources Report 56. FAO (Food and Agriculture Organization of the United Nations), Rome, Italy, pp 31–54

Berhanu F (1991) A survey of earthworms in western Gojam, Dega Damot Awraja. Soil conservation research project research report 20. University of Berne, Switzerland in association with The Ministry of Agriculture and Environmental Protection, Ethiopia, and The United Nations University

Bernoux M, Chevallier T (2014) Carbon in dryland soils: multiple essential functions. In: International CA (ed) Les dossiers thematiques du CSFD. France

Betre A, Tekalign M, Rolf L (2000) Effects of natural forest conversion to plantation and secondary forests on soil chemical properties in Munessa montane forest, Ethiopia. Ethiop J Natl Resour 2:165–187

Beyene D, Fassil A, Zebene A (2016) Response of selected perennial crops and trees to AMF inoculation and NPK application in pot culture. Aca J Biotech 4:325–336

Birhane E, Kuyper TW, Sterck FJ, Bongers F (2010) Arbuscular mycorrhizal associations in Boswellia papyrifera (frankincense-tree) dominated dry deciduous woodlands of northern Ethiopia. For Ecol Manage 260:2160–2169

Birhane E, Sterck FJ, Fetene M, Bongers F, Kuyper TW (2012) Arbuscular mycorrhizal fungi enhance photosynthesis, water use efficiency, and growth of frankincense seedlings under pulsed water availability conditions. Oecologia 169:895–904

Bossuyt H, Six J, Hendrix PF (2006) Interactive effects of functionally different earthworm species on aggregation and incorporation and decomposition of newly added residue carbon. Geoderma 130:14–25

Brady MV, Hedlund K, Cong RG, Hemerik L, Hotes S, Machado S, Mattsson L, Schulz E, Thomsen IK (2015) Valuing supporting soil ecosystem services in agriculture: a natural capital approach. Agron J 107:1809–1821. https://doi.org/10.2134/agronj14.0597

Brady NC, Weil RR (2015) The nature and properties of soils, 15th edn. Pearson Education Inc., Upper Saddle River, New Jersey, USA

Brown TT, Koenig RT, Huggins DR, Harsh JB, Rossi RE (2008) Lime effects on soil acidity, crop yield, and aluminium chemistry in direct-seeded cropping systems. Soil Sci Soc Am J 72:634–640. https://doi.org/10.2136/sssaj2007.00614

Brussaard L, de Ruiter PC, Brown GG (2007) Soil biodiversity for agricultural sustainability. Agric Ecosys Environ 121:233–244

Cagauan B, Uehara G (1965) Soil anisotropy and its relation to aggregate stability. Soil Sci Soc Am Proc 29:198–200

Chapman HD (1965) Cation-exchange capacity. Methods of soil analysis-chemical and microbiological properties. Agronomy 9:891–901

Chestworth W (ed) (2008) Encyclopedia of soil science. Springer, 3300 AA Dordrecht, The Netherlands

Childs EC (1940) The use of soil moisture characteristics in soil studies. Soil Sci 50:239–252

Cowie RH, Wood TG, Barnett EA, Sands WA, Black HIJ (1990) A checklist of the termites of Ethiopia with a review of their biology, distribution and pest status. Afr J Ecol 28(1):21–33. https://doi.org/10.1111/j.1365-2028.1990.tb01133.x

Croney D, Coleman JD (1954) Soil structure inrelation to soil suction. J Soil Sci 5:75–85

Daniel G (2018) Faunal survey of the termites of the genus Macrotermes (Isoptera: Termitidae) of Ethiopia. J Entomol Nematol 10(7):50–64

Delgado-Baquerizo M, Eldridge DJ, Ochoa V, Gozalo B, Singh BK, Maestre FT (2017) Soil microbial communities drive the resistance of ecosystem multifunctionality to global change in drylands across the globe. Ecol Lett 20:1295–1305. https://doi.org/10.1111/ele.12826

Demisachew TA, Wondimu TA, Jaldesa DL, Abiyot Lelisa D, Amsalu TF (2018) Study on community perception of termite expansion and control in Borana plateau: case study of southern Oromia, Ethiopia. Int J Biodivers Conserv 10(9):365–371. https://doi.org/10.5897/IJBC2018.1190

Demissie N, Degefu T, Ergena A, Ojiewo C (2018) Phenotypic characteristics of rhizobial and non-rhizobial isolates recovered from root nodules of chickpea (CicerarietinumL.) grown in Ethiopia. Afr J Microbiol Res 12:73–85

Dereje T, Fasil A, Heluf G, Gemechu K (2015) Nutritional, ecophysiological and symbiotic characteristics of rhizobia nodulating faba bean (Vicia faba L.) collected from acidic soils of Ethiopia. Afr J Environ Sci Technol 9(7):646–654

DeWispelaere L, Marcelino V, Alemayehu R, De Grave E, Dumon M, Mees F, Van Ranst E (2015) Revisiting nitic horizon properties of Nitisols in SW Ethiopia. Geoderma 243–244:69–79

Diriba M, Fassil A, Granhall U (2007) In vitro Antagonism of rhizobacteria isolated from Coffea arabica L. against emerging fungal coffee pathogens. Eng Life Sci 7(6):577–586

Diriba M, Fassil A, Karin H, Stefan R, Ulf G (2009) Characterization of rhizobacteria isolated from Wild Coffea arabica L. Eng Life Sci 9 (2):100–108

Diriba M, Fassil A, Elisabet B, Ulf, G (2013) Phosphate-solubilising rhizobacteria associated with Coffee arabica L. in natural forests of South Western Ethiopia. J Saudi Soc Agr Sci 12:73–84

Diriba T (2017) Genetic and functional diversity of rhizobia and rhizobacteria from soybean: implication for the commercial production and application to enhance soybean production under low input agriculture in Ethiopia. PhD dissertation thesis. Addis Ababa University, Ethiopia

Dinku D, Sheleme B, Nand R, Fran W, Tekleab G (2014) Effects of topography and land use on soil characteristics along the toposequence of Ele watershed in southern Ethiopia. Catena 115:47–54

Diriba M, Fassil A, Sileshi N, Granhall U (2008) Distribution of arbuscular mycorrhizal fungi spores in soils of smallholder agroforestry and monocultural coffee systems in southwestern Ethiopia. Biol Fertil Soils 44:653–659

Doran JW, Zeiss MR (2000) Soil health and sustainability: managing the biotic component of soil quality. Appl Soil Ecol 15:3–11

Dumon M, Alemayehu R, Capon B, Detavernier C, Van Ranst E (2014) Quantitative clay mineralogy of a Vertic Planosol in southwestern Ethiopia: impact on soil formation hypotheses. Geoderma 241–215:184–196

El Ashkar MA, Bodman GB, Peters DB (1956) Sodium hyposulfite-soluble iron oxides and water retention by soils. Soil Sci Soc Am Proc 20:352–365

Ekern PC (1963) The disposition of net radiation over bare low humic Latosols in Hawaii. (Abstr.) Bull Am Metor Soc 44:89

Emiru B, Fatumah N, Kidane G, Amanuel Z, Mohammed S (2018) Vegetation cover density and disturbance affected arbuscular mycorrhiza fungi spore density and root colonization in a dry Afromontane forest, northern Ethiopia. J for Res 29:675–686

Emiru B, Kide M, Tewodros T, Mengsteab HM, Negasi S (2017) Exclosures restored the density and root colonization of arbuscular mycorrhizal fungi in Tigray, northern Ethiopia. Ecol Proc 6:1–12

Emiru B, Sterck FJ, Bongers F, Kuyper TW (2014) Arbuscular mycorrhizal impacts on competitive interactions between Acacia etbaica and Boswellia papyrifera seedlings under drought stress. J Plan Ecol 1:298–308

Endalkachew F, Kibebew K, Bobe B, Asmare M (2018a) Characterization and classification of soils of Yikalo Sub watershed in Lay Gayint District, north-western Highlands of Ethiopia. Eurasian J Soil Sci 7(2):151–166

Endalkachew WM, Tulu D, Berhanu G, Asnake F, Tilahun A, Chris O (2018b) Phenotypic Characteristics and Preliminary Symbiotic Effectiveness of Rhizobia Associated with Haricot Bean Growing in Diverse Locations of southern Ethiopia. Ethiop J Crop Sci 6 (Special Issue, 2)

Ethiopia Institute of Agricultural Research (EIAR) et al. 2011Ethiopia Institute of Agricultural Research (EIAR), TARI (Tigray Agricultural Research Institute) (EIAR and TARI) (2011) Characterization of farming systems and research directions in south Tigray Zone, Tigray Region, Ethiopia. Unpublished document

Eyasu E (2016) Soil of Ethiopian highlands: geomorphology and properties. CASCAPE Project, ALTERA, Wageningen University and Research Centre (Wageningen UR). The Netherlands, 385pp

Eylachew Z (1999) Selected physical, chemical and mineralogical characteristics of major soils occurring in chercher highlands, eastern Ethiopia. Ethiop J Natl Resour 1:173–185

Eylachew Z (2004) Characterization and genesis of some soils of the Ethiopian rift valley showing andic properties. Ethiop J Natl Resour 6:215–235

FAO, Food and Agricultural Organization of the United Nations (2021) Soils portal. http://www.fao.org/soils-portal/data-hub/soil-classification/numerical-systems/chemical-properties/en/. Accessed 12 January 2021

FAO (2006) Guidelines for aoil description, 4th edn, 109p

FAO (1988) Salt-affected soils and their management. Fao Soils Bulletin 39, Rome, Italy. http://www.fao.org/3/x5871e/x5871e05.htm#4.2.2%20Evaluating%20ESP. Accessed 30 August 2021

FAO (1999). World reference base for soil resources. World Soil Resources Rep. Food and Agricultural Organisation of the UN, Rome, Italy, vol 84, 161p

FAO (2011) Soil health: technologies that save and grow. FAO, Rome

FAO (2017) Soil organic carbon: the hidden potential. Food and Agriculture Organization of the United Nations, Rome

Fan S, Guan F, Xu X, Forrester D, Ma W, Tang X (2016) Ecosystem carbon stock loss after land use change in subtropical forests in China. Forests 7:142

Fassil A, Gemechu K, Negash D (2018) Overview of rhizobial inoculants research and biofertilizer production for increased yield of food legumes in Ethiopia. Ethiop J Crop Sci 6(Special Issue, 3). ISSN 2072-8506

Fekadu G, Assefa M, Dagmawi T (2012) Response of yield components and grain yield of groundnut to application of phosphorus and nitrogen in north-western lowlands of Ethiopia. Ethiop J Natl Resour 12(1):63–74

Fentaw Y, Ledin S, Abdu AK (2007) Changes in organic carbon and total nitrogen contents in three adjacent land use types in Bale mountains, southeastern highlands of Ethiopia. For Ecol Manage 242:337–342

Fentaw A (1996) Effect of subsurface drainage system on groundwater table, soil salinity and crop yield in Melka Sadi pilot drainage scheme. In. Proceedings of the conference of the agronomy and crop physiology society of Ethiopia, Addis Abeba (Ethiopia), 30–31 ACPSE

Feyisa B (2021) Survey and identification of plant parasitic nematodes on faba bean crop in Ethiopia. J Plant Pathol Microbiol 12:561

Fikre M (2003) Pedogenesis of major volcanic soils of the southern central Rift Valley region, Ethiopia. MSc thesis. University of Saskatchewan, Saskatoon, Canada

Fisseha A, Tamrat B, Zerihun B (2019) Comparative field survival and growth of selected Ethiopian native tree species and the effect of whole soil AMF inoculation. J Hort For 11:19–31

Fisseha G, Gebrekidan H (2007) Characterization and fertility status of the soils of ayehu research substation, northwestern highlands of Ethiopia. East Afr J Sci 1:160–169

Foster S, Urbanowitz S, Gatzke H, Schultz B (2016) Soil properties, Part 3 of 3: chemical characteristics, extension | University of Nevada, Reno, Fact Sheet FS-16-02

Fuentes M, Govaerts B, De León F, Hidalgo C, Dendooven L, Sayre KD, Etchevers J (2009) Fourteen years of applying zero and conventional tillage, crop rotation and residue management systems and its effect on physical and chemical soil quality. Eur J Agron 30:228–237

Gebremeskel G, Gebremicael TG, Kifle M, Meresa E, Gebremedhin T, Girmay A (2018) Salinization pattern and its spatial distribution in the irrigated agriculture of Northern Ethiopia: an integrated approach of quantitative and spatial analysis. Agr Water Manag 206:147–157

Geering HR, So HB (2006) Texture. In: Lal R (ed) Encyclopedia of soil science, 2nd edn. Taylor and Francis Group

Gelaw AM, Singh BR, Lal R (2014) Soil organic carbon and total nitrogen stocks under different land uses in a semi-arid watershed in Tigray, northern Ethiopia. Agric Ecosyst Environ 188:256–263

Geressu K, Gezaghegne M (2008) Response of some lowland growing sorghum (Sorghum bicolor L. Moench) accessions to salt stress during germination and seedling growth. Afr J Agric Res 3(1):44–48

Getaneh T (2016) Evaluation of rhizobia and phosphate solublizing rhizobacteria for enhancing growth and yield in Phaseolus vulgaris L. via co-inoculation in greenhouse and field. PhD thesis, School of Graduate Studies, Addis Ababa University, Addis Ababa

Getachew A, Amare G (2004) Effects of drainage and genotype on yield and some yield components of faba bean on a highland vertisols in Ethiopia. Ethiop J Natl Resour 10:67–70

Getachew A, Dejene A, Zewdu E, Fisseha I (2013) The state of science and technology in soil fertility and plant nutrient management research in Ethiopia. In: Proceedings of the workshop on the state of agricultural science and technology in Ethiopia. Ethiopian Academy of Sciences, Addis Ababa

Getachew B, Weldesemayat G, Zenebe G, Fantaw Y, Getachew D (2018) Soil macrofaunal diversity and biomass across different land use systems in wondo genet, Ethiopia. Advances in life science and technology, vol 67. ISSN 2224-7181, ISSN 2225-062X

Getu A, Atsushi T, Nigussie H, Taniguchi T, Menale W, Enyew A, Tsugiyuki M, Mitsuru T, Kindiye E, Mulatu LB, Asaminew T (2020) Effects of land use and topographic position on soil organic carbon and total nitrogen stocks in different agro-ecosystems of the upper blue nile basin. Sustainability 12:2425. https://doi.org/10.3390/su12062425

Girma A, Ravishankar H (2004) Dry matter production and nutrient uptake on potato (Solanum tuberosum L) as influenced by N and P application on nitosols of western Ethiopia. Ethiop J Natl Resour 6:183–197

Girmaye K, Fassil A, Mussie Y, Habteselassie (2017) Phenotypic and genotypic characteristics of cowpea rhizobia from soils of Ethiopia. Afr J Biotechnol 17(42):1299–1312. https://doi.org/10.5897/AJB2016.15863

Gizachew A, Selassie YG, Eyasu E, Chiristy B (2015) Soil classification in yigossa watershed, lake tana basin, highlands of northwestern Ethiopia. J Agric Sci 7:106–116

Grohmann C, Oldeland J, Stoyan D, Linsenmair KE (2010) Multi-scale pattern analysis of a mound-building termite species. Insectes Soc 57(4):367–494

Gupta A, Gopal M, Tilak KV (2000) Mechanism of plant growth promotion by rhizobacteria. Review Article. Indian J Exp Biol 38:856–862

Gunnabo AH, Heerwaarden J Van, GeurtsR, Wolde-meskel E, Degefu T, Giller KE (2020b) Symbiotic interactions between chickpea (Cicer arietinum L.) genotypes and Mesorhizobium strains. Symbiosis 82:235–248. https://doi.org/10.1007/s13199-020-00724-6

Gunnabo AH, Heerwaarden J, Van, Geurts R, Woldemeskel E, Degefu T, Giller KE (2020a) Phylogeography and Symbiotic Effectiveness of Rhizobia Nodulating Chickpea (Cicer arietinum L.) in Ethiopia. PLANT MICROBE INTERACTIONS. Microbial Ecology https://doi.org/10.1007/s00248-020-01620-8

Guteta D, Abegaz A (2017) Dynamics of selected soil properties under four land uses in Arsamma watershed, Southwestern Ethiopian Highlands. Phys Geog 38:83–102

Haile T, Tadele GS, Tekelign M (2000) Assessment of salinity/sodicity problems in abaya state farm, southern rift valley of Ethiopia. Ethiop J Natl Resour 2:151–163

Heluf GK (1985) Investigation on salt affected soils and irrigation water quality in melka sedi-amibara plain, rift valley zone of Ethiopia. MSc thesis, School of Graduate Studies, Addis Ababa University. Addis Ababa, Ethiopia, 131p. 93 19

Heluf GK (1987) Problems and prospects of salt affected soils in Ethiopia (Summary Literature Survey Report). AUA, Department of Plant Science, 17p. 20

Heluf GK (1995) Evaluation of the potential use of langbeinite (K2SO4.2MgSO4) as a reclaiming material for sodic and saline sodic soils. PhD Dissertation, Graduate College, The University of Arizona, USA, 166p

Hillel D (2004) Introduction to environmental soil physics. Elsevier Science (USA), 525 B Street, Suite 1900, San Diego, California

Hodgson JM (ed) (1974) Soil survey field hand-book: describing and sampling soil profiles, soil survey of England and Wales technical monograph, no 5, pp 22–25

Huang PM, Li Y, and Sumner ME (eds) (2012) Handbook of soil science-properties and processes, 2nd edn. Taylor & Francis Group 6000 Broken Sound Parkway NW, Suite 300 Boca Raton, FL 33487-2742

IFDC (International Fertilizer Development Center) (2012) Ethiopia fertilizer assessment. The African fertilizer and agribusiness partnership. https://ifdc.org/

Institute of Biodiversity Conservation (IBC) (2009) Convention on Biological Diversity (CBD) Ethiopia's 4th Country Report. Addis Ababa, Ethiopia

International Center for Biosaline Agriculture (ICBA) (2018a) Soil salinity levels and soil types of the afar natoinal regional state, Ethiopia, Addis Ababa, Ethiopia (unpublished)

International Center for Biosaline Agriculture (ICBA) (2018b) Soil salinity levels and soil types of the amhara national regional state, Ethiopia, Addis Ababa, Ethiopia (unpublished)

International Center for Biosaline Agriculture (ICBA) (2018c) Soil salinity levels and soil types of the oromia national regional state, Ethiopia, Addis Ababa, Ethiopia (unpublished)

International Center for Biosaline Agriculture (ICBA) (2018d) Soil salinity levels and soil types of the tigray natoinal regional state, Ethiopia, Addis Ababa, Ethiopia (unpublished)

International Soil Resources Information Center (ISRIC) SoilGrids (2017) FAQ SoilGrids-Global gridded soil information. WUR, The Netherlands. https://www.isric.org/explore/soilgrids. Accessed 31 March 2021

International Union of Soil Science (IUSS) (2015) World reference base for soil resources 2014, update 2015 international soil classification system for naming soils and creating legends for soil maps. Soil Resources Reports No. 106. FAO, Roma IT EU, 192 p

Isreal Z, Tana T, Wogi L, Mohammed A (2018) Soil morphology, physico-chemical properties and classification of typical soils of abelo area masha district south western Ethiopia. Adv Crop Sci Tech 6:341

Jafer D, Gebresilassie H (2017) Application of lime for acid soil amelioration and better soybean performance in south-western Ethiopian journal of biology. Agric Healthc 7(5):95–100

Jenny H (1941) Factors of soil formation: a system of quantitative pedology. McGraw-Hill, New York

Jobbagy EG, Jackson RB (2001) The distribution of soil nutrients with depth: global patterns and the imprint of plants. Biogeochemistry 53:51–77

Jutsi SC (1988) Contribution of vertisols to national food self-sufficiency: actual and potentials. Ethiop J Natl Resour 6:167–182

Jónsson JÖG, Davíðsdóttir B (2016) Classification and valuation of soil ecosystem services. Agric Syst 145:24–38. https://doi.org/10.1016/j.agsy.2016.02.010

Kassa B, Ameha K, Fassil A (2015) Isolation and phenotypic characterization of field pea nodulating rhizobia from Eastern Ethiopia. Soils World Appl Sci J 33(12):1815–1821

Kern JS (1995) Evaluation of soil water retention models based on basic soil physical properties. Soil Sci Soc Am J 59:1134–1141

Kibblewhite M, Chambers B, Goulding K (2016) How good is the evidence to support investment in soil protection? Soil Use Manage 32:172–182. https://doi.org/10.1111/sum.12236

Kidane G, Abebe F, Heluf GK, Fentaw A, Wondimagegne C, Hibstu A, Asegid A, Messele F, Mohammed B (2006) Assessment of salt affected soils in Ethiopia and recommendatiosn on management options for their sustainable utilization. EIAR, Addis Ababa, Ethiopia

Kidd PS, Proctor J (2001) Why plants grow poorly on very acidic soils: are ecologists missing the obvious? J Exp Botany 52:791–799. https://doi.org/10.1093/jexbot/52.357. Laboratory, Addis Ababa, Ethiopia

Kutilek M (1973) The influence of soil minerals and exchangeable cations on soil moisture potential. In: Hadas X et al (eds) Physical aspects of soil water and salts in ecosystems. Springer, New York

Lal R (2010) Beyond Copenhagen: mitigating climate change and achieving food security through soil carbon sequestration. Food Secur 2(2):169–177

Lal R, Shukla M (2004) Principles of soil physics. Marcel Dekker, Inc., 270 Madison Avenue, New York, NY 10016, USA

Lambooy AM (1984) Relationship between cation exchange capacity, clay content and water retention of Highveld soils. S Afr J Plant Soil 1:33–38

Laban P, Graciela M, Jonathan D (2018) Soil biodiversity and soil organic carbon: keeping drylands alive. Gland, Switzerland: IUCN. 8:24p

Landon JR (1991a) Booker tropical soil manual. A handbook for soil survey and agricultural land evaluation in the tropics and sub-tropics. Longman Scientific & Technical Publ, Harlon

Landon JR (1991b) Booker tropical soil manual: a handbook for soil survey and agricultural land evaluation in the tropics and sub-tropics. Longman Scientific and Technical, 474pp

Legass A, Assen M (2012) Characteristics and classification of the soils of gerado catchment, northeastern Ethiopia. Ethiop J Natl Resour 12:1–22

Loveland P, Webb J (2003) Is there a critical level of organic matter in the agricultural soils of temperate regions: a review? Soil Tillage Res 70:1–18.

Lundgre B (1978) Soil conditions and nutrient cycling under natural and plantation forests in Tanzanian highlands. Reports in forest ecology a forest soils No. 31, Department of Forest Soils, Swedish University of Agricultural Sciences, Uppsala, Swede

Mamo T, Dibabe A (1994) Soil microbiology research. Cool-season food legumes of Ethiopia. In: Telaye A, Bejiga G, Saxena MC, Solh MB (eds) Proceedings of the first national cool-season food legumes review conference, 16–20 December 1993, Addis Ababa, Ethiopia. ICARDA/Institute of Agricultural Research, Syria, pp 293–311

Mazid M, Khan TA, Mohammad F (2011) Potential of NO and H2O2 as signaling molecules in tolerance to abiotic stress in plants. J Ind Res Technol 1:56–68

McIntyre DS (1974) Water retention and moisture characteristics. In: Loveday J (ed) 'Methods for analysis of irrigated soils'. Technical communication No. 54. Commonwealth agriculture bureau, farnham royal, England, pp 51–52

Mekasha T, Fassil A, Jemaneh Z (2015) Symbiotic and phenotypic characterization of Rhizobium isolates nodulating fenugreek (Trigonellafoenumgraecum L.) from North and East Shewa, Ethiopia. Int J Agri Agri R 7 (1): 93–104

Meron N (2007). Characterization of salt affected soils in the central Rift Valley and assessing salt tolerance of different plants: a case study at the southwestern shore of Lake Ziway, p 75. MSc Thesis. Addis Abeba University

Meshesha DT, Tsunekawa A, Tsubo M, Ali SA, Haregeweyn N (2014) Land-use change and its socio-environmental impact in Eastern Ethiopia's highland. Reg Environ Change 14:57–768

Miheretu BA, Abegaz A (2017) Spatial variability of selected soil properties in relation to land use and slope position in Gelana sub-watershed, Northern highlands of Ethiopia. Phys Geogr https://doi.org/10.1080/02723646.2017.1380972

Mekete TRA, Sikora S, Kiewnick, Hallmann J (2008) Plant-parasitic nematodes associated with coffee (Coffea arabica L., Rubiaceae) in Ethiopia. Nematropica 38(2):177–186

Mengsteab HM, Emiru B, Zebene A, Solomon Z (2013) Arbuscular mycorrhizal association of indigenous agroforestry tree species and their infective potential with maize in the rift valley, Ethiopia. Agrofor Syst 87:1–14

Mengsteab HM, Emiru B, Girmay GS, Kiros H, Mowo JG (2014) Arbuscular mycorrhizal association and growth response of Faidherbia albida (Del.) A. Chev. as influenced by land use type. A paper presented at World Conference on Agroforestry, 10–14 February, 2014. New Delhi, India

Merga B, Ahmed A (2019) A review on agricultural problems and their management in Ethiopia. Turk J Agric Food Sci Technol 7 (8):1189–1202

Mesfin A (1998) Natures and management of Ethiopian soils. Alemaya University of Agriculture, Ethiopia, p 272

Mesfin A (2007) Nature and management of acid soil in Ethiopia. Haramaya University, Haramaya, Ethiopia, p 99p

Michelsen A (1992) Mycorrhiza and root nodulation in tree seedlings from five nurseries in Ethiopia and Somalia. For Ecol Man 48:335–344

Michelsen A (1993) Growth improvement of Ethiopian acacias by addition of vesicular-arbuscular mycorrhizal fungi or root of native plants to non sterile nursery soil. For Ecol Manage 59:193–206

Minale L, Douglas T, Alemayehu A, Tllahun T (2001) Cropping sequence and fertilizer effects on yield and profitability in s wheat-based crop rotation trial in north-western Ethiopia. Ethiop J Natl Resour 3:195–218

Mitiku H (1987) Genesis, characterization, and classification of soils of the central highlands of Ethiopia. A thesis submitted the State University of Gent, Belgium

Mizota C, Chapelle J (1988) Characterization of some andepts and andic soils in Rwanda, Central Africa. Geoderma 41:193–209

Mizota C, van Reeuwijk LP (1989) Clay mineralogy and chemistry of soils formed in volcanic material in diverse climatic regions. International soil reference and information center, The Netherlands. Soil Monogr 2

MoA (Ministry of Agriculture) (2010) Agricultural inputs and outputs marketing directorate, MoA, Addis Ababa, Ethiopia

Mohammed A, Solomon T (2012) Characteristics and fertility quality of the irrigated soils of Sheneka, Southern Ethiopia. Ethiopian J Nat Res 12(2):116–135

Mohamed S, Tessema G (2013) Evaluation of soil and water salinity for irrigation in north-eastern Ethiopia: case study of Fursa small scale irrigation system in Awash River Basin. Afr J Environ Sci Technol 7(5):167–174

Montañe A (2000) Overview and case studies on biological nitrogen fixation: perspectives and limitations. Food and Agriculture Organization of the United Nations (FAO), Rome

Mulat A, Girma W, Mekiso Y, Solomon Y, Endalkachew WM, Alemayehu C, Jessica GD (2019) Comparison of cyanobacterial bio-fertilizer with urea on three crops and two soils of Ethiopia. Afr J Agric Res 14(10):588–596

Muleta D, Assefa F, Nemomissa S, Granhall U (2007) Composition of coffee shade tree species and density of indigenous arbuscular mycorrhizal fungi (AMF) spores in Bonga natural coffee forest, southwestern Ethiopia. For Ecol Manage 241:145–154

Mulugeta F, Mohd SA, Diriba M, Fikre L (2013) Isolation and characterization of nitrogen deficit rhizobium isolates and their effect on growth of haricot bean. Afr J Agric Res 8(46):5942–5952

Mulisa J, Fassil A (2011) Phenotypic and plant growth promoting characteristics of Rhizobium leguminosarum bv. viciae from lentil growing areas of Ethiopia. Afr J Microbiol Res 5(24): 4133–4142

Mulissa J, Fassil A (2012) Phenotypic diversity and plant growth promoting characteristics of mesorhizobium species isolated from chickpea (Cicer arietinum L.) growing areas of Ethiopia. Afr J Biotechnol 11(29):7483–7493. https://doi.org/10.5897/AJB11.4109

Mulissa JM, Carolin RL, Ruth AS, Fassil A (2015) Characterization of phosphate solubilizing rhizobacteria isolated from lentil growing areas of Ethiopia. Afr J Microbiol Res. 9(25):1637–1648

Mulissa J, Löscher CR, Schmitz RA, Fassil A (2016) Phosphate solubilization and multiple plant growth promoting properties of rhizobacteria isolated from chickpea (Ciceraeritinum L.) producing areas of Ethiopia. Afr J Biotechnol 15(35):1899–1912

Muluneh M, Zinabu W (2013) Effect of cyanobacteria application as biofertilizer on growth, yield and yield components of romaine lettuce (Lactuca sativa L.) on soils of Ethiopia. Am Sci Res J Eng Technol Sci (ASRJETS) 4(1):50–58

Mussa A (2009) Phenotypic and symbiotic characterization of grass pea (Lathyrus sativus) rhizobial isolates from some major growing areas of South Wollo and West Shoa, Ethiopia. MSc. thesis. Addis Ababa University, Addis Ababa, Ethiopia

Nega R, Etisa D, Lule W (2020) Temporal and spatial microalgae diversity in lake bishoftuguda/babugaya, Ethiopia. J Ecol Natl Resour 4(1). https://doi.org/10.23880/jenr-16000184

Negash D, Yohannes B, Waktoli S (2018) Effect of different earthworm feedstocks on vermicompost quality and local earthworm performance. In: Getachew A, Gebreyes G, Tolera A, Daniel M (eds) Soil fertility and plant nutrient management. Ethiopian Institute of Agricultural Research, Addi Ababa, Ethiopia, pp 193–202

Nimmo JR (2004) Porosity and pore size distribution. In: Hillel D (ed) Encyclopedia of soils in the environment, vol 3. Elsevier, London, pp 295–303

Nieder R, Benbi DK (2008) Carbon and Nitrogen in the terrestrial environment. Springer, Dordrecht. https://doi.org/10.1007/978-1-4020-8433-1

Okubay G (2019) Salinity and sodicity hazard characterization in major irrigated areas and irrigation water sources, northern Ethiopia. Cogent Food Agric 5:1–15

Orgiazzi A, Bardgett RD, Barrios E, Behan-Pelletier V, Briones MJ, Chotte JL, De Deyn GB, Eggleton P, Fierer N, Fraser T (2016) Global soil biodiversity atlas. Publications of_ce of the European Union

PGRC (1996) Ethiopia: country report to the FAO international technical conference on plant genetic resources, Addis Ababa, Ethiopia

Pachepsky YA (1989) The effect of the content of soil solutions and exchangeable cations on water retention and hydraulic conductivity of soils. Pochvovedenie 3:53–65

Pal D, Selassie YG (2018) Soil physicochemical properties and their significance for sustainable sugarcane production in Kesem Allaideghe plains irrigation project area, eastern Ethiopia. Ethiop J Sci Technol 11(1):19–28

Paulos D (2002) Present and future trends in natural resources management in agriculture: an overview. Integrated water and land management research and capacity building priorities for Ethiopia

Pekrun C, Kaul H, Claupein W (2003) Soil tillage for sustainable nutrient management. In: El Titi A (ed). Soil tillage in agroecosystems. CRS Press, Florida, USA. 367

Phogat VK, Tomar VS, and Dahiya R (2015) Soil physical properties

Ping CL, Shoji S, Ito T (1988) Properties and classification of three volcanic ash-derived pedons from Aleutian Islands and Alaska Peninsula, Alaska. Soil Sci Soc Am J 52:455–462

Qureshi AS, Tesfaye E, Melese M (2018) Prospects of alternative copping systems for salt affected soils in Ethiopia. J Soil Sci Environ Manage 9(7):98–107

Rawls WL, Gish TJ, Brakensiek DL (1991) Estimating soil water retention from soil physical properties and characteristics. Adv Soil Sci 16:213–234

Reeve MJ and Carter AD (1991) Water release characteristic. In: Smith KA, Mullins CE (eds) Soil analysis: physical methods. Dekker, New York, pp 111–160

Ruffeis D, Loiskandl W, Spendlingwimmer R, Schonerklee M, Awulachew SB, Boelee E, Wallner K (2008) Environmental impact analysis of two large scale irrigation schemes in Ethiopia

Sardo V (2005) Halophytes and salt-tolerant glycophytes: a potential resource. In: Hamdy A (ed) The use of non-conventional water resources. CIHEAM/EU DG Research, Bari, pp 87–98

Seelig, BD (2000) Salinity and sodicity in north Dakota soils. EB-57. North Dakota State University, Fargo, ND

Selamawit A, Kidane, Beira H, Meressa, Solveig H, Trine HE, Christer M, Marjolein C, Wim B, Danny LC (2020) Occurrence of plant-parasitic nematodes on enset (Ensete ventricosum) in Ethiopia with focus on Pratylenchus goodeyi as a key species of the crop. Nematology 1–13https://doi.org/10.1163/15685411-bja10058

Selassie YG (2002) Selected chemical and physical characteristics of soils of adet research center and its testing sites in north-western Ethiopia. Ethiop J Natl Resour 4:199–215

Seid A, Fininsa C, Mekete T, Decraemer W, Wesemael WML (2015) Tomato (Lycopersicon esculentum L.) and root-knot nematodes (Meloidogyne spp)—century old battle. Nematology 17:995–1009

Sharma ML, Uehara G (1968) Influence of soil structure on water relations in low humic latosols. 1. Water retention. Soil Sci Soc Am Proc 32:765–770

Shegena Z, Suryabhagavan KV, Balakrishnan M (2017) Geo-spatial approach for soil salinity mapping in Sego Irrigation Farm, south Ethiopia. J Saudi Soc Agric Sci 16(1):16–24

Sheleme B (2017) Topographic positions and land use impacted soil properties along humbo larena-ofa sere toposequence, southern Ethiopia. J Soil Sci Environ Manage 8:135–147. https://doi.org/10.5897/JSSEM2017.0643

Shimelis D, Mohammed A, Abayneh E (2007) Characteristics and classification of the soils of tenocha-wenchacher micro-catchment, south-west Shewa, Ethiopia. Ethiop J Natl Resour 9:37–62

Sileshi GW, Elias K, Patrick M, Philip ON (2008) Farmers' perceptions of tree mortality, pests and pest management practices in agroforestry in Malawi, Mozambique and Zambia. Agrofor Syst 72:87–101

Sileshi A (2015) Temporal and spatial variations in of salt-affected soils using GIS and remote sensing at Dubti/Tendaho state farm. PhD dissertation, Haramaya University, Ethiopia
Skopp J (2012) Physical properties of primary particles. In: Huang PM et al (ed) Handbook of soil sciences: properties and processes. Taylor &Francis Group, New York
Solomon T (2006) Effects of inoculation by bradyrhizobium japonicum strains on nodulation, nitrogen fixation, and yield of soybean (Glycine max L. Merill) varieties on nitisols of bako, western Ethiopia. MSc thesis. Haramaya University, Haramaya
Solomon Z, Fassil A, Masresha F (2000) Effect of Azospirillum isolates on the growth and nitrogen content of tef (Eragrostis tef (zucc.) Trotter). SINET Ethiop J Sci 23(2):247–262
Solomon L, Fassil A (2013) Symbiotic and phenotypic characteristics of rhizobia nodulating faba bean (Vicia Faba) from Tahtay Koraro, Northwestern Zone of Tigray Regional State, Ethiopia IJEERT 2 (11):15–23
Sun W, Zhu H, Guo S (2015) Soil organic carbon as a function of land use and topography on the Loess Plateau of China. Ecol Eng 83:249–257
Taddese GS (2001) Land degradation: a challenge to Ethiopia. Environ Manage 27(6):815–824. https://doi.org/10.1007/s002670010190
Taddese GS, Sonder K, Peden D (2003) The water of the Awash River Basin a future challenge to Ethiopia International Livestock Research Institute, Addis Ababa
Tadelle GS (1993) Degradation problem of irrigated agriculture, national soil service laboratory, Addis Ababa, Ethiopia
Tadesse C, Fassil A (2013) Arbuscular mycorrhizal fungi associated with shade trees and Coffea arabica L. in a coffee-based agroforestry system in Bonga, southwestern Ethiopia. Afr Foc 26:111–131
Tekalegn M, Killham KS (1987) Effects of soil liming and vescicu-lararbuscular mycorrhiza inoculation on the growth and micronutrient content of the teff plant. Plant Soil 102:257–259
Tekalign M (1987) Vesicular arbuscular mycorrhiza: potential biofertilizers. Soil Science Research Bulletin, ILCA 135–139
Tena W, Woldemeskel E, Walley F (2016a) Symbiotic efficiency of native and exotic rhizobium strains nodulating lentil (Lens culinaris Medik.) in soils of southern Ethiopia. Agronomy 6(11):1–10
Tena W, Woldemeskel E, Walley F (2016b) Response of chickpea (Cicer arietinum L.) to inoculation with native and exotic Mesorhizobium strains in southern Ethiopia. Afr J Biotechnol 15(35):1920–1929
Tenalem A (2007) Water management problems in the Ethiopian rift: challenges for development. J Afr Earth Sc 48(2–3):222–236. https://doi.org/10.1016/j.jafrearsci.2006.05.010
Tesfaye W, Weiß M, Kottke I, Oberwinkler F (2006) Two threatened coexisting indigenous conifer species in the dry Afromontane forests of Ethiopia are associated with distinct arbuscular mycorrhizal communities. Can J Bot 84:1617–1627
Tesfaye W, Kottke I, Demel T, Oberwinkler F (2009) Arbuscular mycorrhizal fungal community structures differ between co-occurring tree species of dry Afromontane tropical forest, and their seedlings exhibit potential to trap isolates suited for reforestation. Mycol Prog 8:317–328
Tulu D, Endalkachew WM, Asa F (2011) Multilocus sequence analyses reveal several unnamed Mesorhizobium genospecies nodulating Acacia species and Sesbania sesban trees in southern regions of EthiopiaInt J Syst Evol Microbiol 55:1439–1452. https://doi.org/10.1099/ijs.0.63534-0
Tulu D, Endalkachew WM, Frank R (2018b) Genetic diversity and symbiotic effectiveness of Bradyrhizobium strains nodulating selected annual grain legumes growing in Ethiopia. Int J Syst Evol Microbiol 68:449–460https://doi.org/10.1099/ijsem.0.002486
Tulu D, Endalkachew WM, Makka A, Asnake F, Tilahun A, Chris O (2018a) Morphophysiological diversity of rhizobia nodulating pigeon pea (Cajanus cajan L. Millsp.) growing in Ethiopia. Afr J Biotechnol 17(6):167–177. https://doi.org/10.5897/AJB2017.16338
Turmel MS, Speratti A, Baudron F, Verhulst N, Govaerts B (2015) Crop residue management and soil health: a systems analysis. Agric Syst 134(6–16). https://doi.org/10.1016/j.agsy.2014.05.009
UNEP, United Nations Environment Programme (2000) Finding alternatives to persistent organic pollutants (POPS) for termite
US Salinity Laboratory Staff (USSLS) (1954) Diagnosis and improvement of saline and alkali soils. US Department of Agriculture Handbook 60, Washington, DC
USDA Natural Resources Conservation Service (USDA-NRCS) (1998) Soil Quality Indicators: pH. National Soil Survey Center in cooperation with the Soil Quality Institute, NRCS, USDA, and the National Soil Tilth Laboratory, Agricultural Research Service, USDA), Washington DC, USA. https://www.nrcs.usda.gov/Internet/FSE_DOCUMENTS/nrcs142p2_052208.pdf. Accessed 30 March 2021
van Beek EE, Selassie YG, Girmay GS, Asrat T, Feyisa H, Mekonnen T, Melmuye M, Gebremeskel Y, Seyoum M (2018) Soil organic matter depletion as a major threat to agricultural intensification in the highlands of Ethiopia. Ethiop J Sci Technol 11(3):271–285
Wang J, Raman H, Zhang G, Mendham N, Zou M (2006) Aluminium tolerance in barely (Horidium vulgarie L.): Physiological mechanisms, genetics and screening methods. J Zhejiang Univ Sci 7:769–787
Wassie H, Shiferaw B (2009) Mitigation of soil acidity and fertility decline challenges for sustainable livelihood improvement: research findings from southern region of Ethiopia and its policy implications
Wegene N (2019) Review on the extent of acid soil in Ethiopia, its impact and management methods. J Biol Agric Healthc 9(15):24–36
Whalen JK, Parmelee R, Subler S (2000) Quantification of nitrogen excretion rates for three lumbricid earthworms using 15N. Biol Fert Soils 32:347–352
Williams J, Ross P, Brlstow K (1990) Prediction of the Campbell water retention function from texture, structure, and organic matter. In: Proceedings of international workshop on indirect methods for estimating the hydraulic properties of unsaturated soils. USDA-ARS/University of California, Riverside, CA
Wolde-Meskel E (2007) Genetic diversity of rhizobia in Ethiopian soils: their potential to enhance biological N fixation (BNF) and soil fertility for sustainable agriculture. Ethiop J Biol Sci 6:77–95. https://doi.org/10.4314/ejbs.v6i1.39042
Wondwosen B, Asresie H, Tesfaye A, Yihenew GS, Tewodros T, Firew T (2015) Effect of liming on yield of barley in Ethiopian highlands. In Eyasu E, van CLB (eds) Scaling innovations and agricultural best practices in Ethiopia: experiences and challenges, pp 78–87
Woldeamlak B, Abebe S (2013) Land-use and land-cover change and its environmental implications in a tropical highland watershed, Ethiopia. Int J Environ Stud 70(1):126–139
World Bank (1995) Staff appraisal report. National Fertilizer Sector Project, Ethiopia, Report No. 13722-ET
Wubet T, Weiß M, Kottke I, Teketay D, Oberwinkler F (2003) Molecular diversity of arbuscular mycorrhizal fungi in Prunus africana, an endangered medicinal tree species in dry Afromontane forests of Ethiopia. New Phytol 161:517–528
Xiao C (2015) Soil organic carbon torage (sequestration) principles and management: potential role for recycled organic materials in agricultural soils of Washington State. https://fortress.wa.gov/ecy/ublications/SummaryPages/1507005.html.
Yeshanew A, Zech W (2013) Litter production and nutrient cycling in two plantations and a Podocarpus falcatus dominated natural forest ecosystems in southeastern highlands of Ethiopia. Afr J Agr Res 8:4810–4818

Yesuf A, Duga D (2000) The effect of rate and time of N application on its uptake by bread wheat varieties and soil characteristics on farmers' fields. Ethiopian J Nat Res 2:137–150

Yimer F, Ledin S, Abdelkadir A (2007) Changes in soil organic carbon and total nitrogen contents in three adjacent land use types in the Bale Mountains, south-eastern highlands of Ethiopia. For Ecol Manag 242:337–342

Yitagesu N, Hayelom B, Abere M (2018) Evaluation of different earthworm species and food sources for their vermiculture and vermicompost outputs in Raya Azebo district, south Tigray. Ethiopian Institute of Agri. Rese. Natural Resources Managment Research Directorate Research proceeding. ISBN 978-99944-66-52-8

Yohannes Y, Assefa F (2007) Phenotypic characteristics of root nodule bacteria and arbuscular mycorrhizal fungi infecting Acacia polyacantha growing in Ghibe wooded grasslands. Ethio J Natu Sci 9 (1):123–139

Yoseph T, Purahong W, Blazevic A, Yitaferu B, Wubet T, Göransson H, Godbold DL (2017) Changes in land use alter soil quality and aggregate stability in the highlands of northern Ethiopia. Sci Rep 7 (13602):1–12

Zaman M, Shabbir A, Shahid Heng L (2018) Guideline for salinity assessment, mitigation and adaptation using nuclear and related techniques. FAO/IAEA

Zebene A (2003) Tree species diversity, top soil conditions and arbuscular mycorrhizal association in the sidama traditional agroforestry land-use, southern Ethiopia. PhD dissertation. Swedish University of Agriculture, Uppsala, Sweden

Zenebe GE (2007) Household fuel and resource use in rural-urban Ethiopia. PhD dissertation, Wageningen University, the Netherlands

Zerihun B, Vestberg M, Fassil A (2013) Diversity and abundance of arbuscular mycorrhizal fungi associated with acacia trees from different land use systems in Ethiopia. Glob J Microbiol 7:24–35

Zerihun B, Vestberg M, Fassil A (2015) Diversity and abundance of Arbuscular Mycorrhizal fungi across different land use types in a humid low land area of Ethiopia. Trop Subtrop Agroecosyst 18:47–69

Zerihun B, Fassil A (2011) Symbiotic and phenotypic diversity of Rhizobium leguminosarum bv. viciae from northern Gondar. Ethiop Afr J Biotechnol 10(21):4372–4379. https://doi.org/10.5897/AJB10.431

Zerihun G, Tigist A, Lejalem A, Bekele A (2018) Vermicompost potential of common Earthworms (Eudrilus eugeniae) and Red Wiggler (Eisenia fetida) worm on the decomposition of various organic wastes. Int J Plant Soil Sci 24(3):1–13. Article no. IJPSS.42986 ISSN 2320-7035

Zewdie E (2013) Properties of major agricultural soils of Ethiopia. LAP LAMBERT Academic Publishing

8 Soil Fertility and Soil Health

Kibebew Kibret, Sheleme Beyene, and Teklu Erkossa

Abstract

Soil is a fundamental resource that supports the production of food and other environmental services. Soil fertility, together with other production factors, is a key soil attribute that determines agricultural productivity and food security, particularly in developing countries. A fertile soil has to provide growth requirements, as determined by its physical, chemical, and biological attributes, at optimum level in order to get optimum benefits. Soil fertility is an indicator of soil health in that fertile soils (in all the three components) are generally healthy. However, the fertility status of most Ethiopian soils is considered to be poor due to severe soil erosion, high levels of nutrient mining, and insignificant use of external inputs, with likely negative consequences on soil health. Deficiency of N and P is the most common problem in almost all the soils in the country. The different forms of degradation to which soils of the country have been prone to have resulted in deterioration of soil quality and soil health. Although empirical evidences on soil sustainability and resilience are none to scant, the decline in soil quality is expected to reflect negatively on soil sustainability and soil resilience. Restoration interventions that improve soil quality, enhance soil sustainability, and build soil resilience are required to ensure a sustainable agriculture in the country. Ensuring soil sustainability and soil resilience requires a shift in policy attention. It is high time that soils receive more respect, value, and policy consideration than before. It is only then that soil resources will be able to support the multiple functions that are central to the survival of human beings, animals, plants, and maintenance of environmental quality. More research should be done on soil quality, sustainability, and resilience as well as their interrelationship for a better understanding of factors and processes that affect their attainment and develop appropriate management interventions.

Keywords

Agroecology • Degradation • Health • Quality • Resilience • Sustainability

8.1 Introduction

Ethiopia, with a population of about 120 million, is an agrarian country. The livelihood of about 80% of its population relies on agriculture, a sector that has been and is serving as the pillar of the country's economy with significant contribution to the national GDP, export earnings, and raw materials to the industry as well as the service sectors. However, low productivity has been the hallmark of the sector since the past several decades. It is besieged by a large number of factors that span across, but to mention a few, environmental, cultural, political, and socio-economic dimensions and their intricate and dynamic interactions. A wealth of literature shows that the success of a country's agriculture depends to a greater extent on the status of a soil resource assessed in terms of conditions or attributes that dictate how it performs its functions. Although it is hard to make a concrete generalization, given the high spatial and temporal variability of soil attributes, high diversity of soils, and the manipulations they are exposed to in the name of 'use', Ethiopian soils have been subject to various forms of heinous degradation. The dominant forms of degradation

K. Kibret (✉)
School of Natural Resources Management and Environmental Sciences, Haramaya University, P.O. Box 138 Dire Dawa, Ethiopia
e-mail: kibebew.kibret@haramaya.edu.et

S. Beyene
School of Plant and Horticultural Sciences, Hawassa University, Hawassa, Ethiopia

T. Erkossa
Duetsche Gesselschaft Fuer Zusammenarbeit, GIZ, Addis Ababa, Ethiopia

S. Beyene et al. (eds.), *The Soils of Ethiopia*, World Soils Book Series,
https://doi.org/10.1007/978-3-031-17012-6_8

include soil erosion, fertility depletion, acidification, and salinization/sodification, all of which resulted in undesirable consequences to soil fertility and quality. Soil erosion by water is particularly an important form of soil degradation in the highlands (Tefera et al. 2002) although wind erosion is not totally absent in the arid regions of the country. Soil erosion removes several tonnes of soil per year with estimates varying across the country. Bewket and Sterk (2003) estimated an average soil loss rate that reaches up to 42 Mg ha^{-1} year^{-1} from cultivated lands across the country, while Hurni et al. (2015) estimated 20 Mg ha^{-1} year^{-1} and 33 Mg ha^{-1} year^{-1} from currently and formerly cultivated lands, respectively.

Nutrient deficiencies, particularly nitrogen and phosphorus, are very common among soils found in different agroecologies across the country and negative nutrient balances are the typical features of most farming systems. Low levels of organic matter are the characteristic features of large proportions of soils, particularly the cultivated ones, in the country. Due to the use of crop residues and manures for other purposes (e.g., as fuel, animal feed, construction, source of cash) and application of other organic inputs such as compost at a rate that is far below the required dose, soil feeding is not commensurate with soil mining. The different degradative processes have resulted in decline in soil quality, which is evidenced by the low productivity. Limited studies conducted across the country indicate the presence of soil quality degradation in relation to physical, chemical, and biological attributes (e.g., Erkossa et al. 2020, 2009; Tefera et al. 2002; Hawando 1997). Deterioration in soil quality is expected to negatively affect soil sustainability and soil resilience, two important concepts introduced to the field of Soil Science very recently. Soil sustainability is the maintenance of soil functionality (Ludwig et al. 2018), while soil resilience is defined as the capacity of a soil to recover from shocks, disturbances, and stresses and as such is associated with the soil's ability to maintain its functional and structural integrity. Soil quality, sustainability, and resilience are interrelated concepts, but with some dissimilarities. The interplay among these attributes determines the soil's capacity to perform its multiple functions. Maintaining good soil quality is the basis for soil sustainability and resilience and in turn resilient soils withstand the impacts of various shocks and/or bounce within the shortest time possible. Periodic evaluation of soil quality, sustainability, and resilience using appropriate indicators is essential for monitoring negative changes and abating undesirable consequences.

Although soils are indisputably one of the most basic natural resources that perform multiple life-supporting functions, the attention given to them at a global level has been at best a lip service. As a consequence, significant proportion of the world's soils are either already exhausted or are on the verge of exhaustion. If this continues unabated, producing food that can feed the continuously increasing world population will be jeopardized. It is, therefore, high time that soils get the recognition that measures up to the multiple ecosystem services that they provide to humans and other living things. Soil quality, sustainability, and resilience should be priority concerns in policies, strategies, interventions, and investments pertinent to soil resources.

8.2 Soil Fertility

8.2.1 Principles

In the past, the concept of soil fertility has generally been most concerned with soil chemical fertility. However, recent literature sources (e.g., Abbott and Murphy 2007) disagree with this perception and claim that soil fertility should be viewed as a sum total of physical, chemical, and biological state of a soil. Its comprehensive definition should take these aspects into account instead of just focusing on supply of certain chemical elements that plants need (SSSA 1997). Recent definitions provided by Abbott and Murphy (2007) and the *Global Soil Partnership* (not dated) fit well into this requirement. Accordingly, soil fertility has been defined by Abbott and Murphy (2007) as "the capacity of soil to provide physical, chemical, and biological requirements for growth of plants for productivity, reproduction and quality relevant to plant type, soil type, land use, and climatic conditions". The *Global Soil Partnership* also provided almost a similar definition. The physical, chemical, and biological fertilities refer to the capacity of a soil to provide physical conditions, chemical and nutritional environment, and biological conditions, respectively, that support the successful and optimum growth and development of plants and animals. Through their interaction, the three components of soil fertility support each other in that the physical fertility creates conditions that enhance the chemical and biological processes and vice versa, with a result that enhances the overall soil fertility.

The status of the three components in a soil influences the overall health of a soil. Soil health is defined as the continued capacity of the soil to function as a vital living system, within ecosystem and land-use boundaries, to sustain biological productivity, maintain or enhance environmental quality, and promote plant, animal, and human health (Doran and Zeiss 2000; Doran et al. 1996, 1998). For a soil to be healthy, it has to be in sound physical, chemical, and biological conditions that enable it to continue providing ecosystem services. The definition shows the link between soil health and human health. Through their effect on food, water, and air qualities, soils influence directly or indirectly the human health. Fertile soils are the foundation for production of healthy food with all the necessary nutrients for a healthy person. Therefore, effective management of soil fertility is essential for maintaining soil health and sustainable and profitable food production that supports healthy human and animal lives.

8.2.2 Soil Chemical Fertility Status Under Agroecologies

The fertility status of soils under different agro-ecologies varies widely due to the climatic factors (mainly temperature and rainfall), parent material, topography, management practices (fertilizer application, soil and water conservation), and so on. The climate, especially drought or erratic rainfall during rainy seasons, strongly impacts and determines the unique characters of the soils. Topography, which is closely related with climate and vegetation distribution also strongly affects soil formation and development. Generally, the fertility status of the soils in the country is considered to be very low due to severe soil erosion, high levels of nutrient mining, low use of external inputs, and hence the soils have low productivity and limited capacity to respond to the environmental shocks. Thus, the country is struggling with challenges such as improving production and productivity for the ever-increasing population, under smallholder and fragmented landholdings, with low input and traditional farming practices.

Ethiopia has a great ecological diversity, ranging from tropical to temperate conditions. Altitude ranges from 126 m below the sea level (Danakil Depression) to 4620 masl in the Ras Dashen Mountains. Agro-ecological zones (AEZ) were determined with a crop suitability approach considering temperature, moisture regime, soil characteristics, and topography differences. The Ministry of Agriculture and Rural Development (MoARD 2005) identified 32 major agro-ecological zones, whereby 51% of the total land area is categorized under arid, semi-arid, and sub-moist zones and the remaining half is moist humid zones (Fig. 8.1). The major crop growing areas are sub-humid, humid, and moist semi-arid climatic zones (Fig. 8.1 and Table 8.1). Nitisols (23%), Cambisols (19%), and Vertisols (18%) comprise more than half of the arable land area in different agro-ecologies of the country (Paulos 2001).

The agricultural practice is mostly exploitative type i.e., mining the natural soil fertility, and the national average nutrient balances are estimated as −41 kg N, −6 kg P and −26 kg K per ha per year, which are among the highest nutrient depletion rates for sub-Saharan Africa (Stoorvogel and Smaling 1993). The decline in soil fertility status is because of increasingly intensive land use without adequate replacement of nutrients removed in harvested materials or loss through erosion and leaching (Agegnehu et al. 2016; Zelleke et al. 2010). The soil nutrient balances for N, P, and K at plot level calculated for 350 farms distributed across the high potential highlands were -23 ± 73, 9 ± 29 and -7 ± 64 kg ha^{-1}, respectively. The situation was most severe for N, where the depletion rate average was 0.2% of the soil total N stock per year, which is about 4.2% of the available soil N pool. Depletion rates were the highest in the relatively intensive farming systems in mountainous areas located in the central and southern parts of the country (Van Beek et al. 2016). Yet, soil nutrient balances can differ considerably between different crops, farming systems, and agro-ecological zones (Sommer et al. 2014). Thus, this review work presents an overview of the soil fertility status of Ethiopia; considering various soil fertility management practices made so far and examines its potential contribution to higher crop yields, as well as its prospects for improvement.

Lowlands

Hot and warm arid lowland plains

The hot and warm arid lowland plains (A1 and A2) cover a total area of 345,610 km^2 (Table 8.1). The soil pH values in these arid AEZs range from 7.0 to 8.8 in A1 and 6.2 to 8.8 in A2 (ATA 2014–19) with mean pH values of 8.1 and 8.0, respectively (Table 8.2). The soil pH values in these AEZs are rated as moderately acidic to strongly alkaline (Fig. 8.2; EthioSIS 2016) which could be the reflections of the parent material and accumulation of basic cations in the surface layers due to limited leaching (Tamirat 1992).

The organic carbon (OC) contents of the soils in the AEZs range from 0.26 to 1.88% in the A1 and 0.27 to 5.16% in A2 (ATA 2014–19) with mean values of 0.76 and 0.97%, respectively (Table 8.2) indicating very low OC level (Fig. 8.3). The low OC in the soils of these AEZs could be attributed to the low supply of organic material due to low vegetation cover, and high decomposition rate of the material under high temperature of the AEZs when there is precipitation (Assefa et al. 2020; Kenea et al. 2017). Similarly, the total nitrogen (TN) content of the soils in the AEZs is very low (Fig. 8.4) ranging from trace (<0.01) to 0.18% in the A1 and trace (<0.01) to 0.73% in the A2 (ATA 2014–19) with mean values of 0.06 and 0.07% in A1 and A2, respectively (Table 8.2). The TN content of the soils in the AEZs is very low in accordance with EthioSIS (2016).

The available P (Mehlich-3) values in the AEZs range between 4.9 and 108.3 mg kg^{-1} in the hot arid (A1) and 1.1 and 329 mg kg^{-1} in the warm arid (A2) soils (ATA 2014–19) with mean values of 20.7 and 19.1 mg kg^{-1} soil, respectively (Table 8.2). The mean values of available P in the soils are low in accordance with EthioSIS (2016). The variability in soil available P content in the AEZs might be the result of differences in the parent material, soil texture,

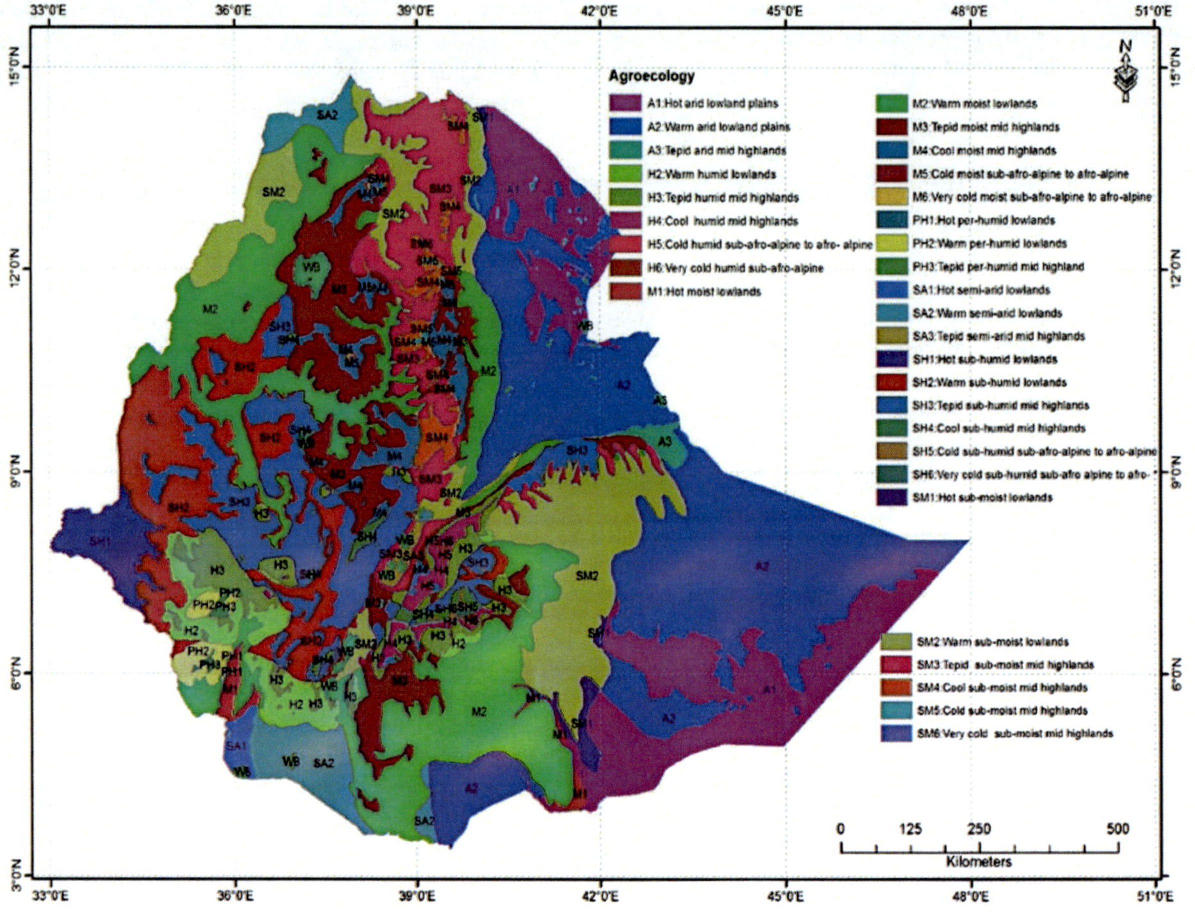

Fig. 8.1 Major Agro-ecological Zones in Ethiopia. *Source* MoARD (2005)

degree of P-fixation, and soil pH level. Soil type, especially clay content, and soil pH influence the amount of available P in the soils (Alemayehu et al. 2017; Sheleme 2016). The available S in the soils of these AEZs ranges from 7.7 to 1,201 mg kg^{-1} soil in A1 and from 5.4 to 2,871 mg kg^{-1} soil in A2 (ATA 2014–19) with mean values of 221.6 and 92.6 mg kg^{-1} soil, respectively (Table 8.2). The very high level of available S in the soils of these arid lowlands could be attributed to the presence of gypsum in the areas.

The exchangeable Ca, Mg, and K contents of the soils in the AEZs are 7,896, 383, and 484 mg kg^{-1} soil. respectively in A1 while the respective contents of the soils in A2 are 6,552, 409, and 535 mg kg^{-1} soil (Table 8.2). The contents of all basic cations in the AEZs are high (Landon 2014) indicating their accumulation in the soils of these arid environments. Additionally, Ca:Mg ratios in A1 and A2 are 16.3 and 16.0, respectively, indicating high Ca:Mg ratio which inhibits the availability of P (Landon 2014). The CEC values of the soils range from 10.4 to 76.4 cmol(+) kg^{-1} soil in A1 and from 10.6 to 67.6 cmol(+) kg^{-1} soil in A2 (ATA 2014–19) with mean values of 37.1 and 31.2 cmol(+) kg^{-1} soil, respectively (Table 8.2). The mean CEC values in the soils of these arid lowland plains are within a high range (Landon 2014) which might be due to high accumulation of cations.

The extractable Fe, Mn, Zn, and Cu contents of the soils in the hot and warm arid lowlands (Table 8.2) show that Fe is high, Mn and Cu are optimum, while Zn is low in accordance with EthioSIS (2016). The low level of Zn is associated with the high pH values of the sols in the environments and application of Zn containing fertilizers is recommended for optimum plant growth.

Hot and warm semi-arid lowland plains

The hot and warm semiarid lowland plains (SA1 and SA2) cover a total area of 35,646 km^2 (Table 8.1). The soil pH values in these semi-arid AEZs range from 6.1 to 9.0 in the hot semiarid (SA1) and from 5.3 to 9.1 in the warm semiarid (SA2) lowland plains (ATA 2014–19) with mean values of

Table 8.1 Major agro-ecologies and total area coverage in the country

No	MajorAgro-ecology	Area (km^{-2})	Area (%)
1	A1(Hotaridlowlandplains)	122,031	10.79
2	A2(Warmaridlowlandplains)	223,579	19.76
3	A3(Tepidaridmidhighlands)	4,882	0.43
4	H2(Warmhumidlowlands)	25,928	2.29
5	H3(Tepidhumidmidhighlands)	30,018	2.65
6	H4(Coolhumidmidhighlands)	9,264	0.82
7	H5(Coldhumidsub-afro-alpinetoafro-alpine)	626	0.06
8	H6(Verycoldhumidsub-afro-alpine)	506	0.04
9	M1(Hotmoistlowlands)	6,722	0.59
10	M2(Warmmoistlowlands)	171,110	15.12
11	M3(Tepidmoistmidhighlands)	91,019	8.05
12	M4(Coolmoistmidhighlands)	19,632	1.74
13	M5(Coldmoistsub-afro-alpinetoafro-alpine)	788	0.07
14	M6(Verycoldmoistsub-afro-alpinetoafro-alpine)	152	0.01
15	PH1(Hotper-humidlowlands)	131	0.01
16	PH2(Warmper-humidlowlands)	7,654	0.68
17	PH3(Tepidper-humidmidhighland)	1,523	0.13
18	SA1(Hotsemi-aridlowlands)	4,498	0.40
19	SA2(Warmsemi-aridlowlands)	31,148	2.75
20	SA3(Tepidsemi-aridmidhighlands)	2,186	0.19
21	SH1(Hotsub-humidlowlands)	18,935	1.67
22	SH2(Warmsub-humidlowlands)	80,474	7.11
23	SH3(Tepidsub-humidmidhighlands)	75,046	6.63
24	SH4(Coolsub-humidmidhighlands)	5,891	0.52
25	SH5(Coldsub-humidsub-afro-alpinetoafro-alpine)	688	0.06
26	SH6(Verycoldsub-humidsub-afrotoafro-alpine)	349	0.03
27	SM1(Hotsub-moistlowlands)	6,373	0.56
28	SM2(Warmsub-moistlowlands)	108,909	9.63
29	SM3(Tepidsub-moistmidhighlands)	58,505	5.17
30	SM4(Coolsub-moistmidhighlands)	13,142	1.16
31	SM5(Coldsub-moistmidhighlands)	768	0.07
32	SM6(Verycoldsub-moistmidhighlands)	180	0.02
33	WB(Waterbody)	8,709	0.77
	Total	**1, 131, 367**	**100.00**

Source MoARD (2005)

7.8 and 7.6, respectively (Table 8.2). The mean soil pH values in both AEZs are within the moderately alkaline range in accordance with EthioSIS (2016). The relatively high soil pH (alkaline condition) could be the reflections of differences in the parent material and its state of weathering, and local landforms that may cause differences in leaching intensities and accumulation of basic cations (Tamirat 1992). Use of quality irrigation water and management practices to reduce the soil pH are required for optimum plant growth and production.

The organic carbon (OC) contents of the soils in the AEZs range from 0.11 to 2.32% in SA1 and 0.24 to 4.79% in SA2 (ATA 2014–19) with mean values of 0.85 and 1.35%, respectively (Table 8.2) indicating very low OC level (Fig. 8.3). The low OC content in the soils of these semi-arid AEZs could be attributed to the low supply of organic material due to low vegetation cover, and high decomposition rate of the material under high temperature of the AEZs when there is precipitation (Assefa et al. 2020; Kenea et al. 2017). Similarly, the total nitrogen (TN) content of the soils

Table 8.2 Mean values of selected soil properties under different agroecologies

Physiographic position	AEZ	pH	OC[a]	TN	Av P	Ex K	Ex Ca	Ex Mg	CEC	Av S	Fe	Mn	Zn	Cu
			%		mg kg^{-1}				cmol (+) kg^{-1}	mg kg^{-1}				
Low-lands	A1	8.1	0.76	0.06	20.7	484	7896	383	26.8	221.6	42.7	107.7	1.15	1.82
	A2	8.0	0.97	0.07	19.1	535	6552	409	34.0	92.6	42.8	126.1	1.18	2.17
	SA1	7.8	0.85	0.06	39.2	637	5143	611	44.1	31.8	101.4	213.9	1.77	2.61
	SA2	7.6	1.35	0.09	15.3	471	6397	782	49.1	18.0	64.9	162.4	1.31	2.86
	SM1	7.9	0.83	0.06	34.8	467	4678	347	30.1	87.6	61.2	146.3	1.23	2.02
	SM2	7.6	1.56	0.11	12.1	424	6029	728	47.6	19.6	62.4	148.7	1.30	2.95
	M1	7.7	1.16	0.08	23.3	545	6162	856	41.6	69.9	78.0	170.0	1.55	2.70
	M2	7.0	1.79	0.13	10.0	369	4775	686	40.5	12.7	83.4	153.0	1.34	2.86
	SH1	6.2	1.68	0.13	20.7	211	2845	687	31.2	16.2	321.0	111.9	2.03	2.92
	SH2	6.3	2.52	0.19	12.7	280	2941	547	28.2	10.3	134.3	127.1	2.07	2.67
	H2	6.6	2.55	0.21	26.4	429	3966	685	33.5	12.5	139.3	172.3	4.19	2.72
	PH1	7.1	1.47	0.10	12.5	503	5263	1307	53.1	19.1	86.8	196.4	1.65	3.52
	PH2	6.3	2.61	0.22	18.4	382	3485	681	32.1	10.7	140.9	176.2	4.65	2.44
Mid-highlands	A3	7.9	1.57	0.10	10.2	599	9756	423	45.6	15.5	39.1	106.9	1.56	1.98
	SA3	7.4	1.64	0.12	12.1	744	4705	399	37.1	10.7	75.4	165.6	3.31	1.62
	SM3	7.2	1.41	0.11	12.7	332	5674	915	45.6	9.4	99.7	126.6	1.52	3.15
	SM4	6.5	1.98	0.17	12.4	240	4523	1058	42.2	9.2	174.0	80.2	1.88	3.01
	SM5	6.4	4.37	0.32	12.1	171	3573	695	36.1	12.0	141.3	29.6	1.47	1.42
	SM6	6.5	4.13	0.28	12.6	161	3544	789	38.0	11.5	143.6	32.8	1.63	1.52
	M3	6.5	2.04	0.16	9.3	357	4297	799	37.6	10.7	128.5	138.3	2.15	3.21
	M4	6.1	2.45	0.21	9.8	269	3811	781	35.3	10.6	184.6	92.6	2.09	2.93
	SH3	6.0	2.76	0.23	7.3	368	2749	425	25.9	11.5	140.7	136.2	3.02	2.63
	SH4	5.7	4.73	0.42	9.0	299	2424	349	23.3	13.2	188.4	80.3	4.42	2.13
	H3	6.0	3.48	0.30	10.0	354	3246	482	27.4	13.4	163.7	145.1	4.56	2.93
	H4	5.7	4.68	0.43	10.1	313	2624	342	24.6	13.7	213.7	92.7	4.19	2.45
	PH3	5.8	3.17	0.28	13.5	304	2618	463	24.4	10.1	173.3	150.3	6.32	1.90
Sub-afro-alpine and afro-alpine	M5	6.3	4.70	0.41	12.5	193	2955	474	29.1	12.5	150.1	31.2	1.61	1.51
	M6	6.2	6.33	0.61	12.3	161	2232	297	25.5	13.9	133.5	18.5	1.57	1.03
	SH5	6.1	7.74	0.68	17.1	269	3369	366	26.0	17.2	199.2	32.8	5.27	1.00
	SH6	6.2	6.44	0.56	21.1	227	4014	435	28.3	17.6	179.1	23.6	3.66	0.61
	H5	5.9	6.00	0.51	11.3	265	2550	305	25.3	14.2	176.5	54.9	3.56	1.54
	H6	6.0	6.41	0.55	17.7	263	3048	364	25.5	15.8	183.6	37.6	4.04	1.04

Source The data are derived from ATA, Agricultural Transformation Agency (2014, 2016, 2017, 2018, 2019)
[a]The values of OM were converted to OC

in the AEZs is also very low (Fig. 8.4; EthioSIS 2016) ranging from trace (<0.01) to 0.21% in the SA1 and from zero to 0.53% in SA2 with mean values of 0.06 and 0.09%, respectively (Table 8.2).

The available P (Mehlich-3) values in the AEZs range between 7.8 and 257.4 mg kg^{-1} soil in SA1 and 0.1 and 346.4 mg kg^{-1} soil in SA2 (ATA 2014–19) with mean values of 39.2 and 15.3 mg kg^{-1} soil, respectively (Table 8.2). The mean available P level in the soils is optimum in SA1 and low in SA2 AEZs in accordance with EthioSIS (2016). The available S ranges from 5.1 to 1,049 mg kg^{-1} soil in SA1, whereas its range is from 2.1 to 3,038 mg kg^{-1} soil in SA2 (ATA 2014–19). The mean values of available S in SA1 and SA2 were 31.8 and

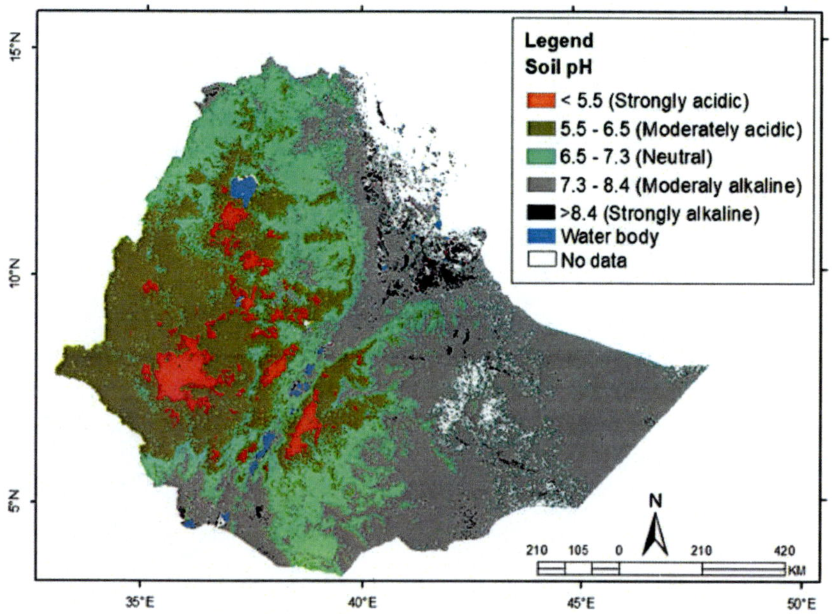

Fig. 8.2 Soil pH ranges and ratings under different agroecologies 2016 (*Source* ATA)

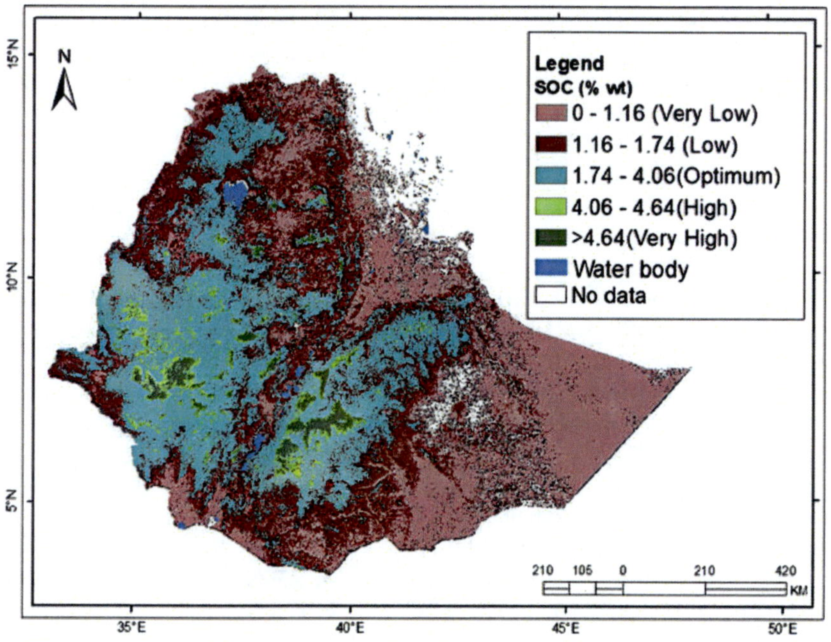

Fig. 8.3 Soil organic carbon contents and ratings under different agroecologies, 20162016 (*Source* ATA, converted to percent OC and the ratings were in accordance with EthioSIS (2016)

18.0 mg kg^{-1} soil (Table 8.2) indicating optimum and low S contents, respectively (EthioSIS 2016).

The exchangeable Ca, Mg, and K contents of the soils in SA1 are 5,143, 611, and 637 mg kg^{-1} soil, respectively, while the respective contents of the soils under SA2 are 6,397, 782, and 471 mg kg^{-1} soil (Table 8.2). The values of exchangeable cations in the soils of both AEZs are high (Landon 2014). The Ca:Mg ratios are 8.4 and 8.2 in SA1 and SA2, respectively, indicating high Ca:Mg in both AEZs which decreases Mg and P availability (Landon 2014). Furthermore, Mg induced K deficiencies are expected in SA2 as K:Mg in the soils of the AEZ is less than 0.7:1 (Loide 2004) calling for the need of K application in soils with low K:Mg ratios.

The CEC of the soils in the AEZs ranges from 16.9 to 73.7 Cmol(+) kg^{-1} soil in SA1, and 13.4 110.6 in SA2 (ATA 2014, 2016) with mean values of 44.1 and 49.1 Cmol (+) kg^{-1} soil, respectively (Table 8.2). The mean CEC

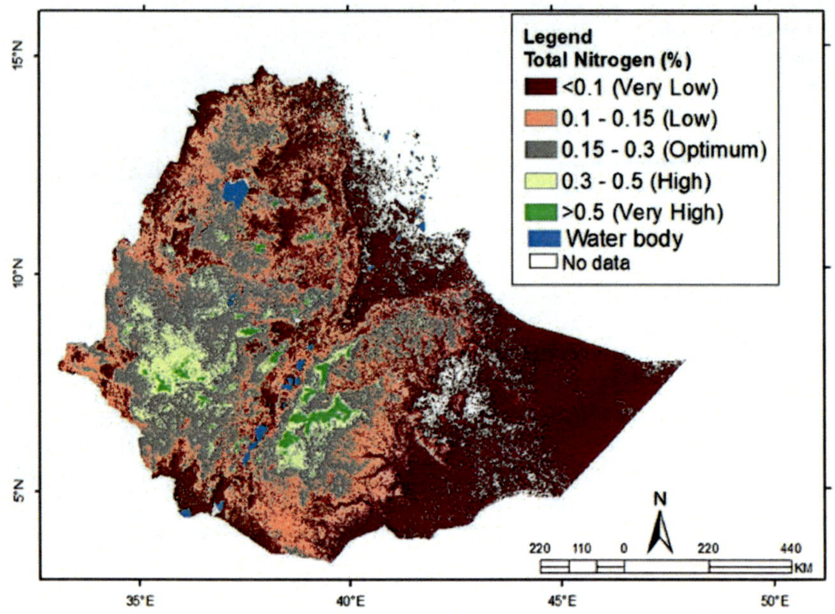

Fig. 8.4 Soil total nitrogen contents and ratings under different agroecologies (*Source* ATA (2016) converted to percent TN and the ratings were in accordance with EthioSIS (2016)

values are within very high range in accordance with Landon (2014) which might be due to high accumulation of basic cations in the surface soil layers. The extractable Fe, Mn, Zn and Cu contents of the soils in the AEZs (Table 8.2) show that Fe is high, whereas Mn, Zn, and Cu are optimum in accordance with EthioSIS (2016).

Hot and warm sub-moist lowlands

The hot and warm sub-moist (SM1 and SM2) lowland AEZs cover a total area of 115,282 km^2 (Table 8.1). The soil pH values in these AEZs range from 7.0 to 8.6 in SM1 and from 4.5 to 8.8 in SM2 (ATA 2014–19) with mean values of 7.0 and 7.6, respectively (Table 8.2). The mean values of soil pH in SM1 and SM2 are neutral and moderately alkaline, respectively (EthioSIS 2016; Fig. 8.2). Similarly, moderately alkaline soil pH value ranging from 7.9 to 8.1 was also reported in the SM2 of Alamata (Table 8.3).

The organic carbon content of the soils in the AEZs ranges from 0.46 to 1.60% in SM1 and from 0.19 to 7.27% in SM2 (ATA 2014–19) with mean values of 0.83 and 1.56%, respectively (Table 8.2). The mean OC levels in the soils of SM1 and SM2 are very low and low, respectively (EthioSIS 2016; Fig. 8.3). Soil OC content ranging from very low to low was recorded in SM2 of Alamata (Table 8.3). Similarly, the TN contents of the soils range from 0.02 to 0.16% in SM1 and from zero to 0.81 in SM2 (ATA 2014–19) with mean values of 0.06 and 0.11%, respectively (Table 8.2). The mean values of TN in SM1 and SM2 are very low and low (EthioSIS 2016; Fig. 8.4). Total N value within the similar range also exists in the soils of Alamata (Table 8.3). Organic C and TN increase with increasing precipitation in the AEZs which could be associated with increase in vegetation and thereby supply of organic materials to the soils.

The available P (Mehlich-3) values in the soils of the AEZs range from 9.1 to 91.9 mg kg^{-1} soil in SM1 and from zero to 575 mg kg^{-1} soil in SM2 (ATA 2014–19) with mean values of 34.8 and 12.1 mg kg^{-1}, soil respectively (Table 8.2). The available P (Mehlich-3) contents of the soils in SM1 and SM2 could be categorized as optimum and very low, respectively, in accordance with EthioSIS (2016). Soil available P (Olsen) ranging from medium to high was also found in soils under SM2 of Alamata (Table 8.3; Landon 2014). The available S contents in the soils of these hot and warm sub-moist lowlands range from 7.8 to 655 mg kg^{-1} soil in SM1 and 0.4 to 3,912 in SM2 (ATA 2014–19) with mean values of 87.6 and 19.6 mg kg^{-1} soil, respectively (Table 8.2). The mean values of available S in SM1 and SM2 are high and low, respectively (EthioSIS 2016). The large variation in the available S in the soils of these hot and warm sub-moist lowlands indicates the existence of different soils with varying gypsum contents.

The exchangeable Ca, Mg, and K contents of the soils in the AEZs are 4,678, 347, and 467 mg kg^{-1} soil Ca, Mg, and K, respectively in SM1 while the respective contents of the soils under SM2 are 6,029, 728, and 424 mg kg^{-1} soil, respectively (Table 8.2). The exchangeable cation contents in the soils of both AEZs are high (Landon 2014). The Ca: Mg in the AEZs is more than 8.3 indicating the possibility of inhibiting P and Mg availability in the soils (Landon 2014). Furthermore, Mg induced K deficiencies are expected as K:

Table 8.3 Range values of selected soil properties under different agroecologies

AEZ	Region	Site	pH	OC	TN	Av. P	Exch. K	Exch. Ca	Exch. Mg	CEC
				%		mg kg^{-1}	Cmol(+) kg^{-1}			
SM2	Tigray	Alamata	7.9–8.1	1.1–1.5	0.13–0.16	11.9–22.2	0.5–0.7	20.4–16.1	7.7–9.2	33.7–40.1
SM3		Endamehone	6.8–7.4	1.0–2.1	0.13–0.26	11.0–15.3	1.0–1.2	24.2–37.0	8.4–12.3	37.4–55.0
		Ofla	6.4–7.8	1.0–2.1	0.10–0.28	9.5–31.7	0.7–1.1	25.3–26.8	8.3–9.7	36.1–45.8
		Raya Azebo	7.2–7.9	1.3–2.8	0.15–0.32	9.6–25.7	0.3–2.0	19.7–35.2	7.0–10.6	38.6–59.2
	Oromiya	Gimbichu	7.0–8.0	0.8–1.6	0.08–0.2	8.8–20.2	0.3–1.0	27.7–38.7	9.5–13.0	41.3–59.7
SM4	Tigray	Ambalage	6.3–7.8	0.9–1.8	0.09–0.22	7.7–24.0	1.0–1.3	10.7–35.5	8.4–14.8	36.1–45.8
M3	Oromiya	Haramaya	6.0–7.9	0.9–1.5	0.10–0.16	11.3–22.9	0.7–1.4	11.5–34.7	4.1–11.7	26.8–54.7
		Kombolcha	6.2–7.9	0.9–1.9	0.10–0.23	7.9–17.5	0.8–1.1	23.9–32.3	8.0–11.2	43.3–50.6
		Becho	5.8–6.8	0.9–2.6	0.09–0.20	14.0–25.1	0.6–1.6	21.2–48.3	7.9–16.4	37.1–68.4
	Amhara	Dera	4.9–5.6	1.4–3.9	0.19–0.35	1.1–12.8	0.1–1.7	4.4–24.6	3.1–12.2	29.8–51.2
		Bure	5.1–5.8	1.6–3.5	0.18–0.32	2.4–10.8	0.1–1.7	11.6–35.2	5.1–16.3	39.2–53.0
		Jebitenan	4.7–7.7	1.6–4.8	0.09–0.22	1.1–7.0	0.1–1.6	10.8–24.6	6.7–20.2	31.0–76.0
		Mecha	4.8–6.0	1.5–4.8	0.15–0.22	2.4–7.4	0.1–2.1	5.9–23.5	4.2–20.4	34.9–56.0
	SNNPR	Cheha	4.7–5.1	1.7–2.6	0.19–0.28	7.3–10.0	1.0–1.2	15.1–19.5	5.0–5.9	32.0–36.8
M4	Oromiya	GirarJerso	6.6–7.3	1.3–1.9	0.14–0.18	20.7–25.9	0.3–0.8	35.4–44.5	12.1–15.4	50.7–63.6
		Munessa	5.1–6.2	2.2–4.0	0.20–0.33	10.7–14.0	0.7–1.5	18.5–35.1	6.2–12.0	36.4–57.2
SH2	Oromiya	Gurawa	6.0–6.1	1.2–2.0	0.15–0.25	14.0–15.5	0.8–0.9	18.3–25.2	6.6–8.4	36.6–50.3
		Bako Tibe	5.3–5.9	1.0–3.1	0.12–0.32	9.8–11.9	0.7–0.9	25.8–29.9	8.3–0.7	39.4–46.9
SH3	Oromiya	Meta (East Hararge)	6.0–7.2	0.4–1.3	0.04–0.16	10.0–24.4	0.6–0.8	11.8–38.5	5.2–12.8	32.4–59.6
		Girawa	5.3–6.4	1.1–2.5	0.10–0.28	10.9–16.1	0.8–1.0	17.9–36.8	6.0–12.8	34.8–54.9
		Habro	5.8–6.8	1.2–1.9	0.12–0.20	12.1–18.6	0.6–1.5	15.0–38.6	4.6–12.9	25.3–58.3
		Bako Tibe	5.8–6.4	1.8–3.5	0.22–0.32	14.4–26.4	0.9–1.2	21.6–30.8	7.9–10.3	41.6–52.5
		Omo Nada	5.0–5.9	1.3–3.7	0.13–0.40	4.2–9.3	1.6–2.1	9.2–21.2	3.4–11.0	30.6–49.3
		Limu Seka	4.9–5.4	2.2–3.9	0.25–0.46	4.3–12.2	1.8–2.2	6.0–12.8	2.6–4.3	32.0–46.1
		Gera	4.3–6.2	2.9–4.0	0.29–0.43	5.5–12.9	0.9–1.8	10.1–18.9	3.4–6.9	39.2–56.6
		Dedessa	4.4–5.4	1.5–5.9	0.13–0.73	7.6–13.2	1.4–1.9	7.7–18.7	3.4–6.8	38.2–51.6
	Amhara	South Achefer	4.7–5.6	2.1–3.6	0.15–0.31	1.3–5.2	0.1–1.6	15.4–29.7	4.2–10.8	41.2–57.4
	Sidama SNNPR	Malga	4.5–6.4	1.3–3.5	0.13–0.31	9.5–17.3	0.3–1.0	14.7–18.6	5.4–6.8	27.4–35.8
		MisrakAzernet Berbere	4.6–5.5	1.1–2.6	0.07–0.25	9.4–11.7	0.8–1.2	13.4–16.0	4.2–5.9	27.4–36.2
		Enamor Ener	4.9–6.0	1.8–4.2	0.16–0.34	8.8–14.3	0.8–1.1	13.5–22.3	4.2–7.7	27.1–37.3
		Cheha	4.5–6.5	1.1–3.8	0.11–0.31	6.9–12.7	1.0–1.2	15.1–19.5	5.0–5.9	32.0–36.8
H3	Oromiya	BedeleZuriya	4.3–4.8	1.4–3.1	0.22–0.31	7.4–12.0	1.2–1.9	9.2–14.3	3.3–5.1	30.8–38.8
H4	SNNPR	Bule	4.9–6.1	2.3–3.3	0.19–0.40	9.8–18.6	0.6–1.0	16.3–20.4	4.9–6.8	31.3–35.5

The data were derived from BENEFIT CASCAPE report (Leenaars et al. 2016) and the ranges were computed by Sheleme Beyene.

Mg in the soils is less than 0.7:1 (Loide 2004) calling for the need of K application. Similarly, high Ca, Mg, and K contents were reported from the soils under SM2 of Alamata (Table 8.3). The mean CEC values in SM1 and SM2 (Table 8.2) are within high and very high ranges, respectively (Landon 2014). Cation exchange capacity values ranging between high and very high were also recorded in SM2 of Alamata (Table 8.3).

The extractable Fe, Mn, Zn, and Cu contents of the soils in SM1 and SM2 AEZs (Table 8.2) show that Fe is high, Mn and Cu are optimum, while Zn is low in accordance with EthioSIS (2016). The low level of Zn is associated due to high pH values of the sols in the environments and application of Zn containing fertilizers is recommended for optimum plant growth. Amanuel et al. (2015) reported that the soils under SM2 of Alamata contain available Fe, Mn,

Zn, and Cu contents of 5.5, 6.9, 0.2, and 1.9 mg kg^{-1} soil, respectively, showing that Fe is low while Mn and Zn contents are very low in soils (EthioSIS 2016). Application of fertilizers containing the micronutrients (Fe, Mn, and Zn) is recommended on the soils of Alamata.

Hot and warm moist lowlands

The hot and warm moist (M1 and M2) lowlands cover a total area of 177,832 km^2 (Table 8.1). The soil pH values in these AEZs range from 6.1 to 8.6 in M1 and 4.9 and 8.9 in M2 (ATA 2014–19) with mean values of 7.7 and 7.0, respectively (Table 8.2). The mean soil pH values in M1 and M2 are moderately alkaline and neutral, respectively (EthioSIS 2016; Fig. 8.2). Soil pH values ranging from moderately acidic (5.9) to moderately alkaline (7.8) were recorded in Vertisols, Fluvisols, and Cambisols of Kobo and Sirinka M2 AEZ (Amare 2015; Wondimu et al. 2006). On the other hand, Kenea et al. (2017) also reported soil pH values ranging from 5.9 to 6.3 (moderately acidic) from Cambisols in warm moist (M2) lowlands of Dida-Hara, Yabello area.

The organic carbon contents of the soils in the AEZs range from 0.13 to 2.73% in M1 and 0.14 to 6.37% in M2 (ATA 2014–19). The mean values of OC in M1 and M2 are 1.16 and 1.79%, respectively (Table 8.2) which could be rated as very low and optimum, respectively, in accordance with EthioSIS (2016; Fig. 8.3). Low level of soil OC content (1.3–1.7%) was also reported from M2 AEZ of Dida-Hara (Kenea et al. 2017) while Vertisols, Fluvisols, and Cambisols in M2 AEZ of Kobo and Sirinka contained low (1.6%) to optimum (2.1%) OC (Amare 2015; Wondimu et al. 2006). The OC content of the M2 soils was higher than that of M1 due to higher precipitation in M2 than M1 which might have resulted in increased in vegetation and thereby relatively high organic material input to the soils. Similarly, the TN contents of the soils range from trace (<0.01) to 0.24% in M1 and zero to 0.72% in M2 (ATA 2014–19). The mean TN values in M1 and M2 are 0.08 and 0.13% (Table 8.3) which could be rated as very low and low, respectively in accordance with EthioSIS (2016). Additionally, TN values within the similar range also exist in the soils of Dida Hara (Kenea et al. 2017) and Vertisols, Fluvisols, and Cambisols of Kobo and Sirinka M2 AEZ (Amare 2015; Wondimu et al. 2006). The TN content followed the same trend with OC in hot and warm moist lowlands.

The available P (Mehlich-3) values in the AEZs range from 2.0 to 107.5 mg kg^{-1} soil in M1 and 0.1 to 417 mg kg^{-1} soil (ATA 2014–16) with mean values of 23.3 and 10.0 mg kg^{-1} soil, respectively (Table 8.2). The available P contents of the soils in M1 and M2 are low and very low, respectively, in accordance with EthioSIS (2016). The P level showed a decreasing trend with increasing moisture which could be attributed to soil development and P fixation. The soils under M2 AEZ of Dida-Hara contained P (Olsen) ranging between low and medium (Kenea et al. 2017); while those of Kobo and Sirinka contained medium (Amare 2015; Wondimu et al. 2006). The available S in the soils of these AEZs ranges from 5.5 to 406 mg kg^{-1} soil in M1 and 0.1 to 4,778 mg kg^{-1} soil in M2 (ATA 2014–19) with mean values of 69.9 and 12.7 mg kg^{-1} soil, respectively (Table 8.2). The average S contents in the soils of M1 and M2 are within the optimum and low range, respectively (EthioSIS 2016).

The exchangeable Ca, Mg. and K in the soils of the AEZs are 6,162, 856, and 545 mg kg^{-1} soil, respectively in M1 while the respective contents of the soils under M2 are 4,775, 686, and 369 (Table 8.2). The Ca:Mg in the AEZs is more than 6.9 indicating the possibility of inhibiting P and Mg availability in the soils (Landon 2014).The soils of Kobo contained exchangeable Ca, Mg, and K of 34.2, 11.4, and 2.3 cmol(+) kg^{-1} soil (Wondimu et al. 2006). The mean values of the cations are within high ranges (Landon 2014). The K:Mg values were below 0.64 in M1 and M2 AEZs, and 0.21 in soils of Kobo indicating the possibility of Mg induced K deficiencies (Loide 2004).

The CEC of the soils ranges from 14.3 to 81.9 cmol(+) kg^{-1} soil in the M1 and 9.1 to 118.7 Cmol(+) kg^{-1} soil in M2 (ATA 2014–19) with mean values of 41.6 and 40.5 cmol (+) kg^{-1} soil, respectively (Table 8.2). The CEC values of the soils in both hot and warm moist lowlands are within a very high range (Landon 2014). Kenea et al. (2017) reported a medium CEC value (15.6 cmol(+) kg^{-1} soil) in M2 AEZ soils of Dida Hara in the Yabelo area.

The extractable Fe contents are high, whereas Mn and Cu contents (Table 8.2) of the soils in both M1 and M2 are optimum (EthioSIS 2016). However, the concentrations of Zn are optimum and low under M1 and M2, respectively calling for external Zn supply on soils under M2 for optimum plant growth.

Hot and warm sub-humid lowlands

The hot and warm sub-humid (SH1 and SH2) lowland AEZs cover a total area of 99,409 km^2 (Table 8.1). The soil pH values in these AEZs ranges from 5.0 to 7.2 in SH1 and 4.8 to 8.5 in SH2 (ATA 2014–19) with mean values of 6.2 and 6.3, respectively (Table 8.2). The mean soil pH values of both AEZs are moderately acidic (EthioSIS 2016; Fig. 8.2). Soil pH values ranging from strongly to moderately acidic were reported in soils of SH2 AEZ at Bako Tibe and Gurawa, respectively (Table 8.3). Additionally, Teshome et al. (2016) also reported soil pH ranging from moderately acidic to neutral in SH2 at Abobo, while Yacob et al. (2014) reported neutral pH values in the soils of the same area.

The organic carbon in the soils of the AEZs ranges from 0.31 to 4.29% in SH1 and 0.33 to 9.11 in SH2 (ATA 2014–19) with mean values of 1.68 and 2.52%, respectively

(Table 8.2). The OC contents of the soils in both SH1 and SH2 AEZs are low in accordance with EthioSIS (2016). Soil OC contents ranging from low to optimum were reported in SH2 AEZs at Gurawa and Bako Tibe (Table 8.3). Additionally, OC content ranging between low and optimum was also found in SH2 AEZ of Abobo (Teshome et al. 2016) while Yacob et al. (2014) reported high levels OC in the soils of the same area. Soil organic C is one of the most important indicators of soil quality (Al-Kaisi et al. 2005; Reeves 1997; Doran and Parkin 1994) because it impacts key soil properties and shows a strong response to land use, land-use change, and land degradation (Vågen and Winowiecki 2013). The OC content of the soil increases with increasing precipitation in the AEZs which could be due increase in vegetation which results in relatively high organic material input to the soils. The organic carbon contents of soils also vary with soil type (Sheleme 2017) land use (Assefa et al. 2020; Alemayehu and Sheleme 2013) climatic conditions (Assefa et al. 2020) soil pH, vegetation cover, and clay content (Assefa et al. 2020).

Similarly, the TN contents of the soils range from trace (<0.01) to 0.48% in SH1 and trace (<0.01) to 1.09% in SH2 (ATA 2014–19) with mean values of 0.13 and 0.19%, respectively (Table 8.2). The mean values of TN in the AEZs are within low to optimum (EthioSIS 2016). Total N values of the soils in both SH1 and SH2 are within low and optimum, respectively (Fig. 8.4). The TN contents of the soils in SH2 AEZs of Gurawa were optimum while that of Bako Tibe were low to high (Table 8.3; EthioSIS 2016). Additionally, TN content ranging between low and optimum was reported from SH2 of Abobo area (Teshome et al. 2016) while Yacob et al. (2014) found values ranging from very low to low in the soils of the same area. Similar to OC content, TN increases with increasing precipitation in the AEZs which could be associated with increase in vegetation and thereby supply of organic materials to the soils.

The available P (Mehlich-3) values in the AEZs range from 3.4 to 295 mg kg^{-1} soil in SH1 and zero to 291 mg kg^{-1} soil in SH2 (ATA 2014–19) with mean values of 20.7 and 12.7 mg kg^{-1} soil, respectively (Table 8.2). The soils of SH1 and SH2 AEZs contain low and very low available P contents (EthioSIS 2016). The available P (Olsen) in the soils of SH2 at Gurawa was medium to high while it was medium at Bako Tibe (Table 8.3; Landon 2014). Teshome et al. (2016) reported high available P (Olsen) in soil of SH2 at Abobo, whereas Yacob et al. (2014) found available P (Olsen) ranging from medium to high in soils of the same area. The available S in the soils of these AEZs ranges from 5.8 to 81.8 mg kg^{-1} soil in SH1 and zero to 954 mg kg^{-1} soil (ATA 2014–19) with mean values of 16.2 and 10.3 mg kg^{-1} soil, respectively (Table 8.2). The available S contents of the soils in both AEZs are low in accordance with EthioSIS (2016) which can be improved by OC management.

The exchangeable Ca, Mg, and K contents of the soils in the AEZs are 2,845, 687, and 211 mg kg^{-1} soil, respectively in SH1 while the respective contents in the soils under SH2 are 2,941, 547, and 280 mg kg^{-1} soil (Table 8.2). The mean values of the cations are within a high range in accordance with Landon (2014). Similarly, high exchangeable Ca, Mg, and K contents were found in the soils under SH2 AEZs of Gurawa and Bako Tibe (Table 8.3). The soils of Abobo (SH2) also contained high levels of exchangeable Ca, Mg, and K (Teshome et al. 2016). Magnesium-induced K deficiencies are expected in soils under both SH1 and SH2 as K: Mg in the soils is less than 0.7:1 (Loide 2004). The CEC of the soils ranges from 10.6 to 67.6 cmol(+) kg^{-1} soil in SH1 and 7.5 and 95.7 Cmol(+) kg^{-1} soil in SH2 (ATA 2014–19) with mean values of 31.2 and 28.2 cmol(+) kg^{-1} soil, respectively (Table 8.2). The CEC of the soils in both AEZs are within a high range (Landon 2014). The soils under SH2 at Abobo also contained high CEC values (Teshome et al. 2016; Yacob et al. 2014).

The extractable micronutrient contents of the soils in the AEZs (Table 8.2) show that Fe is high while Mn, Zn, and Cu are within the optimum range in accordance with EthioSIS (2016). On the other hand, Teshome et al. (2016) reported very low level of Mn (40 mg kg^{-1} soil) from soils under SH2 of Abobo while the levels of the other micronutrients were similar to that of SH2 AEZ in the country.

Warm humid lowlands

The warm humid (H2) lowland AEZ covers a total area of 25,928 km^2 (Table 8.1). The soil pH value in this AEZ ranges from 4.7 to 8.5 (ATA 2014–19) with mean value of 6.6 (Table 8.2) which is neutral (Fig. 8.2). The organic carbon content of the soils in this AEZ ranges from 0.29 to 6.06% with mean value of 2.55 (Table 8.2). The TN content of the soils varies from trace (<0.01) to 0.74% with mean value of 0.25% (Table 8.2). Both the OC and TN levels in the soils of the AEZ are within optimum ranges (Fig. 8.4; EthioSIS 2016). The optimum levels of OC and TN in this humid environment could be due to abundant vegetation cover and inputs of organic materials to the soils.

The available P (Mehlich-3) content of the soils in this AEZ ranges from 1.1 to 476 mg kg^{-1} soil (ATA 2014–16) with a mean value of 26.4 mg kg^{-1} soil (Table 8.2) which is in low range in accordance with EthioSIS (2016). The available S in the soils of AEZ ranges from 3.3 to 747 mg kg^{-1} soil with mean value of 12.5 mg kg^{-1} soil (Table 8.2). Despite optimum levels of the OC in the soils of the AEZ, the available S content is low in accordance with EthioSIS (2016).

The exchangeable Ca, Mg, and K contents of the soils of the AEZ are 3,966, 685, and 429 mg kg^{-1} soil, respectively (Table 8.2). The mean values of the cations in the soils are within the high range in accordance with Landon (2014). The CEC of the soils ranges from 11.7 to 100 cmol(+) kg^{-1} soil with mean value of 33.5 cmol(+) kg^{-1} soil (Table 8.2). The mean CEC value is within a high range in accordance with the ratings of Landon (2014). The extractable Fe, Mn, Zn, and Cu contents of the soils in the AEZ (Table 8.2) show that Fe is high while Mn, Zn, and Cu are within the optimum range (EthioSIS 2016).

Hot and warm per-humid lowlands

The hot and warm per-humid (PH1 and PH2) lowlands cover a total area of 7,785 km^2 (Table 8.1). The soil pH values in these AEZs range from 6.6 to 7.7 in PH1 and 4.7 to 7.7 in PH2 (ATA 2014–16) with mean pH values of 7.1 and 6.3, respectively (Table 8.2). The mean soil pH values in the PH1 and PH2 AEZs are within optimum and moderately acidic ranges, respectively (Fig. 8.2).

The OC content of the soils in these AEZs ranges from 1.01 to 2.48% in PH1 and 0.95 to 6.07% in PH2 (ATA 2014–19) with mean values of 1.47 and 2.61%, respectively (Table 8.2). The OC levels in the soils of PH1 and PH2 are within low and optimum ranges, respectively (Fig. 8.3; EthioSIS 2016). Similarly, the TN contents of the soils vary from 0.07 to 0.19 in PH1 and 0.02 to 0.70% in PH2 (EthioSIS 2014–19) with mean values of 0.10 and 0.22%, respectively (Table 8.2). Accordingly, the TN levels in PH1 and PH2 AEZs are low and optimum, respectively (Fig. 8.4; EthioSIS 2016).

The soil available P (Mehlich-3) contents in these AEZs range from 6.1 to 53.3 mg kg^{-1} soil in PH1 and 1.0 to 171 mg kg^{-1} soil in PH2 (ATA 2014–19) with mean values of 12.5 and 18.4 mg kg^{-1} soil, respectively (Table 8.2). The available P levels are very low and low in the soils of PH1 and PH2, respectively (EthioSIS 2016) which might be due to fixation by relatively weathered and high clay contents of the soils under these environments. Application of P fertilizers to soils of these AEZs is crucial for optimum growth and yield of plants. The available S in the soils of these AEZs ranges from 8.6 to 37 mg kg^{-1} soil in PH1 and 4.5 to 39.4 mg kg^{-1} soil in PH2 (ATA 2014–19) with respective mean values of 19.1 and 10.7 mg kg^{-1} soil (Table 8.2). The mean S contents in the soils of both AEZs are low in accordance with EthioSIS (2016).

The exchangeable Ca, Mg, and K contents of the soils in the AEZs are 5,263, 1307, and 503 mg kg^{-1} soil, respectively in PH1 while the respective contents of the soils under PH2 are 3,485, 681, and 382 mg kg^{-1} soil (Table 8.2). The mean values of the cations in the soils are within a high range in accordance with Landon (2014). The contents of exchangeable cations in the per-humid AEZs are lower as compared to that of the humid environment which could be due to higher precipitation under per-humid AEZs that causes leaching of the cations. Additionally, K:Mg ratios are about 0.38 and 0.56 in the soils under PH1 and PH2, respectively, indicating the possibilities of Mg-induced K deficiencies in the soils (Loide 2004). The CEC of the soils ranges from 32.4 to 70.7 cmol(+) kg^{-1} soil in the soils of PH1 and 12.3 to 68.9 cmol(+) kg^{-1} soil in PH2 (ATA 2014–19) with mean values of 19.1 and 10.7 cmol(+) kg^{-1} soil, respectively (Table 8.2). The CEC values are within medium and low ranges in the soils of PH1 and PH2, respectively (Landon 2014). The extractable Fe, Mn, Zn, and Cu contents of the soils in the AEZs (Table 8.2) show that Fe is high while Mn, Zn, and Cu are within the optimum range (EthioSIS 2016).

Mid-highlands

Tepid arid mid-highland

The tepid arid mid-highland (A3) covers a total area of 4,882 km^2 (Table 8.1). The soil pH value in this arid AEZ ranges from 6.4 to 8.4 (ATA 2014–19) with a mean value of 7.9 (Table 8.2) indicating a moderately alkaline condition (Fig. 8.2; EthioSIS 2016). The OC content of the soils in the AEZ ranges from 0.27 to 5.70% (ATA 2014–19) with a mean value of 1.57% (Table 8.2) indicating low OC level (Fig. 8.3; EthioSIS 2016). However, the OC contents of the soils in these mid-highlands are higher than that of the respective arid lowlands. This could be attributed to the relatively low decomposition rate of the organic material due to low temperatures in the mid-highlands (Assefa et al. 2020). Similarly, the TN content of the soils in the AEZ ranges from trace (<0.01) to 0.69% with a mean value of 0.10% (Table 8.2) which is very low (EthioSIS 2016).

The available P (Mehlich-3) content of the soils in this AEZ ranges from 1.0 to 318 mg kg^{-1} soil (ATA 2014–19) with a mean value of 10.2 mg kg^{-1} soil (Table 8.2) showing very low available P level (EthioSIS 2016). Application of P fertilizers is crucial for optimum plant production in the AEZ. The available S in the soils of AEZs ranges between 6.2 and 76.3 mg kg^{-1} soil with a mean value of 15.5 mg kg^{-1} soil (Table 8.2) which is in low range in accordance with EthioSIS (2016).

The exchangeable Ca, Mg, and K contents of the soils in the AEZ are 9,756, 423, and 599 mg kg^{-1} soil (Table 8.2). The mean values of the cations are high (Landon 2014) and the cations on the exchange complex are in the order of Ca>K>Mg. Additionally, the Ca:Mg ratio ranges from 11.8 to 23.1 in the soils of the AEZs indicating the possibility of Ca-induced Mg deficiencies and inhibition of P availability. The CEC of the soils ranges from 15 to 110 cmol(+) kg^{-1}

soil (ATA 2014–19) with a mean value of 45.6 cmol(+) kg^{-1} soil (Table 8.2) which is within a very high range in accordance with Landon (2014). The extractable Fe, Mn, Zn, and Cu contents of the soils in the AEZs (Table 8.2) show that Fe is high while Mn, Zn, and Cu are within the optimum range (EthioSIS 2016).

Tepid semi-arid mid-highlands

The tepid semi-arid mid-highland (SA3) covers a total area of 2,186 km^2 (Table 8.1). The soil pH value in this semi-arid AEZ ranges from 6.2 to 8.6 (ATA 2014–19) with a mean value of 7.4 (Table 8.2) which is moderately alkaline (Fig. 8.2; EthioSIS 2016). On the other hand, Alemayehu et al. (2016) reported a soil pH value of 10.9 (strongly alkaline) in SA3 of Alage, central Rift Valley. The high soil pH value at Alage is due to the sodic nature of the soil existing in the area which requires management practices to reduce the pH for plant growth.

The organic carbon (OC) content of the soils in the AEZ ranges from 0.35 to 4.41% (ATA 2014–19) with a mean value of 1.64% (Table 8.2) indicating low OC content of the soils (Fig. 8.3; EthioSIS 2016). The OC content of the soils in this mid-highland is higher than that of the respective semi-arid lowlands which could be due to the relatively lower decomposition rate of the organic material in SA3 as compared to A3 due to low temperatures in the mid-highlands (Assefa et al. 2020). Alemayehu et al. (2016) found optimum OC content (2.2%) in SA3 soils of Alage. Similarly, the mean TN content of the soils in the AEZ is low (Fig. 8.4) ranging from trace (<0.01) to 0.49% (ATA 2016) with a mean value of 0.12% (Table 8.2). In SA3 soils of Alage area, however, a TN content of 0.24% was recorded by Alemayehu et al. (2016). The relatively higher OC and TN contents of Alage soils could be due to the strong alkaline condition of the soil that may decrease the decomposition of the organic material.

The available P (Mehlich-3) content of the soils in this AEZ ranges from 2.1 to 301 mg kg^{-1} soil with a mean value of 12.1 mg kg^{-1} soil (Table 8.2) which is very low in accordance with EthioSIS (2016). The Mehlich-3 available P level is higher in the tepid semi-arid (SA3) as compared to the tepid arid (A3) which could be due to the high Ca:Mg (23.1) ratio in A3 that inhibits P availability (Landon 2014). The soils of Alage (SA3) contained 3.7 mg kg^{-1} Olsen available P (Alemayehu et al. 2016) which is also low in accordance with Landon (2014). Application of P fertilizers is crucial for optimum plant production in AEZ. The available S in the soils of this AEZ ranges between 4.5 and 121 mg kg^{-1} soil with a mean value of 10.7 mg kg^{-1} soil (Table 8.2) which is in the low range in accordance with EthioSIS (2016).

The exchangeable Ca, Mg, and K contents of the soils in the AEZs are 4,705, 399, and 744 mg kg^{-1} soil, respectively (Table 8.2). Exchangeable Ca, Mg, and K contents of 15.2, 2.8, and 3.6 cmol(+) kg^{-1} soil were reported in SA3 soils of Alage (Alemayehu et al. 2016). The average values of the cations are high (Landon 2014) and the cations on the exchange complex are in the order of Ca>K>Mg. Additionally, the Ca:Mg ratio in the soils of the AEZ is 11.8 indicating the possibility of Ca-induced Mg deficiencies and inhibition of P availability by high Ca level, except for soils of Alage where the ratio is 5.4. The CEC of the soils ranges from 10.4 to 76.4 cmol(+) kg^{-1} soil with a mean value of 37.2 cmol(+) kg^{-1} soil (Table 8.2) which is high in accordance with Landon (2014). Very high CEC value, 41.4 cmol (+) kg^{-1} soil, was also reported from SA3 soils of Alage (Alemayehu et al. 2016).

The extractable Fe is high, whereas the extractable Mn, Zn, and Cu contents of the soils in the AEZs (Table 8.2) are within the optimum range (EthioSIS 2016). On the other hand, Alemayehu et al. (2016) reported extractable Fe, Mn, Zn, and Cu contents of 4.4, 66.9, 0.28, and 0.37 mg kg^{-1} soil, respectively, in SA3 soils of Alage, central Rift valley. The Fe and Mn contents are low while Zn and Cu are very low in Alage soils in accordance with the ratings of EthioSIS (2016) which could be due to the high pH of the soils. Thus, application of fertilizers containing these micronutrients is needed for soils of Alage for optimum growth of plants.

Tepid and cool sub-moist (SM3, SM4) mid-highlands

The tepid and cool sub-moist (SM3 and SM4) mid-highlands cover a total area of 71,647 km^2 (Table 8.1). The soil pH values in these AEZs range from 5.2 to 9.0 in SM3 and 5.1 to 8.2 in SM4 (ATA 2014–19) with mean values of 6.5 and 6.4, respectively (Table 8.2). The mean pH values of the soils in both SM3 and SM4 are moderately acidic (Fig. 8.2; EthioSIS 2016). The soil pH values in the soils of SM3 at Olfa, Endamehone, Ambalage and Raya Azebo and Gimbichu, and SM4 of Ambalage were in the range from moderately acidic to moderately alkaline (Table 8.3; EthioSIS 2016).

The organic carbon content in the soils of the AEZs ranges from 0.15 to 7.06% in SM3 and 0.22 to 9.91% in SM4 (ATA 2014–19) with mean values of 1.41 and 1.98%, respectively (Table 8.2). The mean OC levels in the soils of both SM3 and SM4 are low (Fig. 8.3; EthioSIS 2016). The OC contents in the soils under SM3 at Endamehone, Ofla, Gimbichu, and in the SM4 at Ambalage were within a very low to optimum range; while that of Raya Azebo (SM3) ranged from low to optimum (Table 8.3; EThioSIS 2016). Similarly, the TN contents of the soils in the AEZs range from zero to 0.81% in SM3 and trace (<0.01) to 1.67% in

SM4 (ATA 2014–19) with mean values of 0.11 and 0.17%, respectively (Table 8.2). The mean TN levels in the soils of SM3 and SM4 are low and optimum, respectively (EthioSIS 2016; Fig. 8.4). Additionally, the TN contents of the soils in the SM3 soils at Endamehone and Ofla ranged from low to optimum; at Raya Azebo optimum to high; while that of Gimbichu and SM4 at Ambalage ranged from very low to optimum (Table 8.3; EThioSIS 2016).

The available P (Mehlich-3) values in the AEZs range from 0.2 to 611 mg kg^{-1} soil in SM3 and 0.8 to 351 mg kg^{-1} soil in SM4 (ATA 2014–19) with mean values of 12.7 and 12.4 mg kg^{-1} soil, respectively (Table 8.1). The mean available P (Mehlich-3) levels in the soils of both AEZs are in the very low range in accordance with EthioSIS (2016) indicating the need for application of P fertilizers for optimum plant production. Additionally, the available P (Olsen) contents of the soils in SM3 AEZ at Endamehone, Ofla, Raya Azebo, Gimbichu, and SM4 at Ambalage (Table 8.3) were from medium to high in accordance with Landon (2014). The available S in the soils of these AEZs ranges from 0.05 to 460 mg kg^{-1} soil in SM3 and 0.04 to 48 mg kg^{-1} soil in SM4 (ATA 2014–19) with mean values of 9.4 and 9.2 mg kg^{-1} soil, respectively (Table 8.2). The available S in the soils of both AEZs is within a very low range in accordance with the ratings of EthioSIS (2016) and there is need to improve it by management of OC in the soils.

The exchangeable Ca, Mg, and K contents in the soils of the AEZs are 5,674, 915, and 332 mg kg^{-1} soil, respectively, in SM3 while the respective contents in SM4 are 4,523, 1,058, and 240 mg kg^{-1} soil (Table 8.2). The mean exchangeable basic cation contents of the soils under both AEZs are high in accordance with Landon (2014). However, Mg-induced K deficiencies are expected in the soils of both AEZs due to low K:Mg ratios (Loide 2004). Additionally, the concentrations of the basic cations were high in the soils of SM3 in Tigray and Oromiya, except for K at raya Azebo and Gimbichu (Table 8.3).

The CEC of the soils in the AEZs ranges from 3.0 to 120 cmol(+) kg^{-1} soil in the SM3 and 8.7 to 77.4 cmol(+) kg^{-1} soil (ATA 2016) with mean values of 45.6 and 42.2 cmol(+) kg^{-1} soil, respectively (Table 8.2). The mean CEC values in the soils of both AEZs are very high in accordance with the ratings of Landon (2014). The CEC values in the soils of SM3 at Endamehone, Ofla, Raya Azebo, and SM4 at Ambalage range from high to very high, whereas that of Gimbichu (SM3) are in the very high range (Table 8.3; Landon 2014).

The extractable Fe is high, whereas the extractable Zn and Cu contents of the soils in the AEZs (Table 8.2) are within the optimum range (EthioSIS 2016). The extractable Mn contents in the soils of SM3 and SM4 are optimum and low, respectively. Amanuel et al. (2015) reported low level of Fe, very low Mn and Zn, and optimum Cu in the soils under SM3 at Olfa and Raya Azebo. Similarly, low level of Fe, very low Mn and Zn, and optimum Cu were recorded in the soils under SM3 at Gimbichu (Engdawork 2015). The variations in the levels of the extractable micronutrients for different soils at different locations of the AEZs might be due to the differences in the parent material (Krauskorf 1972), soil pH (Barghouthi et al. 2012), organic matter, and clay contents. Supplemental application of Zn containing fertilizers is advisable for those soils deficient in the element for optimum crop production.

Cold and very cold sub-moist mid-highlands.

The very cold sub-moist (SM5 and SM6) mid-highlands cover a total area of 948 km^2 (Table 8.1). The soil pH values in these AEZs range from 5.7 to 7.3 in SM5 and 6.0 to 7.3 in the SM6 (ATA 2014–19) with mean values of 6.4 and 6.5, respectively (Table 8.2). The mean soil pH values in the soils of both AEZs is moderately acidic (Fig. 8.2; EthioSIS 2016).

The OC in the soils of the AEZs ranges from 1.02 to 9.80% in SM5 and 1.03 to 9.46% in SM6 (ATA 2014–19) with mean values of 4.37 and 4.13%, respectively (Table 8.2). The mean OC levels in the soils of both AEZs are high (Fig. 8.3; EthioSIS 2016) which could be associated with a low rate of decomposition under such cold and very cold environments (Assefa et al. 2020). Similarly, the TN contents of the soils range from 0.03 to 0.95% in SM5 and 0.08 to 0.86% in SM6 (ATA 2014–19) with mean values of 0.32 and 0.28%, respectively (Table 8.2). The mean TN levels in the soils of SM5 and SM6 are high and optimum (EthioSIS 2016; Fig. 8.3). Similar to the OC content, the high TN levels in the cold and very cold (SM5 and SM6) agroecologies could be associated with the low rate of organic matter decomposition (Assefa et al. 2020).

The available P (Mehlich-3) values in the AEZs range from 1.4 to 75 mg kg^{-1} soil in SM5 and 3.0 to 35 mg kg^{-1} soil in SM6 (ATA 2014–19) with mean values of 12.1 and 12.6 mg kg^{-1} soil, respectively (Table 8.2). The mean available P (Mehlich-3) values in the soils of both AEZs are very low in accordance with EthioSIS (2016) indicating the need for application of P fertilizers for optimum plant production. The available S in the soils of these AEZs ranges from 2.9 to 25.6 mg kg^{-1} soil in SM5 and 3.2 to 19.5 mg kg^{-1} soil in SM6 (ATA 2016) with mean values of 12.0 and 11.5 mg kg^{-1} soil, respectively (Table 8.2). The available S in the soils of the AEZs is low in accordance with the ratings of EthioSIS (2016).

The exchangeable Ca, Mg, and K contents of the soils in the AEZs are 3,573, 695, and 171 mg kg^{-1} soil, respectively, in SM5 while the respective values in SM6 are 3,544, 789, and 161 mg kg^{-1} soil (Table 8.2). The exchangeable K levels in the soils of both AEZs are low in accordance with

EthioSIS (2016). Furthermore, the K:Mg ratios are less than 0.25 indicating Mg-induced K deficiencies in the soils of both AEZs (Loide 2004). The CEC of the soils in the AEZs ranges from 16.3 to 61.5 cmol(+) kg^{-1} soil in SM5 and 15.7 to 64.8 cmol(+) kg^{-1} soil in SM6 (ATA 2014–19) with mean values of 36.1 and 38.0 cmol(+) kg^{-1} soil, respectively (Table 8.2). Accordingly, the CEC of the soils in both AEZs is high in accordance with Landon (2014).

The extractable Fe, Mn, and Cu are high, very low, and optimum, respectively (Table 8.2), in the soils of both AEZs, whereas the extractable Zn is low in SM5 and optimum in SM6 (EthioSIS 2016).

Tepid and cool moist mid-highlands

The tepid and cool moist (M3 and M4) mid-highlands cover a total area of 110,651 km^2 (Table 8.1). The soil pH values in these AEZs range from 4.7 to 9.2 in M3 and 4.5 to 8.8 in M4 (ATA 2014–19) with mean values of 6.5 and 6.1, respectively (Table 8.2). The mean soil pH values in both AEZs are moderately acidic (EthioSIS 2016). The ranges of soil pH values in the soils of tepid moist (M3) mid-highlands are: at Haramaya, Kombolcha, and Becho from moderately acidic to moderately alkaline; at Dera, Bure, Mecha, and Cheha from strongly acidic to moderately acidic; and at Jebitenan from strongly acidic to moderately alkaline (Table 8.3; Fig. 8.2). Soil pH values of 6.2 (Melkamu et al. 2019), 5.5 (Mekonnen and Yihenew 2017), and 6.1 (Abay et al. 2011) were reported in M3 AEZ soils of Dembecha, Fogera, and Hawassa, respectively. In the tepid cool mid-highlands (M4) at Girar Jerso, the soil pH is neutral while at Munessa it ranges from strongly acidic to moderately acidic (Table 8.3; Fig. 8.2). Soil pH values of 5.5 and 6.2 were reported in M4 AEZ of Welmera (Getachew et al. 2015) and Girar Jerso (Engdawork 2015), respectively. Temesgen (2012) also found a soil pH value of 7.6 in M4 AEZ soils of Menz Keye Gebriel.

The organic carbon in the soils of the AEZs ranges from 0.14 to 11.32% in M3 and 0.19 to 11.71% in M4 (ATA 2016) with mean values of 2.04 and 2.45%, respectively (Table 8.2). The mean OC levels in the soils of both AEZs are optimum (Fig. 8.3; EthioSIS 2016). Additionally, OC contents in the soils of M3 AEZ at Haramaya, Kombolcha, and Becho ranged from very low to optimum; at Cheha low to optimum; and at Dera, Bure, Jebitenan, and Mecha low to high (Table 8.3; Fig. 8.3). In the M4 AEZ at Girar Jerso the OC contents of the soils were low to optimum while at Munessa it was optimum (Table 8.3). Melkamu et al. (2019) and Mekonnen and Yihenew (2017) reported optimum levels of OC (2.2%) in the M3 soils of Dembecha and Fogera, whereas Abay et al. (2011) found low OC (1.6%) in the M3 soils of Hawassa. Similarly, low levels of OC with values of 1.5, 1.7, and 1.3% were reported in M4 soils at Welmera (Getachew et al. 2015), Girar Jerso (Engdawork 2015), and Menz Keye Gebriel (Temesgen 2012), respectively.

Similarly, the TN contents of the soils range from zero to 1.31% in M3 and zero to 1.48% in M4 (ATA 2014–19) with mean values of 0.16 and 0.21%, respectively (Table 8.2). The mean TN values in the soils of both AEZs are optimum (Fig. 8.4). Additionally, the TN contents of the soils at Haramaya and Kombolcha were low to optimum; at Becho and Jebitenan very low to optimum; at Mecha and Cheha optimum; and at Dera and Bure optimum to high (Table 8.3; Fig. 8.4). Total nitrogen contents of 0.12, 0.27, and 0.1% were reported from M3 AEZ of Dembecha (Melkamu et al. 2019), Fogera (Mekonnen and Yihenew 2017), and Hawassa (Abay et al. 2011), respectively. In the M4 AEZ at Girar Jerso and Munessa, the TN contents were optimum and optimum to high, respectively (Table 8.3). Total N contents of 0.16, 0.17, and 0.12% were also reported in M4 soils of Welmera (Getachew et al. 2015), Girar Jerso (Engdawork 2015) and Menz Keye Gebriel (Temesgen 2012), respectively.

The available P (Mehlich-3) values in the AEZs range from 0 to 640 mg kg^{-1} soil in M3 and 0 to 479 mg kg^{-1} soil in M4 (ATA 2014–19) with mean values of 9.3 and 9.8 mg kg^{-1} soil, respectively (Table 8.2). The mean values of soil available P (Mehlich-3) in both AEZs are very low in accordance with EthioSIS (2016) indicating the need for application of P fertilizers for optimum plant production. Additionally, available P (Olsen) contents in M3 soils at Jebitenan were low to moderate; at Dera, Bure, Mecha, and Cheha were low to high, whereas at Haramaya, Kombolcha, and Becho the Olsen P contents were high in accordance with Landon (2014). Melkamu et al. (2019), Mekonnen and Yihenew (2017), and Abay et al. (2011) found available P (Olsen) contents of 6.2, 5.1, and 40 mg kg^{-1} soil in M3 AEZ soils of Dembecha, Fogera, and Hawassa, respectively. On the other hand, the soil available P (Olsen) contents were high (Landon 2014) in the soils of M4 AEZ at Girar Jerso and Munessa (Table 8.3). Available P (Olsen) values of 6.7, 9.5, 48.3, and 19.6 mg P kg^{-1}soil were reported in M4 soils of Chelia (Getachew and Hailu 2009), Welmera (Getachew et al. 2015), GirarJerso, (Engdawork 2015) and Menz Keye Gebriel (Temesgen 2012), respectively. The available P (Olsen) contents of the soils in these AEZs range from very low to high in accordance with Landon (2014) and fertilizer response is most likely in soils having < 5 mg P mg kg^{-1}soil. The available S in the soils of these AEZs ranges from 0.03 to 1,625 mg kg^{-1} soil in M3 and 0.03 to 82 mg kg^{-1} soil (ATA 2014–19) with mean values of 10.7 and 10.6 mg kg^{-1} soil, respectively (Table 8.2). The available S in the soils of both AEZs is low (EthioSIS 2016).

The exchangeable Ca, Mg, and K contents of the soils in the AEZs of M3 are 4,297, 799, and 357 mg kg^{-1} soil, respectively, while the respective values in the soils of M4 are 3,811, 781, and 269 mg kg^{-1} soil (Table 8.2). The levels of Ca and Mg are high (Landon 2014) while the K level is optimum (EthioSIS 2016) in the soils of both AEZs. Additionally, the exchangeable Ca and Mg contents of the soils in M3 AEZ were high at Haramaya, Kombolcha, Becho, Cheha, Bure, and Jebitenan; and low to high at Dera (Table 8.3; Landon 2014). The exchangeable K contents were low to high in the Amahara region (Table 8.3); medium to high at Haramaya and Becho; and high at Kombolcha and Cheha. Mulatu (2015) reported exchangeable Ca, Mg, and K contents of 24.8, 7.3, and 2.9 cmol(+) kg^{-1} soil in M3 AEZ at Guzamin, whereas Mekonnen and Yihenew (2017) found 50.2, 22.8, and 0.2 cmol(+) kg^{-1} soil of Ca, Mg, and K, respectively in Fogera soils.

In the M4 AEZ at Girar Jerso and Munessa, the exchangeable Ca and Mg were high in both AEZs, whereas the exchangeable K was medium to high at Girar Jerso and high at Munessa (Table 8.3; Landon 2014). Getachew and Hailu (2009) reported exchangeable Ca, Mg, and K contents of 8.1, 1.9, and 1.7 Cmol(+) kg^{-1} soil in M4 AEZ soils of Chelia while Engdawork (2015) found 31, 1.3, and 0.8 Cmol (+) kg^{-1} soil in similar AEZ soils of Girar Jerso. The average values of the cations are above the critical levels of the individual cations (Landon 2014). Additionally, the Ca:Mg is optimum in majority of soils, except in Girar Jerso (23.8) and soils of Amhara region (1.73) (Table 8.3). Ratio of Ca: Mg above 5:1 may induce Mg deficiency while values lower than 3:1 may result in Ca deficiency while both inhibit P availability. The K:Mg ratios of the soils are below 0.61 in all soils of both AEZs and such low values of K:Mg may result in Mg-induced K deficiencies in soils (Loide 2004).

The CEC of the soils in the AEZs ranges from 0.2 to 111 cmol(+) kg^{-1} soil in M3 and 0.4 to 71.4 mg kg^{-1} soil in M4 (ATA 2014–19) with mean values of 37.6 and 35.3 mg kg^{-1} soil, respectively (Table 8.2). The mean values of the CEC in the soils of both AEZs are in high range (Landon 2014). The CEC levels in M3 soils at Cheha is also high; at Haramaya, Becho, Dera, Bure, Jebitenan, and Mecha high to very high; and at Kombolcha at a very high range (Table 8.3; Landon 2014). The soils in M3 AEZ at Dembecha (Melkamu et al. 2019) and Fogera (Mekonnen and Yihenew 2017) had high and very high (36 and 76 cmol(+) kg^{-1} soil) CEC values, respectively.

In the soils of M4 AEZ at Girar Jerso the CEC value is very high while it is high to very high at Munessa (Table 8.3). Getachew et al. (2015) and Engdawork (2015) also found CEC values of 21.7 and 47.7 cmol(+) kg^{-1} soil in M4 AEZ soils of Welmera and and Girar Jerso, respectively. Generally, the CEC of the soils in these AEZs range from high to very high, except in soils of Welmera which is within the medium range in accordance with Landon (2014).

The extractable Fe is high, whereas the extractable Zn and Cu contents of the soils in the AEZs (Table 8.2) are within the optimum range (EthioSIS 2016). The extractable Mn contents in the soils of M3 and M4 are optimum and low, respectively. Extractable Mn was very low; Cu was optimum; and Fe was high in the soils under M3 at Becho and M4 at Munessa (Engdawork 2015). Kibebew (2015) reported very low level of Mn, low Fe, and optimum Zn and Cu (EthioSIS 2016) in the soils under M3 at Kombolcha and Haramaya. Hussien et al. (2015) also reported 25.9, 27, 3.8, and 0.4 mg kg^{-1} soil of Fe, Mn, Zn, and Cu in the soils under M3 AEZ of Hawassa, respectively. The variations in the levels of the extractable micronutrients for different soils at different locations of the AEZs might be due to the differences in the parent material (Krauskorf 1972), soil pH (Barghouthi et al. 2012), organic matter, and clay contents. Supplemental application of Zn containing fertilizers is advisable for those soils deficient in the element for optimum crop production.

Tepid and cool sub-humid mid-highlands

The tepid and cool sub-humid (SH3 and SH4) mid-highlands cover a total area of 80,937 km^2 (Table 8.1). These AEZs are the main crop production zones in the country. The soil pH values in these AEZs range from 4.5 to 8.7 in SH3 and 4.4 to 7.7 in SH4 (ATA 2014–19) with mean values of 6.0 and 5.7, respectively (Table 8.2). The pH of the soils in both AEZs is moderately acidic (Fig. 8.2). The pH values of the soils of tepid sub-humid mid-highlands (SH3) at Dedessa, Limu Seka, and Misrak Azernet Berbere were strongly acidic; at Girawa, Gera, Omo Nada, South Achefer, Malga, Enemor Ener, and Cheha were strongly acidic to moderately acidic; at Bako Tibe were moderately acidic; and at Meta and Habru were moderately acidic to neutral (Table 8.3). Hillete et al. (2015) found soil pH values ranging between 6.9 and 7.9 in the soils under SH3 at Debre Zeit while Abay et al. (2011) reported a pH value of 4.2 in soils of SH4 at Babicho. The majority of the sites in these AEZs have soil pH values with moderately acidic to moderately alkaline (EthioSIS 2016) that are optimum for plant growth. Generally, the soil pH value decreases with increase in soil moisture content in the AEZs.

The organic carbon in the soils of the AEZs ranges from 0.03 to 11.81% in SH3 and 0.91 to 11.49 in SH4 (ATA 2014–19) with mean values of 2.76 and 4.73%, respectively (Table 8.2). The mean OC levels in the soils of SH3 and SH4 are optimum to high (Fig. 8.3). The higher OC level in SH4 than in SH3 could be associated with higher organic material supply and relatively lower rate of decomposition under the cool sub-humid environment (Assefa et al. 2020). The OC values of the soils in SH3 at Meta were very low to low; at Girawa, Misrak Azernet Berbere, and Cheha were from very low to optimum; at Habru and Omo Nada from low to optimum; at Bako Tibe, Limu Seka, Gera, South

Achefer, and Malga optimum; at Enemor Ener optimum to high; and at Dedessa from low to very high (Table 8.3; Fig. 8.3). Organic C contents of 1.6 and 1.1% were also reported in SH3 soils of Debre Zeit (Hillete et al. 2015) and Babicho (Abay et al. 2011). Generally, the OC contents of the soils in these AEZs range from very low to very high (EthioSIS 2016). The variations in OC contents could be ascribed to the differences in the vegetation cover of the respective locations. Additionally, the low organic carbon contents of the soils could also be attributed to intensive agricultural practices that aggravate organic carbon oxidation (Wakene and Heluf 2003).

Similarly, the TN contents of the soils in the AEZs range from zero to 1.21% in SH3 and trace (<0.01) to 1.23% in SH4 (ATA 2014–19) with mean values of 0.23 and 0.42%, respectively (Table 8.2). The mean TN levels in the soils of SH3 and SH4 are optimum and high, respectively (Fig. 8.4). The TN values in the soils of SH3 at Meta, Cheha, and Misrak Azernet Berbere were very low to optimum; at Girawa and Habro low to optimum; at Omo Nada, Malga, and South Achefer low to high; at Dedessa low to very high; and at Bako Tibe, Limu Seka, and Gera optimum to high (Table 8.3; Fig. 8.4). Total N contents of 0.10 and 0.18% were reported in SH3 soils of Debre Zeit (Hillete et al. 2015) and Babicho (Abay et al. 2011), respectively. Generally, the TN contents of the soils in these AEZs range from very low to very high (EthioSIS 2016) which is in line with the OC contents of the soils.

The available P (Mehlich-3) values in the soils of the AEZs range from 0 to 598 mg kg^{-1} soil in SH3 and 0 to 600 mg kg^{-1} soil in SH4 (ATA 2014–19) with mean values of 7.3 and 9.0 mg kg^{-1} soil, respectively (Table 8.2). The available P (Mehlich-3) level of the soils in both AEZs is very low (EthioSIS 2016). Available P (Olsen) values in the soils of SH3 at South Achefer, Omo Nada, and Limu Seka were low to medium; at Gera, Dedessa, Misrak Azernet Berbere, Enemor Ener, and Cheha were medium; and at Meta, Girawa, Habro, Bako Tibe, and Malga were medium to high (Table 8.3; Landon 2014). Fertilizer response is most likely in soils having <5 mg P (Olsen) mg kg^{-1}soil. The low available P in acidic soils such as South Achefer could be due to fixation, and hence liming together with P application should be implemented for optimum plant growth. The available S in the soils of these AEZs ranges from 0.03 to 871 mg kg^{-1} soil in SH3 and 0.03 to 37.3 mg kg^{-1} soil in SH4 (ATA 2014–19) with mean values of 11.5 and 13.2 mg kg^{-1} soil, respectively (Table 8.2). The available S in the soils of the AEZs is low in accordance with the ratings of EthioSIS (2016).

The exchangeable Ca, Mg, and K contents of the soils in the AEZs are 2,749, 425, and 368 mg kg^{-1} soil, respectively in SH3 while the respective contents under SH4 are 2,424, 349, and 299 mg kg^{-1} soil (Table 8.2). The levels of Ca and Mg are high (Landon 2014) while the K level is optimum (EthioSIS 2016) in the soils of both AEZs. Generally, the exchangeable Ca and Mg contents decrease with increase in soil moisture content in the AEZs which could be due increased precipitation that leaches the cations. The Ca:Mg ratios in the soils are above 5:1 indicating the possibilities of limiting Mg and P availabilities. The exchangeable Ca contents of the soils under SH3 of Amhara, Sidama, SNNPR, and Oromiya were high, except at Omo Nada, Limu Seka, and Dedessa (Oromiya) where it ranged from medium to high (Table 8.3; Landon 2014). Similarly, the exchangeable Mg contents of these soils were high, except for the soils of Omo Nada, Limu Seka, Gera, Dedessa, and South Achefer (Table 8.3). The soils under SH3 of the regions also contained high levels of exchangeable K, except for those at South Achefer and Malga (Table 8.3).

The CEC of the soils in the AEZs ranges from 4.3 to 88.1 cmol(+) kg^{-1} soil in SH3 and 7.1 to 59 cmol(+) kg^{-1} soil in SH4 (ATA 2016) with mean values of 25.9 and 23.3 cmol (+) $kg^{-}1$, respectively (Table 8.2). The mean values of CEC in the SH3 and SH4 AEZs are high and medium, respectively (Landon 2014). The CEC values of the soils in SH3 at Meta, Girawa, Habro, Omo Nada, Limu Seka, Gera, Dedessa, Malga, Misrak Azernet Berbere, Enamor Ener, and Cheha were high to very high while that of Bako Tibe and south Achefer were very high (Table 8.3; Landon 2014).

The extractable Fe is high, whereas the extractable Zn and Cu contents of the soils in the AEZs (Table 8.2) are within the optimum range (EthioSIS 2016). The extractable Mn contents in the soils of SH3 and SH4 are optimum and low, respectively. The extractable Mn contents were very low in the soils under SH3 AEZ of Oromiya at Dedessa, Omo Nada, Limu Seka, and Gera (Alemayehu 2015b). However, Fe was high and Zn and Cu were optimum in the soils at these sites, except for Zn which was high in the soils of Omo Nada. Similarly, high Fe levels, low Mn, and optimum Cu contents were found in the soils under SH3 at Enemor Ener and Misrak Azernet Berbere (Alemayehu 2015a). The Zn content of the soils in Enemor Ener was high. Kibebew (2015) also reported very low levels of Mn, optimum Zn and Cu, and high Fe levels in the soils under SH3 AEZ at Gurawa and Habro. On the other hand, Hussien et al. (2015) found very low levels of Mn and Zn, low Fe, and optimum Cu in SH3 soils at Debre Zeit. The variations in the levels of the extractable micronutrients for different soils at different locations of the AEZs might be due to the differences in the parent material (Krauskorf 1972), soil pH (Barghouthi et al. 2012), organic matter, and clay contents. Supplemental application of Zn containing fertilizers is advisable for those soils deficient in the element for optimum crop production.

Tepid and cool humid mid-highlands

The tepid and cool humid (H3 and H4) mid-highlands cover a total area of 39,282 km^2 (Table 8.1). The soil pH values in these AEZs range from 4.2 to 8.4 in H3 and 4.2 to 8.2 in H4 (ATA 2014–19) with mean values of 6.0 and 5.7, respectively (Table 8.2). The mean pH values of the soils in both AEZs are moderately acidic in accordance with EthioSIS (2016). The pH values of the soils in tepid humid mid-highlands (H3) of Bedele Zuriya were strongly acidic while the soils of the cool humid highland (H4) of Bule had strongly acidic to moderately acidic pH values (Table 8.3). Soil acidity problem is the major challenge to crop production in these agroecologies and liming together with the application of limiting elements is required for optimum crop production.

The OC contents of the soils in these AEZs range from 0.50 to 12.57% in H3 and 0.82 to 11.48 in H4 (ATA 2014–19) with mean values of 3.48 and 4.68%, respectively (Table 8.2). The mean OC levels in the soils of H3 and H4 AEZs are optimum and very high, respectively (Fig. 8.3; EthioSIS 2016). Organic carbon contents ranging from low to optimum (Fig. 8.3) were reported in soils of Bedele Zuriya (H3) while the H4 soils of Bule had OC contents within the optimum range (Table 8.3). The high levels of OC in these AEZs could be associated with high vegetation cover in these humid environments. Similarly, the TN contents of the soils range from trace (<0.01) to 1.06% in H3 and < 0.01 to 1.23% in H4 (ATA 2014–19) with mean values of 0.30 and 0.43%, respectively (Table 8.2). The mean TN values in the soils of H3 and H4 are optimum and high, respectively (Fig. 8.4; EthioSIS 2016) which might also be due to abundant vegetation cover and inputs of organic materials in these humid environments. The soils of Bedele Zuriya (H3) and Bule (H4) also contained TN ranging from optimum to high values(Table 8.3).

The available P (Mehlich-3) content in the soils of these AEZs ranges from 0.03 to 307 mg kg^{-1} soil in H3 and 0.2 to 456 mg kg^{-1} soil in H4 (ATA 2014–19) with mean values of 10.0 and 10.1 mg kg^{-1} soil, respectively (Table 8.2). The available P (Mehlich-3) values in the soils of both AEZs are very low in accordance with the ratings of EthioSIS (2016) which might be due to the acidic conditions of the soils that enhances P fixation. On the other hand, the available P (Olsen) content in H3 soils of Bedele Zuriya was medium while the soils under H4 AEZ of Bule contained medium to high P values (Landon 2014). The available S in the soils of these AEZs ranges from 1.2 to 268 mg kg^{-1} soil in H3 and 4.1 to 51 mg kg^{-1} soil in H4 (ATA 2014–19) with respective mean values of 13.4 and 13.7 mg kg^{-1} soil (Table 8.2). The average S contents of the soils in the AEZs are within the low range (EthioSIS 2016).

The exchangeable Ca, Mg, and K contents of the soils of the AEZs are 3,246, 482, and 354 mg kg^{-1} soil, respectively, in H3 while the respective values in H4 are 2,624, 342, and 313 mg kg^{-1} soil (Table 8.1). The average values of the cations are above the critical levels of the individual cations (Landon 2014). However, the Ca:Mg ratios are above 5 in the soils of both AEZs indicating the potential of inhibiting Mg and P availability (Landon 2014). The soils under H3 of Bedele Zuriya contained exchangeable Ca ranging between medium and high levels while their Mg and K contents were high in accordance with Landon (2014). On the other hand, the soils under H4 AEZ at Bule had high levels of Ca, Mg, and K (Table 8.3; Landon 2014). However, the K:Mg ratio in the soils of the two AEZs are in the range of 0.12–0.37 indicating a potential of Mg-induced K deficiencies in the soils (Loide 2004).

The CEC of the soils ranges from 3.5 to 108 cmol(+) kg^{-1} soil in H3 and 6.4 to 79.1 cmol(+) kg^{-1} soil in H4 (ATA 2014–19) with mean values of 27.4 and 24.6 cmol(+) kg^{-1} soil, respectively (Table 8.2). The soils in H3 of Bedele Zuriya and H4 of Bule contained CEC values (Table 8.3) within a high range in accordance with Landon (2014).

The extractable Fe is high, whereas the extractable Zn and Cu contents of the soils in the AEZs (Table 8.2) are within the optimum range (EthioSIS 2016). The extractable Mn contents in the soils of H3 and H4 are optimum and low, respectively. High level of Fe, optimum Zn and Cu, and very low level of Mn were recorded in the soils of H3 at Bedele Zuriya (Alemayehu 2015b), whereas the soils under H4 at Bule had high level of Fe and Zn, optimum Cu, and very low level of Mn (EthioSIS 2016; Alemayehu 2015a).

Tepid per-humid mid-highland

The tepid per-humid (PH3) mid-highland covers a total area of 1,523 km^2 (Table 8.1). The soil pH value in this AEZ ranges from 4.3 to 6.9 (ATA 2014–19) with a mean value of 5.8 which is moderately acidic (Table 8.2; Fig. 8.2). The OC content of the soils in this AEZ ranges from 1.61 to 6.86% (ATA 2014–19) with a mean value of 3.17% (Table 8.2) which is within the optimum range (Fig. 8.3; EthioSIS 2016). Similarly, the TN content of the soils ranges from 0.06 to 0.60% with a mean value of 0.28% (Table 8.2) which is optimum (Fig. 8.4).

The available P (Mehlich-3) content of the soils in this AEZ ranges from 0.9 to 453 mg kg^{-1} soil (ATA 2014–19) with mean value of 13.5 mg kg^{-1} soil (Table 8.2) which is very low in accordance with EthioSIS (2016). The low level of available P (Mehlich-3) in the soils might be due to the acidic conditions of the soils that enhances P fixation. The available S in the soils of this AEZ ranges from 5.4 to 49.5 mg kg^{-1} soil with a mean value of 24.4 mg kg^{-1} soil

(Table 8.2). The average S contents of the soils in the AEZ is within the optimum range (EthioSIS 2016).

The exchangeable Ca, Mg, and K contents of the soils in the AEZ are 2,618, 463, and 304 mg kg^{-1} soil, respectively (Table 8.2). The exchangeable cations are lower in the per-humid as compared to the humid environments which might be due to increased leaching of the cations owing to the relatively higher rainfall in the per-humid AEZ. The average values of the cations are above the critical levels of the individual cations (Landon 2014). The Ca:Mg ratio is about 5.7 indicating the potential of inhibiting Mg and P availability in soils (Landon 2014). The CEC of the soils ranges from 11.4 to 49.5 cmol(+) kg^{-1} soil (ATA 2014–19) with a mean value of 24.4 cmol(+) kg^{-1} soil (Table 8.2) which is medium in accordance with Landon (2014). The extractable Fe level is high while Mn, Zn, and Cu contents of the soils (Table 8.2) under this AEZ are optimum in accordance with the ratings of EthioSIS (2016).

Sub-afro-alpine to afro-alpine

Cold and very cold moist sub-afro alpine to afro-alpine

The cold and very cold moist (M5 and M6) sub afro-alpine to afro-alpine AEZs cover a total area of 941 km^2 (Table 8.1). The soil pH values in these AEZs range from 5.5 to 7.3 in M5 and 5.6 to 6.8 in M6 (ATA 2014–16) with mean values of 6.3 and 6.2, respectively (Table 8.2). The mean soil pH values in both AEZs are moderately acidic (Fig. 8.2; EthioSIS 2016).

The OC contents of the soils in the AEZs range from 1.04 to 10.54% in M5 and 2.04 to 10.07% in M6 (ATA 2014–19) with mean values of 4.70 and 6.33%, respectively (Table 8.2) indicating very high OC levels (Fig. 8.3: EthioSIS Ethiopian soils information system 2016). High levels of OC could be attributed to the low decomposition rate of the organic material under low temperatures in the environments. Similarly, the total nitrogen (TN) contents of the soils in the AEZs range from 0.04 to 1.25% in M5 and 0.05 to 1.15% in M6 (ATA 2016) with mean values of 0.41 and 0.61%, respectively (Table 8.2). The respective mean TN levels in the soils of M5 and M6 AEZs are high and very high (Fig. 8.4; EthioSIS 2016).

The available P (Mehlich-3) values of the soils in the AEZs range between 1.5 and 328 mg kg^{-1} soil in M5 and 4.7 to 21.8 mg kg^{-1} soil in M6 (ATA 2014–19) with mean values of 12.5 and 12.3 mg kg^{-1} soil, respectively (Table 8.2). The available P (Mehlich-3) level in the soils of both AEZs is very low (EthioSIS 2016) indicating that the application of P fertilizers is crucial for optimum crop growth. The available S in the soils of these AEZs ranges from 3.3 to 20.1 mg kg^{-1} soil in M5 and 5.8 to 18.8 mg kg^{-1} soil in M6 with mean values of 12.5 and 13.9 mg kg^{-1} soil, respectively (Table 8.2). The mean available S in the soils of the AEZs is within the low range in accordance with EthioSIS (2016).

The exchangeable Ca, Mg, and K contents of the soils in the AEZs are 2,955, 474, and 193 mg kg^{-1} soil, respectively in M5 while the respective contents in M6 are 2,232, 297, and 161 mg kg^{-1} soil (Table 8.2). The average values of the cations are above the critical values for the respective cations, except for K (Mulugeta et al. 2020). Additionally, the Ca:Mg, K:Mg ratios are 6.2–7.5 and 0.41–0.54, respectively indicating the potential of inhibiting Mg and P availability in soils with Ca:Mg values > 5.1 (Landon 2014), and Mg-induced K deficiencies in the soils (Loide 2004). The CEC of the soils ranges from 15.8 to 57.2 cmol(+) kg^{-1} soil in M5 and 15.6 to 48.1 cmol(+) kg^{-1} soil in M6 (ATA 2014–19) with mean values of 29.1 and 25.5 cmol(+) kg^{-1} soil, respectively (Table 8.2). The CEC values in the soils of AEZs are in the high range (Landon 2014). The levels of extractable Fe and Mn contents are high and very low, respectively; whereas the levels of Zn and Cu are optimum (Table 8.2) in the soils of both AEZs in accordance with EthioSIS (2016).

Cold and very cold sub-humid sub-afro alpine to afro-alpine

The cold and very cold sub-humid (SH5 and SH6) sub afro-alpine to afro-alpine AEZs cover a total area of 1,037 km^2 (Table 8.1). The soil pH values in these AEZs range from 5.3 to 6.9 in SH5 and 5.7 to 6.6 in SH6 (ATA 2014–19) with mean values of 6.1 and 6.2, respectively (Table 8.2). The mean pH values of the soils in both AEZs are moderately acidic (Fig. 8.2).

The OC contents of the soils in the AEZs range from 2.59 to 11.34% in SH5 and 2.57 to 11.34% in SH6 (ATA 2014–19) with mean values of 7.74 and 6.44%, respectively (Table 8.2). The OC levels in the soils of both AEZs are very high (Fig. 8.3) which could be attributed to the low decomposition rate of the organic material under low temperatures of the environments. Similarly, the TN contents of the soils range from 0.16 to 1.01% in SH5 and 0.12 to 0.93% in SH6 (ATA 2014–19) with mean values of 0.68 and 0.56%, respectively (Table 8.2). The mean TN contents of the soils under both AEZs are very high (Fig. 8.4).

The available P (Mehlich-3) values in the soils of the AEZs range between 5.9 and 337 mg kg^{-1} soil in SH5 and 6.8 to 58.5 mg kg^{-1} soil in SH6 (ATA 2014–19) with mean values of 17.1 and 21.1 mg kg^{-1} soil, respectively (Table 8.2). The available P (Mehlich-3) level in the soils of both AEZs is low (EthioSIS 2016) indicating the need for application of P fertilizers for optimum crop growth. The available S in the soils of these AEZs ranges from 9.5 to 37.7 mg kg^{-1} soil in SH5 and 6.8 to 26.9 mg kg^{-1} soil in

SH6 with mean values of 17.2 and 17.6 mg kg^{-1} soil, respectively (Table 8.2). The available S in the soils of the AEZs is within the low range in accordance with EthioSIS (2016).

The exchangeable Ca, Mg, and K contents of the soils in the SH5 AEZs are 3,369, 366, and 269 mg kg^{-1} soil, respectively, while the respective contents in SH6 are 4,014, 435, and 227 mg kg^{-1} soil (Table 8.2). The average values of the cations are above the critical values for the respective cations. However, the Ca:Mg ratio is about 9.2 in the soils of both AEZs indicating the potential of inhibiting Mg and P availability (Landon 2014). The CEC of the soils ranges from 8.4 to 38.2 cmol(+) kg^{-1} soil in SH5 and 10.8 to 39.3 cmol(+) kg^{-1} soil in SH6 (ATA 2014–19) with mean values of 26.0 and 28.3 cmol(+) kg^{-1} soil, respectively (Table 8.2). The CEC values in the soils of AEZs are at a high range (Landon 2014). The extractable micronutrient contents (Table 8.2) show that Fe is high; Zn is optimum; Mn is very low in the soils of both AEZs, whereas the levels of Cu are optimum and low in the soils of SH5 and SH6, respectively, in accordance with EthioSIS (2016).

Cold humid sub-afro-alpine to afro-alpine

The cold humid sub afro-alpine to afro-alpine (H5) covers a total area of 626 km^2 (Table 8.1). The soil pH value in this AEZ ranges from 5.3 to 6.7 with a mean value of 5.9 (Table 8.2) which is moderately acidic (Fig. 8.2). The OC content of the soils in the AEZ ranges from 2.04 to 11.65% with a mean value of 6.00% (Table 8.2) indicating a very high OC content of the soils (Fig. 8.3). Similarly, the mean TN content of the soils in the AEZ is very high (Fig. 8.4). It ranges from 0.02 to 1.16% with a mean value of 0.51% (Table 8.1). The very high OC and TN contents of the soils in the AEZ could be attributed to the low decomposition rate of the organic material owing to the low temperatures in the environments.

The available P (Mehlich-3) content of the soils in the AEZ ranges from 1.8 to 153 mg kg^{-1} soil with a mean value of 11.3 mg kg^{-1} soil (Table 8.2) which is very low in accordance with EthioSIS (2016). The available S in the soils under the AEZ ranges from 7.1 to 28.9 mg kg^{-1} soil with a mean value of 14.2 mg kg^{-1} soil (Table 8.2) indicating a low level of available S in the soils (EthioSIS 2016).

The exchangeable Ca, Mg, and K contents of the soil in the AEZ are 2,550, 305 and 265 mg kg^{-1} soil, respectively (Table 8.2). The mean values of the cations are above the critical their respective values. However, the Ca:Mg ratio is about 8.4 in soils indicating the potential of inhibiting Mg and P availability (Landon 2014) while the K:Mg ratio could be considered as optimum. The CEC of the soils ranges from 14.0 to 42.0 cmol(+) kg^{-1} soil with a mean value of 25.3 cmol(+) kg^{-1} soil (Table 8.2) which is high in accordance with EthioSIS (2016). The extractable Fe and Mn levels are high and very low, respectively, while Zn and Cu levels are optimum in the soils of the AEZ (Table 8.2; EthioSIS 2016).

Very cold humid sub afro-alpine

The very cold humid sub afro-alpine (H6) covers a total area of 506 km^2 (Table 8.1). The soil pH value in this AEZ ranges from 5.4 to 6.8 with a mean value of 6.0 (Table 8.2) which is moderately acidic (Fig. 8.2). The OC content of the soils in the AEZ ranges from 2.39 to 11.65% (ATA 2014–19) with mean value of 6.41% (Table 8.1) indicating a very high OC content of the soils (Fig. 8.3). Similarly, the mean TN content of the soils in the AEZ is very high (Fig. 8.4). It ranges from 0.03 to 0.97% with a mean value of 0.55% (Table 8.2). The very high OC and TN contents of the soils in the AEZ could be attributed to the low decomposition rate of the organic material owing to the low temperatures in the environments.

The available P (Mehlich-3) content of the soils in the AEZ ranges from 3.8 to 89.6 mg kg^{-1} soil (ATA 2014–19) with a mean value of 17.7 mg kg^{-1} soil (Table 8.2) which is low in accordance with EthioSIS (2016). The available S in the soils under the AEZ ranges from 5.9 to 27.7 mg kg^{-1} soil with a mean value of 15.8 mg kg^{-1} soil (Table 8.2) indicating low level of available S in the soils (EthioSIS 2016).

The exchangeable Ca, Mg, and K contents of the soils in the AEZ are 3,048, 364, and 263 mg kg^{-1} soil, respectively (Table 8.2). The mean values of the cations are above the critical their respective values. However, the Ca:Mg ratio is about 8.4 in soils indicating the potential of inhibiting Mg and P availability (Landon 2014) while the K:Mg ratio could be considered as optimum. The CEC of the soils ranges from 13.7 to 38.5 cmol(+) kg^{-1} soil with a mean value of 25.5 cmol(+) kg^{-1} soil (Table 8.2) which is high in accordance with EthioSIS (2016). The extractable Fe and Mn levels are high and very low, respectively, while Zn and Cu levels are optimum in the soils of the AEZ (Table 8.2; EthioSIS 2016).

8.3 Soil Quality and Soil Health

8.3.1 Principles

Soil quality (SQ) and soil health is a concept that has increased in popularity over the past decades, especially since the early 1990s. It continued to gain traction by soil managers, agricultural scientists, extension specialists, and other groups that work with soil. Soil quality and soil health are vital for sustainable agroecosystem management and survival of life on planet Earth. While awareness of SQ and health is increasing, it is important to have a good understanding of what SQ and health entail, how they are measured, and how to manage them for optimal and sustainable delivery of the ecosystem services that soils provide.

Soil quality is the capacity of the soil to function, within its natural or managed ecosystems, to sustain productivity, enhance water and air quality, support human and animal health, and habitation. Soil quality is a holistic concept, which recognizes soil as part of a dynamic and diverse production system with biological, chemical, and physical attributes (Sanchez et al. 2003; Swift 1999). It is related to the concept of soil capability, which is as old as the civilization itself (Carter et al. 2004). Although there has been an interest in soil and land quality from the beginning of agriculture (Carter et al. 2004), the SQ concept is continuously evolving (Warkentin 1995). The concept is not limited to agriculture, but most work and evaluation have occurred on agricultural lands. Deterioration of SQ due to high rates of soil erosion, losses of organic matter (OM), reductions in fertility and productivity, chemical and heavy metal contamination, and degradation of air and water quality have sparked interest in the concept of SQ and its assessment (Karlen et al. 2003; Doran and Parkin 1994; Larson and Pierce 1991).

In its current context, however, SQ is relatively new (Doran and Parkin 1994; Arshad and Cohen 1992) as it differs from the traditional technical approaches that focus solely on productive functions of the soil (Sanchez et al. 2003; Swift 1999). The concept was formally initiated when Alexander (1971) suggested developing 'Soil quality criteria 'about agriculture's role in environmental improvement. Subsequently, Warkentin and Fletcher (1977) introduced the concept at an international seminar on soil environment and fertility management for intensive agriculture. The concept was needed to facilitate better land-use planning, because of the increasing number of functions that the soil resources provide (Karlen et al. 2003).

To understand SQ in its modern context, the complexity of soil and its functions must be recognized. Soil should be recognized as a complex system of minerals, organic compounds, soil solution, soil gases, and living organisms that interact continuously in response to natural and imposed biological, chemical, and physical forces. Besides, the soil should be accepted as a living system, which represents a finite resource vital to life on earth (SSSA 1995).

8.3.2 Soil Quality and Soil Functions

Bouma et al. (1998) argue that SQ must be linked to certain types of land use and the associated management. Therefore, 'the capacity of the soil to function' is the central idea in most SQ definitions (Karlen et al. 1997). In other words, the term quality implies value judgment or degree of excellence for a specific purpose (Schjonning et al. 2004). Therefore, SQ can be defined in a more elaborated manner as the capacity of a specific kind of soil to function, within natural or managed ecosystem boundaries, to sustain plant and animal productivity, maintain or enhance water and air quality, and support human health and habitation (Larson and Pierce 1994).

The biophysical soil functions include nutrient cycling, water dynamics, filtering and buffering, physical stability and support of plant systems and human structures, and promotion of biodiversity and habitat. These broad categories of soil functions require some expansion to fully understand the vital nature of soils in the food, energy, water nexus. There are several soil properties related to soil functionality (Fig. 8.5). However, the ability of the soil to provide these functions is more complex than merely listing the soil properties. Our understanding of the linkage between soil properties and soil functions and the resultant ecosystem services is incomplete (Adhikari and Hartemink 2016; Swinton et al. 2006). The ability of the soil to provide these functions depends upon the state of the soil properties. For example, soil water holding capacity is a soil property related to water dynamics and filtering and buffering meaning that sandy soil with a low water holding capacity will not provide those functions as well as clay loam soil with a high water holding capacity.

Soils provide the products and environmental services indicated in Fig. 8.5 by performing five essential functions (Karlen et al. 1997) which include:

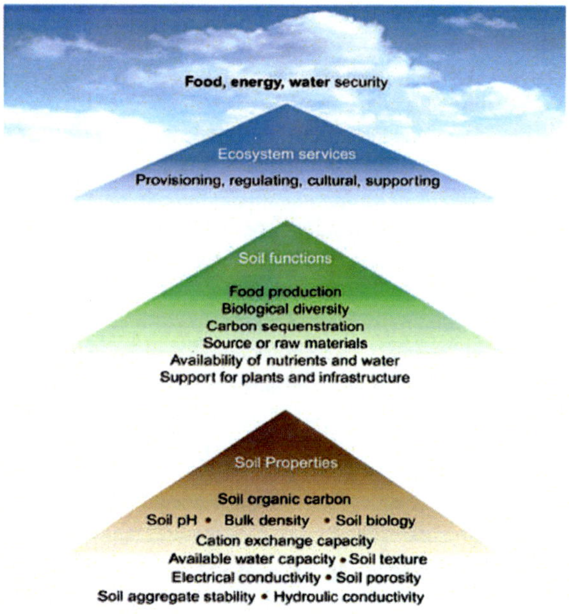

Fig. 8.5 Interface of soil properties relative to soil functions and ecosystem services (*Source* Hatfield et al. 2017)

I. *Regulating water movement:* Soil helps to control the whereabouts of precipitation and irrigation water. Water and dissolved solutes flow over the land surface or into and through the soil depending on the quality of the soil.
II. *Sustaining plant and animal life:* The diversity and productivity of living things depend on soil functions.
III. *Filtering and transforming potential pollutants*: The minerals and microbes in the soil are responsible for filtering, buffering, degrading, immobilizing, and detoxifying organic and inorganic materials in the soil.
IV. *Cycling nutrients:* Carbon, nitrogen, phosphorus, and many other nutrients are stored, transformed, and cycled through soils.
V. *Supporting structures:* Buildings need stable soil for support, and archaeological treasures associated with human habitation are protected in soils.

The relative importance of the functions depends on the socio-economic and development status of the society. For instance, for the substance farmers in developing countries, soil functions that are directly related to their basic needs such as sustaining plant and animal life (diversity and productivity) are prioritized while for advanced communities in developed countries to environment and leisure related functions may also get greater importance. According to Erkossa et al. (2005), for the farmers in the central highlands of Ethiopia the production-related functions such as crop, forest, and grazing received high ranking although other functions such as raw materials for pottery and construction of house and grain storage structures (Silo), source of water, sink for waste including burial of coffins, and fire control were also recognized. Therefore, although universally recognized functions are essential, due attention needs to be given to the locally recognized functions in SQ assessment.

Agricultural productivity is the primary recognized production function of soils. Soil quality affects crop productivity through its important functions such as nutrient cycling, physical stability and support, resistance and resilience, and water relations (Andrews et al. 2004). Good quality soil stores and cycles nutrients and allows crops to grow and use nutrients efficiently (Andrews et al. 2004; USDA-NRCS 1998). In such soils, nutrients become available when the plants need them, thus reducing the chance of nutrients being lost from the root zone through leaching, from the surface by runoff or above the crop canopy by volatilization.

Among the important soil parameters related to nutrient cycling, soil pH, potentially mineralizable nitrogen and microbial biomass are often considered as indicators of SQ (Sparling 1997; Karlen et al. 1996) for agricultural productivity. Nutrients and soil organic matter contained in the topsoil are vulnerable to erosion or runoff water loss (USDA-NRCS 1998). Nutrient loss increases in agricultural production costs due to the additional nutrient while the risk of water pollution that leads to higher societal costs increase. Soil compaction is among the key physical constraint concerning SQ for agricultural productivity and regulating functions of the soil. Compact soils restrict the movement of roots and nutrients in the soil, and hence reduce nutrient uptake and restrict air movement and gas exchange in the root zone, which leads to nutrient loss (USDA-NRCS 1998; Arshad et al. 1996). Often, soil aggregate stability and bulk density are considered indicators for soil physical stability and support (Arshad et al. 1996; Karlen et al. 1996; Doran and Parkin 1994). As soil moisture is an important attribute determining soil productivity function, plant available water capacity is often considered as an indicator of water relations (Lowery et al. 1996; Smith and Doran 1996).

8.3.3 Soil Quality Indicators

Bünemann et al. (2018) assert that assessments of the suitability of soil for crop growth might have been applied even before the evidence of written records. Documentation can be found in ancient Chinese books such as "Yugong" and "Zhouli", written during the Xia (2070–1600 BCE) and Zhou (1048–256 BCE) dynasty, respectively (Harrison et al. 2010), and in the work of Roman authors such as Columella (Warkentin 1995). Bünemann et al. (2018) described several concepts related to SQ assessment which include soil fertility, land quality, soil capability, SQ, and health. According to Patzel et al. (2000), the suitability of soil for agricultural production is captured in the concept of soil fertility, originating from the German literature on "Bodenfruchtbarkeit" that is predominantly aligned to crop yields. Accordingly, the FAO describes soil fertility as "the ability of the soil to supply essential plant nutrients and soil water in adequate amounts and proportions for plant growth and reproduction in the absence of toxic substances which may inhibit plant growth" (www.fao.org).

The obvious purpose of SQ assessment is to monitor the effects of management systems on SQ attributes and to avoid practices that negatively affect its capacity to function (Gary et al. 2000). Soil quality assessment involves spatial or temporal monitoring of management effects on soil functions. There are no universally accepted methods and tools for SQ assessment. The Soil Management Assessment Framework (SMAF) is among the tools widely used (Erkossa et al. 2007; Andrews et al. 2004). SMAF was designed to follow three basic steps (Andrews 1998):

indicator selection, indicator interpretation, and integration of the indicator values into an index.

Relevant and reliable indicators for specific functions under agro-climatic and socio-economic circumstances are essential since SQ cannot be directly measured. Indicators are the key attributes of the soil system that have the greatest sensitivity to changes in soil function (Andrews et al.2004). They are measurable physical, chemical, and biological attributes (Doran and Perkin 1996) or morphological and visual features (USDA-NRCS 2001) of soils or plants. Useful indicators are those that can be easily assessed by qualitative or quantitative methods (USDA-NRCS 2001), and able to evaluate changes in soil functions, assessed in a reasonable duration, sensitive to variations in climate and management. Gary et al. (2000) suggest that SQ indicators must well correlate with quantifiable soil functions, respond to external changes in a measurable way, be adaptable for use by individuals with a range of background and skills, be found in the existing database that is accessible and of value to SQ assessment, and be easily integrated into larger ecosystem-scale models, including socio-economic models. The best SQ indicators are those soil characteristics that show a significant change in one to three years. Readers are referred to Andrews et al. (2004) or Erkossa et al. (2007) for an elaborated example of SQ assessment.

8.3.4 Soil Quality and Soil Health Status in Ethiopia

The major threat to SQ is its degradation, which is a change in the soil health status resulting in a diminished capacity of the soil to provide the expected functions within the economically and environmentally sustainable management systems. Soil quality degradation is the reduction or loss of the biological or economic productivity functions of the soil. Globally, both the crop and grazing lands are degrading at an increasing rate, due to natural and anthropogenic factors. Human-induced SQ degradation is a serious threat, especially in marginal agro-ecosystems characterized by low and variable rainfall and steep slopes where the soil fertility and organic matter content are depleted resulting in low productivity (Erkossa et al. 2020; Scherr and Yadav 1995).

In Ethiopia, due to the combined effects of anthropogenic activities like poor farming practices, overgrazing, deforestation, soil erosion, salinity and alkalinity, and the use of livestock manure and crop residues as fuel, SQ degradation is hastening desertification (Erkossa et al. 2009; Hawando 1997; Cesen 1986). The major physical agent for SQ degradation in the highlands of Ethiopia is soil erosion (Tefera et al. 2002). The average soil erosion rate from cultivated lands across the country was estimated at 42 Mg ha^{-1} year^{-1} (Bewket and Sterk 2003). Another study estimated the rates of soil erosion from the currently cultivated and formerly cultivated lands at 20 Mg ha^{-1} year^{-1} and 33 Mg ha^{-1} year^{-1}, respectively (Hurni et al. 2015). Soil erosion selectively removes the finest and fertile topsoil rich in organic matter and clay content resulting in enriched nutrients and OM in the eroded sediment as compared to the surface soil from which it is eroded. The deterioration of SQ and health due to natural and anthropogenic factors can be depicted in biological, chemical, and physical aspects.

(a) Biological aspect

Biological activities in soil are widely recognized as playing a vital part in nutrient cycling and availability to plants and in developing and maintaining soil structure and contributing to SQ and soil health. Biological SQ degradation refers to the process that leads to a decline in the dead organic matter (OM) content mainly through mineralization or erosion (Solomon 1994) or that of the essential living organisms. Because it concentrates on the surface and has a low density, OM is easily removed by erosion. According to Okolo et al. (2019), unsustainable land management practices including deforestation, extractive harvesting, uncontrolled grazing, continuous cultivation without fallowing, and bush burning that are practiced in many locations in Ethiopia are the key factors in the depletion of organic matter. Due to increases in pressure on the cultivated land, landholdings have continuously shrunk leading to a short or no fallow period (Erkossa et al. 2020). Based on a study conducted in western Ethiopia, Solomon (1994) reported that soil OM content dropped from 20 to 7% in less than three years of continuous cultivation due to mineralization. A recent study conducted in the Northern and Eastern parts of the country showed an average annual removal of 3 Mg C ha^{-1}, with the lowest depletion in the dry areas (van Beek et al. 2018). Depending on their intensity, timing, and type of implements used, tillage practices increase the loss of OM by enhancing the mineralization through aggregates breakdown (Erkossa et al. 2007). The stability of soil aggregates is dependent on microbial biomass. Thus, the elimination of soil microorganisms or limitation of their activities causes physical damage to the soil ecosystem, including increased soil erosion and OM depletion.

(b) Chemical aspect

The major chemical aspects of SQ degradation of agricultural soils are macro- and micro-nutrient depletion, acidification, salinization/sodification, most of which are induced or aggravated by human activities. Nutrient depletion is arguably the greatest management-induced chemical SQ degradation in Ethiopia in terms of its extent and effect on crop productivity. Various independent studies have shown that the soil nutrient balance in the country is negative, with an average annual depletion per hectare of 122 kg N, 13 kg P, and 82 K due to soil erosion and crop uptake (van Beek et al. 2016; Stoorvogel and Smaling 1993). The suboptimal use of only N and P fertilizers led to a multi-nutrient deficiency of the soils. A recent national soil fertility survey revealed a widespread deficiency of Sulphur (S), Boron (B), and Zinc (Z) in addition to N and P (EthioSIS 2013). Some research reports also indicated a localized deficiency of potassium (K), copper (Cu), manganese (Mn), and iron (Fe) (Fanuel et al. 2018; Yifru and Sofia 2017). In addition to crop uptake, the plant nutrients are lost in soluble form with runoff and with eroded sediments, which are dominated by finer soil fractions that are rich in their nutrient concentration (Wudneh et al. 2014; Erkossa et al. 2005).

The other management-related chemical SQ degradation in Ethiopia is soil acidification and salinization or sodification. Soil acidity is increasingly challenging crop productivity in the high rainfall areas of Ethiopia. According to Getachew et al. (2019), about 43% of cultivated land in humid and sub-humid highlands of the country is affected by soil acidity, of which about 28% are strongly acidic (ATA 2014). Soil acidity limits the availability of essential nutrients such as P, K, Calcium, and Magnesium, and affects the activities of essential soil organisms.

The soil acidification is accelerated through burning and clearing of vegetation, continued use of acid-containing fertilizers, and excessive irrigation. Soil acidification has gotten attention as a major problem affecting crop productivity in the high rainfall areas of the country. Soil acidity and associated low nutrient availability are key constraints to crop production in acidic soils, mainly Nitisols of Ethiopian highlands (Zeleke et al. 2010).

According to Getachew et al. (2019), continuous application of inorganic and organic fertilizers can increase soil acidity. For instance, hydrogen is added to the soils in the form of ammonia-based (NH_4), urea-based [(CO $(NH_2)_2$], and proteins (amino acid) in organic fertilizers. The transformation of N fertilizers into nitrate (NO_3^-) releases hydrogen ions (H^+) which contribute to soil acidity. In addition, nitrogen fertilizers increase soil acidity by increasing crop yields, which increases the uptake of the basic elements from the soil solution.

According to Qureshi et al. (2018), about 11 million ha of land in Ethiopia has been affected by different levels of salinity and sodicity, which is mainly concentrated in the rift valley. The salinity and sodicity levels are related to poor drainage conditions in most areas of the valley, and the problems are more pronounced in the irrigated areas. In the middle Awash Valley, the large state-owned irrigated farms are fast going out of production due to increasing soil salinity (Fantaw 2007). For instance, Qureshi et al. (2018) reported a significant increase in the extent of salt-affected soils from 1972 to 2014 at Tendaho state farm due to poor irrigation practices, where normal soil was reduced from 35 to 20% of the total farm. This degradation was attributed to the use of poor-quality irrigation water and lack of drainage facilities. Given the policy-driven fast increasing trend in areas equipped for irrigation in response to increased demand for agricultural production both for domestic consumption and export (Gebul 2021), there is a paramount risk of irrigation-induced salinization due to the poor irrigation water management practices.

(c) Physical aspect

The manifestation of physical SQ degradation in Ethiopia includes soil compaction and hardpan formation due to repeated cross-ploughing, use of heavy agricultural implements, and cattle grazing (Tebebu et al. 2017; Tefera et al. 2002; Raper et al. 1998; Mwendera and Saleem 1997), which is exacerbated due to organic matter depletion. Soil compaction occurs when soil particles are pressed together, reducing the pore space between them. In the conventional teff cultivation system, the land undergoes more than four tillage operations (Fekremariam et al. 2021) using an oxen-drawn plough and sometimes up to 9 tillage passes (Erkossa and Ayele 2003), followed by soil trampling using a large number of cattle and donkeys immediately before sowing the crop. A study conducted around Hawassa (Feto 2016) concluded that agricultural soils may be compacted up to 0.3 m depth with heavy machinery farm operations, where soil bulk density, air-filled porosity, and saturated hydraulic conductivity are negatively affected.

Overstocking and overgrazing including grazing of leftover residues on cropland after harvesting cause soil compaction due to heavy and continuous trampling by livestock. Watering points and cattle routes are particularly vulnerable to soil compaction, which leads to excessive runoff and reduced water infiltration. Revegetation in these areas is, therefore, impeded. Unimpeded water flowing down slopes causes rills and gullies. The bulk density of grazing land in the Illubabor area was measured to be 1.34 g/cm^3. This is a high figure compared to 0.83, 0.79, and 1.12 g/cm^3 for a coffee forest, ungrazed grass fallow, and cropland, respectively (Solomon 1994).

(d) Land use and land management effects on soil quality and health

Land use in Ethiopia has been traditional with little regard to the land's best potential use and without due consideration for conservation of natural resources and safeguarding the environment (Zemen et al. 2017). Demand for agricultural products is burgeoning due to the ever-increasing human population. However, the agricultural productivity increase could not cope with the demand due to inappropriate land use, poor land and crop management practices, and limited use of inputs such as fertilizers. The dominant strategy to bridge the gap between demand and supply is an expansion of the cultivated land (Abate and Lemenih 2014). As the prime agricultural lands, especially in the settled highlands are used up, expansion of cultivation is directed toward lands with different land uses such as forest (Bishaw 2001) or grazing lands (Olson and Maitima 2006) which are generally marginal for agriculture given the prevailing rudimentary agricultural practices. According to Erkossa et al. (2018), the expansion of agriculture to the new lands usually involves either moving into the areas adjacent to the current cultivated lands or migration to other areas, including to those in different agro-ecological settings. Despite the quality of the new land and the agro-ecological settings, farmers often attempt to produce the same traditional crops using techniques that are not necessarily suitable for the new site (Erkossa et al. 2015). Coupled with the undulating terrain and fragile nature of the lands and the heavy seasonal rains, the rudimentary land and crop management adopted exacerbate soil erosion and deterioration of SQ (Fanuel et al. 2016; Erkossa et al. 2015).

Reducing or reversing the degradation of SQ and health and sustaining the productivity and ecological functions of the soils require the adoption of innovative management practices including those that enhance soil organic matter content, reduce soil disturbance and keep the soil surface covered and prevent soil compaction. Research reports from different parts of the country show that the use of physical and biological soil and water conservation measures and improved agronomic practices such as intercropping, row seeding/planting, and fertilization increase crop yields while enhancing SQ (Erkossa et al. 2018; Araya et al. 2016). In addition, agricultural inputs such as liming of acidic soils (e.g. Agegnehu et al. 2021), application of gypsum in salt-affected areas (eg. Sisay et al. 2021), biochar and other biological (Bikila 2019; Tariku et al. 2017), physical, and chemical amendments are needed depending on the dominant constraints. The key management strategy for enhancing SQ is increasing biomass and crop productivity, which leads to ecological and economic resilience while reducing the need for further expansion of the agricultural land.

Several studies revealed that farmlands with physical and biological soil and water conservation measures significantly improved physical and chemical SQ indicators (e.g., Melku et al. 2019; Erkossa et al. 2018). Integrated soil and water conservation combined with soil fertility management practices can enhance SQ and increase crop yield and biomass productivity. Recent studies in Ethiopia showed that, when properly implemented, conservation agriculture (CA) which reduces disturbance of the soil and increases surface cover can enhance biomass productivity and crop yield while improving SQ in 3–5 years (Dejene et al. 2020). Other studies show that the use of soil bunds reduce surface runoff, increase infiltration, and improve the availability of water and nutrient to plants (Schmidt and Zemadim 2015; Tadele et al. 2013, Tireza et al. 2013) and consequently contribute to higher biomass and crop yield (Tadele et al. 2013; Soomro et al. 2009), especially in areas where soil moisture is a key constraint (Kassie et al. 2008). In the humid highlands of Ethiopia, the use of level soil bund (Fig. 8.6) and integrated soil fertility management including fertilization, crop rotation, and line seeding have increased biomass, crop yield, and economic return while improving biological, chemical, and physical SQ indicators (Erkossa et al. 2018). Therefore, the use of physical soil and water conservation in tandem with appropriate agronomic practices which increase crop yield and biomass productivity may be a judicious and cost-effective strategy to enhance or maintain SQ and health under the smallholder farming systems like those in Ethiopia. Appropriate use of external inputs such as fertilizers, agricultural lime, gypsum, living and nonliving organic matter may lead to rapid improvement of soil quality and increased productivity to encourage farmers to invest, which can be sustained through other agroecological strategies.

8.4 Soil Sustainability and Soil Resilience

8.4.1 Principles

In an era of Anthropocene (Steffen et al. 2015) where there is a major swing in the functioning of the Earth's System resulting in surpassing of many planetary boundaries, it is no surprise if experts in different fields of study talk about concepts like sustainability and resilience of systems and resources. Soil is one of the resources to which these concepts have been attached to since recently. The burden of feeding the ever increasing world population (UN 2015) and meeting the associated escalating demands for energy and clean water (IRENA 2015), for which soil is the link (Jónsson et al. 2016), rests heavily on soil resources. The state of soil resources is, therefore, fundamental to ensuring food security and maintaining environmental quality including but not limited to fighting against climate change through at least carbon sequestration. Thus, soil management holds the key to, to what extent the soil fulfils these commitments including sustainability of agriculture itself (Baveye et al. 2016). Hence, introduction of concepts like soil sustainability and resilience to the field of soil science is not a mere coincidence, but a necessity. Soil management has to consider the multiple uses and functions of soil and its effects on soil sustainability and resilience needs to be monitored periodically.

What is soil sustainability? Although there have been large number of definitions, the most comprehensive definition of soil sustainability is derived from the Brundtland Commission's definition of sustainable development, which was coined in 1987. On the basis of this, Abbott and Murphy (2007) defined soil sustainability as "soil management that meets the needs of the present without compromising the ability of future generations to meet their own needs from that soil". This definition places more emphasis on management rather than the natural entity, the soil, its self. However, the definition of soil sustainability should focus more on the soil's overall capacity to continue performing its ecosystem functions not only under managed ecosystems but also natural ones today and tomorrow. If management is the center of the definition for soil sustainability, then, it becomes very difficult to define management that brings about soil sustainability universally, for soil is a dynamic entity requiring different management interventions. Buol (1995) claims that soil sustainability is achieved if and only if management options be matched with specific soil types and its uses in a given geographical environment. The management interventions should work toward building the soil's resilience, a concept closely related to sustainability. On the other hand, Sullivan (2004) described the term "Sustainable" in light of soils as capable of being maintained at length without interruption, weakening, or losing in power or quality. Another important concept that is introduced into the discipline of soil science and contributes to soil sustainability is soil resilience.

Many experts (e.g., Lal 2015; Barrow 1991; Peters 1991) view *resilience*, in a very general way, as an ecological concept that dictates a given entity's responses to disturbances of different nature. More specifically, Lal (2015) defined *soil resilience* as "the ability of a given soil to resist or recover from an anthropogenic or natural perturbations". Others defined it in a more elaborated way as "the capacity to recover from a disturbance while retaining its structure, function, identity, and feedbacks (Brand and Jax 2007; Walker and Salt 2006; Holling et al. 2002; Lal 1997)". More recently, Ludwig et al. (2018) described *soil resilience* very briefly as "The potential to recover functional and structural integrity after a disturbance". Soil resilience is fundamental to soil sustainability. A soil's resilience is influenced by both internal and external conditions that can be broadly categorized into processes, factors, and causes of resilience (Blanco and Lal 2010). Lal (2015) and Blanco and Lal (2010) have compiled long lists of processes, causes, and factors for soil resilience. Few examples include the following: processes (new soil formation, aggregation, soil organic carbon accumulation, nutrient cycling and transformation, increase in biodiversity); factors (landscape position, climate, parent material, soil quality, water balance, vegetation); and causes (socio-economic and political forces that govern land use, land rights, institutional support, and income).

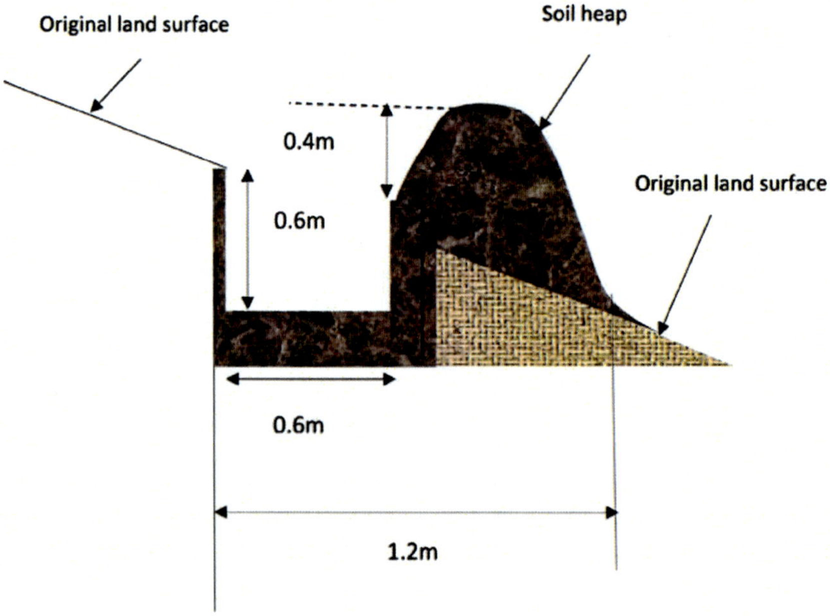

Fig. 8.6 Schematic sketch and dimension of a typical soil bund used in the humid tropical areas of Ethiopia (*Source* Erkossa et al. 2018)

8.4.2 Soil Sustainability

(a) Concepts of soil sustainability

Most definitions of sustainability are concerned with performance of an entity at a certain acceptable or profitable level over a given time span under any given management or set of management practices (Eswaran 1994) and as such is related to the productivity, economic, social, and environmental aspects of a land use system, i.e., agriculture (Smyth and Dumanski 1995). By similar analogue, soil sustainability refers to the soil's ability to function at an optimal level at present and continue to do the same in the future. Undoubtedly, maintaining this optimal level on a sustainable basis requires management interventions. However, the debate in this regard remains providing a universally accepted understanding of what sustainable practices are and for which purpose or land use (Ludwig et al. 2018; Rist et al. 2014; Wezel et al. 2014).

Whether a given management leads to sustainability or not has to be evaluated in terms of its effects on soil functioning. In this regard, the most important question remains "how can the soil functions be evaluated?" given the fact that soil performs multiple functions. The recommendation is that multifunctionality effects should be considered in the evaluation process, which requires a holistic approach. Therefore, defining sustainability requires selection and matching of sustainability indicators with soil functions. It also requires defining optimum levels of the indicators that lead to optimum functioning of the soil. However, we have to bear in mind that not all systems have one sustainable 'optimal' state (Walker et al. 2012).

Despite the many controversial issues surrounding the concept of sustainability in general and in view of soils, the importance of its concept in safeguarding soil resources is indisputable. This is particularly important in developing countries like Ethiopia for which agriculture is the backbone of their economy and source of livelihood for high proportion of their population. In Ethiopia, the soil resource is the victim of various forms of degradative processes such as nutrient mining, soil fertility depletion, consistent soil organic matter decline, soil erosion, soil acidification, and salinization/sodification due mainly to the use of inappropriate management practices and land use decisions. These heinous degradation processes have resulted, over the years, in sustained decline in the productivity of the soils, among others. The issue of soil sustainability has always been an ever present concern in the country and studies on soil sustainability are non-existent.

(b) Soil sustainability indicators

Whether a given soil is performing all functions it is performing at an optimum level or at least at a minimally acceptable level needs to be evaluated. Doing this helps to understand the status of the soil's properties which are necessary for proper functioning and development and implementation of appropriate management interventions when required. It is, therefore, important to select a minimum set of soil attributes that are sensitive to disturbances in order to evaluate the soil sustainability via soil function. In a broader sense, indicators are required to evaluate and report how sustainably soils are managed in fulfilling different goals. Selection of appropriate indicators is a prerequisite for evaluating soil sustainability, providing appropriate feedback to decision makers, and suggest feasible and sound interventions in the form of management. However, given the multiple functions soils perform, their diversity, and the myriad purposes for which they are put, providing a universally accepted list of indicators could be almost a daunting task (Jónsson et al. 2016; Moebius-Clune et al. 2016; Singh et al. 2011). Furthermore, the emergence of concepts like sustainability, which require multidisciplinary approach, has led to change in approach towards selection of indicators.

Jónsson et al. (2016) added two dimensions (social well-being and economic) to the commonly used nature dimension for selection of indicators to evaluate soil sustainability. The nature dimension focuses mainly on soil attributes that include the physical, chemical, and biological properties (Marchese et al. 2017; Jónsson et al. 2016; Moebius-Clune et al. 2016; Singh et al. 2011). Ludwig et al. (2018) argues that selection of bioindicators for sustainability requires a holistic concept. Extensive literature sources on indicators selected for different purposes in the nature dimension can be found in Jónsson et al. (2016). Most of the indicators from the nature dimension are also those commonly selected for assessing soil quality, which shows the link between soil quality, soil function, and soil sustainability. The requirement in the selection of indicators from the nature dimension is that the indicators proposed must be linked to a certain soil function. Commonly selected indicators include: physical (aggregate stability, bulk density, soil depth, water retention and transmission, compaction, susceptibility to erosion, and aeration status), chemical (soil reaction-pH, cation exchange capacity, nutrient reserve/retention and release in the right amount and form and at the right time, elemental balance that does not cause antagonism), and biological (soil biodiversity, soil organic carbon pool, microbial biomass, pests and pathogens).

Not much information is available on the list of indicators used in assessing sustainability in terms of social and well-being, and economy dimensions. Very recently, Jónsson et al. (2016), using the Delphi survey technique in which stakeholders from different sectors were involved in the selection of indicators, assessed large number of indicators from the two dimensions. They used three stakeholder groups categorized as scientists, soil practitioners, and policy makers. From the society and well-being dimension, government policies, expenditure on soil related research and development, education on soil sustainability, public awareness of the value of soil, public participation, bioavailability of essential major and trace elements, and population growth were accepted as indicators of sustainability by the stakeholder groups. Similarly, from the economy dimension, economic value of soil ecosystem services, change in land use diversity, yield under conditions of no change in fertilization, chemical fertilizer use intensity, pesticide use intensity, and soil salinity due to irrigation were selected by the different stakeholder groups as indicators to define soil sustainability. The fact that not much has been done on the two dimensions other than the nature dimension calls for the dire need to do more research to come-up with reasonable number of indicators that can be helpful in assessing sustainability comprehensively.

(c) Opportunities for soil sustainability

There are a number of compelling reasons and opportunities for why soil sustainability should be taken more seriously than ever before. The prevalence of heinous soil degradation worldwide (FAO 2010) that has undermined the multiple functions of soils that range from serving as a medium for food production to maintaining environmental quality including climate change mitigation (Jónsson et al. 2016; Keesstra et al. 2016; Bindraban et al. 2012; Nellemann 2009) is among the persuasive reasons for ensuring soil sustainability. Schulte et al. (2014) summarize the five key functions that arable soils provide as biomass production, water purification, carbon sequestration, serving as habitat for biodiversity, and recycling of nutrients and (agro) chemicals. Cognizant of these functions and services of soils and the role they play in human life, they have been included in many of the sustainable development goals (SDGs) (e.g., SDGs 2, 7, 13, and 15) of the United Nations (General Assembly 2016; Jónsson et al. 2016; Keesstra et al. 2016; Bouma 2014; Godfray et al. 2010), which can be viewed as an opportunity for ascertaining soil sustainability. The link between soil processes and SDGs is usually conceptualized via soil functions (Keesstra et al. 2016), which is an indicator of soil sustainability. In the coming many years, therefore, the protection and sustainable management of soil resources will be an even more important priority to the world (Jónsson et al. 2016). Added to this, recognition is the availability of enumerable technological options to overcome soil degradation and enhance opportunities for soil sustainability.

Soil is one of the most important strategic resources affecting the agriculture sector in Ethiopia. It, therefore, goes without saying that the status of the soil resource plays a pivotal role in ensuring food security for the rapidly growing population, harnessing economic development, and maintaining environmental health. In view of these, government has started giving more attention to soil resources (e.g., inclusion of soil health issues in policies, establishing soil information system-EthioSIS, national soil database, etc.) in a bid to ensure soil sustainability. Therefore, opportunities for soil sustainability are highly promising in Ethiopia.

8.4.3 Soil Resilience

(a) Concepts of soil resilience

The term resilience has been used in different fields, such as Ecology (Seybold et al. 1999), Psychology (Olsson et al. 2015), and unspecified many different fields (Ludwig et al. 2018), before it was introduced into the field of soil science as a new concept (Blum 1994). In these different fields, it was given different meanings.

After its concept has been introduced to soil science, different definitions have been provided by different scientists. Blanco and Lal (2010) defined soil resilience as the inherent capability of a soil to recover from degradation and return to a new equilibrium similar but not identical to the original state. Seybold et al. (1999) and Herrick and Wander (1998) defined it, in a more comprehensive way, as the ability of the system to recover its "functional and structural integrity" following shocks, perturbations, or stresses. According to this source, functional integrity refers to the capability of the soil to perform its multiple functions at optimum level. On the other hand, the structural integrity refers to the soil's ability to improve its structural properties (e.g., soil aggregation and porosity) to a level that is close to the initial status following disturbance. Rozanov (1994) described soil resilience as the capacity to resist change caused by disturbance. Disturbance is defined by Forman and Gordon (1986) as any event or process that causes a significant change from the normal pattern or functioning of an ecosystem.

Soil resilience measures the extent of resistance to forces driving soil degradation and gives insight into mechanisms that endow soils with favorable properties and processes for overcoming destructive forces. Beinroth et al. (1994) and

Table 8.4 An example of soil attributes used for monitoring soil resilience

Soil properties		
Physical	Chemical	Biological
Horizonation	Soil pH	Microbial biomass
Color and depth	Essential nutrients	Microbial activity
Clay content	Total organic C	Earthworm population and activity
Bulk density and porosity	Total N content	Above- and below-ground biomass
Water-stable aggregates	C:N ratio	Rooting depth
Soil temperature and air permeability	Particulate organic matter	Root abundance
Water infiltration	Particulate P	Biodiversity
Water retention capacity	Cation exchange capacity	Plant growth and population
Cone index and tensile strength	Oxidation and reduction	
Coefficient of linear extensibility and plastic limit	Sodicity	

Eswaran (1992) indicated that the concept of resilience is relatively new in soil science literature and it is closely related to the sustainability of agricultural systems (Szablocs 1994). However, Lal (1994) underlined that the concept of resilience remains inexactly expressed and quantified albeit its high relevance for understanding and management of soil ecosystems. Blanco and Lal (2010) argue that the dynamic, variable, and heterogeneous nature of the soil system make it difficult to provide an accurate definition of resilience. Not all soils have the same resilience implying the inadequacy of a single definition to fully express the resilience of diverse soils. Blanco and Lal (2010) recommend that the appropriate and comprehensive definition must be based on the soil's ability to recover from perturbation to perform a specific process or function.

In Ethiopia, literature on soil resilience is almost non-existent. However, many interventions geared toward rehabilitation and/or restoration of soils through watershed works and others are meant to improve soil resilience, signifying that the concept of soil resilience exists.

(b) Indicators of soil resilience

Identifying soil attributes or processes that elucidate soil resilience is necessary in identifying management options and monitor changes. In general, selected soil physical, chemical, and biological properties are used as indicators. The state of these indicators should be monitored before and after intervention in order to quantify the rate and magnitude of change. In the process of monitoring, distinction should be made between those properties that change rapidly and those that require relatively longer time to change. Requirements for measuring resilience include the use of long-term data derived from replicated experiments and developing standard sensitive indicators focusing on a small but complete set of sensitive properties that are enough to detect significant differences in soil function. Examples of soil properties often used as indicators of soil resilience are summarized in Table 8.4 (Adapted from Blanco and Lal 2010).

(c) Soil quality and soil resilience

Soil quality is defined as the fitness of a specific kind of soil, to function within its capacity and within natural or managed ecosystem boundaries, to sustain plant and animal productivity, maintain or enhance environmental quality, and support human health and habitation (Arshad and Martin 2002; Karlen et al. 1997; Doran and Parkin 1994). There is a strong link between soil quality and soil resilience. Seybold et al. (1999) wrote an article with the heading "Soil Resilience: A Fundamental Component of Soil Quality" to elucidate the strong relationship between the two soil attributes. However, they indicated that the relationship between soil resilience and soil quality has not been well defined or developed albeit the concept of soil resilience was introduced to soil science in order to address issues related to sustainability of soil resources and their degradation. In general, the two terms are inter-related; however, they are different (Lal 1997). In brief, soil quality pertains to ecosystem functions and services, while resilience is related to a soil's capability to restore itself following disturbance or even resist the effects of the disturbance. Soil resilience and soil resistance ["the capacity of a soil to continue to function without a change throughout

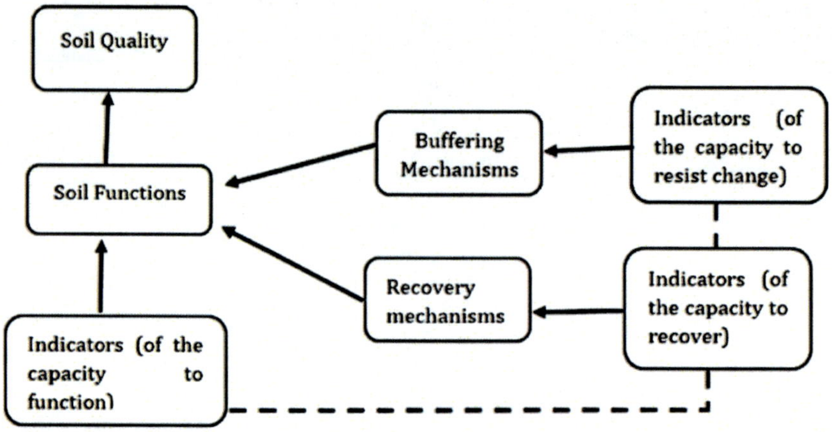

Fig. 8.7 "The concept of how soil resilience and resistance relate to soil quality through soil functions" (Adapted from Seybold et al. 1999)

a disturbance (Herrick and Wander 1998; Pimm 1984)"] are important components of soil quality, which is a key component of soil sustainability.

Soil quality can be categorized into two as inherent (derived from the intrinsic soil properties determined by the five soil-forming factors) and dynamic quality (change in soil function contingent upon those soil attributes that are affected by natural and anthropogenic disturbances including management). Seybold et al. (1999) presented an interesting schematic concept that shows the relationship between soil resistance and soil resilience with soil quality through soil functions as follows (Fig. 8.7):

(d) Enhancing soil resilience

Soil resilience is important for ensuring sustainable use of soil resources. Building the resilience of soils in an era where there is increasing pressure on soil resources is of paramount importance. Degradation due to anthropogenic activities is a serious problem challenging soil resources and their capacity to perform essential functions. Soils should be able to recover from various shocks in the shortest time possible to continue feeding the increasing world population and maintain environmental quality. Before interventions are identified, it is important to identify factors that degrade soil resilience and enhance it. The factors that reduce soil resilience are related to degradation and are generally termed as disturbances. Disturbances can be natural or anthropogenic. Blanco and Lal (2010) provided a long list of disturbances under two categories. Of the two disturbances, the anthropogenic ones are of major concern because of the magnitude of disturbance they result in within a short period of time. In general, these disturbances influence soil attributes (physical, chemical, and biological) that are important for facilitating or enhancing a number of soil processes that affect soil functions.

In addition to soil characteristics (parent material, physical, chemical, and biological properties), landscape characteristics and biota (e.g., slope, aspect, flora and fauna), and climate (e.g., precipitation, radiation, temperature, wind) and time (rate of soil formation and degradation) are identified as factors affecting soil resilience (Lal 1997). On the basis of this, Blanco and Lal (2010) indicated that soil resilience is affected by the same factors that affect soil formation, for it is these factors that influence the initial state of different soil characteristics. Although there are many soil properties that are related to soil resilience, it is the intrinsic textural and dynamic structural properties and their interaction that have the biggest control over soil resilience since these salient properties affect most other soil attributes. Thus, the soil resilience can be enhanced by maintaining these and other soil properties at an optimum level. In fact, some of the intrinsic properties such as texture and soil depth cannot be managed easily; however, their effects can be modified by managing other dynamic properties such as soil structure.

Adopting practices that result in minimum disturbance of the soil and increasing organic matter input into the soil can help in enhancing soil resilience. The choice of appropriate practices, however, could depend on specific local conditions including socioeconomic aspects. The key to enhancing soil resilience is maintaining the level of organic matter in the soil at an optimum level. Organic matter enhances soil resilience through improving the soil pore structure, increasing water infiltration, and reducing soil compaction, runoff and soil erosion. Conservation tillage, green manuring, leaving residues on soil surfaces, incorporation of crops that produce large biomass into the cropping system, appropriate and complex crop rotation, application of animal and poultry manures, compost, and other organic amendments increase organic matter input into the soil and, thus, enhance soil resilience.

8.4.4 Relationship Between Soil Resilience and Sustainability

Resilience is a measure of the (pre-) adaptive potential to cope with future disturbances while at the same time representing the multifunctionality of the system. Resilience is a product of the past and the present as well as a prospect of the system's future. Hence, it reflects all-important indication levels for evaluating the soil state: it is derived from the soil's [management] past as part of the soil memory, representing the soil's present status over the affectedness under the given pressures and disturbances as well as providing possible recommendations for future improvement of holistic soil management by evaluating the intrinsic adaptive potential. Hence, in many concepts resilience is seen as a crucial part of sustainability or even equated with the term of sustainability (Marchese et al. 2017). Therefore, soil resilience can be used as an appropriate measure of soil management sustainability.

8.4.5 Policy Implications of Soil Sustainability and Resilience

Given the importance of soils in fulfilling most of the UN's Sustainable Development Goals, directly or indirectly, and the forecasted increases in food production in the future which requires the presence of good soils and their role in climate change mitigation, soil resources deserve more attention than ever by the global community. People in the profession, soil scientists, claim, which is actually true, that soil is the most basic of all natural resources that supports all terrestrial life (serving as the substrate for life) and most living things on planet Earth are directly or indirectly derived from soil. However, the value given to this important resource has never been nearly equal to what it serves the world. As a result, it has been exposed to different forms of degradation that continuously undermined its ability to perform its multiple functions at the desired level. Since recent times, this has alerted the global community since soil degradation has jeopardized food security and environmental safety. The various degradation processes have resulted in negative trends of soil quality, which in turn has put attainment of soil sustainability and soil resilience in balance. In the years to come, meeting the globally increasing the demand for food, fiber, and bio-based products for the continuously increasing world population will not be possible without ensuring soil sustainability and soil resilience. Furthermore, fighting climate change through carbon sequestration in soils will not be possible without soils that have favorable conditions for such processes. Soils cannot perform their task of serving as a 'geologic kidney' in which they purify water resources from different harmful substances and even help in cleaning the air that we inhale unless they are healthy. They can't maintain their rich biodiversity unless they have conditions within them that are conducive to the survival of the millions of biological organisms living in them. They can't support production of adequate food that is safe and nutritious if they are misused or abused. At global level, soil resources are manifesting their tiredness through different symptoms, some visible and some latent.

Without sustainable and resilient soils, the challenges the entire world will face are quite enumerable. It is therefore high time that soil resources get the kind of attention they deserve in policies, investments (in research, education, infrastructure and others), and protection before they are exhausted beyond repair.

8.5 Conclusion

Soil performs multiple functions that are fundamental to the well-being of human beings and other biological organisms as well as maintenance of environmental quality. Soil fertility is one of the key attributes of soils affecting food production, particularly in developing countries like Ethiopia. Ethiopian soils have major fertility issues related to deficiency of major nutrients such as N and P, low organic matter level, and strongly acidic and strongly alkaline reactions, among others, with significant variations across agroecologies. These limitations are affecting the quality and/or health of soils negatively. The deterioration in soil quality is affecting soil sustainability and soil resilience negatively. Current and future interventions must give attention to restoration of the already lost soil qualities and maintenance of the better ones in order to create resilient soils that are resistant to disturbances or bounce back within a short period following shocks. It is such interventions that are likely to ensure soil sustainability. Soil sustainability and soil resilience should appear as core issues in policies dealing with natural resources.

References

Abay A, Kelsa K, Tesfaye D (2011) Application of NP fertilizers for better production of Teff (Eragrostis tef (Zucc.) Trotter) on different types of soils in Southern Ethiopia. J Nat Sci Res 1(1):6–15

Abate A, Lemenih M (2014) Detecting and quantifying land use/ land cover dynamics in nadda asendabo watershed, South Western Ethiopia. Int J Environ Sci 3(1):45–50

Abbott LK, Murphy DV (2007) Soil biological fertility: a key to sustainable land use in agriculture. 1007/978-1-4020-6619-1. ISBN:978-1-4020-6618-4

Abera Y, Kassa S (2017) Status of soil micronutrients in Ethiopian Soils: a review. J Environ Earth Sci 7(4)

Adhikari K,Hartemink AE (2016) Linking soils to ecosystem services—a global review. Geoderma 262:101–111. https://doi.org/10.1016/j.geoderma.2015.08.009

Agegnehu G, Nelson PN, Bird MI (2016) Crop yield, plant nutrient uptake and soil physicochemical properties under organic soil amendments and nitrogen fertilization on Nitisols. Soil Tillage Res 160:1–13

Agegnehu G, Amede T, Erkossa T, Yirga C, Henry C, Tyler R, Matthew GN, Sheleme B, Gudeta WS (2021) Extent and management of acid soils for the sustainable crop production system in the tropical agroecosystems: a review. Acta Agric Scand Sect B Soil Plant Sci. https://doi.org/10.1080/09064710.2021.1954239

Agegnehu G, Yirga C, Erkossa T (2019) Soil acidity management. Ethiopian Institute of Agricultural Research (EIAR). Addis Ababa, Ethiopia

Alemayehu K (2015a) Characterization of agricultural soils in CASCAPE intervention Woredas in Southern Nations Nationalities People Regional State. CASCAPE report. Hawassa University, Ethiopia, p 511. library.wur.nl/WebQuery/isric/2259398

Alemayehu K, Sheleme B (2013) Effects of different land use systems on selected soil properties in South Ethiopia. J Soil Sci Environ Manag 4(5):100–107

Alemayehu K, Sheleme B, Schoenau J (2016) Characterization of problem soils in and around the southcentral Ethiopian Rift Valley. J Soil Sci Environ Manag 7(11):191–203

Alemayehu K, Sheleme B, Schoenau J (2017) Fractionation and availability of phosphorus in acid soils of Hagereselam, Southern Ethiopia, under defferent rates of lime. Chem Biol Technol Agric 4:21. https://doi.org/10.1186/s40538-017-0105-9

Alemayehu R (2015b) Characterization of agricultural soils in CASCAPE intervention Woredas in Western Oromia. CASCAPE report. Jimma University, Ethiopia, p 117. library.wur.nl/WebQuery/isric/2259397

Alexander M (1971) Agriculture's responsibility in establishing soil quality criteria. In: Environmental improvement—agriculture's challenge in the seventies. National Academy of Sciences, Washington, DC, pp 66–71

Al-Kaisi MM, Yin X, Licht MA (2005) Soil carbon and nitrogen changes as influenced by tillage and cropping systems in some Iowa soils. Agr Ecosyst Environ 105:635–647

Amanuel Z, Girmay GS, Atkilt G (2015) Characterization of agricultural soils in CASCAPE intervention Woredas in Tigray, Ethiopia. CASCAPE report. Mekelle University, Ethiopia, p 284. library.wur.nl/WebQuery/ isric/2259410

Amare A (2015) Yield response of local long maturing sorghum varieties to timing of nitrogen fertilizer application in eastern Amhara Region, Ethiopia. J Biol Agric Healthc 5(3):184–189

Andrews SS (1998) Creating a minimum data set to compare soil quality under poultry litter management alternatives. In: Sustainable agriculture alternatives: ecological and managerial implications of poultry litter management alternatives applied to agronomic soils. University of Georgia, USA, pp 68–109

Andrews SS, Karlen DL, Cambardella CA (2004) The soil management assessment framework: a quantitative soil quality evaluation method. Soil Sci Soc Am J 68:1945–1962

Araya T, Nyssen J, Govaerts B, Deckers J, Sommer R, Bauer H et al (2016) Seven years resource-conserving agriculture effect on soil quality and crop productivity in the Ethiopian drylands. Soil Tillage Res 163:99–109

Arshad MA, Coen GM (1992) Characterization of soil quality: physical and chemical criteria. Am J Altern Agric 7: 25–31

Arshad MA, Martin S (2002) Identifying critical limits for soil quality indicators in agro-ecosystem. Agr Ecosyst Environ 88:153–160. https://doi.org/10.1016/S0167-8809(01)00252-3

Arshad MC, Lowery B, Grossman B (1996) Physical tests for monitoring soil quality. In: Doran JW, Jones AJ (eds) Methods for assessing soil quality. Soil Science Society of America Special Publication No. 49. SSSA, Madison, WI, pp 123–141

Assefa A, Lulseged T, Wuletaw A, Tesfaye Y, Habtamu H, Niyawira SS, Mayesse DS (2020) Sommer R (2020) Soil organic carbon dynamics along chrono-sequence land-use systems in the highlands of Ethiopia. Agr Ecosyst Environ 300:106997

ATA, Agricultural Transformation Agency (2014) Soil fertility status and fertilizer recommendation Atlas for Tigray Regional State, Addis Ababa, Ethiopia

ATA, Agricultural Transformation Agency (2016) Soil fertility status and fertilizer recommendation Atlas of Amhara, Oromiya and SNNP Regional States, Addis Ababa, Ethiopia

ATA, Agricultural Transformation Agency (2017) Soil fertility status and fertilizer recommendation Atlas of Dire Dawa administrative region, Addis Ababa, Ethiopia

ATA, Agricultural Transformation Agency (2017) Soil fertility status and fertilizer recommendation Atlas of Harari national regional state, Addis Ababa, Ethiopia

ATA, Agricultural Transformation Agency (2018) Soil fertility status and fertilizer recommendation Atlas of Benshangule national regional state, Addis Ababa, Ethiopia

ATA, Agricultural Transformation Agency (2018) Soil fertility status and fertilizer recommendation Atlas of Gambela national regional state, Addis Ababa, Ethiopia

ATA, Agricultural Transformation Agency (2019) Soil fertility status and fertilizer recommendation Atlas of Somali national regional state, Addis Ababa, Ethiopia

ATA, Agricultural Transformation Agency (2019) Soil fertility status and fertilizer recommendation Atlas of Afar national regional state, Addis Ababa, Ethiopia

Barghouthi Z, Sameer A, Natsheh B, Salman M (2012) Analysis of macro and micronutrients in soils from Palestine using ion exchange membrane technology. Open J Soil Sci 2:44–49

Barrow CJ (1991) Land degradation: development and breakdown of terrestrial environments. Cambridge University Press, pp 295

Baveye PC, Baveye J, Gowdy J (2016) Soil "ecosystem" services and natural capital: critical appraisal of research on uncertain ground. Front Environ Sci 4:1–49

Beinroth FH, Eswaran H, Reich PF, Van Den Berg E (1994) Land related stresses in agroecosystems. In: Virmani SM, Katyal JC, Eswaran H, Abrol IP (eds) Stressed ecosystems and sustainable agriculture. Oxford and IBH, New Delhi

Bewket W, Sterk G (2003) Farmers' participation in soil and water conservation activities in the Chemoga watershed, Blue Nile basin, Ethiopia. Land Degrad Dev 13:189–200

Bikila TA (2019) Soil acidity and its management options in Western Ethiopia: review. J Environ Earth Sci 9:10

Bindraban PSV, Ye M, van den Berg LM, Materechera M, Kiba S, Tamene DI, Ragnarsdottir L, Jongschaap KV, Hoogmoed R, Hoogmoed M, van Beek W, van Lynden CG (2012) Assessing the impact of soil degradation on food production. Curr Opin Environ Sust 4:478–488

Bishaw B (2001) Deforestation and land degradation on the Ethiopian Highlands: a strategy for physical recovery. Int Conf Contemp Devel Issues Ethiopia, August 16–18 Kalamazoo, Michigan

Blanco H, Lal R (2010) Principles of Soil Conservation and Management. https://doi.org/10.1007/978-1-4020-8709-7

Blum WEH (1994) General approaches and definition. XVth International Congress of Soil Science, Acapulco, Mexico, 10–16 July, vol 29, pp 233–237

Bouma J (2014) Soil science contributions towards sustainable development goals and their implementation: linking soil functions with ecosystem services. J Plant Nutrit Soil Sci 177:111–120

Brand FS, Jax K (2007) Focusing the meaning(s) of resilience: resilience as a descriptive concept and a boundary object. Ecol Soc 12(1):23. http://www.ecologyandsociety.org/vol12/iss1/art23/

Bünemann EK, Bongiorno G, Bai Z, Creamer RE, De Deyn G, Goede R et al (2018) Soil quality–a critical review. Soil Biol Biochem 120:105–125. https://doi.org/10.1016/j.soilbio.2018.01.030

Buol SW (1995) Sustainability of soil use. Ann Rev Ecol Syst 26:25–44

Carter MR, Andrews SS, Drinkwater LE (2004) Systems approaches for improving soil quality. In: Schjonning P, Elmholt S, Christensen BT (eds) Managing soil quality: challenges, pp 261–281

Cesen (1986) Biomass energy resources. Technical Report. Ministry of Mines and Energy, Addis Ababa. Ministry of mines, Addis Ababa, Ethiopia, pp 34–41

Dejene A, Bedru B, Feyera ML (2020) The role of conservation agriculture on soil quality improvement: a review Ethiop J Agric Sci 30(4):197–222

Doran JW and Parkin TB (1994) Defining and assessing soil quality. In: Doran JW, Coleman DC, Bezdicek DF, Stewart BA (eds) Defining soil quality for a sustainable environment, pp 3–21. https://doi.org/10.2136/sssaspecpub35.c1. (Soil Science Society of America Journal)

Doran JW, Liebig MA, Santana DP (1998) Soil health and global sustainability: In: Proceedings of the 16th world congress of soil science, Montepellier, France. Paper 1923.

Doran JW, Parkin TB (1996) Quantitative indicators of soil quality: a minimum data set. In: Doran JW, Jones AJ (eds) Methods for assessing soil quality. SSSA Special Publications, vol 49. SSSA, Madison, WI, pp 25–37

Doran JW, Sarantonio M, Leibig M (1996) Soil health and sustainability. Adv Agron 56:1–54

Doran JW, Zeiss MR (2000) Soil health and sustainability; managing the biotic component of soil quality. Appl Soil Ecol 15:2–11

Engdawork A (2015) Characterization and classification of the major agricultural soils in CASCAPE intervention Woredas in the central highlands of Oromia region, Ethiopia. Addis Ababa University, Ethiopia. library.wur.nl/ WebQuery/isric/2259411

Erkossa T, Awulachew SB, Hagos F (2009) Characterization and productivity assessment of the farming systems in the upper part of the Nile Basin. Ethiop J Nat Res 11(2):149–167

Erkossa T, Ayele G (2003) Indigenous knowledge and practices for soil and water management in East Wollega, Ethiopia. In: Wollny C, Deininger A, Bhandari N, Maass B, Manig W, Muuss U, Brodbeck F, Howe I (eds) Technological and Institutional Innovations for Sustainable Rural Development. Deutscher Tropentag. www.tropentag.de

Erkossa T, Fisseha I, Stahr K (2007) Indexing soil quality: a new paradigm in soil science research. Aust J Soil Res 45:1–9

Erkossa T, Geleti D, WilliamsTO, Laekemariam F, Haileslassie A (2020) Restoration of grazing land to increase biomass production and improve soil properties in the Blue Nile basin: effects of infiltration trenches and Chloris Gayanareseeding. Renew Agric Food Syst 1–9. https://doi.org/10.1017/S1742170519000425

Erkossa T, Stahr K, Gaiser T (2005) Effect of methods of land preparation on runoff, and soil and nutrient loss from a Vertisol in the Ethiopian highlands. Soil Use Manag 21:253–259

Erkossa T, Wudneh A, Desalegn B, Taye G (2015) Linking soil erosion to onsite financial cost: lessons from watersheds in the Blue Nile basin. Solid Earth Discuss 6:1–30

Erkossa T, Williams TO, Laekemariam F (2018) Integrated soil, water and agronomic management effects on crop productivity and selected soil properties in Western Ethiopia. Int Soil Water Conserv Res. https://doi.org/10.1016/j.iswcr.2018.06.001

Eswaran H (1992) Role of soil information in meeting the challenges of sustainable land management (18th Dr.R.V. Tamhane memorial lecture). J Indian Soc Soil Sci 40:6–24

Eswaran H (1994) Soil resilience and sustainable land management in the context of Agenda 21. In: Greenland DJ, Szabolcs I (eds) Soil resilience and sustainable land use. CAB International Publisher, Wallingford, England, pp 21–32

Ethiopian Soil Information System (EthioSIS) (2013) Towards improved fertilizer recommendations in Ethiopia–nutrient indices for categorization of fertilizer blends from EthioSIS Woreda Soil Inventory Data. Addis Ababa, Ethiopia

Ethiopian Soil Information System (EthioSIS) (2016) Soil fertility status and fertilizer recommendations Atlas of Ethiopia, Addis Ababa.

Fantaw A (2007) An overview of Salt-affected soils and their management status in Ethiopia. In: A paper presented in the 3rd International Workshop on Water Management project, Haramaya University, Ethiopia. https://www.researchgate.net/publication/326352427_Prospects_of_alternative_copping_systems_for_salt-affected_soils_in_Ethiopia. Accessed 02 Nov 2021

Fanuel, L, Kibebew K, Tekalign M, Karltun E, Gebrekidan H (2016) Physiographic characteristics of agricultural lands and farmers' soil fertility management practices in Wolaita zone, Southern Ethiopia. Environ Syst Res 5(24):1–16

Fanuel L, Kibret K, Shiferaw H (2018) Potassium (K)-to-magnesium (Mg) ratio, its spatial variability and implications to potential Mg-induced K deficiency in Nitisols of Southern Ethiopia. Agric Food Secur 7:13

FAO (2010) Statistical yearbook. Food and agricultural organization.

Fekremariam AM, Atsushi T, Nigussie H, EnyewA, Mitsuru T, Kindiye E, Tsugiyuki M, Birhanu K, Derege T, Meshesha W, Tsujih M, Bayable M, Liyew B (2021) Tillage and crop management impacts on soil loss and crop yields in northwestern Ethiopia.

Feto E (2016) Soil physical properties as affected by repeated Wheeling. https://www.researchgate.net/publication/313023323_Soil_physical_properties_as_affected_by_repeated_Wheeling. Accessed 12 Nov 2021

Forman RTT, Godron M (1986) Landscape Ecology. John Wiley and Sons Ltd., New York

Gary MP, Thomas J, Vance GF (2000) Soil and environmental quality, 2nd ed, pp 315–336

Gebul MA (2021) Trend, status, and challenges of irrigation development in Ethiopia—a review. Sustainability 13:5646. https://doi.org/10.3390/su13105646

General Assembly (2016) Transforming our world: the 2030 agenda for sustainable development, A/RES/70/1. http://www.un.org/en/ga/search/view_doc.asp?symbol=A/RES/70/1. Accessed 5 Feb 2022

Getachew A, Hailu B (2009) Response of yield and yield components of field pea to tillage frequency, phosphorus fertilization and weed control on Nitisols of central Ethiopian highlands. East Afr J Sci 3 (2):161–169

Getachew A, Nelson PN, Bird MI, Beek C (2015) Phosphorus response and fertilizer recommendations for wheat grown on Nitisols in the central Ethiopian Highlands. Commun Soil Sci Plant Anal. https://doi.org/10.1080/00103624.2015.1081922

Godfray HCJ, Beddington JR, Crute IR, Haddad L, Lawrence D, Muir JF, Pretty J, Robinson S, Thomas SM, Toulmin C (2010) Food security: the challenge of feeding 9 billion people. Science 327:812–818

Harrison R, Strahm B, Yi X (2010) Soil education and public awareness. In: Verheye WH (ed) Soils, plant growth and crop production. Encyclopaedia of Life Support Systems (EOLSS), vol III. EOLSS Publishers/UNESCO. UK, pp 196–218

Hatfield JL, Wacha KM, Dold C, Cambardella CA, Olk DC, Trabue SL (2017) Enhancing soil begins with soil biology and a stable soil microclimate. International Erosion Control Association. 2017 Environmental Connection. 20–24 Feb 2017

Hawando T (1997) Desertification in Ethiopian highlands, Norwegian Church AID Addis Ababa Ethiopia. RALA Report No. 200.

Herrick JF, Wander MM (1998) Relationships between soil organic carbon and soil quality in cropped and rangeland soils: the importance of distribution, composition and soil biological activity. In: Lal R et al (eds) Soil Processes and the Carbon Cycle

Hillete H, Tekalign M, Riikka K, Karltun E, Heluf GK, Taye B (2015) Soil fertility status and wheat nutrient content in Vertisol cropping systems of central highlands of Ethiopia. Agric Food Secur 4(19). https://doi.org/10.1186/s40066-015-0038

Holling CS, Gunderson LH, Peterson GD (2002) Sustainability and panarchies. In: Gunderson LH, Holling CS (eds) Panarchy: understanding transformations in human and natural systems. Island Press, Washington D.C., USA, pp 63–102

Hurni H, Zeleke G, Kassie M (2015) Economics of land degradation in Ethiopia, soil degradation and sustainable land management in the rainfed agricultural areas of Ethiopia: an assessment of the economic implications. In: Report for the economics of land degradation initiative, vol 94

Hussien M, Walley FL, Bueckert R, Tar'an B (2015) Response of snap bean cultivars to Rhizobium inoculation under dryland agriculture in Ethiopia. Agronomy 5(3):291–308

IRENA (2015) Renewable energy in the water, energy & food nexus. http://www.irena.org/menu/index.aspx?mnu=Subcat&PriMenuID=36&CatID=141&SubcatID=496. Accessed 20 Jan 2022

Jónssona JÖG, Davíðsdóttira B, Jónsdóttirb EM, Kristinsdóttirc SM, Ragnarsdóttir KV (2016) Soil indicators for sustainable development: a transdisciplinary approach for indicator development using expert stakeholders. Agr Ecosyst Environ 232:179–189

Karlen DL, Andrews SS, Weinhold BJ, Doran JW (2003) Soil quality: Humankind's foundation for survival. Research editorial. J Soil Water Conserv 58:171–178

Karlen DL, Mausbach MJ, Doran JW, Cline RG, Harris RF, Schuman GE (1997) Soil quality: a concept, definition, and framework for evaluation. Soil Sci Soc Am J 61(1):4–10

Karlen DL, Parkin TB, Eash NS (1996) Use of soil quality indicators to evaluate conservation reserve program sites in Iowa. In 'Methods for assessing soil quality. In: Doran JW, Jones AJ (eds) Soil Science Society of America Special Publication No. 49. Madison, WI, SSSA, Inc., pp 345–356

Kassie M, Pender J, Yesuf M, Kohlin G, Bluffstone R, Mulugeta M (2008) Estimating returns to soil conservation adoption in the northern Ethiopian highlands. J Int Agric Econo 38(2):213–232

Keesstra SD, Bouma J, Walling J, Tittonell P, Smith P, Cerdà A, Montanarella L, Quinton JN, Pachepsky Y, van der Putten WH, Bardgett RD, Moolenaar S, Mol G, Jansen B, Fresco LO (2016) The significance of soils and soil science towards realization of the United Nations Sustainable Development Goals. SOIL 2:111–128

Kenea F, Sheleme B, Ayana A, Mohammed YS, Leeuw J, Aster A, Bekele M (2017) Effects of enclosure management on carbon sequestration, soil properties and vegetation attributes in East African rangelands. CATENA 159(2017):9–19

Kibebew K (2015) Characterization of agricultural soils in CASCAPE intervention woredas in eastern region. Haramaya University, Ethiopia, p 263. library.wur.nl/WebQuery/isric/2259414

Krauskorf KB (1972) Geochemistry of micronutrients. In: Mortvedt JJ, Giordano PM, Lindsay WL (eds) Micronutrients in agriculture. Madison, WI, pp 7–40. (Soil Sci Soc Am J)

Lal R (1994) Tillage effects on soil degradation, soil resilience, soil quality, and sustainability. Soil Tillage Res 27:1–8

Lal R (1997) Degradation and resilience of soils. Philos Trans R Soc Biol Sci 352:997–1010. https://doi.org/10.1098/rstb.1997.0078

Lal R (2015) Restoring soil quality to mitigate soil degradation. Sustainability 7:5875–5895. https://doi.org/10.3390/su7055875

Landon JR (2014) Booker tropical soil manual: a handbook for soil survey and agricultural land evaluation in the tropics and subtropics. Long man scientific and technical. Booker Tate Ltd. Wiley, New York

Larson WE, Pierce FJ (1991) Conservation and enhancement of soil quality. Evaluation for sustainable land management in the developing world. In: Proceedings of the international workshop on evaluation for sustainable land management in the developing world. Chiang Rai, Thailand, 15–21 Sept 1991. International Board of Soil Research and Management, Bangkok, pp 175–203

Larson WE, Pierce FJ (1994) The dynamics of soil quality as a measure of sustainable management. In: Doran JW et al (eds) Defining soil quality for a sustainable environment. Soil Science Society of America Special Publication No. 35. Madison, WI, SSSA, pp 37–52

Leenaars JGB, Elias E, Wösten H, Ruiperez G, Kempen M, Ashenafi A, Brouwer F (2016) Major soil-landscape resources of the CASCAPE intervention woredas, Ethiopia. Soil information in support to scaling up of evidence-based best practices in agricultural production (with dataset). CASCAPE Project, ALTERRA, Wageningen UR, The Netherlands. library.wur.nl/WebQuery/isric/2259349

Loide V (2004) About the effects of the contents and the ratios of soil's available calcium, potassium and magnesium in liming of acid soils. Agron Res 2(1):71–82

Lowery B, Arshad MA, Lal R, Hickey WJ (1996) Soil water parameters and soil quality. In: Doran JW, Jones AJ (eds) Methods for assessing soil quality. Soil Science Society of America Special Publication No. 49. SSSA, Inc., Madison, WI, pp 143–155

Ludwig M, Wilmes P, Schrader S (2018) Measuring soil sustainability via soil resilience. Sci Total Environ 626:1484–1493

Marchese D, Reynolds E, Bates ME, Morgan H, Clark SS, Linkov I (2017) Resilience and sustainability: similarities and differences in environmental management applications. Sci Total Environ 613–614:1275–1283. https://doi.org/10.1016/j.scitotenv.2017.09.086

Mekonnen G, Yihenew GS (2017) Characterization, classification and mapping of soils of agricultural landscape in Tana Basin, Amhara National Regional State, Ethiopia. In: Social and ecological system dynamics. Springer, Cham, pp 93–115

Melkamu D, Asmare M, Wondwosen T (2019) Effects of soil conservation practice and crop rotation on selected soil physico-chemical properties: the case of Dembecha district, Northwestern Ethiopia. Applied and Environmental Soil Science 2019

Melku D, Awdenegest M, Asfaw K, Adane A (2019) Effects of soil and water conservation measures on soil quality indicators: the case of Geshy Sub catchment, Gojeb River Catchment, Ethiopia Research Article, vol 2020, Article ID 1868792.https://doi.org/10.1155/2020/1868792

MoARD, Ministry of Agriculture and Rural Development (2005) Major agro-ecological zones of Ethiopia. Forestry, Land Use and Soil Conservation Department. Addis Ababa, Ethiopia

Moebius-Clune BN, Moebius-Clune DJ, Gugino BK, Idowu OJ, Schindelbeck RR, Ristow AJ, van Es HM, Thies JE, Shayler HA, McBride MB, Kurtz KSM, Wolfe DW, Abawi GS (2016) Comprehensive assessment of soil health-the cornell framework, 3.2 edn. Cornell University, Geneva, NY

Mulatu K (2015) Effect of mineral NP and organic fertilizers on the productivity and nitrogen use efficiency of bread wheat (*Triticum aestivum*L.) in Gozamin district, North-Western Ethiopia. PhD Dissertation, Haramaya University, Haramaya, Ethiopia, p 108

Mulugeta D, Tekalign M, Sheleme B, Selamyihun K (2020) Effects of potassium level on tef (*Eragrostis tef (Zucc.) Troter*) growth and yield in central highland Vertisos of Ethiopia. Eurasian J Soil Sci 9 (2):105–118

Mwendera E, Saleem MM (1997) Hydrologic response to cattle grazing in the Ethiopian highlands. Agric Ecosyst Environ 64:33–41. (Google Scholar, Cross Ref)

Nellemann C (2009) The environmental food crisis: the environment's role in averting future food crises: a UNEP rapid response assessment. UNEP

Okolo CC, Gebresamuel G, Retta AN (2019) Advances in quantifying soil organic carbon under different land uses in Ethiopia: a review and synthesis. Bull Natl Res Cent 43:99. https://doi.org/10.1186/s42269-019-0120-z

Olsson L, Jerneck A, Thoren H, Persson J, O'Byrne D (2015) Why resilience is unappealing to social science: theoretical and empirical investigations of the scientific use of resilience. Sci Adv 1: e1400217

Olson JM, Maitima JM (2006) Sustainable intensification of mixed crop-livestock systems, land use change impacts and dynamics 110 (LUCID) Policy Brief # 1. International livestock research institute, Nairobi, Kenya

Patzel N, Sticher H, Karlen DL (2000) Soil fertility - Phenomenon and concept. J Plant Nutr Soil Sci-Zeitschrift Fur Pflanzenernahrung Und Bodenkunde 163(2):129–142

Paulos D (2001) Soil and water resources and degradation factors affecting their productivity in the Ethiopian Highland Agro-ecosystems. In: International Conference on the Contemporary Development Issues in Ethiopia, CADPR, and Western Michigan University

Peters RH (1991) A critique for ecology. Cambridge University Press

Pimm SL (1984) The complexity and stability of ecosystems. Nature 307:321–326

Qureshi AS, Ertebo T, Mehansiwala M (2018) Prospects of alternative cropping systems for salt-affected soils in Ethiopia. J Soil Sci Environ Manag 9(7): 98–107

Raper R, Schwab E, Dabney S (1998) Measurement and variation of site-specific hardpans. In: 2001 ASAE annual meeting; American society of agricultural and biological engineers. St. Joseph, MI, USA. (Google Scholar)

Reeves DW (1997) The role of soil organic matter in maintaining soil quality in continuous cropping systems. Soil Tillage Res 43:131–167

Rist L, Felton A, Nystro¨m M, Troell M, Sponseller RA, Bengtsson J, O¨sterblom H, Lindborg R, Tida˚ker P, Angeler DG, Milestad R, Moen J (2014) Applying resilience thinking to production ecosystems. Ecosphere 5(6):73. https://doi.org/10.1890/ES13-00330.1

Rozanov BG (1994) Stressed soil systems and soil resilience in drylands. In: 15th International congress of soil science, Acapulco, Mexico, 10–16 July, 1994, vol 29, pp 238–245

Sanchez PA, Palm CA, Buol SW (2003) Fertility capability soil classification: a tool to help assess soil quality in the tropics. In: Sposito G, Zabel A (eds) The assessment of soil quality, vol 114, pp 157–186. (Int J Soil Sci)

Scherr SJ, Yadav S (1995) Land degradation in the developing world: implications for food, agriculture, and the environment to 2020. Food, Agriculture and the Environment Discussion Paper 14, IFPRJ. Washington, OC

Schmidt E, Zemadim B (2015) Expanding sustainable land management in Ethiopia: scenarios for improved agricultural water management in the Blue Nile. Agric Water Manag 158:166–178

Schjonning P, Elmholt S, Christensen BT (2004) Soil quality management concepts and terms. In: Schjonning PS, Elmholt S, Christensen BT (eds) Managing soil quality challenges in modern agriculture. CABI Publishing, pp 1–15

Schulte RPO, Creamer RE, Donnellan T, Farrelly N, Fealy R, O'Donoghue C, O'hUallachain D (2014) Functional land management: a framework for managing soil-based ecosystem services for the sustainable intensification of agriculture. Environ Sci Policy 38:45–58

Seybold CA, Herrick JE, Bredja JJ (1999) Soil resilience: component of soil quality. Soil Sci 164:224–234

Sheleme B (2016) Effects of lime and phosphorus application on soil chemical composition, growth and nutrient uptake of maize (*Zea mays* L.) in an acid soil of Gununo, Sothern Ethiopia. J Sci Dev 4 (1):1–9

Sheleme B (2017) Topographic position and land use impacted soil properties along HumboLarena-Ofa Sere toposequence, southern Ethiopia. J Soil Sci Environ Manag 8(8):135–147

Singh JS, Pandey VC, Singh DP (2011) Efficient soil microorganisms: a new dimension for sustainable agriculture and environmental development. Agric Ecosyst Environ 140:339–353

Sisay A, Moltote B, Fantahun A, Manuel TM (2021) Reclamation and amelioration of saline-sodic soil using gypsum and halophytic grasses: Case of Golina-Addisalem irrigation scheme, Raya Kobo Valley, Ethiopis. Cogent Food Agric 7:1. https://doi.org/10.1080/23311932.2020.1859847

Smith JL, Doran JW (1996) Measurement and use of pH and electrical conductivity for soil quality analysis. In: Doran JW, Jones AJ (eds) Methods for assessing soil quality. Soil Science Society of America Special Publication No. 49. SSSA, Inc., Madison, WI, pp 169–186

Smyth AJ, Dumanski G (1995) A framework for evaluating sustainable land management. Can J Soil Sci 75:40–406. https://doi.org/10.4141/cjss95-059

Soil Science Society of America (SSSA) (1995) SSSA statement on soil quality, Agron. News, June 1995: 7, Soil Science Society of America, Madison, WI

Solomon A (1994) Land use dynamics, soil degradation and potential for sustainable use in Metu area, Illubabor Region, Ethiopia. University of Berne, Switzerland, p 135

Sommer R, Thierfelder C, Tittonell P, Hove L, Mureithi J, Mkomwa S (2014) Fertilizer use should not be a fourth principle to define conservation agriculture: response to the opinion paper of Vanlauwe et al. (2014) 'A fourth principle is required to define conservation agriculture in sub-Saharan Africa: the appropriate use of fertilizer to enhance crop productivity. Fields Crops Res 169:145–148

Soomro UA, Rahman MU, Odhano EA, Gul S, Tareen AQ (2009) Effects of sowing method and seed rate on growth and yield of wheat (Triticum aestivum). World J Agric Sci 5(2):159–162

Sparling GP (1997) Soil microbial biomass, activity and nutrient cycling as indicators of soil health. In: Pankhurst C, Doube BM, Gupta VVSR (eds) Biological indicators of soil health. CAB International, New York, pp 97–119

SSSA (1997) In Glossary of Soil Science Terms; SSSA, Madison

Steffen W, Broadgate W, Deutsch L, Gaffney O, Ludwig C (2015) The trajectory of the Anthropocene: the great acceleration. Anthr Rev 2:81–98

Stoorvogel JJ, Smaling EMA (1993) Assessment of soil nutrient depletion in Sub-Saharan Africa: 1983–2000. Technical Report Winand Staring Centrum, Wageningen

Sullivan P (2004) Sustainable soil management. Soil Systems Guide, National Sustainable Agriculture Information Service, ATTRA. www.attra.ncat.org

Swift MJ (1999) Integrating soils, systems and society. Nature and Resources 35(4)

Swinton S, Lupi F, Robertson G, Landis D (2006) Ecosystem Services from Agriculture: lookin beyond the Usual Suspects. Am J Agric Econ 1160–1166

Szabolcs I (1994) Concepts of soil resilience. In: Greenland DJ, Szabolcs I (eds) Soil resilience and sustainable land use. CAB International, Wallingford, UK

Tamirat T (1992) Vertisols of central highlands of Ethiopia: characterization and evaluation of the phosphorus status. MSc Thesis, Alemaya University of Agriculture, Alemaya, Ethiopia, p 144

Tariku B, Muluken T, Firew K (2017) The application of biochar on soil acidity and other Physico-chemical properties of soils in southern Ethiopia. J Plant Nutr Soil Sci 180(3):269–420. https://doi.org/10.1002/jpln.201600343

Tadele A, Aemro A, Yihenew G, Birru Y, Wolfgramm B, Hurni H (2013) Soil properties and crop yields along the Terraces and Toposequece of Anjeni watershed, central highlands of Ethiopia. J Agric Sci 5(2):134–144

Tebebu TY, Bayabil HK, Stoof CR, Giri SK, Gessess AA, Tilahun SA, Steenhuis TS (2017) Characterization of degraded soils in the humid Ethiopian highlands. Land Degrad Dev 28:1891–1901. (Google Scholar)

Tefera B, Ayele G, Atnafe Y, Jabbar MA, Dubale P (2002) Nature and causes of land degradation in Oromia Region, Ethiopia—a review; socio-economics and policy research working paper 36; ILRI (International Livestock Research Institute): Nairobi, Kenya, 2002, vol 82, p 6. (Google Scholar)

Temesgen K (2012) Response of tef [Eragrostis tef (zucc.) trotter] cultivars to nitrogen and phosphorus fertilizer rates at Menzkeya district, North Shewa, Ethiopia. MSc Thesis, Haramaya University, Haramaya, Ethiopia. 66p

Teshome Y, Sheleme B, Kibebew K (2016) Characterization and classification of soils of Abobo Area, Western Ethiopia. Applied and Environmental Soil Science.

Tireza N, Eyasu Y, Nata T (2013) Quantification of the impact of integrated soil and water conservation measures on groundwater availability in Mendae catchment, Abraha We-Atsebaha, eastern Tigray, Ethiopia. Momona Ethiop J Sci (MEJS) 5(2):117–136

UN (2015) World population prospects: the 2015 revision. In: Nations U (ed) Population division of the department of economic and social affairs of the United Nations Secretariat https://esa.un.org/unpd/wpp/publications/files/key_findings_wpp_2015.pdf. Accessed 22 Jan 2022

USDA-NRCS (1998) Soil quality. Agronomy Technical Note No. 6, March 1998

USDA-NRCS (2001) Soil quality test kit guide. USDA-NRCS Soil Quality Institute, Ames, IA, p 82

Vagen TG, Winowiecki LA (2013) Mapping of soil organic carbon stocks for spatially explicit assessments of climate change mitigation potential. Environ Res Lett 8(1):1–9

Van Beek CL, Elias E, Yihenew GS, Heesmans H, Tsegaye A, Feyisa H, Mengist S (2016) Soil nutrient balances under diverse agro-ecological settings in Ethiopia. Nutr Cycl Agroecosyst 106 (3):257–274

Van Beek CLC, EliasE SelassieYG, GebresamuelG TsegayeA, HundessaF TollaM, MunayeM YemaneG, Mengistu S (2018) Soil organic matter depletion as a major threat to agricultural intensification in the highlands of Ethiopia. Ethiopian J Sci Technol 11:271–285

Wakene N, Heluf GK (2003) Influence of land management on morphological, physical and chemical properties of some soils of Bako, Western Ethiopia. Agropedology 13(2):1–9

Walker B, Salt D (2006) Resilience thinking: sustaining ecosystems and people in a changing world. Island Press

Walker B, Salt D, Reid W (2012) Resilience thinking: sustaining ecosystems and people in a changing world. Island Press

Warkentin BP (1995) The changing concept of soil quality. J Soil Water Conserv 50(3):226–228

Warkentin BP, Fletcher HF (1977) Soil quality for intensive agriculture. In: Intensive agriculture. Proceedings of the international seminar on soil environment and fertility management. Society of Soil Science and Manure, National Institute of Agriculture Science, Tokylu, pp 594–598

Wezel A, Casagrande M, Celette F, Vian JF, Ferrer A, Peigné J (2014) Agroecological practices for sustainable agriculture. A review. Agron Sustain Dev 34:1–20. https://doi.org/10.1007/s13593-013-0180-7

Wondimu B, Rethman NFG, Hammes PS, Getachew A (2006) Effects of farmyard manure and inorganic fertilizers on sorghum growth, yield, and nitrogen use in semi-arid areas of Ethiopia. J Plant Nutr 29(2):391–407

Wudneh A, Erkossa T, Devi P (2014) Sediment and nutrient lost by runoff from two watersheds, Diga district in Blue Nile basin, Ethiopia. Afr J Environ Sci Technol 8:498–510

Yacob A, Heluf GK, Sheleme B (2014) Pedological characteristics and classification of soils along landscapes at Abobo, southwestern lowlands of Ethiopia. J Soil Sci Environ Manag 5(6):72–82

Zelleke G, Agegnehu G, Abera D, Rashid S (2010). Fertilizer and soil fertility potential in Ethiopia: constraints and opportunities for enhancing the system. https://doi.org/10.13140/RG.2.1.3632.5846.

Zemen H, Solomon BW, Alehegne DB, Tigistu GA, Belete TD (2017) Ethiopia's move to a national integrated land use policy and land use plan, paper prepared for presentation at the "2017 World Bank conference on land and poverty" the World Bank-Washington Dc, March 20–24, 2017

9 Soil Management

Bobe Bedadi, Sheleme Beyene, Teklu Erkossa, and Endalkachew Fekadu

Abstract

Ethiopia is characterized by diverse climate, topography, geology, soils, hydrology, vegetation, and culture. The diversity in these soil forming factors is responsible for the formation/occurrence, distribution, level of degradation and management of natural resources including soils. The major soil types occupying more than 97% of the area of the country include Leptosols, Nitisols, Vertisols, Cambisols, Calcisols, Luvisols, Gypsisols, Fluvisols, Alisols, Solonchacks, Acrisols, and Regosols. Other soil types including Andosols, Arenosol, Pheozems, Lixisols, and Chernozems also occur widely. The diversity of soil types and their problems across the various landscapes and associated factors call for the need to develop and implement soil management practices that are relevant to the specific problems of the soils. Although soil management may not directly correspond to the types of soils as they appear in specific classification systems like the WRB, soils with similar characteristics can be clustered together and be treated with similar management practices. For instance, acid soils like Nitisols, Acrisols, Lixisols, and Alisols can primary be managed through liming and fertilization while salt-affected soils like Calcisols, Gypsisols, Solonchacks, and Solonetz can be managed by leaching soluble salts out of the rootzone. On the other hand, Leptosols, Cambisols, and Regosols that are characterized by shallow depths, primarily require soil and water conservation practices in addition to other specific management practices. Soils with impeded drainage like Vertisols require drainage management. On top of these, Integrated Soil Fertility Management (ISFM) has been proven to be beneficial for improving soil health, crop yields and the livelihood of farmers. Various attempts have been made to manage the restoration of soil health in the country. However, managing the restoration of soil health is subject to different challenges that could be soil level or systemic or both. The soil level challenges include: depletion of organic matter, soil fauna and flora, nutrient; biomass coverage removal; soil erosion; salinity and sodicity; acidity; waterlogging; low moisture availability; physical land degradation; soil structural deterioration; and soil pollution, while the systemic challenges include soil information management; technology generation, dissemination and linkage; input value chain; strategic and regulatory framework; and organization and management systems. On the otherhand, there have been also a lot of opportunities for managing the restoration of soil health in Ethiopia. These include development of various policies by the governments of Ethiopia from the Minilik time to the current regime including development of the Ministry of Agriculture (MoA) during the Minilik regime, land use planning and regulatory department during the Hailese-lasie regime followed by various policies during the existing regime such as, climate resilient green economy strategy, agricultural-led industrialization, sustainable development and poverty reduction program, participatory and accelerated sustainable development, and GTPI and II. Moreover, the development of community-based participatory watershed management guidelines, introduction of environmental impact assessment proclamation, and strategic investment framework for SLM have contributed towards the same. In order to reduce degradation, it is of paramount importance to use lands

B. Bedadi (✉)
School of Natural Resources Management and Environmental Sciences, Haramaya University, P.O. Box 138 Dire Dawa, Ethiopia
e-mail: bobe.bedadi@haramaya.edu.et

S. Beyene
School of Plant and Horticultural Sciences, Hawassa University, Hawassa, Ethiopia

T. Erkossa
Deutsche Gesellschaft Fuer Internationale Zusammenarbeit, GIZ, Addis Ababa, Ethiopia

E. Fekadu
College of Agriculture, Woldia University, Woldia, Ethiopia

S. Beyene et al. (eds.), *The Soils of Ethiopia*, World Soils Book Series,
https://doi.org/10.1007/978-3-031-17012-6_9

according to their capacities based on the principles of land evaluation and land use planning.

Keywords

Challenges • Integrated soil fertility management • Major soil types • Restoration • Soil health • Sustainable land management

9.1 Introduction

Ethiopia is endowed with various natural resources (Gete et al. 2010) owing to its diverse climate, topography, geology, soils, hydrology, vegetation, and culture. Some of these natural resources have been degraded over centuries due to severe soil loss, deforestation, low vegetative cover and unbalanced crop and livestock production (Girma 2001), overexploitation, misuse and mismanagement that are mainly associated with rapid population growth and the associated increase in demand for more resources. These resulted in changes in land uses mainly from forests and grasslands to agricultural and degraded lands. Farmers are forced to cultivate marginal lands through cutting of forests in order to feed the growing population. The expansion in agricultural lands is not adequately supported with appropriate land management practices. Hence, lands are subjected to various types of degradation including physical, chemical, and biological degradation processes depending on the context of a specific area. In the highlands, the main land degradation processes are mainly associated with soil erosion, leaching, nutrient depletion and acidification. In the lowlands, however, the main land degradation processes include salinization/alkalinization, low organic matter and associated depletion of plant nutrients, water shortage, and shallow soil depth.

The government has tried to respond to the calamities of land degradation processes by developing various land management and regulatory policies and guidelines. Soil conservation in Ethiopia is considered to be of top priority, which is the long-term objective of the agricultural development program (Martin 2001). For instance, there had been massive soil and water conservation programs (including afforestation, construction of physical soil and water conservation practices like terraces, bunds, ditches, etc.) through mass mobilization using various platforms such as the Food for Work. A great deal of attention has been paid to soil management practices that promote sustainable soil quality and productivity (Magdoff and Van Es 2000). Moreover, International and National initiatives like REDD+ , Sustainable Development Goals (SDG) of the United Nations, Green Legacy of Ethiopia also significantly contribute towards reducing soil degradation through protecting forests and other natural resources. In an attempt to treat acid soils, the government is producing agricultural limes by developing lime factories. Chemical fertilizers mainly urea and DAP are also imported by the government and distributed to farmers. Blended fertilizers such as NPS are also made available by the government to farmers. Initiatives like integrated watershed management and integrated soil fertility management/integrated nutrient management also play paramount importance in managing soil degradation. All these efforts contribute to the reduction of soil degradation and improvement of soil productivity. Attempts of Vertisols management and limited soil salinity/sodicity management are also witnessed especially in the middle Awash Irrigation areas though the drainage structures that were established for salinity control have not been maintained properly.

The main problem in addressing soil degradation is that most of the technologies that are made available to farmers are not site and context-specific. For instance, blanket fertilizer recommendations are widely practiced in the country regardless of the soil type/fertility status or crop type. Besides, the types of fertilizers used in Ethiopia are very few (mainly Urea and DAP with recent inclusion of NPS). However, under natural conditions, given the diversity of the soils in different parts of the country, the management approach should have been specific to the real problems on the ground.

A wide range of soils occur in Ethiopia in various extents. The most dominant ones that occur on more than 75% of the area include Leptosols, Nitisols, Vertisols Cambisols, Calcisols and Luvisols in decreasing order of magnitude. Management practices on these soils vary based on the specific problem of the soils. Although similar management practices may be implemented on different types of soils, there could be cases where the same type of soil may require different sets of management under different conditions. Hence, soil management should be based on the actual problem of a specific soil occurring in a specific locality. With this understanding, the use and management of soils have been discussed for all the major soil types and for those soils requiring special management attentions. ISFM practices implemented in Ethiopia are also presented. Besides, the paper provides an analysis of the challenges and opportunities of soil restoration in the country.

9.2 Use and Management of Major Soils

According to the Reference (blended) Soil Map (Source ISRIC 2017 + SOTER soil maps) (IUSS Working Group WRB 2015), Leptosols (29.64%), Nitisols (12.11%), Vertisols (10.62%), Cambisols (9.44%), Calcisols (9.38%), Luvisols (7.78), Gypsisols (7.66%), Fluvisols (4.19%), Alisols (2.67%), Solonchacks (1.72%), Acrisols (1.3) and Regosols (1.13%) cover more than 97% of the area. Other soil types including Andosols (0.62%), Arenosol (0.57%), Pheozems (0.41%), Lixisols (0.18%), and Chernozems (0.12%) also occur widely (Fig. 6.8), The agricultural potential, limitation, use and management aspects of the major reference soils groups are briefly discussed.

9.2.1 Leptosols, Cambisols, and Regosols

Leptosols, Cambisols, and Regosols have limited agricultural potential due to their occurrence mostly on hillslopes and dry lowlands and their susceptibility to erosion. The soil depth is too shallow to cater for most of the agricultural crops. These soils are suitable for naturally adapted plants and traditional herbs. These lands are used for grazing, but overgrazing may degrade them seriously.

Leptosols do not support good biological activities and have low organic carbon and total nitrogen. Since the soils are at low stage of weathering and development, most of the ions are held in the form of primary minerals. Therefore, they have low exchangeable cations and exchangeable capacity to sustain crop production. It is preferable not to disturb the natural vegetation grown on Leptosols or use for forest. However, if there is no other option, Leptosols on steep slopes with shallow surface soil can be transformed into crop land through using intensive soil conservation structures preferably stone terrace.

Cambisols show variation in physical, chemical, and mineralogical characteristics based on climatic zones, and parent material. Most Cambisols have good drainage, good water retention and aeration, neutral to weakly acid soil reaction, bearing a satisfactory chemical fertility and an active soil fauna.

Regosols located on fluvitile materials tend to contain better levels of soil organic matter, total nitrogen, and available phosphorus. Those occurring in areas receiving reliable rainfall are used for crop production.

The major limitation of Leptosols is their shallow soil depth, which is exacerbated by their occurrence on steep slopes that are highly vulnerable to erosion. The shallow depth restricts plant root development. Other types of Leptosols also occur in dry lowlands where the rate of weathering of parent materials and soil formation is slowed down by the low moisture. The shallower and coarse textured Regosols, in most cases, have low water holding capacities, which is a major limiting factor of crop farming. Chemically, Regosols have very low to low OC, N, and available P contents and the values also show a decreasing trend with depth (Rabia et al. 2013).

Leptosols of the mountainous areas are commonly reserved for forests and grazing lands. However, in the highly populated highlands, the farmers use these soils for subsistence farming with or without soil and water conservation practices. This aggravates the degradation of these soils in the highlands of Ethiopia.

Depending on the parent materials on which these soils are formed, some Leptosols are chemically rich while others are poor. For example, Calcaric Rendzic Leptosol in Abergelle, Tigray, occurs on steep slopes, and has shallow soil depth with high topsoil stoniness, the area is used for (marginal) cultivation due to its high content of Ca^{2+}, Mg^{2+}, and P (Nyssen et al. 2019). In Adigrat sandstone cliff, because of the very limited soil depth, this area is used as rangeland but the vegetation cover is sparse. Leptosols in Delanta District, north central highlands of Ethiopia are found in a strongly dissected topography of continuous rock with limited soil depth (Abate et al. 2014). Most of the lands in these areas are covered by highland cereals such as barley.

Eutric Cambisols have a wider use for cereal and fruit cultivation, grass and forest production. The Dytstric Cambisols though less fertile, are used for (mixed) arable farming and as grazing land. Vertic and Calcaric Cambisols in the lowlands are intensively used for the production of cotton and other crops such as maize, sorghum and fruits and vegetable crops. Cambisols in steep slopes of Ethiopian highlands are used for crop production. Those on undulating or hilly terrain are used for a variety of annual and perennial crops or as grazing lands whereas, on flat and alluvial lands, they are mainly devoted for food crop production. Their productivity is limited where there are skeletals, shallow depths, and low bases status. These soils must be under soil and conservation program. Otherwise, they lose their productivity through time. Kebede et al. (2015) suggested that amending the Cambisols with lime, essential and deficient nutrients will be vital for producing food and feed crops.

The land use practices on Regosols vary widely. Except the Regosols in moisture deficit areas, in the remaining parts of Ethiopia, they are extensively/intensively used for crop production. These soils have low moisture holding capacity and especially those located in an area with 500–700 mm rainfall range need supplementary irrigation for good crop production. Regosols that occur in the lowlands of Ethiopia are left for grazing and along the banks of major rivers they are under small- or big-scale irrigation and used for crop production. Irrigating these soils in lowlands where evapo-transpiration exceeds precipitation make them liable for salinization as it is observed in the Middle Awash area. Thus, it requires the identification of salt tolerant crops and use of appropriate irrigation methods.

Soil conservation practices should be implemented to reduce the vulnerability of these soils to erosion. Either or both physical (such as terraces, bunds, ditches) and biological (such as contour farming, crop rotation, strip copping, and mulching) soil conservation practices should be implemented to protect the shallow soils from being lost. Agroforestry (combination of crop and forest) could be a promising approach to retain the ecosystem stable and the approach could also help to retain excessive internal drainage.

Farmers usually practice a number of soil and water management practices on Regosols including hillside ditches, terraces, bunds, fanyajuus, grass strips, contour cultivation, and permeable barriers like loose stone bunds. On non-agricultural lands, these soils are managed through protecting the natural vegetation and/or plantations.

9.2.2 Vertisols

Vertisols have diverse importance in agriculture depending on climate, soil properties, population density, and other socioeconomic factors. Their use in Ethiopia range from very extensive grazing in arid areas to a wider range of food crop production such as teff, barley, faba beans, wheat, field pea, oats, lentils, linseed, niger seed, chickpeas, and sorghum in rainfed and irrigated areas, but the crop yields are quite low unless specific management practices are put in place to give sustainable production (Berhanu Debele 1985a, b).

Vertisols are considered to be among the most fertile soils and rich in base cations. However, the chance for accumulation of organic matter in these soils is extremely low as they are intensively utilized and not fallowed.

The major agricultural constraint of Vertisols is associated with the poor drainage conditions that cause excess water ponding (Fig. 9.1) on the land surface and root zone of crops which exposes the soils for plant nutrient deficiency and erosion. In response to their high clay content, vertisols have slow infiltration and more commonly, water flows slowly on the surface into depressions where it accumulates as swamps or marsh from which the water is lost mainly by evaporation and, to a lesser extent, drainage (Mesfin 1998). Vertisols are generally heavy and difficult to work with traditional oxen-drawn implements. Their cohesiveness when wet can cause problems of workability under both hand tillage and mechanized farming. They pose drainage difficulties when wet and expanded, root damage when dry contracted and cracked. They have deficiency of nitrogen and phosphorous and thus, both have to be applied in form of inorganic fertilizers. Besides, they are also reported to have deficiencies of K, S, Zn, B, and Mo (Hailu et al. 2015).

The management of Vertisols comprises several alternative practices and technologies such as improved land and water management, improved cropping systems, and appropriate fertility management. Several researchers in Ethiopia investigated various land preparation and drainage methods including flatbed planting, drainage furrows, ridges and furrows, broadbeds and furrows, green manure, reduced tillage, post rainy season planting and soil burning (Erkossa et al. 2006). For instance, at Inewari, all crops except tef are grown on manually constructed broad beds and furrows (BBF). The crops grow on the beds while the excess water drains out through the furrows (Jutzi 1988a, b; Abebe and Jutzi 1989).

As most of the broadbed and furrow makers that were developed to solve the waterlogging problem on Vertisols (Fig. 9.2) were too heavy for animal traction, a research consortium has developed an animal-drawn cultivator called a broadbed maker (BBM) (Fig. 9.2) by modifying the local implement known as '*maresha*' to make raised beds and furrows more efficiently and effectively, thus reducing water

Fig. 9.1 Water-logged soil (Vertisol) in Fogera Plains (*Source*: Dawit and John 2020)

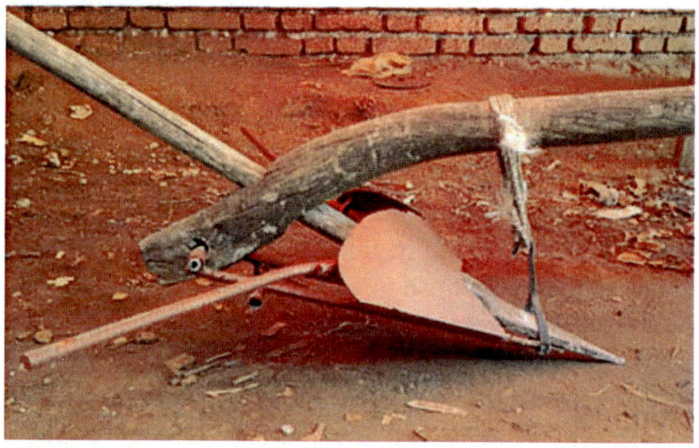

Fig. 9.2 Broad Bed Maker (BBM) Technology package innovations in Ethiopian farming systems (*Source*: Rutherford 2008; www.aybareng.com)

logging and encourage early planting of cereals (Rutherford 2008). The BBM creates 0.8 m wide beds separated by 0.4 m wide furrows that allow excess water during heavy rains to be discharged to a main drain or other outlet at the bottom end of a plot. This technology allows early planting to take advantage of a longer growing period, resulting in higher yields and less erosion as there is adequate vegetative cover during the rainy season (Astatke and Saleem 1998). The implement is manufactured by Aybar Engineering PLC in Ethiopia and is called Aybar BBM.

Efficient cropping systems play important roles in making use of available resources such as soil moisture, labor, and time in managing Vertisols. Double cropping involves growing crops on a given land for two seasons within a year and helps in using the land for production for up to 8 months in summer instead of four. Areas of Vertisols with annual rainfall above 750 mm could produce two short-duration crops in sequence without irrigation. Cropping systems involving integration of forage legumes is another alternative to improve the fodder yield and quality. Results of fieldwork conducted over two seasons showed that intercropping wheat with clover or sequential cropping of an oat/vetch mixture followed by chickpea provided high-quality fodder; the effect was greater under fertilized conditions (Tedla et al. 1999). Tillage practices that incorporate conservation agriculture (CA) principles are effectively increasing green water in the root zone available for crops and thus, improve crop productivity and yields substantially on Vertisols in drylands without other inputs (Araya et al. 2015).

9.2.3 Luvisols

Luvisols are among the high potential agricultural soils. They are among the most intensively used and most productive soils throughout the country. They are generally considered as fertile soils that are suitable for a wide range of agricultural uses because they are characterized by high base saturation.

The major limitation of Luvisols is associated with the subsurface clay accumulation layer that may inhibit water and air movement as well as root penetration depending on the degree of cementation of the Argic layer. Moreover, the surface soils with higher silt content are usually susceptible to erosion. Luvisols that occur on steep slopes are vulnerable to soil erosion and hence shall be protected by implementing soil and water conservation practices.

Luvisols show variability in soil chemical properties. According to the reports of studies in CASCAPE intervention districts of Ethiopia (e.g. Mekonnen 2015), most Luvisols are characterized by slightly to strongly acid, low to medium available P, low to medium N, low to very low OM, and high to very high CEC. Such variability in soil chemical properties of Luvisols was also reported by several authors in Ethiopia (Fekadu et al. 2018a, b, c; Yitbarek et al. 2016). Yihenew et al. 2014) suggested that Luvisols are variably suitable for maize (*Zea mays* L.), finger millet (*Eleusine coracana* L.), teff (*Eragrostis tef Zucc.*), and rice (*Oryza sativa* L.) production. For optimum yield of crops, addition of experimentally proven rates of OM and N, P, and K fertilizers on the farm fields is recommended (Rabia et al. 2013).

Their high porosity and deep profile permit deep rooting. They have stable soil structure that makes them less susceptible to erosion as compared to other soils on the same catena. They have good internal drainage, water holding capacity, and workability. Those in the southern and southwestern part of the country (Jima, Sidama, Gamu Gofa, Wollega and Illubabor) are under coffee, pineapple, and tea plantations as well as maize/sorghum production where as those in the northern part (Gojjam) are under wheat and Teff production. This disparity in cropping system has resulted organic matter content differences. Luvisols in southern parts have relatively higher organic matter compared with the northern ones and respond well to fertilizer applications. Like the other soils in the country, these soils must be put under soil and water conservation program if they are under crop production. Use of agroforestry system is reported to give a maximum return (Eylachew 1999). Farmers usually practice a number of soil and water management practices including hillside ditches, terraces, bunds, fanya juus, grass strips, contour cultivation, permeable barriers like loose bunds made of stones collected from the farm lands, soil or weeds and residues.

9.2.4 Nitisols, Alisols, Acrisols, and Lixisols

Nitisols and Acrisols are among the high potential agricultural soils especially if proper acidity management practices are implemented. The major crops grown on these soils include tea, coffee, and pine apple. Acrisols are suitable for production of rainfed and irrigated crops only after liming and full fertilization. Rotation of annual crops with improved pasture maintains the organic matter content. Alisols, on the other hand, are considered as less productive soils.

The major limitation of Nitisols and Acrisols is their strong acidity and low basic cations. Toxicity of acidic cations such as aluminum and iron retards the growth of crops that are sensitive to these ions. Besides, deficiency of basic cations such as Ca, Mg, and K also limits crop growth and development. The strong acidity also induces high P fixation resulting in P deficiency to the plants. Microbial activity will also be retarded due to the strong acidity, which will in turn decrease decomposition of organic matter and availability of nutrients like nitrogen and sulfur.

Alisols occur predominantly on old land surfaces with hilly or undulating topography. The generally unstable surface soil of cultivated Alisols makes them susceptible to erosion; truncated soils are quite common. Toxic levels of aluminum at shallow depth and poor natural soil fertility are added constraints. As a consequence, many Alisols allow only cultivation of shallow-rooting crops and these suffer from drought stress during the dry season.

Lixisols could be described as one of the highly leached soils of Ethiopia. Hence, they have generally low CEC and nutrient retention capacity.

Nitisols, Alisols, and Acrisols are usually used to grow acid-tolerant crops such as tea, coffee, pineapple, etc. With appropriate management of the soil acidity problems through liming and fertilization, they can be used for a wide range of crops.

They support the bulk of the cereal and livestock production in the Ethiopian highlands. More importantly, the production of coffee (*Cofea arabica*), the most important export commodity in Ethiopia, relies almost exclusively on Nitisols. In addition, the large proportion of tea production comes from strongly acidic Nitisols in the western part of the country (Elias 2002). However, due to over cultivation of the land for cereals and inappropriate land use, the soils are depleted of nutrients and OM. Most Nitisols have become degraded and acidic with high P fixation capacity (Elias 2017).

Nitisols are planted with coffee and tea in most highlands of the country and are also widely used for food crop production on smallholdings. High P-sorption calls for application of P-fertilizer, liming, and OM (Ayenew et al. 2018; Fekadu et al. 2018a, b, c). Elias (2017) suggested the integrated use of N, P, K, Zn, and liming with sufficient rates to raise the productivity of Nitisols. Besides, use of organic fertilizers sources such as crop residues, and manure could improve soil OM content and water retention capacity. On steep lands, suitable physical soil and water conservation structures reinforced with adaptable biological and agronomic measures can reduce soil erosion and nutrient depletion.

Acrisols are also widely used for crop production, and partly they are left for grazing. Complete fertilization and careful management are required for sustainable farming on Acrisols. Liming and use of acid tolerant crop varieties are important management options for most Acrisols affected by Al toxicity.

The use Alisols is generally restricted to acidity-tolerant crops or low volume grazing. The productivity of Alisols in subsistence agriculture is generally low as these soils have a limited capacity to recover from chemical exhaustion. If fully limed and fertilized, crops on Alisols may benefit from the considerable cation exchange capacity and rather good water holding capacity. Alisols are increasingly planted to aluminum-tolerant estate crops such as tea and rubber but also to oil palm and in places to coffee and sugar cane (IUSS Working Group WRB 2015).

Lixisols have a relatively higher content of organic matter, clay, base saturation, and pH but low aluminum toxicity problems than Acrisols. They, however, have generally low CEC and nutrient retention capacity and hence requiring fertilization and liming a prerequisite for crop production on these soils. Areas with Lixisols that are still under natural savanna or open woodland vegetation are widely used for low-volume grazing.

9.2.5 Fluvisols

Fluvisols have dominantly loamy sand and sandy loam textures. They have granular and simple grain structures and very rapid drainage characteristics. Their depth depends on seasonal transportation and deposition of materials. Chemically, Fluvisols in most lowlands are characterized by medium to high CEC, high to very high $CaCO_3$, and slight to moderate alkalinity (Nyssen et al. 2019), low to medium OC and N, low to medium available P, and significant amounts of exchangeable Ca and Mg that suggests that they are potentially fertile soils (Rabia et al. 2013). They usually occur on gentle slopes ranging from 1 to 5%. Hence, they are considered as fertile and productive soils especially with irrigation because they regularly receive deposition of fresh sediments containing high organic matter and exchangeable cations.

The major limitations of fluvisols are their susceptibility to flood due to their occurrence on lower landscapes. Most of the Fluvisols in the flood plains in Ethiopia are used for both rainfed and irrigated agriculture. These soils are widely used for production of rice, cotton, and other vegetables (Habtegebrial et al. 2013).

9.2.6 Calcisols, Gypsisols, and Solonchak

Calcisols, Gypsisols, and Solonchaks are usually rich in basic cations and can be used for agricultural production where sufficient water is available for irrigation. However, water is the main limiting resource in the drier parts of the country where these soils commonly occur.

Most Calcisols are alkaline with pH values between 8.0 and 8.5, low in OM, N, and available P (Rabia et al. 2013). Gypsisols also have similar chemical properties like the Calcisols except that the former has low CEC due to the dominance of gypsum granules with low exchange sites in the soil particles. Another limitation of these soils is fixation of P by calcium.

The soils of the Melka Sedi-Amibara Plain of the Middle Awash Valley are highly saline with ECe ranging from 16 to 18 dS m^{-1} (Table 9.1). Soluble Na^+, Ca^{2+}, Cl^-, and SO_4^{2-} are the dominant soluble salt constituents throughout the depths of the profile. Accordingly, chloride and sulfate salts of sodium and calcium (mainly NaCl and $CaSO_4$) are assumed to be the major soluble salts contributing to the very high salinity level of these soils (Auge et al. 2018). The high evaporation rate coupled with lack of drainage facilities to remove soluble salts from the root zone has contributed to the formation of saline soils in these areas. Salinity affects crop growth directly through ion toxicity and indirectly through its effects on soil water potential, which cause soil/plant osmotic imbalances (Endris and Mohammed 2007). Besides, due to the dry climate under which these soils usually occur, they are characterized by low organic matter and associated plant nutrients like nitrogen and phosphorous.

The productivity of Calcisols, Gypsisols, and Solonchaks can be improved with application of organic and inorganic fertilizers. As these soils are highly rich in calcium, P fixation is expected. Hence, application of high amount of P sources is required where these soils occur. Large areas of Calcisols in the country are mainly covered with shrubs, grasses, and herbs and are used for extensive grazing. Extensive areas of Calcisols in the lowlands of Afar region are used for the production of irrigated wheat, melons, and cotton.

Solonchaks in these areas have developed from use of poor-quality water for irrigation, injudicious on-farm water management and inadequate drainage schemes (Gebremeskel et al. 2018). Under the prevailing system of irrigation, and water quality, it is expected that cultivated lands will get out of production in the near future.

Restoration of salt-affected lands into productive lands and protection of newly developed areas from the spread of salinity through improved irrigation and crop management is therefore of paramount importance. In the high salinity areas where growth of normal field crops is restricted, the use of bioremediation methods including planting halophytic forages could bring these soils back into production.

The incorporation of salt-tolerant crops in the cropping system is another alternative strategy, which is recently introduced through research and encouraging results have been observed. Based on results reported by Ethiopian Agricultural Research Institute (EIAR 2015), four forage crop species (*Cenchrus sp.*, *Panicum antidotale*, Sudan grass (*Sorghum sudanense*), and *Chloris gayana*) and three legume species (*Desmodium triflorum*, *Sesbania sesban*, and *Medicago sativa* (alfalfa)) were identified as salt tolerant crops.

Although barley is among the commonly grown cereal crops and has been well described for its potential ability to tolerate stress induced by salinity, its introduction and potential use in marginal environments is not common in the country. However, farmers in the Ziway Dugda area in Ethiopia had practiced growing barley instead of maize and other horticultural crops when the soils became saline (Muruts and Haileselass 2019). The yield and growth of barley under saline soil were improved with high doses of potassium, which apparently demonstrated the positive contribution of potassium nutrition to plants exposed to high soil salt levels (Endris and Mohammed 2007).

Irrigated solonchaks are usually used for the production of monoculture banana, mango, or maize, while those occurring far from the irrigation watercourses are predominantly used for agroforestry and extensive pastoralism.

Table 9.1 Chemical composition of soils in Melka Sedi-Amibara Plain of the Middle Awash valley *Source*: (Qureshi et al. 2018)

Depth (cm)	pH (H_2O)	EC (dS m^{-1})	Soluble cations (me l^{-1})				SAR	HCO_3^-	Cl^-	SO_4^{2-}
			Ca^{2+}	Mg^{2+}	K^+	Na^+				
0–25	7.2	18.6	59.4	3.7	208.1	15.4	37.0	1.1	165.8	77.1
25–50	7.2	17.8	55.5	3.3	197.6	14.4	36.4	1.0	160.2	71.8
50–70	7.2	17.5	53.1	3.7	197.5	14.1	37.0	1.0	157.0	68.5
70–90	7.2	17.2	51.5	3.4	120.0	13.7	22.9	1.4	147.0	64.3
90–120	7.2	16.6	47.3	3.1	132.5	10.6	26.4	1.0	132.4	59.9

9.3 Soils with Special Management Requirement

9.3.1 Soils on Sloppy Lands

(a) **Characteristics, extent, and occurrence**

Most of the highlands (altitude >1500 m.a.s.l) constituting about 43% of the total area of the country (Hurni 1988), and which are considered to be the source of livelihood for the majority of the population, are characterized by sloppy landscapes (Fig. 9.3). Hence, these lands are much more vulnerable to soil erosion by water than the low lands. The average annual loss on agricultural lands is about 137t/ha/year, which is equivalent to an annual soil depth loss of 10–13 mm (Spielman and Pandya-Lorch 2009), which will take more than 200 years to replenish. Some sources list Ethiopia as one of the most severely erosion-affected countries in the world (IFPRI 2010). As compared to East African and other developing countries, Ethiopia is among countries with the highest soil erosion which is estimated to be 342,000 km^2 (31%) of eroded lands of the total land area.

The degradation and loss of the soils resulting from erosion, depletion of organic matter and nutrients are much faster than they can be replaced (Hurni 1993). This resulted in significant yield losses where soil degradation is estimated to cost Ethiopia a huge amount of money. Nearly 80% of the yield losses are attributed to reduced crop production and the rest to reduced livestock production.

(b) **Agricultural potentials**

In Ethiopia, cultivation of lands on steep slopes is a common practice due to the high demand for arable lands especially in areas with high population pressure. However, since soils on steep slopes are more prone to erosion and shallow in depth, their cultivation exacerbates the soil degradation. Hence, these soils need to be protected through afforestation and area closures.

(c) **Major limitations**

The main problem on steep slopes comes with the increased difficulty in detaining or slowing down run-off to non-erosive rates as slopes increase. On top of this, as slopes increase, the soils become shallower and their capacity to hold water decreases. The shallow soils also suffer from nutrient depletion due to increased erosion aggravated by the steep slopes. These soils require expensive soil conservation measures.

Soils that occur on sloppy lands are highly vulnerable to soil erosion by water. Soil erosion results in loss of fertile topsoil, meaning that the base on which inputs applied and crops grown is increasingly depleted and thinly-spread. This leads to reduced water-holding capacity of soil (making it more susceptible to extreme conditions such as drought) and limited crop emergence, growth, yield, and rooting depth. This in turn contributes to a vicious cycle of increased rate of loss of organic matter–caused by a combination of

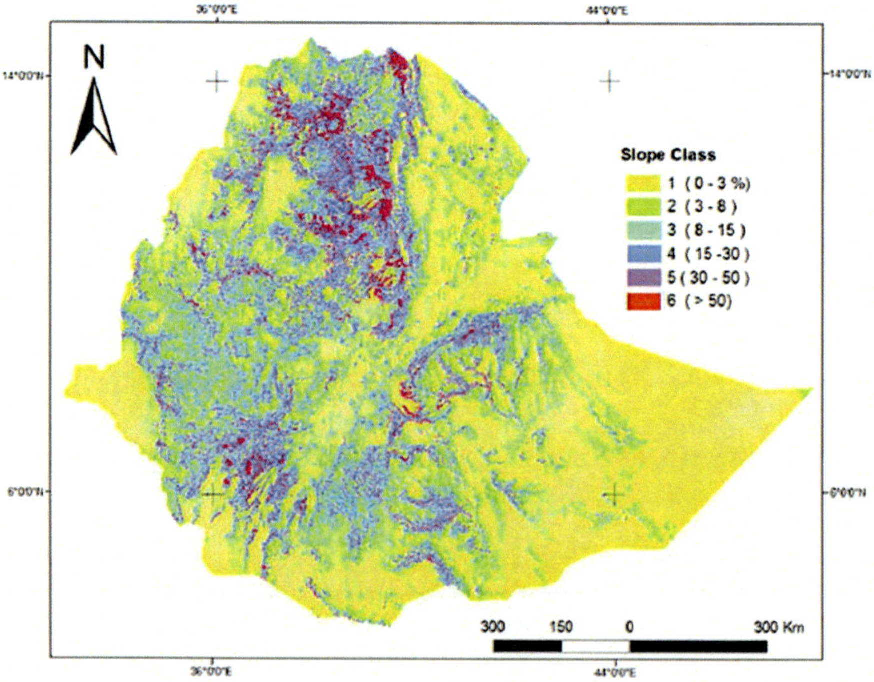

Fig. 9.3 Slope classes of Ethiopia (*Source*: Belete et al. 2013)

cultivation of slopes with poor management, high rainfall and inappropriate drainage (water erosion), and significant loss of vegetation cover (deforestation, overstocking, and overgrazing).

(d) **Use and management**

In order to manage erosion-induced soil degradation on slopy lands, farmers implement both biological (such as crop rotation, contour cultivation, strip cropping, mulching, etc.) and Mechanical (such as terraces, soil bund, stone bund, fanya juu, micro basins, etc) soil and water conservation measures. However, most of the soil and water conservation measures used in Ethiopia are old traditional practices. For instance, among the mechanical soil conservation measures, terracing has been developed under traditional agriculture in the Highlands of Tigray (Virgo and Muno 1977), North Showa, in the Chercher Highlands and in Konso (Westphal 1975; Huffnagel 1961). In the Chercher Highlands, soil and stone bunds were built primarily for the cultivation of coffee (Coffee arabica) and chat (Kata edulis). In Konso and many other places, bunds were built for water harvesting as well as conservation for annual crop production. It is still possible to see contour lines and hedges on many hill slopes indicating former terracing specially in Konso (Fig. 9.4), North Shoa and the Hararge highlands. However, the traditional soil conservation practices were limited to few areas and rendered ineffective due to reduced vegetative cover and inappropriate land use systems such as overgrazing and destructive farm management practices (Hurni 1988). Plant residues are usually removed after harvesting to use them as animal feed, fuel wood or sale which leaves the fields totally barren and exposing it to all forces including rainfall, trembling of the soil by livestock and humans that cause aggregate disturbances. Moreover, animal dung, which otherwise could have contributed to soil fertility management and reduce soil erodibility, is not brought back to the field, but rather sold on the markets for income generation in addition to its use as fuel. In light of all these facts, a need to intervene with the traditional conservation system by building on the existing knowledge and introducing new technologies was recognized by the government and the government started mass mobilization for soil and water conservation.

Fig. 9.4 Traditional soil and water conservation practices in Konso, Ethiopia. *Source*: McCausland (2010)

In fact, Hurni et al. (2016) recommended different soil and water conservation measures to be applied on different land use types in the country based on predefined 12 agroecological zones namely; Moist High Dega, Wet High Dega, Dry Dega, Moist Dega, Wet Dega, Dry Weyna Dega, Moist Weyna Dega, Wet Weyna Dega, Dry Kolla, Moist Kolla, Wet Kolla and Moist Berha (Table 9.2).

9.3.2 Acid Soils

(a) **Extent and distribution**

Acid soils in Ethiopia occupy some 40.9% of the country of which 27.7% are moderate to weakly acidic (pH of 5.5–6.7); 13.2% are strong to moderately acidic (pH < 5.5) and nearly one-third have aluminum toxicity problem (Mesfin 2007; Schlede 1989).

The most productive highland parts of Ethiopia are the most prone to acidity in the country. Significant areas of Jimma, Wollega, Ilubabor, and Gojjam, (Fig. 7.10) are adversely influenced due to soil acidity. These areas produce cereals (such as maize, wheat, and teff) and pulses, which are the major food crops of the country (Gurmessa 2021). The productivity of these crops has been reported very low when no soil acidity management practices are put in place.

(b) **Agricultural potentials**

The well-drained acid soils are very productive if they are properly treated. The main crops raised by farmers on acidic soils are barley, wheat, teff, rapeseed, faba bean and field pea (Regassa and Agegnehu 2011). Commercial crops such as coffee, tea, and pine apple are widely grown.

(c) **Major limitations**

The major limitations of acid soils include deficiency of basic cations such as Calcium (Ca), Magnesium (Mg), and Potassium (K) and toxicity of aluminum. The strongly acidic soil pH and associated aluminum toxicity reduce microbial activity which in turn results in reduction in the rate of decomposition of organic matter. This leads to deficiency of

Table 9.2 Soil and water conservation measures recommended for different land uses in Ethiopia. *Source*: (Adapted from Hurni et al. 2016)

Cultivated land	Grasslands	Forest land	Common to all land uses
Alley cropping	Controlled grazing	Hillside terrace	Hillside terrace
Bench terrace	Cut and carry	Microbasin	Microbasin
Broadbed and furrow	Grassland improvement	Tree planting	Tree planting
Conservation tillage		Trench	Trench
Graded bund			Area closure
Graded Fanya Juu			Checkdam
Grass strip			Cutoff drain
Level bund			Gully rehabilitation
Level Fanya Juu			Revegetation
Mulch			Water harvesting
Trash line			Waterways
Vetiver			

some essential plant nutrients like S, nitrogen and phosphorous. P fixation occurs when P reacts with other elements like Al and Fe to form insoluble compounds and becomes unavailable to crops.

(d) **Use and management**

Acid soils are usually managed through conservation-based sound soil management with application of appropriate technologies that promote their better productivity and production since they occur in all but the very driest climate.

Soil acidity management researches and practices have been promoted in Ethiopia to improve crop production and productivity. The most notable practices include liming, organic amendments, development of acid tolerant crop varieties as reviewed in the following subsections.

Liming Acid Soils

The ministry of agriculture has made continuous progress in designing and disseminating packages to acid soil management, and launched trials across the country to extend the practices and adoption of liming (Tamene et al. 2017). However, no significant changes have been observed in acid-prone areas. Some of the reasons include limited lime factories, high cost of lime and its transportation as well as sole application of lime alone.

Acid soil management through liming (Fig. 9.5) improves the biological, chemical, and physical properties of soils and subsequently increases crop yield and efficiency of nutrient use (Fageria and Baligar 2005). Activity, diversity, and population of microorganisms that are involved in the process of biological N_2 fixation, OM mineralization, P solubilization are suppressed under acid conditions (Ibekwe et al. 1997; Condron et al. 1993). Thus, liming improves soil conditions for mineralization to occur, increases P availability, increases soil pH and enhances N availability in the soil. The improvement of soil pH creates suitable environment for microbial activity, nutrient uptake, and thereby increasing yield of crops (Didaand Etisa 2019). Most reports of liming experiments conducted in Ethiopia (Fekadu et al. 2018c; Bore and Bedadi 2016; Dejene et al. 2016; Boke and Fekadu 2014; Kidanemariam et al. 2013) showed yield improvement in different crops with lime. Application of lime in acid soils reduces acidity due to chemical properties of the cations (Ca^{2+} and Mg^{2+}) and anions (CO_3^{2-}) in liming materials. The divalent cations in the liming materials replace Al^{3+} from exchange sites on soil colloids and make it precipitate as insoluble $Al(OH)_3$. Anion of the liming material (CO_3^{2-}) reduces active acidity by consuming H^+ in soil solution (Yadesa et al. 2019). Other similar reports (Fekadu et al. 2019; Shanka et al. 2018; Bekere and Dawud 2013) also revealed increased dry matter production and grain yields, nodule number and dry weight of soybeans, common bean and faba bean, respectively due to liming of acid soils. The improvement was attributed to lowering soil acidity thereby enhancing proliferation of bacteria and increasing availability of P, Ca, and Mo (Bambara and Ndakidemi 2010). Cells of N_2 fixing bacterial strains such as *R. leguminosarum* are swollen and lack rigidity in acid soils where there is Ca deficiency. Liming can increase soil Ca level that plays the main role in cytoplasmic pH maintenance, P mobilization and ion transport, which are caused mostly by changes in membrane properties (Basak and Rakshit 2015). Nodulation, N2 fixation, nitrogenase and specific nodule activities are directly related to the P supply.

P availability is usually poor in acid soils due to its fixation resulting from either chemisorption of P to Fe/Al-oxides and clay minerals or forming precipitates of insoluble Fe/Al-phosphates (Nurlaeny et al. 1996).

Table 9.3 Effect of lime on P availability of acid soils in Ethiopia

District	Soil pH	Rate of lime applied (tons ha-1)	Initial soil available P (mg kg^{-1})	Final soil available P (mg kg^{-1})	Change in available P (mg kg^{-1})	Sources
Bedi	4.80	2.2	4.5	5.9	1.4	Yadesa et al. (2019)
Ebantu	4.83	4.0	4.5	6.3	1.8	Bekele et al. (2018)
Lay Gayint	4.85	10.0	5.8	9.3	3.5	Fekadu et al. (2018b)
Gununo	5.0	0.4	0.6	2.9	2.3	Mesfin et al. (2014)
Farta	4.89	11.2	7.0	8.7	1.7	Melese and Yli Halla (2016)
Nedjo	4.82	5.4	1.5	2.8	1.3	Yadesa et al. (2019)

Experiments conducted in different areas of Ethiopia (Fekadu et al. 2018b; Mesfin et al. 2014; Melese and Yli Halla 2016; Bekele et al. 2018) obtained significant P improvement due to liming only, while others (Alemu et al. 2017; Yadesa et al. 2019) observed relatively small increases in labile P concentrations (Table 9.3).

The observed changes in available P as a result of liming are related to increased pH, reduced exchangeable Al and a resulting addition of phosphate ions into soil solution (Opala et al. 2010). Lime application can also stimulate mineralization of soil organic phosphorus to increase phosphate availability (Haynes 1982) but measuring P mineralization rates to explain its practical significance requires further investigation (Bolan et al. 2003). The smaller rise in available P is related to the relatively low inherent P content of the soil (Yadesa et al. 2019). This shows that soils having high P fixing potential and inherently poor in available P need optimum lime and high rate of P fertilizers to saturate fixation sites (Opala et al. 2010).

In addition to N and P, the other primary macronutrient affected by liming is K. Exchangeable and reserve form of K are also influenced by liming. Potassium uptake may decrease up on liming acid soils. This could be due to high lime rate that creates antagonistic effects of Ca and Mg (Fageria and Zimmermann 1995). Therefore, consideration of the contents and ratios of Ca, Mg, and K in the soil is important to maintain nutrient balance and (Loide 2004).

Fig. 9.5 Management of acid soils in Ethiopia by Liming. *Source*: Tilahun (2019)

(i) **Organic materials as acid soil amendment**

As an option to manage soil acidity, integration of organic amendments such as crop residues, manures, compost, and biochar with lime are advisable to smallholders in Ethiopia whose purchasing power is low. Experimental results (Adisu et al. 2019; Fekadu 2018; Shanka et al. 2018; Demissie et al. 2017; Agegnehu et al. 2005) obtained in different districts of the country showed a marked increase in yield of different crops (Table 9.4 and Fig. 9.6). Despite the promising results registered, these practices have not been widely disseminated in various parts the country due to competition of these organic resources for fuel, construction materials, and animal feed. Industrial waste products, such as precipitated calcium carbonate from pulp and paper industry are alternative sources of lime that can be used to amend acid soils after they are checked in the greenhouse for their effectiveness (Poykio and Nurmesniemi 2008).

The increase in yield could be ascribed to an increase in CEC and nutrients such as N, P, Ca, and Mg upon decomposition of the added organic substances (Ewulo 2005; Kheyrodin and Antoun 2011). Furthermore, the acid neutralization and lowering the toxicity of Al as a result of complex formation with organic materials, which also enhances the release P from fixation could contribute to the improvement of crop yields (Haynes and Mokolobate 2001;

Table 9.4 Effects of different treatment combinations on yield of crops

District	Rate of treatments applied (tons ha-1)	Yield (kg ha^{-1})	Yield advantage of combination over lime/control (%)	Test crop	Sources
Lay Gayint	7.2 lime ha^{-1}	1153	25.5	Faba bean	Fekadu et al. (2018c)
	Lime ha-1 + 8 t compost ha-1	1447			
Welmera	0.6 t lime ha-1	1682	3.7	Barley	Demissie et al. (2017)
	0.6 t lime ha^{-1} + 5 t compost ha-1	1744			
Areka	1.28 t lime ha^{-1}	3864	62.7	Common bean	Shanka et al. (2018)
	1.28 t lime ha^{-1} + 10 t compost ha-1	6287			
Assosa	4.9 t lime ha^{-1}	1800	6.1	Soya bean	Adisu et al. (2019)
	4.9 t lime ha^{-1} + 1.5 t ha^{-1} Vermicompost ha^{-1}	1910			
Holetta	Control	606	208	Faba bean	Agegnehu et al. (2005)
	8 t farmyard manure ha^{-1} + 39 kg P ha^{-1}	1867			

Fig. 9.6 Effects of FYM, P (TSP), compost, and lime on growth of faba bean. (*Source*: Fekadu et al. 2018c)

Agegnehu and Ademe 2017). Most researchers in Ethiopia also suggest the addition of inorganic fertilizers in the integrated acid soil management systems since acid soils are low in essential elements (Fekadu et al. 2018c). For instance, according to Fekadu et al. (2018c), application of 4 t FYM ha^{-1} + 15 kg P ha^{-1} + 7.2 t lime ha^{-1} increased grain yield of faba bean by 102% over the control, and by 53% over lime alone. The same researchers reported improved growth of faba bean due to integrated application of 4t FYM·ha^{-1} + 15 kg P·ha^{-1} + 5 t lime·ha^{-1} over the control (Fig. 9.6).

(ii) **Use of acid soil tolerant varieties**

Utilization of acid soil/Al-toxicity tolerant crop varieties along with other management methods is a common practice in acid soil-prone areas of the world. Many literatures worldwide indicate the possibility of developing varieties to adapt acid soils (Chen et al. 2002; Nguyen et al. 2002). In Ethiopia, however, development of varieties adapted to acid soils has not yet obtained adequate research attention (Tamene et al. 2017; Abate et al. 2013). However, selection of varieties that can give maximum yield under toxic Al concentration in the soil has been practiced in Ethiopia. For example, to evaluate acid-tolerant barley varieties, five released barley varieties were tested with lime and without lime in acid soils at Endibir (Agegnehu et al. 2019; AGP 2011). Barley variety (HB-1307) was superior under both limed and unlimed conditions (Fig. 9.7). Likewise, Alemu and Lule (2018) reported that Natoli and DZ-2012-CK-20113-2-0042 were better chickpea genotypesin tolerating soil acidity. Recently, acid-tolerant cultivars such as Walala (sweet lupin), 79 Ab 382 (Tx) 80 SA 94 (food oat), ETBW—6785 (bread wheat) and ETCL-161 (triticale) are identified to perform well on hot spot areas (Fekadu et al. 2018b).

The mechanisms for Al tolerance by plants include exclusion of Al from root apex and detoxification of Al in the root and shoot symplasm (Brunner and Sperisen 2013; Ermias et al. 2013). Aluminum exclusion is related to the ability of Al-tolerant plants to release organic acids (mainly citric acid and oxalate in maize) and phenolics from the root apex (Pinieros et al. 2005). The secreted organic anions (OA) bind with Al to form a complex (Al–OA), which protects the root apex, thus allowing it to continue growing (Yang et al. 2011). Detailed review has been presented by other researches on the mechanisms of Al detoxification by different crops (Ngoune et al. 2018; Abate et al. 2013; Kochian et al. 2004).

9.3.3 Salt-Affected Soils

Salts are composed of positively charged ions (cations) and negatively charged ions (anions). They can be dissolved in water (soluble salts) or be present as solids. Salts in soil can originate from soil parent material; irrigation water; or fertilizers, manures, composts, or even other amendments (Horneck et al. 2007).

The predominant salts that accumulate in soils are salts of calcium, magnesium, sodium, potassium, sulfate, chloride, carbonate, and bicarbonate. Any salt that accumulates in excessive amounts in soil can cause problems on plant growth. There are three categories of salt-affected soils, namely, saline, sodic, and saline-sodic based on their ECe, SAR/ESP, and pH. Saline soils have ECe of >4 dS m^{-1} at 25 °C for soil saturation extracts and ESP or SAR of <15%

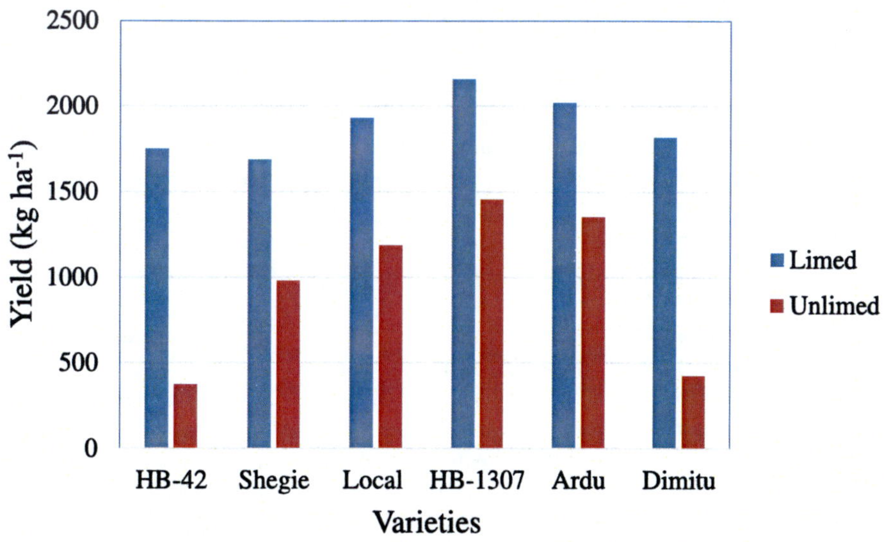

Fig. 9.7 Effect of lime and barley varieties on grain yield. (*Source*: modified from Agegnehu et al. (2019)

Table 9.5 Soil salinity indicator properties of surface soils at different locations

Locations	Salinity/sodicity indicator parameter				Type of salt-affected soils	Source
	EC (dS/m)	pH (H_2O)	ESP (%)	SAR		
Alage	4.6	10.9	8.7	6.9	Saline	Alemayehu et al. (2016)
Zeway	4.4	7.9	42.9		Sodic	
Babile (site 1)	5.2	7.7	14.5	6.7	Saline	Adane et al. (2019)
Babile (site 2)	4.5	8.5	20.3	13.4	Saline sodic	
Babile (site 3)	4.7	8.4	20.9	15.1	Saline sodic	
Babile (site 4)	5.0	7.9	11.7	7.0	Saline	
Tendaho	7.4	8.2	3.5	0.4	Saline	Asmamaw et al. (2018)
Shewarobit	5.1	8.6	9.3	3.7	Saline	Giday (2019)
Melka Sadi	12.3	8.4	33.8	—	Sodic	Ashenafi et al. (2016)

or 13, respectively. Sodic soils have ECe of <4 dS m^{-1} at 25 °C for soil saturation extracts and ESP or SAR of >15% or 13, respectively. These soils are characterized by dispersion of colloidal clays leading to low permeability and infiltration rates, poor aeration, surface soil crusting and difficult to till and for plant roots to penetrate through (Saied et al. 2017; Worku and Bedadi 2016). Saline sodic soils have ECe of >4 dS m^{-1} at 25 °C for soil saturation extracts and ESP or SAR of <15% or 13, respectively.

(a) **Extent, occurrence, and distribution**

Ethiopia ranks first among African countries in the extent of salt-affected soils caused by natural and anthropogenic causes. Although no systematic nationwide data on the extent of salinity problem in Ethiopia is available, many researchers have reported that about 11 million ha of land is exposed to salinity (Worku and Bedadi 2016; Abebe 2012; Gedion 2009; Fantaw 2007). These soils are concentrated in the Rift Valley, Wabi Shebelle River Basin, the Denakil Plains, and various other lowlands and valleys of the country, where about 9% of the population lives (Ayenew et al. 2013; Sileshi 2016). Some reports on types and properties of salt-affected soils occurring in Ethiopia are given in Table 9.5.

(b) **Agricultural potentials**

Salt-affected soils can be of good agricultural potential if good quality water is adequately available and soluble salts are properly managed. The soils are rich in basic cations and hence have high base saturation.

Salt-affected soils are marginally productive for most cereals and pulses grown in the country, demanding special management practices for sustainable agricultural development. Given the low amount of water available to satisfy crop growth with competition for domestic and industrial consumption, economic use of available water and the use of salt-tolerant crops are strategies to increase farm productivity and return in areas where these soils occur. As these soils occur on flat lowlands, they can be of good potential for large-scale production of commercial/industrial crops like cotton and sugarcane with proper implementation of salt and water management practices.

(c) **Major limitations**

The major limitations of salt-affected soils include their low osmotic potential due to the high soluble salt contents; low contents of micronutrients such as Fe, Mn, Zn, and Cu. Soil compaction due to surface sealing and crusting are also common problems in sodic soils. This is associated with the dispersion effect of sodium on soil aggregates. As a result, the sodic soils are usually characterized by low infiltration rate and hydraulic conductivity.

Salt-affected soils adversely affect soil fertility and productivity due to changes in their physical, chemical, and biological properties. A continuous osmotic phase inhibits water uptake by plants due to osmotic pressure of saline soil solution lowering its potential energy; and a slower ionic phase when the accumulation of specific ions in the plant over a period of time leads to ion toxicity or ion imbalance (Munns and Tester 2008).

(d) **Management and Use**

Most salt-affected soils in Ethiopia especially those in the rift valley are used for large-scale production of commercial crops such as sugarcane, cotton, citrus fruits, banana, and vegetables. The soluble salts that accumulate due to irrigation water applications can be managed by installing appropriate drainage structures to leach soluble salts out of the root zones. Once salinity is built up, reclamation of saline soils is made through leaching the soluble salts using good quality water. For sodic and saline sodic soils, the excess sodium should be replaced by a suitable cation like calcium through application of amendments like gypsum, which is usually followed by leaching. Application of organic matter will also improve soil structure and reduce soil compaction.

There have been experiences of saline soil reclamation through leaching, leaching plus artificial drainage, and managed land shaping to reduce salt accumulation in the drylands of Ethiopia (Muruts and Haileselass 2019). Leaching trial conducted by Fentaw and Girma (1996) demonstrated a decrease in salinity level on the surface of a soil profile. Gelaye et al. (2019) observed a decline in soil salinity level following leaching in plots with deep groundwater table. The areas of low to moderate salinity levels can be restored by introducing improved irrigation and crop management practices. However, in areas where increased salinity levels have restricted the growth of normal field crops, use of biosaline approach could be a potential solution. This approach is based on adaptable technology packages composed of salt-tolerant fodders and halophytes integrated with livestock and appropriate management systems (Qureshi et al. 2018).

Studies conducted by the Werer Agricultural Research Center in Ethiopia during 2011–2014 (EIAR 2015) showed promising results in terms of salinity tolerance, biomass yield, and ameliorative effects for four forage crop species (*Cenchrus sp*., *Panicum antidotale*, Sudan grass (*Sorghum sudanense*), and *Chloris gayana),* and three legume species (*Desmodium triflorum*, *Sesbania sesban*, and *Medicago sativa* (alfalfa).

Barley (*Hordeum vulgare* L.), sorghum, wheat, mustard, and oilseeds (safflower, sunflower, and sesame) are among economically important crops with diverse genetic diversity for better adaptation under saline soil conditions. Barley is one of the highland cereal crops with important contributions for food, malt, and feed. However, its production has been limited in areas where salinity is not a major production problem (Qureshi et al. 2018). Recent evidences indicate that barley is expanding to midlands where there is salinity problem. For instance, farmers at Zeway Dugda area grow barley instead of maize and other horticultural crops when the soil gets more salinized. Safflower is moderately salt-tolerant and cultivation on salt-affected lands can prove beneficial to farmers (ICBA 2014).

Studies have also shown that Karnal grass (*Diplachne fusca*), Rhodes grass (*Chloral gayana*), Para grass (*Brachiaria mutica*), and Bermuda grass (*Cynodon dactylon*) are highly salt-tolerant and can be successfully grown in saline and sodic soils. Karnal grass performed well in highly sodic soils even when no amendments were applied. A combined application of gypsum with Bermuda grass (*Cynodon dactylon*) and Rhodes grasses (*Chloral gayana*) showed relatively faster effectiveness in decreasing soil pHe, ECe, SAR, and ESP for all soil layers (Abate et al. 2021).

Emerging results in Ethiopia indicate that the use of bio-drainage is promising to control soil salinity and water logging, which are the major problems in most of irrigation schemes. Plantation of trees such as *Eucalyptus hybrid, Prosopis juliflora,* and *Acacia nilotica* is found to be effective in removing excess water and controlling groundwater rise to root depth. The advantages of bio-drainage as an eco-friendly technique for combating waterlogging and salinity are cost-effective strategies as compared to the expensive conventional drainage systems (Qureshi 2017).

For management of sodic and saline sodic soils, integrated application of gypsum, farmyard manure, and commercial humic acid play significant role in reducing soil pH, electrical conductivity, sodium adsorption ratio, and increasing yield of various crops (Shaaban et al. 2013). The decomposed organic matter from the added manure increases soil CO_2 concentrations due to microbial respiration and releases H^+ when it dissolves in water. The released H^+ enhances $CaCO_3$ dissolution and liberates more calcium (Ca) to replace sodium (Na) on the exchange site (Ghafoor et al. 2008). In another study, combined application of compost and poultry manure accelerated sodium leaching and reduced EC and exchangeable sodium percentage (ESP), while water-holding capacity and soil aggregate stability were improved (Tejada et al. 2006). Addition of organic matter enhances multiplication of the microbial population, which promotes soil aggregation, structure, and stability and provides a steady supply of nutrients into the soil. Moreover, organic materials improve the soil physicochemical properties that accelerate exchange of cations on soil solids and leaching of salts from the root zone (Clark et al. 2007), hence preventing roots from salt injuries and promoting growth of roots more smoothly.

9.3.4 Sandy Soils

(a) Extent and occurrence

Sandy soils occur in all parts of the world. When these soils have >70% sand and <15% clay, they are classified as Arenosols in the World Reference Base (FAO 2001) and in Soil Taxonomy as sandy Entisols: Psamments when well-drained, or as Psammaquents when in tidal marshes, deltas, and wetlands (Soil Survey Staff 1999). In the World Reference Base, sandy soils may also occur in the reference groups Regosols, Leptosols, and Fluvisols. Arenosols may have developed in residual sands, in the weathering products of quartz-rich rock or in recently deposited sands common to deserts and beaches (FAO 2001; Hartemink and Huting 2008).

Sandy soils are represented well under Arenosols according to the IUSS Working Group WRB (2015). Arenosols comprise deep sandy soils. This includes soils in residual sands after in situ weathering of usually quartz-rich sediments or rock, and soils in recently deposited sands such as dunes in deserts and beach lands. Corresponding soils according to the Soil Taxonomy of the United States of America include Psamments. The essential characteristic of Psamments includes coarse textured unconsolidated materials with a large proportion of quartz derived exclusive from recent alluvial deposits (Abebe 1998).

(b) **Agricultural potentials**

Sandy soils occur in widely differing environments and possibilities to use them for agriculture vary accordingly. Their coarse texture accounts for their generally high permeability and low water and nutrient storage capacity.

On the other hand, sandy soils offer ease of cultivation, rooting, and harvesting of root and tuber crops. In arid and semi-arid lands with annual rainfall of less than 300 mm, they are predominantly used for extensive (nomadic) grazing. Dry farming is possible where the annual rainfall exceeds 300 mm.

(c) **Major limitations**

Sandy soils have low coherence, low nutrient and water storage capacity, and high sensitivity to erosion. They are structure less or show weak differentiation, low in cation exchange capacity, and low in organic matter content and their consequent characteristics. As a result, they are usually excessively drained.

Uncontrolled grazing and clearing for cultivation without appropriate soil conservation measures can easily destabilize these soils, reverting them back to shifting dunes.

Sandy soils in the humid and sub-humid temperate zone have limitations similar to those of the dry zone, albeit that drought is a less serious constraint. In some instances, e.g., in horticulture, the low water storage of these soils is considered advantageous because the soils warm up early in the season.

(d) **Use and management**

Good yields of small grains, melons, pulses, and fodder crops have been realized on irrigated sandy soils, but high percolation losses may make surface irrigation impracticable. Drip or trickle irrigation, possibly combined with careful dosage of fertilizers, may remedy the situation. In mixed farming systems (which are much more common) with cereals, fodder crops and grassland, supplementary sprinkler irrigation is applied during dry spells.

Permanent cultivation of annual crops would require management inputs that are usually not economically justifiable. Root and tuber crops benefit from the ease of harvesting, notably cassava with its tolerance of low nutrient levels. In some cases, sandy soils may be prone to develop water-repellency, typically caused by hydrophobic exudates of soil fungi that coat sand grains.

9.3.5 Volcanic Ash Soils

(a) **Extent and occurrence**

Volcanic ash soils are usually represented by the term Andosols or Andisols (US Soil Taxonomy) and occur in volcanic regions all over the world. The total area with volcanic ash soils (Andosols) is estimated at some 110 million ha or less than 1% of the global land surface (IUSS Working Group WRB 2015). In Africa, major occurrences of these soils are along the East African Rift Valley in Kenya, Rwanda, and Ethiopia but also in Cameroon and in Madagascar. In Ethiopia, these soils mainly occur in the Rift valley as well as in the mountain areas of the country such as in the northwest, and northeast, and southeastern Ethiopian highlands.

The soils of Shewa, Eastern Gojjam and Gonder, Arsi, and Northern Bale and Sidamo are dominantly developed on trap series volcanos. Soils of Eastern Shewa and Central Wello have developed almost exclusively on volcanic materials. Much of the area forms the so-called north and eastern escarpment of Ethiopia, i.e. the northeastern extreme of the central plateau formed by the western horse arm of the Rift system, which has strong structural influence on landform and soil development. The soils in these regions are strongly influenced by the chemical and mineralogical composition of the volcanic rocks that were deposited during eruptive phases.

Andosols are located in the central Ethiopian Rift Valley of quaternary volcanos largely between Lake Abaya and Metahara, and around Mount Ras Dashen in northern Ethiopia. Andosols accommodate soils that develop in glass-rich volcanic ejecta under almost any climate whose colloidal fraction is dominated by short range-order minerals, and/or Al-humus complexes, allophane, and imogolite.

(b) **Agricultural potentials**

Andosols have a high potential for agricultural production, but many of them are not used to their potential. They are generally fertile and are continually renewed by the high proportion of weatherable minerals characteristic of young soils. These soils are developed from volcanic ash deposits of andesitic to basaltic nature that could furnish high cation exchange capacity contributing to its high potential for agricultural production.

The free drainage, high infiltration rates and available water capacity (Nanzyo et al. 1993b) simplify management of irrigation water, particularly if leaching is required. High available water capacity minimizes the implied irregularities in moisture supply under rainfed conditions. In the drier parts of the Rift Valley where they are predominantly used by pastoralists, Andosols are used for livestock grazing. Fine textured Andosols, on the other hand, are among the most productive soils in Ethiopia given an adequate supply of water, in the absence of special limitations that are not general to these soils.

(c) **Major limitations**

Andosols have high proportion of amorphous clay minerals that lead to strong phosphate fixation caused by active Al and Fe (Nanzyo et al. 1993a). Phosphorus applied to these soils will be immediately adsorbed to the surface of amorphous clay minerals to form complex iron and aluminum phosphates, which is unavailable to plants. Their major "limitation" to agricultural use, however, is a demand for high applications of phosphate fertilizers to overcome phosphorous fixation (Komiyama et al. 2014; Mohammed and Belay 2008; Tekalign et al. 1991).

Accumulation of salts occurs in Halaquepts, mainly as sodium chloride. They occur quite extensively in the drier climates under aridic and ustic soil moisture regimes in Ethiopia as in the Awash Valley, the Afars, the Somali, and parts of Oromiya. In addition, the predominant occurrence of Andepts, Halaquepts and other Inceptisols in semi-arid areas not only resulted in many sodic, and saline phases, but phosphorus fixation is also a common phenomenon. Then, the reactions of P with calcium, aluminum, and iron as part of different solid compounds or complexes, or being in the soil solution assumes great importance.

(d) **Use and management**

Volcanic ash soils have favorable properties for cultivation, plant roots, and water storage. The soils are planted with a wide variety of crops including sugar cane, tobacco, sweet potato (tolerant of low phosphate levels), tea, vegetables, wheat, and orchard crops. Volcanic ash soils on steep slopes are perhaps best kept under forest. Paddy rice cultivation is a major land use on Andosols in lowlands with shallow groundwater (IUSS Working Group WRB 2015). Volcanic ash soils that occur in the mountain areas of Ethiopia are cultivated with potato and barley, mainly conditioned by the local climatic characteristics while those occurring in the rift valley are cultivated with maize and sorghum. On the other hand, the uncultivated volcanic ash soils, both in the mountains and rift valley areas, are used as grazing and pasture lands (Mohammed and Belay 2008).

The management of Andosols in the Ethiopian Rift Valley, being fragile to erosion need wind break, and moisture conservation and planting trees in a row in the agriculture field must be encouraged. Besides, soils must be covered by intensive cropping system including crop residue mulching. In order to reduce runoff velocity and energy, grass strips and earth bunds should be managed each 10–20 m depending of slope steepness and soil erodibility. These strips must be productive of forage, firewood and poles for building to be acceptable for poor farmers. Zeleke et al. (2004) suggested incorporating crop residues, especially in conjunction with the use of inorganic fertilizers, can improve rainwater use efficiency and soil tilth of Andosols. This will also have a direct effect in minimizing the rate of soil erosion in the area.

Ameliorative measures to reduce phosphate fixation effects include application of lime, silica, organic matter and phosphate fertilizer. Manure compost application with supplemental N fertilizer to Andosols is reported to improve P accumulation and nutrition to increased maize production (Komiyama et al. 2014). The organic manures seemed to have lowered the capacity of the soil to fix P and increased the availability of nutrients. Increased P availability was observed with liming of Andosols (Guadalix and Pardo 1994). Organic material amendments could improve P availability through enhancing organic P mineralization. For example, addition of cover crop such as rye and rapeseed in the cropping system (Takeda et al. 2009), colonization of subsequent crops with arbuscular-mycorrhizal (AM) increased P uptake and P use efficiency (Karasawa et al. 2002).

9.3.6 Soil and Water Management in Drylands

Drylands are lands characterized by the ratio of annual precipitation to potential evapotranspiration (P/PET) ranging between 0.05 and 0.65 (UNCCD 2000). They are also defined as those areas with a growing season length of 1–179 days and a climatic classification of arid, semi-arid and dry sub-humid (FAO 2000). Drylands in Ethiopia cover about 75% of the total land mass of the country (Georgis 2014). They consist of a wide range of agroecologies, including arid, semi-arid and dry sub-humid and are most prevalent in the north, east, central Rift Valley areas, south and southeastern parts of the country. The lowland dryland areas in Ethiopia cover about 61% of the land mass of the country. The altitude of lowlands ranges from −124 to 1500 m above sea level (m.a.s.l.) and average annual rainfall varies between 200 and 700 mm. The length of the growing period is 90–180 days. The drylands are naturally rich in natural resources. The details of resource bases in the dryland areas are summarized as follows.

Water resources: Drylands are endowed with water resources much of which have been untapped and have valuable potential. For example, they have substantive river systems and groundwater, which are cradles of important watersheds and riverside ecologies, diversity of fish and have high values for hydropower and irrigated commercial crop agriculture. There are three categories of water basins in the area (EPA 1998).

Surface water—lakes: There are seven major lakes of the Rift Valley, namely: Ziway, Langano, Abiyata, Shalla, Awassa, Abaya, and Chamo. These lakes are used for commercial fisheries, irrigation, recreation & industrial purposes.

Surface water—rivers: There are 12 river basins with a total surface run-off of about 110 billion m^3. Major river basins are Awash, Genale-Dawa, Wabi Shebele, Baro-Akobo, Tekeze, Merb, Fafem and Abay. Some of these rivers are used for commercial fisheries, irrigation, recreation and industrial purposes.

Groundwater: There is also a high potential of groundwater in many pastoral areas, including the Rift Valley areas of Oromia, Eastern Afar and Eastern Tigray, and Somali. These areas have large quantities of groundwater along the valleys that can irrigate millions of hectares for food and cash crop production.

Soil resources: If the land is well managed, particularly in pastoral areas and valley bottoms, there is a high agricultural potential, with nutrient-rich soils, and providing possibilities for surpluses.

Despite the availability of natural resources in the drylands, production is usually constrained by several factors mainly water stress, low soil fertility, prevalence of pests and disease. Therefore, the livelihoods of farmers in semi-arid farming systems should be improved through application of integrated genetic and natural resource management strategies. There is much evidence that appropriate agronomic management practices, such as improved soil and water, fertility conditions, pest and disease control, are more critical factors for determining crop yield than improved variety alone.

Soil and water management: Soil and water management practices such as mulching, contour farming, in-situ water harvesting techniques such as tied ridges are practiced to conserve moisture in drylands. Tied ridges have proven to be very effective for soil and water conservation in Ethiopia and many other African countries. The technique has been extensively tested and evaluated with smallholder farmers in Kenya, Ethiopia, Eritrea, Zimbabwe, Uganda, Tanzania, Burkina Faso, Nigeria, and other areas in Africa (Georgis 2003). The experimental evidence of the effectiveness of tied ridges for soil and water conservation and its impact on yield increase and water use efficiency is well demonstrated. Its use increased grain and straw yield of crops by 150% and 90%, respectively, compared to traditional methods of planting in the flat seedbed. Most soils in semi-arid areas have the problem of compaction or surface sealing or crusting which leads to low water infiltration and high runoff. Mulching is traditionally used to alleviate these problems and research results in the central Rift Valley indicated that use of mulches at the rate of three tons increased yield by 30% compared to without mulching. The mulching materials are obtained from pigeon peas and sesbania sesban, which are drought resistant, and can easily be produced locally in many drought-prone areas. The use of sesbania sesban mulches was also found to increase grain and stover yield substantially in wheat compared to the control without mulching (Gebrekidan 2005).

Improving water productivity: Improving rainwater management alone cannot deliver increased productivity. Water management must form part of a farming system that includes a whole range of inputs, such as fertilizer, pesticides, improved seeds and adequate farm power. This concept is well described in integrated soil fertility management.

9.4 Integrated Soil Fertility Management

9.4.1 Principles and Practices

Several definitions for Integrated Soil Fertility Management (ISFM) are proposed in global literature, but most of them have focused on the farming systems in Sub-Saharan Africa (SSA). Most of the proposed definitions attempted to capture the full set of principles that are required to sustainably increase crop productivity in smallholder farming systems in the Sub-Saharan Africa (SSA) context (Vanlauwe and Zingore 2011). In this context, they underlined that at the regional scale, overall agro-ecological and soil conditions have led to a diverse population and livestock densities across SSA and formed a wide range of farming systems. Each of these systems has different crops, cropping patterns, soil management considerations, and access to inputs and commodity markets. Within farming communities, a wide diversity of farmer wealth classes, inequality, and production activities may be distinguished. For instance, the use of cattle manure and more fertilizer by the wealthier farmers results in higher farm-level productivity than on poorer farms. At the individual farm level, it is important to consider the variability between the soil fertility status of individual fields. Variability arises due to farmer preference or capacity to apply limited fertilizers and organic nutrient resources to small areas of the farms. They then suggest that any definition of ISFM must consider these attributes.

Consequently, they have provided an operational definition of ISFM as 'A set of soil fertility management practices that necessarily include the use of mineral fertilizer, organic inputs, and improved germplasm combined with the knowledge on how to adapt these practices to local conditions, aiming at maximizing agronomic use efficiency of the applied nutrients and improving crop productivity'.

ISFM consists of a set of site-specific practices to increase soil fertility and reduce soil degradation, reduce input cost and thereby improve yields, incomes, and ultimately livelihoods of farm households (Giller 2001; Vanlauwe et al. 2011). ISFM can also be described as a set of good soil management technologies that can be applied in an integrated fashion. While each component can have a positive contribution to soil fertility and crop productivity, the aim is to integrate multiple technologies to exploit complementarities among the different technologies (According to Vanlauwe (2013), the definition ISFM must include several concepts (Fig. 9.8) described below:

1. **Focus on agronomic use efficiency**. Mineral fertilizers and organic inputs are both scarce resources in the areas where agricultural intensification is needed. Consequently, the definition focuses on maximizing their use efficiency. Agronomic efficiency (AE) is defined as the extra produce generated (kg) per unit of nutrients applied.
2. **Fertilizer and improved germplasm**. In terms of management response, two general classes of soils are distinguished:
   i. soils that show acceptable responses to fertilizer (Step A – blue line, Fig. 9.8) and
   ii. soils that show minimal or no response to fertilizer due to other constraints besides the nutrients contained in the fertilizer (Step B—green line, Fig. 9.8).

These soils are classified as 'responsive soils' and 'poor, less-responsive soils' respectively. In some cases, where land is newly opened, or where fields are close to homesteads and receive large amounts of organic inputs each year, the third category of soil exists where crops respond little to fertilizer as the soils are fertile. These soils need only maintenance fertilization and are termed *fertile, less responsive soils*.

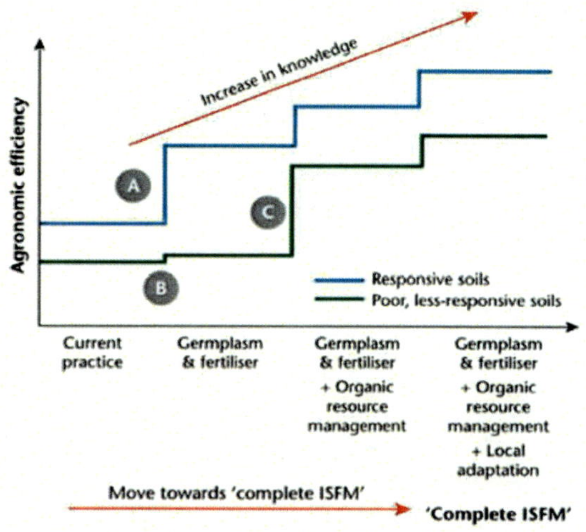

Fig. 9.8 Conceptual relationship between the agronomic efficiency (AE) of fertilizers and organic resources and the implementation of various components of ISFM, culminating in complete ISFM towards the right side of the graph. (*Source*: Adapted from Vanlauwe et al. (2011)

The definition proposes that the application of fertilizers to improved germplasm on responsive soils will boost crop yield and improve the *agronomic efficiency* relative to current farmer practice, characterized by traditional varieties receiving too little and insufficiently managed nutrient inputs (Step A—blue line, Fig. 9.8). Major requirements for achieving production gains on 'responsive fields' within Step A include:

(i) *the use of disease-resistant and improved germplasm*
(ii) *the use of the correct fertilizers sources and rates*
(iii) *appropriate fertilizers use in terms of placement and timing, and*
(iv) *crop and water management practices.*

The definition focuses on *maximizing the efficiency* with which fertilizer and organic inputs are used since both are scarce resources in the areas where agricultural intensification is needed. This involves the use of carefully calculated combinations of inorganic and organic fertilizers in association with complementary agronomic practices including tillage, rotation and sequencing of the crop as well as soil and moisture conservation.

9.4.2 Components of ISFM and Their Integration

Ruganzu et al. (2015) provided a detailed description of ISFM components that they broadly categorized into land assessment, agricultural practices, and soil conservation. The authors of this chapter borrowed the description with some modifications in this section.

Land Assessment

Land assessment is concerned with the evaluation of land performance when used for specified purposes. It involves the execution and interpretation of basic surveys of climate, soils, vegetation and other aspects of land in terms of the requirements of alternative uses of the land. Land assessment may be concerned with present land performance, but often involves change and its effects with change in the use of land and in some cases change in the land itself. The assessment considers the economics of the proposed enterprises, the local and national social consequences, and the beneficial or adverse environmental effects.

Agricultural Practices

Agricultural practices are a set of activities combining farming systems and soil conservation practices for promoting sustainable agriculture and enhancing land productivity. *A farming system (FS)* is the population of individual farm systems that have broadly similar resource bases, enterprise patterns, household livelihoods, and constraints and for which similar development strategies and interventions would be appropriate. FS consists of cropping system and crop management including land preparation, planting pattern, and timing. The following paragraphs provide a brief elaboration of the common land and crop management practices.

Land Preparation

Land preparation (LP) refers to the activities that make the land suitable for crop establishment, release soil nutrients, reduce weeds and pests and improve the soil and plant contact.

Cropping Pattern

Cropping pattern (CP) refers to the proportion of land under cultivation of different crops at different points in time. Any change in the cropping pattern would cause:

- *change in the proportion of land under different crops*
- *change in space sequence and time of crops*

Intercropping

Intercropping system (IC) is cultivating two or more crops in a field simultaneously. In addition to cash crops, sometimes cover crops are used in intercropping. It is a sustainable practice that can improve resource-use efficiency, such as nutrients and water, allowing low input agricultural practices. The increased coverage of the soil surface reduces soil and nutrient losses through erosion. Mostly, the intercropping system makes use of legumes in cereal growing farms to enable the latter to benefit from nitrogen that is fixed by legumes.

Cover Crops

Cover crops (CC) are plants that grow rapidly and close to the ground. CC are effective at reducing soil erosion by maintaining the soil surface covered which reduces soil detachment associated with the impact of raindrops hitting soil particles and by reducing the volume and velocity of runoff over the soil surface.

Mulching

Mulching is maintaining a protective cover, usually of organic matter such as leaves, straw on the soil surface around plants. Mulching reduces the splash effect of rainfall, controls weeds, conserves soil moisture, increases soil water permeability, reduces runoff, moderates soil temperatures.

Agroforestry

Agroforestry is a collective name for land-use systems and technologies, where woody perennials are deliberately used on the same land management unit as crops in some form of spatial arrangement or temporal sequences.

Crop Rotation

Crop rotation (CR) refers to the recurrent succession of crops so that soil health is not impaired. CR helps to balance the fertility demands of various crops and reduces soil nutrient depletion and avoids the build-up of pathogens and pests that often occur when one species is continuously cropped on the same piece of land.

Improved Fallow

A fallow period is a period when farmers decide to let their land rest to regain fertility; it can be traditional or improved. The improved fallow mainly consists of growing fast-growing leguminous shrubs and nitrogen-fixing trees for soil amelioration.

Soil Conservation

Soil conservation (SC) is the prevention of loss of the topmost layer of the soil mainly by erosion or prevention of reduced fertility caused by nutrient mining, acidification, salinization, or other chemical soil contamination. The basic principles to reduce soil erosion and maintain soil quality are to keep the soil covered and minimize soil disturbance. Among the widely used soil conservation practices are.

Conservation Tillage

Conservation Tillage (CT) involves the aspects of reduced soil disturbance such as zero tillage and reduced tillage and soil cultivation that leaves crop residue on fields before and after planting crops. It involves maintaining 30% of the soil surface covered with residue after planting crops.

Contour Tillage

Contour tillage is the practice of ploughing across a slope following its elevation contour lines. The rows formed following the contour have the effect of slowing water runoff and increasing infiltration during rainstorms so that the soil is not washed away.

Strip Cropping

Strip cropping (SC) is a system whereby strips of row crops and closely growing crops, planted on the contour, are alternated, commonly used when a slope is too steep or too long, or when other types of farming may not prevent soil erosion. Erosion is largely limited by the row-crop strips and soil removed from upslope is trapped in the next strip down the slope, which is generally planted with a leguminous or grass crop.

Terraces

A terrace is a piece of sloped plane that has been cut into a series of successively receding flat surfaces or platforms, which resemble steps, for more effective farming. It is designed as a method of soil conservation to slow down or prevent the rapid surface runoff of rainwater and irrigation.

Organic Resource Management

One of the key organic resources is organic matter (OM), which is a material composed of organic compounds that have come from the remains of plants and animals and their waste products. When OM decays to the point at which it is no longer recognizable is called soil organic matter (SOM). The sources of SOM include compost, farmyard manure, green manure, and bio-fertilizers. Green manuring is a practice of enriching the soil by turning a crop into the soil, whether originally intended or not, irrespective of its state of maturity, for the beneficial effects to the soil and subsequent crop improvement. Legumes are the most used as green manure due to their nitrogen-fixing capacity coupled with their capacity to grow fast. Bio-fertilizer is a substance that contains living microorganisms that, when applied to seeds, plant surfaces, or soil colonizes the rhizosphere or interior of the plant and promotes growth by increasing the availability of primary nutrients to the host plant. The most used soil inoculants are rhizobacteria that live symbiotically with legumes.

9.4.3 Application of ISFM Within the Ethiopian Context

Traditionally, the use of elements of ISFM including intercropping, crop rotation, manuring have been well established since antiquity. However, mineral fertilizers, row planting, and several improved crop varieties were introduced to the major cereal-growing farming systems during the last decades. The cereal-pulse rotation systems with strategic application of P fertilizer to legumes to maximize biological N-fixation to benefit the subsequent cereals (Amanuel et al. 2000) are widely used in conjunction with complementary agronomic practices. The recent introduction of liming acidic soils and the use of bio-fertilizers have further augmented the integrated soil fertility management efforts in Ethiopia.

Since recently, the ISFM-related efforts that were largely limited to research and small-scale field demonstrations are augmented by initiatives that involved international partners and external funding. Several projects including the Wageningen-EIAR Integrated Nutrient Management project, MoARD/SG-2000 Conservation Agriculture demonstrations, AGRA-EIAR ISFM project, N2Africa, GIZ-ISFM project have engaged in the demonstration and promotion of ISFM options into the agricultural system of Ethiopia (GIZ 2020).

9.4.4 The ISFM+ Project

The Implementation Approaches

The Integrated Soil Fertility Management (ISFM+) project is funded by the German Federal Ministry of Economic Cooperation and Development (BMZ) through its special initiative 'one world—no hunger (SEWOH).' It is programmed to run for 9 years (January 2015–December 2024) and has the objective of promoting integrated soil fertility management technologies and approaches in Amhara, Oromia, Tigray, and SNNP regions. The project aims to achieve the following outputs: (1) the concept of integrated soil fertility management is incorporated into the national system, (2) existing and newly acquired knowledge of soil fertility improvement has been processed for training purposes, (3) small and medium-sized enterprises offer ISFM-specific production inputs for which there is demand from farmers and (4) ISFM is mainstreamed into the relevant national institutions.

The project has hypothesized that the implementation of ISFM+ enhances biomass production, which increases food for humans and feed for livestock while improving soil organic matter content. Initially, the project was introduced as a quick-win solution to increasing both yield and biomass production through varied but complementary technology packages (Leta et al. 2020). In its strategy, the project defined the quick-win technologies as baskets of options for the simultaneous use of improved crop varieties, organic and inorganic fertilizers, line seeding, increased use of legumes and rhizobia, improved agronomic management practices and, in acid soil areas, the application of agricultural lime.

The project has followed a stepwise approach to transform the existing situation dominated by organic matter depletion, nutrient mining and low yield and biomass production toward the envisaged complete ISFM characterized by high on-farm biomass production and wealthy and healthy farmers and sustainably productive soil (Fig. 9.9) as outlined by Vanlauwe (2013). The Community Level Participatory Planning (CLPP) process that is guided by the MoA's USAID-funded Agricultural Growth Program (AGP) was used as a planning tool. The process involves communities and their institutions prioritizing their soil fertility and productivity challenges, preparing their plans, mobilizing resources, allocating budgets and identifying ways in which to implement activities and monitor progress.

Awareness creation and capacity buildings play a key role in mobilizing stakeholders including farmers. Accordingly, the project trains and exposes farmers to organic fertilizer use, compost production with effective microorganisms (EMOs), vermicompost and bioslurry, use of technologies and practices such as biofertilizers, agronomic practices such as intercropping of large cereal crops with legumes, and small-scale mechanization for minimum tillage and other farming activities. The project has developed and promoted locally adapted ISFM practices via a group-based learning approach, like farmers' field schools (e.g., Davis et al. 2012). The project woredas Demonstrations sites were set up for showcasing these technologies in project woredas. The sites were used to increase awareness of the farmers, development workers, and policymakers by showcasing the enhanced efficiency and productivity of fertilizers due to the increased biomass yield and soil organic matter accumulation.

Following the successful introduction of ISFM+ 'quick win' technologies, the project pushes for the more advanced step 2 and 3 technologies, which includes agroforestry, bioslurry, conservation agriculture, green manuring, urine collection, and water harvesting or water conservation. An

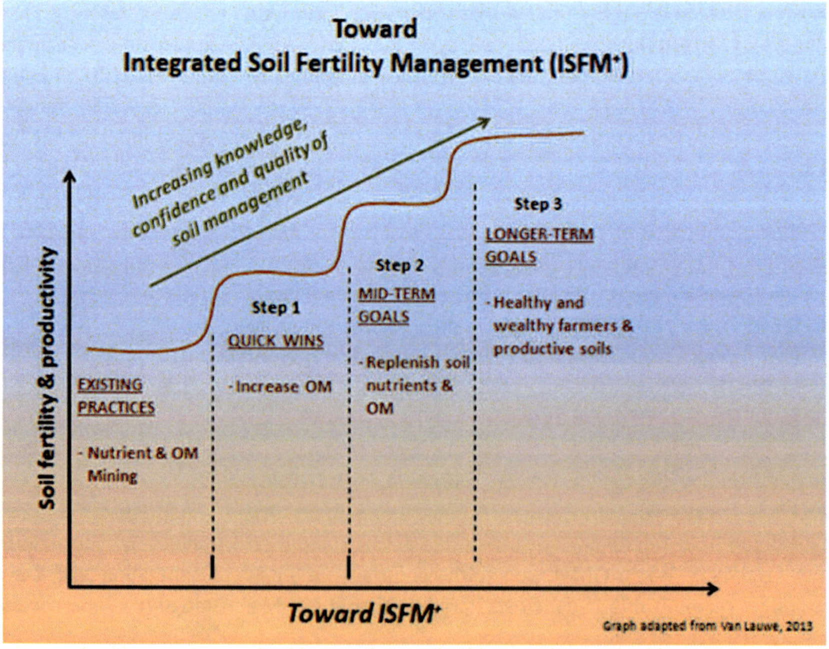

Fig. 9.9 A stepwise implementation of integrated soil fertility management practices in the Ethiopian highlands. (*Source*: Adapted from Vanlauwe 2013)

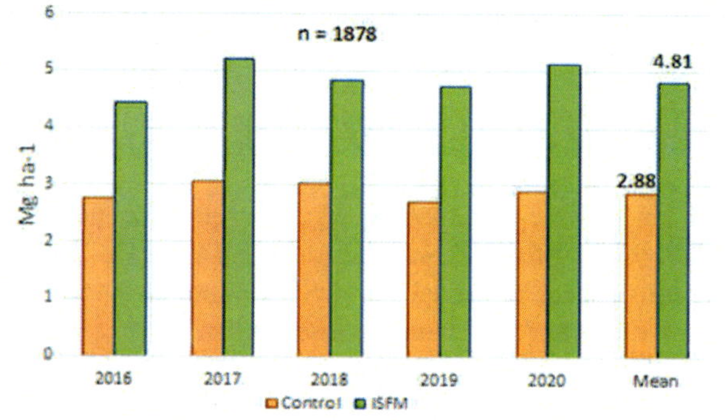

Fig. 9.10 Effect of ISFM technologies on average grain yield of cereal crops in the Ethiopian highlands. (*Source*: GIZ 2020)

attempt is also underway to promote the application of the ISFM technologies at the farm level within an agro-ecological setting to ensure socio-economic and environmental resilience.

9.4.5 Effects of Adopting ISFM in Ethiopia

Effects on Crop yield

The results from the field demonstration from 2016 to 2019 by the project have shown a yield advantage of 60% in 2016 and above 100% in 2019, as compared to the farmer's practice (GIZ 2020) (Fig. 9.10). Despite the proven economic benefits through increased yield and ecological roles including enhanced soil quality of using ISFM were established by the research system, the widespread adoption of a fully-fledged ISFM practice was limited to the project-based interventions.

A long-term (5 years) continuous application of ISFM resulted in stable crop productivity while the control plots (farmer's practice) have shown declining trends, leading to a yield difference of about 70–80% to more than 150% in 2020 (Fig. 9.11). The ISFM project report indicates that the effects of ISFM practices on grain yields are even greater when they are used on acidic soils (with a 76% increase) than on non-acidic soils (with 58% increase) as compared to the farmer's practices (GIZ 2020).

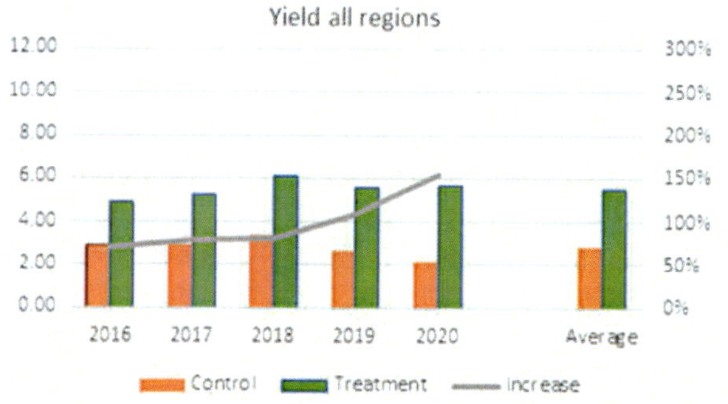

Fig. 9.11 Long-term demonstration yields development over 5 years (n = 481 observations from 103 farmers, only 69 in 2020). (*Source*: GIZ 2020)

Effects on Income and Food Security

Hörner and Wollni (2020) conducted a comprehensive analysis of ISFM effects on net crop value, labor demand, labor productivity and returns to unpaid labor using survey data from over 6000 tef, maize, and wheat plots and 2000 households in Ethiopia, based on the data from the GIZ-ISFM+ project's demonstration plots. Accordingly, data from Amhara and Oromia regions showed that adoption of ISFM leads to a significant increase of around 32% in total household income per capita. This increase is likely to stem from higher per capita income achieved from the production of three cereal crops (maize, wheat or tef), for which average increases of approximately 38 and 37% due to partial or full, respectively ISFM adoption. The estimated income per hectare obtained from these crops suggests that both adopting partial or full ISFM increases productivity.

In Tigray, by contrast, the per capita household income has a negative sign, though they are not statistically significant. As in the other two regions, adopting ISFM for one of the three main portions of cereal seems to be associated with a significant increase in income generated from these crops of about 20–21% when measured per capita. In Tigray, adopting full ISFM goes along with a significant decrease in the average probability of households generating off-farm income by 12%. As opposed to the other regions, in Tigray, no indication for the food security-enhancing effect of adopting ISFM for the three cereal crops. Even though ISFM adoption increases income from the three cereal crops in both agroecological zones, it is only related to an improvement in food security in areas where it also goes along with an increase in household income.

Effects on Labor Demand

The study revealed that adoption of ISFM resulted in a significant increase in total labor demand, both when measured in labor days per hectare and absolute labor days. The estimates suggest that in Amhara and Oromia, this additional labor demand is primarily absorbed by adult males and to some extent, adult females in the household, increasing their seasonal labor input on average by around 10–11, respectively. By contrast, in Tigray, ISFM adoption appears to increase the labor input of adult females and children in the household. Moreover, in Tigray, partial or full, as well as full ISFM adoption for the three cereal crops, seems to significantly increase the probability of school-aged children to work to produce these crops by 13% points on average.

Effects on Children's Education

Estimates for Amhara and Oromia suggest a positive effect of adopting partial or full as well as full ISFM on enrollment of primary school-aged children, increasing their average likelihood to be enrolled by 15 and 18% points. In Tigray, by contrast, there is no evidence of a significant effect of ISFM adoption on school enrollment. The adoption of ISFM has not significantly affected the average number of missed school days. Therefore, adopting ISFM has some positive effects on school enrollment in Amhara and Oromia, possibly a consequence of higher household income in these regions.

Institutionalization and Scaling

Through the collaborative efforts with the Soil Fertility and Health Directorate of MoA, ISFM has been part of the national 'Soil Health and Fertility Improvement Strategy' of MoA to sustainably enhance soil fertility, productivity, and livelihoods of the rural population (MoANR 2017). Building on that the ISFM+ project facilitated the federal and regional partners who have developed a strategy for the institutionalization of ISFM (GIZ 2020). The strategy outlines the concrete activities and timelines to incorporate ISFM+ approaches and lessons into learning institutions and the agricultural extension system. In addition, the strategy promotes enhancing knowledge management such as sharing lessons learned including through stakeholders' platforms and social media. The project also focuses on creating an enabling environment at the policy level.

Key lessons from ISFM woredas were disseminated outside the 60 ISFM+ woredas through the annual field days. Motivated by the success observed in the project woredas, the regional Bureaus of Agriculture (BoA) have facilitated the implementation of ISFM outside the project woredas aka 'out scaling woredas', where ISFM is now part of the regular planning and implementation process with only minimum support and backstopping from the ISFM project. In 2021, 45 woredas in Amhara and 25 in Oromia have implemented ISFM by the BoAs who have provided training of trainers at their respective regions. The training is cascaded to woreda and kebele levels, and ISFM practices are implemented and further disseminations are done through field days.

9.5 Challenges and Opportunities for Managing Restoration of Soil Health

Soil Health

Soil health is the continued capacity of a soil to function as a vital living system, within ecosystem and land use boundaries, to sustain biological productivity, promote the quality of air and water environments, and maintain plant, animal, and human health (Pankhurst et al. 1997). It provides essential nutrients, water, oxygen, and support to the roots, all elements that favor the growth and development of plants for food production. Furthermore, healthy soils emit less greenhouse gasses and actually sequester carbon while having the added benefit of greater agricultural productivity (ISD 2016).

Soil health consists of five principles: soil armor, minimizing soil disturbance, plant diversity, continual live plant/root, and livestock integration. Less healthy soils emit more carbon into the atmosphere, whereas in healthy soils the carbon can actually be sequestered through different biological, physical, and chemical processes, which prevent escape into the atmosphere. Therefore, soil has a vital role to play with regard to the global climate issue (ISD 2016).

Soil health provides an environment for plants to grow in, by anchoring roots and storing nutrients. It filters and cleans our water and helps prevent natural hazards such as flooding. Soil health best practices minimize surface disruption (e.g., less tilling or no tilling), keep the soil covered as much as possible to limit erosion, and provide plant diversity to naturally feed the soil and support greater biological diversity (Kihara et al. 2020).

9.5.1 Challenges for Managing Restoration of Soil Health

There are two categories of challenges in managing restoration of soil health in the country. These are the soil-level and systemic challenges

(a) **Soil-level Challenges**

The soil-level challenges for managing restoration of soil health include physical, chemical, and biological issues within the country's soils in different agroecologies. The issues of these challenges are found within the soils, both on farmer's fields as well as within the surrounding landscapes around agricultural areas that have direct consequences on agricultural soils and their productivity. The challenges have negative impact on soil health—in terms of physical, chemical, and biological conditions—and on soil fertility. Assessment of the challenges on soil health (MoA 2013) identified 12 key soil-level challenges in various parts of the country:

- Organic matter depletion
- Nutrient depletion
- Soil erosion
- Acidity
- Low moisture availability
- Soil structural deterioration
- Soil fauna and flora depletion
- Biomass coverage removal
- Salinity and sodicity
- Waterlogging
- Physical land degradation
- Soil pollution

Organic Matter Depletion

The soil organic carbon content is extremely low in the country, far below the absolute standards for a productive soil environment (Braun and Muchugu 1997). The declining productivity of the soils is associated with loss of soil organic matter (Gelaw et al. 2014). The use of dung as a fuel instead of as a fertilizer reduced Ethiopia's agricultural GDP by 7% (Tamene et al. 2017). There is also strong competition in the use of crop residues as animal feed and cooking fuel that leaves very little for the soil (Zenebe 2007).

The depletion of organic matter is also caused by poor farming practices with insufficient use of organic inputs, excessive tillage, over grazing, and deforestation. Organic residues are rarely utilized for the purpose of conserving and improving organic matter; instead, they are mainly used for feed, building materials, and fuel, and therefore are not being returned to the soil. The traditional application of livestock dung has been decreasing overtime due to the competing use of dung as an energy source. The complete removal of crop residues from agricultural lands deprives the soil of its cover and exposes it to erosion.

Soil Fauna/Flora Depletion

The soil biota, including soil microbial biomass (soil flora) and soil fauna provide the means and regulate the transformation of organically bound nutrients into plant available forms through decomposition and mineralization. They also contribute to nutrient cycling and symbiotic processes in the rhizosphere. Furthermore, the process of litter decomposition is critical to maintain the proper functioning of natural and managed ecosystems. Beneficial soil organisms, if properly and effectively utilized, are the cheapest and most readily available options for the peasant farmers to increase agricultural productivity, through sustainable maintenance of soil fertility. However, there has been a decline in activity and diversity of soil fauna and flora, mainly due to organic matter and plant nutrient depletion, soil acidity, and topsoil erosion. Lack of attention on conservation of microbial biodiversity also contributes to declining soil fauna and flora.

Nutrient Depletion

Soil nutrient mining coupled with low fertilizer use is the main cause of soil fertility decline and consequently, the nutrient balances in the Ethiopian farming systems are generally negative (Haileslassie et al. 2005). Accordingly, the nutrient balances for nitrogen, phosphorus and potassium are –122, –13, and –82 kg ha^{-1} per year respectively signifying the large rates of macronutrient depletion.

Low soil OC; total N; available P, K, S; and micronutrients (B and Cu) are the major limiting factors to plant growth in some part of the country (Laekemariam and Kibret 2020). The most frequently reported and highly ranked limiting factors that contribute to poor crop yields in Ethiopia are nitrogen and phosphorous deficiencies, acidity, and low soil organic C content (Stewart et al. 2020). Micronutrient deficiencies are also part of the top limiting factors. The estimated annual nation-wide loss of nutrients through crop residue removals and dung burning is estimated to be equivalent of 200,000 MT of urea and 350,000 MT of DAP (MoARD 2005).

The reasons for multiple nutrient deficiency are primarily the agronomically unbalanced fertilizer use, whereby only N and P in the form of DAP (18–46% $N\text{-}P_2O_5$) and urea (46% N) are supplied until the introduction of compound and blended fertilizers based on soil fertility assessment by Ethiopian Agricultural Transformation Agency (ATA 2014, 2016). The application of N and P alone is believed to have exhausted the soil nutrient stocks since crops remove many more nutrients than just N and P. In particular, the uptake and deficiency of potassium and micronutrients (mainly Zn, Cu, B) can be accentuated due to continued application of N and P fertilizers alone (FAO 1984).

The decline in total organic matter content of the soils and inadequate levels of organic inputs result in nutrient exhaustion and lower crop yields. Additionally, macronutrient depletion by crop nutrient uptake, topsoil erosion, and crop residue removal also result in nutrient depletion from the soil. The nutrient loss due to the burning of crop residues and livestock manure is high in different regions (Tamene et al. 2017).

Fig. 9.12 Removal of maize stover for chickpea growth (relay cropping) in Kebena, near Wolkitie. *Source:* (Photo courtesy of Sheleme Beyene)

Removal of Organic Residues and Other Biomass

Ethiopia is among the top 10 countries in the world with high woodland removal from the total forest and wood-covered areas (FAO 2010). Decrease of biomass coverage results in limited alternative livestock feed and fuel, other than crop residue, leading to reduced crop residue usage for organic matter in farming fields. The coverage of grasslands is more than halved from 1980 to 2000, mainly due to uncontrolled grazing. On the other hand, the number of livestock has increased significantly from 58 million in 1995/6 to 107 million in 2009/10, which is almost double in 14 years (CSA 2011).

Severe organic matter depletion is driven by complete removal of crop residues from fields as livestock feed or burning for household energy. Most cultivated soil of Ethiopia is poor in organic matter content due to low amounts of organic materials applied to the soils and complete removal of the biomass from the fields (Yihenew 2002; Figs. 1.1 and 9.12). Animal dung is made into dung 'cake' and used for household energy. This deprives the soil of an important source of organic matter and nutrients. The removal of crop residues without sufficient replacement is a major reason for nutrient mining, causing nutrient deficiency and imbalance and low productivity of crops, particularly in erosion-prone regions (Tamene et al. 2017). There is also strong competition for biomass, with about 63, 20, 10, and 7% of cereal straws being used for feed, fuel, construction, and bedding purposes, respectively (Tamene et al. 2017).

Burning of dung cake and crop residues is common in Ethiopia due to a lack of widely-available and affordable fuel wood; dung cake has been reported to account for about 50% of households' fuel supply, particularly in the north and central highland cereal zones, and in some cases, manure is used as a source of supplementary cash income as a result (Bojo and Cassells 1995). It was also reported that 63% of cereal straws are used for feed, 20% for fuel, 10% for construction, and 7% for bedding (Gete et al. 2010).

Soil Structure Deterioration

Soil structure deterioration is a form of physical degradation of soils that is often difficult to observe directly. Formation of a surface crust when the soils are dry and the presence of a cultivation pan are often indicators of soil structural deterioration. These prevent seedling emergence and affect root growth and development. They also reduce the rate at which water can infiltrate the soil, and increase the amount of work that is required to cultivate the soil. Nowadays, soil structural deterioration (often called soil compaction) is increasingly becoming a critical problem in some parts of the country, particularly in acidic and waterlogged (Vertisols) areas. It leads to lower root penetration and reduced water entry and storage in the soil. This particular problem can be addressed using organic matter input applications and management methods and through the use of deep ploughs such as 'Tenkara Kende.'

Soil Erosion

Land degradation is a severe environmental problem across sub-Saharan Africa, and Ethiopia is among the most affected countries (Abiy 2008). Soil erosion is a major part of land degradation (Figs. 9.13 and 9.14) that affects the physical, chemical, and biological properties of soils and results in on-site nutrient loss and off-site sedimentation of water resources in Ethiopia (Huni 1993). The rate of soil erosion losses, 130 tons ha^{-1} for cultivated fields, is considered to be one of the highest in Africa (FAO 1986). The productive

Fig. 9.13 Eroded site at Konso, southern Ethiopia. *Source:* (Photo courtesy of Sheleme Beyene)

Fig. 9.14 Erosion affected Andosols at Toga, between Shashemane and Hawassa. *Source:* (Photo courtesy of Tesfaye Abebe)

land has been exposed to degradation and menace both economic and survival of the people (Genene and Abby 2014). The soil loss due to erosion in Ethiopian highlands was high, varying between 42 and 175.5 t/ha/yr (Adimassu et al. 2017).

Estimates indicate that 1.9–3.5 billion tons of soil is lost annually (Soil and Water Conservation 1992), and the resultant loss in soil nutrients reduced land, crop and livestock productivity, costing an estimated 2–3% of the country's annual agricultural GDP (WLE 2017). Additional estimate also indicates that out of the cultivated range and pasture lands in the country, which is close to 780,000 km^2, annual soil loss ranges from 1.3 to 7.8 billion metric tons (Tamirie 2005). This amounts to topsoil loss of 10–300 tonnes/hectare per year, which is among the highest rate in the world (Abebe et al. 2013). The degraded area on the highlands is about 27 million hectares, of which 14 million hectares are very seriously eroded, with 2 million hectares having reached 'a point of no return' (FAO 1984). Compared to some other East African and developing countries with similar topography and agricultural practices, Ethiopia ranked as the top fourth, with 31% of its total land eroded (FAO 2000). Soil erosion is the major cause of on-site topsoil loss (Figs. 9.13 and 9.14) and significant amounts of fertile topsoil are eroded, mainly due to water and wind erosion, leading to severe depletion of nutrients and organic matter in upstream areas and over-deposition in downstream ones.

High population pressure, continuous and steep slope cultivation, low vegetation cover, deforestation, and inadequate soil conservation practices cause annual soil loss of about 1.5 billion metric tons in the highlands of Ethiopia (Girma 2001). The ever-increasing land use change is aggravating the rates of soil erosion, soil fertility reduction, crop yield decline, and food insecurity (Tsegaye et al. 2010). Under agricultural conditions, it takes approximately 200 years to replenish 10 mm loss of topsoil. Ethiopia is listed as one of the most severely erosion-affected countries in the world (Gete et al. 2010).

The rugged topography in different parts of the country contributes to soil loss by erosion. The running water, if the site is unprotected, may erode soils on slopes and form thinner surface layer because the surface layer is consistently removed by erosion, whereas the increment in thickness of the surface layer at lower slopes can be attributed to the soil deposition at lower landscape position (Mulugeta and Sheleme 2010; Dinku et al. 2014; Sheleme 2017).

Salinity and Sodicity

Salt-affected soils cover, at least, 11 m ha that is nearly 10% of total land, and are dominantly found in Rift Valley Zone, Wabi Shebelle River basin, and various other lowlands and valley bottoms (Kefyalew and Kibebew 2016; Abebe et al. 2015; Meron 2007). Specifically, the salt-affected soils are concentrated in Melkassa, Melka Sadi, Melka Werer, Abaya state farm, and Dams in Mekelle plateau (Habtamu 2013). Also, the surface and sub-surface soils of the Meki-Ogolcha area that were irrigated by water sources of Lake Zeway and groundwater were found to be saline-sodic and sodic soils (Kefyalew and Kibebew 2016). This may be due to massive expansion of irrigated agriculture by saline water. For instance, it has been reported that of 4000 ha of irrigated lands at Melka Sedi, about 40, 16.9, and 0.02% are saline, saline-sodic and sodic soils, respectively, and 39% of the Abaya state farm is also salt-affected.

These areas are characterized by low levels of annual rainfall and high daily temperatures that led to high water evaporation rates and consequently contributed to high concentrations of soluble salts (Sileshi 2016). Most saline soils are concentrated in the plain lands of the Ogaden lowlands, the Denkil Plains and the Red Sea Coast, and various valley bottoms. At present, most of the irrigated large state farms producing export crops such as cotton, sugarcane, fruits, and vegetables are situated in these zones. However, following poor practice of irrigation management, salinity has emerged as a major problem responsible for reduction of land productivity and natural resources degradation (Wondimu et al. 2020). This problem has deleterious impact on soil fertility, which in turns reduces the crop production and soil productivity (Farifteh et al. 2006).

Saline-sodic soil contains both soluble salts and exchangeable sodium on soil colloidal complex, which affect the water infiltration, air movement, water holding capacity, root penetration, and seedling emergence (Ghafoor et al. 2008). Also, bulk density of the soil increases with increasing sodicity due to dispersive action of exchangeable Na^+ ions on soil colloids that result in soil compaction and altering the pore size distribution and decrease the total volume of soil (Berek et al. 2011).

Acidity

Vast areas of land in the western, southern, southwestern, northwestern, and even the central highlands of the country (which receive high rainfall) are affected by soil acidity (Mesfin 2007). Soil acidity affected about 41% of the cultivated land in the high rainfall areas (Agegnehu et al. 2021) whereby 13% is strongly acidic (pH < 4.5) and 28% is moderately to weakly acidic (pH 4.5–5.5) (Mesfin 2007).

Acidity could be the result of high rainfall accompanied with leaching, cation mining due to growing high-yielding crops, acidic soil parent material, and continuous application of acidifying fertilizers and a combination of those factors (Goulding 2016; Pearson 1975). Soil acidity affects crop production in many ways: (i) acid soils commonly fix phosphorus, reduce cation exchange capacity and cause the loss of basic cations including Ca, Mg, and Mo, making it unavailable for plant growth (Goulding 2016); (ii) crop yield is commonly low in strongly acidic soils due to the possible Al and Mn toxicities (Barber et al. 1988); (iii) soil acidity restricts choice of crops that could grow in the farming system, thereby affecting crop rotation and diversification of crops; and (iv) severe acidification could also cause non-reversible clay mineral dissolution and structural deterioration.

Crop yield loss due to acidity ranges from 20 to 80% depending on the degree of acidity, agronomic practices and agro-ecologies (Amede et al. 2019). Most of the areas affected by soil acidity occupy the highlands where wheat, maize, and teff are grown. Some barley and wheat-growing farmers in the central and southern highlands have shifted to producing oats, which is more tolerant to soil acidity than the usually grown crops (Wassie and Shiferaw 2009). The most acid soil-affected areas are found in the high rainfall parts of the west, northwest, southwest and southern parts of the country and are limited by the eastern escarpments of the Rift Valley (Mesfin 2007; Gete et al. 2010). The southwestern and northwestern parts of the country, where Nitisols and Alfisols dominate, are acidic to a level that could limit productivity (Tamene et al. 2017). In the soils of northwestern highlands, the pH values are below 5.0, with significant Al concentration (Abate et al. 2021). The problem is threatening given the area extent of acidic soils, particularly in the north-central and south-western highlands (Mesfin 2007).

Low Moisture Availability

The Association for Strengthening Agricultural Research in Eastern and Central Africa (ASARECA) in collaboration with USAID and ILRI reported that about 50% of Ethiopia's land is arid or semi-arid in 2012, characterized by low, erratic, and highly inconsistent rainfall levels with poor soil moisture and soil quality. Most rain-fed regions are exposed to low moisture content in soil during the dry season, lacking sufficient water supply during crop growth. Areas with rainfall of 100–800 mm, which are arid and semi-arid areas, are dry lands presenting difficulties for rain-fed agriculture. Also, the hyper-arid areas (called deserts) have annual rainfall of less than 100 mm. More than 40% of the country's total land falls under hyper-arid and dryland areas, which are going through severe water availability issues, much worse than most other surrounding countries (FAO 2000).

Surface cover with crop residues is necessary to achieve acceptable yield along with minimum tillage, particularly in low moisture stress areas. Reduced tillage and maintenance of surface cover with crop residues commonly improve soil water availability and increased grain and straw yields in semiarid areas (Mesfine et al. 2005). Incorporating crop residues has a positive effect, particularly when using inorganic fertilizers in improving rainwater use efficiency and soil tilth, and in minimizing the rate of soil erosion (Bationo et al. 1993).

Soil Pollution

Soil pollution or soil contamination is caused by the presence of xenobiotic (human-made) chemicals or other alterations in the natural soil environment. Specifically, it is often caused by industrial activity, agricultural chemicals, and/or municipal waste disposals. The rate and extent of application of these chemicals on agricultural land was not a major concern until recently. However, looking at the current rate of urbanization, industrialization, and the extent of agro-chemical use in Ethiopia, soil pollution will be a major concern in the future.

The status of soil pollution in Addis Ababa showed that the concentration of heavy metals, such as zinc (Zn), chromium (Cr), nickel (Ni), cobalt (Co), and lead (Pb), in the soil samples of the dumpsite and nearby open land were higher than the internationally acceptable limit for the soil (AGP 2011). From this report, it could be deduced that continuous application of all categories of solid waste on land results in degraded quality of the soil and stream water, accumulation of metals, and release of concentrated leachate to the environment, which further become a potential source of entry into food chain.

Although the government introduced the Environmental Impact Assessment (EIA) Proclamation, a variety of key problems hinder the effective implementation of the proclamation, including: lack of awareness and widespread misconception about EIA by the general public, lack of human and resource capacity to implement EIA, lack of regulations and rules to support implementation, and lack of coordination among sectorial and regional offices. Thus, lack of capacity and mechanisms for soil pollution control enforcement is the primary systemic bottleneck hindering the implementation of holistic soil pollution control mechanisms (Mellese and Mesfin 2008).

Waterlogging

In most parts of the level planes and gently sloping hills of the central highlands, where soils are heavy and dominated by smectite clay minerals, internal drainage is very low, causing a serious waterlogging problem during the main rainy season. Such soils cannot be used for crop production during the normal growth period unless they are properly drained. Instead, they are mostly left for grazing during the rainy season.

Waterlogged lands cannot be easily managed with the traditional plows owned by most farmers due to their workability problems. In some areas, they are used for growing some short-season crops that are planted at the end of the rainy season; particularly crops that can make use of the reserve moisture in the soil. Consequently, a large acreage of land is either out of production or gives very low yields. Vertisols are very vulnerable to the issue of waterlogging (CSA 2007) and one of the problematic soil types, covering about 12.6 million ha (10.3% of the land in the country), with 7.6 million ha of this in the highlands and a large proportion in agricultural areas (Kebede and Yamoah 2009). Although Vertisols are relatively fertile, they are underutilized in the country due to waterlogging problem in the highland areas.

Physical Land Degradation

Landslides, earth movement, and deep gullies are major issues related to physical land degradation. The rugged topography, the slopping and rolling land forms, overgrazed grasslands and farmlands after harvest time, de-vegetation and cultivation on steep lands, and heavy rains have created conditions ripe for excessive soil erosion on cultivated fields. Moreover, flooding and sediment deposits in low-lying areas and riverbanks also made some degradation problems, due to heavy rainfalls and uncontrolled runoff from the highlands (FAO 2000). About 50% of the highland area is subjected to severe-to-moderate erosion. This is about 28% of the total land that is severely degraded (mostly physical land degradation), which is a much higher compared to other countries. Of this, about 20% of the land is very severely degraded indicating the biotic function is completely destroyed and the land is non-reclaimable, whereas the remaining 8% is severely degraded and non-reclaimable at farm level.

(b) **Systemic Challenges**

The bottlenecks affecting the soil system exist not only at the soil level but also in the surrounding regulatory system. 'Systemic' bottlenecks refer to issues within the set of entities and institutions that form the country's soil system. These include aspects related to knowledge, capability, organization, and management systems. There are five major components in the system, which contain systemic bottlenecks hindering achievement of the vision. They are: soil information management; technology generation, dissemination and linkage; input value chain; strategic and regulatory framework; and organization and management systems.

Soil Information Management

The bottlenecks under soil information management include: lack of up-to-date information on soil resource data and exchange system, and absence of shared soil resource information database. Comprehensive, up-to-date data and knowledge base of the soil status across the nation is lacking. However, the Agricultural Transformation Agency (ATA 2014–16) presented soil fertility maps of the different regions of the country, whereby different nutrient blends were recommended based on soil testing. Given the fact that soil fertility condition undergoes through dynamic physical, chemical, and biological changes that directly or indirectly influence crop production and productivity, it will be misleading to base soil fertility inputs and management practice recommendations on this soil information data without periodical revision (IFPRI 2010). The understanding of the physical, chemical, and biological characteristic of soils in different geographies is critical for soil fertility and productivity improvement, as the lack of nationwide soil information inhibits the recommendation of local specific inputs/practices for soil health and fertility.

Another overarching constraint is the absence of up-to-date, comprehensive, and actionable soil data. The soil resource data available at ATA are not widely available. Although research and testing institutes and facilities exist under EIAR, Regional Agricultural Research Institutes (RARIs) and the Ministry of Agriculture and Rural Development (MoARD), they face challenges in maintaining adequate skilled staff as well as equipment and chemicals. This lack of a basic fact base—up-to-date, localized data covering all aspects of soil health—is a major constraint to intervention effectiveness (Gete et al. 2010).

There also exist a lack of institutions, mechanisms, management systems, and capacity to manage a database for sharing the vast amount of information generated in different parts of the country. The research institutes, soil testing laboratories, higher learning institutions, and various development partners have generated valuable soil data across. However, the data are not centrally compiled, and hence not accessible among the various actors causing a significant waste of resources and effort in the sector.

Technology Generation, Dissemination, and Linkage

Until recently, there were only two fertilizers, urea and diammonium phosphate (DAP), readily available in Ethiopia. These fertilizers contain only nitrogen and phosphorus and have been applied based on nationwide recommendations (normally 100 kg/ha of each) for all crops and soil types, irrespective of the diverse agroecological characteristics of the country (IFPRI 2010). It does not also consider the variations and fertility gradients encountered within a given field. The main reason for the continuation of this blanket application rate is the lack of soil test-based fertilizer recommendations. Investigations on the amount, type, method, and time of fertilizer application have been conducted by different institutions, including the MoA, EIAR, RBOAs, RARIs, development partners, and higher learning institutions. However, there have been three underlying problems in these early trials, which are: (1) limited soil testing to support the results of the fertilizer trials, (2) use of experimental designs which were not suitable for generation of soil analysis/crop yield response relationships, and (3) limited central coordination of efforts; resulting in the lack of development of a national recommendation or implementation strategy.

Despite significant increase in fertilizer use, the average fertilizer application rate in Ethiopia is in general lower than the recommended rate. This is due to various reasons including: low fertilizer/nutrient use efficiency; high price of fertilizer; constrained knowledge of farmers on how to use the fertilizers; acid soils in the highly-weathered soils; water logging in Vertisols; nutrient imbalance in alkaline and saline soils; and blanket fertilizer recommendation irrespective of soil type and crop variety (Tamene et al. 2017). Low levels of fertilizer application result in mining the soil nutrient stocks.

A soil technology registry for validation of the effectiveness of the technologies and their release has never been conducted. There are also no accompanying guidelines as to when, how, and where the technology can best be used as a method of releasing technologies. The absence of this system resulted in 1) a lack of farmer confidence in technologies that are intended to improve farm-level productivity, 2) duplication of efforts that spends time and resource to understand technologies that have already been tested, and 3) discouragement and reluctance by researchers.

Another problem is with respect to the emphasis given to soil fertility extension system. The natural resource management wing, which is responsible for guiding natural resource and soil health and fertility management practices, is not significant in size in the federal and regional extension structure. Additionally, limited research emphasis is given to soil health and fertility. These limitations on soil research led to limited generation of technology and practices that are fundamental to improving the soil health and fertility conditions. In addition, the facilities across the different research centers and laboratories are also insufficient, in terms of both variety and quality. The existing soil laboratories do not include the ideal conditions, in terms of their infrastructure, human capability, and management systems. The laboratories lack the required infrastructure, human capability, and management systems to provide state-of-the-art soil laboratory services and up-to-date recommendations.

The weak connection between research and extension, particularly research-extension-academia is also a problem in agriculture. The two core reasons for the problem are the gap in the management structure and the relatively weaker capacity of the extension system. Although the Research-Extension Liaison Committees (RELCs) were formed in 1986, both at the national and zonal levels, the RELCs had limited impact and did not last long enough to be of practical use. This was due to (1) lack of research centers and lacked the capacity to steer the extension role through staff development, support, and reward, (2) poor technical know-how and skills in monitoring and evaluating research and extension activities by local government officials, (3) funding constraints and absence of decision-making power of RELCs, (4) absence of farmers representations, (5) frequent changes in organizational structure of the Ministry, and (6) shortage of relevant technologies that were proven to provide directly measurable results or perceived benefits.

Input Value Chain

The issues of fertilizer value chain (e.g., weak distribution/handling system, capital provision, demand projection) are limiting the accessibility and the affordability of fertilizers for farmers. Fertilizer access to very remote areas is limited; and fertilizer is not available on time, or it is simply unavailable in these hard-to-reach areas (IFPRI 2010), which might be due to poor quality of feeder roads. Moreover, the domestic transportation system capacity is limited, making it difficult to deliver fertilizer before planting time ends (IFPRI 2012).

The dissemination of rhizobium as a bio-fertilizer is also at its infant stage, whereby only few strains of rhizobium, developed by the National Soil Testing Laboratory (NSTC) and Menegesha Biotech Industry, are being produced and distributed. There is no linkage between Research centers and registered private enterprises to promote dissemination of newly released strains. Strain management and handling at the woreda and FTC levels suffer from lack of facilities, such as refrigeration, and limited knowledge of best practices. This potentially leads to reduced effectiveness due to the death of strains prior to application.

The lime distribution system is still inefficient resulting in price build-up from the production site to the farmers' fields, mainly due to price of transportation the farm gate price is high for the farmers. Additionally, the distributed lime will remain unused by farmers turning the lime into unusable form, whereby the powder limes will be resource wastage for farmers.

There are also other inefficiencies that exist in the input value chain, especially for fertilizers. One example is the gap between demand and consumption. The amount of fertilizer carryover is considerable, especially over the last 3 years, accounting for nearly 50% of the imported volume. In addition, there are also lack of sufficient transportation infrastructure, strategic location of central warehousing, and weak last mile distribution.

Capital is the major resource that is constrained, and the underlying bottlenecks are the difficulty in accessing the capital, and the limited risk aversion mechanism for farmers. Credit and insurance services are critical in helping often cash-constrained farmers to purchase the required quantities of lime as well as complementary inputs (WLE 2017). The issue of lack of input credit for fertilizer for smallholders is a relatively recent problem that was caused by channeling input credit through 'regular' cooperatives. Although cooperatives have access to fertilizer using government credit guarantees, fertilizer access for farmers is primarily cash-based turning into a major issue should the quality of harvests decrease.

Organization and Management Systems

Lack of focus on soil research and inefficient coordination across enabling factor drivers has led to less soil research activities and less prime soil technology service provision. Soil resource information generation initiatives in Ethiopia have been fragmented and often redundant. Each activity tries to address issues separately, without implementing representative site selection across the country in a coordinated manner. Because of the lack of center of excellence for soil research data in the country, the available information generated by various institutions is scattered among various institutions (Belay 2003).

The soil testing laboratories across the nation are administered by different organizational systems and lack centralized management to solve cross-cutting issues. In addition to the human resources and infrastructure problems, a lack of coherence in the management system poses a significant problem on the laboratories. Specifically, some regional laboratories are under the management of their respective Regional Bureau of Agriculture (RBoAs) while others are under the management of the RARIs. Some of the reasons identified for this include: inadequate infrastructural services, difficulty in retaining experienced staff, no staff development plan in place, instruments have limited exposure to maintenance and spare parts, and reagents/consumables are in short supply (Mohammed 2011).

Strategic and Regulatory Framework

The growing demand for agricultural inputs in the country, particularly for inorganic fertilizers, and the increasing number of business societies engaged in the sector necessitates setting up laws and appropriate quality control mechanisms. Though the government recognized the need for ensuring fertilizer quality and fair distribution, which led to the Fertilizer Trade and Manufacturing Proclamation No. 137/98 on November 24, 1998, the enforcement mechanism was not established at the district level. Additionally, strategy and framework for proper land use planning is absent. The productive arable land is encroached by competing land use, especially through urbanization and industrialization as the economies of countries develop and the industrial focus shifts more to manufacturing from agriculture. This leads to loss of fertile soils and negatively affects the overall soil fertility of the nation, by losing the fertile areas and migrating to inferior soil for agricultural production.

Land use is a human-driven activity that is considered as one of the land's major characteristics (Lambin et al. 2003). The type and expansion rate of a given land use are relied on biophysical conditions, production preference, socio-economic status, demographic variables, political strategies, and cultural manners (Tadesse et al. 2014; Schmook and Vance 2009; Lambin et al. 2001; Duguma et al. 2010; De Fries et al. 2004) that operating across space and time (Kolb et al. 2013). Whereas, land use change (LUC) is an unspecified conversion between natural and artificial land use systems; and, it is among key drivers of environmental quality degradation (Kolb and Galicia 2018; Fearnside 2000) due to its direct relation with expansion and intensification of agriculture, grazing and settlement processes (Lambin et al. 2003; Lambin and Geist 2006).

Generally, soil fertility status varies spatially and is influenced by both land use and soil management practices (Sun et al. 2003). Change in land use is another factor affecting the soil chemical properties (Sheleme 2017; Dinku et al. 2014; Alemayehu and Sheleme 2013). Additionally, severe deforestation, steep relief conditions and excessive erosion hazards also result in low organic matter content of the soils (Eylachew 1999). Significant changes in soil quality are found following deforestation and the subsequent conversion of the land into different land uses (Islam and Weil 2000; Campos et al. 2007).

Environmental quality degradation caused by inappropriate land use is a problem whereby the conversion of forestland to cropland has got a big concern in specific (Matson et al. 1997) as it speeded up the land use variability for having sustainable agricultural production systems (Ayoubi et al. 2011) with sufficient land availability (Lambin and Meyfroidt 2011). Inappropriate land use change and agricultural management practices, such as imprudent cultivation pattern on steep and fragile soils that mainly characterized by extensive deforestation and erosion, are still the major causes of land degradation in the country (Mulugeta 2004; Solomon et al. 2002; Hurni 1988; Belay 2003). The extensive land use change happening now in the country is attributed to many causes such as drought, resettlement, land tenure system and governmental policies and strategies (Reid et al. 2000; Urgesa et al. 2016).

9.5.2 Opportunities for Managing and Restoration of Soil Health

Soil health is an important consideration in the Climate Resilient Green Economy (CRGE) Strategy, adopted by Ethiopia in 2011, setting the ambitious targets for Ethiopia to achieve both carbon neutral and middle income status by 2025 (ISD 2016). To rehabilitate acid soil, different options are available and these practices can help to return the unproductive soils to productive. There have been four lime crushers engaged in the production of lime, though the capacity of the crushers to meet the required amount is by far below the required level. Given an annual demand of 1,919,619 quintals, the capacity of these crushers is limited to only 98,280 quintals annually and lime use is restricted to about 5,100 ha. In GTP II, the Ethiopian government plans to rehabilitate about 226,000 ha of agricultural land by expanding the production, distribution, and promotion of lime by smallholder farmers (MoANR 2015).

As an alternative to lime, the application of biochar was also found to improve soil pH. Application of biochar and lime as a soil amendment significantly increased yield, even in the absence of fertilizer (Abewa et al. 2014; Agegnehu et al. 2016). Application of 12 t ha^{-1} biochar and 2 t ha^{-1} lime without fertilizer exceeds the full fertilizer rate without amendment in grain yield. However, increasing soil pH by the same magnitude would require application of more biochar than lime (Abewa et al. 2014). Integrating lime application with biochar can be beneficial. A combined application of moderate rates of biochar (e.g. 2 to 4%) with lime (an equivalent of exchangeable acidity or about 2 t ha^{-1}) could significantly improve soil quality and increase crop growth (Berek et al. 2011).

The annual production of crop residues in the country was significantly increased, from 6.3 million t in 1980 to about 19 million t, mainly due to the expansion of cultivated land, and hence increasing the availability of crop residue for soil amendment (CSA 2008). The addition of organic matter plays a key role in nutrient availability, soil water content and nutrient recycling by adding nutrients to the soil, influence mineralization-immobilization patterns, serving as an energy source for microbial activities and as precursors to soil organic matter, reducing the P adsorption of the soil, and reducing leaching of nutrients and making them available to crops over a longer period of time (Tamene et al. 2017).

Researchers developed technologies such as camber bed and broad-based furrow (BBF) maker to remove excess moisture and improve productivity of the soil. The use of BBF increased seed yield by 204 and 7.2% compared to Ridge-Furrow (RF) and Flat Bed (FB), respectively (Regassa 2008).

(a) Agricultural Policies of the Imperial, Derg, and Current Governments

A crude attempt was made to evaluate and measure land in connection with the allocation of land to soldiers during the reign of Emperor Tewodros (1855–1868). Land was categorized into four major groups from 'Highly productive' to 'Least productive'. The massive deforestation that took place during the time of Emperor Menelik (1889–1913) led to acute shortage of fuel wood and forced Menelik to shift his capital city from Ankober to Addis Alem, Holetta, Entoto and Addis Ababa successively. This situation also prompted him to issue some regulations with relation to land use and forest conservation. In order to implement these regulations, the Ministry of Agriculture was established in 1890.

During Emperor Haile Selassie's time, various catchment studies were undertaken by foreign consulting firms but no land use planning unit was established. When the Derg displaced Haile Selassie in 1974, it issued a proclamation that abolished the feudal land ownership system. To enforce the proclamation, it established the Land Distribution Department under the Ministry of Land Reforms and Settlement. The Ministry of Land Reform and Administration (MLRA) was established with land reform issues. In 1977, the MLRA and the Ministry of Agriculture (MoA) were amalgamated and the 'Land Use Planning and Distribution Department' was formed under the MoA. In 1979, agreement was reached between the Government of Ethiopia and UNDP/FAO to support land use planning and the Land Use Planning and Distribution Department of the MoA was renamed as Land Use Planning and Regulatory Department (LUPRD). Support from UNDP/FAO lasted from 1979 to 1990 in three phases through the Assistance to Land Use Planning Project (ALUP).

The Federal Democratic Republic of Ethiopia Constitution was enacted on December 8, 1994. This Constitution, under article 89 (5), underlines the Ethiopian government's duty to hold, on behalf of the People, land and other natural resources and to deploy them for their common benefit and development. Likewise, the Constitution, under Article 92, further provides that the design and implementation of development programs and projects shall not damage or destroy the environment. The Constitution also stipulates, under article 40(3), that the right to the ownership of rural and urban land as well as of all natural resources is to be exclusively vested in the state and in the peoples of Ethiopia. Thus, land is a common property of the Nations, Nationalities and Peoples of Ethiopia and shall not be subject to sale or to other means of exchange. This was done for the reasons of national interest and sustainability.

Since 1991, the government employed a number of agricultural policies to increase productivity and production in agriculture, efficiency in the processing and marketing chain that have a substantial positive effect on rural households' welfare. The policies include Market Liberalization, Structural Adjustment, Climate Resilient Green Economy Strategy, Agricultural-Led Industrialization, Sustainable Development and Poverty Reduction Program, Participatory and Accelerated Sustainable Development to Eradicate Poverty and successive Growth and Transformation Plans I and II to raise productivity in agriculture between 1991 and 2016. All subsidies and price support measures to agriculture were abolished. A structural adjustment program reduces the role of the government and increases the role of demand and supply forces in the allocation of resources in the Ethiopian economy. All these policy interventions have been implemented to increase agricultural productivity and production which, in turn, reduce poverty and food insecurity. Agricultural research and development, irrigation, access to credit and price support policies have great impacts on agricultural transformation if they are implemented jointly and successfully (Eicher 1995; Smale 1995).

The government has pledged to restore 15 million hectares of degraded lands by 2020 through investments in a range of land, soil, and water management practices. These investments improved farm incomes by 50% on average, doubled crop productivity and fodder availability and halved the risk of crop failure due to moisture stress and climate shocks in some watersheds located in Tigray, Oromia and Amhara regional states (Gebregziabher et al. 2016). Soil and water conservation efforts reduced stormwater runoff and sediment yield, and increased infiltration and groundwater recharge (WLE 2015).

The Agricultural Development Led Industrialization (ADLI) strategy set out agricultural intensification as a major pillar to increase food production and achieve food security at each household level. In this connection, the Growth and Transformation Plan (GTP) has led to land use changes due to agricultural landscape intensification (FDRE 2010) in the country. Currently, this strategy becomes the major driver of land use changes throughout the country. The land use changes induced by government policies and strategies have now, at least, partially been taken up by the farming communities (Nyssen et al. 2009).

The government recognized the need for ensuring fertilizer quality and fair distribution, which led to the Fertilizer Trade and Manufacturing Proclamation No. 137/98 on November 24, 1998. However, the enforcement mechanism was not established at the district level. As a result, there are cases where underweight and low-quality fertilizers (unwholesome in physical properties) are being distributed to farmers.

The establishment of Research-Extension Farmer Advisory Councils (REFACs) at three levels: the federal, regional, and zonal level (research center-based) followed in the late 90 s was an enabling forum for knowledge transfer. However, there remains significant room for improvement in coordinating stakeholders from research and higher learning institutions, governmental and non-governmental organizations, the private sector, regional and international organizations at all levels. Research findings and conclusions largely remain within journals and not in the hands of extension workers or smallholder farmers (Gete et al. 2010).

In 2008, an Agricultural and Rural Development Partners Linkage Advisory Council (ARDPLAC) was established at the federal, regional, zonal, and district levels. The higher learning institutes should also be actively engaged into the chain, supporting both the research and the extension aspects. Research-Extension Liaison Committees (RELCs) were also formed in 1986, both at the national and zonal levels, although the available evidence shows that both the national and zonal level RELCs had limited impact and did not last long enough to be of practical use. The current field-level extension service in Ethiopia has a strong foundation of Farmers Training Centres (FTCs) with 62,764 trained Development Agents (DAs) and 45,812 staff. Each Kebele has at least one FTC, staffed by three DAs, supported by an itinerant DA covering three FTCs. Each FTC also has one specialist each in the areas of livestock, crops, and natural resource management.

The government established the project 'The Ethiopian Soil Information System (EthioSIS)' under the overall Soil Fertility Road Map of MoA mainly to conduct soil fertility mapping to reformulate fertilizer recommendations in 2010 to address many of the soil fertility management and fertilizer use-related problems. Huge efforts and investments were made to prepare digital maps of soil fertility parameters from data collected through intensive soil sampling campaigns in cultivated and arable lands. The approach was limited to auger-based composite sampling mainly in the top 20 cm for annual crops and in some cases at 20–50 cm.

The digital soil fertility maps are predicted soil property maps developed using Digital Soil Mapping (DSM)-Soil Spatial Prediction Models (SSPM). The ATA/EthioSIS project finalized and released the digital soil fertility and fertilizer formulation maps/Atlas for almost all regions and has collected approximately 100,000 soil datasets for about 13 soil nutrients (N, P, S, K, Ca, Mg, Na, Fe, Cu, Zn, B, Mn, Si) and 5 soil properties (pH, EC, CEC, OC, $CaCO_3$) in the top 20 cm soil depth (ATA 2016).

(b) **Community-based Participatory Watershed Development Guidelines**

The absence of an overarching national land use policy resulted in the overall degradation of the nation's land use resources, jeopardizing their ability to sustain the livelihood of current and future generations including unprecedented damages to the environment. The Ministry of Agriculture and Rural Development developed a community-based participatory watershed guideline in 2005 (MoARD 2005).

The Guideline aims to build upon existing community-based participatory watershed efforts to harmonize and consolidate planning procedures at the grass-roots level. The intent is to provide Development Agents (DAs) and rural communities with a workable and adaptable planning tool, as well as, to provide practical guidance on the correct selection of technologies under different conditions and their sequentially correct implementation. Local Level Participatory Planning Approach (LLPPA) was developed for DAs, as a practical approach focusing mostly on integrated NRM interventions, productivity intensification measures, and small-scale community infrastructure such as water ponds and feeder roads. Several NGOs and bilateral organizations adopted participatory land use-planning approach in the last decade in their respective areas of intervention and in close collaboration with government partners. These gave birth to the current community-based participatory watershed development guidelines.

(c) **Natural Resources and Environmental Policy**

The Environmental Protection Authority in collaboration with the Ministry of Economic Development and Cooperation formulated Environment Protection Policy to (1) improve and enhance the health and quality of life of the Ethiopians and (2) promote sustainable social and economic development through the sound management and use of natural, human-made and cultural resources and the environment as a whole so as to meet the needs of the present generation without compromising the ability of future generation to meet their own needs. The Policy considers the adoption of 'conservation culture' in environmental matters and integration of natural resource and environmental management activities laterally across all sectors.

The sectoral environmental policies also include promoting the use of appropriate organic matter and nutrient management for improving soil structure, nutrient status and microbiology in improving soil conservation and land husbandry; and safeguarding the integrity of the soil and protecting its physical and biological properties, through management practices for the production of crops and livestock which pay particular attention to the proper balance in amounts of chemical and organic fertilizers, including green manures, farmyard manures, and compost. Ensuring that planning for agricultural development incorporates economic cost–benefit analysis and the potential costs of soil degradation through erosion and salinization as well as soil and water pollution; ensuring that inputs shall be as diverse and complementing as the physical, chemical and biological components of the soil requirement, and shall not focus solely on a quick and transitory increase in plant nutrients to the long-term determent of soil structure and microbiology.

Control of hazardous materials and pollution from industrial waste, atmospheric pollution and climate change, environmental economics and impact assessment and monitoring were also included. The implementation of the policy was enacted by the Proclamation No. 295/2002 and No. 299/2002. Following the provision of the environment policy, the government introduced the Environmental Impact Assessment (EIA) Proclamation. The proclamation envisages the issuance of specific directives and guidance that further specify the implementation process requiring the Environmental Protection Authority (EPA) to identify categories of projects likely to have negative impact, and thus require an EIA.

(d) **Strategic Investment Framework for Sustainable Land Management**

The Ethiopian Sustainable Land Management Investment Framework (ESIF) provides a holistic and integrated strategic planning framework under which government and civil society stakeholders can work together to remove the barriers, and overcome the bottlenecks, to promoting and scaling up sustainable land management (SLM). The ESIF has been formulated with the goal of serving as a national-level strategic planning framework that is to be used to guide the prioritization, planning and implementation, by both the public and private sector, of current and future investments in SLM with the aim of addressing the interlinked problems of poverty, vulnerability and land degradation at the rural community level (Daniel 2008).

The ESIF is planned to be implemented in three phases, over a 15-year period (phase 1: 2009–2013, Phase 2: 2014–2018, and Phase 3: 2019–2023). Activities to be implemented under the auspices of the ESIF would fit within one, or more, of six broad (and interrelated) component areas, namely: (i) investment in field-based projects and programs for promoting and scaling up SLM; (ii) improving land administration and certification system; (iii) building the capacity of public and private sector SLM advisory and other support services providers; (iv) improving the enabling policy, legal, institutional and financial environment for SLM; (v) building the ESIF SLM Knowledge Base; and (vi) management and implementation of the ESIF (MoA 2019), and these need to be implemented in integration.

9.6 Conclusions

Ethiopia has diverse climate, topography, geology, soils, hydrology, and culture, which affects the formation/occurrence, distribution, level of degradation and management of natural resources including soils. These natural resources have been subjected to degradation over centuries due to severe deforestation, low vegetative cover and associated soil loss, and unbalanced crop, and livestock production, overexploitation, misuse and mismanagement resulting in reduction in soil health. The diversity of soil types and their problems across the various landscapes and associated factors reveal the need to develop and implement soil management practices that are relevant to context-specific problems. It is necessary to move the soil management practice from the ordinary blanket application of few fertilizer and related technology packages to a management that is based on the site-specific problems. However, given the complexity of soil problems, although soil management may not directly correspond to the types of soils as they appear in various classification systems like the WRB, the similarity in the characteristics of soil types can be considered in clustering soils that can be managed in similar ways. For instance, acidic soils such as Nitisols, Acrisols, Lixisols, and Alisolscan principally be managed through liming and fertilization while salt-affected soils like Calcisols, Gypsisols, Solonchack need to be leached the to remove dominant soluble salts responsible for the problem. On the other hand, soils that are characterized by shallow depth like Leptosols, Cambisols and Regosols require soil and water conservation practices in addition to other site-specific management practices. Soils with drainage problems like Vertisols are principally managed through drainage of excess water by using implements like BBM. In addition to these cluster-based context-specific management practices, it is worth implementing practices like integrated soil fertility management (ISFM), which has proven beneficial especially in Amhara and Oromia regions where average yield increase of around 38% has been reported with even more benefits in acidic soils.

Although there are a number of soil levels and/or systemic challenges in managing the restoration of soil health in Ethiopia, there are also opportunities involving the development of various land management and regulatory policies and guidelines that contribute to restoration of soil health from the time of Minilik regime up to the current regime. It is, therefore, necessary to properly exploit the existing opportunities in tackling the challenges of soil health restoration. For instance, in order to reduce degradation, it is of paramount importance to use lands according to their capacities based on the principles of land use planning. Particularly, the land use patterns in the mountainous parts of the country need to be changed and agricultural development activities need to move to more favorable soils.

References

Abate N, Kibret K, Gebrekidan H, Esayas A (2014) Characterization and classification of soils along the toposequence at the Wadla Delanta Massif, north central highlands of Ethiopia. J Ecol Nat Environ 6(9):304–320

Abate E, Hussein S, Laing M, Mengistu F (2017) Soil acidity under multiple land-uses: assessment of perceived causes and indicators, and nutrient dynamics in small-holders mixed-farming system of northwest Ethiopia. Acta Agriculturae Scandinavica, Section B Soil& Plant Sci 67(2):134–147. https://doi.org/10.1080/09064710.2016.1230227

Abate S, Belayneh M, Ahmed F (2021) Reclamation and amelioration of saline-sodic soil using gypsum and halophytic grasses: Case of Golina-Addisalem irrigation scheme, Raya Kobo Valley, Ethiopia. Cogent Food Agricult 7(1):1859847

Abebe M (1998) Nature and management of Ethiopian soils. Alemaya University of Agriculture, Ethiopia p 272

Abebe S, Hurni H, Gete Z (2013) A review on soil carbon sequestration in Ethiopia to mitigate land degradation and climate change. J Environ Earth Sci 12:1–15

Abebe F, Alamirew T, Abegaz F (2015) Appraisal and mapping of soil salinity problem in Amibara irrigation Farms, middle awash basin, Ethiopia. Int J Innov Sci Res 13:298–314

Abebe M, Jutzi SC (1989) The joint project on improved management and utilization of dark clay soils in Ethiopia: retrospects and prospects. In: Tekalign et al (1993) Improved management of vertisols for sustainable crop-livestock production in the Ethiopian highlands: Synthesis Report 1986–92

Abebe F (2012). Appraisal and mapping of soil salinity problems in Amibara area of middle awash basin. M.Sc thesis, School of Graduate Studies, Haramaya University. Haramaya, Ethiopia

Abewa A, Yitaferu B, Yihenew GS, Tadele Amare T (2014) The role of biochar on acid soil reclamation and yield of teff (eragrostis tef [Zucc] trotter) in Northwestern Ethiopia. J Agric Sci 6(1):1–12. https://doi.org/10.5539/jas.v6n1p1

Abiy T (2008) Area closure as a strategy for land management: a case study at Kelala Dalacha enclosure in the Central Rift Valley of Ethiopia. M.Sc thesis submitted to School of Graduate Studies &Environmental Science, Addis Ababa University, Ethiopia

Adane A, Gebrekidan H, Kibret K (2019) Assessment of the pysicochemical status soils and irrigation water qualities in eastern Ethiopia. Forest Res Eng Int J 3(3):91–104

Adimassu Z, Langan S, Johnston R, Mekuria W, Amede T (2017) Impacts of soil and water conservation practices on crop yield, run-off, soil loss and nutrient loss in Ethiopia: review and synthesis. Environ Manage 59(1):87–101. https://doi.org/10.1007/s00267-016-0776-1

Adisu T, Wogi L, Feyisa T (2019) Soybean (Glycine max L.) response to lime and vermicompost amelioration of acid Nitosols of Assosa, North Western Ethiopia. International Journal of Plant $ Soil Science, 27:1–18

Agegnehu G, Amede T (2017) Integrated soil fertility and plant nutrient management in tropical agro-ecosystems: a review. Pedosphere 27 (4):662–680

Agegnehu G, Bekele T, Tesfaye A (2005) Phosphorus fertilizer and farmyard manure effects on the growth and yield of faba bean and some soil chemical properties in acid Nitisols of the central highlands of Ethiopia. Ethiop J Nat Resourc 7(1):23–39

Agegnehu G, Nelson PN, Bird MI (2016) Crop yield, plant nutrient uptake and soil physicochemical properties under organic soil amendments and nitrogen fertilization on Nitisols. Soil Tillage Res 160:1–13

Agegnehu G, Tilahun A, Teklu E, Chilot Y, Henry C, Tyler R, Nosworthy MG, Sheleme B, Gudeta WS (2021) Extent and management of acid soils for sustainable crop production system in tropical agrosystem: a review. Acta Agriculturae Scandinavica, Section B—Soil Plant Sci 71. https://doi.org/10.1080/09064710.1954239

AgegnehuG, Yirga C, Erkossa, T (2019) Soil acidity management. Ethiopian Institute of Agricultural Research (EIAR). Addis Ababa, Ethiopia

AGP (2011) Agricultural growth program (AGP) of Ethiopia: baseline report, Addis Ababa

Alemayehu K, Sheleme B, Schoenau J (2016) Characterization of problem soils in and around the south central Ethiopian Rift Valley. J Soil Sci Environ Manag 7(11):191–203

Alemayehu K, Sheleme B (2013) Effects of different land use systems on selected soil properties in south Ethiopia. J Soil Sci Environ Manag 4(5):100–106. https://doi.org/10.5897/JSSEM12.20

Alemu B, Lule D (2018) Yield and agronomic performances of desi type chickpea genotypes against acidic soil of Western Ethiopia. J Agricult Biotechnol Sustain Dev 10(6):116–121

Alemu G, Desalegn T, Debele T, Adela A, Taye G, Yirga C (2017) Effect of lime and phosphorus fertilizer on acid soil properties and barley grain yield at Bedi in Western Ethiopia. Afr J Agric Res 12:3005–3012

Amanuel G, Kuhene RF, Tanner DG, Vlek PLG (2000) Biological nitrogen fixation in faba bean (Vicia faba L.) in the Ethiopian highlands as affected by P fertilization and inoculation. Biol Fert Soils 32:353–359. https://doi.org/10.1007/s003740000258

Amede T, Schulz S, Warner J, Tolera S (2019) Managing acid soils for reclaiming livelihoods in Ethiopia. ICRISAT-GIZ, Addis Ababa

Araya T, Nyssen J, Govaerts B, Deckers J, Cornelis WM (2015) Impacts of conservation agriculture-based farming systems on optimizing seasonal rainfall partitioning and productivity on vertisols in the Ethiopian drylands. Soil Tillage Res 148:1–13

Asamenew G, Beyene H, Negatu W, Ayele G (1993) A survey of the farming systems of Vertisol areas of the Ethiopian highlands

Asfaw E, Nebiyu A, Bekele E, Ahmed M, Astatkie T (2019) Coffee-husk biochar application increased AMF root colonization, P accumulation, N2 fixation, and yield of soybean grown in a tropical Nitisol, southwest Ethiopia. J Plant Nutr Soil Sci 182 (3):419–428

Worku A, Bedadi B (2016) Studies on soil physical properties of salt-affected soil in Amibara area, Central Rift Valley of Ethiopia. Int J Agricult Sci Nat Resourc 3(2):8–17

Ashenafi W, Melese M, Kidan HG (2016) Impact of gypsum and sulfuric acid application on cotton yield under saline sodic soil condition in Melka Sadi irrigated farm. Academia J Agricult Res 4 (2):91–95

Asmamaw M, Haile A, Abera G (2018) Characterization and classification of salt affected soils and irrigation water in Tendaho sugarcane production farm, North-Eastern Rift Valley of Ethiopia. Afr J Agric Res 13(9):403–411

Assefa E (2002) Characteristics, classification and potentials of soils in Werkaryia area, South Welo, Ethiopia. SINET: Ethiop J Sci 25 (1):45–70

Assen M, Tegene B (2008) Characteristics and classification of the soils of the plateau of SIMEN Mountains National Park (SMNP), Ethiopia. SINET: Ethiop J Sci 31(2):89–102. https://doi.org/10.4314/sinet.v31i2.66544

Assen M, Yilma S (2010) Characteristics and classification of the soils of Gonde micro-catchment, Arsi highlands, Ethiopia. SINET: Ethiop J Sci 33(2):101–116

Astatke A, Saleem MM (1998) Effect of different cropping options on plant-available water of surface-drained Vertisols in the Ethiopian highlands. Agric Water Manag 36(2):111–120

ATA, Agricultural Transformation Agency (2014) Soil fertility status and fertilizer recommendation atlas for Tigray Regional State, Addis Ababa, Ethiopia

ATA, Agricultural Transformation Agency (2016) Soil fertility status and fertilizer recommendation atlas of Amhara, Oromiya and SNNP Regional States, Addis Ababa, Ethiopia

Auge KD, Assefa TM, Woldeyohannes WH, Asfaw BT (2018) Potassium dynamics under enset (Ensete ventricosom cheesman) farming systems of Sidama zone, Southern Ethiopia. J Soil Sci Environ Manag 9(4):47–58

Ayalew A (2011) The influence of applying lime and NPK fertilizers on yield of maize and soil properties on acid soil of Areka, Southern Region of Ethiopia. Innov Syst Des Eng 2(7):33–42

Ayenew T, GebreEgziabher M, Kebede S, Mamo S (2013) Integrated assessment of hydrogeology and water quality for groundwater-based irrigation development in the Raya Valley, northern Ethiopia. Water Int 38(4):480–492

Ayenew B, Tadesse AM, Kibret K, Melese A (2018) Phosphorous status and adsorption characteristics of acid soils from Cheha and Dinsho districts, southern highlands of Ethiopia. Environ Syst Res 7 (1):1–14

Balba AM (2018) Management of problem soils in arid ecosystems. CRC Press (1995)

Bambara S, Ndakidemi PA (2010) The potential roles of lime and molybdenum on the growth, nitrogen fixation and assimilation of metabolites in nodulated legume: a special reference to Phaseolus vulgaris L. Afr J Biotech 9(17):2482–2489

Barber S, Mackay A, Kuchenbuch R, Barraclough P (1988) Effects of soil temperature and water on maize root growth. Plant Soil 111:267–269

Basak BB, Rakshit R (2015) Chemistry of macronutrients fixation in acid soils

Basga SD, Madi OP, Balna J, AbibFC TD, Njiemoun A (2018) Sandy soil fertility restoration and crops yields after conversion of long-term Acacia Senegal planted fallows in North Cameroon. Afr J Agric Res 13(40):2154–2162

Basso AS, Miguez FE, Laird DA, Horton R, Westgate M (2013) Assessing potential of biochar for increasing water-holding capacity of sandy soils. Gcb Bioenergy 5(2):132–143

Bationo A, Christianson CB, Klaij MC (1993) The effect of crop residue and fertilizer use on pearl millet yields in Niger. Fert Res 34 (3):251–258

Bedada W, Lemenih M, Karltun E (2016) Soil nutrient build-up, input interaction effects and plot level N and P balances under long-term addition of compost and NP fertilizer. Agr Ecosyst Environ 218:220–231. https://doi.org/10.1016/j.agee.2015.11.024

Bekele A, Kibret K, Bedadi B, Yli-Halla M, Balemi T (2018) Effects of lime, vermicompost, and chemical P fertilizer on selected properties of acid soils of Ebantu District, Western Highlands of Ethiopia. Appl Environ Soil Sci, Article ID 8178305 13 p

BekereW, Kebede T, Dawud J (2013) Growth and nodulation response of soybean (Glycine max L.) to lime, Bradyrhizobium japonicum and nitrogen fertilizer in acid soil at Melko, South western Ethiopia. Int J Soil Sci 8(1):25

Belay K (2003) Agricultural extension in Ethiopia: the case of participatory demonstration and training extension system. J Soc Dev Afr 18(1):49–83

Berek AK, Hue N, Ahmad A (2011) Beneficial use of Biochar to correct soil acidity. Hānai'Ai/The Food Provider

Berhanu B, Assefa M, Melesse B, Yilma S (2013) GIS-based hydrological zones and soil geo-database of Ethiopia. CATENA 104:21–31

Beyene S (2017) Topographic positions and land use impacted soil properties along Humbo Larena-Ofa Sere toposequence, Southern Ethiopia. J Soil Sci Environ Manag 8(8):135–147

Bojo J, Cassells D (1995) Land degradation and rehabilitation in Ethiopia: a reassessment.The World Bank

Boke S, Fekadu A (2014) Lime and NPK effect on soil acidity and yield of barley in different acid soils of southern region, Ethiopia. Int J Nat Sci Res 2(7):113–122

Bolan NS, Adriano DC, Curtin D (2003) Soil acidification and liming interactions with nutrient and heavy metal transformation and bioavailability. Adv Agron 78(21):5–272

Bore G, Bedadi B (2016) Response wheat (Triticum aestivum L.) to liming of acid soils under different land use systems of Loma Woreda, Dawuro Zone, Southern Ethiopia. J Environ Earth Sci 6:99–108

Braun AX, Muchugu EL (1997) Maintenance and improvement of soil productivity in the highlands of Ethiopia, Kenya, Madagascar and Uganda, Geomorphology and Soils ICRAF (International Center for Agricultural Forestry)

Bruand A, Hartmann C, Lesturgez G (2005) Physical properties of tropical sandy soils: A large range of behaviours. In: Management of tropical sandy soils for sustainable agriculture. A holistic approach for sustainable development of problem soils in the tropics

Brunner I, Sperisen C (2013) Aluminum exclusion and aluminum tolerance in woody plants. Front Plant Sci 4:172

Campos AC, Oleschko KL, Etchevers JB, Hidalgo CM (2007) Exploring the effect of changes in land use on soil quality on the eastern slope of the Cofre de Perote Volcano (Mexico). For Ecol Manag 248(3):174–182. https://doi.org/10.1016/j.foreco.2007.05.004

CGIAR Research Program on Water, Land and Ecosystem (WLE) (2017) Healthy soils for productive and resilient agricultural. International Water Management Institute (IWMI). https://doi.org/10.5337/2017.211

Chekol W, Mnalku A (2012) Selected physical and chemical characteristics of soils of the middle awash irrigated farm lands, Ethiopia. Ethiop J Agricult Sci 22(1):127–142

Chen ZC, Yamaji N, Motoyama R, Nagamura Y, Ma JF (2012) Up-regulation of a magnesium transporter gene OsMGT1 is required for conferring aluminum tolerance in rice. Plant Physiol 159(4):1624–1633

CIAT (2016) Big win: Soil Organic Carbon: absorbing carbon back into the soil

Clark GJ, Dodgshun N, Sale PWG, Tang C (2007) Changes in chemical and biological properties of a sodic clay subsoil with addition of organic amendments. Soil Biol Biochem 39(11):2806–2817

Condron LM, Tiessen H, Trasar-Cepeda C, Moir JO, Stewart JWB (1993) Effects of liming on organic matter decomposition and phosphorus extractability in an acid humic Ranker soil from northwest Spain. Biol Fertil Soils 15(4):279–284

CSA, Central Statistical Agency (2007) International Food Policy Research Institute (IFPRI) and Ethiopian Development Research Institute (EDRI) Atlas of Ethiopian Rural Economy. CSA, IFPRI and EDRI. Addis Ababa

CSA, Central Statistical Agency (2008) Statistical abstracts. Federal Democratic Republic of Ethiopia, Addis Ababa

CSA, Central Statistical Agency (2011) Statistical abstracts. Federal Democratic Republic of Ethiopia, Addis Ababa

Daniel D (2008) Sustainable land management and Ethiopian strategic investment framework, Nile BDC Workshop, Addis Ababa, April 8, 2011

Das R, Saha D (2014) Effect of liming on the changes of different forms of potassium in an acid soil treated with N and K fertilizers. J Indian Chem Soc 91:1–7

Davis K, Nkonya E, Kato E, Mekonnen DA, Odendo M, Miiro R, Nkuba J (2012) Impact of farmer field schools on agricultural productivity and poverty in East Africa. World Dev 40:402–413. https://doi.org/10.1016/j.worlddev.2011.05.019

Dawit A, John T (2020) What is the future for rice as a strategic crop in Ethiopia? APRA Working Paper 44. Future Agricultures. https://www.future-agricultures.org/blog/what-is-the-future-for-rice-as-a-strategic-crop-in-ethiopia/

De Fries RS, Foley JA, Asner GP (2004) Land-use choices: balancing human needs and ecosystem function. Front Ecol Environ 2(5):249–257

Berhanu Debele (1985a) The Vertisols of Ethiopia: Their properties, classification and management. In: Proceedings of the fifth meeting of the Eastern African soil correlation and land evaluation

sub-committee, F.A.O., (Ed.), World Soil Resources Report, vol 56, Rome, pp 31–54
Debele B (1985b) The vertisols of Ethiopia: their properties, classification and management
Dejene T, Tana T, Urage E (2016) Response of common bean (Phaseolus vulgaris l.) to application of lime and phosphorus on acid soil of Areka, southern Ethiopia. J Nat Sci Res 16(19):90–100. ISSN 2224-3186 (Paper), ISSN 2225-0921
Demiss M, Beyene S (2010) Characterization and classification of soils along the toposequence of kindo koye watershed in Southern Ethiopia East African. J Sci 4(2):65–77
Demissie W, Kidanu S, Cherukuri VR (2017) Effect of integrated use of lime, blended fertilizer and compost on productivity, nutrient removal and economics of barley (Hordeum vulgare L.) on acid soils of high lands in West Showa Zone of Ethiopia. Int J Life Sci 5 (3):311–322
Dida G, Etisa D (2019) Effect of lime and compost application on the growth and yield of common bean (Phaseolus Vulgaris L.): a review. Adv Oceanogr Marine Biol 1(3):2019. AOMB. MS. ID, 512
Dinku D, Sheleme B, Nand R, Walley F, Teleab SG (2014) Effects of topography and land use on soil characteristics along a toposequence of ele subwatershed in southern Ethiopia. CATENA 115:47–54
Duguma L, Hager H, Sieghardt M (2010) Effects of land use types on soil chemical properties in smallholder farmers of central highland Ethiopia. Ekologia 29(1):1–14
EIAR Ethiopian Institute of Agricultural Research (2015) Annual Research Report. Ethiopian Institute of Agricultural Research, Addis Ababa, Ethiopia
Eicher CK (1995) Zimbabwe's maize-based green revolution: preconditions for replication. World Dev 23(5):805–810
Elias E (2016) Soils of the Ethiopian highlands: geomorphology and properties. CASCAPE Project, ALTERA, Wageningen University and Research Center (Wageningen UR). The Netherlands. 385 pp
Elias E (2017) Characteristics of Nitisol profiles as affected by land use type and slope class in some Ethiopian highlands. Environ Syst Res 6(1):1–15
Elias E, Agegnehu G (2020) Mineralogical and physicochemical properties of Nitisols in the Ethiopian highlands. Ethiop J Agricult Sci 30(4):97–119
Endris S, Mohammad MJ (2007) Nutrient acquisition and yield response of Barley exposed to salt stress under different levels of potassium nutrition. Int J Environ Sci Technol 4(3):323–330
EPA, Environmental Protection Authority (1998) Background information on drought and desertification in Ethiopia. Environmental-protection authority of Ethiopia, Addis Ababa P9
Erkossa T, Stahr K, Gaiser T (2006) Soil tillage and crop productivity on a vertisol in Ethiopian highlands. Soil Tillage Res 85(1–2):200–211
Ermias A, Hussein S, Laing M, Mengistu F (2013) Aluminum tolerance in cereals: a potential component of integrated acid soils management in Ethiopia. Ethiop J Nat Resourc 13(1):43–66
Ewulo BS (2005) Effect of poultry dung and cattle manure on chemical properties of clay and sandy clay loam soil. J Anim Vet Adv 4 (10):839–841
Eylachew Z (1999) Selected physical, chemical and mineralogical characteristics of major soils occurring in Chercher highlands, eastern Ethiopia. Ethiop J Nat Resourc 1(2):173–185
Fageria NK, Baligar VC (2005) Enhancing nitrogen use efficiency in crop plants. Adv Agron 88:97–185
Fageria NK, ZimmermannFJP BVC (1995) Lime and phosphorus interactions on growth and nutrient uptake by upland rice, wheat, common bean, and corn in an Oxisols. J Plant Nutr 18(11):2519–2532
Fantaw A (2007) An overview of Salt-affected soils and their Management status in Ethiopia. A paper presented in the 3rd International Workshop on Water Management project, Haramaya University, Ethiopia
Farifteh J, Farshad A, George RJ (2006) Assessing salt-affected soils using remote sensing, solute modelling, and geophysics. Geoderma 130(3–4):191–206. https://doi.org/10.1016/j.geoderma.2005.02.003
FAO, Food and Agriculture Organization of the United Nations, (1984) Geomorphology and soils of Ethiopia. In: Assistance to land use planning and provisional soil association map of Ethiopia
FAO, Food and Agriculture Organization of the United Nations (1986) Ethiopian Highlands Reclamation Study, Final Report, vol 1
FAO, Food and Agriculture Organization of the United Nations (2000) Land resource potential and constraints at regional and country levels. World Soil Resources Reports 90. Land and Water Development Division, Food and Agriculture Organization of the United Nations, Rome
FAO, Food and Agriculture Organization of the United Nations, (2001) Lectore notes on the major soils of the world. World Soil Resources Reports 94, FAO, Rome
FAO, Food and Agriculture Organization of the United Nations (2010) Global forest resources assessment 2010 main report. FAOUN, Rome, p 2010
FDRE, Federal Democratic Republic of Ethiopia (2010). Growth and transformation plan (2010/11–2014/15). Addis Ababa
Fearnside PM (2000) Global warming and tropical landuse change: greenhouse gas emissions from biomass burning, decomposition and soils in forest conversion, shifting cultivation and secondary vegetation. Clim Change 46(1–2):115–158. https://doi.org/10.1023/A:1005569915357
Fekadu M (2018) Progress of soil acidity management research in Ethiopia. Adv Crop Sci Tech 6(4):1–8
Fekadu E, Kibret K, Bedadi B, Melese A (2018a) Characterization and classification of soils of yikalo subwatershed in lay gayint district, northwestern highlands of Ethiopia. Euras J Soil Sci 7(2):151–166
Fekadu E, Kibret K, Bedadi B, Melese A, Yitaferu B (2018b) Organic and inorganic amendments on soil chemical properties at different period of incubation of acidic soil. Euras J Soil Sci 7(3):273–283
Fekadu E, Kibret K, Melese A (2019) Integrated acid soil management for growth, nodulation, and nutrient uptake of Faba Bean (Vicia faba L.) in Lay Gayint District, Northwestern Highlands of Ethiopia. Int J Agron 2019:1–10
Fekadu E, Kibret K, Melese A, Bedadi B (2018c) Yield of faba bean (Vicia faba L.) as affected by lime, mineral P, farmyard manure, compost and rhizobium in acid soil of Lay Gayint District, northwestern highlands of Ethiopia. Agricult Food Secur 7(1):16
Fentaw A, Girma T (1996) Reclamation of salt affected soils in the Middle Awash region. In: Proceedings of the 3rd conference of the Ethiopian Society of Soil Science (ESSS), Addis Ababa, 28–29 February 1996, pp 49–59
Gebregziabher G, Abera D, Gebresamuel G, Giordano M, Langan S (2016) An assessment of integrated watershed management in Ethiopia. International Water Management Institute (IWMI), Colombo (IWMI Working Paper 170)
Gebrekidan H (2005) "Soil and water conservation practices (tied ridges and planting methods) on cultivated lands of the sub-humid and semiarid highlands of eastern Ethiopia." Haramaya University publication
Gebremeskel G, Gebremicael TG, Kifle M, Meresa E, Gebremedhin T, Girmay A (2018) Salinization pattern and its spatial distribution in the irrigated agriculture of Northern Ethiopia: an integrated approach of quantitative and spatial analysis. Agric Water Manag 206:147–157

Gedion T (2009) Surface water-groundwater interactions and effects of irrigation on water and soil resources in the Awash Valley. M.Sc thesis, Addis Ababa University. Addis Ababa, Ethiopia
Gelaw AM, Singh BR, Lal R (2014) Soil organic carbon and total nitrogen stocks under different land uses in a semi-arid watershed in Tigray, Northern Ethiopia. Agr Ecosyst Environ 188:256–263. https://doi.org/10.1016/j.agee.2014.02.035
Gelaye KK, Zehetner F, Loiskandl W, Klik A (2019) Effects of soil texture and groundwater level on leaching of salt from saline fields in Kesem irrigation scheme, Ethiopia. Soil Water Res 14(4):221–228
Genene T, Abiy GM (2014) Review on overall status of soil and water conservation system and its constraints in different agro ecology of Southern Ethiopia. J Nat Sci Res 4(7):59–69
Gete Z, Getachew A, Dejene A, Shahidur R (2010) Fertilizer and soil fertility potential in Ethiopia: constraints and opportunities for enhancing the system. IFPRI Working paper, p 42
Georgis K (2003) Land degradation adoption, low soil fertility and water stress: the major issues for improving crop production and food security in the dryland areas of Ethiopia. proceedings of the conference (Ed 13-15) Challenges and prospects of food security in Ethiopia, 3–15 August 2003, United Nations Conference Centre, Addis Ababa, Ethiopia pp 201–216.
Ghafoor A, Murtaza G, Ahmad B, Boers THM (2008) Evaluation of amelioration treatments and economic aspects of using saline-sodic water for rice and wheat production on salt-affected soils under arid land conditions. J Soil Irrig Drainage 57:424–434
Giday AO (2019) Salinity and sodicity hazard characterization in major irrigated areas and irrigation water sources, Northern Ethiopia. Cogent Food Agricult 5(1):1673110
Giller KE (2001) Nitrogen fixation in tropical cropping systems, 2nd edn. CAB International, Wallingford
Girma T (2001) Land degradation: a challenge to Ethiopia. International Livestock Research Institute, Addis Ababa, Ethiopia, pp 815–823
GIZ (Deutsche Gesellschaft für Internationale Zusammenarbeit). (2020) Integrated Soil Fertility Management Project (ISFM+) 9th Progress Report (1 July 2019–30 June 2020), Addis Ababa
Guadalix ME, Pardo MT (1994) Effects of liming on soil properties and P solubility in some Spanish Andosols. Geoderma 63(1):53–61
Gurmessa B (2021) Soil acidity challenges and the significance of liming and organic amendments in tropical agricultural lands with reference to Ethiopia. Environ Dev Sustain 23:77–99
Goulding (2016) Soil acidification and the importance of liming agricultural soils with particular reference to the United Kingdom. Soil Use Manag 32(3):390–399. https://doi.org/10.1111/sum.12270
Habtamu AA (2013) Impact of salinity on tolerance, vigor, and seedling relative water content of haricot bean (*Phaseolus vulgaris* L.) cultivars. J Plant Scie 1(3): 22–27
Habtegebrial K, Mersha S, Habtu S (2013) Nitrogen and sulphur fertilizers effects on yield, nitrogen uptake and nitrogen use efficiency of upland rice variety on irrigated Fulvisols of the Afar region, Ethiopia. J Soil Sci Environ Manag 4(3):62–70
Haileslassie A, Priess J, Veldkamp E, Teketay D, Lesschen JP (2005) Assessment of soil nutrient depletion and its spatial variability on smallholders' mixed farming systems in Ethiopia using partial versus full nutrient balances. Agr Ecosyst Environ 108(1):1–16. https://doi.org/10.1016/j.agee.2004.12.010
Hailu H, Mamo T, Keskinen R, Karltun E, Gebrekidan H, Bekele T (2015) Soil fertility status and wheat nutrient content in Vertisol cropping systems of central highlands of Ethiopia. Agricult Food Secur 4(1):1–10
Hailu R, Getachew A (2011) Potentials and limitations of acid soils in the highlands of Ethiopia: a review. Barley Research and Development in Ethiopia, Ethiopian Institute of Agricultural Research, International Center for Agricultural Research in the Dry Areas
Hartemink AE, Huting J (2008) Land cover, extent and properties of Arenosols in Southern Africa. Arid Land Res Manag 22(2):134–147
Haynes RJ (1982) Effects of liming on phosphate availability in acid soils. Plant Soil 68(3):289–308
Haynes RJ, Mokolobate MS (2001) Amelioration of Al toxicity and P deficiency in acid soils by additions of organic residues: a critical review of the phenomenon and the mechanisms involved. Nutrient Cycl Agroecosyst 59(1):47–63
Horneck DA, Ellsworth JW, Hopkins BG, Sullivan DM, Stevens RG (2007) Managing salt-affected soils for crop production
Hörner D, Wollni M (2020) Integrated soil fertility management and household welfare in Ethiopia. Food Policy 100. https://doi.org/10.1016/j.foodpol.2020.102022
Huffnagel HP (1961) Agriculture in Ethiopia. Rome 484 pp
Hurni H (1988) Degradation of and conservation of soil resources in the Ethiopian highlands. Mt Res Dev 8(2/3):123–130
Hurni H, Berhe WA, Chadhokar P, Daniel D, Gete Z, Grunder M, Kassaye G (2016) Soil and water conservation in Ethiopia: guidelines for development agents, 2nd revised edn. Centre for Development and Environment (CDE), University of Bern, with Bern Open Publishing (BOP), Bern, Switzerland, 134 pp
Hurni H (1993) Land degradation, famine, and land resource scenarios in Ethiopia. In: Pimental D (ed) World soil erosion and conservation, Cambridge, 27–61 pp
Ibekwe AM, Angle JS, Chaney RL,Van Berkum P (1997) Enumeration and N_2 fixation potential of Rhizobium leguminosarum biovar trifolii grown in soil with varying pH values and heavy metal concentrations. Agricult Ecosyst Environ 61(2–3):103–111
IFPRI, International Food Policy Research Institute (2012) Fertilizer market in Ethiopia
IFPRI, International Food Policy Research Institute (2010) Fertilizer and Soil Fertility Potential in Ethiopia constraints and opportunities for enhancing the system. https://www.researchgate.net/profile/Getachew_Agegnehu/publication/277813347
International Center for Biosaline Agriculture (ICBA) (2014) Highlights and Achievements 2009–2013. International Center for Biosaline Agriculture, United Arab Emirates, Dubai
ISFM Project Annual Report (2020) Integrated Soil Fertility Management Project (ISFM+) 9th Progress Report (01 July 2019 to 30 June 2020). Addis Ababa
ISD, Intervention for Sustainable Development (2016) Soil health insights from the 2015 Ethiopian soil campaign. Retrieved from www.agriprofocus.com
Islam KR, Weil RR (2000) Land use effects on soil quality in a tropical forest ecosystem of Bangladesh. Agricult Ecosyst Environ 79(1):9–16
IUSS Working Group WRB (2015) World Reference Base for Soil Resources 2014, update 2015 International soil classification system for naming soils and creating legends for soil maps. World Soil Resources Reports No. 106. FAO, Rome
Jutzi S (1988a) Deep black clay soils (Vertisols): management options for the Ethiopian highlands. Mountain Res Dev 153–156
Jutzi SC (ed) (1988b) Management of vertisols in sub-Saharan Africa: proceedings of a conference held at ILCA, Addis Ababa, Ethiopia, 31 August–4 September 1987. ILRI (aka ILCA and ILRAD)
Karasawa T, Kasahara Y, Takebe M (2002) Differences in growth responses of maize to preceding cropping caused by fluctuation in the population of indigenous arbuscular mycorrhizal fungi. Soil Biol Biochem 34(6):851–857
Kebede W, Habitamu T, Efrem G, Fantaw Y (2015) Soil erosion risk assessment in the Chaleleka wetland watershed, Central rift valley of Ethiopia. Environ Syst Res 4(5):1–12

Kebede F, Yamoah C (2009) Soil fertility status and NuMaSS fertilizer recommendation of typic hapluusterts in the northern highlands of Ethiopia. World Appl Sci J 6(11):1473–1480. http://www.idosi.org/wasj/wasj6(11)/4.pdf

Kefyalew A, Kibebew K (2016) Evaluation of the effect of salt affected soil on selected hydraulic properties of soils in Meki Ogolcha Area in East Showa Zone of Oromia Region, Ethiopia. J Biol Agricult Healthcare 6(21)

Kheyrodin H, Antoun H (2011) Tillage and manure effect on soil physical and chemical properties and on carbon and nitrogen mineralization potentials. Afr J Biotech 10(48):9824–9830

Kidanemariam A, Gebrekidan H, MamoT, Fantaye KT (2013) Wheat crop response to liming materials and N and P fertilizers in acid soils of Tsegede highlands, northern Ethiopia. Agricult Forest Fisheries 2(3):126–135

Kihara J, Bolo P, Kinyua M, Nyawira SS (2019) Sommer R (2020) Soil health and ecosystem services: lessons from sub-Sahara Africa (SSA). Geoderma 370:114342. https://doi.org/10.1016/j.geoderma.2020.114342

Kochian LV, Hoekenga OA, Pineros MA (2004) How do crop plants tolerate acid soils? Mechanisms of aluminum tolerance and phosphorous efficiency. Ann Rev Plant Biol 55:459–493

Kolb M, Galicia L (2018) Scenarios and story lines: drivers of land use change in southern Mexico. Environ Dev Sustain 20(2):681–702. https://doi.org/10.1007/s10668-016-9905-5

Kolb M, Mas JF, Galicia L (2013) Evaluating drivers of land-use change and transition potential models in a complex landscape in Southern Mexico. Int J Geogr Inf Sci 27(9):1804–1827. https://doi.org/10.1080/13658816.2013.770517

Komiyama T, Ito T, Saigusa M (2014) Effects of phosphorus-based application of animal manure compost on the yield of silage corn and on soil phosphorus accumulation in an upland Andosol in Japan. Soil Sci Plant Nutr 60(6):863–873

Laekemariam F, Kibret K (2020) Explaining soil fertility heterogeneity in smallholder farms of Southern Ethiopia. Appl Environ Soil Sci 2020:16. https://doi.org/10.1155/2020/6161059

Lambin EF, Geist HJ (2006) Land-use and land-cover change: local processes and global impact. Springer, Berlin Heideberg

Lambin EF, Turner BL, Geist HJ, Agbola SB, Angelsen A, Bruce JW, Folke C (2001) The causes of land-use and land-cover change: moving beyond the myths. Glob Environ Chang 11(4):261–269. https://doi.org/10.1016/S0959-3780(01)00007-3

Lambin EF, Geist HJ, Lepers E (2003) Dynamics of land-use and land-coverchange in tropical regions. Ann Rev Environ Resour 28 (1):205–241

Leta G, Schulz S, Alemu GG (2020) Agricultural extension approach: evidence from an Integrated Soil Fertility Management Project in Ethiopia. Front Agr Sci Eng 7(4):427–439. https://doi.org/10.15302/J-FASE-2020331

Loide V (2004) About the effect of the contents and ratios of soil's available calcium, potassium and magnesium in liming of acid soils. Agron Res 2(1):71–82

Magdoff F, Van Es H (2000) Building soils for better crops (No. 631.584/M188b). Sustainable Agriculture Network. Accessed at: https://www.soilandhealth.org/wp-content/uploads/0302hsted/030218bettersoils.pdf [Google Scholar]

Martin M (2001) The impact of community animal health services on farmers in low-income countries: a literature review. VETAID, UK

Matson PA, Parton WJ, Power AG, Swift MJ (1997) Agriculture intensification and ecosystem properties. Sci 277:504–509

McCausland A (2010) Konso-Ethiopia-agriculture and culture. Permaculture Research Institute. https://www.permaculturenews.org/2010/12/22/konso-ethiopia-agriculture-and-culture/

Melese A, Yli-Halla M (2016) Effects of applications of lime, wood ash, manure and mineral P fertilizer on the inorganic P fractions and other selected soil chemical properties on acid soil of Farta District, Northwestern highland of Ethiopia. Afr J Agric Res 11(2):87–99

Melesse D, Mesfin B (2008) Overview of environmental impact assessment in Ethiopia, gaps and challenges

Mekonnen G (2015) Characterization of agricultural soils in CASCAPE intervention woredas of Amhara region, Ethiopia. BDU-CASCAPE working paper 12: pp90

Merga B, Ahmed A (2019) A review on agricultural problems and their management in Ethiopia. Turkish J Agricult Food Sci Technol 7 (8):1189. https://doi.org/10.24925/turjaf.v7i8.1189-1202.2626

Meron N (2007) Characterization of salt affected soils in the central Rift Valley and assessing salt tolerance of different plants: a case study at the southwestern shore of Lake Ziway. MSc thesis, Addis Ababa University, Ethiopia, 75 p

Mesfin A (2007) Nature and management of acid soils in Ethiopia. Haramaya University, Haramaya, Ethiopia, 82 p

Mesfin K, Belay Y, Abera H (2014) Effects of liming and phosphorus levels on yield and yield components of haricot bean (Pharsalus vulgarism L.) varieties on nitosols at Wolaita Zone, Ethiopia. Asian J Crop Sci 6(3):245–253

Mesfine T, Abebe G, Al-Tawaha A (2005) Effect of reduced tillage and crop residue ground cover on yield and water use efficiency of sorghum (*Sorghum bicolor* L. Moench) under semi-arid conditions. World J Agricult Sci 1(2):152–160. http://www.idosi.org/wjas/wjas1(2)/9.pdf

MoA (2013) Five-year strategy for the transformation of soil health and fertility in Ethiopia. A working paper, 97 p

MoA (2019) Ethiopian Strategic Investment Framework for Sustainable Land Management (ESIF). A revised working paper, 51 p

MoARD, Ministry of Agriculture and Rural Development (2005) Agricultural Sector Policy and Investment Framework (PIF). In: Handbook of federal countries, 2005

MoANR, Ministry of Agriculture and Natural Resource (2015) Assessment of Fertilizer Consumption and Use by Crop in Ethiopia. IFDC, Addis Ababa

MoANR, Ministry of Agriculture and Natural Resource (2017) Soil Health and Fertility Improvement Strategy Document. Federal Democratic Republic of Ethiopia, Addis Ababa

Mohammed A, Belay T (2008) Characteristics and classification of the soils of the plateau of SIMEN Mountains National Park (SMNP), Ethiopia. SINET: Ethiop J Sci 31(2):89–102

Mulugeta D, Sheleme B (2010) Characterization and classification of soils along the toposequence of Kindo Koye watershed in Southern Ethiopia. East Afric J Sci 4(2):65–77

Mulugeta L (2004) Effects of land use changes on soil quality and native flora degradation and restoration in the highlands of Ethiopia: Implication for sustainable land management. PhD thesis presented to Swedish University of Agricultural Sciences, pp 64

Munns R, Tester M (2008) Mechanisms of salinity tolerance. Ann Rev Plant Biol 59:651–681

Muruts H, Haileselass DY (2019) Reclamation of salt-affected soils in the drylands of Ethiopia: a review. Res Dev Saline Agricult 579–589

Nanzyo M, Shoji S, Dahlgren R (1993a) Chemical characteristics of volcanic ash soils. Dev Soil Sci 21:145–187

Nanzyo M, Shoji S, Dahlgren R (1993b) Physical characteristics of volcanic ash soils. Dev Soil Sci 21:189–207

Ngoune Tandzi L, Mutengwa CS, Ngonkeu ELM, Gracen V (2018) Breeding maize for tolerance to acidic soils: a review. Agronomy 8 (6):84

Nguyen V, Nguyen B, Sarkarung S, Martinez C, Paterson A, Nguyen H (2002) Mapping of genes controlling aluminum tolerance in rice: comparison of different genetic backgrounds. Mol Genet Genomics 267(6):772–780

Nurlaeny N, Marschner H, George E (1996) Effects of liming and mycorrhizal colonization on soil phosphate depletion and phosphate uptake by maize (Zea mays L.) and soybean (Glycine max L.) grown in two tropical acid soils. Plant and Soil 181(2):275–285

Nyssen J, Poesen J, Deckers J (2009) Land degradation and soil and water conservation in tropical highlands. Soil Tillage Res 103 (2):197–202. https://doi.org/10.1016/j.still.2008.08.002

Nyssen J, Tielens S, Gebreyohannes T, Araya T, Teka K, Van de Wauw J, Deckers J (2019) Understanding spatial patterns of soils for sustainable agriculture in northern Ethiopia's tropical mountains. PLoS One 14(10):e0224041

Opala PA, Okalebo JR, Othieno CO, Kisinyo P (2010) Effect of organic and inorganic phosphorus sources on maize yields in an acid soil in western Kenya. Nutr Cycl Agroecosyst 86(3):317–329

Pankhurst CE, Double MD, Gupta VVSR (1997) Biological indicators of soil health: Synthesis. In: Pankhurst CE (ed) Gupta VVSR biological indicators of soil health. CaABI International, Oxon, UK, 436 p

Pearson RW (1975) Soil acidity and liming in the humid tropics. Cornell Int Agric Bull 30. Cornell University, Ithaca, New York

Pineros MA, Shaff JE, Manslank HS, Carvalho Alves VM, Kochian LV (2005) Aluminum resistance in maize cannot be solely explained by root organic acid exudation A comparative physiological study. Plant Physiol 137(1):231–241

Poykio R, Nurmesniemi (2008) Calcium Carbonate waste from an Integrated pulp and paper mill as a potential liming agent. Environ Chem Lett 6:47–51. https://doi.org/10.1007/s10311-007-0110-5

Qureshi AS, Ertebo T, Mehansiwala M (2018) Prospects of alternative copping systems for salt-affected soils in Ethiopia. J Soil Sci Environ Manag 9(7):98–107

Qureshi AS (2017) Sustainable use of marginal lands to improve food security in the United Arab Emirates. J Exp Biol Agricult Sci 5(Spl 1-SAFSAW)

Rabia AH, Afifi RR, Gelaw AM, Bianchi S, Figueredo H, HuongTL AV (2013) Soil mapping and classification: a case study in the Tigray Region, Ethiopia. J Agricult Environ Int Dev (JAEID) 107(1):73–99

Regassa A (2008) Response of Chickpea (*Cicer arietinum L*) to Land preparation methods and sowing dates on vertisols of Taba Watershed, Southern Ethiopia. MSc thesis, College of Agriculture, Hawassa University

Regassa H, Agegnehu G (2011) Potentials and limitations of acid soils in the highlands of Ethiopia: a review In: Mulatu B, Grando S Barley research and development in Ethiopia. ICARDA. Aleppo Syria pp 103–112

Reid RS, Kruska RL, Muthui N, Taye A, Wotton S, Wilson C, Mulatu W (2000) Land-use and land-cover dynamics in response to changes in climatic, biological and socio-political forces: the case of southwestern Ethiopia. Landsc Ecol 15:339–355

Ruganzu V, Rukangantambara H, Fashaho A, Sriramulu A, Nduwumuremyi A, James (2015) Integrated Soil Fertility Management (ISFM) training module for teachers in Agriculture Secondary Schools. Rwanda Soil Health Consortium, RAB, Kigali-Rwanda

Rutherford AS (2008) Broad bed maker technology package innovations in Ethiopian farming systems: an ex post impact assessment. ILRI (aka ILCA and ILRAD)

Saied M, Elsanat GM, Talha N, El Barbary SM (2017) On-farm soil management practices for improving soil properties and productivity of rice and wheat under salt-affected soils at North Delta, Egypt. Egypt J Soil Sci 57(4):445–453

Schlede H (1989) Distribution of acid soils and liming materials in Ethiopia. Ethiopian Institute of Geological Surveys, Ministry of Mines and Energy. Addis Ababa, Ethiopia

Schmook B, Vance C (2009) Agricultural policy, market barriers, and deforestation: the case of Mexico's southern Yucatán. World Dev 37(5):1015–1025. https://doi.org/10.1016/j.worlddev.2008.09.006

Shaaban M, Abid M, Abou-Shanab RAI (2013) Amelioration of salt affected soils in rice paddy system by application of organic and inorganic amendments. Plant Soil Environ 59(5):227–233

Shanka D, Dechassa N, Gebeyehu S, Elias E (2018) Dry matter yield and nodulation of common bean as influenced by phosphorus, lime and compost application at Southern Ethiopia. Open Agricult 3 (1):500–509

Sheleme B (2017) Topographic positions and land use impacted soil properties along Humbo Larena-Ofa Sere toposequence, Southern Ethiopia. J Soil Sci Environ Manag 8(8). https://doi.org/10.5897/JSSEM2017.0643

Sileshi A (2016) Status of salt affected soils, irrigation water quality and land suitability of Dubti/Tendaho area, North Eastern Ethiopia. Doctoral dissertation, Haramaya University, Ethiopia

Smale M (1995) Maize is life: Malawi's delayed green revolution. World Dev 23(5):819–831

Solomon D, Fritzsche F, Tekalign M, Lehmann J, Zech W (2002) Phosphorus forms and dynamics as influenced by land use changes in the sub-humid Ethiopian highlands. Geoderma 105:21–98

Soil and Water Conservation (1992) Soil and Water Conservation working paper 15. Ethiopian Forestry Action Progress, Addis Ababa

Soil Survey Staff (1999) Soil Taxonomy, 2nd edition, Agriculture Handbook Number 436, USDA. National Resources Conservation Services, Washington DCVanlauwe B, Bat

Spielman DJ, Pandya-Lorch R (eds) (2009) Millions fed proven successes in agricultural development. IFPRI

Stewart ZP, Pierzynski GM, Middendorf BJ, Prasad PV, Dhankher O (2020) Approaches to improve soil fertility in sub-Saharan Africa. J Exp Bot 71(2):632–641. https://doi.org/10.1093/jxb/erz446

Sun B, Zhou SL, Zhao (2003) Evaluation of spatial and temporal changes of soil quality based on geostatistical analysis in the hill region of subtropical China. Geoderma 115:85–99

Tadesse G, Zavaleta E, Shennan C, Fitz Simmons M (2014) Prospects for forest-based ecosystem services in forest-coffee mosaics as forest loss continues in southwestern Ethiopia. Appl Geogr 50:144–151. https://doi.org/10.1016/j.apgeog.2014.03.004

Takeda M, Nakamoto T, Miyazawa K, Murayama T, OkadaH (2009) Phosphorus availability and soil biological activity in an Andosol under compost application and winter cover cropping. Appl Soil Ecol 42(2):86–95

Tamene L, Amede T, Kihara J, Tibebe D, Schulz S (2017) A review of soil fertility management and crop response to fertilizer application in Ethiopia: towards development of site- and context-specifc fertilizer recommendation. CIAT Publication No. 443. International Center for Tropical Agriculture (CIAT), vol 443, p 86

Tamirie H (2005) Cost of land degradation in Ethiopia: a critical review of past studies. IFPRI, EEPFE

Tedla A, Mamo T, Klaij MC, Diedhiou ML (1999) Effects of cropping system, seed bed management and fertility interactions on biomass of crops grown on a Vertisol in the central highlands of Ethiopia. J Agron Crop Sci 183(3):205–211

Tejada M, Garcia C, Gonzalez JL, Hernandez MT (2006) Use of organic amendment as a strategy for saline soil remediation: influence on the physical, chemical and biological properties of soil. Soil Biol Biochem 38(6):1413–1421

Tilahun A (2019) Managing acid soils for reclaiming livelihoods in Ethiopia. ICRISAT. http://oar.icrisat.org/11409/1/Managing%20acid%20soils_Ethiopia_02%20May%202019_low.pdf

Tsegaye D, Moe S, Vedeld P, Aynekulu E (2010) Land use/cover dynamics in Northern Afar rangelands, Ethiopia. Agr Ecosyst Environ 139:174–180

UNCCD, United Nations Convention to Combat Desertification (2000) An introduction to the United Nations convention to combat desertification

Urgesa AA, Abegaz A, Bahir AL, Antille DL (2016) Population growth and other factors affecting land-use and land cover changes in north-eastern Wollega, Ethiopia. Trop Agricult 93(4):298–311

Vanlauwe B, Bationo A, Chianu J et al (2010) Integrated soil fertility management: operational definition and consequences for implementation and dissemination. Outlook Agricul 17 39:17–24. https://doi.org/10.5367/000000010791169998

Vanlauwe B, Zingore S (2011) Integrated soil fertility management: an operational definition and consequences for implementation and dissemination. Better Crops 95(3):4–7

Vanlauwe B, Bationo A, Chianu J, Giller KE, Merckx R, Mokwunye U, Ohiokpehai O, Pypers P, Tabo R, Shepherd K, Smaling E, Woomer PL, Sanginga N (2011) Integrated soil fertility management: operational definition and consequences for implementation and dissemination. Outlook Agricult 39:17–24. https://doi.org/10.5367/000000010791169998

Vanlauwe B (2013). Report I: Investment options for adoption of integrated soil fertility management (ISFM). Prepared for Bill and Melinda Gates Foundation

Virgo KJ, Muno RN (1977) Soil and erosion features of the Central Plateau of Tigray. Geoderma 20:131–157

Wassie H, Shiferaw B (2009) Mitigation of soil acidity and fertility decline: Challenges for sustainable livelihood improvement: evidence from the southern region of Ethiopia. In: Sustainable land management and poverty alleviation. Ethiopian Development Research Institute, Addis Ababa, pp 131–143

Water, Land and Ecosystems (WLE) (2017) CGIAR research program on water, land and ecosystems. Healthy soils for productive and resilient agricultural landscapes. (WLE Towards Sustainable Intensification: Insights and Solutions Brief 2). International Water Management Institute (IWMI), Colombo, Sri Lanka Colombo, Sri Lanka, 12 p. https://doi.org/10.5337/2017.211

Westphal E (1975) Agricultural systems in Ethiopia, Wageningen, p 278

WLE, Water Land and Ecosystem (2015) Sustaining the benefits of soil and water conservation in the Highlands of Ethiopia. Int Water Manag Inst 2014(03):2013–2016. https://doi.org/10.3389/feart.2015.00049

Wondimu G, Daba A, Qureshi A (2020) Effects of salinity on producers' livelihoods and socio-economic conditions; the case of Afar region, Northeastern Ethiopia. J Sustain Agricult Sci 46:35–46

Yadesa W, Tadesse A, Kibret K, Dechassa N (2019) Effect of liming and applied phosphorus on growth and P uptake of maize (Zea mays subsp.) plant grown in acid soils of West Wollega, Ethiopia. J Plant Nutr 42(5):477–490

Yang Y, Wang QL, Geng MJ, Guo ZH, Zhao Z (2011) Rhizosphere pH difference regulated by plasma membrane H+-ATPase is related to differential Al tolerance of two wheat cultivars. Plant Soil Environ 57(5):201–206

Yihenew GS, Gizachew A, Eyasu E, Mekonnen G (2014) Soil characterization and land suitability evaluation to cereal crops in Yigossa Watershed, Northwestern Ethiopia. J Agric Sci 6(5):109

Yihennew GS (2002) Selected chemical and physical charateristics of soils of adet research center and its testing sites in northwestern Ethiopia. Ethiop J Nat Resourc 4:199–215

Yitbarek T, Beyene S, Kibret K (2016) Characterization and classification of soils of Abobo Area, Western Ethiopia. Appl Environ Soil Sci 2016:1–16

Zeleke TB, Grevers MCJ, Si BC, Mermut AR, Beyene S (2004) Effect of residue incorporation on physical properties of the surface soil in the southcentral Rift Valley of Ethiopia. Soil Tillage Res 77(1):35–46

Zenebe (2007) Household fuel consumption and resource use in rural-urban Ethiopia. Retrieved from http://www.efdinitiative.org/publications/household-fuel-consumption-and-resource-use-rural-urban-ethiopia

10 Land Evaluation and Land Use Planning

Berhan Gessesse, Ashenafi Ali, and Alemayehu Regassa

Abstract

Land degradation harms sustainable economic growth and food security. Accordingly, different forms of soil, water, and land management practices are required to reverse the degradation challenge and food insecurity trends and eliminate fundamental barriers to Ethiopia's goal of achieving green economic growth. In this connection, land evaluation and land use planning can be used as strategic instruments to manage farming landscape dynamics in Ethiopia, where urgent interventions are needed to enhance agricultural landscape sustainability. In doing so, we collected scientific documents such as policies, regulations, published literature, and unpublished reports produced from 1980 onwards to analyze and synthesize the nature and status of land evaluation practices and land use planning efforts. As a result, the general principles and the historical development of land evaluation and land use planning approaches and practices are reviewed and documented. The findings of this review showed that land evaluation is the cornerstone of land use planning, and it can make it easier to allocate the land to the users bringing them long-term benefits. However, lack of awareness; up-to-date technical standards and systematized capacity; poor linkages among the concerned institutions; financial and incentive limitations; policy and strategy implementation restraints and limited and inconsistent datasets availability are the key constraints of land evaluation and land use planning development. On the other hand, the existence of environmental and land use policies, involvement of community-based participatory watershed development initiatives, rich experience in land evaluation practice, existence of education and training institutes in agriculture and land administration are identified as opportunities to enhance land evaluation and land use planning in Ethiopia. In general, the review provided the baseline information regarding the practices and status of land evaluation and land use planning and acquired and synthesize information on land suitability analysis for crop production. This chapter will also contribute a lot in assisting decision-makers and land users to identify and implement land use that will best fulfill people's needs while protecting natural resources and ecosystem services for current and future generations.

Keywords

Land evaluation • Land use planning • Ethiopia

10.1 Introduction

Most agrarian nations rely on land resources (water, soil, and vegetation) for survival and economic growth (FAO 2007; Mcrae and Burnham 1981; Dent and Young 1981). Although owing to population pressure and declining land productivity, agriculture, which is highly dependent on the efficient uses of rural land uses, is one of the essential economic sectors in most developing countries supporting human existence (FAO 2000). In Ethiopia, for example, highlands (above 1,500 m a.s.l) account for over 90% of regularly cultivated land and sustain nearly 90% of the country's population (MoARD and WB 2007; Hurni 1993). Accordingly, the country is one of the largest grain producer nations in Africa. Despite the large-scale output, the farming system is characterized by high subsistence and reliance on

B. Gessesse (✉)
Entoto Observatory and Research Center (EORC), Ethiopian Space Science and Technology Institute, Addis Ababa, Ethiopia
e-mail: berhanga@essti.gov.et

A. Ali
Department of Geography and Environmental Studies, Addis Ababa University, Addis Ababa, Ethiopia

A. Regassa
Department of Natural Resource Management, Jimma University, Jimma, Ethiopia

S. Beyene et al. (eds.), *The Soils of Ethiopia*, World Soils Book Series,
https://doi.org/10.1007/978-3-031-17012-6_10

rainfall, aggravated by the world's lowest yield (Ewnetu and Bliss 2010; Yizengaw and Verheye 1995).

Besides, land degradation is thought to be affecting farmlands, resulting in huge soil erosion and fertility loss. As a result, agricultural production and productivity have been below the potential and insufficient to meet the country's population demands (Deressa et al. 2008; MoA and SLMS 2008; Hurni 1993). Declining the agricultural land use productivity has negative consequences for food production, poverty, and related social dysfunction (Shiferaw and Holden 1999), and Ethiopia has been a net importer of grains since 1981/82 (Deressa et al. 2008; MoA and SLMS 2008). According to Gebeyehu et al. (2017), the land is being exploited without respect for its environmental, social, and economic appropriateness, as well as natural resource protection. Land also used for different purposes it is also used without adequate planning and management across numerous economic sectors including in agriculture, forest, livestock, water development, wildlife, and tourism as well as leading to severe deterioration of natural resources and creating a low level of economic development, environmental degradation, and poverty trap.

In this regard, land evaluation and land use planning are solutions for analyzing and developing the best land management options to hold back land resource degradation and the different biophysical limiting factors of agricultural production. Land evaluation and land use planning have positive implications on sustainable land management programs and crop production, which are essential to devise strategies to check out land degradation and enhance agricultural production and productivity. According to Calzolari (2009), sustainable land use is a form of practice that can be executed indefinitely without causing the degradation or impoverishment of different land resources mostly in developing countries.

As a result, the land evaluation exercise is an essential aspect of tackling Ethiopia's rural land use problem, the optimization of agricultural production, and the realization of sustainable land management practices. Conceptually, land evaluation is a classification system of land that assesses the best use for a given portion of land and determines the fitness of this fraction of land for a specified and defined use while evidencing the existing limitations for one or more minor specific types of land use requirements (Calzolari 2009; Sys et al. 1991a; FAO 1983). However, in their seminal book, Sys et al. (1991a) argued that a full-fledged and critical land evaluation considers two primary components of the land: physical resources such as soil, topography, and climate, as well as socioeconomic resources like farm size, management level, labor availability, market, and other land-related economics issues.

Historically, FAO (1984a) and Assen (2003) claimed that land evaluation has been carried out from reconnaissance scale to plot levels in Ethiopia since 1982. A land evaluation exercise has been carried out on the physical parameters as part of the project activities for the formulation of a national master land use plan to halt land degradation. The evaluation process of the land was established by assessing the fundamental surveys of natural resources such as geomorphology, climate, soils, and vegetation in Ethiopia, for both the actual and potential land use systems. Meanwhile, these days, land evaluation exercises in the country serve as a connection between basic resource surveys and agricultural and other land development plans. However, most of the land evaluation practices in Ethiopia were not possible to completely combine economic and social analysis with physical land evaluation (Gessesse 2013; Kassa et al. 2010; Assen 2003; Yizengaw and Verheye 1995; FAO 1984a).

10.2 The Need for Land Evaluation

The amount of cultivable land in Ethiopia is limited, and land degradation in the form of soil erosion and nutrient loss, biodiversity loss, land use transformation as well as vegetation, water, and forest resources degradation, is nearly irreversible in most parts of the highlands of Ethiopia (Gessesse 2013; MoARD and WB 2007; Lakew et al. 2000; Fistum et al. 1999). Besides, complex processes and factors trigger land use changes as well as land degradation in Ethiopia (Zeleke et al. 2006; Lakew et al. 2000). Land degradation is primarily a result of extraordinary population pressure and a lack of adequate innovative farming practices by smallholder farmers (MoRAD 2008; Woldemariam 1991). However, the expansion of crop cultivation into marginal lands; erratic and erosive rainfall patterns; declining use of fallow farming practices; inadequate recycling of dung and crop residues to the soil; limited application of external inputs of plant nutrients as well as deforestation and overgrazing are the direct causes of land degradation (MoARD and WB 2007).

On the other hand, population pressure, poverty, high costs of and limited access to agricultural inputs and credit, low profitability of agricultural production, fragmented landholdings, and a lack of information about appropriate alternative technologies for farmers are socio-economic and institutional factors that exacerbate land degradation in the country (Zeleke et al. 2006; Lakew et al. 2000). Furthermore, government policies have mostly failed to address appropriate infrastructure and market development, input and credit supply, land tenure, agricultural research and extension development program, and land conservation enhancement programs for sustainable land management practices (Gessesse 2010; Lakew et al. 2000). At the federal, regional, and community levels, the absence, as well as lack of implementation of integrated and comprehensive land administration policies, proclamations, regulations, and land use master plans developed in a participatory manner, are major factors contributing to the massive land degradation process in the country (FAO 2014; Zeleke et al. 2006). In general, when the natural, socioeconomic, and institutional factors are coupled with poor land management practices, the result is a creeping soil degradation characterized by loss of soil nutrients from agricultural and range land landscapes.

In agrarian countries like Ethiopia, the quest for arable land necessitates full-scale spatial–temporal rearrangement, as well as a more logical use of the existing land, which is unquestionable. This is only possible if the land's traits and attributes are properly identified, a full inventory of natural resources is completed, and a thorough evaluation of the land's relative appropriateness for various uses is performed. In this context, land evaluation is needed for describing, assessing, and presenting as well as mapping various biophysical factors for acceptable land use plans and relevant decision-making of crop cultivation recommendations. Based on the foregoing argument, it is obvious that land use decision-making is a socio-economic and political arrangement with significant social, economic, and environmental negative consequences. Therefore, because of the increased public interest in planning, sound land evaluation tools for identifying diverse land uses alternatives must be devised and applied.

10.3 History of Land Evaluation

According to studies conducted in Ethiopia, a primitive attempt was made to evaluate and measure the quality of land in conjunction with the transfer of land to the soldiers during the reign of Emperor *Tewodros* (1855–1868) (FAO 1984a, 2014), and the quality of land was divided into four categories ranges from "Highly productive" to "Least productive" land. On the other hand, during the reign of Emperor Menelik II, the ministry of agriculture was founded in 1890 to issue land use and forest protection regulations. Besides, in 1974, under Emperor Haile Selassie's rule, international consulting firms conducted numerous catchment-based characterization comprising diverse land suitability studies. Between late 1950s and 1974, various institutions made fragmented efforts to generate biophysical baseline information and land suitability studies through either foreign technical aid agencies or private foreign consultants for evaluating the agricultural development potential land in Ethiopia. As a result, extensive land evaluation studies were done, however, the scope, scale, and technique of these studies were different greatly (Esayas and Debele 2006; Debele 1980).

However, with the formation of the first soil survey and land evaluation division in 1971 under the then Institute of Agricultural Research (IAR), land suitability studies have garnered substantial institutional legitimacy. Soil surveys and land evaluation studies were required for agricultural research and development, regional planning, and land and water resource conservation. In 1973, the unit was moved to the Ministry of Agriculture's Land Use Planning and Regulatory Department (LUPRD) (Esayas 2001; Debele 1980). Accordingly, the MoA's LUPRD conducted countrywide and regional land suitability evaluations for the preparation of land use plans in various geographic regions. The studies included the national level reconnaissance land suitability assessments; three semi-detailed land use planning studies (i.e., conducted in *Borkena, Bichena*, and *Hossaina* areas); three area-based reconnaissance land use plans in *Menagesha, Hayikoch,* and *Butajira* administrative areas; and three area-based reconnaissance land use plans in *Yerer and Kereyu* (FAO 1984; Debele 1980).

In the later stage, river basin-based land resource assessments and land evaluation began in the 1980s under the Ministry of Water Resources for drafting an integrated development master plan (IDMP). It has conducted several reconnaissance land suitability studies for major river basins of the country to prepare for the development of plans including Abay in 1998 (MoWR 1998), Omo-Gibe in 1996 (MoWR 1996), Wabeshebele in 2007 (MoWR 2007), Genale Dawa in 2006 (MoWR 2006), and Rift Valley Lakes Regions in 2007 and 2010 (MoWR 2010). The river basin master plans are quite extensive and deal with the utilization of various natural resources as well as socio-economic elements of land-based on-site suitability assessments, even though the primary goal of these studies was to develop water resources.

Recently, mainly as of the 2000s, various land evaluation and use planning studies have been completed or are currently being conducted in some regional states, with land evaluation studies being one of the key components. The Ethiopian Construction Design and Supervision Works Corporation (ECDSWCo), the Oromia Water Works Design Supervision Enterprise (OWWDSE), and the Amhara Design and Supervision Works Enterprise (ADSWE) conducted the majority of these studies (Ali et al. 2020). The planning efforts are aimed at "Integrated Land Use Planning" or "Land Suitability Evaluation for Various Irrigated Cropping", which includes all sectoral activity in the study region and results in a multi-sectoral and multi-disciplinary plan.

The plans are based on bio-physical, economic, social, and environmental land suitability assessments and have been undertaken at various scales, for instance, reconnaissance surveys at the basin level (1:250,000); semi-detailed surveys at the sub-basin level (e.g. Borena, Hararge, East Amhara, scale 1:50,000); semi-detailed and detailed surveys (e.g. Oromia Special Zone, scale 1:25,000); Kuraz irrigated sugarcane development project, scale 1:15,000 to 25,000); intensive surveys (e.g. Local Development Plan Fentale and Tibila, scale 1:5,000). Furthermore, land evaluation studies were carried out in various locations of Ethiopia for the partial fulfillment of graduate-level university degrees, and many of them are remarkable instances of large-scale level land evaluation exercises in the country (Gessesse 2013; Assen 2003; Kassa et al. 2010; Yizengaw and Verheye 1995).

10.4 Basics of Land Evaluation

Ethiopia, being an agrarian state, has land demands for arable farming, grazing, forestry, wildlife habitat, tourism development, urban and rural infrastructures development, and related activities that are more significant than the available land resources. These demands become more pressing every year in Ethiopia due to population growth and pressure. Even where land is still plentiful, many people may have inadequate access to the land or benefits from its use. Thus, planning guides are necessary in case of conflicts between different land users where they act by indicating the specific areas of the land that are most valuable for specific land use systems. Accordingly, land evaluation is perceived as the process of assessment of land performance when used for specified purposes involving the execution and interpretation of biophysical and socio-economic surveys and other aspects of land to identify and make a comparison of good kinds of land use in terms applicable to the objectives of the evaluation (Gessesse 2013; Woubet et al. 2013; Kassa et al. 2010; Assen 2003; Yizengaw and Verheye 1995). In general, the basics of land evaluation exercise in Ethiopia consist of land mapping units, key categories of land uses, land utilization types, land characteristics, land qualities, diagnostic elements, land use requirements, limiting factors, as well as land evaluation outcomes categorization and spatialization.

10.4.1 Spatial Entity for Land Evaluation

The spatial entity under evaluation refers to the land-mapping units (LMUs) with specified land characteristics (FAO 1976, 2007; Sys et al. 1991a). Land mapping units are portions of the landscape that have similar characteristics and qualities mapped. They are mapping units in which land suitability evaluation, analysis, and description will be made for a specified purpose (FAO 1983). The boundary of LMUs can be defined by a natural resource survey and the final output of the evaluation process is a map that generally derives directly from LMUs. Depending on the objective of the analysis, different types of terrain features, slope classes, land use types, land cover units, soil units and types, farming units, landform classes, geological and geomorphic units, well-defined landscape units, climate types, and agro-ecological zones can be considered as LMUs for land evaluation purpose as well as land use planning exercises in Ethiopia (Gessesse 2013; Assen 2003; Yizengaw and Verheye 1995; Miteku 1996).

On the other hand, according to Calzolari (2009), a single land-mapping unit may contain two or more distinct types of land. For example, a single river flood plain land-mapping unit is mapped as a single unit but is known to contain both well-drained alluvial flood plains and swampy depressions flood plains. The degree of homogeneity or internal variance of land mapping units, on the other hand, varies depending on the size and breadth of the study. In this regard, natural resource surveyors, environmentalists, pedologists, geologists, agronomists, land evaluators, land use planning professionals, foresters, and geospatial and remote sensing experts, to mention a few, can define and map these land mapping units. As a result, land suitability evaluation entails assigning land-mapping units to certain land uses, and land suitability evaluation involves relating land-mapping units to specified types of land use. The types of use considered are limited to those, which appear to be relevant under general physical, economic and social conditions prevailing in areas and these kinds of land use serve as the subject of land evaluation.

10.4.2 Major Kinds of Land Use and Land Utilization Types (LUTs)

A major kind of land use is a subdivision of rural land use such as rainfed agriculture, irrigated agriculture, grassland, forestry, or recreation (FAO 1976, 2007). In this regard, the most common major land use types in Ethiopia are agricultural lands; urban or rural built-up lands; range or grazing lands; forest land use; water and wetlands, livestock grazing lands; geothermal, wind, and hydroelectric power production areas; tourist attraction sites; wildlife conservation and management sites; mining sites and manufacturing sites just to mention a few (Gebeyehu et al. 2017). On the other hand, LUTs are kinds of land uses described in detail greater than that of a major kind of land use in the form of a crop or crop combination with a specified technical and socioeconomic setting (FAO 1983). For example, residential, commercial, and services; transportation and communication utilities; industrial and commercial complexes; as well as mixed urban or built-up lands are detailed land utilization types under urban or built-up major land-use types. Similarly, the agricultural land use types consist of different major kinds of land uses including rainfed cropland, irrigated cropland, pasturelands, horticulture land uses, and many more detailed land utilization types. More specifically, from rainfed agricultural practices, about 102 different land utilization types (LUTs) were identified for physical land evaluation purposes in Ethiopia (FAO 1984).

10.4.3 Land Characteristics, Land Qualities, and Diagnostic Factors

Land use characteristics (LCs) are observable physical qualities of the environment that are linked to land use features. The major land characteristics such as climate, topography, wetness (drainage and flooding), texture, soil depth, pH level, cation exchange capacity, base saturation, organic matter content, salinity, and alkalinity are some examples of LCs that can be used to characterize land-mapping units to conduct land evaluation analysis (Sys et al.1991a; FAO 1976, 1985; Van Diepen et al. 1991). On the other hand, LQs are calculable attributes of the land representing the immediate requirements of land utilization types. The value of LQs is mostly quantified utilizing land characteristics (FAO 1983); however, they can sometimes be estimated or measured directly from the physical environment. More than two dozen land qualities were identified for land evaluation (FAO 1983). However, temperature regime, moisture availability, oxygen availability, nutrient availability, rooting conditions, soil workability, absence of salinity/alkalinity, flood and erosion hazards are the most common LQs considered during rainfed-based land suitability analysis and evaluation in Ethiopia (Gessesse 2013; Kassa et al. 2010; Assen 2003; Yizengaw and Verheye 1995; FAO 1984; Debele and Ochtman 1974). The purpose of quantifying the value of LQs or LCs is to determine the limitations of LMUs. These limitations are known as diagnostic factors; and there will be critical values for each diagnostic factor, which are used to define suitability class limits (FAO 1983).

10.4.4 Land Use Requirements and Limitations

LURs are sets of agro-climatic and edaphic factors that determine production and management conditions of a specified land use considering the crop requirements, management requirements, and conservation requirements (Calzolari et al. 2009; Sys et al. 1991a). From the land-use mathematical equation perspective, LQs are the 'supply' side of the land (what the land cn offer to LUTs), whereas LURs are the 'demand' side of the land, i.e., what the LUTs require from the land (Rossitor 1996). On the other hand, limiting factors are land qualities or land characteristics that adversely affect the potential of the land for a specified kind of major land use or LUTs (FAO 1983, 2007). Limiting factors (limitations), according to Sys et al. (1991a), are deviations from the ideal circumstances of land characteristics/land qualities that negatively influence a kind of land use.

10.4.5 Suitability Classification and Cartographic Spatialization

The term "land suitability for a certain application" refers to how well the LMUs fit the LUTs' criteria. The land suitability categorization system, according to Calzolari et al. (2009), Sys et al. (1991b) and FAO (1983, 2007) consists of four hierarchical levels: (i) orders, (ii) classes, (iii) subclasses, and (iv) units. There are only two suitability orders, one showing whether a certain LMU is suitable (S) and the other indicating if it is unsuitable (N). The suitability (S) order designates a land area where a specific LMU can be used without endangering the environment. The unsuitable (N) order, on the other hand, indicates that a particular LMU is not suited for long-term economic and environmental usage. The suitable (S) order has three suitability classifications [very suitable (S1), moderately suitable (S2), and marginally suitable (S3)]. In contrast, the N order has two suitability classes [actually unsuitable but perhaps suitable in the future with the treatment of the land with different land management options (N1) and actually and potentially unsuitable (N2).

On the other hand, suitability subclasses specify the types of land quality constraints that exist and the primary types of improvement actions necessary within each class of LMU (Fig. 10.1). Lowercase letters are used to represent subclasses. Finally, the suitability units are used to indicate the relative importance of land improvement works, and as Arabic numbers present standard units, separated by a hyphen. For example, from the land suitability classification S3e-1 structure: 'S' stands for suitability Order', '3' stands for suitability class of marginally suitable, 'e' stands for' suitability Subclass' of erosion hazard and 1' stands for' land management options. Standard subclasses (limitations) are presented in lower case letters (FAO 1985). As recommended by FAO (1976) subclasses such as 'c' (climatic limitations), '*t*' (topographic limitations), 'w' (witness limitations), 's' (physical soil limitations; i.e. influencing soil/water relationships and management), 'e' (erosion hazard), 'd' (soil degradation hazard), 'f' (soil fertility limitation not readily to be corrected) and 'n' (salinity and/or alkalinity limitations) are some of the identified subclasses in the framework. Finally, land suitability units are used to indicate the relative importance of land improvement works or agronomic management practice, and as Arabic numbers present standard units, separated by a hyphen.

In Ethiopia, varieties of land suitability evaluation classification and outcomes are already available at the national level. These contain studies and maps on land suitability for rain-fed and irrigated crop farming, rangeland and pasture development, livestock, and forestry production systems.

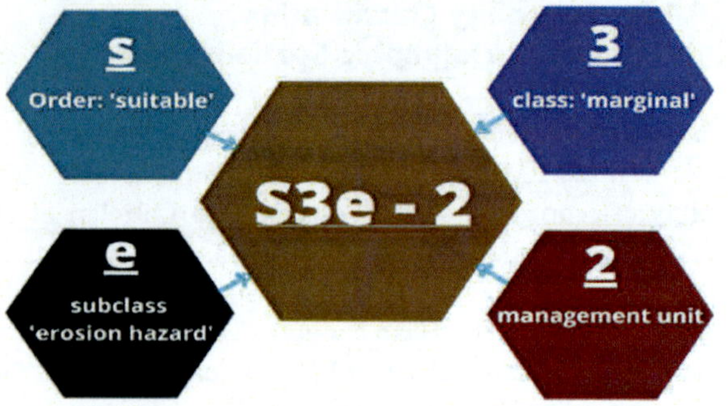

Fig. 10.1 Structure of land suitability classification classes (*Source* Modified by the authors, after FAO 1976)

For example, in the 1980s, the then land use program of rural development (LUPRD) conducted a national-level land suitability evaluation, but only three suitability classes (S1, S2, N) were used, with suitability displayed for both "low input" and "high input" land-use types (FAO 1984). Besides, land suitability studies undertaken by the Ethiopian Institute for Agricultural Research (EIAR) in the 2010s are expressed in four classes (S1, S2, S3, N) (EIAR 2014). However, international scientists in the field of land evaluation devised five suitability classes namely S1, S2, S3, N1, and N2 (FAO 2007; 1976). In general, the studies vary in scope, approach, and methods and are mainly based on qualitative biophysical land suability evaluation techniques with little consideration and integration of socioeconomic and economic-based assessments.

10.5 Principles, Scales, Data Sources, and Approaches of Land Evaluation

10.5.1 Principles of Land Evaluation

Land evaluation is principally focused on collecting, interpreting, and analyzing biophysical and socioeconomic datasets about the land, i.e. its soil, climate, vegetation, and realistic alternatives for improving the land-use system (FAO 2007). In the early stage, six fundamental principles of land evaluation were framed under rainfed, irrigation, forestry, and economic land evaluation exercises to assess the particular land uses for distinct land utilization types based on the initial idea of land terminology and its applications (FAO 1976, 1983, 2007). In light of this, since the 1980s, these six principles have been considered in several land evaluation studies in Ethiopia. However, concerns of biodiversity, agro-ecosystem functions, stakeholder engagement, and agro-environmental monitoring have been included in land evaluation exercises due to the UN's (1995) most recently updated concept and definition of the totality of land. Although most of the original principles (FAO 1976) have been preserved or slightly amended; land evaluation principles incorporate the evaluation of not just the products but also services provided by the land, and two new principles have been introduced (FAO 2007).

As a result, eight standardized land evaluation principles were framed and all of these principles shall be included in every land use exercise in its landmark of the updated land evaluation guideline. According to the FAO (2007) recommendation, the revised land evaluation basic principles are: (i) land suitability evaluation should be assessed and classified with specific types of land use and services; (ii) land evaluation necessitates a comparison of benefits obtained and inputs required on various types of land to assess productive potential, environmental services, and long-term livelihood; (iii) land evaluation necessitates a multidisciplinary and cross-sectoral approach; (iv) land evaluation should take into account the biophysical, economic, social and political context as well as the environmental concerns; (v) land evaluation suitability relates to long-term usage of the land, while sustainability encompasses productivity, social equality, and environmental considerations; (vii) all stakeholders needs, preferences, and views must be considered while evaluating land; and (viii) before the land evaluation process, the scale and level of decision-making need to be clearly defined.

10.5.2 Scale of Land Evaluation

The process of land evaluation takes place at all levels and geographical scales, i.e., from farm size (large scale) to more generic ones (reconnaissance survey level). In this sense, the Food and Agricultural Organization first established the map's scale standards for land evaluation in Ethiopia (FAO 1984). According to the same source, the land evaluation map scale is clustered into four classes in which land evaluation and overall land use-planning procedures are carried out. The standardized four classes of mapping scale are: (i) fast reconnaissance (scale of 1:1,000 to 1:2,000); (ii) reconnaissance (scale of 1:250,000, occasionally 1:500,000); (iii) semi-detailed (scales of 1:50,000); and (iv) detailed (scale of 1:50,000) (at scale of 1:25,000 and larger than this scale). The reconnaissance, semi-detailed, and detailed land assessment scales may be used to find an approximate development to precisely determine the limits of agricultural projects; conduct land use planning, and give detailed farm layouts and conservation works, all in the same sequence.

10.5.3 Sources and Datasets for Land Evaluation

The fundamental purpose of land evaluation would be to find the optimal land use for each specified land unit, taking into consideration physical and socioeconomic variables as well as natural resource protection for future use. To discover the key limiting variables affecting land production, the evaluation process includes surveying and assessing landforms, soils, vegetation, climate, and other land features. The integration of land evaluation exercise and agricultural systems analysis may significantly enhance the present land use planning procedures as a tool for sustainable land use and rural development in Ethiopia.

(i) Physiographic and Agro-climatic Datasets: Land qualities of topographic parameters (elevation, slope inclination, and aspect), geology, geomorphology, and regional landform characteristics are the most essential physiographic factors that would be examined for a land evaluation exercise. Besides, flood and erosion hazard-prone area datasets and information are important for land evaluation exercises. Most of the time, in Ethiopia, slop inclination, elevation, aspect, and contour information are extracted from ground survey data, 3D based stereoscopic aerial photo interpretation, and available satellite mission-based 3D image products of digital elevation models (DEM) (Getahun et al. 2018; FAO 2014; Gessesse 2013; Assen 2003; FAO 1984).

Similarly, study area boundary and land mapping units, drainage networks, related micro-relief (landform) characteristics, geological and geomorphological information as well as land use/land cover units information could be derived from both ground survey data (FAO 1983), remote sensing products interpretation (Rossiter 1994; Gessesse 2013) and available geological and geomorphological maps of Ethiopia (FAO 1984). More specifically, remotely sensed imageries are used for land evaluation and land use planning to produce land use/land cover maps, identify land mapping units and ecological zones, update base maps without conducting a full-fledged field survey, locate specific areas of interest to the land evaluation, and provide time series spatial, temporary, or seasonal phenomena such as crop growth, vegetation intensity, and land use variation over time (Gessesse 2013; FAO 2007; Rossiter 1994).

Major land quality attributes employed for land evaluation, on the other hand, include agro-ecological zones, moisture availability, and temperature regime, length of the growing period, good ripening and harvesting period, which are very essential for crop production in rainfed and irrigated agricultural systems (FAO 1976, 1983, 2007; Sys et al. 1991a). Different climatic elements such as temperature, potential evaporation, and the temporal and geographical variability of rainfall, all of which are peculiar to a certain region or localities, are used as input for estimating the growth season under the land evaluation exercise. Specifically, temperature, rainfall, wind velocity, radiation, relative humidity, evapotranspiration, and soil moisture are the minimal climatic data needed for land evaluation study in Ethiopia, and most of the time these information are extracted from observed climate data using both conventional and automatic weather stations. Furthermore, most of these climate and soil hydrological-related parameters, information can be calculated using satellite-based forecast weather information (Tesfaye et al. 2018; Gebrehiwot et al. 2019; Mengistu et al. 2019), and the products can be used for land evaluation exercise in metrological measured data scarce environment in Ethiopia.

(ii) Soil Datasets: According to Getahun et al. (2018), FAO (1984, 2014), Gessesse (2013), Assen (2003), the key datasets for land evaluation research are soil information, including its geographical distribution, psychochemical qualities and quantities, and a logical explanation of legends based on dominating, related, and inclusion soil units. Soil data and information are usually obtained from ground surveys and laboratory analyses or existing soil legacy datasets. In Ethiopia, different soil parameters such as soil texture, pH, electrical conductivity (EC), organic carbon (OC) concentration, Total nitrogen (TN) content, soil available phosphorus (P), sodium bicarbonate extraction, exchangeable bases, and cation exchange capacity (CEC), bulk density (BD), field capacity (FC), and permanent wilting thresholds (PWP) are factors usually considered for land evaluation exercise. On the other hand, standard methods can be used to calculate land characteristic parameters such as soil organic matter (SOM), percentage base saturation (PBS), cation/cation exchange capacity ratio (C: CEC), available water holding capacity (AWC), and carbon/nitrogen ratio (C: N) are extracted from laboratory analysis results (FAO 2014). Besides, in Ethiopia, different soil qualities and characteristics can be extracted from remotely sensed products interpretation.

(iii) Socioeconomic Data Sources and Types: Although the fourth principle of the 1976 Framework did emphasize the importance of economic land evaluation (FAO 1976), there are hardly available any published results on economic land evaluations so far (FAO 2007). However, the FAO (1983, 2007) suggested that the physical survey is able to locate the areas where the land is suitable for land utilization types exists; when combined with socio-economic inputs the resulting land evaluation shows where the land is suitable for profit and is socially relevant. As a result, socioeconomic criteria must be defined and included in any land evaluation exercise to assess the overall viability of land mapping units for various land use categories.

Accordingly, socio-economic datasets such as demographic variables, type and size of labor costs to be involved, land tenure systems, transport and infrastructure accessibility for agricultural inputs (fertilizer and other chemicals), capital and level of technology investment, credit and saving availability, costs of production, productivity (yield), prices gross and net returns, accessibility to markets and food products, shape, size and accessibility of the farm affecting management are some examples of socioeconomic data sources for economic land evaluation practices in Ethiopia (FAO 1983; Young 1980). Besides, socioeconomic datasets for land evaluation exercise in Ethiopia would be acquired from surveys and censuses conducted by the Ethiopian Central Statistical Agency, published reports, agricultural experimental plots information and sample surveys, and to some extent from remote sensing image analysis and interpretation.

10.5.4 Approaches to Land Evaluation

Land evaluation is always applied for analyzing the land for cost-effective purposes, while also safeguards the environment's diverse ecological functions from any degrading agents and processes (Calzolari et al. 2009). As a result, multiple qualitative and quantitative-based land evaluation approaches have been established in different countries in general and in Ethiopia, in particular, to evaluate the land for various applications, both before and after the "FAO's Land Evaluation Framework" were developed in 1976, and scientists and organizations working in the land evaluation field, such as Calzolari et al. (2009); FAO (2007), Van Diepen et al. (1991), Sys et al. (1991a, b), and Rossiter (2009), provide in-depth analyses and exhaustive discussion for these different approaches.

However, many of them share major fundamental concepts, and the extent of these techniques are tied to the FAO's land evaluation framework principles (FAO 1976, 1983, 2007). Subsequently, the FAO framework is considered the standard reference in land evaluation (Van Diepen el al. 1991). Considering the FAO (1976, 1985, 2007) and Calzolari et al.'s (2009) argument, three historical periods can be distinguished in the development of land evaluation on the global scale. These are land evaluation techniques that were produced before the FAO framework (FAO 1976), the FAO framework (FAO 1976), and recent land evaluation approaches that were developed following the Framework.

(i) Land Evaluation Approach before the FAO Framework

Land Capability Classification (LCC): According to FAO (1976, 1985, 2007) and Calzolari et al. (2009), the United States Department of Agriculture's Soil Conservation Service developed the LCC approach. Although being designed to support specific soil conservation projects, this land evaluation approach has been widely adopted around the world with numerous adaptations. LCC's mission was to assist farmers in making the most use of their land. Soil mapping units were divided into eight categories based on their capacity to sustain a wide range of land uses without degrading the environment or causing substantial off-site consequences. The first four classifications are arable land, with increasing restrictions on usage and the need for conservation measures and careful management (FAO 2007). The remaining four groups are not suited for crops but may be ideal for grassland, woods, grazing, wildlife, recreation, and other uses. Subclasses within the broad classes denote specific limits such as erosion, excessive moisture, root zone difficulties, and climate constraints.

Land Classification System for Irrigation: The United States Bureau of Reclamation (USBR) classification of land evaluation for irrigation framework was designed in the United States in the 1950s (Calzolari et al. 2009; FAO 2007), to classify land properties based on their capacity to support agricultural irrigation. The classes in the USBR framework are classified based on their relative ability to generate a profit, or "capacity for return" in financial terms. They are categorized based on the entrepreneur's net income after deducting operating and maintenance expenditures. It is an example of land valuation mainly reliant on economic criteria in this respect. The system is divided into six categories, each of which is defined by the features of the region being investigated. The first three are irrigable, with a decreasing capacity to recoup expenditures as time goes on. However, it is not possible to irrigate the sixth class.

Fertility Capability Classification (FCC): According to Calzolari et al. (2009), the Soil FCC that was developed in 1975 and later modified in 1982, has been examined and revised (Sanchez et al. 2003). It is a system for reclassifying soils into homogenous groups for the sake of fertility management, especially chemical fertility. This system envisions three levels: the first is defined by the texture of the surface horizon; the second is defined by the texture of the subsurface horizon (within 50 cm); and the third is defined by "specifiers" that represent chemical and physical conditions that can negatively affect soil fertility, such as moisture regime, toxic element presence, salinity, and other factors. The FCC has been given as a semi-quantitative tool for assessing the soil quality since its revision in 1982.

(ii) The FAO Framework for Land Evaluation

For more than 45 years, the FAO has developed and used land-evaluation guidelines that have been effectively implemented in many regions of the world including Ethiopia. In land evaluation exercise, the FAO (1976) framework is regarded as a standard reference (Van Diepen et al. 1991; Dent and Young 1981). However, this approach is essentially a framework that may be tweaked to fit certain circumstances, intending to establish a theoretical framework for the evaluation procedure (Davidson 1980). There are four hierarchical levels in the categorization structure: order, class, subclass, and unit (See the details in Sect. 10.4.5). The FAO framework consists of five primary components. The framework is used to document the nature and principles of land assessment (Sect. 10.1), fundamental concepts of land evaluation (Sect. 10.2), land suitability classifications (Sect. 10.3), land evaluation techniques (Sect. 10.4), and use cases of land evaluation findings from Brazil, Surinam, and Kenya (Sect. 10.5).

In the past, improvements to existing approaches have been triggered by the evolving technologies and increased awareness of the important influence of a wider range of factors on the sustainable use of land and the livelihood of land users. Subsequently, according to Rossiter (2009, 1996), the 1976 Framework is not deficient in these areas and is remarkably forward-looking. He further added that unsatisfactory results of land evaluation exercises can largely be attributed to the misguided or mechanical application of the Framework and Guideline, in particular the lack of creative thinking and difficulties forming and working in multidisciplinary teams.

Because of this, FAO issued the document "Land Evaluation: Towards a New Framework" in 2007, which mostly confirms the notions presented in the 1976 framework. According to Calzolari et al. (2009) and Rossiter (2009), one of the most significant advancements in the land evaluation process is the requirement for stakeholders to have a more active and substantial role. Both the starting stage, when the objectives, issues, and relative causes are identified, and the final stage, when the outcomes are evaluated, should be addressed with this in mind. The contribution that local communities may make with soil maps, as well as how this might be integrated into international categorization, are also highlighted.

The revised framework begins with an executive summary and Chap. 1 explains the rationale behind this work and the perceived need for revision of the framework. A chapter follows this on historical development, which gives a fair summary of the many approaches that have been taken toward land evaluation approaches advancement. The 3rd Chapter addresses the "expansion of concepts and definitions" of the land evaluation approach. The 4th Chapter "Revised principles and procedures" summarizes the previous chapter by proposing revisions to the original Framework's principles. Following the four main chapters is an outline of a proposed revised Framework to replace the 1976 FAO's Land Evaluation Framework. This is followed by an extensive annex: a glossary of terms, a useful catalog of data sources, a catalog of tools (stakeholder analysis, land resource survey, eliciting local knowledge, environmental characterization, interviews, GIS, remote sensing and modeling), and some case studies showing where these data sources and tools have been applied.

(iii) Recent Evolution of Land Evaluation Approaches

Since 1976, the framework for land evaluation has influenced several land evaluations approaches. The majority of these are based on agro-ecological, computer simulation, remote sensing, and GIS. Land evaluation processes have recently advanced significantly from methodologies utilized in the early 1970s, thanks to the expanded availability of more sophisticated computation devices and tools, such as more powerful computational infrastructures for data processing, GIS, remote sensing, terrain modeling, process models; extensive public data sources complement these (FAO 2007). Besides, experience over the past decades has shown the value of participatory approaches with multiple stakeholders, and of demand-driven approaches to identify land use options. A few approaches that were developed and applied for the land evaluation process since 1976 are briefly presented in the following section and details of all these approaches have been exhaustively outlined and discussed in Calzolari et al. (2009); Rossiter (1996) and FAO (1976).

Among others, productivity indices, soil potential ratings, land evaluation and site assessment (LESA), agro-ecological zoning, land evaluation, and farming systems analysis sequence (LEFSA), sustainable land management (SLM), and the framework for evaluating sustainable land management (FESLM) are typical land evaluation methodologies developed after the FAO framework. Besides, computerized land evaluation systems; multi-criteria analyses using geographic information systems; automated land evaluation system (ALES); land evaluation decision support system and intelligent system for land evaluation land-use capability investigation and evaluation; crop yield simulation and land assessment model for Botswana; land evaluation using earth observation; physico-mathematical modeling and driving forces and drivers, pressures, state, impacts, and response

(DPSIR) conceptual model are some of the recently additional developed land evaluation approaches.

In addition, in recent times, geoinformation technologies such as remote sensing, geographic information systems (GIS), and global positioning systems (GPS) have offered many rewards to fulfill the demand for both qualitative and quantitative spatial information for land evaluation exercises (FAO 2007). These integrated technologies have an advantage during land evaluation and land-use modeling application for data extraction (Remote sensing), field inventory purpose (GPS) as well as data storage, management, visualization, and automation (GIS). Different land evaluation models such as 'Automated Land Evaluation System' (ALES), Intelligent System for Land Evaluation (ISLE), and 'Land Evaluation and Farming Systems Analysis' (LEFSA) have been fully integrated with the GIS platform to store, manage, and map the outputs. Although various land evaluation approaches have been applied for different land evaluation exercises, the FAO land evaluation framework (Gessesse 2013; MoWR 2006, 2007, 2010; FAO 1984), automated land evaluation system (ALES) (Assen 2003; Yizengaw and Verheye 1995), integrated remote sensing, GIS, and geospatial modeling approaches are the most commonly used in many of the land evaluation exercises in Ethiopia.

10.6 Land Evaluation Practices

10.6.1 Overview

In Ethiopia, it is in the mid-1960s that some project-based catchment/basin level land evaluation studies were undertaken by foreign consulting firms as part of river basin development studies (FAO 1984, 2014; Debele 1980). However, land evaluation exercises have gained institutional form in 1971 under the then Institute of Agricultural Research in which the first soil survey and land evaluation section was established (Debele 1980). Many studies were conducted on detailed land evaluation studies of the experimental agricultural research stations and their surrounding areas to support agricultural technology scaling up efforts. Moreover, land evaluation studies were conducted, in about 4,500 km^2 areas, in *Gode, Holeta, Jima,* and *Bako* agricultural research stations at a 1:5,000 scale and *Wollenkomi-Addis Ababa* and *Sendafa-Debre Zeit* regions at a scale of 1:50,000 (FAO 1984; Debele 1980). The land evaluation exercise was based on mapping units constructed and described the biophysical factors and the unit of analysis evaluated against various land utilization types.

In addition, the Ministry of Agriculture (MoA) established the `Land Use Planning and Distribution Department well ahead renamed as the Land Use Planning and Regulatory Department. This department with development partners was engaged in conducting various land evaluation exercises between 1979 and 1990. According to Debele (1980), the department's major land evaluation efforts included: (i) national-level land evaluation study for land use master plan preparation at a 1:2,000,000 scale; (ii) three reconnaissance land evaluation studies and land use plans at a 1:250,000 scale (*Menagesha, Hayikoch and Butajira, Yerer and Kereyu* administrative zones) covering a total area of 3,245,790 ha, and (iii) three semi-detailed land evaluation and land use planning studies at 1:50,000 scale covering about 1,000,000 ha in *Borkena, Bichena, and Hossaina* project areas. The scopes of the 1984 national land evaluation mission were both for major kinds of land uses and specific land utilization types for specific and/or a combination of crops, forest, and livestock development. Further, the land suitability evaluation was carried out for two land utilization types ("low input" and "high input" rainfed and irrigated crop farming) at three suitability sub-classes (S1, S2, N) (FAO 1984).

10.6.2 Land Evaluation Practices at River Basin Level

In the later stage, integrated development master plans for major river basins of the country were being developed by the Ministry of Water, Irrigation, and Energy in a coordinated and successive way in the mid-1990s (Table 10.1). The ministry commissioned various public and/or private national and international consulting firms to conduct land evaluation, as a basis for land-use master plan development, which is based on basins' biophysical and socio-economic data generated for such purpose. Reconnaissance level (1:250,000) land evaluations for major land uses and specific land utilization types have been conducted for major river basins of Ethiopia (Table 10.1) including Omo-Gibe, Baro-Akobo, Abay, Tekeze and Mereb, Wabishebele, Genale Dawa, and Rift Valley Lakes basin (MoWR 1996, 1998, 2006, 2007, 2010).

The Ethiopia Institute of Agricultural Research (EAIR) undertook land evaluation studies at a national level in 2014 (FAO 2014). Accordingly, land suitability evaluation for specific land utilization types for rained and irrigated crop farming was conducted and a land suitability atlas was produced for specific crops including faba bean (Vicia faba), field pea, noug (Guizotia abyssinica), castor, coffee, highland sorghum, mid-land sorghum, lowland sorghum, rain-fed rice, cotton, and jatropha (EAIR 2014). The land suitability classification was expressed in four classes (S1, S2, S3 and N) following updated FAO land evaluation guidelines for rain-fed and irrigated agriculture.

Table 10.1 Basin, sub-basin, and irrigation project-based land evaluation studies (*Source* Compiled after Ali et al. 2020)

ID	Author	Project name	Scale	Area (km^2)
1	OWWDSE (2017a)	Omo-Gebe sub-basin irrigation potential assessment: A land evaluation study	1:50,000	20,180.96
2	OWWDSE (2017b)	Baro sub-basin irrigation potential assessment: A land evaluation study	1:50,000	24,956.62
3	OWWDSE (2018a)	Awash sub-basin irrigation potential assessment: A land evaluation study	1:50,000	25,986.45
4	OWWDSE (2018b)	Wabi-Shebele sub-basin irrigation potential assessment: A land evaluation study	1:50,000	76,241.70
5	OWWDSE (2014a)	Dhihidhessa sub-basin Integrated Land Use Planning Study Project: A land evaluation study	1:50,000	36,182.05
6	OWWDSE (2014b)	Upper Awash Sub-basin/Upper Awash Sub-basin Integrated Land Use Planning Study Project: A land evaluation study	1:50,000	19,373.23
7	OWWDSE (2010a)	Lege Wata sub-basin Integrated Land Use Plan Project: A land evaluation study	1:50,000	22,363.90
8	OWWDSE (2010b)	Dawa Sub-basin Integrated Land Use Plan Project	1:50,000	17,403.70
9	ECDSWC (2009)	Land Evaluation of Mille and Dirma Sub-basin	1:50,000	57.19
10	ADSWE (2012)	Land evaluation of Terru Sub-basin	1:50,000	12,159.00
11	ADSWE (2011)	East Amhara Development Corridor Land Use Planning and Environmental Study project: A land evaluation study	1:50,000	18,772.78
12	ADSWE (2011)	North Western Amhara Development Corridor Land Use Planning Study Project: A land evaluation study	1:50,000	30,866.20
18	ADSWE (2015)	Tekeze Sub-basin Land Use Planning and Environmental Study Project: A land evaluation study	1:50,000	29,094.60
13	ADSWE (2014)	Tana Sub-basin Land Use Planning and Environmental Study Project: A land evaluation study	1:50,000	15,896.55
14	OWWDSE (2012)	Gilo Sub-basin/Gambela Integrated Land Use Planning Project; A land evaluation study	1:50,000	6,520.11

10.6.3 Land Evaluation for Integrated Land Use Planning

Various land-use planning studies have been done in many regional states in which land evaluation studies are integral components. Most of these studies are at basin or sub-basin (watershed or sub-watershed) and irrigation project command areas and are mainly executed by public consulting firms such as ECDSWCo, ADSWE, and OWWDSE (Table 10.2). The land evaluation efforts were aimed at "Integrated Land Use Planning (ILUP)" or "Land Suitability Evaluation for Various Irrigated Cropping", which includes all sectoral studies to generate biophysical and socio-economic data as inputs for land evaluation exercises. The ILUP activities aim at incorporating all sectoral activity in the study area, resulting in a multi-sectoral and multi-disciplinary plan in which the plans were based on four major land suitability assessments i.e., biophysical, economic, social, and environmental.

The studies have been undertaken at various scales including (i) reconnaissance surveys at basin level, scale 1:250,000; (ii) semi-detailed surveys of sub-basins at 1:50,000 scale; (iii) detailed surveys at <1:25,000 scale, and (iv) intensive surveys at 1:5,000 scale. The scope of most land evaluation studies for ILUP and irrigated crop farming projects included biophysical land evaluation for major land uses (rain-fed, irrigated, livestock production, and multi-purpose forestry) and a specific land utilization defined in more detail, according to a set of technical specifications in a given socio-economic setting. Most of the land suitability exercises (Table 10.2) employ a combination and overlay of various land resources input data that result in the delineation of relatively uniform "land units" or agro-ecological cells. Accordingly, land suitability is usually expressed in

Table 10.2 Irrigation project-based land evaluation studies (*Source* Compiled after Ali et al. 2020

ID	Source	Project name	Scale	Area (km^2)
1	ECDSWCo (2016)	Land Suitability Evaluation of Lower Awash Multi-Purpose Project	1:10,000	150.00
2	ECDSWCo (2015)	Land Suitability Evaluation of Kuraz Sugar Development Project	1:10,000	2,529.54
3	ECDSWCo (2014)	Land Suitability Evaluation of Lower Awash Flood Protection Multi-Purpose Dam Project	1:10,000	120.00
4	ECDSWCo (2013)	Land Suitability Evaluation of Kesem-Bolahamo Sugar Development Project	1:10,000	150.00
5	ECDSWCo (2010a)	Land Suitability Evaluation of Rate Irrigation Development Project	1:10,000	134.52
6	ECDSWCo (2010b)	Land Suitability Evaluation of Kelafo/Wabishebele Multi-purpose (WS-18) Project	1:10,000	412.25
7	ECDSWCo (2009a)	Land Suitability Evaluation of Welmel Irrigation Development Project	1:10,000	123.89
8	ECDSWCo (2009b)	Land Suitability Evaluation of Humera Irrigation Project	110,000	960.85
9	ECDSWCo (2008a)	Land Suitability Evaluation of Welkayit Irrigation Project	1:10,000	451.28
10	ECDSWCo (2008b)	Land Suitability Evaluation of Raya Valley Pressurized Irrigation Project	1:10,000	225.30
11	ECDSWCo (2007a)	Land Suitability Evaluation of Welenchiti Bofa Irrigation Expansion Project	1:10,000	226.01
12	ECDSWCo (2007b)	Land Suitability Evaluation of Gumera Irrigation Project	1:10,000	139.76
13	ECDSWCo (2007c)	Land Suitability Evaluation of Gidabo Irrigation Development Project	1:10,000	286.21
14	ECDSWCo (2006)	Land Suitability Evaluation of Tendaho Sugar Development Dam and Irrigation Project	1:10,000	600.00
15	ECDSWCo (2005)	Land Suitability Evaluation of Kessem-Kebena Dam and Irrigation Project	1:10,000	218.23

four classes (S1, S2, S3, N) and several sub-classes are defined by the most limiting factor. Further, land suitability results are presented in land suitability maps and/or land suitability matrix in such a way that can depict the extent, spatial distribution, and percent share of the gross study area that is suitable for major or specific land utilization type.

10.7 Land-Use Planning

10.7.1 Concepts and Applications

Inappropriate land use practices aggravate unwise natural resource exploitation, land degradation, and deterioration of land resources as well as massive poverty and other socio-economic crises. In this regard, Pratibha et al. (2020) argued that the initial step toward sustainability is to develop and apply land use planning as well as using land according to its capabilities as well as competencies. In this connection, land evaluation is part of the process of land use planning and is concerned with the assessment of land performance when used for specified purposes sustainably. Land evaluation is also an important element of land use planning, according to UNCCD (2017), which is a process that answers questions like "Which areas of land are most suited for any given sort of land use?" and "Which kind of usage is best suited for any given area of land?".

The procedure is used to describe promising land-use types, determine requirements for each land-use type, conduct surveys to map land units and describe their physical properties, and compare the requirements of land-use types to the properties of land units to arrive at a land suitability classification (FAO 1985). Thus, land evaluation is seen as a means of guiding land use planning at the national and local levels. In broader terms, land use planning requires an understanding of the extent of land usage for different purposes. The optimal level of exploitation of natural resources based on their potential and limits is critical for the long-term growth and development of agrarian nations like Ethiopia. This is achieved through the application of land evaluation for enhancing land use planning to gain immediate and long-term benefits, determine priorities, and meet society's diversified requirements while also protecting vulnerable ecosystems and the planet earth's genetic legacy.

Land use planning, according to UNCCD (2017), is the systematic evaluation of land and water potential, land-use possibilities, and economic and social circumstances to choose and implement the optimal land-use options. Its goal is to identify and implement land uses that will best fulfill people's needs while also conserving resources for the future. In both developing and developed countries, land use planning is widely used for rural, regional, and local land-use applications. However, the process requires gathering a wide range of data, including physical, technological, economic, and institutional information, and methodologically integrating them to establish a feasible plan and program of action (FAO 1984). Land resources utilized for land use planning, in particular, are made up of a variety of components that fall into two categories and resources These land resources are natural land resources and artificial land resources. Climate, topography, geology, hydrology, soils, and vegetation are among the subgroups of the former, while cultivated land, settlement areas, roads, canals, and terraces fall under the second category.

Land-use planning should consist of the nature and purpose of a given tract of land: what it is, why it is needed, who benefits from it, and who carries it out. Similarly, it identifies the persons involved and specifies the various levels or scales at which planning is carried out by land users, decision-makers, and the planning team. On top of this, land-use planning involves establishing goals and terms of reference, organizing work, identifying opportunities for change, evaluating land suitability, evaluating alternatives, choosing the best option, preparing the land-use plan, implementing the plan, monitoring, and revising the plan (FAO 1985).

10.7.2 Land Use System

The major types of uses and sub-classes of applications to which the land is placed can describe the land-use system in Ethiopia. The most distinguishing kinds of uses of productive land are grazing, cropping (annuals as well as perennials), and woody vegetation divided between natural forests and other woody vegetation. The proportion of land put to these major kinds of uses to the productive land potential (total geographic area less un-utilizable land for agricultural purposes) is a good measure to assess the status of land use in the country. Utilizable land in agriculture including cropping, grazing, and forestry is a significant category in many parts of the country.

The function of the land is determined by natural circumstances and human activity, and it is referred to as a land-use system. Cropland, forestland, grazing land, nature reserve, aquaculture, non-agricultural land, and multifunctional uses, for example, help categorize land-use systems based on the status and service of the land. Ethiopia has over 12 primary land use systems as well as 30 sub-level land use and land cover classifications. Urban and built-up land, cultivated land, afro-alpine/sub-afro-alpine vegetation, forest land, woodland, riparian woodland, bush land, scrubland, grassland, wetland, bare land, and water bodies are the principal land use land cover groups in Ethiopia (Karra 2021; EMA 2013; FAO 1984a).

10.7.3 Land-Use Planning Practices at National Level

Since prehistoric times, Ethiopians have used some form of land use planning (FAO 1984a). Similarly, the FAO (2014) outlined that pastoralists want to transfer their herd from one location to another depending on the seasons for securing forage and water availability in Ethiopia. Similarly, sedentary farmers have developed farm planning to cultivate diverse crops in distinct agro-ecological zones. As a result, cohesive communities often make land-use plans for how they will use the land they own. In some places of the country, forest and grazing land uses are subject to various unwritten restrictions, and some trees or forests may be protected or have restricted usage. On the other hand, authorities, neighborhoods, and plot owners all have a role in land use planning in cities. Due to several drivers and influences such as unprecedented population pressure, other demographic changes, severe droughts or other natural calamities, as well as rapidly changing needs and ambitions of the resident population, traditional land use planning practices are no longer applicable, effective, or sustainable in Ethiopia. As a result, when traditional land use management approaches fail or prove inefficient, land use planning that is more systematic and institutionalized is required to boost land productivity and capacity.

Although several attempts were done to introduce land use planning, systematic and national level efforts were introduced at the beginning of the 'Dergue Regime' of Ethiopia (FAO 1984a, 2014). During this time, the feudal land-use ownership system was abolished by a land-use planning proclamation issued in 1974. Subsequently, the 'Land Distribution Department' under the Ministry of Land Reforms and Settlement was established to enforce the decree. Later on, the Ministry of Land Reform and Administration was founded to cope with land reform challenges. However, the MLRA and the MoA merged, and the MoA created the "Land Use Planning and Distribution Department" in 1977. Consequently, an agreement was reached between the Government of Ethiopia and UNDP/FAO to introduce and support land use planning in Ethiopia. The Land Use Planning and Distribution Department of the MoA was re-named as Land Use Planning and Regulatory Department.

This department was in charge of performing land evaluation studies, designing, and implementing land use planning programs in Ethiopia, and it was supported by UNDP/FAO in three phases from 1979 to 1990 under the Assistance to Land Use Planning Project. The main achievements of the ALUP project include: institutional strengthening of the LUPRD; reconnaissance of land suitability assessments; adaptation of FAO methodologies on agro-ecological zoning and land evaluation; preparation of draft guidelines for land use planning; semi-detailed land use planning studies for *Borkena, Bichena and Hossaina areas;* reconnaissance land use plans *for Menagesha, Hayikoch and Butajira, Yerer, and Kereyu* administrative zones; master land use plan preparation with land development potential map for different kinds of land uses at a scale of 1:2,000,000. The master land use plan's goal was to "provide planners and higher-level decision-makers with an objective assessment of Ethiopia's agricultural resources major constraints and potentials, to identify appropriate policies and interventions to stimulate food production for the next 25 years." Public consulting firms in some regional administrative states are currently conducting various land use-planning studies. The majority of these studies are at the basin or sub-basin level, with planning activities geared at ILUP implementation. In conjunction, the Ethiopia Strategic Investment Framework for SLM established the SLM initiative in 2008. The project allows the government, development partners, and civil society to work together to promote SLM. The MoA and development partners' SLM project primarily focuses on land administration and registration, but it also includes land use components such as participatory forest management and participatory management planning for a designated watershed/catchment.

10.7.4 Land Use Planning Practices at the Local Level

Various forms of local-level land use planning have been employed, with varying success (FAO 2014). These include:

i. "Mass Mobilization for Natural Resource Management (NRM)" was initiated in the early 1990s by the Tigray Regional State. In this program, every farming household is obliged to take part in NRM activities for a specific number of days in a year. Every household is organized into Working Groups, many of which are managed by an Agriculture Cadre. Higher institutional levels are the Kebele and the Woreda Natural Resources Conservation Commissions, respectively.
ii. The "Participatory Land Use Planning and Implementation" approach was developed in the 1990s for the *Meket Woreda* Development Project of SOS-Sahel. Planning was done at the village *("Got")* level, which was considered the core unit for community-based natural resource management. Village Development Committees, *Kebele* Land Use Committees, and *Woreda* Rural Development Committee were established;
iii. The Ministry of Agriculture in collaboration with the World Food Program (WFP) developed the "Local Level Participatory Planning Approach" in the 1990s. A manual was produced and about 900 plans covering 500,000 ha were prepared.
iv. "Participatory Forest Management" was piloted in several sites in the Oromiya Regional State in the late 1990s by regional forest authorities with support from international development agencies such as GTZ, Farm Africa, and SOS-Sahel.
v. 'The Land Use Planning and Resource Management Project' in Oromia Region introduced "Participatory Land Use Planning" with technical assistance from GTZ. This project lasted from 1997 to 2003 and involved Participatory LUP at Village, Kebele, Woreda, and Zone levels and the formulation of Community Action Plans. The MoA introduced "Community-based Participatory Watershed Development" in 2005 to standardize local level LUP methodology. Watersheds covering 200 to 500 ha were adopted as planning units and Watershed Teams were established at Woreda, Kebele, and Community levels, respectively.

In general, both the national and municipal master land-use plans were not adequately executed or enforced, in part due to major political shifts in the years following and weaknesses in the planning process. Besides, lack of coordination between ministries, departments, and institutions involved in various aspects of land-use and rural development programs; insufficient legal framework, and institutional capacity to support land-use plan implementation and low awareness among decision-makers regarding the use of land use planning were some of the barriers.

10.8 Challenges and Opportunities

Land evaluation and land-use planning are critical tools for tracking land resource dynamics and degradation concerns, where immediate action is needed to enhance environmental sustainability. Subsequently, land evaluation and land use planning confront similar issues to address unplanned growth of land use demand trends and for improving strategic planning and the sustainability of farmed areas. The key challenges and opportunities of land evaluation and land use planning for the environmental resource sustainability of farmlands in the country are discussed in this section of the chapter.

10.8.1 Constraints of Land Evaluation

In Ethiopia, rapid population growth continued reliance on agriculture, and expanding urbanization have resulted in inefficient land use and degradation of environmental resources (Admassie 2000). Land-use conflicts and disputes have arisen resulting from increased land demand in rural and urban areas, as well as in and around metropolitan areas, between diverse interest groups such as the community and the government. As a result, since the 1980s, the Ethiopian government has included the development of national and regional integrated land use plans in its national development strategy, as well as established the national integrated land use planning and policy office, which is responsible for cross-sectoral coordination of land use planning processes at all levels of the government (Gebeyehu et al. 2017). However, failure of appropriate land evaluation and land use planning policy development and implementation has been observed across the country. In this regard, identifying the major constraints of land evaluation and land use planning development and implementation in the country could help the policymakers and implementers to find an alternative solution for successful implementation. Some of the major constraints of land evaluation and land use planning development and implementation program in Ethiopia are low levels of awareness among experts and policymakers, lack of technical standards and capacity constraints, poor linkages among the concerned institutions, and policy and strategy application constraints. Below, we go through each of the primary restrictions in depth.

(i) Lack of Awareness, Technical Standards, and Limited Capacity Constraints

Even though land evaluation and land use planning procedures have been used in Ethiopia for a long time, there is a crucial gap between experts and decision-makers in terms of understanding the nature, methodologies, and technical standards of these processes. In addition, in Ethiopia, there is a serious shortage of technical specialists that can design and implement land evaluation, land use planning standards, and methods (FAO 1984a, 2014). Furthermore, experts and decision-makers at all levels lack competency in technical issues and procedures connected to land appraisal and planning programs. The lack of focused and in-depth training; poor quality of training modalities; well-experienced and professional staff turnover from agricultural and land-use-related sectors; and a lack of appropriate and up-to-date guidelines, standards, and reference materials to assist professionals in improving the quality of their interventions and guiding communities in better directions are all contributing to these challenges. Furthermore, there are major limits in the research system competence in terms of natural resource management, land management, land governance, land assessment, and land use planning programs at the national level.

(ii) Poor Linkages among the Concerned Institutions

It is very natural that land evaluation and land use planning should be conducted in terms of the biophysical, economic, social, and political demission of a country. In this context, they need the involvement as well as engagement of multi-disciplinary professionals, experts, and even decision-makers. According to the FAO (1993, 2007), land evaluation and land use planning processes require a multi-disciplinary approach, i.e., the involvement of a range of specialists from the fields of natural science, remote sensing and geospatial sciences, the technology aspect of land use, economics and sociology and even decision-makers. The different specialists may work in a team or successively. However, presently in Ethiopia, there are no multi-stakeholder or inter-ministerial level compositions on land evaluation or land use planning program implementation. Therefore, there should be a multi-stake-holder level team composed of different fields, departments, bureaus, and ministries at national, regional, and local levels to run land evaluation and land use planning successfully. This team will develop and update land evaluation guidelines and standards, land use policy, coordinate land use planning activities, manage training at the national and regional levels, advise the government on issues concerning the use and management of land resources, and make final decisions in the event of conflicting land-use objectives.

(iii) Policy and Strategy Implementation Constraints

Although several laws, policies, and guidelines help with full land evaluation and land use planning programs such as the current land evaluation guideline of Ethiopia (FAO 1984a), Rural Land Administration and Land Use Proclamation (FDRE 1997), Ethiopia's Strategic Investment Framework for Sustainable Land Management (MoARD 2008), Ethiopia's Agricultural Sector Policy and Investment Framework (MoARD 2010), Environmental Policy of Ethiopia (EPA and MoEDC 1997), and River basins Councils and Authorities proclamation (MoWR 2007) are formulated and available in Ethiopia, however, proper implementation has not met with success. Thus, to realize land evaluation and land use planning that contribute to sustainable development, the government of Ethiopia should design land use policy and guideline enforcement, and implementation strategies at all levels.

(iv) Limited and Inconsistent Datasets Availability for Land Evaluation:

One of the key obstacles to land assessment studies in Ethiopia is the lack of spatially precise datasets and information on the country's biophysical and socioeconomic datasets at all sizes and levels of research (FAO 2014). Furthermore, because it is such a huge and diverse country, the challenges today facing thorough land suitability assessments are significant. To fix this condition and create current, quantitative, comprehensive, and geographically consistent biophysical and socio-economic information at diverse spatial scales with appropriate precision with low level of uncertainty, a significant and strategically deployed resource investment will be required.

10.8.2 Opportunities to Promote and Realize Land Evaluation

(i) Existence of Environmental and Land Use Policies

Ethiopia has made impressive legislative and strategic initiatives to address environmental degradation and land resource management (USAID 2004). The country's Environmental Policy, which is one of the most important environmental-related umbrella policies, was designed and adopted in 1997 (EPA 1997). This policy is well-targeted to address a variety of sectoral and cross-sectoral environmental issues. The goal was to ensure that natural, human-made, and cultural resources, as well as the environment, were appropriately utilized and conserved. Following that, to execute the aforementioned umbrella policy, some sectoral policies and plans were drafted and authorized. Furthermore, the Federal Democratic Republic of Ethiopia Rural Land Administration and Land Use Proclamation (No. 456/2005) was enacted to provide a legal framework for the administration and use of rural land (FDRE 2005).

Following the formulation of the Federal Government of Ethiopia's "Rural Land Administration Proclamation," the Amhara, Oromia, Tigray, and Sothern Peoples, Nations, and Nationalities regional states formulated regulations and procedures for the implementation of the "Rural Land Administration and Land Utilization Proclamation," which will be implemented in each respective region (USAID 2004). Despite efforts being made to develop land-use planning and natural resource management initiatives, enforcement and regulatory mechanisms were lacking in their execution. Besides, a lack of legally enforceable regulatory instruments impeded the efforts of several Regions, Zones, and Woredas to undertake rural land use planning of varied sizes and methodologies. In addition, studies reveal that the country's natural resources are fast depleting, highlighting the need for a national land-use strategy that maximizes resource distribution while simultaneously encouraging conservation and long-term usage (EPA 1997). In general, the country's existing environmental proclamations and policies will help to improve land evaluation and land use planning operations significantly. If the land evaluation exercise and implementation of land use planning are expanded across Ethiopia, land resource use efficiency will improve to support the growing population, realization of land use planning, reverse natural environment degradation, and provide equitable and efficient access to the economic benefits of the land.

(ii) The Practice of Community-based Participatory Watershed Development: The planning and implementation of Ethiopia's watershed development program began in the 1980s (Chimdesa 2016; Worku and Sangharsh 2015; Birhanu 2014; Gete et al. 2006; Lakew et al. 2005; Admassie 2000). The main goal was to put natural resource protection and development plans in place. However, large-scale efforts remained mostly unsatisfactory due to a lack of effective community participation, a limited sense of responsibility for assets created, and unmanageable planning units. The lessons learned from this experience encouraged MoA and support agencies like the FAO and international development partners to initiate watershed planning approaches on a bottom-up basis, using smaller units and following community-based approaches.

Meanwhile, since 2005, a guideline for community-based participatory watershed development has been established and implemented to enhance watershed management in Ethiopia (FAO 2014; Lakew et al. 2005). There are two components to this guideline. The first portion, divided into three sections, describes the procedures to be taken, interventions to be made, and technologies to be used in Ethiopia's community-based participatory watershed development initiative. The annex, includes participatory mapping and understanding of the target area, participatory planning and socioeconomic survey, biophysical survey and mapping, sample survey methods, interventions and their suitability, a summary of national work norms, a list of useful plant species, community-based solidarity efforts, and planning formats-samples, is the second part of the guideline. As a result, conducting a complete biophysical parameter assessment for watershed management and development gives useful information to local land evaluators, land-use planners, and managers, which is used to implement land evaluation and land use planning programs successfully in Ethiopia.

(iii) Rich Experience in Land Evaluation Practice since the 1980s: The availability of extensive experience in land evaluation and land use planning (Debele 1980; FAO 1984, 2014; Esayas 2001; Esayas and Debele 2006) provides a good prospect for Ethiopia to scale up land evaluation exercises. Basic surveys of climates, soils, vegetation, and other features of the land by the Federal, Regional, Zonal, and 'Woreda' levels, as well as research institution experts and professionals, have been interpreted in terms of the requirements of various types of land use in land evaluation exercises. These might be broad categories of land use such as rainfed agriculture, animal production, forestry, and so on, or more specific land-use types.

Besides, many stakeholders have been involved in land evaluation tasks and the genuine participation of stakeholders and communities at all levels of the decision-making process is one of the key requirements of successful land evaluation and land use planning undertakings. Accordingly, there are very good land evaluation and land use planning experiences in the country as mentioned in Sect. 4.2; however, there are many issues for different approaches, which need careful inquiry. In this regard, the government has recognized the need for participatory land evaluation and land use planning and recently. Gebeyehu et al. (2017) recommended a national guideline known as Ethiopia's move to a national integrated land use policy and land use plan, by taking the potential procedures from selected approaches in the country. This is one of the most important steps taken by the government for effecting land evaluation and land use planning implementation in the country.

(iv) Initiatives of Donor Support and Development Partners Program: A variety of multilateral and bilateral international development collaborative organizations and donors have provided, and continue to offer, financial and technical assistance to Ethiopia's Federal and Regional governments to enhance land resources management in the country. The projects and programs of these international development partners will provide most of the necessary resources and logistics for sustainable land management practices (MoARD and WB 2007). Furthermore, in Ethiopia, there are more than 300 non-governmental and international organizations (NGOs) that are actively involved in activities relating to sustainable land management and monitoring areas in the country (Gessesse 2010). Land resource conservation, management, and development are typically significant components of international and local NGO projects. Land evaluation and land use planning techniques for sustainable land resource management are also being promoted by NGOs, including the integration of soil, water, and forest management with income creation, credit, value-added chains connected to marketing, and so on. Integrating land evaluation and land use planning procedures with non-governmental organizations (NGOs) are, therefore, a great way to boost efforts.

(v) Expansion of Agricultural and Land Administration Education Institutes: Agricultural land evaluation, and land use planning, as well as natural resource management professionals with various degrees, are trained by higher learning institutes that integrate agriculture, natural resource management, and environmental education into their curriculum. The number of institutions and graduates has recently risen dramatically (MoARD and WB 2007). This is a significant step forward for Ethiopia's efforts to produce skilled full, knowledgeable land evaluation and land use planning professionals as well as to promote and implement effective land evaluation and land use planning practices and resource management techniques.

10.9 Conclusions

Land degradation and irresponsible land management practices are harmfull to soil resource and consequently to all food and water security practices as well as lead to diminished agricultural production and jeopardizes farmers' and the rural economy's ability. Sustainable land management (SLM) practices are required to reverse these trends and tackle fundamental roadblocks to Ethiopia's goal of achieving green and resilient economic growth. Land evaluation and land use planning are considered strategic instruments to manage agricultural landscape dynamics in most developing nations, including Ethiopia, where urgent interventions are needed to enhance environmental sustainability. They confront similar issues to address unplanned growth of land use demand trends and for improving strategic planning and the sustainability of cultivated lands. This review looked into and identified the need for, and the history, basics, principles, scale, data sources, approaches, practices status, key challenges, and opportunities of land evaluation and land use planning in Ethiopia. Land evaluation is always should be applied for analyzing the land for cost-effective purposes, while also safeguarding the environment's diverse ecological functions from any degrading agents and processes. The quest for arable land necessitates a full-scale spatio-temporal rearrangement, and the more logical use of the existing land is unquestionable in Ethiopia. This is only possible if the land's performance is properly identified and a thorough evaluation of the land's relative appropriateness for various uses is performed. In this context, although land evaluation practices were started in the nineteenth century, it is also needed for describing, assessing, and presenting as well as mapping various biophysical factors and acceptable land-use plans for optimal and relevant decision-making for crop cultivation and other useful recommendations in Ethiopia.

Besides, the basics of land evaluation such as the land-mapping unit, major kinds of land use and land utilization types, land characteristics, land qualities, diagnostic factors, land use requirements, limitations, suitability classification, and cartographic spatialization are reviewed and documented. Multiple qualitative and quantitative-based land evaluation approaches have been established so far to evaluate the land for various applications. As a result, the historical development of land evaluation approaches has been clustered into three phases. These are land evaluation techniques that were produced before the FAO framework (FAO 1976), the FAO land evaluation framework, and recent land evaluation approaches that were developed following the FAO Framework.

Since the 1970s, when Ethiopia established the first soil survey and land evaluation department, land evaluation exercises have been carried out. Many extensive land evaluation studies have been done since then on the experimental agricultural research station sites, sub-watershed, watershed, sub-basin, and basin scales. Following the introduction of the land evaluation concept in Ethiopia, the necessity of introducing the fundamental concept of land use planning was coined in many land resource conservation practices in the country. It can be concluded that the role of land use planning in Ethiopia is critical in guiding land-use decisions in such a manner that environmental resources are put to the best productive use for man while also safeguarding those resources for the future. The various land use exercises in Ethiopia revealed that land use planning was based on a thorough grasp of both the natural environment and the types of land use that were envisioned. Land use planning is a multi-sectoral, multi-stakeholder, and scale-dependent participatory process that evaluates land potential and options for optimal land uses and better economic and social circumstances. Besides, it is helpful to assist decision-makers and land users in identifying and implementing land use that best satisfies people's needs while also protecting natural resources and ecosystem services for current and future generations.

It can be concluded that some of the major constraints of land evaluation and land use planning development and implementation program in Ethiopia are at low levels of awareness among experts and policymakers, lack of technical standards and capacity constraints, poor linkages among the concerned institutions, and policy application constraints. On the other hand, the existence of environmental and land use policies, ironic familiarity with the community-based participatory watershed development program, rich experience in land evaluation practice since the 1980s, presence of donor support and development partners in land management activities, expansion of agricultural and land administration education and training institutes are identified opportunities to enhance land evaluation and land use planning in Ethiopia. In general, the chapter provided the baseline information regarding the practices and status of land evaluation and land use planning and acquired and synthesized information on land suitability analysis for crop production. In the future, land use planning tools and procedures at suitable scales should encourage and aid different and frequently competing land use and management alternatives that boost productivity, support sustainable agriculture and food systems, promote land and water resource management and governance, and satisfy societal demands. Consequently, expanding the current land evaluation and land use planning practices will help to protect soil degrdation, vegetation clearing, natural habitat destruction, urban and peri-urban encroachment to agricultural land, groundwater quality and quantity from pollution, and the rehabilitation of degraded and contaminated sites.

References

Admassie Y (2000) Twenty years to nowhere: property rights, land management and conservation in Ethiopia. Red Sea Press

Ali A, Tamene L, Erkossa T (2020) Identifying, cataloguing, and mapping soil and agronomic data in Ethiopia. CIAT Publication No. 506. International Center for Tropical Agriculture (CIAT). Addis Ababa, Ethiopia, p 42

Amhara Design and Supervision Works Enterprise (ADSWE) (2011a) East Amhara development corridor land use planning and environmental study project: a land evaluation study. Bahir Dar, Ethiopia

Amhara Design and Supervision Works Enterprise (ADSWE) (2011b) North Western Amhara development corridor land use planning study project: a land evaluation study. Bahir Dar. Ethiopia

Amhara Design and Supervision Works Enterprise (ADSWE) (2012) Land evaluation of Terru Sub Basin. Bahir Dar, Ethiopia

Amhara Design and Supervision Works Enterprise (ADSWE) (2014) Tana sub basin land use planning and environmental study project: a land evaluation study. Bahir Dar, Ethiopia

Amhara Design and Supervision Works Enterprise (ADSWE) (2015) Tekeze sub basin land use planning and environmental study project: a land evaluation study. Bahir Dar, Ethiopia

Assen M (2003) Land suitability evaluation in the Jelo catchment, Chercher highlands, Ethiopia. Thesis (PhD). The University of the Free State

Birhanu A (2014) Environmental degradation and management in Ethiopian highlands: review of lessons learned. J Environ Protect Policy 2:24–34

Calzolari C, Costantini EAC, Venuti UF, L, (2009) Soil and land evaluation:history, definitions and concepts. In: Costantini EAC (ed) Manual of methods for soiland land evaluation. Science Publishers, Enfield (NH), Jersey, Plymouth, pp 3–6

Chimdesa G (2016) Historical perspectives and present scenarios of watershed management in Ethiopia. Int J Natl Resour Ecol Manage 1(3):115–127

Davidson DA (1980) Soils and land use planning. longman, London, p 129

Debele B, Ochtman LHJ (1974) Soil resource appraisal and evaluation studies for ruraldevelopment in Ethiopia. A country report presented at the East African soilCorrelation Committee Nairobi, 11–16 March 1974

Debele B (1980) Survey of studies conducted about soil resources appraisal and evaluation for rural development in Ethiopia. Institute of Agricultural Research. Addis Ababa. Ethiopia

Debele B (1994) The soils of Ethiopia: annotated bibliography. Technical Handbook no. 9. Regional Soil Conservation Unit, Swedish International Development Authority. Nairobi, Kenya

Dent D, Young A (1981) Soil survey and land evaluation. George Allen and Unwin, London

Deressa T, Hassan RM, Alemu T, Yesuf M, Ringler C (2008) Analyzing the determinants of farmers' choice of adaptation methods and perceptions of climate change in the Nile Basin of Ethiopia. IFPRI Discussion Paper 00798: Environment and Production Technology Division.

Ethiopian Construction Design and Supervision Works Corporation (ECDSWCo) (2005) Land suitability evaluation of Kessem-Kebena dam and irrigationproject. Addis Ababa. Ethiopia

Ethiopian Construction Design and Supervision Works Corporation (ECDSWCo) (2006b) Land suitability evaluation of Tendaho sugar development dam andirrigation project. Addis Ababa. Ethiopia

Ethiopian Construction Design and Supervision Works Corporation (ECDSWCo) (2007a) Land suitability evaluation of Welenchiti Bofa irrigation expansion project. Addis Ababa. Ethiopia

Ethiopian Construction Design and Supervision Works Corporation (ECDSWCo) (2007b) Land suitability evaluation of Gumera irrigation project. AddisAbaba, Ethiopia

ECDSWCo (2007c) Land suitability evaluation of Gidabo Irrigation Development Project, Addis Ababa, Ethiopia

Ethiopian Construction Design and Supervision Works Corporation (ECDSWCo) (2008a) Land suitability evaluation of Welkayit irrigation project. AddisAbaba, Ethiopia

Ethiopian Construction Design and Supervision Works Corporation (ECDSWCo) (2008b) Land suitability evaluation of Raya Valley pressurized irrigationproject. Addis Ababa. Ethiopia

Ethiopian Construction Design and Supervision Works Corporation (ECDSWCo) (2009a) Land suitability evaluation of Welmel irrigation development project. AddisAbaba. Ethiopia

Ethiopian Construction Design and Supervision Works Corporation (ECDSWCo) (2009b) Land suitability evaluation of Humera irrigation project. Addis Ababa, Ethiopia

Ethiopian Construction Design and Supervision Works Corporation (ECDSWCo) (2010a) Land suitability evaluation of rate irrigation development project. AddisAbaba. Ethiopia

Ethiopian Construction Design and Supervision Works Corporation (ECDSWCo) (2010b) Land suitability evaluation of Kelafo/ Wabishebele multi-purpose (WS-18)project. Addis Ababa. Ethiopia

Ethiopian Construction Design and Supervision Works Corporation (ECDSWCo) (2013) Land suitability evaluation of Kesem-Bolahamo sugar development project. Addis Ababa. Ethiopia

Ethiopian Construction Design and Supervision Works Corporation (ECDSWCo) (2014) Land suitability evaluation of lower awash flood protection multi-purposedam project. Addis Ababa. Ethiopia

Ethiopian Construction Design and Supervision Works Corporation (ECDSWCo) (2015) Land suitability evaluation of Kuraz sugar development project. AddisAbaba. Ethiopia

Ethiopian Construction Design and Supervision Works Corporation (ECDSWCo) (2016) Land suitability evaluation of lower awash multi-purpose project .AddisAbaba. Ethiopia

Environmental Protection Authority (1997). The environmental policy of Ethiopia. Addis Ababa, Ethiopia

Environmental Protection Authority (EPA) and Ministry of Economic Development and Cooperation (MoEDC) (1997) Environmental Policy, Addis Ababa, Ethiopia

Esayas A, Debele B (2006) Soil survey in Ethiopia: the past, present and future. In: Proceedingsof the seventh conference of the Ethiopian society of soil science on soils for sustainabledevelopment, 27–28, April 2006, Addis Ababa, Ethiopia, pp 61–79

Esayas A (2001) Application of Geographic Information System (GIS) for soil resource study inEthiopia. In: Proceedings of the national sensitization workshop on agro metrology and GIS,17–18 December 2001. Addis Ababa, Ethiopia, pp 162–169

Ethiopian Institute of Agricultural Research (EIAR) (2014) Land suitability atlas for selected crops of Ethiopia. Addis Ababa, Ethiopia

Ethiopian Mapping Agency (2013) National Atlas of Ethiopia. Addis Ababa, Ethiopia

Ewnetu Z, Bliss JC (2010) Tree growing by smallholder farmers in the Ethiopian highlands. In: Medved M (ed) Small-scale forestry in a changing world: opportunities and challenges and the role of extension and technology transfer, 06–12 June 2010 Bled, Slovenia. Ljubljana: Slovenia Forestry institute and Slovenia Forest Service, pp 166–187

FAO (1976) A framework for land evaluation. Soils Bulletin 32. Rome: Food and Agriculture Organization of the United Nations

FAO (1983) Guidelines: land evaluation for rainfed agriculture. Soils Bulletin 52. Rome: Food and Agriculture Organization of the United Nations

FAO (1984a) Assistance to land-use planning, Ethiopia. Land-use, production regions, and farming systems inventory, based on the work of V.A.O. Odenyo. Rome. 121 p 16 maps. AG: DP/ETH/78/003, Technical Report 3

FAO (1985) Guidelines: land evaluation for irrigated agriculture. Soils Bulletin 55. Rome: Food and Agriculture Organization of the United Nations

FAO (1993) Guidelines for land-use planning. Development series 1. Rome: Food and Agriculture Organization of the United Nations

FAO (2000) Manual on integrated soil management and conservation practices. Land and Water Bulletin 8. Rome: Food and Agriculture Organization of the United Nations

FAO (2007) Land evaluation: towards a revised framework. Land and water discussion paper 6. Rome: Food and Agriculture Organization of the United Nations

FAO (2014) Land use planning guidelines - Ethiopia. FAO/MoA Technical Cooperation Project TCP/ETH/3402. Technical document 1. Rural Land Administration and Utilization Directorate of the Ministry of Agriculture. Addis Ababa

FAO (2017) Land resource planning for sustainable land management: Current and emerging needs in land resource planning for food security, sustainable livelihoods, integrated landscape management, and restoration, Food and Agricultural Organization, United Nations

Federal Democratic Republic of Ethiopia (FDRE) (1997) (995) Proclamation no. 89/1997: Ethiopia's federal rural land administration and land use planning. Addis Ababa. Ethiopia

Federal Democratic Republic of Ethiopia (FDRE) (2005) Rural land administration and land use proclamation (no. 456/2005), Addis Ababa, Ethiopia

Federal Democratic Republic of Ethiopia (FDRE) (2007) River basin councils and authorities proclamation, proclamation no. 534/2007

Gebeyehu Z, Woldegiorgis SB, Belete AD, Abza TG, Desta BT (2017) Ethiopia's move to a national integrated land use policy and land use plan. In: Paper prepared for presentation at the "2017 World Bank conference on land and poverty". The World Bank - Washington DC, March 20–24, (2017)

Gebrehiwot G, Melgani F, Gessesse B (2019) Rainfall retrieval and drought monitoring skill of satellite rainfall estimates in the Ethiopian Rift Valley Lakes Basin. J Appl Remote Sens 13 (1):014522. https://doi.org/10.1117/1.JRS.13.014522

Gessesse B (2010) Review of land degradation and land management in Ethiopia up to 2008/09. In: Edwards S (ed) Ethiopian environment review. Forum for environment (FfE), Addis Ababa, Ethiopia, pp 187–214

Gessesse B (2013) Characterization and modelling of landscape transformation for optimizing agricultural land-use in the ethiopian highlands: a case study of Modjo Watershed. PhD Dissertation, Addis Ababa, Ethiopia

Getahun Z, Bedada TB, Gessesse B, Vermeer M (2018) Positional accuracy test of digital orthophoto mapping: case Bahir Dar City, Ethiopia. Nordic J Surv Real Estate Res 13(1):7–17

Hc H (1993) Land degradation, famines and resource scenarios in Ethiopia. In: Pimentel D (ed) World Soil erosion and conservation. Cambridge University Press, Cambridge, pp 27–62

Karra K (2021) Global land use/land cover with Sentinel-2 and deep learning. In: IGARSS 2021–2021 IEEE international geoscience and remote sensing symposium. IEEE (2021)

Kassa T, Van Rompaey A, Poesen J (2010) Land suitability assessment for different irrigation methods in Korir Watershed, Northern Ethiopia. J Drylands 3(2):214–219 (2010)

Lakew D, Kassie M, Benin S, Pender J (2000) Land degradation in the highlands of Amhara Region and strategies for sustainable land management. Livestock policy analysis program. Working paper No. 32. Nairobi: International Livestock Research Institute (ILRI)

McRae SG, Burnham CP (1981) Land evaluation. Clarendon press, London

Mengistu M, Gessesse B, Bedada TB, Tibebe D (2019) Evaluation of long-term satellite-based retrieved precipitation estimates and spatio-temporal rainfall variability: the case study of Awash Basin, Ethiopia. In: Melesse AM, Abtew W, Senay G (eds). Extreme hydrology and climate variability: monitoring, modeling, adaptation and mitigation. Elsevier Inc., Amsterdam, pp 23–35

Ministry of Agriculture and Rural Development (MoARD) (2010) Ethiopia's Agricultural Sector Policy and Investment Framework (PIF) 2010–2020. Draft Final Report. 15 September 2010. Addis Ababa

Ministry of Agriculture and Rural Development and the World Bank (2007) Ethiopia: thematic papers on land degradation in Ethiopia. Addis Ababa: Ministry of Agriculture and Rural Development and World Bank

Ministry of Agriculture and Rural Development (2008) Ethiopian strategic investment framework for sustainable land management. Addis Ababa: Ministry of Agriculture and Rural Development and Sustainable Land Management Secretariat (Unpublished)

Ministry of Finance and Economic Development (MoFED) (2003) Rural development policy and strategies. Addis Ababa

Ministry of Water Resources (MoWR) (1996) Integrated development of Omo-Gibe river basin master plan study, vol XI F1, F2, F3, Addis Ababa, Ethiopia

Ministry of Water Resources (MoWR) (1998) Abay river basin integrated development master plan project phase 2: Land Resources Development. Addis Ababa

Ministry of Water Resources (MoWR) (2006) Genale Dawa river basin integrated resource development master plan study project. Master plan report: land evaluation/land suitability. Addis Ababa

Ministry of Water Resources (MoWR) (2007) Wabeshebele Gennale-Laga Wata/sure, rivers basin Integrated development master plan study. Addis Ababa

Ministry of Water Resources (MoWR) (2010) Rift valley lakes basin integrated resources development master plan study. Phase 3. Final report. Western Lake Ziway Sub-basin integrated watershed management feasibility study. Volume 1: Main Report

Mitiku H (1996) Genesis, characteristic and classification of the Central Highland soils of Ethiopia. PhD thesis, State University of Ghent, Belgium

Oromia Water Works Design and Supervision Enterprise (OWWDSE) (2010a) Lege Wata sub-basin integrated land use plan project: a land evaluationstudy. AddisAbaba. Ethiopia

Oromia Water Works Design and Supervision Enterprise (OWWDSE) (2010b) Dawa sub basin integrated land use plan project. Addis Ababa. Ethiopia

Oromia Water Works Design and Supervision Enterprise (OWWDSE) (2012) Gilo sub basin-Gambela integrated land use planning project: a landevaluation study. Addis Ababa. Ethiopia

Oromia Water Works Design and Supervision Enterprise (OWWDSE) (2014a) Dhihidhessa sub basin integrated land use planning study project: a landevaluation study. Addis Ababa. Ethiopia

Oromia Water Works Design and Supervision Enterprise (OWWDSE) (2014b) Upper Awash sub basin integrated land use planning study project: a landevaluation study. Addis Ababa. Ethiopia

Oromia Water Works Design and Supervision Enterprise (OWWDSE) (2017a) Omo Gebe sub basin irrigation potential assessment: a land evaluation study. AddisAbaba. Ethiopia

Oromia Water Works Design and Supervision Enterprise (OWWDSE) (2017b) Baro sub basin irrigation potential assessment: a land evaluation study. AddisAbaba. Ethiopia

Oromia Water Works Design and Supervision Enterprise (OWWDSE) (2018a) Awash sub basin irrigation potential assessment: a land evaluation study. Addis Ababa. Ethiopia

Oromia Water Works Design and Supervision Enterprise (OWWDSE) (2018b) Wabi Shebele sub basin irrigation potential assessment: a land evaluationstudy. Addis Ababa. Ethiopia

Pratibha TD, Saikia B, Raju PLN (2020) Land use planning using geospatial technology and soil health card data for a micro watershed in sub tropical humid region of Meghalaya. Agri Res Tech: Open Access J 25(2):556301. https://doi.org/10.19080/ARTOAJ.2020.25.556301

FAO (1984b) Assistance to land use planning project. Ethiopia. Land evaluation. Part 1, Introduction and Procedures. R.B. Ridgway. Addis Ababa (1982), p 62. AGOA: ETH/78/003. Technical Report 5, Part 1

Rossiter DG (1996) A theoretical framework for land evaluation. A Discussion Paper. Geoderma 72:165–202

Rossiter DG (1994) Lecture notes: "Land Evaluation". http://www.css.cornell.edu/landeval/le_notes/lecnot.htm

Rossiter DG (2009) Land evaluation: towards a revised framework; Land and water discussion paper 6, FAO. FAO, Rome (2007), p 107, ISSN: 1729–0554; Geoderma 148:428–429

Sanchez PA, Palma CA, Buol SW (2003) Fertility capability soil classification: a tool to help assess soil quality in the tropics. Geoderma 114:157–185

Shiferaw B, Holden S (1998) Resource degradation and adoption of land conservation technologies in the Ethiopian highlands: a case study in Andit Tid. North Shewa Agricult Econ 18(3):233–247

Sys IC, Van RE, Debaveye IR (1991a) Land evaluation: principles in land evaluation and crop production calculations, part I. Agricultural publication no 7. Ghent: ITC, University of Ghent

Sys IC, Van RE, Debaveye IJ (1991b) Land evaluation: methods in land evaluation, part II. Agricultural publication no 7. Ghent: ITC, University of Ghent

Sys IC, Van RE, Debaveye, IR (1993) Land evaluation: crop requirements, part III. Agricultural Publication No 7. Ghent: ITC, University of Ghent

Tesfaye G, Tadesse T, Gessesse B, Denku T (2018) Validation of new satellite rainfall products over the Upper Blue Nile Basin, Ethiopia. J Atmosp Meas Tech 11:1–16. https://doi.org/10.5194/amt-11-1-2018

United Nations (UNs) (1995) Chapter 40: Information for decision-making and Earthwatch. Commission on Sustainable

Development, Economic and Social Council E/CN.17/1995/7. United Nations, New York
United Nations Convention to Combat Desertification (UNCCD) (2017) Land use planning: global land outlook working paper. Sydney, Australia
United States Aid International Development (USAID) (2004) Ethiopia land policy and administration assessment final report with appendices. Addis Ababa, Ethiopia
Van Diepen VA, Van Keulen H, Wolf J, Berkhout JA (1991) Land evaluation: from intuition to quantification. Adv Soil Sci 15:139–204
Woldemariam M (1991) Suffering under God's environment: a vertical study of the predicament of peasants in north-central Ethiopia. Berne: African Mountains Association and Geographica Bernensia
Worku T, Sangharsh KT (2015) Watershed Management in highlands of Ethiopia: a review watershed management in highlands of Ethiopia. Open Access Library J 2:e1481
Woubet GA, Tadele A, Birru Y, Yihenew G, Bettina W, Hans H (2013) Impacts of soil and water conservation on land suitability to crops: the case of Anjeni Watershed, Northwest Ethiopia. J Agric Sci 5 (2):95–103
Yizengaw T, Verheye W (1995) Application of computer captured knowledge in land evaluation, using ALES in Central Ethiopia. Geoderma 66(3–4):297–311
Young A (1980) Current land evaluation activities in the Land USE Planning Department of the Ministry of Agriculture and on the future activities and staffing of a national Land Use Services Institution for Ethiopia. Consultant's report no. 2. AGOA: ETH/78/003. FAO, Rome
Zeleke G Menale K, Pender J, Mahmud Y (2006) Stakeholder analysis for sustainable land management (SLM) in Ethiopia: assessment of opportunities, strategic constraints, information needs, and knowledge gaps. Environmental Economics Policy Forum for Ethiopia (EEPFE), Addis Ababa, Ethiopia

Soils and Society

11

Kibebew Kibret, Girma Abera, and Sheleme Beyene

Abstract

Humans and soils have been intricately linked since antiquity, for soil touches people's lives in many ways, including serving as the source of food and clothing, and for its multiple ecological services, such as filtering drinking water and maintaining environmental health (for which it is often called 'a geologic kidney' by environmentalists). It has influenced the rise and decline of many civilizations across the world. However, improper use and management of this basic natural resource have led to its gradual degradation with consequent loss in capacity to function. It influences human life in many ways that include, but not limited to, wealth, nutrition (quantity as well as quality of food), and health. It affects human health directly or indirectly and positively or negatively. The indirect effects come from the food produced on soils since the nutritional value of many foods is markedly influenced by the soil's ability to supply essential nutrients to food systems and most of the elements that are essential for plants are also essential for human health. The direct effects could come from the exposure of humans to soil contaminated by various chemicals and pathogens through ingestion, inhalation or absorption. Geophagy, the habit of eating soil, is often one way through which humans ingest soil. The harmful substances and pathogens as well as deficiencies of nutrients could be causes of many diseases of complex nature. The soil materials to which humans could be exposed include heavy metals, organic pollutants, toxic materials in fertilizers and other agro-chemicals such as pesticides and herbicides, radioactive materials, pathogens, and polluted water. On the positive side, soils have been the origins of a large number of medicines (e.g., antibiotics) that are used for curing or treating a large number of human health problems. In Ethiopia, where agriculture is the main stay of the economy and supports the livelihood of about 79% of the population, soil is a strategic resource. However, unwise use and improper management have exposed it to degradation of intense nature, reducing its capacity to support life systems and exposing the country to persistent struggles to ensure food security. Fertility depletion is among the challenges the country has been facing since many decades, leading to the production of food that is both insufficient and nutrient-poor. Most of the beneficial physical, chemical, and biological soil attributes are below their expected threshold levels and thus the soils are unhealthy. The country carries the greatest burden of many of the soil-transmitted diseases such as helminths and podoconiosis. Most of the cultivated soils require restoration interventions that help them regain their quality.

Keywords

Civilization • Degradation • Geophagy • Health • Nutrients • Wealth

11.1 Introduction

The global demand for considerable increases in agricultural production is ever-growing to meet food and energy requirements, together with access to the necessary resources. Soil is one of the most important resources over most of human history as people got most of their food from plants grown in soils (McNeill and Winiwater 2010). Humans interact with soils in the process of crop and livestock production through continued plowing, overgrazing,

K. Kibret (✉)
School of Natural Resources Management and Environmental Sciences, Haramaya University, P.O. Box 138 Dire Dawa, Ethiopia
e-mail: kibebew.kibret@haramaya.edu.et

G. Abera · S. Beyene
School of Plant and Horticultural Sciences, Hawassa University, Hawassa, Ethiopia

S. Beyene et al. (eds.), *The Soils of Ethiopia*, World Soils Book Series,
https://doi.org/10.1007/978-3-031-17012-6_11

deforestation, erosion, desertification and use of various agro-chemicals: fertilizers, herbicides, insecticides and fungicides that contaminate and pollute the environment.

Humans and soils have been intricately linked since antiquity, as soil touches people's lives in many ways. Soils have served as foundations for many ancient civilizations and are still playing a pivotal role in modern-day civilizations. Many earlier civilizations began where farming was productive and that depended on the availability of good soil. Historically, highly developed political systems existed around Aksum in the northern part of Ethiopia on productive soils, and the Axumite Kingdom declined probably due to land degradation (Butzer 1981). Wylde (1901) expressed Ethiopia's agriculture as "producing everything that man wants in this world", giving a long list of crops cultivated at Yeju village in the then Wag. Gradual decline in soil fertility, together with a devastating drought, exposed the country to recurrent famine.

Decades of rapid population growth have contributed to soil degradation through over-farming and deforestation (Haile and Teller 2005). As a result, many previously habitable areas have now been transformed into dry lands and deserts (Haile and Teller 2005). Continuously tilled soils coupled with the removal of plant residues depletes OC, total N, available P, K, S and other nutrients and also limits the soil microbial diversity, abundance and activity. A farmer in Alaba district, southern Ethiopia, expressed "we are responsible for making our soils continuously poorer and poorer through removing plant residues from our farms" (Sheleme 2003). The regions with the greatest damage due to soil degradation are also the ones most affected by food insecurity. Certainly, soil degradation results in a higher vulnerability to famine. Soil fertility and wealth are also closely related. Societies that inhabit the fertile parts of the country are relatively better economically and in terms of food security than those that leave on degraded lands of the country.

Sizeable portion of Ethiopian highlands soils are either strongly or moderately acidic, which could be attributed to the inherent soil properties, climate and management practices. Currently, about 43% of Ethiopian soils were classified as acidic soils (EthioSIS 2014), of these about 28.1% of soils in Ethiopia are dominated by strong acid soils (pH 4.1–5.5). On the other hand, poor irrigation water management and operation coupled with the absence of drainage system cause salinization and considerable losses in crop yields.

Soils that provide a healthy, nutrient-rich growth medium for plants will result in adequate concentrations of elements required for human life in plant tissues. Soil type also influences the micronutrient concentration in food items (e.g., Joy et al. 2015; Chilimba et al. 2011). Most Ethiopian soils have fertility issues that can affect food quantity and quality. The soils are generally deficient in major nutrients like nitrogen and phosphorus. Besides, about 94% of the soils are deficient in Zn (EthioSIS 2014). Protein malnutrition, stunting and wasting in children were observed in young children around Hawassa Zuriya, southern Ethiopia, (Abebe et al. 2007) which could be attributed to N and Zn deficiencies in the soils. Application of Zn increased the concentration of Zn in seeds of chickpea (Legesse et al. 2017) and haricot bean (Abay et al. 2015) which in turn resulted in mean weight gain of young children consuming the seeds.

Soils can affect human health positively as well as negatively through their influence on food quantity and quality (Lal 2009). Soils are the origin of many medicines that are used to treat many health problems. Medicines derived from soils include mostly antibiotics, prescription drugs, and cancer drugs (Pepper et al. 2009). On the other hand, there are so many diseases that are associated directly or indirectly with soils. The country carries the greatest burden of the disease originating from soils (Deribe et al. 2017; Davey 2010) with more than one million people affected. Leta et al. (2020) indicated that soil-transmitted helminths are also almost endemic to Ethiopia and are posing serious health problems. Very common among these diseases is Podoconiosis, which is common among barefoot individuals who are in long-term contact with irritant soils of volcanic origin (Deribe et al. 2013a, b; Davey et al. 2007; Price 1990, 1976).

Potential health concerns related to soils are many and may include cancers, respiratory diseases, neurological disorders, diseases of the excretory system, skin diseases and secondary diseases like heart failure, increased susceptibility to infections and others (Nieder et al. 2018). Geophagia practiced by pregnant women on contaminated soil resulted in exposure to toxic minerals, pathogenic microbes, and helminthic infections (Steinbaum et al. 2016). Humans can also be exposed to contaminated soils through ingestion, respiration, and skin absorption or penetration (Brevik 2013). Important radioactive materials in soils that could harm human health upon exposure include radon, isotopes of cesium, cobalt, curium, neptunium, strontium, plutonium, uranium, technetium, tritium, thorium, americium, radium, and iodine, the sources of which can be natural or anthropogenic.

11.2 Soil Resource and Human Interactions

11.2.1 Principles

Human community survival has always depended, and still depends, on certain environmental goods and services. Soil is one of the most important resources over most of human history as people got most of their food from plants grown in soils (McNeill and Winiwater 2010). As human history goes on, its direct dependence on soil obviously increased; at the same time, its impact on soil significantly increased perhaps through eroding, degrading and enriching. Thus, the survival, prosperity and power of almost any given farming community rested on its success in resolving the problem of long-term nutrient depletion/loss. Egypt was the chief exception. There the waters of the annual Nile flood brought a regular nutrient subsidy left behind by receding flood waters on the banks of the Nile (McNeill and Winiwater 2010). Historically, highly developed political systems existed around Aksum in the northern part of Ethiopia until the seventh century. The Axumite Kingdom then declined probably due to land degradation (Butzer 1981). Everywhere, long-term economic trajectories, the ebb and flow of political power and the waxing and waning of populations rested on the successful management of soils.

Soil is so fundamental to human life that it has been reflected for millennia in our languages. However, soil is a very slowly renewable resource, and the use we make of them today as in much of the past-history, is often quite unsustainable. Humans must manage and manipulate soils to utilize them successfully in a sustainable manner. Ethiopia has mosaics of soil resources attributed to the diverse physiographic and climatic conditions, which has made possible the presence of diverse faunal and floral resources (Simachew 2020). Ethiopia's economy primarily depends on agriculture, which in turn is based on soil health and its sustainable use. However, mismanagement of soils leads to widespread environmental problems, soil loss, nutrient decline, acidification, salinization/solidification and overall soil degradation.

Soil degradation is a human-induced or natural process which impairs the capacity of soil to function. For instance, in 3000 BC, the Sumerians built large cities in the deserts of Southern Mesopotamia. Using irrigation, they farmed the desert soils and created large food surpluses that made their civilization possible. But around 2200 BC, the civilization collapsed due to the mismanagement of irrigation water and soils. Irrigating in dry climates can cause a build-up of salt through a process of salinization. When unregulated/excess irrigation water is applied to soils, soluble slats rise from the sub-surface horizon to the surface layer, resulting in a build-up of salinity due to high evaporation in semi-arid and arid areas. This could be a potential problem in irrigated farms of semi-arid and arid parts of Ethiopia, particularly in the vast Rift Valley region.

Humans interact with natural resources and soils in the process of crop and livestock production through continued plowing, overgrazing, deforestation, erosion, desertification and use of various agro-chemicals: fertilizers, herbicides, insecticides and fungicides that contaminate/pollute the environment. Recent studies have shown that rapid land use change and soil fertility degradation have been observed in various agro-ecologies of Ethiopia as human-soil interaction progresses. The shift from a natural ecosystem to a managed ecosystem is the main direction of change (Daniel 2020). The outcomes of these changes are deterioration of soil physicochemical properties, increased soil erosion or soil compaction (Rao and Pant 2001) and land degradation (Khresat et al. 2008; Woldeamlak and Stroosnijder 2003). These land use changes can lead to soil quality deterioration that affects the productivity of the soils and its sustainable use in the future. Overall, despite the close connections we have with soils, many fail to preserve the health of soils for generations to come.

11.2.2 Population Growth Nexus Soil Degradation

Most of the population of Ethiopia are making their livelihoods with soil. This population is growing rapidly over the last several years, on average by about 2 million every year. Decades of rapid population growth have contributed to over-farming and deforestation, which have degraded the environment and undermined development (Haile and Teller 2005). Most researchers believe that the fast-growing population of Ethiopia is playing a significant role in hastening land degradation in a way that the increasing population abused land through deforestation and overgrazing for more cropland and grazing areas, resulting in a loss of soil productivity (Temesgen et al. 2014; Berry 2003; Paulos 2001). Thus, population growth exacerbates land degradation, in line with the classical theory of Malthus. The central tenet of Malthus is "the growth of human population always tends to outstrip the productive capabilities of land resources".

As population increases, more forest lands are converted to agricultural farms. People must travel long distances to find firewood, the principal fuel, which reduces time spent for farming and related activities. Without sufficient firewood in the vicinity, many farmers resort to burning animal dung, instead of using it to fertilize their depleted soil. Without trees to help hold soil in place, the soil erodes from the steep highlands. As a result, many previously habitable areas have now been transformed into dry lands and deserts (Haile and Teller 2005). Pressure from the growing population forces the farmers to cultivate marginal lands and discontinue the use of crop residues to maintain soil fertility (Tilahun et al. 2001). Moreover, the impact of population increase on landholding size led to limiting soil fertility restoration practices such as fallowing, crop diversification and rotation, which in turn resulted in depletion of soil nutrients (Corbeels et al. 2002).

Soil degradation manifests itself through soil erosion, nutrient depletion and loss of organic matter, acidification and salinization (Haile and Fetene 2012; Bewket and Teferi 2009). The current population of Ethiopia is nearly 120 million and is expected to increase at an alarming rate in the years to come. Available evidence shows that, over the last several years, the national land holding size per household has declined. In addition, agricultural soils are moderately to severely degraded, mostly in the Ethiopian highlands, where more than 88% of the country's population lives (Birhanu 2014). Considering the increasing population, land degradation in Ethiopia is bound to proceed at aggravated rates unless significant measures are made in conservation and rehabilitation. Currently, one-third of rural households cannot produce adequate amounts of food to meet their subsistence needs as they cultivate less than half hectares of land per capita (Demise et al. 2010). On top of this, future trends in population growth, corresponding increases in food and energy demand, climate change, and globalization add challenges to agriculturists to develop innovative production systems that are highly productive and environmentally sound (Hanson et al. 2007). Overall, agriculture is known to be extractive and a resource-intensive dynamic enterprise that requires investment, innovation and contemporary knowledge for sustainable production. Therefore, the economic and societal health of any nation is largely based on soil health and sustainable ecosystem function.

11.2.3 Health Cares

Soil is one of the spectacular products of nature, serving as a medium for food, feed, fiber and medicinal crop production; regulation of water resources and health of the environment. It is a habitat for diverse soil microorganisms and regulates atmospheric concentrations. The quality of soil determines the nature of plant ecosystems, the types of wild and domestic animals and the overall capacity of the land to support human society. In addition, any process that takes place in soil has far reaching impact on earth's biosphere and beyond.

The soil is the critical element of life support systems because it delivers several ecosystem goods and services such as carbon storage, water regulation, soil fertility, and food production, which have significant effects on human well-being (FAO 2015; Jones et al. 2013). These ecosystem goods and services are broadly categorized as provisioning, supporting, regulating, and cultural services (Millennium Ecosystem Assessment 2005). Soil is so complex, there are still knowledge gaps, and fundamental research is still needed to better understand the relationships between different facets of soils and the array of ecosystem services they underpin, although enough is known to implement best practices (Smith et al. 2015).

Ethiopia's arable land soils are estimated to be 113 million hectares. The proportion of land under agriculture has been slightly increasing in recent years, with an average of 33.6% in 2018. Major grain crops: cereals, pulses and oilseeds are produced on about 13 million hectares of land in Ethiopia. Small grain cereals and oil seed crops are commonly produced on highland agro-ecologies, where there is severe erosion and high soil fertility degradation. Whereas vegetables, root crops, fruit crops, coffee, khat, hops and sugar cane are produced on about 2 million hectares of land across various agro-ecologies of Ethiopia. Some segments of arable land soils are also allocated for the production of forage and feed crops, although the estimates vary from year to year. Overall major crop production and productivity have been increasing over the last three decades by conversion of other land uses to agriculture, besides the use of improved technologies (fertilizers, improved seeds, etc.). Chemical fertilizers use has been increasing over the last four decades in Ethiopia, from a minimum of 0.1 kg ha^{-1} of arable land in 1961 to a maximum of 36.2 kg ha^{-1} of arable land in 2018. However, still, soil fertility has been steadily declining because smallholder farmers use very low rates of chemical and organic fertilizers.

Soil will continue to supply nearly all foods. All grain crops (cereals, pulses, and oil seeds), forest trees and livestock feeds (grasses, legumes, forage trees and shrubs) grow on soils. Most of the fiber we use for lumber, paper and clothing has its roots in the soils of forests and farmlands. In addition, biomass-feed stocks for fuel manufacturing grow on soils. Furthermore, in the twenty-first century as the population increases, demand for all these products may increase, while the amount of available soil remains constant. Indeed, the resource base has been and will be shrinking because of soil degradation and urbanization. Hence, our understanding and management of soil resources must improve.

Most medicinal plants (trees, shrubs and herbs) grow on diverse types of soils in natural forests. Ethiopia is the origin and/or center of diversity for many medicinal plant species (Endashaw 2007) attributed to the diverse soil landscapes and climates. Most medicinal plants have been produced on fertile soils in forest ecosystems, but currently they are under threat due to soil degradation. The proportion of consumers who rely on harvesting medicinal plants is the highest in the rural area, since collecting them from natural forests is most accessible and cost effective. The greater diversity of medicinal plants found in south and southwestern Ethiopia is a reflection of the biological and cultural diversity (Edwards 2001), high rain fall, dense vegetation cover, and fertile soils. As land degradation becomes the hallmark of Ethiopia, correspondingly medicinal plant species become endangered. In nutshell, soil is the greatest reservoir and the last frontier of biodiversity. It may harbor a wide diversity of organisms that function in the production of various medicines. Most known antibiotics are produced from organisms that were isolated from the soil. For example, Penicillin was a medicine produced from *Penicillium*, a fungus found in soil, while Vancomycin was produced from a bacterium isolated from dirt soils.

11.2.4 Impact of Land Use Types on Soil Fertility Degradation

Land use and land cover (LULC) change is the human modification of Earth's terrestrial surface from existing management of the land or land cover to new management of land or new land cover type (Hailemariam et al. 2016). Land use and land cover change and its management have a significant impact on various soil chemical and physical properties. In Ethiopia, LULC change is mainly dominated by the conversion of natural vegetation to agricultural activities (Fasika et al. 2019; Gashaw et al. 2018). According to FAO (2020), the forest cover in Ethiopia has decreased from 13.3% of the total area of the country (14.69 million ha) in 1993 to 11.4% of the total area (12.54 million ha) in 2016 with an estimated annual rate of change of 0.8% (104,600 ha·year^{-1}). On the contrary, the agricultural land of the country has increased from 27.66% (30.54 million ha) in 1993 to 32.83% (36.26 million ha) in 2016 (FAO 2020). Land use and land cover change in Ethiopia is triggered by the interaction of various demographic, socioeconomic, institutional, and biophysical factors (Birhanu et al. 2019). A plethora of studies have shown that population pressure, widespread agricultural expansion and settlement, rural poverty, inadequate management of common property resources, and land tenure insecurity resulted in land use and land cover change (Ajanaw 2021; Berihun et al. 2019; Fasika et al. 2019; Alemu et al. 2015; Ariti et al. 2015; Bewket and Abebe 2013).

Several studies conducted across various agro-ecologies of Ethiopia revealed that soil pH declined in cultivated/crop lands as compared to forest and grass lands (Daniel 2020; Yifru and Taye 2011). The reasons could be cation nutrients removal by grains and biomasses, as a result, exchangeable acidity and exchangeable aluminum increased in cultivated lands. Moreover, the use of acid forming chemical fertilizers contributes to the decline of soil pH in cultivated soils, and continuous cultivation also enhanced organic matter decomposition and soil acidification. Similarly, the conversion of native forest to crop land and the subsequent cultivation resulted in a distinct decrease in OC and total N content of surface soils (Daniel 2020; Okolo et al. 2019; Yifru and Taye 2011). However, the absence of long-term field trials and the non-existence of OC databases are among the major drawbacks of land use change studies in Ethiopia (Okolo et al. 2019). Soil OC and total N contents of surface soil crop lands declined as compared to grass land and native forest lands in selected Bale highlands (Yifru and Taye 2011; Fantaw et al. 2007) (Tables 11.1 and 11.2, respectively). A study conducted in northwestern Ethiopia showed that OC content of the cultivated land was low as compared to other land uses (Yihenew et al. 2015). The OC content followed the order grassland > cultivated land > *Eucalyptus* woodlot in the upper layer soils of Jaldu area, central Ethiopia (Daniel 2020). *Enset* land use also showed significantly higher OC content as compared to crop and grazing land uses in southern Ethiopia (Bahilu2014). In Kersa sub-watershed, east Harareghe of Oromia region, the highest OC content was recorded under grazing land (1.87%), while the lowest was observed in soils under the fallow land (Table 11.3) (Mulat et al. 2021). The results imply that under the cultivated land use system, losses of OC were not fully compensated by organic matter inputs from the crop residues and any other sources.

11.2.5 Management-Induced Degradation of Soils

Land degradation is a broader concept that includes degradation of soil, vegetation and water resources, and biodiversity/soil organisms degradation therein. According to Paulos (2001) land degradation is a temporary or permanent decline in the productive capacity of the land, or its potential for environmental management. In Ethiopia, natural resources including soils are under the influence of various interconnected factors. These factors include population pressure, agricultural expansion, deforestation, overgrazing, rapid

Table 11.1 Effects of land use types on carbon and total nitrogen at Sinana Dinsho, Bale

Land uses	Total nitrogen (%)				Organic carbon (%)			
	Soil depth (cm)				Soil depth (cm)			
	0–5	5–15	15–30	30–60	0–5	5–15	15–30	30–60
Forest land	0.80	0.50	0.34	0.19	12.95	7.76	4.54	2.86
Grassland	0.44	0.31	0.25	0.17	7.03	4.65	2.31	2.39
Fallow land	0.21	0.13	0.12	0.09	5.71	3.33	2.64	1.72
Cultivated	0.21	0.19	0.17	0.16	2.75	2.45	2.18	1.97
LSD (0.05)	0.07	0.05	0.05	0.03	1.44	0.91	0.69	0.65

Source Yifru and Taye (2011)

Table 11.2 Effects of land use on soil pH, organic carbon (%) and total N (%) (Mean + SE) at Bale highlands

Variables	Depth (m)	Land uses			
		Cropland	Grazing	Native forest	Overall
pH	0.0–0.2	6.03(0.14)	5.63(0.14)	5.59(0.14)	5.75(0.14)
	0.2–0.4	6.00(0.14)	5.55(0.14)	5.10(0.14)	5.55(0.14)
	0.4–1.0	6.11(0.14)	5.53(0.15)	4.91 (0.14)	5.52 (0.15)
	Overall	6.04(0.08)a	5.57(0.08)b	5.20 (0.08)c	
OC	0.0–0.2	5.04(0.43)	6.33(0.43)	6.65(0.43)	6.00(0.25)a
	0.2–0.4	3.67(0.43)	5.17(0.43)	5.08(0.43)	4.64(0.25)b
	0.4–1.0	2.29(0.43)	4.57(0.45)	4.19(0.43)	3.68(0.25)c
	Overall	3.67(0.25)a	5.35(0.25)b	5.31(0.25)b	
Total N	0.0–0.2	0.56(0.04)	0.66(0.04)	0.72(0.04)	0.64(0.02)a
	0.2–0.4	0.38(0.04)	0.54(0.04)	0.55(0.04)	0.49(0.02)b
	0.4–1.0	0.27(0.04)	0.39(0.04)	0.51(0.04)	0.39(0.04)c
	Overall	0.40(0.02)b	0.53(0.02)a	0.59(0.02)a	

Source Fantaw et al. (2007)
Means followed by the same letters across rows and columns (last column) show non-significant difference at $p < 0.05$

Table 11.3 Means of soil quality parameters' scores in the different land use types in Kersa sub-watershed, East Harareghe, Oromia

Land use type	Clay	Sand	WSA	OC	CEC	pH	BD	Av. P	SQI
Grassland	0.93	0.91	1.00a	0.24a	0.84a	0.51b	0.86a	0.29a	0.69a
Cropland	0.90	0.91	0.98b	0.11b	0.67b	0.50b	0.78ab	0.15b	0.62b
Fallow	0.92	0.89	0.87b	0.03c	0.63c	0.59a	0.75b	0.12b	0.59b
LSD (0.05)	NS	NS	0.03	0.02	0.04	0.08	0.09	0.11	0.03
CV (%)	14.9	10.4	1.72	2.03	2.16	8.17	6.16	29.2	2.24

Source Mulat et al. (2021)
Means followed by the same letters within a column are not statistically different at p lt; 0.05

urbanization, resettlement, climate change, and environmental pollution. Population pressure has been putting a great burden on the sustainability of almost all types of natural resources. Thus, the degradation of land, water, forest, rangeland, and wildlife resources appear to feed off each other. These result in severe soil loss, low vegetative cover, unsustainable farming practice, continuous use of dung and crop residues for fuel, overgrazing, and destruction and/or migration of wildlife, which again intensify the degradation of available resources in a vicious circle (Simachew 2020).

Soil is a natural entity which is formed as a result of both natural and managed processes and varies greatly in space and time. The rate and extent of soil formation depend on the types of rocks, climate, vegetation, organisms, topography and time. Ethiopia is marked with a great variation in these soil forming factors. Soil is very important for Ethiopia where most of the economic activities are dependent on agriculture (Engdawork 2015). Managed lands (crop and livestock production) share the largest proportion of arable lands in Ethiopia. Correspondingly, average crop production makes up 60% of the sector's outputs, whereas livestock accounts for 27% and other areas contribute 13% of the total agricultural value added (Gebre-Selassie and Bekele 2010). These contributions can be sustained when we are able to maintain healthy and fertile soil conditions.

Soil degradation is the loss of the intrinsic physical, chemical, and/or biological qualities of soil either by natural or anthropogenic processes, which result in the diminution or annihilation of important ecosystem functions (Nunes et al. 2020). Land use and land cover change, and agricultural soil management practices significantly affect the status of soil health. For instance, various researchers from different corners of Ethiopia reported that the conversion of native forest/grasslands to croplands and grazing lands resulted in substantial soil degradation (Fasik et al. 2019; Mengistu et al. 2017; Fantaw et al. 2007). Historically, highly developed political systems existed around Aksum in the northern part of Ethiopia until the seventh century, when the soil was not much degraded. The Axumite Kingdom then declined probably due to soil degradation subsequently (Butzer 1981). Soil degradation can be regarded as a direct result of the past agricultural practices. Soil erosion by water must be considered the most important of all degradation processes in the Ethiopian highlands (Hurni 1988).

Table 11.4 Ideal and root restricting bulk densities

Soil texture	Ideal bulk density (g/cm^3)	Bulk density restricts root growth (g/cm^3)
Sand, loamy sand	<1.60	>1.80
Sandy loam, loam	<1.40	>1.80
Sandy clay loam, clay loam	<1.40	>1.75
Silt, silt loam	<1.30	>1.75
Silty clay loam	<1.40	>1.65
Sandy clay, silty clay	<1.10	>1.58
Clay	<1.10	>1.47

SDA (1999). Soil quality test kit guide. USDA Soil Quality Institute. Washington, D.C

The regions with the greatest damage due to soil degradation are also the ones most affected by food insecurity. Certainly, soil degradation results in a higher vulnerability to famine. Generally, the northern and eastern regions have much thinner soil than the central, western, and southern regions of Ethiopia. The differences, apart from topography and pedogenic arguments, can be explained by the history of the land mismanagement, which proceeded from the north to the center and then toward the south and west of the country (Hurni 1988). Indeed, land degradation and soil fertility decline have posed tremendous challenges to increasing agricultural productivity and economic growth in Ethiopia. Nutrient depletion in Ethiopia has several causes. The major ones are nutrient export via harvested products (grain, stover), soil erosion, leaching and acidification. The crop productivity decline attributed to the loss of soil fertility may limit Ethiopia's opportunities in striving for food security, development and self-reliance.

Agriculture is highly extractive and exploitative of soil resources. It depletes OC, total N, available P, K, S and other nutrients. It also limits the soil microbial diversity, abundance and activity. The situations can be nastiest under subsistence smallholder, and low input farming systems. Globally, several long-term agricultural experiments revealed a large decline in OC, total N and other essential nutrients. In Ethiopia, the impact of agriculture on soil nutrient depletion has been documented by various researchers. Conversion of forest to farmland contributed greatly to enhanced erosion rates over a large part of Ethiopia. Excessive tillage for some crops, e.g., tef, wheat, maize (the main grain crops), tilling sloping land, reduction of fallow period and crop rotation practices, and overgrazing are some of the agricultural practices that might have enhanced erosion (Bezuayehu et al. 2002). Continuous cropping without adequate crop nutrition is also causing soil nutrient mining and erosion (Zhang et al. 2018; Tittonell et al. 2010).

(a) **Physical soil degradation**

Physical soil degradation comprises very different processes and morphometric forms, mainly through the deformation of the inner soil structure by compaction, caused by tracking with heavy agricultural machinery (Blum 2011). Physical soil degradation reflects structural decline (compaction and surface sealing or crusting) and mass movement. Soil compaction is the reduction of soil volume due to external factors; this reduction lowers soil productivity and environmental quality. This physical deformation may affect total soil porosity, its water holding capacity, bulk density, nutrient retention, microbial diversity and microbial activity of soils. Soil compaction leads to low water infiltration, water ponding, high surface runoff, and soil erosion after heavy rains. Physical degradation also includes soil erosion by water and wind as well as the formation of crusts at the soil surface (soil crusting). The most direct effect of soil compaction is an increase in the bulk density of soil. However, optimum bulk densities for soils depend on the soil texture (Table 11.4).

As a large mass of Ethiopian soil is tilled by oxen power, compaction may not be a serious problem except in selected areas such as Arsi-Bale mechanized farms and large-scale irrigated farms in the central Rift Valley of Ethiopia. Moreover, knowledge and scientific evidences on soil compaction are limited in Ethiopia. Nevertheless, soil compaction may be expected in teff farmlands as land preparation often include pressing/trampling the seedbeds by driving livestock on fields. Continuously tilled soils over the last many decades and low organic matter containing soils may be prone to physical compaction. Currently, Ethiopia is heading towards mechanization of agricultural practices, tillage and harvesting activities. Therefore, soil compaction may be a potential threat to Ethiopian soils unless proper management are crafted along with the introduction of mechanized technologies. There must be a precaution to avoid the potential challenges of soil compaction. Indeed, soil compaction becomes a serious problem when soils are tilled at very dry and very wet conditions. In general, soil scientists, practitioners and farmers don't give due attention to the impact of tillage on soil compaction and its consequences, as the problem is perceived to be insignificant.

Soil erosion is the greatest challenge to Ethiopia. It is the main driver of land degradation in Ethiopian highlands, and in the whole region of East Africa (Adugna et al. 2015). However, the soil losses have shown spatio-temporal variations. The soil loss rate by water ranges from 16 to over 300 Mg ha^{-1} yr^{-1} in Ethiopia, mainly depending on the degree of slope gradient, intensity and type of land cover and nature of rainfall intensities (Tesfaye et al. 2014; Tamrie 1995). According to Adugna et al. (2015), soil loss estimate made by means of the Revised Universal Soil Loss Equation (RUSLE) showed ranges of 4.5 Mg ha^{-1} yr^{-1} in forest to 65.9 Mg ha^{-1} yr^{-1} in cropland in Northeast Wollega. Based on field assessment of rill and inter-rill erosion, Bewket and Teferi (2009) estimated an annual soil loss of 93 Mg ha^{-1} yr^{-1} for the entire Chemago watershed, Blue Nile Basin, Ethiopia. Whereas, about 97% of Kilie catchment, East Shewa recorded an estimated soil loss of 0–10 Mg ha^{-1} yr^{-1} (Haile and Fetene 2012). In the Borena district of south Wello, the rate of soil loss was estimated to be between 10 and 80 Mg ha^{-1} yr^{-1} 10 (Abate 2011). Approximately, 75% of the total area of the Gerado catchment, Northeastern Ethiopia, was found to have rates of soil losses which were above 25 Mg ha^{-1} yr^{-1}. Berhan and Mekonnen (2009) estimated a soil loss of 35.4 Mg ha^{-1} yr^{-1} at the Medego watershed with steep mountains (slope 30–50%). A similar study in the highlands of Ethiopia showed a soil loss by water erosion ranging from 3.4 to 84.5 4 Mg ha^{-1} yr^{-1} with an average of 32.0 4 Mg ha^{-1} yr^{-1} (Berry 2003). All exceeded both the suggested soil loss tolerance limit of 18 Mg ha^{-1} yr^{-1} (Hurni 1983a) and the estimated soil formation rate ranging from 2 to 22 Mg ha^{-1} yr^{-1} (Hurni 1983b).

Soil erosion is taking place all over the country but because of the effect of overpopulation on land that is already fragile (steep and mountainous), and mismanagement of the land in the northern and central highlands are the worst affected (Paulos 2001). The most common types of soil erosion include sheet, rill and gully erosion by water and wind energy. Deforestation, population growth, overgrazing and use of marginal lands intensify erosion, and the intensification of agriculture production also results in high erosion rates. About 43% of the country is classified as highland (above 1500 m a.s.l.), where most of the population (about 88%) carry out mixed crop-livestock agriculture activities (Bewket and Teferi 2009). Generally, soil erosion by water is the most pressing environmental problem in Ethiopia, particularly in the highlands where the topography is highly rugged, population pressure is high, steeplands are cultivated and rainfall is erosive (Bewket and Teferi 2009).

Although indigenous soil conservation techniques have been applied for centuries in Ethiopia (Ciampalinia et al. 2012; Beshah 2003), institutionalized soil and water conservation (SWC) programs have been significant only since the 1970s (Osman and Sauerborn 2001). Indigenous soil and water conservation techniques date back to 400 BC in Ethiopia. In order to reduce/mitigate the anthropogenic soil erosion, institutionalized soil water conservation (SWC) activities have begun since 1970s (Haregeweyn et al. 2015). Various nationwide SWC initiatives have been undertaken, especially since the 1980s supported by multiple donors, and then have undergone a series of changes in terms of approach, technologies, and technical standards. These initiatives include food-for-fork (FFW) (1973–2002), managing environmental resources to enable the transition to more sustainable livelihoods (MERET 2003–2015), productive safety net programs (PSNP 2005–present), community mobilization through free-labor days (1998–present), and the national sustainable land management project (SLMP 2008–2018). The FFW program started in the form of food aid and gradually shifted in the 1980s to a development-oriented program through engaging the community in the rehabilitation of degraded lands (Devereux et al. 2006).

(b) **Chemical soil degradation**

Chemical soil degradation is manifested by nutrient depletion, acidification, salinization and soil pollution. The large proportions of Ethiopian soils are nutrient depleted due to mismanagement and continuous cultivation with less chemical and organic fertilizer inputs. Nutrient depletion has caused serious nutrient imbalances in soils under low-nutrient input agriculture systems (Osman 2014). Few earlier nutrient balance studies of agricultural soils of Ethiopia exhibited negative NPK balances (Abegaz et al. 2007; Haileslassie et al. 2005). A country-wide analysis of nutrient balance indicated a depletion rate of 122 kg N ha^{-1} yr^{-1}, 13 kg P ha^{-1} yr^{-1} and 82 kg K ha^{-1} yr^{-1} (Haileslassie et al. 2005). They further stated that soil stocks in all regional states of Ethiopia were decreasing with the exception of areas under permanent and vegetable crops. They accounted soil erosion as the major cause for nutrients depletion in Ethiopia, although the model result has high uncertainty. Recent nutrient balance studies of agricultural soils of Ethiopia exhibited negative NK, but positive P balance in northern parts of Ethiopia (van Beek et al. 2016; Shimbahri et al. 2020; Girmay et al. 2021). This positive P balance observed in most agricultural soils of Ethiopia may be related with P fertilization and P fixation in soils. According to Abebayehu et al. (2011), a comparative analysis of soil nutrient balance in the Jimma Zone revealed an average depletion of 55.61 kg N ha^{-1} yr^{-1}, 9.7 kg P ha^{-1} yr^{-1} and 49.81 kg K ha^{-1} yr^{-1} at the highlands and 35.73 kg N ha^{-1} yr^{-1}, 9.1 kg P ha^{-1} yr^{-1} and 29.69 kg K ha^{-1} yr^{-1} at lowlands. The landscape position (highlands and lowlands) showed large differences for N and K contents. The P depletion was observed to be low, thus attributed to the high P fixation and building up of P in soils when P fertilizer was regularly applied to soils.

Most land use studies conducted across various agro-ecologies of Ethiopia also exhibited soil fertility degradation in crop lands as compared to native forests and grasslands (Mulat et al. 2021; Mengistu et al. 2017; Yifru and Taye 2011; Fantaw et al. 2007). Recent records show that about 43% of the agricultural lands of Ethiopia are acidic. These acidic soils are sparingly suitable for crop production, albeit the degree of acidity varies across agro-ecologies of Ethiopia. Liming is an ancient method of reclaiming acid soils. Moreover, the modeling approach developed for estimation of soil nutrient depletion in current management practice showed 16% reduction of OC in Cambisols and 32% reduction in Luvisols in the northern Ethiopian highlands. The depletion rates of soil N are similar to those in OC under current management conditions (Abegaz and Keulen 2009).

Soil acidification is one of the most important soil chemical degradation processes attributed to the inherent soil properties, climate and management practices. A sizeable portion of Ethiopian highlands are either strongly or moderately acidic (Fig. 3.9). Currently, about 43% of Ethiopian soils were classified as acidic soils (Ethio SIS 2014), of these about 28.1% of soils in Ethiopia are dominated by strong acid soils (pH 4.1–5.5). The soil acidity is increasing in scope and magnitude in Ethiopia. Thus, soil acidity is a critical agricultural problem in certain agro-climatic zones of Ethiopia. However, the extent of soil acidity varies from semi/arid lowlands of eastern Ethiopia to the high rainfall areas of south, southwestern and western Ethiopia. For instance, the high rainfall areas of western Oromia have been adversely affected by soil acidity and associated soil quality problems; as a result, farmers' livelihood has been significantly affected (Bulti and Abdulatif 2019). Soil acidity and associated low-nutrient availability, and toxicity of certain nutrients (Al and Mn) are some of the major constraints to crop production on acid soils. For instance, a shortage of available K, Ca, P and Mo on the one hand, and an excess of soluble Al, and Mn are effects of high acidity problems in soils (Agegnehu and Sommer 2000; Somani 1996). When the pH of a soil is less than 5.5 phosphates can readily be rendered unavailable to plant roots as it is the most immobile of the major plant nutrients (Agegnehu and Sommer 2000; Sanchez 1977), and the yields of crops grown in such soils are very low. On the other hand, in a soil pH between 5.5 and 7, P fixation is low and its availability to plants is higher. Toxicity and deficiency of Fe and Mn may be avoided if the soil reaction is held within a soil pH range of 5.5 to 7; this pH range seems to promote the readiest availability of plant nutrients (Somani 1996). It is important to note that most crops' performance was observed to be deleteriously affected by soil acidity in the highlands of Ethiopia. Liming is one of the best soil acidity management strategies in various parts of the world. In Ethiopia, currently, it is widely practiced through establishing lime producing factories in different parts of Ethiopia such as in Amhara, Oromia and SNNP regions. Moreover, indigenous soil acidity management practices appears to be critically important in Ethiopia. This indigenous practice includes keeping livestock overnight for certain days on a fenced plot (barn) for manure production. It is an environmentally resilient and locally adapted indigenous practice used by farmers to ameliorate acidic soil with organic manure, and thereby improve soil quality and crop productivity (Bulti and Abdulatif 2019).

(c) **Biological soil degradation**

Soil biology is one of the most unexplored frontiers associated with understanding the dynamics of soil resources' health or quality (Lehman et al. 2015). However, soil represents one of the most important reservoirs of biodiversity. Soil biology is an interesting area of soil sciences and has yielded considerable information that is used in soil fertility management. In Ethiopia, soil biological teaching and studies have been very limited to biological N_2 fixation, due to limited laboratory facilities, knowledge and skills. Few experiments have been conducted on carbon and nitrogen mineralization studies in Ethiopia. These include the C and N mineralization studies with glutamate and legume residues amendment in laboratory, greenhouse and field conditions. Thus, the results are a good indicator of soil biological properties and a proxy indicator of soil health (Girma et al. 2012). Even when fertilizers are applied to soils, it become useable by the action of soil microorganisms. For instance, urea (NH_2CONH_2) and diammonium phosphate $(NH_4)_2HPO_4$ can only be mineralized into usable mineral nitrogen and phosphorous with the presence of significant microorganisms including nitrifying bacteria.

Few nitrogen mineralization studies from soil organic matter (SOM) were conducted in situ and in laboratory incubation under different soil moisture regimes with coffee and crop land use soils (Awassa, Yirgalem, Wondogent and Ziway) in southern Ethiopia. The results showed that extractable NO_3^--N and mineral N were strongly increased, while NH_4^+-N declined in response to soil moisture increase (from air dry to 100% FC, field capacity) during laboratory incubation. In situ extractable NO_3^--N and mineral N were strikingly low in December (dry) in both coffee and crop lands. However, they were consistently greater during March to August due to more rain flushes, suggesting better N release during the cropping season. The nitrogen mineralization of coffee land use was about double of crop land attributed to its higher soil organic matter content. Accordingly, the soil biological processes, the microbiota may be higher in coffee land than in crop land. Generally, the knowledge of annual patterns of nitrogen mineralization in relation with soil moisture is necessary to synchronize crop N demands with plant-available N in the soil (Girma et al. 2012).

Biological degradation of soil occurs due to the impairment or elimination of one or more "significant" populations of microorganisms in soil, often with a resulting change in biogeochemical processing within the associated ecosystem. "Significant" microorganisms are those for which an ecologically significant role is understood (Sims 1990). Increases in SOM, particularly in biologically-available forms, are intimately linked to changes in the size, activity and composition of the soil microbial community, enhanced cycling and retention of nutrients, improved aggregate stability, and increased water-holding capacity of soils (Lehman et al. 2015). Soil management practices that increase SOM and enhance soil health create expanded habitats and greater niche diversity for soil biological communities. Inputs of organic matter from plant residues and exudates provide carbon and energy sources for soil organisms.

11.2.6 Rain-Fed Agriculture

Over 95% of Ethiopia's agriculture is rain fed, subsistence smallholder agriculture. Soil resources in Ethiopia are considered as an asset but its management is treated as a challenge (Engdawork 2015). The severe rain fed agricultural soil degradation in Ethiopia is attributed to mismanagement, overexploitation (over-cultivation, overgrazing), removal of crop residues, use of manure (dung) for fuel, limited use of chemical fertilizers and erosion which causes billions of tons of topsoil removal every year and, and reduction/loss of the functions and services that soils provide.

Soil degradation can be regarded as a direct result of the past agricultural practices. Soil erosion by water must be considered the most important of all soil degradation processes in the Ethiopian highlands (Hurni 1988). Very few estimates are available about the overall soil loss rates at regional or national scale (Haregeweyn et al. 2015; Hurni 1988) estimated a nationwide annual gross soil loss of 1.5×10^9 tons, extrapolating data obtained from six soil conservation research project stations in which the highest loss was from croplands (42 t ha^{-1} yr^{-1}). Sonneveld et al. (2011) provided a tentative nationwide mean annual soil loss map combining the results of different model estimates. They stated that soil loss varies remarkably from 0 t ha^{-1} yr^{-1} in the eastern and southeastern parts of Ethiopia to more than 100 t ha^{-1} yr^{-1} in the northwestern part of the country. Moreover, various studies have also been reported on chemical soil degradations across agro-ecologies of Ethiopia (Tamene et al. 2017). These include soil fertility decline as observed by depletion of OC, total N, available P and acidification, and which is also substantiated by crop response to chemical fertilizers.

Agriculture is highly extractive and exploitative of soil resources. It depletes soil OC, total N, available P, S and other nutrients. It also limits the soil microbial diversity, abundance and activity, and thereby limits soil biological processes. The situations can be nastiest under subsistence, smallholder and low input farming systems. According to Gete et al. (2010), declining soil fertility is one of the most significant constraints to increased food production in Ethiopia. Globally several long-term experiments revealed a large decline in soil OC, total N and other essential nutrients due to agriculture practiced. In Ethiopia, the impact of agriculture on soil nutrient depletion has been documented by various researchers. For instance, Ethiopia is one of the 14 SSA countries with the highest nutrient deficit (−41 N, −26 P and −6 K y^{-1}) in agricultural fields.

A meta-data analysis on fertilizer application across various agro-ecologies of Ethiopia showed substantial positive crop response in most highlands of Ethiopia. The lowlands areas also had limited crop response (Tamene et al. 2017). This data shows a good representation of Ethiopian highlands that were generated by FAO and others. The meta-data results show that there is a limited response to N and P in the absence of the other, while combining N and P results in large increases in yield. Not including P in crop nutrition has a greater effect on attainable yield in areas with low than with high (>4 t ha^{-1}) unfertilized/control yield. Therefore, recommendations for fertilizer must likely include both N and P fertilizer sources. With fertilizer application, many observations indicate elevated yields beyond the national averages. There was a positive response to N, P and S with the test crops (wheat, maize, teff and rice) although some of the observations show no response/ negative responses to the applied nutrient (Tamene et al. 2017). This is perhaps related with the nutrient balance issue, as some nutrients increase may create an antagonistic relationship with non-applied essential nutrients.

11.2.7 Degradation of Soils in Irrigated Agriculture

Ethiopia has about 4–5 million hectares of irrigable land area. Out of these about 400,000 to 500, 000 hectares were already developed. Traditional irrigated agriculture is widely distributed across wider agro-ecologies, while commercial and large-scale irrigated agriculture are limited to semi-arid and arid agro-ecologies of Ethiopia. Overall, irrigated agriculture is very much limited in geographic coverage and technologies. Nevertheless, commercial large-scale irrigated agriculture is practiced in some Sugar Estat farms such as Wonji, Metahara, Fincha, Tendaho, Kuraz, Didessa and Wolikayti areas. Some of these sugar estate farms are young, while others are more than three decades old. In addition, there exist commercial irrigated vegetables, fruits and flower farms in middle Awash, Upper Awash and central rift valley areas, where potential salinity development is expected if not properly managed.

Over 11 million ha of land in Ethiopia are known to be salt affected soils (Ruffeis et al. 2008; Taddese 2001). The soil salinity problems in Ethiopia stems from the use of poor-quality water coupled with the intensive use of soils for irrigation, poor on-farm water management practices and lack of adequate drainage facilities (Gebremeskel et al. 2018). The arid and semi-arid lowlands prevalent in the rift valley and other areas that are characterized by higher evapotranspiration rates are salt affected soils in Ethiopia (Asfaw and Itanna 2009; Geressu and Gezaghegne 2008; Dubale et al. 2002). The salt affected soils have increased from 6 to 16% of the total land area of Ethiopia (Abraha and Yohannes 2013). According to Zewdu et al. (2014), in Sego Irrigation Farm around Arbaminch, southern Ethiopia, the coverage of moderately and strongly saline areas has increased at an average annual rate of 4.1% and 5.5% respectively from 1984 to 2010. Earlier estimate showed that over 44 million ha (36% of the country's total land area) is potentially susceptible to salinity problems (Hawando 1995). These soils are currently at risk to salinity development due to mismanagement of irrigation water and poor drainage. The Awash basin can be considered as a typical example where salinization has been a critical problem in its many large and medium scale irrigation schemes including the Amibara irrigation project in the Middle Awash and the Metahara sugar plantation in the Upper Awash (Ayenew 2007).

The soils of irrigated farms in Ethiopia are challenged by poor drainage and salinity problems. Waterlogging is the main drainage problem in the small-scale irrigation schemes in the Vertisols dominated highland areas while salinity and salinization is a common phenomena in the large and medium scale irrigation schemes located in the lowlands of the country's major river basins with predominantly salt affected soils (Bulti and Abdulatif 2019). Large areas of the middle and lower parts of the Awash basin are also saline or sodic and thus potentially exposed to salinization and sodicity (Ruffeis et al. 2008). Salinization is more spreading in irrigated lands because of inappropriate management of irrigation and drainage. The major sources and/or causes of salinity are shallow groundwater tables and natural saline seeps. Poor drainage and lack of appropriate irrigation water management are also known to facilitate secondary salinization development in Ethiopia (Abebe et al. 2015). Improperly planned irrigation projects not supported by improved irrigation and drainage management technologies had invited serious degradation causing salinity and sodicity problems in the Awash basin which accounts for about one-third of the total irrigated area of the country (Dubale et al. 2002; Ruffeis et al. 2008). This high salinity problem is also related to uncontrolled irrigation practices and lack of knowledge on crop water requirements and water management leading to increased saline groundwater levels or capillary rise (Ayenew 2007). Climate is also a key factor in the salinization process. The high temperature of the Middle Awash (annual average 26.7 °C) and low annual rainfall (500 mm) and the high free evaporation of water have aggravated the salinization process (Ayenew 2007; Bekele 2005).

Salt affected soils are characterized with excess concentrations of calcium (Ca^{2+}), sodium (Na^{+}) and chloride (Cl^{-}) which are easily soluble (Bekele 2005). This has an adverse effect on the seedling growth of several crops, by creating an osmotic deficit in the rhizosphere of the plant. Thus, the soluble salts also inhibit the absorption of water or create toxicity effects due to Na^{+} and Cl^{-} to the roots and the whole crop (Singh 2015; Abraha and Yohannes 2013). When salt affected soils are intensively cultivated without proper caution for the gradual accumulation of salts and soluble substances, it may result in severe land degradation. Poor irrigation water management and operation coupled with the absence of drainage system can cause groundwater rise (waterlogging), salinization and considerable losses in crop yields which ultimately led to the abandonment of substantial irrigable areas. The problems of salinity and waterlogging persist in many regions where farmers apply excessive irrigation water, and where farmers and irrigation departments fail to invest in its proper management.

Considering the country's agriculture dependent economy, increasing food demand due to increasing population and insufficient rain fed agriculture, it is evident that the country should plan to promote irrigated agriculture. Provided that the likely target of planned expansion would not be out of the highland or lowland areas which are being affected or vulnerable to waterlogging and salinity problems respectively; drainage technologies are inevitable. The literatures referred in this paper signals the importance of drainage in boosting crop yields by controlling the above-mentioned drainage problems in irrigated agriculture.

It can be concluded that drainage is as important as irrigation for a productive and profitable irrigated agriculture that could help the country achieve its planned development goal. The unavoidable challenges that might be faced in the process are related to costs and technology in the design, implementation, operation and maintenance of drainage systems. But this might be copped by introducing low cost technologies like BBF for small-scale irrigators and the modern surface and sub-surface drainage systems.

11.2.8 Soil Pollution

Soil pollution refers to the contamination of soil with anomalous concentrations of toxic substances. It is a serious environmental concern since it harbors many health hazards. It is important to understand that all soils contain compounds that are harmful/toxic to human beings and other living organisms. However, the concentration of such substances in unpolluted soil is low enough that they do not pose any threat to the surrounding ecosystem. When the concentration of one or more such toxic substances is high enough to cause damage to living organisms, the soil is said to be contaminated. The root cause of soil pollution is often one of the following:

- Excessive/improper use of pesticides in agriculture
- Excessive industrial activity
- Poor management or inefficient disposal of waste

The extensive use of pesticides in agricultural production can degrade and damage the community of microorganisms living in the soil, particularly when these chemicals are overused or misused as chemical compounds build-up in the soil. It harms and destroys the beneficial organisms in the soils and affects their natural fertility and pest resistance. Although many studies have found deleterious effects of pesticides on soil microorganisms and biochemical processes, the full impact of pesticides on soil microorganisms is still not entirely understood.

Generally, there are very limited studies in Ethiopia with respect to soil pollution. In recently flourished flower farms, soil pollution is perceived to be a critical danger to soils and aquatic ecosystems, for example in rift valley areas. In large cities like Addis Ababa, soils are expected to be polluted due to the use of polluted water as a source of irrigation water for vegetable production. Another source of soil pollution can be improper disposal of agricultural as well as agro-processing/industrial wastes. Therefore, poor waste collection, control, storage, transfer, process, and disposal may be another good reason to expect soil and environmental pollution in Ethiopia. For instance, some Ethiopian soils are polluted by misuse and mismanagement of soils. These polluted soils are largely concentrated in the rift valley area and in Upper Awash State Farm associated with residues of floriculture and horticulture industry, coffee processing byproducts and gold mining residues across various parts of Ethiopia. According to Degytun (2011), the floriculture industry's inappropriate choice of cultivation methods and wide range of use of chemicals and fertilizers have negatively impacted soil and water condition. Upper Awash Agro Industry Enterprises is one of the major state farms in Ethiopia with known large-scale pesticide use. The major contaminants identified comprised of previously used persistent organic pollutants (POPs) and currently used insecticides. Low concentrations or non-detectable levels of certain POPs (aldrin, dieldrin, endrin, and heptachlor) indicate a positive phasing out of these persistent organic pollutants (POPs). Similarly, HCHs were found in soils at low concentrations. Endosulfans and DDTs were detected in substantial amounts in the soils with endosulfans up to 56,000 and DDTs up to 230ngg^{-1}dry weight, which is a threat to the surrounding and downstream ecosystems. Moreover, considering the investigated POPs constituted 29,000L of the 63000L of pesticides applied annually on the fields. Additional apprehension must be raised concerning the synergistic effects of all pesticides added.

A total heavy metal (Cr, Cd, Pb, As, Cu, Ni, Zn, Co) concentration study was performed on 33 soil samples taken from different profiles and soil types in a highly urbanized and industrial sector of Addis Ababa, central Ethiopia. The results show a relatively high content of the analyzed trace metals in the soil attributed to anthropogenic and geogenic sources. According to the heavy metal SE analysis, the major heavy metal contribution is from the residual followed by the hydroxide phases (Molla and Stefan 2006). Another potential pollutants of soil and water in Ethiopia are the coffee processing byproducts, pulp and husk, which are often damped to soils or streams and rivers within the vicinity of the processing plants. A report showed that in western Ethiopia large quantities are either dumped into streams or burnt in big piles, whic contributes to environmental hazards (Bikila 2019). However, these byproducts can be good organic fertilizer sources if properly managed and utilized as a soil amendment. Gold mining is a tremendously important economic activity in rural districts of Ethiopia. Gold mining removed colossal volumes of soil from the mining landscape. In addition, various chemicals used in the gold mining process can also be sources for soil and aquatic ecosystem pollution in various parts of Ethiopia.

11.3 Civilization, Soil Fertility and Wealth

11.3.1 Soils and Civilization

Humans and soils have been intricately linked since antiquity, for soil touches people's lives in many ways, including serving as the source of food and clothing, and for its multiple ecological services, such as filtering drinking water and maintaining environmental health (often called 'a geologic kidney' by environmentalists) (McNeill and Winiwarter 2004). Some scholars associate the origin of the word *human* with soil claiming that the Latin root of the word *human* itself is similar to the root of the word *humus*, which means earth. Hillel (1980) wrote, "The primeval association of man with soil is manifested most strongly in the name *Adam*, derived directly from *adama*, a Hebrew word with the composite connotation of earth, land, and soil (*pp xiv*)." Similarly, many other historical accounts in many traditions educe the presence of strong associations between humans and soils, which helped people to recognize the value of soil at least since the dawn of cropped agriculture. Since then, soils have found their way into societies' cultures, religion, history, art, literature, thought, civilization, and livelihood (Minami 2009).

Soils have served as foundations for many ancient civilizations and are still playing a pivotal role in modern-day civilizations. Many earlier civilizations began where farming was productive and that depended on the availability of good soil. At some stage, ancient civilizations have gone to the extent of worshiping soils as a foundry of their life itself since soils provided most of their food and nutrients (McNeill and Winiwarter 2004) through agriculture, arguably the first systematic use of soils (Brevik 2005). This triggered people to make an attempt to organize, preserve, and impart knowledge about soils, leading to the writing of documents in the form of agricultural manuals. Because of these earlier documents, civilizations all around the world showed fairly advanced soil knowledge as early as the fourth century AD including irrigation, the use of terraces to control soil erosion, various ways of improving soil fertility, and ways to create productive artificial soils (Brevik 2005).

When farm productivity declined, usually as a result of soil mismanagement leading to soil degradation, civilizations also declined—and occasionally vanished entirely. Testimonies to this claim include the collapse of the 1700-year-old Mayan civilization in South America around 900 AD (due to soil loss caused by erosion), the demise of civilizations at Mesopotamia (due to salinity and waterlogging) (Essington 2004), Sumerians (due to salinity), and Babylonians (due to siltation caused by sediment washed from the surrounding hillsides which were left barren). On the other extreme, successful civilizations that are worth mentioning in relation to soils include those of the Greeks, Romans, The Mediterranean, Northern Europe, Asia, and many others although the level of soil knowledge of the Greeks and Romans was more refined than the others.

Agriculture has been practiced in Ethiopia, a country with one of the oldest civilizations in Africa, for centuries. Its people are dominantly agrarian where the agricultural productivity relies heavily on soil resources. As such, the contribution of soils to the civilization of the country has been laudable. Butzer (1981) related the rise and fall of the Axum civilization with soils/natural resources. An Egyptian noble, Ibn Fadl Allah Al-Omari, mentioned the cultivation of many cereal crops like wheat, barely, sorghum, *teff*, chickpeas, and lentils and mentioned the possibility of two harvests a year in his geographical work 'MasalikelAbsar' (1342–1349), which was devoted to Ethiopia. Other early writers also gave a good account of Ethiopia's agriculture mentioning the different types of crops (including fruits and vegetables of different kinds) grown, the yield they used to obtain, the good quality of the soil, and irrigation practices around water sources (e.g., Crawford 1958; Alvares 1540; Zorzi 1522-all cited in Westphal 1975). Writers such as Manuel Almeida described Ethiopia to be a very fertile country (Westphal 1975). Ludolphus, writing about Ethiopia in 1684, said: 'The fertility of the Soil of Habessinia is to be admired; for the land where it admits of Tillage, abounds in all sorts of Fruits (Westphal 1975). Burton (1856) described the environs of the walled city of Harar mentioning that the soil on both sides of the path is rich and red.

Ben (1896) described the soil and water conservation experiences of Ethiopians based on his observation that all the surrounding hills in a valley near Yeha are terraced, comparing the appearance of the terraces in Greece and Asia Minor, but with a much greater extent. He further concluded that "Hundreds and thousands of acres must have been under the most careful cultivation, right up almost to the tops of the mountains." Wylde (1901) wrote extensively about Ethiopia's agriculture describing it as "produced everything that man wants in this world", giving a long list of crops cultivated at Yeju village in the then Wag. He also described some of the soil management practices that the farmers were practicing (e.g., burning of roots and weeds on cultivated lands). Bunting (1963) described the farming in Ethiopia as more advanced in terms of management of crops and soils than any other indigenous system in tropical Africa. However, a gradual decline in soil fertility, together with a devastating drought, exposed the country to recurrent famine that claimed the lives of millions of Ethiopians. Soil degradation (through soil erosion, soil fertility depletion, salinization, soil acidity, and others) stands among the top factors that led to a decline in agricultural productivity and persistent food insecurity. This, together with other contributing factors, has been undermining its civilization and economic development.

11.3.2 Fertile Soils and Wealth

Soil fertility and wealth are closely related. The United Nations (2013) declared that soils form the basis for agricultural development, ecosystem functions, and food security and thus are key to sustaining life on earth. This is particularly true in the least developed agrarian countries like Ethiopia whose income is solely dependent on agricultural commodities. Fertile soils contribute towards economic growth, biodiversity, agricultural sustainability, and food security, all of which are fundamental to eradicating poverty. However, until very recently, our dependence on natural resources in general and soils, in particular, seems to have been largely undermined by intellectuals, cultured people, economists, policy makers, and even the society at large. Inauspiciously, many view soil as a 'dirt', which has to be removed or cleaned from a place if possible. Even the neo-classical economics excludes natural resources, including soils, from its theories, erroneously assuming that these entities are nearly free and infinite, and thus should not be a matter of concern (Gomiero 2016). It is only very recently that the UN General Assembly declared 2015 as the International Year of Soils. Emphasizing on the special place and attention soils deserve in the global agenda, one FAO Director-General stated that "the multiple roles of soils often go unnoticed. Soils don't have a voice, and few people speak out for them. They are our silent ally in food production." (Da Silva 2014). Jack et al. (2009) argue that soil should be viewed as an important source of greater wealth; however, establishing a direct relationship between wealth and soil does not lend itself to simplicity (Burras et al. 2013).

As vividly pointed out by Gomiero (2015), there can be no agriculture at a scale without fertile soils and, hence, it is important that people comprehend the tight and delicate link soil health (which includes fertility status) has to land use, food production, people's health, the use of inputs (e.g., fertilizers), and many other environmental and socioeconomic issues. It is through this link that the status of our soils affects society's wealth creation and accumulation. Evidences from different sources (e.g., Montgomery 2007; McNeill and Winiwarter 2006; Diamond 2005; Troeh et al. 2004; Hillel 1991; Dale and Carter 1955) indicate the role soils played in the rise and fall of the wealth of early civilizations claiming that soil degradation (e.g., soil exhaustion, soil fertility depletion, soil erosion, and salinization) contributed to the demise of notable early civilizations that prevailed in the Middle East, Greece, and the Roman Empire, among others. Narrations in many literatures converge towards one reality, i.e., a society without food security can't be viewed as a wealthy society. With soils at the heart of the requirements for ensuring food security, the role they play in wealth creation speaks by itself. Their contribution is direct as well as indirect.

In Ethiopia, largely an agrarian nation, the standard of living of the farming community in particular is very much dependent on the quality and size of land (which includes soils) they own. In general, those societies that inhabit the fertile parts of the country are relatively better economically and in terms of food security than those that leave on degraded lands of the country. This generalization, however, does not preclude the existence of individual differences among those inhabiting fertile lands. Apart from individuals, the country's economy and its growth are heavily dependent on the agriculture sector whose productivity is tellingly influenced by the quality of soil resources, among others. The contribution of agriculture to the country's gross domestic product and foreign export earnings is significant. The other economic sectors, such as industry and service, are heavily dependent on the agriculture sector for inputs. By and large, the contribution of soil fertility towards food security, income generation, and economic development of the county is extraordinarily significant. Equally, the country is losing a huge amount of wealth due to soil degradation, particularly soil erosion. However, the contribution of natural resources including soils to the national economy has always been greatly undermined or underestimated.

11.4 Soils in Relation to Human Nutrition and Health

11.4.1 Principles

Soil, which provides the substrate for almost all terrestrial life, is a basic natural resource for food production, with the major proportion of food that we consume either directly or indirectly coming from it (Silver et al. 2021; Menta 2012). Soil quality, the continued ability of the soil as a vital living system to support important ecosystem functions (Karlen et al. 1997; Pankhurst et al. 1997), determines the quantity (calories) and the quality (nutritional value and safety) of the foods grown on it (Zhu 2009). Soil quality is dictated by the status of physical, chemical, and biological attributes which affect one or more of the many soil functions. Maintaining these soil attributes and their integration at an optimum level is therefore of vital importance in safeguarding global food security.

Soil can also affect human health directly or indirectly. The direct influences could come from the ingestion of soil material accidentally or deliberately through practices such as geophagia. The indirect effects could come from the food produced on soils and consumed by human beings. The effects on human health could be positive or negative. Healthy soils that support the production of healthy and nutritious food will enhance human health, while food produced on unhealthy soils (e.g., polluted soils) could hurt human health through some contents of the food produced on such soils. The influences could range from simple nutritional deficiencies, which can be easily corrected, to complicated diseases such as cancer. Similarly, ingested soil material could be beneficial when it contains essential components such as nutrients or harmful if it contains dangerous substances such as minerals (e.g., radioactive nucleotides, heavy metals) and pathogens. These components can be a source of concern to human health since they can be the cause of many diseases and disorders. The link between soil health/quality and human health should be understood properly and comprehensively so that management scenarios that assist in maintaining soil health/quality, which enhances the production of adequate food that is safe and nutritious, are put in place.

11.4.2 Soils and Human Nutrition

(a) **Soil and food quality**

It was as early as 1921 that McCarrison (1921) was able to link the fertility of a soil with the nutrient content of food crops stating that the fertility of a soil determines the nutrient content of food crops, and therefore the health of humans who ate the crops. The nutritional value of many foods is markedly influenced by the soil's ability to supply essential nutrients to food systems (Allaway 1986), particularly where food is produced and consumed locally, influencing to what extent the nutritional requirements of people are met from the crops grown on these soils. Nevertheless, in more complex food chains where there are different food sources and processed foods are supplemented with essential minerals and vitamins, the contribution of soils in meeting nutritional requirements are less evident (Oliver and Gregory 2015). Naturally, fertile soils are expected to produce healthy and nutrient-rich crops that nourish people and animals (IFOAM 2011). However, a large number of evidences indicate the existence of macro- and micronutrient deficiencies in significant proportions of agriculturally important soils of the world. These deficiencies not only affect crop productivity but also could lead to the poor nutritional quality of crops and animals, often leading to malnutrition in humans. On the other hand, under certain soil conditions, some nutrient elements occur in high concentrations that may diminish crop yield, with potentially negative consequences on human health (Hodson and Donner 2013). Soil conditions like acidity and sodicity dictate as to which elements may accumulate in the soil in toxic concentrations.

Soil acidity favors the accumulation of toxic concentrations of manganese, aluminum, and iron, while salinity and sodicity could create suitable conditions for toxic concentrations of sodium, boron, and chloride, which could reduce crop production and affect food quality. Anthropogenic activities, such as mining and manufacturing could lead to toxic accumulation of arsenic, cadmium, mercury, and lead, which are carcinogenic and thus could affect the quality of food that we consume (Hodson and Donner 2013). Therefore, from the foregoing discussion, it can be inferred that soil quality affects food quality in at least in two ways. One of these is through the essential nutrient elements which it supplies to plants that make our food. The second way is through its chemical composition which may not be nutrient elements but absorbed by plants indiscriminately. The former affects the quantity and nutritional quality of the food produced on a soil, while the latter influences food safety. Food that is produced on polluted soils (soils containing excessive concentrations of toxic chemical elements and even microorganisms), is unsafe for human consumption since the toxic substances can be transferred to humans via the food we eat.

Mineral or nutrient malnutrition results from crops produced on soils with poor phytoavailability of the elements essential to human nutrition. For example, alkaline and calcareous soils (25–30% of all agricultural land) have small availabilities of Fe, Zn and Cu (White and Greenwood 2013; Broadley et al. 2007), and coarse-textured, calcareous or strongly acidic soils contain little Mg, mid-continental regions have little I and soil derived mostly from igneous rocks contains little Se (Gregory 2012). Consequently, crops also have inherently small concentrations of certain elements (Karley and White 2009; White and Broadley 2009).

Most Ethiopian soils have fertility issues that can affect food quantity and quality. The soils of the country are generally deficient in major nutrients like nitrogen and phosphorus. Besides, the organic matter content of almost all agricultural soils is in the range of very low to low. Most soils in the high rainfall areas of the country are severely affected by soil acidity, while those in vast lowlands are affected by various levels of salinity and sodicity, soil conditions that might result in deficiency of some nutrients and toxic concentration of others-particularly micronutrients. In line with this, a recent study by Gashu et al. (2021) identified soil pH and organic matter as covariates of grain micronutrient concentration in staple cereal grains grown in Ethiopia. The same study reported with-in species variation in cereal grain Ca, Fe, Se, and Zn concentration and attributed it to the spatial variation in soil and landscape factors. Tessema et al. (2019) also reported a significant correlation between soil Zn and serum Zn. Similarly, De Groote et al. (2021) reported that serum Zn was correlated to soil Zn for children and the prevalence of a high level of human deficiency throughout

Ethiopia, while a significant reduction in the prevalence of Zn deficiency in young children was only observed at high soil Zn levels. Contrary to this, Berkhout et al. (2019) found weak associations between soil nutrients, including Zn, and some health indicators, such as child mortality and stunting in sub-Saharan Africa. Studies conducted elsewhere (e.g., Joy et al. 2015; Chilimba et al. 2011) demonstrated the influence of soil type on micronutrient concentration in food items. Crops growing on localized soil types (e.g., Vertisols) have greater micronutrient concentrations than crops growing on more weathered, acidic soils (Ligowe et al. 2020; Joy et al. 2015; Chilimba et al. 2011).

(b) **Essential soil elements for plants and humans**

Tisdale et al. (1993) define an element considered to be essential to plant growth and development as "that involved in plant metabolic functions without which the plant cannot complete its life cycle." Nieder et al. (2018) and Tisdale et al. (1993) recognize 16 elements (carbon, hydrogen, oxygen, nitrogen, phosphorus, potassium, calcium, magnesium, sulphur, chlorine, boron, iron, manganese, copper, zinc, and molybdenum,) as essential to plant growth and development, while Marschner (2012) and Havlin et al. (2005) add nickel to the list. In addition to these elements, sodium, selenium, cobalt, aluminium, bromine, vanadium, and silicon are identified as beneficial or *quasi-essential* elements, which are elements needed by some but not all plants for optimum growth and production (Nieder et al. 2018; Marschner 2012; Havlin et al. 2005). Based on their relative abundance in plant tissues, the essential elements are further divided into macro (carbon, hydrogen, oxygen, nitrogen, phosphorus, potassium, calcium, magnesium, sulphur) and micronutrients (chlorine, boron, iron, manganese, copper, zinc, molybdenum, and nickel). The macronutrients, which constitute greater than 0.1% of dry plant tissue, are used in relatively large quantities to form constituents of organic compounds that act as building blocks for cells or act as osmotica, while the micronutrients are used in relatively small amounts, constituting less than 0.1% of dry plant tissue (Nieder et al. 2018). However, the essentiality criteria applies equally to both macro- and micronutrients. Carbon, oxygen, and hydrogen, the most abundant elements in plants (making up 94–99.5% of the plant tissue), are obtained from air and water and they hardly limit plant growth. It is the other essential elements, called mineral nutrients, which most often limit plant growth and development and the quality or nutrient composition of food products (Nieder et al. 2018; Havlin et al. 2005). These mineral nutrients are obtained from the soil and/or artificially added to soils through fertilizers and manures. The mineral nutrients are taken up by plants in ionic forms. Among the macronutrients, N, P and K are termed major or fertilizer nutrients, while Ca, Mg and S are called secondary nutrients.

Most of the elements that are essential for plants are also essential for human health (Leitzmann 2009; Klasing et al. 2005) although humans may require several others (White and Brown 2010). Around 29 elements are considered essential for human life, of which 13 are essential plant nutrients obtained from the soil and another 5 are beneficial elements obtained from the soil. Only eleven elements comprise 99.9% of the atoms found in the human body, with H, O, C, and N making up about 99% and Na, K, Ca, Mg, P, S, and Cl making up about 0.9% (Combs 2005). Hydrogen (H), O, C and N are called major nutrient elements, while Na, K, Ca, Mg, P, S, and Cl are called minor elements required for human life. Although there is no agreement on the number and identity of the elements by human health experts, many workers have added 18 elements to the above list as essential trace elements (e.g., Leitzmann 2009; Deckers and Steinnes 2004; Abrahams 2002). These trace elements are lithium (Li), vanadium (V), chromium (Cr), manganese (Mn), iron (Fe), cobalt (Co), nickel (Ni), copper (Cu), zinc (Zn), tungsten (W), molybdenum (Mo), silicon (Si), selenium (Se), fluorine (F), iodine (I), arsenic (As), bromine (Br), and tin (Sn) (Combs 2005).

Therefore, soils that provide a healthy, nutrient-rich growth medium for plants will result in plant tissues that contain many of the elements required for human life. In fact, most of the elements necessary for human life are obtained from either plant or animal tissues (Leitzmann 2009; Shetty 2009). Plant tissues are among the most important sources of Ca, P, Mg, K, Cu, Zn, Se, Mn, and Mo in the human diet and these elements are obtained by plants from the soil. Furthermore, the trace elements that are required for healthy human life cannot be synthesized (McMillian 2002) but obtained from plant tissues, underpinning the strong and direct link between soil, food, and human nutrition. Soil mineral concentration is important not only for crop yields but also for the mineral concentration of the edible portion of crops (Allaway 1986). In Ethiopia, a similar connection has been found between soil zinc deficiency and human zinc deficiency, where the latter was measured in country-wide blood samples (Tessema et al. 2019). A different study in Ethiopia found that soil organic matter influenced the zinc status of crops, though the implications for humans were not directly examined (Wood and Baudron 2018).

11.4.3 Soils and Human Health

If the medium where primary producers of our food are growing does not have a connection with human health, what else can have? Evidences indicate that the connection between soils and human health has a long history with some of them dating back to 1400 BC in the Bible in the book of Numbers (Steffan et al. 2018; Brevik and Hartemink 2010). Soils can affect human health positively as well as negatively through their influence on food quantity and quality (Lal 2009), and human exposure to various chemicals and pathogens present in the soil albeit making a strong scientific connection between the two could be delicate and complex (Brevik and Sauer 2015). Oliver and Gregory (2015) classify the effects of soils on human health into direct or indirect, which can be beneficial or harmful (Pepper et al. 2009). The direct influences come through ingestion, inhalation and absorption of soil or its constituents, while the indirect influences come through dynamic interaction between the pedosphere, biosphere, atmosphere and hydrosphere and are related to the quantity and quality of food that is derived from soil-based agriculture. Ingested soil as much as it potentially supplies essential nutrients, can also expose humans to various dangerous chemicals and pathogens that can cause various diseases. Different workers have reported intestinal obstruction (Henry and Cring 2013), coccidioidomycosis (Stockamp and Thompson 2016) and chronic bronchitis (Zosky et al. 2014) from inhaled soil dust, and podoconiosis (endemicnon-filarial elephantiasis) (Deribe et al. 2013a, b), among others. Potential health concerns related to soils are many and may include cancers, respiratory diseases, neurological disorders, diseases of the excretory system, skin diseases and secondary diseases like heart failure, increased susceptibility to infections and others (Nieder et al. 2018).

In the following sections, soil effects on human health through exposure to soil materials, geophagy, and as sources/origins of different medicines are discussed.

(a) **Human exposure to soil material**

Soils can be a source of materials that are harmful to human health. Humans can be exposed to contaminated soils through ingestion, respiration, and skin absorption or penetration (Brevik 2013). The soil materials to which humans could be exposed include heavy metals, organic pollutants, toxic materials in fertilizers and other agro-chemicals such as pesticides and herbicides, radioactive materials, pathogens, and polluted water. Some of the heavy metals known to affect, especially when they are present in their ionic form and bound to organic molecules, human health are mercury (Hg), cadmium (Cd), chromium (Cr), strontium (Sn) and arsenic (As) (Baird and Cann 2005; Sparks 2003). These heavy metals can get into the human body through various routes, such as inhalation of contaminated soil dust, consumption of crops grown in contaminated soils (Handschumacher and Schwartz 2010) and purposeful or incidental consumption of contaminated soil (Abrahams 2005). The heavy metals form bonds with sulfhydryl groups on enzymes so that the enzyme cannot function properly (Baird and Cann 2005), causing damage to the central nervous system that in turn leads to many health problems that include, but a few, hypertension, gastrointestinal damage, lowering IQ, bone deterioration, and increased cancer rate (Deckers and Steinnes 2004; Sparks 2003). Owing to its many different sources that could cause widespread exposure, lead has been identified as the most problematic heavy metal on a global basis (Balabanova et al. 2017; Baird and Cann 2005).

The most important groups of organic pollutants that are of major concern to human health are the Persistent Organic Pollutants (POPs) (Brevik 2013), which are either recalcitrant to decomposition or bioaccumulate through the food web (Lee et al. 2003). Commonly known POPs in this regard include organochlorines, organophosphates, carbamates, chloroacetamides, glyphosate, and phenoxy herbicides coming from various sources such as pesticides and entering the human body through dermal contact with contaminated soil and ingestion (Peterson et al. 2006). Besides, environmental estrogens, which interfere with the endocrine system and are suspected to cause infertility and increased cancer rates in reproductive organs (Safe 2000), and antibiotic residues, which are feared to result in the development of antibiotic-resistant bacteria (Wellington et al. 2013; Chee-Sanford et al. 2009), are additional organic compounds found in the soil and might present some risks to human health. Veterinary pharmaceuticals and agricultural animal wastes are likely sources of environmental estrogens (Bradford et al. 2008), while the antibiotic residues in the soil could come from animal manures and sewage sludge, implying that the use of these substances as organic fertilizers needs serious health risk considerations before application to soils.

Some chemical fertilizers, usually applied to replenish deficient nutrients, are also known to contain impurities that could be dangerous to human health when accumulated in the soil to toxic levels through repeated applications (Keller et al. 2001). According to different sources, heavy metals are among the toxic materials commonly found in fertilizers, particularly in phosphate and nitrate fertilizers (Fuge 2005), micronutrient fertilizers (Bourennane et al. 2010; Chen et al. 2008), and sewage sludge (Mbila et al. 2001). Lack of stringent regulation on the filler content of fertilizers has allowed some companies to use fertilizers as an inexpensive way to dispose of hazardous wastes.

Important radioactive materials in soils that could harm human health upon exposure include radon, isotopes of cesium, cobalt, curium, neptunium, strontium, plutonium, uranium, technetium, tritium, thorium, americium, radium, and iodine. The sources can be natural or anthropogenic. The natural sources are rocks and minerals that contain these radioactive materials. For instance, granites, felsic metamorphic rocks, organic-rich shales, and phosphatic rocks contain a high amount of uranium, which upon decaying produces radon. Important anthropogenic sources of radioactive materials in the environment include nuclear weapons manufacturing and testing, accidental release from nuclear facilities, the burning of coal, smelting of nonferrous metals, mining activities, and medical wastes (Hu et al. 2010). Different forms of cancer, including lung cancer, and genetic mutations are human health risks often associated with exposure to radioactive materials in the soil (Cheever 2002).

Soil is often called a 'living entity' because of the number and diversity of biological organisms it contains. It is rich in biodiversity. There is no wonder if some of these biological organisms are posing health risks to humans upon exposure to contaminated soils although most of the organisms found in soils are not harmful to humans. The harmful organism can affect human health directly or indirectly through affecting crop production. According to Bultman et al. (2005) and Abrahams (2002), organisms get access to the human body through ingestion, respiration, and skin penetration. Ingestion is arguably the most common way that humans are infected by soil pathogens (Bultman et al. 2005). Groups of organisms that can cause disease include Helminths (hookworms, roundworms, and tapeworms), Protozoa, Fungi, Bacteria, Actinomycetes, and viruses. Recent study by Mestawet et al. (2021) revealed the prevalence of *Ascaris lumbricoides* (56.2%), *Trichuris trichiura* (23.8%), and hookworms 12 (15%) among pregnant women who practiced geophagy.

Water that passes through the soil and gets its way to surface or groundwater sources that could be developed for utilization can pose serious threats to human health by picking up heavy metals, organic pollutants, and soil pathogens that are present in the soil. Counting on soil's ability to filter toxic substances, humans also intentionally dispose of contaminated water into the soil system through septic systems and the application of manure as fertilizers, among others (Kresic 2009; Fetter 2001).

(b) Geophagy

Geophagy, a habit of intentionally eating or ingesting soil, is common among animals and humans (Young et al. 2011; Abrahams 2005), particularly children (von Lindern et al. 2016), pregnant women (Steffan et al. 2018) and people of low socioeconomic status (Henry and Kwong 2003; Oliver 1997). It can be beneficial as well as harmful. The benefits come from the soil material ingested being a source of mineral nutrients such as Ca, Fe, Cu, Mg, and Mn (Abrahams 2006; Abrahams 2002), used as a medicine, helpful as a way to detoxify foods and sometimes conciliate hunger (Sing and Sing 2010; Abrahams 2005; Oliver 1997). Pregnant women, who practice geophagia the most, claimed a number of benefits that come from geophagia, which include relief from morning sickness related nausea and stomach upsets (Diko and Diko 2014) and protection of gastrointestinal tract disturbance and/or infections by clay fractions through directly adsorbing distress causing agents, reinforcing the luminal epithelium by absorbing liquids, and lysing bacterial cells (Kambunga et al. 2019).

On the other hand, geophagy can also be a source of human health concerns depending on the nature of the soil ingested and its chemical as well as biological compositions. Deficiency of certain nutrients (e.g., Fe, Zn, and K) due to the high cation exchange capacity of some clays (Abrahams 2005; Oliver 1997), carcinogenic substances like heavy metal toxicity (Singh and Singh 2010; Calabrese et al. 1997), disorders associated with deficiency of iodine and infection by soil pathogens (Singh and Singh 2010; Hough 2007) have been cited as potential disadvantages attributed to geophagy. Soil pathogens such as bacteria and, fungi when ingested through geophagy can cause disorders in big organs such as liver, while the soil ingested can also be a source of many infectious diseases (e.g., hookworm, podoconiosis, geohelminth). Other health risks associated with geophagy include constipation due to accumulation of soil material in the gastrointestinal tract, abdominal pain that reduced absorption of food, perforation of the colon (Abrahams 2005), and intestinal obstruction (Henry and Cring 2013). In pregnant women, geophagia practiced on contaminated soil resulted in exposure to toxic minerals, pathogenic microbes, and helminthic infections (Steinbaum et al. 2016). Even humus in the soil can cause anemia and malnutrition through chelating essential nutrients such as iron (Henry and Kwong 2003).

Similar to many other developing countries, geophagy is practiced among different members of the society in Ethiopia, particularly pregnant women (Mestawet et al. 2021). However, there are no statistics that shows what percent of the society practices geophagia at the national level. Nevertheless, some studies conducted on pregnant women at some places reported that it varies from 20.3% (Sidama zone) (Handiso 2015) to 30.4% (Addis Ababa) (Kuma et al. 2013).

(c) **Effects of soil on human health**

Soil can influence human health positively or negatively as well as directly or indirectly (Steffan et al. 2018). Provision of important nutrients through food produced on soils and/ or incidental or intentional ingestion of soil material (e.g., through geophagy) and medicines extracted from soil organisms could be taken as positive contributions, while health risks posed by dangerous chemicals and harmful biological organisms (pathogens) and nutrient imbalances can be viewed as negative effects. However, the connection between soils and human health entails complex interactions, demanding further research. In this section, the contribution of soils in the production of a variety of medicines is emphasized.

Soils have been used as origins of a large number of medicines that are used for curing or treating myriad human health problems. However, this value of soils has not been acknowledged widely although the first antibiotic was isolated from soil actinomycetes in 1940 (Ginsberg 2011). Yet, most of the currently relevant antibiotics are those that are extracted from the actinomycetes (Pepper et al. 2009). Available evidences reveal that many beneficial soil organisms, such as actinomycetes and fungi, can make antibacterial molecules (Wolf and Snyder 2003). Medicines derived from soils include mostly antibiotics, prescription drugs, and cancer drugs (Pepper et al. 2009). Significant number of these medicines had their origin in the soil. Important groups of antibiotics derived from actinomycetes include aminoglycosides, glycopeptides, and tetracyclines, while the cephalosporin group are the antibiotics derived from soil fungi. By and large, though not fairly acknowledged, soils seem to be playing a commendable role in helping humans fight against many diseases.

In addition to the varied antibiotics and other drugs extracted from soil organisms, the soil material itself has been used in enhancing human health in different ways. Typical examples include the use of clay mineral kaolin for treating various health problems. This clay mineral has been used as a digestive aid, for making, in combination with pectin, anti-diarrheal medicine Kaopectate (Allport 2002), included in some toothpaste formulas (EPA 1999), used for treating diaper rash and as an emollient and drying agent in treating poison ivy, poison oak, and poison sumac cases (AJN 1989). Another clay mineral, montmorillonite, has been used for treating poisoning by herbicides paraquat and diguat (Abrahams 2005; Clark 1971). A work by Handschumacher and Schwartz (2010) indicated that pharmaceuticals and cosmetics industries use clays in products that are developed to prevent wrinkles and skin ageing.

In addition to the effects of soil on human health discussed above, soils can also put human health under treat indirectly through nutritional deficiencies in food produced locally. Many studies (e.g., Oliver and Gregory 2015; Black et al. 2008) have documented the multiple effects of under-nourishment on human health leading to various disorders and diseases that range from stunted growth and disability through more serious diseases such as diabetes, cardiovascular diseases, anaemia, and mental retardation. Furthermore, dangerous substances taken up by plants indiscriminately from the soil (e.g., antibiotics, antibiotic resistant genes or bacteria) can pose health risks to humans (Wellington et al. 2013).

In Ethiopia, there are so many diseases that are associated directly or indirectly with soils. Very common among these diseases is Podoconiosis, which is common among barefoot individuals who are in long-term contact with irritant soils of volcanic origin (Deribe et al. 2013a, b; Davey et al. 2007; Price 1990, 1976). The country carries the greatest burden of the disease (Deribe et al. 2017; Davey 2010) with more than one million people affected. It is also one of the most extensively studied diseases in the country (Deribe et al. 2017; Deribe et al. 2015a, b; Tekola Ayele et al. 2013; Geshere Oli et al. 2012; Molla et al. 2012; Alemu et al. 2011; Desta et al. 2003). Different studies have found associations of the disease with soil attributes, together with other environmental factors, such as particle size distribution, clay mineralogy (Deribe et al. 2017; Le Blond et al. 2017; Molla et al. 2014; Kloos et al. 1997) and organic matter content (Frommel et al. 1993).

The results of a recent study by Leta et al. (2020) revealed that soil-transmitted helminths are also almost endemic to Ethiopia and are posing serious health problems. According to this study, the prevalence of any soil-transmitted helminths infection across the study population was 21.7%, with *Ascaris lumbricoides* (12.8%) being the most prevalent, followed by hookworms (7.6%) and *Trichuristrichiura* (5.9%).

11.5 Conclusion

Irrefutably, soil is one of the natural resources that are exploited the most by human beings. It is essentially a living entity that provides a substrate for all terrestrial life. It had been and is playing a central role in the rise as well as the demise of many notable civilizations that this world has witnessed. It is, directly or indirectly, associated with almost everything about humans that include their economic status, nutrition and health. It produces not only food and is involved in many environmental regulatory functions, but also is the origin of many medicines that are used to treat many health problems. However, it has been prone to various intensities and forms of degradation that greatly undermine its capacity to perform multiple functions. This has put the production of adequate and nutritious food for the ever increasing world population in jeopardy. The importance of soil in Ethiopia is ever more pronounced given the number of people that rely on this resource for their livelihood is high and the level of awareness is increasing. It is, therefore, imperative that soils be managed properly and used according to their capacity so that they continue to provide goods and services at the required level, which ensures food security, human health, and economic prosperity.

References

Abay A, Sheleme B, Walley F (2015) Response of haricot bean varieties to different level of zinc application in selected areas of Ethiopia. Food Sci Quality Manag 45:78–90

Abate S (2011) Estimating soil loss rates for soil conservation planning in the Borena Woreda of South Wollo highlands, Ethiopia. J Sustain Dev Afr 13(3):1–20

Abebayehu A, Eyasu E, Diels J (2011) Comparative analysis of soil nutrient balance at farm level: a case study in Jimma zone, Ethiopia. Int J Soil Sci 6(4):259–266

Abebe TF, Alamirew T, Abegaz F (2015) Appraisal and mapping of soil salinity problem in Amibara irrigation farms, Middle Awash basin, Ethiopia. Int J Innovat Sci Res 13(1):298–314

Abebe Y, Bogale K, Hambidge M, Stoecker BJ, Bailey K, Gibson RS (2007) Phytate, zinc, iron and calcium content of selected raw and prepared foods consumed in rural Sidama, Southern Ethiopia, and implications for bioavailability. J Food Compos Anal 20:161–168

Abegaz A, Keulen H (2009) Modelling soil nutrient dynamics under alternative farm management practices in the Northern Highlands of Ethiopia. Soil and Tillage Res 103(2):103–115

Abegaz A, Keulen H, Haile M, Oosting SJ (2007) Nutrient dynamics on smallholder farms in Teghane, Northern Highlands Of Ethiopia. In: Bationo A, Waswa B, Kihara J, Kimetu J (eds) Advances in integrated soil fertility management in sub-Saharan Africa: challenges and opportunities. Springer, Dordrecht. https://doi.org/10.1007/978-1-4020-5760-1_34

Abraha B, Yohannes G (2013) The role of seed priming in improving seedling growth of maize (*Zea mays* L.) under salt stress at field conditions. Agricult Sci 4(12):666–672. https://doi.org/10.4236/as.2013.412089

Abrahams PW (2002) Soils: their implications to human health. Sci Total Environ 291:1–32

Abrahams PW (2005) Geophagy and the involuntary ingestion of soil. In: Selinus O, Alloway B, Centeno JA, Finkelman RB, Fuge R, Lindh U, Smedley P (eds) Essentials of medical geology. Elsevier, Amsterdam, The Netherlands, pp 435–458

Abrahams PW (2006) Soil, geography and human disease: a critical review of the importance of medical cartography. Prog Phys Geogr 30:490–512

Adugna A, Abegaz A, Cerdà A (2015) Soil erosion assessment and control in Northeast Wollega, Ethiopia. Solid Earth Discuss 7:3511–3540

Agegnehu G, Sommer K (2000) Optimization of the efficiency of phosphate fertilizers in acidic-ferralitic soils of the humid tropics. Ethiopian J Nat Resour 2(1):63–77

Ajanaw N (2021) Impacts of land use and land cover change on soil erosion and hydrological responses in Ethiopia. Appl Environ Soil Sci 1–10

AJN (1989) Nurses' drug alert. Am J Nurs 89(8):1063–1070

Alemu B, Garedew E, Eshetu Z, Kassa H (2015) Land use and land cover changes and asociated driving forces in north western lowlands of Ethiopia. Int Res J Agricult Soil Sci 5(1):28–44

Alemu G, TekolaAyele F, Daniel T, Ahrens C, Davey G (2011) Burden of podoconiosis in poor rural communities in Gullisoworeda, west Ethiopia. PLoSNegl Trop 5:e1184

Allaway WH (1986) Soil-plant-animal and human intrrelationships in trace element nutrition. p. 465–488. In Mertz W (ed.) Trace elements in human and animal nutrition. Academic Press, Orlando, San Diego, New York, Austin, London, Montreal, sydeney, Tokyo, Toronto

Allport S (2002) Women who eat dirt. Gastronomica J Food Culture 2 (2):28–37

Alvares FF (1540) The Prester John of the Indies. A true relation of the lands of the Prester John, being the narrative of the Portuguese Embassy to Ethiopia in 1520, 1 and 2. Edited by Beckingham, C. F. &G. W. B. Huntingford, 1961. Ben (1896)

Ariti AT, van Vliet J, Verburg PH (2015) Land-use and land-cover changes in the Central Rift Valley of Ethiopia: assessment of perception and adaptation of stakeholders. Appl Geogr 65:28–37

Asfaw KG, Itanna F (2009) Screening some tef [*Eragrostis tef* (Zucc.) Trotter] accessions / varieties for salt tolerance during germination and seedling stage. Momona Ethiopian J Sci 1(2). https://doi.org/10.4314/mejs.v1i2.46046

Ayenew T (2007) Water management problems in the Ethiopian rift: challenges for development. J Afr Earth Sc 48(2–3):222–236. https://doi.org/10.1016/j.jafrearsci.2006.05.010

Bahilu B (2014) Soil fertility status as affected by different land use types and topographic positions: a case of delta Sub-Watershed, Southwestern Ethiopia. Biol Agricult Health Care 4(27)

Baird C, Cann M (2005) Environmental chemistry, 3rd edn. W.H. Freeman, New York, NY

Balabanova B, Stafilov T, Šajn R, Andonovska KB (2017) Quantitative assessment of metal elements using moss species as biomonitors in downwind area of lead-zinc mine. J Environ Sci Health 52:290–301

Bekele G (2005) The Nature and properties of salt affected soils in middle Awash Valley of Ethiopia. J Soil Sci 8:23–30

Beshah T (2003) Understanding farmers: explaining soil and water conservation in Konso, Wolaita, and Wollo, Ethiopia. PhD Thesis, Wageningen University and Research Center, Wageningen

Bekele E (2007) Study on actual situation of medicinal plants in Ethiop. http://jaicaf.or.jp/publications/ethiopia_ac.pdf. Accessed 16 September 2021

Berhan G, Mekonnen K (2009) Estimating soil loss using Universal Soil Loss Equation for soil conservation planning at Medego watershed, northern Ethiopia. J Am Sci 5:58–69

Berihun ML, Tsunekawa A, Haregeweyn N et al (2019) Exploring land use/land cover changes, drivers and their implications in contrasting agro-ecological environments of Ethiopia. Land Use Policy 87

Berkhout ED, Malan M, Kram T (2019) Better soils for healthier lives? An econometric assessment of the link between soil nutrients and malnutrition in Sub-Saharan Africa. PLoS ONE 14(1):e0210642. https://doi.org/10.1371/journal.pone.0210642

Berry L (2003) Land degradation in Ethiopia: its impact and extent. In Berry L, Olson J, Campbell D (eds), Assessing the extent, cost and impact of land degradation at the national level: findings and lessons learned from seven pilot case studies. Commissioned by global mechanism with support from the World Bank

Bewket W, Stroosnijder L (2003) Effects of agro-ecological land use succession on soil properties in the Chemoga watershed, Blue Nile Basin, Ethiopia. Geoderma 111:85–95

Bewket W, Abebe S (2013) Land-use and land-cover change and its environmental implications in a tropical highland watershed, Ethiopia. Int J Environ Stud 70(1):126–139

Bewket W, Teferi E (2009) Assessment of soil erosion hazard and prioritization for treatment at the watershed level: case study in the Chemoga watershed, Blue Nile basin, Ethiopia. Land Degrad Dev 20:609–622

Tefera B, Ayele G, Atnafe Y, Dubaleb P, Jabbarc MA (2002) Nature and causes of land degradation in the Oromiya Region: a review. OCio-economic and Policy Research Worki

Birhanu A, Masih L, van der Zaag P, Nyssen J, Cai X (2019) Impacts of land use and land cover changes on hydrology of the Gumara catchment, Ethiopia. Phys Chem Earth 112:165–174

Birhanu A (2014) Environmental degradation and management in Ethiopian highlands: review of lessons learned. J Environ Protect Policy 2:24–34. https://doi.org/10.11648/j.ijepp.20140201.14

Bikila T (2019) Soil acidity and its management options in Western Ethiopia: review. J Environ Earth Sci 9(10): 27–35

Black RE, Allen LH, Bhutta ZA, Caulfield LE, de Onis M, Ezzati M, Mathers C, Rivera J (2008) Maternal and child undernutrition: global and regional exposures and health consequences. Lancet 371:243–260

Blum WEH (2011) Physical degradation of soils, risks and threats. In: Gliński J, Horabik J, Lipiec J (eds) Encyclopedia of agrophysics. Encyclopedia of earth sciences series. Springer, Dordrecht. https://doi.org/10.1007/978-90-481-3585-1_111

Bourennane H, Douay F, Sterckeman T, Villanneau E, Ciesielski H, King D, Baize D (2010) Mapping of anthropogenic trace elements inputs in agricultural topsoil from northern France using enrichment factors. Geoderma 157:165–174

Bradford SA, Segal E, Zheng W, Wang Q, Hutchins SR (2008) Reuse of concentrated animal feeding operation wastewater on agricultural lands. J Environ Qual 37:S97–S115

Brevik EC (2005) A brief history of soil science. Land use, land cover and soil sciences. vol VI

Brevik EC (2013) Soils and human health: an overview. In: E.C. Brevik&L.C.Burgess, (ed) Soils and human health. CRCPress, BocaRaton, FL, pp 29–56

Brevik EC, Hartemink AE (2010) Early soil knowledge and the birth and development of soil science. CATENA 83:23–33

Brevik EC, Sauer TJ (2015) The past, present, and future of soils and human health studies. Soil 1:35–46

Broadley MR, White PJ, Hammond JP, Zelko I, Lux A (2007) Zin in plants. New Phytol 173:677–702. https://doi.org/10.1111/j.1469-8137.2007.01996.x

Bultman MW, Fisher ES, Pappagianis D (2005) The ecology of soil-borne human pathogens. In: Selinus O, Alloway B, Centeno JA, Finkelman RB, Fuge R, Lindh U et al (eds) Essentials of medical geology. Elsevier, Amsterdam, pp 481–511

Bulti M, Abdulatif A (2019) A review on agricultural problems and their management in Ethiopia. Turkish J Agric Food SciTechnol 7 (8):1189–1202

Bunting AH (1963) A plan for agricultural research and specialist services in Ethiopia. Reading

Burras CL, Nyasimi M, Butler L (2013) Soils, human health and wealth. In: Brevik EC, Burgess LC (eds), = Soils and human health. Taylor & Francis Group, 6000 Broken Sound Parkway NW, Suite 300, Boca Raton, FL 33487–2742

Burton R (1856) First footsteps in East Africa. Waterfield G (ed), 1966. London

Butzer KW (1981) Rise and fall of Axum, Ethiopia: a geo-archaeological interpretation. Am Antiq 46(3):471–495

Calabrese EJ, Stanek EJ, Pekow P, Barnes RM (1997) Soil ingestion estimates for children residing on a superfund site. Ecotoxicol Environ Saf 36:258–268

Chee-Sanford JC, Mackie RI, Koike S, Krapac IG, Lin YF, Yannarell AC, Maxwell S, Aminov RI (2009) Fate and transport of antibiotic residues and antibiotic resistance genes following land application of manure waste. J Environ Qual 38:1086–1108

Cheever CL (2002) Ionizing radiation. In: Plog BA, Quinlan PJ (eds) Fundamentals of industrial hygiene. National Safety Council Press, Itasca, IL, pp 257–280

Chen W, Krage N, Wu L, Pan G, Khosrivafard M, Chang AC (2008) Arsenic, cadmium, and lead in California cropland soils: role of phosphate and micronutrient fertilizers. J Environ Qual 37:689–695

Chilimba ADC, Young SD, Black CR, Rogerson KB, Ander EL, Watts MJ, Lammel J, Broadley MR (2011) Maize grain and soil surveys reveal suboptimal dietary selenium intake is widespread in Malawi. Sci Rep 1:72

Clark DG (1971) Inhibition of the absorption of paraquat from the gastrointestinal tract by adsorbents. Br J Ind Med 28:186–188

Combs GF (2005) Geological impacts on nutrition. In: Selinus O, Alloway B, Centeno JA, Finkelman RB, Fuge R, Lindh U, Smedley P (eds) Essentials of medical geology. Elsevier, Amsterdam, The Netherlands, pp 161–177

Corbeels M, Abebe S, Mitiku H (2002) Farmers' knowledge of soil fertility and local management strategies in Tigray, Ethiopia. Managing Africa's Soils No 10

Crawford OGS (1958) Ethiopian itineraries circa 1400–1524. London

Da Silva HG (2014) Soils are the foundations of family farming, FAO, Rome, Italy

Daniel JN (2020) Effects of land use types on selected soil properties in central highlands of Ethiopia. Appl Environ Soil Sci Volume 2020 Article ID 7026929. https://doi.org/10.1155/2020/7026929

Dale T, Carter VG (1955) Topsoil and civilization. University of Oklahoma Press, Norman, OK, USA

Davey G (2010) Podoconiosis, non-filarial elephantiasis, andlymphology. Lymphology 43:168–177

Davey G, GebreHanna E, Adeyemo A, Rotimi C, Newport M, Desta K (2007) Podoconiosis: a tropical model for gene-environment interactions? Trans R Soc Trop Med Hyg 101:91–96

De Groote H, Tessema M, Gameda S, Gunaratna NS (2021) Soil zinc, serum zinc, and the potential for agronomic biofortification to reduce human zinc deficiency in Ethiopia. Sci Rep 11:8770. https://doi.org/10.1038/s41598-021-88304-6

Deckers J, Steinnes E (2004) State of the art on soil-related geo-medical issues in the world. In: Sparks DL (ed) Advances in agronomy, vol 84. Elsevier, Amsterdam, The Netherlands, pp 1–35

Degytun GT (2011) OCio-economic and environmental impact of floriculture in Ethiopia. MSc thesis. 63 P. Humboldt University of Berlin

Demise C, Berhanu A, John M (2010) Ethiopia's agricultural sector policy and investment framework (PIF). In: Ministry of Agriculture and Rural Development. Addis Ababa, Ethiopia

Deribe K, Brooker SJ, Pullan RL, Sime H, Gebretsadik A etc (2015) Epidemiology and individual, household and geographical risk factors of Podoconiosisin Ethiopia: results from the First Nationwide Mapping. Am J Trop Med Hyg 92(1):148–158. https://doi.org/10.4269/ajtmh.14-0446

Deribe K, Cano J, Newport MJ, Golding N, Pullan RL, Sime H, et al (2015b) Mapping and modelling the geographical distribution and environmental limits of podoconiosis in Ethiopia. PLoSNegl Trop Dis 9(7):e0003946. https://doi.org/10.1371/journal.pntd.0003946

Deribe K, Tomczyk S, Tekola AF (2013a) Ten years of podoconiosis research in Ethiopia. PLoS Negl Trop Dis 7:e2301

Deribe K, Tomczyk S, Tekola-Ayele F (2013b) Ten years of podoconiosis research in Ethiopia. PLoS NTD

Deribe K, Cano J, Newport JM, Golding N et al (2017) Mapping and modelling the geographical distribution and environmental limits of Podoconiosis in Ethiopia. PLoS Negl Trop Dis 9(7):e0003946. https://doi.org/10.1371/journal.pntd.0003946

Desta K, Ashine M, Davey G (2003) Prevalence of podoconiosis (endemic non-filarial elephantiasis) in Wolaitta, southern Ethiopia. Trop Doct 32:217–220

Demlie M, Stefan W (2006) Soil and groundwater pollution of an urban catchment by trace metals: case study of the Addis Ababa region, central Ethiopia. Environ Geol 51:421–431

Devereux S, Sabates-Wheeler R, Tefera M, et al (2006) Trends in PSNP transfers within targeted households: Final report. Institute of Development Studies, Sussex, UK, Indak International Pvt. L. C., Addis Ababa, Ethiopia

Diamond J (2005) Collapse-How OCieties choose to fail or succeed. Viking, New York, NY

Diko M, Diko S (2014) Physico-chemistry of geophagic soils ingested to relieve nausea and vomiting during pregnancy. Afr J Tradit Complement Altern Med 4(11):21–24

Dubale P, Menkir M, Zewdie M, Zeray L (2002) Agriculture, irrigation and drainage research in the past and the future Integrated water and land management research and capacity building priorities for Ethiopia, 38

EthioSIS (Ethiopia Soil Information System) (2014) Soil fertility status and fertilizer recommendation Atlas for Tigray Regional State, Ethiopia. Ethio SIS, Addis Ababa

Edwards S (2001) The ecology and conservation status of medicinal plants on Ethiopia. What do we know? pp 46–55, In: Zewdu M, Demissie A (eds) Conservation and sustainable use of medicinal plants in Ethiopia, Proceedings of national workshop on biodiversity conservation and sustainable use of medicinal plants in Ethiopia, Institute of Biodiversity Conservation and Research, Addis Ababa

Engdawork A (2015) Characterization and classification of the major agricultural soils in CASCAPE intervention woredas in the central highlands of Oromia region, Ethiopia

EPA (1999) Kaolin (100104) Fact sheet. Environmental Protection Agency, Washington, DC

Essington M (2004) Soil and water chemistry: an integrative approach. CRC Press, Boca Raton

Endashaw B (2007) Study on actual situation of medicinal plants in Ethiop. http://jaicaf.or.jp/publications/ethiopia_ac.pdf retrieved on September 16, 2021

Fantaw Y, Ledin S, Abdu A (2007) Changes in soil organic carbon and total nitrogen contents in three adjacent land use types in the Bale Mountains, south-eastern highlands of Ethiopia. Forest Ecol Manage 242:337–342

FAO (2015) Revised world soil charter, food and agriculture organization. Rome, Italy

FAO (2020) Food and Agriculture Organization Corporate Statistical Database (FAOSTAT). http://www.fao.org/faostat/en/#data

Fasika A, Motuma T, Gizaw T (2019) Land use land cover change trend and its drivers in Somodo watershed Southwestern, Ethiopia. Afr J Agricult Res 14(2):102–117

Fetter CW (2001) Applied hydrogeology, 4th edn. Prentice-Hall, Upper Saddle River, NJ

Frommel D, Ayranci B, Pfeifer HR, Sanchez A, Frommel A, Mengistu G (1993) Podoconiosis in the EthiopianRift Valley. Role of beryllium and zirconium. Trop Geogr Med 45(4):165–167. PMID: 8236466

Fuge R (2005) Anthropogenic sources. In: Selinus O, Alloway B, Centeno JA, Finkelman RB, Fuge R, Lindh U, Smedley P (eds) Essentials of medical geology. Elsevier, Amsterdam, The Netherlands, pp 43–60

Gashaw T, Tulu T, Argaw M, Worqlul AW (2018) Modeling the hydrological impacts of land use/land cover changes in the Andassa watershed, Blue Nile Basin, Ethiopia. Sci Total Environ 619:1394–1408

Getachew M, Yeshigeta R, AbebawTiruneh YA, Dereje E, Mekonnen Z (2021) Soil-transmitted helminthic infections and geophagia among pregnant women in Jimma town health institutions, Southwest Ethiopia. Ethiop J Health Sci 31(5):1033. https://doi.org/10.4314/ejhs.v31i5.16

Gashu D, Nalivata PC, Amede T, Ander EL, Bailey EH, Botoman L, Chagumaira C, Gameda S et al (2021) The nutritional quality of cereals varies geospatially in Ethiopia and Malawi. Nature 594:71–76

Gebremeskel G, GebremicaelTG KM, Meresa E, Gebremedhin T, Girmay A (2018) Salinization pattern and its spatial distribution in the irrigated agriculture of Northern Ethiopia: an integrated approach of quantitative and spatial analysis. Agric Water Manag 206:147–157. https://doi.org/10.1016/j.agwat.2018.05.007

Gebresamuel G, Opazo-Salazar D, Corral-Núnez G, van Beek C, Elias E, Okolo CC (2021) Nutrient balance of farming systems in Tigray, Northern Ethiopia. J Soil Sci Plant Nutr 21:315–328. https://doi.org/10.1007/s42729-020-00362-3

Gebresamuel G, Opazo-Salazar D, Corral-Núnez GC, van Beek C, Elias E, Okolo CC (2021) Nutrient balance of farming systems in Tigray, Northern Ethiopia. J Soil Sci Plant Nutrit https://doi.org/10.1007/s42729-020-00362-3

Gebre-Selassie A, Bekele T (2010) A review of Ethiopian agriculture: roles, policy and small-scale farming systems. 36 Global Growing Casebook. Country Analyses: Ethiopia and D. R. Congo 37

Geressu K, Gezaghegne M (2008) Response of some lowland growing sorghum (Sorghum bicolor L. Moench) accessions to salt stress during germination and seedling growth. Afr J Agricult Res 3 (1):044–48

Geshere Oli G, TekolaAyele F, Petros B (2012) Parasitological, serological, and clinical evidence for high prevalence ofpodoconiosis (non-filarial elephantiasis) in Midakegn district, central Ethiopia. Trop Med Int Health 17:722–726

Gete Z, Getachew A, Dejene A, Shahid R (2010). Fertilizer and soil fertility potential in Ethiopia: Constraints and opportunities for enhancing the system. Technical Report in Review July 2010. https://doi.org/10.13140/RG.2.1.3632.5846

Girma A, endalkahew W, Sheleme B, Bakken LR, (2012) Nitrogen mineralization dynamics undes different moisture regimes in Tropical soils. Intrenat J Soil Sci 7:132–145

Ginsberg J (2011) Selman Waksman and Antibiotics. American Chemical society

Gomiero T (2016) Soil degradation, land scarcity and food security: reviewing a complex challenge. Sustainability 8, 281. https://doi.org/10.3390/su8030281

Gregory PJ (2012) Soils and food security: challenges and opportunities. In: Hester RE, Harrison RM (eds) Soils and food security. RSC Publishing, London, pp 1–30

Haile GW, Fetene M (2012) Assessment of soil erosion hazard in Kilie catchment, East Shoa, Ethiopia. Land Degrad Dev 23:293–306

Haile S, Teller C (2005) Population, development, and environment in ethiopia. environmental change and secuirty program. https://www.wilsoncenter.org/event/population-development-and-environment-ethiopia. Accessed 17 September 2021

Hailemariam SN, Teshome S, Demel T (2016) Land use and land cover change in the bale mountain eco-region of Ethiopia during 1985 to 2015. Land 5:41. https://doi.org/10.3390/land5040041

Haileslassie A, Priess J, Veldkamp E, Teketay D, Lesschen JP (2005) Assessment of soil nutrient depletion and its spatial variability on smallholders' mixed farming systems in Ethiopia using partial versus full nutrient balances. Agric Ecosyst Environ 108:1–16

Handiso YH (2015) Prevalence of food aversions, cravings and pica during pregnancy and their asOCiation with nutritional status of pregnant women in Dale Woreda, Sidama zone, SNNPRS, Ethiopia. Int J NutrMetab 7:1–14

Handschumacher P, Schwartz D (2010) Do pedo-epidemiological systems exist? In Landa ER, Feller C (eds) Soil and culture 355–368. Springer, New York, NY

Hanson JD, Liebig MA, Merrill SD, Tanaka DL, Krupinsky JM, Stott DE (2007) Dynamic cropping systems: increasing adaptability amid an uncertain future. Agron J 99:939–943

Haregeweyn N, Tsunekawa A, Nyssen J, Poesen J, Tsubo M, Derege TM, Schu B, Enyew A, Firew T (2015) Soil erosion and conservation in Ethiopia: A review. Progress in Physical Geography 1–25

Havlin JL, Beaton JD, Tisdale SL, Nelson WL (2005) Soil fertility and fertilizers: an introduction to nutrient management, 7th edn. Pearson Prentice Hall, Upper Saddle River, NJ

Hawando T (1995) The survey of the soil and water resources of Ethiopia. UNU/Toko

Henry J, Kwong AM (2003) Why is geophagy treated like dirt? Deviant Behav 24(4):353–371

Henry JM, Cring FD (2013) Geophagy:ananthropologicalperspective. In: Brevik EC, Burgess LC (eds) Soils and human health. CRCPress, Boca Raton, FL, pp 179–198

Hillel D (1991) Out of the earth: civilization and the life of the soil. University of California Press, Berkeley, CA, USA

Hillel H (1980) Applications of soil physics. Academic Press, INC. 111 Fifth Avenue, New York, New York 10003

Hodson ME, Donner E (2013) Managing adverse soil chemical environments. In: Gregory PJ, Nortcliff S (eds) Soil conditions and plant growth, 12th edn. Wiley-Blackwell, Chichester, pp 195–237

Hough RL (2007) Soil and human health: an epidemiological review. Eur J Soil Sci 58:1200–1212

Hu QH, Weng JQ, Wang JS (2010) Sources of anthropogenic radionuclides in the environment: a review. J Environ Radioact 101:426–437

Hurni H (1983a) Soil formation rates in Ethiopia, soil conservation project. Working paper 2, Ministry of Agriculture, Addis Ababa, Ethiopia

Hurni H (1983b) Soil erosion and formation in agricultural ecosystems: Ethiopia and Northern Thailand. Mt Res Dev 3:131–142

Hurni H (1988) Degradation and conservation of the resources in the Ethiopian highlands. Mt Res Dev 8:123–130

IFOAM (2011) Definition of organic farming. http://ifoam.org/growing_organic/definitions/doa/index.html

Jack CG, Moss JE, Wallace MT (2009) Is growth in land-based wealth sustaining part-time farming? Euro Choices 8:29–35

Jaleta Negasa D (2020) Effects of land use types on selected soil properties in central highlands of Ethiopia. Appl Environ Soil Sci 2020, Article ID 7026929. https://doi.org/10.1155/2020/7026929

Jones A, Breuning-Madsen H, Brossard M et al (2013) Soil Atlas of Africa. Publications Office of the European Union, Brussels, Belgium, European Commission

Joy EJM, Stein AJ, Young SD, Ander EL, Watts MJ, Broadley MR (2015) Zinc-enriched fertilizers as a potential public health intervention in Africa. Plant Soil 389:1–24

Kambunga SN, Candeias C, Hasheela I et al (2019) Review of the nature of some geophagic materials and their potential health effects on pregnant women: some examples from Africa. Environ Geochem Health 41:2949–2975

Karlen DL, Mausbach MJ, Doran JW, Cline RG, Harris RF, Schuman GE (1997) Soil quality: a concept, definition, and framework for evaluation. Soil Sci OCiety Am J 61:4–10

Karley AJ, White PJ (2009) Moving cationic minerals to edible tissues: potassium, magnesium, calcium. Curr Opin Plant Biol 12:291–298

Keller A, von Steiger B, van der Zee SEATM, Schulin R (2001) A stochastic empirical model for regional heavy-metal balances in agroecosystems. J Environ Qual 30:1976–1989

Khresat S, Al-Bakri J, Al-Tahhan R (2008) Impacts of land use/cover change on soil properties in the Mediterranean region of northwestern Jordan. Land Degrad Dev 19:397–407

Klasing KC, Goff JP, Greger JL, King JC, Lall SP, Lei XG, Linn JG, Nielsen FH, Spears JW (2005) Mineral tolerance of animals. The National Academies Press, Washington, DC

Kloos WE, Ballard DN, Webster JA, Hubner RJ, Tomasz A, Couto I, Sloan GL, Dehart HP, Fiedler F, Schubert K, de Lencastre H, Sanches IS, Heath HE, Leblanc PA, Ljungh A (1997) Ribotype delineation and description of Staphylococcus sciuri subspecies and their potential as reservoirs of methicillin resistance and staphylolytic enzyme genes. Int J Syst Bacteriol 47(2):313–23. https://doi.org/10.1099/00207713-47-2-313

Kresic N (2009) Groundwater resources: sustainability, management, and restoration. McGraw Hill, New York, NY

Kuma T, Yusuf L, Negash S (2013) Assessment of nausea and vomiting of pregnancy on antenatal clients of Addis Ababa. Ethiop J Health Dev 27:200–207

Lal R (2009) Soil degradation as a reason for inadequate human nutrition. Food Sci 1:45–57

Le Blond JS, Baxter PJ, Bello D, Raftis J, Molla YB, Cuadros J et al (2017) Haemolyticactivity of soil from areas of varying podoconiosisendemicity in Ethiopia. PLoS ONE 12(5):e0177219

Lee WY, Iannucci-Berger W, Eitzer BD, White JC, Mattina MI (2003) Persistent organic pollutants in the environment: chlordane residues in compost. J Environ Qual 32:224–231

Legesse H, Walelign W, Hussein M, Bunyamin T (2017) Effects of zinc application strategy on zinc content and productivity of chickpea grown under zinc deficient soils. J Soil Sci Plant Nutr 17 (1):112–126

Lehman RM, Cambardella CA, Stott DE, Acosta-Martinez V, Manter DK, Buyer JS, Maul JE, Smith JL, Collins HP, Halvorson JJ, Kremer RJ, Lundgren JG, Ducey TF, Jin VL, Karlen DL (2015) Understanding and enhancing soil biological health: the solution for reversing soil degradation. Sustainability 7:988–1027. https://doi.org/10.3390/su7010988

Leitzmann C (2009) Adequate diet of essential nutrients for healthy people. In Squires V (ed) The role of food, agriculture, forestry and fisheries in human nutrition. Encyclopedia of Life Support Systems (EOLSS), Developed under the Auspices of the UNESCO, EOLSS Publishers, Oxford

Leta1 GT, Mekete K, Wuletaw Y, et al (2020) National mapping of soil-transmitted helminth and schistosome infections in Ethiopia. Parasites Vectors 13:437.https://doi.org/10.1186/s13071-020-04317-6

Ligowe IS, Phiri FP, Ander EL, Bailey EH, Chilimba ADC, Gashu D, Joy EJM, Lark RM, Kabambe V, Kalimbira AA, Kumssa DB, Nalivata PC, Young SD, Broadley MR (2020) Selenium (Se) deficiency risks in sub-Saharan African food systems and their geospatial linkages. Proceedings of the Nutrition OCiety 1–11. https://doi.org/10.1017/S0029665120006904

Marschner P (ed) (2012) Mineral nutrition of higher plants, 3rd edn. Academic Press, London

Mbila MO, Thompson ML, Mbagwu JSC, Laird DA (2001) Distribution and movement of sludge-derived trace metals in selected Nigerian soils. J Environ Qual 30:1667–1674

McMillian RJ (2002) Malnutrition. In: Katz SH, Weaver WW (eds) Encyclopedia of food and culture. Scribner, New York, pp 431–435

McCarrison R (1921) Observations on the effects of fat excess on the growth and metamorphosis of Tadpoles. Proceed Royal Soc B 92 (647):295–303 Bibcode:1921RSPSB..92..295M, https://doi.org/10.1098/rspb.1921.0026, JSTOR 81036

McNeill JR, Winiwarter V (2004) Breaking the sod: humankind, history, and soil. Science 34:1627–1629

McNeill JR, Winiwarter V (eds) (2006) Soils and OCieties—perspectives from environmental history. The White Horse Press, Isle of Harris

McNeill, John R, Verena Winiwater, eds (2010) Soils and OCieties: perspectives from environmental history, pp 1–21. The White Horse Press, Cambridge

Merga B, Ahmed A (2019) A review on agricultural problems and their management in Ethiopia. Turkish J Agricult Food Sci Technol 7 (8):1189–1202

Getachew M, Yeshigeta R, Tiruneh A, Alemu Y, Dereje E, Mekonnen Z (2021) Soil-Transmitted helminthic infections and geophagia among pregnant women in Jimma Town Health Institutions, Southwest Ethiopia. Ethiop J Health Sci 31(5):1033. https://doi.org/10.4314/ejhs.v31i5.16.18

Mengistu C, Kibebew K, Tarekegn F (2017) Influence of different land use types and soil depths on selected soil properties related to soil fertility in Warandhab Area, Horo GuduruWallaga Zone, Oromiya, Ethiopia. Int J Environ Sci Nat Res 4(2):1–12

Menta C (2012) Soil fauna diversity–function, soil degradation, biological indices, soil restoration. In: Lameed GA (ed) Biodiversity conservation and utilization in a diverse world. Rijeka, Croatia: In Tech, pp 59–94

Millennium Ecosystem Assessment (2005) Ecosystems and Human Well-Being: Desertification Synthesis, World Resources Institute, Washington, DC, USA TANGO. 2012. MERET Impact Evaluation Report. Technical Assistance to NGOs. Tucson

Minami K (2009) Soil and humanity: culture, civilization, livelihood and health (a review article). Soil Sci Plant Nutrit 55:603–615

Molla YB, Tomczyk S, Amberbir T, Tamiru A, Davey G (2012) Podoconiosis in east and west Gojam Zones, northern Ethiopia. PLoSNegl Trop Dis 6:e1744

Molla YB, Wardrop NA, Le Blond JS, Baxter P, Newport MJ, Atkinson PM, Davey G (2014) Modeling environmental factorscorrelated with podoconiosis. Int J Health Geogr 13:24

Montgomery DR (2007) Dirt: the erosion of civilization. University of California Press, Berkeley, CA, USA

Mulat Y, Kibret K, Bedadi B, Mohammed M (2021) Soil quality evaluation under different land use types in Kersa sub-watershed, eastern Ethiopia. Environ Syst Res 10:19. https://doi.org/10.1186/s40068-021-00224-6

Nieder R, Benbi DK, Reichl FX (2018) Soil components and human health. Springer Nature, The Netherlands

Nunes FC, AlvesLJ, CarvalhoCCN, Gross E, SoaresTM, Prasad MNV (2020) Soil as a complex ecological system for meeting food and nutritional security. In: Climate change and soil interactions, pp 229–269

Okolo CC, Gebresamuel G, Abeba NR, Amanuel Z, Mitiku H (2019) Advances in quantifying soil organic carbon under different land uses in Ethiopia: a review and synthesis. Bullet Nat Res Centre 43:99. https://doi.org/10.1186/s42269-019-0120-z

Oliver MA (1997) Soil and human health: a review. Eur J Soil Sci 48:573–592

Oliver MA, Gregory PJ (2015) Soil, food security and human health: a review. Eur J Soil Sci 66:257–276

Osman KT (2014) Chemical soil degradation. In: Soil degradation, conservation and remediation. Springer, Dordrecht. https://doi.org/10.1007/978-94-007-7590-9_5

Osman M, Sauerborn P (2001) Soil and water conservation in Ethiopia. J soils sediment 1(2):117–123

Pankhurst CE, Doube BM, Gupta VVSR (1997) Biological indicators of soil health: synthesis. In: Pakhurst CE et al (ed) Biological indicators of soil health. CABI, Wallingford, Oxfordshire, England, pp 419–435

Paulos D (2001) Soil and water resources and degradation factors affecting their productivity in the Ethiopian highland agro-ecosystems. Michigan State University Press 8(1):1–18

Pepper IL, Gerba CP, Newby DT, Rice CW (2009) Soil: a public health threat or savior? Crit Rev Environ Sci Technol 39:416–432

Peterson RKD, Macedo PA, Davis RS (2006) A human-health risk assessment for West Nile virus and insecticides used in mosquito management. Environ Health Perspect 114(3):366–372

Price E (1990) Podoconiosis: non-filarial elephantiasis. Oxford Medical Publications, Oxford

Price EW (1976) The association of endemic elephantiasis of the lower legs in East Africa with soil derived from volcanic rocks. T Roy Soc Trop Med H.70:288–295

Rao KS, Pant H (2001) Land use dynamics and landscape change pattern in a typical micro watershed in the middle elevation zone of central Himalaya, India. Agric Ecosyst Environ 86:113–123

Ruffeis D, Loiskandl W, Spendlingwimmer R, Schonerklee M, Awulachew SB, Boelee E, Wallner K (2008) Environmental impact analysis of two large scale irrigation schemes in Ethiopia

Safe SH (2000) Endocrine disruptors and human health: is there a problem? An update. Environ Health Perspect 108(6):487–493

Sanchez PA (1977) Properties and management of soils in the tropics. Soil Sci 124(3):187

SDA (1999) Soil quality test kit guide. USDA Soil Quality Institute. Washington, DC

Setffan JJ, Brevik LLS, Burgess LC, Cerdà A (2018) The effect of soil on human health: an overview. Eur J Soil Sci 69:159–171

Sheleme B (2003) Low Input Sustainable Agriculture in southern Rift Valley Project final technical report (Canadian International Development Agency funded project)

Shetty P (2009) Fundamentals of human health and nutrition. In Squires V (ed) The role of food, agriculture, forestry and fisheries in human nutrition. Encyclopedia of Life Support Systems (EOLSS), Developed under the Auspices of the UNESCO, EOLSS Publishers, Oxford. http://www.eolss.net

Shimbahri M, Girmay G, Amanuel Z, Mitiku H (2020) Nutrient balances in smallholder farms in northern Ethiopia. Soil Use Manag 37:468–478

Silver WL, Perez T, Mayer A, Jones AR (2021) The role of soil in the contribution of food and feed. Phil Trans r OC B 376:20200181. https://doi.org/10.1098/rstb.2020.0181

Simachew B (2020) Natural resource degradation tendencies in Ethiopia: a review. Environ Syst Res 9:33. https://doi.org/10.1186/s40068-020-00194-1

Sims GK (1990) Biological degradation of soil. In: Lal R, Stewart BA (eds) Advances in soil science. Advances in Soil Science 11. Springer, New York, NY. https://doi.org/10.1007/978-1-4612-3322-0_8

Sing D, Sing C (2010) Impact of direct soil exposures from airborne dust and geophagy on human health. Int J Environ Res Public Health 7:1205–1223

Singh A (2015) Soil salinization and waterlogging: a threat to environment and agricultural sustainability. Ecol Ind 57:128–130. https://doi.org/10.1016/j.ecolind.2015.04.027

Smith P, Cotrufo MF, Rumpel C, Paustian K, Kuikman PJ, Elliott JA, McDowell R et al (2015) Biogeochemical cycles and biodiversity as key drivers of ecosystem services provided by soils. Soil 1:665–685

Somani LL (1996) Crop production in acid soils. Agrotech Publishing Academy, 1st edition/Ed. Agrotech Publishing Academy, New Delhi

Sonneveld BGJS, Keyze MA, Stroosnijder L (2011) Evaluating quantitative and qualitative models: an application for nationwide water erosion assessment in Ethiopia. Environ Model Softw 26:1161–1170

Sparks DL (2003) Environmental soil chemistry, 2nd edn. Academic Press, Amsterdam, The Netherlands

Steinbaum L, Njenga SM, Kihara J et al (2016) Soil-transmitted helminth eggs are present in soil at multiple locations within households in rural Kenya. PLoS One 11:e0157780

Stockamp NW, Thompson GR (2016) Coccidioidomycosis. Infect Dis Clin North Am 30:229–246

Taddese G (2001) Land degradation: a challenge to Ethiopia. Environ Manage 27(6):815–824. https://doi.org/10.1007/s002670010190

Tamene L, Amede T, Kihara J, Tibebe D, Schulz S (eds) (2017) A review of soil fertility management and crop response to fertilizer application in Ethiopia: towards development of site- and context-specific fertilizer recommendation. CIAT Publication No. 443. International Center for Tropical Agriculture (CIAT), Addis Ababa, Ethiopia, p 86. http://hdl.handle.net/10568/82996

Tamirie H (1995) The survey of the soil and water resources of Ethiopia. United Nations University, Tokyo, p 1995

Takala B (2019) Soil acidity and its management options in Western Ethiopia: review. J Environ Earth Sci 9(10):27–35

Tefera B, Ayele G, Atnafe Y, Jabbar MA, Dubale P (2002) Nature and causes of land degradation in the Oromiya Region: a review. OCio-economic and Policy Research Worki

TekolaAyele F, Alemu G, Davey G, Ahrens C (2013) Community-based survey of podoconiosis in BedeleZuria woreda, west Ethiopia. In Health 5:119–125

Temesgen G, Amare B, Abraham M (2014) Population dynamics and land use/land cover changes in Dera District, Ethiopia. Global J Biol. Agricult Health Sci 3(1):137–140

Tesfaye A, Negatu W, Brouwer R, Van Der Zaag P (2014) Understanding soil conservation decision of farmers in the Gedeb watershed, Ethiopia. Land Degrad Dev 25:71–79

Tessema M, De Groote H, Brouwer ID, Feskens EDJM, Tefera B, Zerfu D, Belay A, Demelash Y, Gunaratna NS (2019) Soil zinc is asociated with serum zinc but not with linear growth of children in Ethiopia. Nutrients 11:221. https://doi.org/10.3390/nu11020221

Tilahun A, Takele B, Endrias G (2001) Reversing the degradation of arable land in the Ethiopian highlands. Managing Africa's soils No. 23. International Center for Research in Agro forestry 1–20

Tisdale SL, Nelson WN, Beaton JD, Havlin JL (1993) Soil fertility and fertilizers, 5th edn. Macmillan Publishing Company, New York, NY

Tittonell P, Muriuki A, Shepherd KD, Mugendi D, Kaizzi KC, Okeyo J, Vanlauwe B (2010) The diversity of rural livelihoods and their influence on soil fertility in agricultural systems of East Africa-a typology of smallholder farms. Agric Syst 103:83–97. https://doi.org/10.1016/j.agsy.2009.10.001

Troeh FR, Hobbs JA, Donahue RL (2004) Soil and water conservation for productivity and environmental protection, 4th edn. Prentice Hall, Upper Saddle River, NJ

UN (United Nations) (2013) Draft Resolution submitted by the vice-chair of the committee, Ms. Farrah Brown (Jamaica), on the Basis of Informal Consultations on Draft Resolution A/C.2/68/L.21" (PDF). United Nations General Assembly, Washington, DC, USA, 2013

van Beek CL, Elias E, Yihenew GS, Heesmans H, Tsegaye A, Feyisa H, Tolla M, Melmuye M, Gebremeskel Y, Mengist S (2016) Soil nutrient balances under diverse agro-ecological settings in Ethiopia. Nutr Cycl Agroecosyst 106:257–274

Von Lindern I, Spalinger S, Stifelman ML, Stanek LW, Bartrem C (2016) Estimating children's soil/dust ingestion rates through retrospective analyses of blood lead biomonitoring from the Bunker Hill Superfund Site in Idaho. Environ Health Perspect 124:1462–1470

Wellington EMH, Boxall ABA, Cross P, Feil EJ, Gaze WH, Hawkey PM, Johnson-Rollings AS, Jones DL, Lee NM, Otten W, Thomas CM, Prysor Williams AP (2013) The role of the natural environment in the emergence of antibiotic resistance in Gram-negative bacteria. Lancet Infect Dis 13(2):155–165

Westphal E (1975) Agricultural systems in Ethiopia. Agric Res Rep (Versl. landbouwk. Onderz)

White PJ, Broadley MR (2009) Biofortification of crops with seven mineral elements often lacking in human diets–iron, zinc, copper, calcium, magnesium, selenium and iodine. New Phytol 182:49–84

White PJ, Brown PH (2010) Plant nutrition for sustainable development and global health. Ann Bot 105:1073–1080. https://doi.org/10.1093/aob/mcq085

White PJ, Greenwood DJ (2013) Properties and management of cationic elements for crop growth. In: Gregory PJ, Nortcliff S (eds) Soil conditions and plant growth. Oxford: Blackwell Publishing, pp 160–194

Wolf B, Snyder GH (2003) Sustainable soils. Food Products Press, Binghamton, NY

Woldeamlak B, Stroosnijder L (2003) Effects of agro-ecological land use succession on soil properties in the Chemoga watershed, Blue Nile Basin, Ethiopia. Geoderma111: 85–95

Wood SA, Baudron F (2018) Soil organic matter underlies crop nutritional quality and productivity in smallholder agriculture. Agr Ecosyst Environ 266:100–108. https://doi.org/10.1016/j.agee.2018.07.025

Wylde AB (1901) Modern Abyssinia. London

Yifru A, Taye B (2011) Effects of land use on soil organic carbon and nitrogen in soils of bale, Southeastern Ethiopia. Tropical Subtropical Agroecosyst 14:229–235

Yihenew G, Fentanesh A, Solomon A (2015) The effects of land use types, management practices and slope classes on selected soil physico-chemical properties in Zikre watershed, North-Western Ethiopia. Environ Syst Res 4:3. https://doi.org/10.1186/s40068-015-0027-0

Young SL, Sherman PW, Lucks JB, Pelto GH (2011) Why on earth? Evaluating hypotheses about the physiological functions of human geophagy. Q Rev Biol 86:97–120

Zewdu S, Suryabhagavan KV, Balakrishnan M (2014) Geo-spatial approach for soil salinity mapping in Sego irrigation farm, South Ethiopia. J Saudi Soc Agric Sci (2015) https://doi.org/10.1016/j.jssas.2014.12.00

Zhang W, Kato E, Bianchi FJJA, Bhandary P, Gort G, van der Werf W (2018) Farmers' perceptions of crop pest severity in Nigeria are associated with landscape, agronomic and ocio-economic factors. Agr Ecosyst Environ 259:159–167. https://doi.org/10.1016/j.agee.2018.03.004

Zhu YG (2009) Soil science in the understanding of the security of food systems for health. Asia Pac J Clin Nutr 18(4):516–519

Zosky GR, Boylen CE, Wong RS, Smirk MN, Gutiérrez L, Wood-ward RC, Shan Siah W, Devine B, Maley F, Cook A (2014) Variability and consistency in lung inflammatory responses to particles with a geogenic origin. Respirology 19:58–66

12 Soil and Industry

Bipin B. Mishra, Alemayehu Regassa, and Endalkachew Fekadu

Abstract

Ethiopia is gifted with diversified types of soils. Being the foundation of agricultural industries, soil is the driving force to sustain safe and profitable production. Soil is not only meant for production functions but is directly linked to many other functions and services of industrial and socio-economic relevance. Soil is vital for our shelter in mud houses or houses made of clay bricks and tiles. Similarly, clay ceramics are popularly used in pottery, household utensils, mud or clay stoves and bins. Soils through sand and clays have tremendous contributions to the construction industries, roads, and even embankments. Besides, Ethiopia has zeolite, bentonite, volcanic material and kaolin deposits either in soil or in their weathered states and are of high industrial values. Clay rich soils, having a high potential of organic carbon sequestration, open opportunities for carbon trading in days to come, especially in forest soils. However, some emerging threats to soils are alarming which need to be minimized by sustainable soil and land conservation and management options following the nexus approach.

Keywords

Soil based Industry • Construction Industry • Carbon trading • Mud house • Bricks • Tiles • Ceramics • Zeolite • Kaolin

B. B. Mishra (✉)
Department of Soil Science, Bihar Agricultural University, Sabour, Bhagalpur, 813210, India
e-mail: bbmsoil@rediffmail.com

A. Regassa
Department of Natural Resource Management, Jimma University, P.O. Box 307 Jimma, Ethiopia
e-mail: alemrega@yahoo.com

E. Fekadu
College of Agriculture, Woldia University, P.O. Box 400 Woldia, Ethiopia
e-mail: endalkf@gmail.com

12.1 Introduction

Soils are an essential natural resource both for the survival and prosperity of humans and serve as the basic income source within a sense of socio-economic and livelihood security among farming communities (Bedasa and Hussein 2018). Accordingly, the soils have a direct linkage to industries in agricultural and other sectors including soil based many small as well as large industries, although Ethiopia's major industries include agriculture, construction, food processing, resources & energy, tourism and manufacturing. Soils play a major role in the area of quality food production and security. With agricultural progress, there is extension and expansion of the market and industrial relevance of soils. Agriculture is the country's most promising resource and economic backbone. Soils of the arable land are suitable for agriculture and allow water and air to move through and finally get to roots, soils have a huge but diverse population of microorganisms, but they do have a considerable amount of organic matter through decayed or dead plants and animals, and they contain readily available nutrient elements. But such arable land is by and large shrinking, while a part is being lost to soil degradation caused by salinization-desalinization, desertification, erosion, chemical spills, and so on.

The soil resources of Ethiopia are under the pressure of interconnected components and inputs like high population pressure, agricultural expansion and migration followed by rapid urbanization, resettlement, climate change, and environmental pollution collectively imposing acute challenges to sustainability (Wassie 2020). Soil erosion by water and nutrient export are the most important ones resulting in low agricultural productivity. Ethiopia has been described as one of the countries in the world with the most serious soil erosion, with an estimated total annual soil loss ranging from 16 to 179 t ha^{-1} $year^{-1}$ in croplands as reviewed by Adimassu et al. (2020), who also observed that overgrazing by livestock has been considered as one of the most important

S. Beyene et al. (eds.), *The Soils of Ethiopia*, World Soils Book Series,
https://doi.org/10.1007/978-3-031-17012-6_12

causes of soil erosion and nutrient export in Ethiopia. However, all runoff, soil, and nutrient losses are highly variable due to variations in land use, topographic, edaphic, and climatic factors (Adimassu et al. 2020). Besides, an appreciable land area is being covered up by buildings, roads and other construction and engineering works.

Agro-industries cover only 5% of the GDP in Ethiopia, wherein 50% of the total manufacturing production rests in food and beverage. The country has altogether 13 sugar manufacturing factories owned by the state. Recently as a part of PM Abiy Ahmed's economic reform program, the transfer of sugar factories to private ownership is underway. The initial plan is to privatize six out of the 13 sugar plants to local or international private investors. Establishment of effective supply chains, including cold chain, may increase agro-processor access to local producers. With the establishment of agro-industrial parks, the integration of smallholder farmers and processors into the industries as part of the commercial value chain could improve the local economy (EIC 2021).

12.2 Multiple Soil Functions

Soils have multiple and diversified functions, and this is well indicated by FAO (Fig. 12.1). The soil functions refer to as benefits that are derived from soils, not just by human populations, but more generally also by plants and animals. Blum (1988) however organized soil functions into five categories viz. (a) the extraction of raw materials and water, (b) physically supporting buildings and other man-made structures, (c) the production of biomass, (d) filtration, buffering, storage, and chemical/biochemical transformations, and (e) the preservation of biodiversity or potentially useful genetic material, as well as of geogenic and cultural heritage. The socio-economic functions include the supply of water by soils, and different raw materials like clay, silt, sand, gravel, or peat, which are used in industrial, and manufacturing operations including embankments, dams, roads, buildings etc. Daily et al. (1997) reviewed in detail on services provided by soils, such as buffering and moderation of the hydrological cycle, mechanical or physical support of plants, retention or uptake and delivery of nutrients to plants, disposal and degradation of wastes including dead organic matter, renewal of soil fertility, and regulation of element cycles. Pimentel et al. (1997) proposed the economic and environmental benefits of biodiversity.

They focus on the vital services that are provided by all biota (biodiversity), including their genes and biomass, to humans and to the environment, and they refer explicitly to the role of soil biota in terms of topsoil formation, nitrogen fixation by soil-borne diazotrophic bacteria, and the bioremediation or biotreatment of highly polluted soils. Schlichting (1972) introduced the concept of multi-functionality of soils and described how the buffering capacity of soils in different respects is important to a multitude of organisms and to other parts of the landscape. Brümmer (1978) was the first to propose a classification of soil functions. Using soil to make houses could interfere with other usages in construction, medicine, or in the production of wool and sugar. Simonson (1966) discussed the fact that soil resources not only produce food and fiber, but are also important as construction materials for highways and dams, as the foundation for homes, and for waste disposal. The photopedogenesis as the most fundamental soil forming process may open an academic passage to move to lunar and mars surfaces even in order to learn and discover if pedogenesis is operative there in outer space, since pedology is neither dead nor buried but under a functional state with visual signatures of horizonation (Mishra and Roy 2019).

In Ethiopia, it is believed that about half the population is currently living in houses made of dried clay or mud. If one counts fired bricks, the proportion of people whose houses are made of soil material might even be higher (Staubach 2005). Importantly, soils after being sealed support directly in the construction of buildings, roads, parks, play grounds and sports facilities besides landfills. The ecological functions of soils are directly related to the production of all types of biomasses including crop and plantation production and physical buffers in the global water cycle, besides fostering biological and geobiochemical transformations of toxic organic constituents. Additional ecological function includes the preservation of genetic diversity such as the production of antibiotics and carbon sequestration, a powerful tool to mitigate climate change.

The Government of Ethiopia (GOE) aims to boost exports and trade in agro-processing industrial parks to make Ethiopia a top manufacturing hub in the African continent. Industrial parks are basically a key focus of Ethiopia's economic development strategy. Importantly through the Ministry of Trade and Industry (MoTI) and the Industrial Parks Development Corporation (IPDC), 17 agro-industrial growth corridors (AIGC) are under execution for development in all ten regional states. Commodities intended for processing include coffee, sorghum, maize, sesame, horticulture products, meat and dairy, and cereals, among others. The IAIPs will include companies that export value-added agricultural products as well as those products for domestic consumption. Major agriculture processing potential includes livestock feed manufacturing, wheat-based food production (e.g., pasta, biscuits), sesame processing (e.g., tahini), soybean crushing (e.g., soybean oil and feed), sugar production and processing as well as juice and dairy manufacturing (EIC 2021).

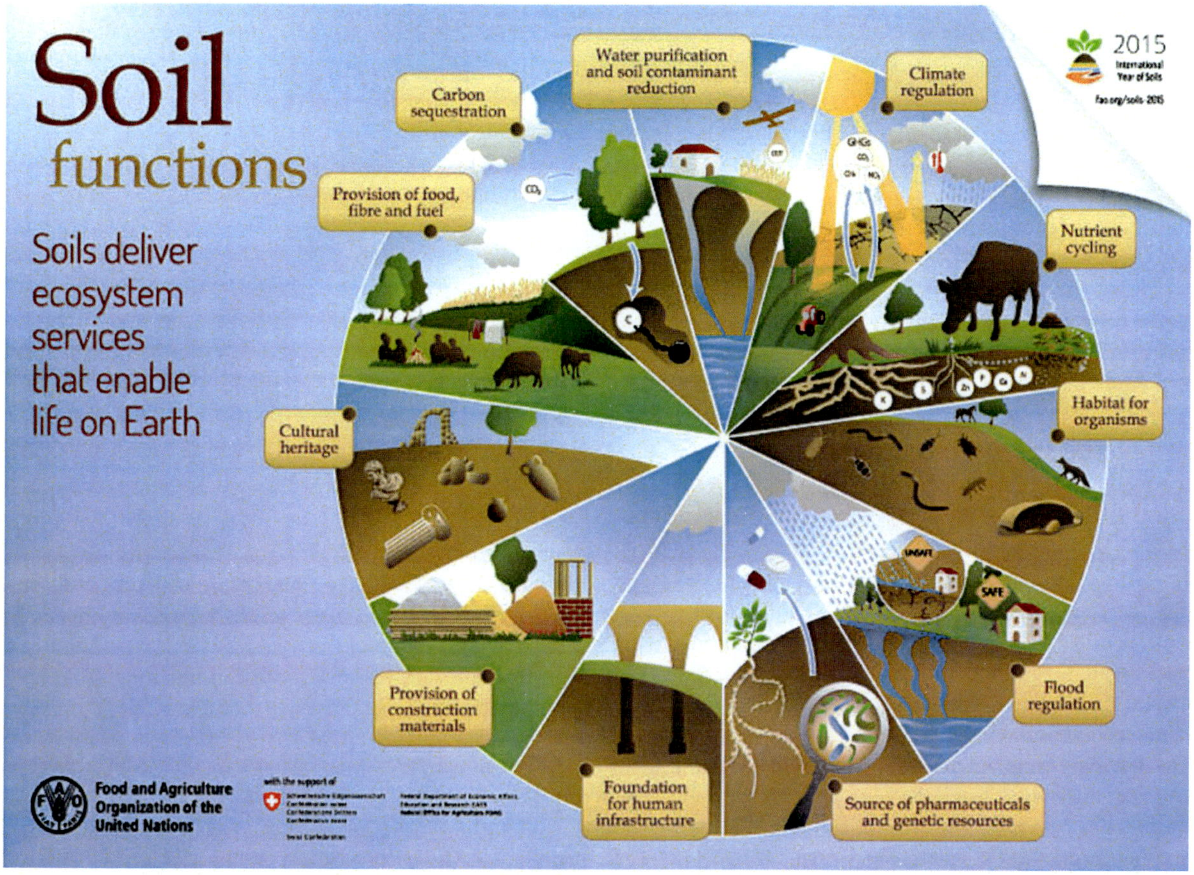

Fig. 12.1 Soil functions (*Source* Food and Agriculture Organization)

12.3 Soil Stabilization with Pumice in Road Construction

Pumice is a light-colored highly vesicular glass composed normally of 60–70% SiO_2, 12–14% Al_2O_3, 1–2% Fe_2O_3 and alkali oxides, with a specific gravity of <1 g cm^{-3} (Day 1990). The chemical analysis of pumice normally shows the total contents of Silicon Dioxide (SiO_2), Aluminum Oxide (Al_2O_3) and Iron Oxide (Fe_2O_3) of 82.68%, which was above the minimum of 70% and it resembles a good Pozzolan (Table 12.1). Thus, the Pumice sample resembles a cementitious compound containing calcium oxide, alumina, and iron oxide with a total of about 15.90% limit (Mesfin et al. 2019). Swell-shrink soils are widely occurring in Ethiopia and their presence in road is often disastrous due mainly to wide cracking. However, commonly recommended methods of stabilizing weak swell-shrink soils (Vertisols) are mechanical compaction or chemical treatments (Mesfin et al. 2019). Ancient Romans used to make specific cement just by mixing pozzolanic materials with lime to build structures and a few of which are still in existence to use (Zhang et al. 1996). In fact, pozzolanic reactions are associated with silica reactions in presence of calcium hydroxide and water in order to produce "calcium silicate hydrates" or C–S–H (Hewlett 2004; Jackson et al. 2003) which develops a dense microstructure that enhances the material strength, reduces the permeability of concrete, and improves its resistance to chemical attack. Besides, such mixing of pozzolana reduces the pore sizes and porosity leading to an increase of strength (Mindess et al. 2003). In Ethiopia, raw materials of the pozzolanic materials are available over a long stretch and wide area of the Rift Valley. Importantly, on the other hand, pumice is a kind of glass formed and not a mixture of minerals.

12.4 Soil Based Products

Construction industry is one of the three sectors of the economy identified by the Ethiopian Government for special consideration to accelerate the country's economic development. The construction industry in Ethiopia is a major economic driving sector and as reported by Gashaw and

Table 12.1 The chemical composition of clay samples collected from Ethio-bricks factory

Components	White soil (%)	Red soil (%)
CaO	<0.01	<0.01
SiO_2	66.61	55.30
Al_2O_3	11.56	16.40
Fe_2O_3	8.16	9.62
MgO	0.16	0.60
Na_2O	2.60	1.16
K_2O	3.06	1.48
MnO	0.32	0.34
P_2O_5	0.09	0.15
TiO_2	0.57	1.12
H_2O	2.17	4.32
LOI	4.04	8.15

Aklilu (2020), the National Bank of Ethiopia (NBE) observed that the construction industry accounted for 71.4% of the nation's industrial output in 2018, and signifying expanded by 15.7% from its previous share, and is playing the leading role of the construction sector. In fact, the rate of urbanization is increasing, and this created a huge need for improved infrastructure systems and a big housing project. The low-cost building construction from mud as the main building material is common in Ethiopian villages, often called stick-and-mud construction and is the traditional housing method found in the country. The thatched roofs of huts have dirt floors and are amazingly cool, but the cooking is usually done outdoors. The lifespan of a thatched roof stick-and-mud house is around 10–15 years.

Based on the material used for construction; Ethiopia construction industry is mainly classified into modern and traditional building material-based buildings. The traditional building material-based buildings are Chika-bet, mud block house, stone house and brick house while the modern building material-based buildings are HCB building, glass, steel, and modern building material-based buildings. When we compare this housing typology based on the construction materials; the traditional building material-based are more sustainable in the case of economy, energy efficiency, material availability, local labor, environmental adaptability and construction technique. The mud blocks-based buildings are constructed both in the traditional and modern construction industry in some parts of the country (Ann-Charlotte 2008).

The most common soil-based source in construction industries is sand. For central and southeastern parts of Ethiopia, the preferred choice for sands is from Meki, Langano, Sodere, Koka, Metehara, and Minjar. Construction of roads is one of the major focus areas to fast-track economic growth (Uge 2017). The road density and number of vehicles per 1000 populations are low compared to other African countries. When the Ethiopian Roads Authority (ERA) was established in 1951, the total road network was only 6400 km but increased many folds. All roads use naturally occurring soils and rocks as the basic foundation and construction materials. However, road construction engineers do face multiple challenges in using soils and rocks of unknown and variable quality (Uge 2017). Besides, construction activities in road sector encounter variable swelling clays or expansive soil types due to its wide occurrence covering around 40% of surface areas (Geremew et al. 2016; Berhane 2010; Arega et al. 2009; Sisay 2004).

The sand found in Debre Markos town and its surroundings contains silt and clay content and organic impurities that exceed the allowable limits and it resulted in a significant reduction in; strength (Gashahun 2020). Construction industry is one of the booming economic sectors in the country. The governmental policies support infrastructure development in order to realize the transformation from agriculture to industry (Gashahun 2020). The main natural cheapest sources of sand are riverbeds and natural deposits. The parameters that determine the quality of the fine aggregate are the silt, clay and organic impurities contents and gradation of it (Alexander and Mindess 2005). Silt materials are fine aggregate particles having non-cohesive properties due to which they do not react with cement. However, silt starts reacting by shrinking and swelling concrete but still exists in concrete which causes unwanted major cracks in the concrete (Cho 2013; Olanitori and Olotuah 2005). Most of the sand sources are natural deposits or riverbeds, wherein silt content is much higher than the standard. On the other hand, enforcing regulation and timely supervision is weak in Ethiopia's construction industry (Negash 2014).

Clays are composed of clay minerals like kaolinite, halloysite, illite, bentonite/montmorillonite, attapulgite, chlorite, vermiculite and other aluminosilicates as well as ingredients like quartz grains, zeolite, apatite, granite, iron hydroxide, etc. However, natural clays are mixtures of different types. With differing specific properties like surface tension, capillarity, rheological properties, hardening, plasticity, thixotropy and high swell-shrink features, clays are widely used in various industries. The swelling feature allows clays to occupy an important role in mining and geology while the high swelling degree of clays is exploited in the construction industry. In contact with freshly poured concrete, clays swell and create a waterproof barrier that is used during the construction of underground structures, sewerage systems, water tanks, nuclear and other waste storages, etc. Clays are widely used in the steelmaking industry as well as binding agents in the manufacturing of iron ore pellets. One of the important properties of clays is adsorption and the ability to interact with the metal ions from the surrounding environment. Therefore, clays take a prominent place in the list of natural sorbents and can be used to remove heavy metal ions and organic compounds from industrial water that represents one of the major challenges of modern civilization. Besides being used in environmental protection, the use of clays as adsorbents are found in the food industry for clarification of wine. In addition, clays are also very important in medicine and cosmetics when mixed with water create a colloidal solution, which when consumed acts as a natural laxative absorbing both organic and inorganic contaminants, heavy metals and free radicals. Clays and clay minerals are ingredients in a large number of skin care products such as creams but are also independently applied to eczema and different types of rashes.

Huge amounts of silica sand deposits are found in North Shewa of Abay basin around Mugher valley, Haro Genda and Alem Ketema. Recent demand for silica sand in industries like glass, cement, paint, chemical, smelting and others is increasing and thus compelling even for import. In particular, Mugher Cement Factory and Ethio Glass Factory are mining small quantities of silica sand only for their own consumption. Quartz is the most common mineral of sand followed by feldspar and often a small quantity of white mica (muscovite). However, red, brown, and yellow sands contain iron compounds, while red desert sands are usually made of quartz coated with iron oxides. Major industrial uses of sands are (i) the Construction industry in making bricks and tiles, filters to be extensively used in filtering municipal water supply, swimming pools and in sewage treatment plants and for testing sand to determine the strength of cement. (ii) Glass industry for manufacturing glass containers, flat glass (sheet glass), safety glass, pressed and blown glass, fiber glass, optical glass and industrial glass (iii) Metallurgy and abrasives. Any sand that is supposed to be used as standard sand may be free from clay and organic impurities.

The construction industry is growing at a faster rate to provide houses for the increasing urban population. Much of the constructional materials are heavily dependent on soils as row materials. Among others, tiles which are used for floors and roofs, and walkways are largely produced in cities such as Adiss Ababa, Bahir Dar, Jimma, and Debre Tabor. The frequently used industrial products in Ethiopia include bricks and tiles, pottery and ceramics besides earthenwares, stone wares and porcelain. Cosmetics, soil made cooking furnaces and mud houses are other products essentially demanded particularly in rural areas. Coffee celebration pot sets with the furnace as well as Injera baking stove and pan are integral to the Ethiopian lifestyle.

12.4.1 Bricks and Tiles

The construction industry is growing at a faster rate to provide houses for the increasing urban population. Much of the constructional materials are heavily dependent on soils as raw materials. Among others, tiles which are used for floors and roofs, and walkways are largely produced in cities such as Adiss Ababa, Bahir Dar, Jimma, and Debre Tabor. The frequently used industrial products in Ethiopia include bricks and tiles, pottery and ceramics, besides earthenware, stoneware and porcelain. Cosmetics, soil made cooking furnaces and mud houses are other products essentially demanded particularly in rural areas. Coffee celebration pot sets with furnace as well as Injera baking stove and pan are integral to the Ethiopian lifestyle.

Brick is one of the oldest manufactured building materials in the world and it is extensively used even at present because of its durability, strength, reliability, low-cost, easy availability etc. Burnt clay bricks are commonly been used as a solid matrix mainly due to their characteristics, such as good mechanical resistance and satisfactory stability (Jahagirdar et al. 2013). In many areas of the world, conventional bricks are produced from clay as a basic building material for the construction of houses with high temperature kiln firing which resulted in the scarcity of natural resource materials for the production of the conventional bricks. Clays used for brick making vary broadly in their composition and are dependent on the locality from which the soil originates. Different proportions of clays are composed mainly of silica, alumina, lime, iron, manganese, sulfur and phosphates. For example, the chemical composition of Ethio-bricks factory clay samples were shown in Table 12.1.

The worldwide annual production of bricks is currently in billion units and the demand for bricks is expected to be continuously rising. Quarrying operations for obtaining the clay are energy intensive, adversely affect the landscape, and generate high levels of wastes. It is also noted that there is a shortage of clay in many parts of the world. To protect the clay resource and the environment, some countries such as China have started to limit the use of bricks made from clay. The increase in demand for construction materials in recent years as a result of the development has called for an alternative way to develop or derive construction materials from other sources. In order to meet the increased demand, attention has been given to the development of sustainable construction materials. The usage of improved construction materials in the construction industry has been on increasing daily which has led to the investigation of its environmental impact and meeting required standards when waste is used in developing sustainable construction materials. The treatability studies using solidification/stabilization indicate that chemical sludge generated from the treatment of textile dyeing waste water has the possibility to be used as the construction material (Patel and Pandey 2009).

In the Gilgel Gibe catchment area, southwestern highlands of Ethiopia, there is massive brick production for local house construction by the communities (Fig. 12.2). The bricks are made from the bleached top layer of Vertic Planosols. In these areas, the average thickness of the bleached layer is about 45 cm (Van Ranst et al. 2011). There are over 15 legally organized small scale brickmaking entrepreneurs each with an average production capacity of 33,000 bricks per year (Abebe et al. 2013; Mertens 2013).

The process of brick making involves six consecutive steps (Fig. 12.3): First, the bleached top layer is chopped and piled up. This material is mixed with a bucket of water using hands and feet. Wooden molds are used to mold this moistened material in the desired shape. Molds are covered in ash to ease the removal of the molded brick. Any excess material is removed using a metal wire. Molded bricks are then placed on a flat surface to dry for one to two days.

The quality of bricks especially their strength and stability could be improved with the application of enzymes (Mitikie et al. 2017). Maximum compressive strength value was noted for the brick units formulated with lateritic soil of 15% replacing clay (Geremewe and Mamuye 2019). Nowadays, clay bricks are used to construct improved buildings everywhere in the country. Clay roof tiles are the best material that one can use to cover the roof of a building. There are advantages of clay roof tiles viz. easy installation and longevity, such roof tiles are the most environmentally friendly and much better even than concrete roof tiles. Clay tiles may be handmade or machine made. In Ethiopia, clay tiles are often used to cover the building roof.

12.4.2 Household Utensils and Earthenware

According to the study by Jessie et al. (2015) on technical traditions and pottery craftsmanship among Wolayta and Oromo groups in Ethiopia, pottery craftsmanship is a female activity, although the Wolayta men and young boys actively participate in the extraction of clay materials, as well as in the firing and sale of pottery. In both ethnic groups, pottery is considered as a domestic activity, i.e., the work is carried out at home in workshops or in an annex to the home environment. For both groups, clay sources are located near the dwelling places, i.e., between 500 m and 5 km away. Both groups use a mixture of different types of clay (Wolayta) or a mixture of clay and different types of temper (Oromo). Raw-material sources are not the same for both

Fig. 12.2 Brick making from top layer of Planosols in the Gilgel Gibe catchment, southwest Ethiopia (*Source* Photo by Alemayehu Regassa)

Fig. 12.3 The local brick making process in the Gilgel Gibe catchment, Southwestern Ethiopia (*Source* Photo by Alemayehu Regassa)

groups. The type of pottery produced by these two different ethnic groups is not the same. While the Wolayta produce 18 different types of pottery (Fig. 12.4) the Oromo produce 22 (Fig. 12.5).

There are 15 different shaped pottery types but they are used for the same purposes by both groups (making coffee, carrying water and milk, food conservation, cooking meat, cereal processing, distilling local alcohol and making homemade beer, cooking bread and patties, etc.). The majority of the productions are meant for domestic use while surplus productions are sold at local markets. The method of extraction and processing of the raw material, the clay varies among the two ethnic groups. Wolayta raw material extraction takes place by pit extraction or quarrying. they use a mixture of red and white clays. They derive from the ferralitic alteration of ignimbrites; truncated by erosion and cut into by the hydrographic network. The red clays colored by iron oxides, are made up of kaolinites and contain a considerable proportion of sand corresponding to residual sands from the original rock. The white clay is located in the lower part of the alteration profile. It coats blocks of ignimbrite in the process of deterioration. The Oromo extract their clay using tunnel extraction or quarrying. Unlike the Wolayta, the Oromo do not use the second type of clay but incorporate this very white temper extracted from the ignimbrite substratum into the clay, when it is available (Jessie et al. 2015).

Earthenware is often used by the Ari people of southwestern Ethiopia when they cook. Potters, who are predominantly women, make cooking vessels using locally available soil clays and then distribute the completed products throughout the area (Kaneko 2009). In the same way, Kechene Women's Pottery Cooperative functioning in the surrounding area of Addis Ababa is an independent organization consisting entirely of women working for the last 15 years on the creation of artistic, traditional and functional pottery products. Among the products, there are traditional coffee-sets and tea-sets (including pots, cups and small plates), candle holders, incense burners, lamps, bowls of various forms and dimensions, sculptures of flowers and animals (lions, monkeys, frogs, etc.), frames, jewellery cases, icons, small sculptures (Fig. 12.6).

In Ethiopia, coffee plays a significant role in establishing social networking and enhancing the participation of people at all levels to share their views, feelings, and concerns related to their life. People from their neighbors call each other at intervals, gather and drink cups of coffee while discussing happiness and sorrow. Coffee is a symbol of peace-making among the Ethiopians not only because its ceremonies bring together different communities but also because quarreled individuals are reconciled at the ceremony. *Jebena* (coffee pot) is one of the useful traditional pottery vessels in the coffee ritual (Fig. 12.7a–c). The powdered coffee is added to boiled water in a clay pot

Fig. 12.4 The different shapes of utensils made by Wolyta ethnic group (*Source* Jessie et al. 2015)

Fig. 12.5 The different shapes of utensils made by Oromo ethnic group (*Source* Jessie et al. 2015)

Fig. 12.6 Household utensils and earthenwares (*Source* Kechene Women's Pottery Cooperative)

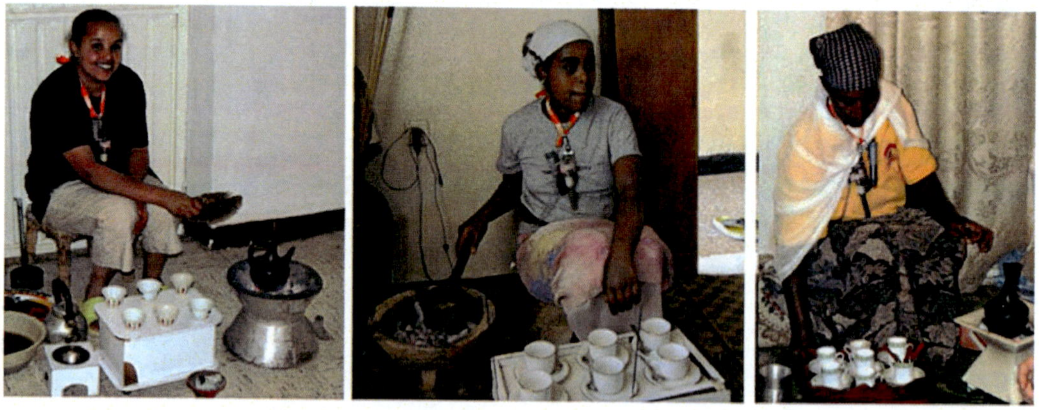

(a) (b) (c)

Fig. 12.7 a–c Ethiopian coffee ceremony (*Source* Keil et al. 2010)

(*Jebena*), which is over the fire on hearthstones. The use of *Jebena* for making coffee has continued in both rural and urban areas.

Injera is very common with a special sauce (wot) is the major staple food for Ethiopians with people from Somalia and Sudan (Asfaw et al. 2014). In fact, *Injera* is prepared from the local indigenous grain mostly of teff and sometimes of corn, sorghum, millet, and a mix of two or three of these grains. In Ethiopia, the people bake *Injera* using traditional biomass on open-fire stoves. *Injera* baking process starts from the preparation of dough and ends in baking on the plate made of clay called *Eelee* (Afaan Oromo or *Mitad* (Amharic) (Fig. 12.8). Other common pottery products used by many people include *Desit* (cooking pots), *Insera* (water pot) *Gan* (water or beverage storage), and *Jebena* (coffee pot). Drinks and food prepared with a clay material have special test and delicious flavor. Potters in Wellega often dig five meters below the surface, until they access clay which they perceive to be of high quality for making different pots. It is believed that in most cases quality clay is waterlogged, and for this reason, potters dig down below the water table.

For example, potters from Haroji Agamsa (located in *aanaa* Lalo Assabi) mine clay from two sites: the Huwa and Suphee River Valleys. Clay mined from Golbe Valley has a medium texture and is dark-brown in color. Because they believe that the Golbe Valley yields good-quality clay, potters travel up to 30 min to reach the site. On the other hand, clay at the Suphee site, which is only 15 min away, is fine textured, dark in color, and assumed to be poor in quality because pots shaped from this clay often crack (Wayessa 2011). Almost in all rural areas, clay furnace or oven or kitchen hearth is common for making food of any type. It is of different types and often represents the requirements and local culture even. Accordingly, the cost varies in the markets. Traditionally developed enclosed *Injera* baking stoves are commonly used in the northern part of Ethiopia, mostly in Tigray (Tigray Regional State) and Wollo (a zone in Amhara Regional State), and some parts of semen Shewa (Oromia Regional State). They are named after the area where they are most popular in Tigray and hence, it is commonly referred to as Tigrian stove (Fig. 12.9). These stoves, unlike three-stone *Injera* baking stoves, are permanently built on the ground or on a raised platform made up of mud and stones. The raw materials used for making Tehesh *Injera* baking stove are mud, stones, and barley straw. A small amount of fresh dung is mixed together with mud to increase its adhesion and strength. This composition is smeared over the vertically stacked stone from inside and outside. The straw is used as insulation by placing it between the outer and inner walls of the stove and under the combustion chamber (Tadesse 2020).

Fig. 12.8 Injera Plate (*Eelee*) locally made from clays in the Gilgel Gibe catchment, Southwest Ethiopia (*Source* Photo by Alemayehu Regassa)

Soil and clay may be in a wide range of colors depending on the parent materials and the associated organic matter.

12.5 Industrial Use of Soils and Clays

12.5.1 Zeolite in Soils

Sidama highlands of Ethiopia are dominantly sesquioxidic reddish soils suitable for major agricultural crops but are intensively cultivated for coffee. Here, feldspar and ferromagnesian are the dominant minerals in the basalt, whereas quartz, feldspar and minor amounts of hornblende phenocrysts embedded in a groundmass of volcanic ash and glass are dominant in the ignimbrites and rhyolite. Zeolites (clinoptilolite and mordenite) were discovered in substantial amounts (Fig. 12.10) in the ignimbrites (https://agris.fao.org/agrissearch/search.do?recordID=ET2005OOOO73).
Typical crystal sizes for most of the zeolite minerals are up to 5 mm for chabazite and 15 to 25 mm for analcime and phillipsite. Zeolite formation in lake and soil environments requires conditions with high alkalinity and Si concentrations and elevated (Na + K + Ca)/H^+ ratios (Sheppard and Hay 2001). Such situations are met in geological settings with volcanic activities as common in rift valley lakes with zeolite bearing deposits (Hay 1966). Zeolites deposits in Ethiopia are scattered and no efforts are virtually been made in exploring the presence and potential industrial interest of zeolites. On a survey of zeolites for industrial uses in Ethiopia, it was observed that the main demand of zeolites was as adsorbents or ion exchangers in water and wastewater treatments (Diaz 2017).

Zeolites are aluminosilicate minerals with extensive pores and channels that enable soils to absorb moisture and slowly avail it to growing crops and horticultural plants. Zeolites are one such naturally occurring phyllosilicate minerals capable

Fig. 12.9 Injera backing stove in Tigray region, Northern Ethiopia (*Source* https://www.internationaljournalssrg.org/IJME/2020/Volume7-Issue1/IJME-V7I1P102.pdf)

Fig. 12.10 Natural zeolite in the cavity of basalt (*Source* https://www.dreamstime.com/natural-zeoliteminerals-black-basalt-rock-zeolite-minerals-black-basalt-rock-image131748112)

of holding water as much as 60% of their weights which will be gradually availed to growing plants. They have several pores and channels which look like honey bee comb and they are also negatively charged particles like most clays. These are the two characteristics that enable zeolites to adsorb/absorb water and cations of several types which can be exchanged with materials in the environment at any time (Ramesh et al. 2011). Moisture conserving ability of zeolite minerals frequently in basaltic soils can be exploited to enhance crop and tree production, especially in drought prone areas of Ethiopia.

Huge deposits of zeolites have been identified in Ethiopia that can be used to conserve soil moisture. Steps in testing this mineral for its ability to conserve soil moisture include experimenting first in the greenhouse by amending soils with different amounts of zeolites to which various moisture regimes will be applied. Then test crops/trees of choice will be planted and all the necessary parameters showing the performance of zeolite amendment will be collected. The resulting data from all participating countries will be analyzed from which the appropriate dose of zeolite that should be applied in different soils for optimum moisture conservation will be known. Some 10 years ago or so, huge deposits of zeolites minerals have been identified in different parts of Ethiopia and several million tons of high-grade zeolite deposits (Clinoptilolite and Mordenite) were discovered by geologists of the Ethiopia-Canada Agroecology project in the central rift valley region of the country (Van Straaten 2002).

Following are the most outstanding benefits of using zeolites for moisture conservation purposes. (i) Amending soil with zeolite will increase moisture availability to crops growing in moisture stress areas as a result production and productivity of crops and livestock will be increased in drought prone areas of Africa; (ii) Increased resilience/adaptation of vulnerable societies to drought through increased food and feed availability; (iii) Decreased risk of losing crop and livestock due to water shortage increased feed availability; (iv) Decreased in numbers and frequencies of human migration due to drought; (v) Increased forest cover in semi-arid, arid and help to combat desertification.; (vi) Increased carbon sequestration due to increasing vegetation cover; and (vii) Creates job opportunity for youths (zeolite mining, processing etc.). It was observed that adding zeolite on the sandy soil increases the water retention capacity of the soil (Desta et al. 2019). Zeolites have desalinization properties thus, increases the quality of drinking and irrigation water. It is also important to note that zeolites are highly stable minerals unlike other soil minerals like clay minerals even. So, once zeolite is applied to soils it is functional for a very long period of time. It is believed that once zeolite is applied to the soil, there is normally no need to apply it again. As huge deposits of zeolites were identified in different parts of Ethiopia, publicizing zeolites for soil moisture conservation purposes is important and timely.

12.5.2 Bentonite Clays

Bentonite is a term that was first used to designate particular, highly colloidal, plastic clay found near Fort Benton in the Cretaceous beds of Wyoming, USA, which possesses the unique characteristic of swelling to several times its original volume when placed in water and that forms thixotropic gels. It is a smectite clay formed from the alteration of siliceous, glass-rich volcanic rocks such as tuffs and ash deposits (Nadežda et al. 2011). Later the term is used to describe plastic clays, generated by alteration of volcanic tuff and ash, with the dominant content of smectite minerals, usually montmorillonite, named after the deposits in Montmorillon, France (Kutlic et al. 2012).

According to the Ethiopian Geological survey, Ethiopia has a huge amount of Bentonite clay deposits/resources which are found in the Afar and Oromia regions. The main occurrences in Afar are located at Ledi, Gewane, Hadar, and Warseiso (Mesfin 2012). The bentonitic beds are part of a thick sequence of lacustrine sediments which consist of clays, silts, sands, calcareous grits, gravels, conglomerates, basaltic flows, and ashes. The main constituent, which is the determinant factor in the clay's properties, is the clay mineral montmorillonite. They are easily accessible, as they are located near the main road. The total resource in the Afar region is estimated to be170 million tons. The bentonite beds are well exposed and the overburden consists of loamy gravel and sandy clay. Tests conducted so far confirm that some beds could be used for the preparation of drilling mud and iron ore palletization, and if upgraded, may have found dry applications as well (Mesfin 2012).

Bentonite, primarily composed of montmorillonite, is used as light- drilling mud; and is also used in ceramic industry. Arerti Ceramics Manufacturing, a Chinese company in Ethiopia, signed an agreement with the Ministry of Mines and Petroleum to start mining bentonite minerals in Chacha, Angolola area in north Shewa of Amhara Region. The company has allocated a total investment of 41 million Ethiopian Birr. The Ministry of Mines and Petroleum indicated that when the company goes fully operational, it will create 3000 jobs. In addition to the Amhara region, bentonitic clay resources are found in the Afar and Oromia regions. Bentonite is used in the decolorizing of edible oils, although smaller amounts are used in well-drilling, for wine clarification and, possibly, foundry use. Edible oil processing accounts for 90% of bentonite consumption. There are numerous small scale edible oil producers in Ethiopia which do not bleach their products, but forthcoming government regulations may make this mandatory. This should lead to an increase in demand for bleaching-grade bentonite. Although there is no domestic production of bentonite, it is recommended locally available quality Bentonite occurrence with a standardized requirement as a raw material input for various industrial applications based on laboratory test results.

12.5.3 Kaolin

Kaoline is a 1:1 type clay mineral composed of $Al_2Si_2O_5(OH)_4$. Kaolin clay is a promising adsorbent and attracts attention due to the alternative low-cost, eco-friendly, and highly abundant. The China clays (reference kaolin) and Ethiopian kaolin are highly kaolinitic and are similar in physico-chemical and mineralogical properties. Treated Ethiopian kaolin (with mechanical and thermal treatment) is found to have properties closer to that of China kaolin, even by far it is better compared to the reference kaolin based on their particle property and pH values (Aragaw and Kura 2019). Ansho and Bombowha kaolin are produced in Ethiopia and are efficient in synthesizing zeolite A, serving for tannery waste water treatment (Ayele et al. 2018). Several factories in Ethiopia including Melkasa Aluminum Sulfate and Sulfuric Acid Factory, and Nazret Aluminum Sulfate Industries are potential consumers of Bombowha kaolin. The Alemtena kaolin deposit located within the main Ethiopian rift is characterized by comparable grain size distribution, mainly white color, high Al_2O_3 kaolinite, relatively low concentrations of Fe_2O_3, TiO_2 and alkali elements that make it favorable for various industrial applications including paper, filler, ceramics, pharmaceuticals, and agricultural industries (Getnet and Worash 2020). The kaolin deposit of Alemtena is related to the acidic volcanic rocks of pumice, tuffs, rhyolite and trachy-andesites. Debre Tabor kaolin deposit is formed from the weathering of felsic rocks mainly trachyte and tuff units (Alemu et al. 2021).

12.6 Nanoclay Technology

Salissou (2021) reviewed the efforts being made in Ethiopia on nanoclays in food packaging. Nanoclays are phyllosilicates and are usually produced on natural bentonite, which consists of around 60–80% montmorillonite, consisting of about 1 nm thick aluminosilicate layers in about 10 μm large stacks. Such layered structure favors nanoclay in swelling and/or shrinking depending upon the availability of moisture. Zeolites, on the other hand, indicate a rigid structure, wherein liquid may move freely through the pores without changing the structure of the zeolites. Nanoclays have recently become of increasing interest for many industrial applications and uses. Among nano products, important ones are nano carbon, quantum dots, fertilizer and pesticide, metal nano powder, clay nano powder, grapheme nano powder etc. However, it is yet to be seen how far Ethiopian regulation, ethical engagement, socio-economic and environmental governance is stretching to address issues that the desired rise of nanotechnology is pushing to main stream. Nanotechnology for nanofilters in water purification in Ethiopia is broadening the scope of ongoing approaches.

Hatefi et al. (2016), based on their research on the impact of nanoclay to minimize erosion of sandy soils in Iran, observed that the rate of soil erosion significantly decreased with the increase of nanoclay concentration and 97.4 and 100% decreased in 0.5 and 1.5 g L^{-1}nanoclay as compared with control, respectively. The results indicated that the mean weight diameter significantly increased in 0.5 g L^{-1} (0.403 mm) and 1.5 g L^{-1} (0.481 mm) as compared to the control (0.345 mm). Besides, the proportion of aggregates >1 mm significantly increased with the increase of nanoclay concentration. The Norwegian company "The Desert Control" currently based in UAE is using "Liquid nano clay (LNC) technology" with a noble mission to transform deserts into fertile land (https://www.iamrenew.com/product-and-startup-reviews/this-company-is-transforming-deserts-into-fertile-land/). Here, a mixture of water and LNC may be sprayed directly with irrigation system being followed. The nanoparticles seep into the sand to give spongy and hollow structures that retain water about 40–60 cm underground (normal depth of plant roots). This requires around 40 L of water and 1 kg of clay per square meter. This approach supports sand particles to be nanostructured clay coating, completely changing their physical properties and allowing them to bind to water. Such technology needs appreciation and adoption under desert control in Ethiopia using LNC. Kumar et al. (2017) at Dire Dawa Technical Institute, observed that nanoclay is an excellent tool in mechanical engineering work too.

12.7 Impact of Industry on Soils

Soil pollution by heavy metals has become a severe global problem (Danica and Juraj 2020; Fawen et al. 2020), especially for African countries like Ethiopia where waste disposal is a major problem. Contaminations of soil by heavy metals become a serious concern in many developing countries due to intense industrialization and urbanization (Mireles et al. 2012). Although heavy metals occur naturally in soil, anthropogenic activities such as agriculture (fertilizers and pesticide application), urbanization, industrialization and mining significantly raise the levels of heavy metals. Pollution of heavy metals, directly and indirectly, affects human health. These substances adversely affect the productivity of soils, plants, animals and the entire environment if exceed certain limits (Farid et al. 2015). As a result of rapid industrialization and urbanization, the disposal of untreated and/or partially treated effluents from various industries, urban wastes, and the use of agrochemicals, the pollution level has reached alarming situation in Ethiopia with increasing metal levels and deterioration of agricultural soil quality (Fitsum and Abraha 2018). Especially, the problem is more severe in Addis Ababa, the capital, where most of the industrial establishments of the country are taking place.

According to Gebeyehu and Bayissa (2020), the dominant heavy metal pollutant sources in Ethiopian agricultural soils are irrigation with rivers/streams laden with industrial effluents and application of fertilizers and pesticides present, agricultural soils in Ethiopia are becoming increasingly polluted with heavy metals, especially, in urban centers.

The chemical fertilizer based intensive production causes soil acidification, organic matter depletion, greenhouse gas emission and are major contributors of climate change (Johannes 2015). The direct effect of fertilizers on soil living organisms are either short term immediately after application or long term after recurring application in a given cropping season. The indirect effects are resulted from prolonged periods of application due to drastic changes in soil pH, plant productivity, residue inputs and soil organic matter levels (Bunemann and McNeill 2004). These effects significantly affect nutrient availability to crops and consequently brought changes in productivity. In Ethiopia, there is a misperception among extension agents, crop production experts, researchers and particularly political leaders on the issue of chemical fertilizer consumption and crop productivity. They do believe that the application of inorganic fertilizers alone can increase yields and promotes its unwarranted utilization for alleviation of abject poverty in the country. This argument is so powerful than the negative impact of fertilizers on the soils, environment and climate are often suppressed or treated as external costs which simply have to be accepted. Temporarily, it may be a solution to aggressively compensate for the available food, fiber and oil crop demand gaps that a country requires urgently.

12.8 Conclusions

Soil is uniquely multifunctional with human existence although it lies beneath the human's feet. It is the functional key even to the origin of life and is capable to sustain our livelihood in a comfortable socio-economic framework. It has wide scope to support soil-based industries in order to meet the environmentally sustainable needs of humans. Soil is strangely full of wisdom that needs to be captured and discovered by the interconnection and interdependence of surrounding resources in order to develop more and more soil-based industries. Importantly, the soils of Ethiopia in diversity have the potential to mitigate climate change within a framework of planning. Soil-based sustainable agricultural production is by itself a mega industry, while forestry and plantation form another valuable industrial avenue. Besides, the value-added food products cover giant agri-business opportunities. Ethiopia is among a few African countries with sufficient geothermal resources. The future of soil-based industries in Ethiopia is thus brighter than ever before.

References

Abebe S, Hans H, Gete Z (2013) A review on soil carbon sequestration in Ethiopia to mitigate land degradation and climate change. J Environ Earth Sci 3(12). ISSN 2224-3216 (Paper), ISSN 2225-0948 (Online)

Adimassu Z, Tamene L, Degefie DT (2020) The influence of grazing and cultivation on runoff, soil erosion, and soil nutrient export in the central highlands of Ethiopia. Ecol Process 9:23

Alemu M, Teklay G, Tilahun W, Wuletaw M, Tamrat M, Yahya A, Gizachew M, Betelhem T, Mulgeta B (2021) Mineralogical and geochemical characterization of kaolin deposit from Debre Tabor area northwestern Ethiopia. Appl Earth Sci 130(1):42–56

Alexander M, Mindess S (2005) Aggregates in concrete. Taylor and Francis, Oxon

Ann-Charlotte R (2008) Low-cost housing for the Kambaata Region, Ethiopia. Halmstad

Aragaw TA, Kura F (2019) Physico-chemical characterizations of Ethiopian Kaolin for industrial applications: case study WDP propoxur formulations. In: Advances of science and technology. Springer Nature, pp 122–134

Arega F, van der Meer F, van der Werff H (2009) Quantifying engineering parameters of expansive soils from their reflectance spectra. Eng Geol 105:151–160

Asfaw HT, Mulu BK, Ole JN (2014) Design and development of solar thermal Injera baking: steam-based direct baking. Energy Procedia 57(2014):2946–2955

Ayele L, Pérez E, Mayoral Á, Chebude Y, Díaz I (2018) Synthesis of zeolite A using raw kaolin from Ethiopia and its application in removal of Cr (III) from tannery wastewater. J Chem Technol Biotechnol 93(1):146–154

Berhane G (2010) Engineering geological soil and rock characterization in the Mekelle Town, Northern Ethiopia: implications to engineering practice. MEJS 2(2):64–86

Bedasa NA, Hussein JW (2018) Challenges in managing land-related conflicts in East Hararghe Zone of Oromia Regional State, Ethiopia. Soc Nat Resour 31:351–366. https://doi.org/10.1080/08941920.2017.1400626

Blum WH (1988) Problems of soil conservation. In: Nature and environment series, no 39, Strassburg

Brümmer G (1978) Funktionen des Bodens im Stoffhaushalt der Ökosphäre. Deutscher Rat Für Landespflege: Zum Ökologischen Landbau 31:13–20

Bunemann EK, McNeill A (2004) Impact of fertilizers on soil biota. In: Lines R (ed) Proceedings current research into soil biology in agriculture. Kelly, Tamworth, pp 64–71

Cho SW (2013) Effect of silt fines on the durability properties of concrete. J Appl Sci Eng 16(4):425–430

Daily GC, Alexander S, Chrlich PR, Goulder L, Lubchenco J, Matson PA, Mooney HA, Postel S, Schneider SH, Tilman D, Woodwell GM (1997) Ecosystem services: benefits supplied to human societies by natural ecosystems. Ecology 2:1–18

Danica F, Juraj F (2020) Soil quality and heavy metal pollution assessment of iron ore mines in Nizna Slana (Slovakia). Sustainability 12:2549

Day RL (1990) Pozzolans, for use in low-cost housing. A State-of-the-Art Report Prepared for the International Development Research Centre, Ottawa, Canada

Desta Y, Habtu NG and Alemu A (2019) Hydrothermal synthesis of Na-P1 zeolite from Pumice to enhance moisture content and water retention capacity of sandy soil. In: 6th EAI international conference, ICAST 2018, Bahir Dar, Ethiopia; October 5–7, 2018, Proceedings, pp 457–472. https://doi.org/10.1007/978-3-030-15357-1_38

Diaz I (2017) Environmental uses of zeolites in Ethiopia. Catal Today 285:29–38

EIC, Ethiopian Investment Commission (2021) International Trade Administration U.S. Department of Commerce, 1401 Constitution Ave NW, Washington, DC, 20230. http://www.investethiopia.gov.et/

Farid G, Sarwar N, Saifullah AA, Ghafoor A (2015) Heavy metals (Cd, Ni and Pb) contamination of soils, plants and waters in Madina Town of Faisalabad metropolitan and preparation of Gis based maps. Adv Crop Sci Tech 4:199

Fawen Z, Yulong H, Changmin Z, Yuanbo K, Kai H (2020) Heavy metals pollution characteristics and health risk assessment of farmland soils and agricultural products in a mining area of Henan province, China. Pol J Environ Stud 29(5):3929–3941

Fitsum G, Abraha G (2018) Health risk assessment of heavy metals via consumption of Spinach vegetable grown in Elalla River. Bull Chem Soc Ethiop 32(1):65–75

Gashahun A (2020) Investigating sand quality effect on concrete strength: a case of Debre Markos and its vicinities. Int J Construct Manag. https://doi.org/10.1080/15623599.2020.1774838

Gashaw Assefa and Aklilu Gebregziabher (2020) Environmental impact and sustainability of aggregate production in Ethiopia. Sandy Mater Civil Eng. https://doi.org/10.5772/intechopen.90845

Gebeyehu HR, Bayissa LD (2020) Levels of heavy metals in soil and vegetables and associated health risks in Mojo area, Ethiopia. PLoS One 15(1)

Geremew A, Quezon ET, Kebede G (2016) Influence of subsurface courses materials on pavement performance: a case study in Yebu-Agaro road. Int J Sci Eng Res 7(10):1252–1260

Geremewe A, Mamuye Y (2019) Improving the properties of clay soil by using laterite soil for production ofbricks. Sciendo 15(2):134–141. https://doi.org/10.2478/cee-2019-0017

Getnet G, Worash G (2020) Economic geology and genesis of kaolin resources in the East African Rift system: the case of Alemtena kaolin deposit, Ethiopia. Appl Earth Sci. https://doi.org/10.1080/25726838.2020.1809315

Hatefi O, Jalalian A, Padidar M, Fallahzade J (2016) Effect of nanoclay on Wind Erosion a sandy loam soil in Segzi Region, Isfahan, Iran. J Environ Sci Technol 9:296300. https://doi.org/10.3923/jest.2016.296.300

Hay RL (1966) Zeolites and zeolitic reactions in sedimentary rocks. Geological Society of America Special Paper 85, 130 pp

Hewlett P (2004) Lea's chemistry of cement and concrete, 4th edn. Elsevier Science & Technology Books

Jackson M, Sagnotti L, Rochette P, Sølheid P (2003) The IRM Quarterly 13(3):112

Jahagirdar SS, Shrihari S, Manu B (2013) Reuse of textile mill sludge in burnt clay bricks. Int J Adv Technol Civil Eng 2(1):10

Jessie C, Claire M, Vincent A, Joséphine C, Ayed B, Amara AB, Laurent B, Nadia C, Xavier S, Fabien Victoria B (2015) Technical traditions and pottery craftsmanship among the Woloyta and Oromo groups in Ethiopia. In: Proceedings of the workshop of Namur (Belgium), 29–30 May 2015

Johannes K (2015). A soiled reputation: adverse impacts of mineralfertilizers in tropical agriculture. Association for Agriculture andEcology, Christine Chemnitz (Heinrich Böll Foundation), WWF,Germany

Kaneko M (2009) Variations in pottery making in southwestern Ethiopia. In: Proceedings of the 16th international conference of Ethiopian studies, vol 1, pp 383–394

Keil C, Kassa H, Tefera W (2010) Inhalation exposures to particulate matter and carbon monoxide duringEthiopian coffee ceremonies in Addis Ababa: a pilot study. J Environ Public Health. 213960. https://doi.org/10.1155/2010/213960

Kumar MSS, Adugna A, Balasundaram K, Kumar A (2017) Vibration analysis of nanoclay reinforced glass fiber/epoxy nanocomposite. J Recent Res Eng Technol 4(5):2252–2349

Kutlic A, Bedekovic G, Sobota I (2012) Bentonite processing, professional paper 24, pp 61–65

Mertens K (2013) Land use dynamics in the planosol belt of the gilgel gibe catchment, South-West Ethiopia. MSc thesis, Ghent University, Ghent

Mesfin T (2012) The occurrence of bentonite in Ethiopia: Bentonite Promotion document. Ethiopian geological survey

Mesfin K G (2019) Decade of South Ethiopian coffee improvement activities at Awada coffee research center.Adv Crop Sci Technol 7 (1):420. https://doi.org/10.4172/2329-8863.1000420

Mindess S, Young JF, Darwin D (2003) Concrete, 2nd edn. Prentice-Hall, London

Mireles F, Davila JI, Pinedo JL, Reyes E, Speakman RJ, Glascock MD (2012) Assessing urban soil pollution in the cities of Zacatecas and Guadalupe, Mexico by instrumental neutron activation analysis. Microchem J 103

Mishra BB, Roy R (2019) Photopedogenesis: a fundamental soil forming process in rock weathering and rhizospheric stability on earth and lunar surface. Agricult Res Technol: Open Access J 23 (3):556231. https://doi.org/10.19080/ARTOAJ.2019.23.556231

Mitikie BB, Lee J, Lee TS (2017) The impact of risk in Ethiopian construction project performance.Open Access Lib J 4:e4233. https://doi.org/10.4236/oalib.1104233

Nadežda S, Mihovil L, Jelena L, Jelena P, Miljana M, Biljana B, Radosavljević M (2011) Characterization of bentonite clay from "Greda" deposit. Process Appl Ceramics 5(2):97–101

Negash B (2014) The impact of standardization of aggregate in Ethiopian construction industry [MSC thesis]. AA

Olanitori LM, Olotuah AO (2005) The effect of clayey impurities in sand on the crushing strength of concrete (A case study of sand in Akure Metropolis, Ondo State, *Nigeria). In: 30th conference on our world in concrete and structures, Singapore, August 23–24, 2005

Patel H, Pandey S (2009) Exploring the reuse potential of chemical sludge from textile wastewater treatment plants in India-A hazardous waste. Am J Environ Sci 5(1):106–110

Pimentel D, Wilson C, McCullum-Gomez C, Huang R (1997) Economic and environmental benefits of biodiversity. BioSci 47 (11):747–757. https://doi.org/10.2307/1313097

Ramesh K, Biswas AK, Patrar A (2011) Zeolite farming. Indian Journal of Agronomy 60(2):185–191

Salissou A (2021) Nanoclays in food packaging. Afric J Food Sci Technol 12:01–02. https://doi.org/10.14303/ajfst.2021.003

Schlichting E (1972) Böden puffern Umwelteinflüsse ab. Umschau in Wissenschaft Und Technik 72:50–52

Sheppard RA, Hay RL (2001) Occurrences of zeolites in sedimentary rocks. Bish DL, Ming DW (eds) In:natural zeolites: occurrence, properties, applications. Reviews in Mineralogy &Geochemistry, An overview 45:217–234. Mineralogical Society of America & Geochemical Society, Washington DC

Simonson RW (1966) Shifts in the usefulness of soil resources in the USA. Agriculture 23:11–15

Sisay A (2004) Assessment of damage of buildings constructed in expansive soil areas of Addis Ababa, MSc thesis, Addis Ababa University, Ethiopia

Staubach S (2005) Clay: the history and evolution of humankind's relationship with earth's most primal element. Berkley Books, New York, NY

Tadesse M (2020) The developmental patterns of Injera Baking Stoves: review on the efficiency, and energyconsumption in Ethiopia. SSRG International J Mech Eng 7(1):7–16. ISSN:2348–8360. https://doi.org/10.14445/23488360/IJME-V7I1P102, http://creativecommons.org/licenses/by-nc-nd/4.0/

Uge BU (2017) Performance, problems and remedial measures for roads constructed on expansive soil in Ethiopia–a Review. Civil Environ Res 9(5)

Van Straaten P (2002) Rocks for crops: agrominerals of sub-Saharan Africa. ICRAF, Nairobi, Kenya, 338 pp. www.uoguelph.ca/lrs

Van Ranst E, Dumon M, Tolossa AR, Cornelis JT, Stoops G, Vandenberghe RE, Deckers J (2011) Revisiting ferrolysis processes in the formation of Planosols for rationalizing the soils with stagnic properties in WRB. Geoderma 163:265–274

Wassie SB (2020) Natural resource degradation tendencies in Ethiopia: a review. Environ Syst Res 9:33

Wayessa B (2011) The technical style of Wallagga pottery-making: an ethnoarchaeological study of Oromo potters in southwest highland Ethiopia. Africa Archaeol Rev 28(4):301–326

Zhang MH, Lastra R, Malhotra VM (1996) Rice-husk ash paste and concrete: some aspects of hydration and the microstructure of the interfacial zone between the aggregate and paste. Cem Concr Res 26(6):963–977

Future/Emerging Soil Issues

13

Sheleme Beyene, Kibebew Kibret, and Teklu Erkossa

Abstract

Soils are basic prerequisite for meeting human needs but human pressure on soil resource is reaching critical limits. Anthropogenic effects including the depletion of soil organic matter, erosion, acidity, salinity and/or sodicity, and pollution cause major challenges on future advances in soils. The reversal of soil degradation through the build-up of soil organic matter and the sustainable management of soils offers large potential to contribute to climate change mitigation by sequestering atmospheric C into the soil. Soils are also subjected to indirect impacts arising from human activity, such as heavy metal pollution and acid deposition. Globally, the advancement in data storage, processing, and sharing capacity as well as analytics and modeling competency are optimizing agricultural decision-making by predicting the various complexities that influence agricultural production processes. Pedometrics is also playing major role in advancing the role of land evaluation and, the quantification of soil quality for land management, and sustainable use of land resources. Multivariate methods are available in the country for the analysis of high-dimensional data such as those obtained from hyperspectral sensors in conjunction with data obtained directly from the field, for calibration and validation of the remote sensing data. A combined application of the interactive decision support tools within the framework of the Agronomic Response Unit and the socioeconomic circumstances including investment capacity of the farmers is believed to enhance resource use efficiency. Soil science education produced qualified scientists in the field, and research in soil science has been also instrumental in the generation of new knowledge and technologies that helped solving many societal problems. However, multiple challenges exist for future advances in soil science. Thus, future education in soil science must respond to emerging societal challenges and sees soils beyond agriculture. The education must ensure that soil continues to be among the most important strategic resources that are central to solving societal challenges and make sure that the multiple functions of soils in an ecosystem are understood comprehensively. Thus, there should be courses dealing with soil and society, economics of soil resources, soils and policy, soil politics, and legal aspects pertinent to soil resources.

Keywords

Anthropogenic effects • Agronomic response unit • Challenges • Data access • Education • Multivariate methods • Research

S. Beyene (✉)
School of Plant and Horticultural Sciences, Hawassa University, P.O. Box 5 Hawassa, Ethiopia
e-mail: shelemeb@gmail.com

K. Kibret
School of Natural Resources Management and Environmental Sciences, Haramaya University, P.O. Box 138 Dire Dawa, Ethiopia
e-mail: kibebew.kibret@haramaya.edu.et

T. Erkossa
Deutsche Gesellschaft Für Internationale Zusammenarbeit (GIZ), P.O. Box 10009 Addis Ababa, Ethiopia
e-mail: teklu.erkossa@giz.de

13.1 Introduction

Healthy soils are basic prerequisite in meeting various human needs including food, biomass energy, fiber, fodder, and other products, and ensuring the provision of essential ecosystem services in all regions of the world. However, human pressures on soil resources are reaching critical limits (FAO 2015). The main causes of soil degradation and thereby threats to its ecological functions are erosion, organic matter decline, loss of biodiversity, compaction, sealing, diffused contamination, pollution, and salinization. Soil erosion is a major environmental issue with a worldwide

S. Beyene et al. (eds.), *The Soils of Ethiopia*, World Soils Book Series,
https://doi.org/10.1007/978-3-031-17012-6_13

impact and direct and indirect effects on soil productivity and consequently on human survival. It is significantly increased due to human intervention such as intensive agriculture, overgrazing, deforestation, and industrial and mining activities. Presently, soil erosion and degradation promoted by human action have reached extreme levels, necessitating urgent measures to promote soil conservation and rehabilitation.

The depletion of soil organic matter causes the vicious cycle of decline in crop yield–food insecurity–soil and environmental degradation. The tranny of hunger as a result of this cycle can be broken by improving soil fertility through enhancement of the soil organic matter, which requires the use of sustainable agricultural technologies for water and nutrients management, including no-till farming, composts and mulching, leguminous cover crops, water harvesting, agroforestry, and integrated farming systems, along with judicious use of agrochemicals (Lal 2004). The reversal of soil degradation, through the build-up of soil organic matter and the sustainable management of soils, therefore, offers large potential to contribute to climate change mitigation by sequestering atmospheric C into the soil. Additionally, these processes would increase the capacity of soils to act as a buffer against climate change, which in turn would improve the resilience of agricultural systems to the impacts of climate change (Amanullah and Hidyatullah 2016; Amanullah and Khalid 2016). Soil organic matter, therefore, regulates the resilience of the agricultural system to climate change.

Soils are also subjected to indirect impacts arising from human activity, such as heavy metal pollution and acid deposition. Soil pollution has been internationally recognized as a major threat to soil health, and it affects the ability of the soils to provide ecosystem services, including the production of safe and sufficient food, compromising global food security. Contaminated soils can have an impact on ecosystem services including loss of productivity, contaminating the surrounding environment, and affecting the health of humans or other organisms, which can occur through direct contact with the soil, or via food-chain transfer of contaminants. In developed countries, the extent of soil contamination is likely to only slowly increase, or perhaps even improve as environmental regulations reduce the amounts of contaminants released into the environment, and remediation of existing contamination progresses. In developing countries, however, the extent of contamination is likely to increase, due to the absence of required regulatory framework or the ability to enforce it. Land use land cover changes have been one of the most critical threats to soil resources, for instance, soil erosion after deforestation, increased soil salinity following the expansion of irrigation, and transformation of rainfed lands into weakly-managed irrigated lands (Darwish et al. 2005; Dregne and Chou 1992).

Soil acidity is another threat that can be accelerated by anthropogenic activities, such as the combustion of fossil fuels and subsequent acidic deposition. Wet and dry acid deposition occurs because of the combustion of fuels that leads to the emission of sulfur and nitrogen oxides, which are then returned to the soil as acid precipitation, or acid-forming dry deposition. Ammonia emissions to the atmosphere, primarily from fertilizer use and intensive livestock production also contribute to soil acidity as the ammonia returns to the soil via precipitation and undergoes nitrification. Increases in the global use of inorganic fertilizers in developing countries can be attributed to greater demands for food from rapidly increasing populations and desires for a higher standard of living. Positive nutrient balances will become more pronounced or prevalent in such areas (Pretty and Bharucha 2014). However, many of the world's most common inorganic fertilizers including anhydrous ammonia, urea, ammonium sulfate, diammonium phosphate, and monoammonium phosphate contribute to soil acidification through the nitrification process that converts ammonium to nitrate.

Globally, the advancement in data storage, processing, and sharing capacity as well as analytics and modeling competency are optimizing agricultural decision-making by predicting the various complexities that influence agricultural production processes. Untapped opportunities include artificial intelligence and machine learning, drone technology, and robotics. Mathematical, statistical, and numerical methods are being applied to resolve the uncertainty and complexity inherent in a soil system model, including numerical approaches to classification, which deals with deterministic variation (Webster 2006). Pedometrics is playing a major role in advancing the role of land evaluation and, more recently, the quantification of soil quality for land management and sustainable use of land resources (Rossiter et al. 2018). It also addresses other soil questions including digital soil mapping, soil monitoring, quantitative pedogenesis and soil utility, soil functions and security and to diversify the tools and techniques used for such research: soil spectroscopy, machine learning, spatio-temporal soil inventories, and quantitative process-based soil-landscape models (Wadoux et al 2022).

With its current human population exceeding 110 million, Ethiopia is grappling with the daunting challenges of producing food for its fast-growing population on low fertility soils managed by poor smallholder farmers. Some of the major challenges for future advances in soils include the depletion of organic matter, erosion, acidity, salinity and/or sodicity, and pollution. Additionally, systemic challenges such as lack of up-to-date information on soil resource data and exchange system; absence of soil technology registry; weak connection between research, extension, and academia; inefficient input value chain and lack of credit; limited human resources and infrastructure; and absence of proper land use planning are drawbacks in advancing soil issues (ATA 2013).

The emergence of a large soil survey and agronomic research legacy datasets and recent development in data analytical approaches such as data mining and machine learning may allow the development of better-informed decisions across geographic and social scales. Machine learning (ML) and artificial intelligence (AI) algorithms are proved to have substantial potential to predict site-specific optimal agronomic inputs such as fertilizer recommendations. In recent years, cheaper and more accurate methods of generating site-specific data are developed and used around the globe. For instance, multivariate methods are available for the analysis of high-dimensional data such as those obtained from hyperspectral sensors in conjunction with data obtained directly from the field, including for calibration and validation of the remote sensing data (e.g., Vohland et al. 2017). For instance, Optimizing Fertilizer Recommendations for Africa (OFRA) developed an agro-ecological zone (AEZ)-based fertilizer recommendation using data from a few sites (Kaizzi et al. 2017).

Following the recent developments in real-time data generation capacities and geospatial analytical technologies, the creation of a land unit with similar response to agronomic management interventions, which in conjunction with socio-economic settings can result in effective and affordable recommendations, was suggested (Tamene et al. 2022). This land unit also known as 'Agronomic Response Unit' (ARU) can be developed through mapping areas with similar geomorphological and soil properties and ecological potentials. A combined application of the interactive decision support tools (DST) within the framework of the ARUs (the spatial and temporal biophysical characteristics of the farming systems) and the socioeconomic circumstances including investment capacity of the farmers is believed to enhance resource use efficiency and also empower the farmers that are the ultimate decision-makers.

The contribution of soil science education in the country to produce scientists in the field and its rapid development and transformation could be critical. Research in the field of soil science has been instrumental in the generation of new knowledge and technologies that helped solving many societal problems related to agricultural production and productivity, and more recently environmental quality. Future soil science education must respond to emerging societal challenges and be able to see soils beyond agriculture (Field et al. 2017; Bouma 2014; McBratney et al. 2014). The education must ensure that soil continues to be among the most important strategic resources that are central to solving societal challenges related to food, climate change, water and energy security, enhancing biodiversity and being a potential reservoir for future pharmaceutical (Brevik 2013; Field et al. 2013; Koch et al. 2013; Mol and Keesstra 2012). Thus, the future soil science education or teaching must foster an interdisciplinary approach and instill state-of-the art soil science knowledge and skills that enable graduates to work across disciplines in a bid to address increasingly complex environmental challenges (Field et al. 2013).

Furthermore, future soil science education should make sure that the multiple functions of soils in an ecosystem are understood comprehensively, provides state-of-the art knowledge and skill that helps in developing management interventions that help the soil to continue functioning sustainably. It has to make sure that the currently overlooked roles of soils in human nutrition and health are adequately emphasized. Thus, there should be courses dealing with soil and society, economics of soil resources, soils and policy, soil politics, and legal aspects pertinent to soil resources.

13.2 Direction in Soil Science Education

Soil science is a relatively young science emerging as an independent discipline in the mid-1800s (Hartemink 2015; Sandor et al. 2006) and rapidly developing in the twentieth century (Sandor et al. 2006). Since then, it has been passing through different developmental phases leading to the emergence of many sub-disciplines dealing in depth with specific aspects of soils but having important interfaces among each other. Enormous literature exists on the history of soil science's development as a distinct and independent field of study (Brevik and Hartemink 2010; Churchman 2010; Berthelin et al. 2006; Evtuhov 2006; Feller et al. 2006; Hasegawa and Warkentin 2006; Pachepsky and Rawls 2005; McNeill and Winiwarter 2004; Gong et al. 2003; Dobrovolskii 2001; Yaalon and Berkowicz 1997; Yaalon 1997; Krupenikov 1993; Hendricks and Fry 1930).

The contribution of soil science education in the production of the critical mass of scientists who played a pivotal role in its rapid development and transformation has been central. Research in the field of soil science has been instrumental in the generation of new knowledge and technologies that helped in solving many societal problems related to agricultural production and productivity and more recently environmental quality. Not less is expected from the future soil science education. However, most soil science education hitherto has been focusing dominantly on the use of soils for agricultural purposes (Brevik 2009). Future soil science education must respond to emerging societal challenges and be able to see soils beyond agriculture (Field et al. 2017; Bouma 2014; McBratney et al. 2014). The education must ensure that soil continues to be among the most important strategic resources that are central to solving societal challenges related to, but not limited to, food, climate change, water, and energy security, enhancing biodiversity and being a potential reservoir for future pharmaceutical (Brevik 2013; Field et al. 2013; Koch et al. 2013; Mol and Keesstra 2012). As pointed out by Field et al. (2013), the future soil science education or teaching must foster an interdisciplinary approach and instill state-of-the art soil science knowledge and skills that enable graduates to work across disciplines in a bid to address increasingly complex environmental challenges.

In Ethiopia, most under- and postgraduate level soil science curricula have been focusing more on the use of soils for agricultural purposes, ignoring the other functions and ecosystem services of soils. As a consequence, the current curricula are not able to produce graduates that are well versed with non-agriculture uses of soils and the challenges thereof. On the other hand, environmental challenges such as pollution (e.g., of water and soil itself) are becoming eminent. Nutritional deficiencies are posing a number of human and animal health concerns, while the importance of soils in alleviating these challenges is not well covered in current courses or curricula of soil science. Literature on direct and indirect influences of soils on human health abounds in other parts of the world. However, this has not been integrated well in the current soil science curricula. Although a number of antibiotics and many line drugs have been extracted from soils, the contribution of soils in this regard has not been touched upon altogether. The role soils have been playing and are expected to play in adapting to and mitigation of climate change at least through carbon sequestration has not been boldly considered. Social, economic, cultural, and aesthetic aspects of soils are not adequately addressed in the current soil science education although many sources claim that soil is at the center of these dimensions. There should be courses or at least topics in relevant courses dealing with soil and society, economics of soil resources, soils and policy, soil politics, and legal aspects pertinent to soil resources.

The ongoing environmental changes caused by natural and/or anthropogenic factors are expected to influence soil properties and processes in different directions. This will in turn influence management considerations under changing environment. Therefore, anticipated changes in soil properties and process should be able to be modeled. Most existing models are considering most soil attributes as static or not responding to perturbations, which is not the case. Therefore, future curricula should include analytical courses such as soil modeling, pedometrics and others to guide future interventions and policy directions. Emerging concepts like soil sustainability, soil resilience, and soil security are missing in most of the current soil science education. They should be inculcated in the topics of relevant courses. Soil is a habitat for large number of biodiversity. However, that aspect of the soil is not largely captured. Future curricula should place strong emphasis on biodiversity of soil, which holds immense potential for predicting future changes and responses to the changes. Courses like soil ecology are not being offered in most programs that have soil courses in them. Nanotechnologies have proven to have an important application in the field of soil science including fertilizer use efficiency and increasing the surface area of soils. Such technologies should be used aggressively in a bid to increase resource use efficiency. State-of-the art equipment are being developed and applied in quantification of soil parameters that were not measurable before. Examples of such equipment include Radar technologies for monitoring, for instance soil moisture dynamics near the soil surface. The invention of high magnification microscopes and thermographs has given a promising result in studying pore geometry and soil structure. In general, advances in the field of soil science should be included in the future soil science curriculum. Emerging fields of study such as Hydropedology should not be overlooked.

By and large, future soil science education should make sure that the multiple functions of soils in an ecosystem are understood comprehensively, provides state-of-the art knowledge and skill that helps in developing management interventions that help the soil to continue functioning sustainably. It has to make sure that the currently overlooked roles of soils in human nutrition and health (through supporting the production of nutritious and safe food as well as supporting biodiversity from which a number of medicines are extracted and purifying water and air from harmful substances) is adequately emphasized. It has to impart a mix of basic and interdisciplinary knowledge and skills that enable graduates to view soil as a complex system that has multiple dimensions (social, economic, cultural, political, etc.) and attempt to address complex challenges through a holistic approach. It has to help scientists and experts in the field to do cutting-edge researches that enable them understand how a soil functions as system, interacts with many systems such as the atmosphere and the lithosphere and forms an interface with the hydrosphere and biosphere and how these interactions can be used to create a soil that is of good quality (healthy), sustainable and resilient. It has to equip graduates that are capable of taking soil's issues to decision and policymakers and make sure that soil as a vital life-supporting resource gets the recognition and attention it deserves in national as well as international policies and investments.

13.3 Direction in Soil Science Research

With its current human population exceeding 100 million, Ethiopia is grappling with the daunting challenges of producing food for its fast-growing population on low fertility soils managed by poor smallholder farmers (Tamene et al.

2017). Increasing agricultural productivity to meet food security and agriculture-led economic development needs without compromising the integrity of the agricultural ecosystems requires understanding the spatial and temporal trends of the potentials and constraints of the soils and the agricultural systems at large. Soil research in Ethiopia mainly focused on survey and characterization, plant nutrition and erosion and soil and water conservation.

The formal attempt to understand the soil resource in Ethiopia has begun in the late 1950s when the roadside soil survey expedition was conducted. Prompted by the results of the national soil survey that indicated a widespread and localized deficiency of the major plant nutrients including Phosphorus (P) and Nitrogen (N), soil fertility management research has begun in the 1960s (Fig. 13.1). During the early stages, the studies focused on the use and management of mineral fertilizers for cereal crops including tef (Eragrostis tef), wheat (Triticum aestivum) and maize (*Zea mays*) (Erkossa et al. 2022). The fertilizer trials carried out around 1967, confirmed a widespread and significant response to N, P, and their interaction. This led to the nationwide on-farm trials that were started in the 1970s, the result of which led to the blanket but location-based N and P fertilizers recommendations irrespective of crop and soil types. A more recent study conducted across agro-ecological and edaphic spectrum recommended 30–138 kg N ha^{-1} and 0–50 kg P ha^{-1}, respectively. However, studies show that only 30–40% of Ethiopian smallholder farmers use fertilizers and those that do apply on average 37–40 kg ha^{-1}, which is significantly below the recommended rates.

Erosion and soil and water conservation attracted the attention of soil science researchers in the country since the 1970s, mainly in response to the tragic effect of drought and famine during the early 1970s. According to Haregeweyn et al (2015), at least six national soil and water conservation-related programs have been initiated around the 1970s initially focusing on food relief, which gradually shifted to land conservation and livelihoods. While the soil conservation research and development during the early years focused on the evaluation of physical conservation measures in reducing soil erosion, it has gradually evolved to integrated approaches, including in watershed settings.

As stated in the Federal Negarit Gazeta (1997), the Ethiopian Institute of Agricultural Research (EIAR) is mandated for generating, improving, and adapting technologies and coordinating, encouraging, and assisting agricultural research activities in the country. With its multiple research centers distributed across the different regional

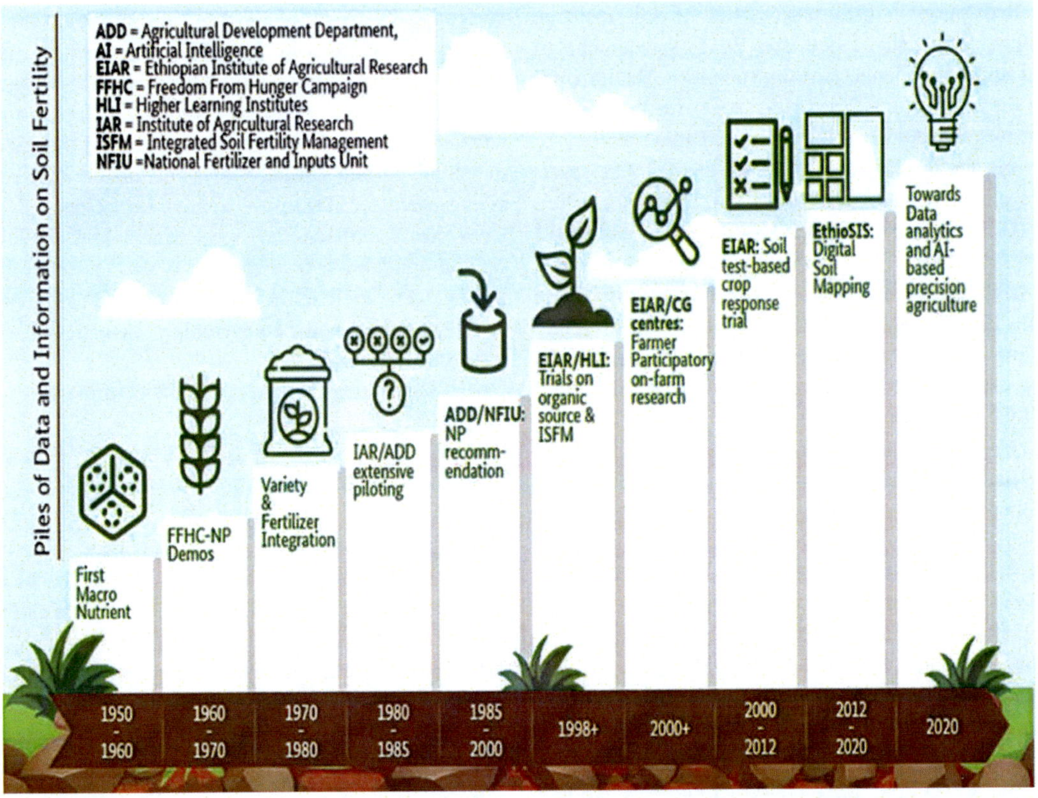

Fig. 13.1 Evolution of soil fertility research and development in Ethiopia: From reconnaissance to data-mining approaches Erkossa et al. (2022)

states and agro-ecologies of the country, the institutes are attempting to fulfill the current and long-term agricultural requirements of the country. In addition, all the regional governments have agricultural research centers with similar mandates within their respective regions. Besides, most of the over 40 government universities have agricultural faculties with research mandates where soil science research is a major component. Furthermore, 17 soil laboratories are established and operationalized by the ministry of agriculture, mainly to support farmers in testing soils, but they also contribute to the research services.

While these have created a huge research capacity in terms of human and infrastructure for soil and other agriculture-related research, and generation of scientific knowledge including recommendations of technologies and management practices, adoption by farmers and agricultural productivity remains low. Consequently, the agriculture-dominated rural livelihood remains subsistent and the agricultural ecosystems remain fragile. The limited impact of soil science and overall agricultural research is partly attributed to a multitude of factors including poor targeting of technological recommendations and agricultural advisories that are not well correlated with positive economic and environmental impact.

However, encouraged by some of the localized successful experiences, the government of Ethiopia has shown a steadfast dedication to the improvement of soil resource management, especially since the 2015 soil campaign organized under the theme 'Healthy soils for a Health Life' as part of the International Year of Soils. During the event, it was emphasized that soil fertility is an essential prerequisite for achieving the targets set forth under the national development agendas, such as the Growth and Transformation Program (GTP), Climate Resilient Green Economy (CRGE) and others, in terms of enhancing agricultural outputs and GDP growth. Other recent initiatives include scaling up of sustainable land management (SLM), reclamation of acidic soils, improving the productivity of waterlogged soils, increasing capacity for climate-smart agriculture, integrated soil fertility management (ISFM), digital mapping of soils in the country's arable lands, and most recently, liming of acid soils. The nationwide initiative to map the soil fertility status of the country based on which site-specific fertilizer recommendations can be developed represents a significant milestone in the nation's soil research. Together with a huge, underutilized legacy soil survey and research data collected for different purposes over decades, this would create a solid foundation on which advanced research methods and approaches can be applied to generate a more robust agricultural advisory that can be better targeted to the local contexts while providing the big picture for supporting policy directions. Despite these concerted efforts, there is a consensus among the soil science community that agricultural land and soil management in the country is not sufficiently benefiting from the emerging scientific methods.

The emergence of a large soil survey and agronomic research legacy datasets and recent development in data analytical approaches such as data mining and machine learning may allow the development of better-informed decisions across geographic and social scales. Globally, the advancement in data storage, processing, and sharing capacity as well as analytics and modeling competency are optimizing agricultural decision-making by predicting the various complexities that influence agricultural production processes. Untapped opportunities include artificial intelligence and machine learning, drone technology, and robotics. Developing countries like Ethiopia can benefit from the opportunities through the adoption of global knowledge and skills and accessing the available open data sources in conjunction with the nationally generated data sets, on which advanced analytical approaches can be applied.

Understanding the potential of data access and sharing and with the knowledge and information about the wealth of soil and agronomy data accumulated over decades, a group of soil and agronomy researchers agreed in 1918 to create a coalition to share their data and to promote the idea of soil and agronomy data sharing in Ethiopia. In a short period, the group has expanded in terms of scope and membership and currently, the Coalition of the Willing (CoW) for soil and agronomy data sharing in Ethiopia is seen as one of the thriving African initiatives toward data-driven agriculture. Through its efforts to collate legacy data, implementation of advanced data analytics, the group realized the need for the development and adoption of standardized data collection, storage, and management approaches. Consequently, the CoW has facilitated the development and publication of several guidelines to standardize data creation and data management approaches, which is believed to pave the way to an advanced and new paradigm in soil and agronomy research in Ethiopia.

13.3.1 Development of Soil and Agronomy Data Portals

Through the work of the CoW and other initiatives, two central data portals are developed; one at the Ministry of Agriculture, mainly for soil survey-related data and the other at the Ethiopian Institute of Agricultural Research, which is dedicated to field research related to soil fertility and agronomy. The two data portals are believed to facilitate the application of the FAIR (Finable, Accessible, Interoperable and Reusable) data principles, which requires that data from different sources follow a certain standard to ensure interoperability. The soil and agronomy data ecosystem mapping conducted in 2019–2020 and collation of the legacy data from data holders that were willing to share their data have revealed that data collection generally followed divergent approaches (Ali et al. 2020). This jeopardized the possibility of combining the data or required arduous harmonization efforts both store in a centralized database and to conduct a combined analysis.

13.3.2 Development of Data Standardization Guidelines

Although a concerted effort is made to harmonize the data using a standard template, a large volume of soil survey data could not be mapped to the standard template, jeopardizing its interoperability and reusability. Therefore, to avoid such constraints in the future, CoW decided to develop guidelines for the standardization of soil and agronomy-related data collection and management as elaborated in the following sections.

Accordingly, a task force was established to facilitate the process, which has identified six thematic areas including soil fertility and agronomy, soil biology, laboratory analysis of soil, plant and water, agricultural water management, watershed management, soil survey and mapping for which guidelines should be developed. A team of technical experts for each of the thematic areas has been established to draft the guidelines, which were then reviewed by local and international scholars before publication. Passing through this rigorous process, three guidelines, including for soil fertility and agronomy, soil biology and laboratory analysis of soil, plant and water analysis have been published and dissemination is underway, while the rest are expected to be published shortly. Although the guidelines are not yet officially adopted by the Institute, the guidelines were used during the 2021–2022 research planning process. In addition, the soil fertility and agronomy guideline were used in the design and implementation of a national fertilizer response trial involving 700 farm plots. Besides, about 100 focal persons of GIZ's Integrated Soil Fertility Management Project (ISFM) from Oromia, Sidama and Southern Nations Nationalities and People (SNNP) regional states were trained on the use of these guidelines in 2021, which they have used in collecting on-farm demonstration data. Currently, the agreement is reached with EIAR to work on the mainstreaming of the guidelines into the national agricultural research system, and a concept note for financing agreement of the process is developed. In the subsequent sections, a brief description of the already published guidelines for field data collection (soil fertility and agronomy, and soil biology) and the laboratory analysis will be highlighted.

13.3.3 Standardization of Field Data Collection

Recently, recognition has been growing of the power of data and information for better decision-making and service provision in agriculture. Standard field research design, field data collection, and data reporting are required for well-informed meta-analyses and syntheses of agricultural research data as well as for making these data more accessible for calibration and evaluation of process-based models. Cognizant of this, the guidelines for field data collection including soil fertility and agronomy (Abera et al. 2020), and soil biology data (Mnalku et al. 2020) were published. The guidelines are aimed at enhancing temporal and spatial data interoperability and reusability, which would enable meta-analysis of different data collected over years and/or space to accumulate evidence and generate new knowledge or insights to facilitate informed decision-making in soil and crop management. The guidelines are developed based on accepted standards and procedures in the field. Nevertheless, they are not exhaustive in terms of covering the soil fertility, agronomy, and soil biology data types. Hence, additions and updates depending on the development of research facilities, the ever-changing focus of agricultural research and production systems, and advances in technology are warranted. The guidelines are intended for use by researchers, academicians, students, and other interested professionals in Ethiopia and beyond.

13.3.4 Standardization of Laboratory Procedures

Generation of data related to soils, plants, water, and fertilizers has been underway ever since the commencement of soil and agronomy research in the country. However, the data generated from the laboratories are not accessible since there is no centralized data repository and data sharing and access policy and guideline are missing. Different methods and procedures of sample collection and analyses are used across the country for analysis of the same soil, plant, water, and fertilizer parameters. This poses a conspicuous impediment to the comparison or to a combined analysis of the data accessed informally from the laboratories or extracted from publications (Wogi et al. 2021).

In some instances, a method designed for extracting certain soil nutrients, such as phosphorus (P), from the soil with specific physical and chemical characteristics, may also be used to extract the nutrient from the soil with different physical and chemical properties, for which a different method is supposed to be used or no method is yet developed (Wogi et al. 2021). This often leads to the use of ill-suited laboratory methods and procedures for determining soil, plant nutrient status, water as well as fertilizer quality indicators. Consequently, interpretation of data obtained through one method of analysis for various soil types could lead to erroneous conclusions and recommendations. Similarly, data on the same soil, plant, water, and fertilizer parameters are often documented and shared in different units of measurement, making it difficult to compare, combine and reuse. To address the issues, the CoW has facilitated the development and publication of a guide to standardized methods of analysis for soil, water, plant, and fertilizer resources for data documentation and sharing (Wogi et al. 2021). The guideline is hoped to bridge information gaps and promote standardized laboratory methods and procedures for soil, plant, water, and fertilizer analyses, thereby facilitating standardized data documentation and sharing to allow advanced data analytics needed for evidence-based decision-making. The guideline should be used with a standard laboratory manual, which needs to be developed based on the standardization guideline.

13.3.5 Advanced Approaches and Tools for Data Collection and Analysis

Field experimentation is almost a rule rather than an exception in the traditional soil science research in Ethiopia. The validity of data obtained this way depends on the appropriateness of the design, data collection facilities and importantly the skill of the data collector. In the era of precision agriculture, which in addition to getting averages and variances of both crop and soil parameters requires enhanced description and understanding of the Spatio-temporal variability using newly developed technologies. In recent years, cheaper and more accurate methods of generating site-specific data are developed and used around the globe. A remote sensing approach focusing on the spatial variability of the soils using spectroscopic techniques from visible to near-infrared light energy reflection is becoming a standard approach in soil science research. A real-time soil spectrophotometer is one of the innovative tools to provide information about multiple soil parameters, such as moisture and organic matter content.

Multivariate methods are available for the analysis of high-dimensional data such as those obtained from hyperspectral sensors in conjunction with data obtained directly from the field, including for calibration and validation of the remote sensing data (e.g., Vohland et al. 2017). The use of remote sensing and real-time data generation approaches in Ethiopia is at an infant stage.

Currently, a two-prong approach is followed to improve soil and agronomy data availability, i.e., collating, cleaning, and making accessible the legacy data and generation of fresh data including through more advanced methods. Regarding the legacy data, the CoW has created a large historical soil and agronomy dataset that can allow for advanced data mining approaches to generate evidence-based agricultural advisories. For instance, the soil fertility-related datasets cover a wide environmental space, which can be used to develop national-level fertilizer recommendations (Erkossa et al. 2022). With data-sharing initiatives, machine learning (ML) and artificial intelligence (AI) algorithms are proved to have substantial potential to predict site-specific optimal agronomic inputs such as fertilizer recommendations. However, developing optimal fertilizer recommendations is challenging because it involves many variables (e.g., weather, soils, land management, genotypes, and crop diseases), which should be used as determinant covariates. With the availability of high-resolution spatial layers of the determinant environmental factors and large agronomic datasets, ML and AI approaches can integrate these large and diverse datasets to develop decision support tools.

13.3.6 Recommendation from DST, Targeted and Area-Based Experiments

Recommendations and decisions of crop management in sub-Saharan Africa (SSA) are often based on traditional field experimentation. In Ethiopia, soil test and field experiments in conjunction with economic analysis are customary approach used to recommend soil and crop management technologies and practices. Such approach seldom takes the agro-ecological, landscape, and socio-economic heterogeneity of the farming systems, which jeopardizes the validity of the recommendations beyond the experimental sites. On the other hand, individual plot or household level recommendations are arduous. To overcome the problem, various initiatives have been implemented. For instance, Optimizing Fertilizer Recommendations for Africa (OFRA) developed an agro-ecological zone (AEZ) based fertilizer recommendation using data from a few sites (Kaizzi et al. 2017). However, the approach did not consider the micro-factors that influence crop response to nutrients, thus resulted in a coarse recommendation. Amede et al (2020) showed the effect of local-scale topography on yield response to fertilizer and developed a landscape-based fertilizer recommendation, but as the approach ignored other important determinant factors it is not suited for holistic, site-specific fertilizer recommendations (Abera et al (2022).

Thanks to the recent developments in real-time data generation capacities and geospatial analytical technologies, researchers have suggested the creation of a land unit with similar response to agronomic management interventions, which in conjunction with socio-economic settings can result in effective and affordable recommendations (Tamene et al.

2022). This land unit also known as 'Agronomic Response Unit' (ARU) can be developed through mapping areas with similar geomorphological and soil properties and ecological potentials. One of the key components to creating similar ARU is understanding yield-controlling factors and generating yield-response differential maps. Based on defined ARU, it will be possible to design management plans geared towards tackling constraints and harnessing potentials. A recent work by Tamene et al. (in press) provided detail insights into how similar ARUs can be developed and applied for fertilizer recommendation targeting and scaling in a complex context like Ethiopia.

Several studies reported on increased fertilizer use efficiency and reduced crop production risks with the use of decision support tools (DST). In the last one decade, the application of ML to guide agronomic management decisions has been increasing in many parts of the world (Chlingaryan et al. 2018). A combined application of the interactive decision support tools (DST) within the framework of the ARUs (the spatial and temporal biophysical characteristics of the farming systems) and the socioeconomic circumstances including investment capacity of the farmers is believed to enhance resource use efficiency but also to empower the farmers that are the ultimate decision-makers. Development, validation, refinement, and most importantly simplification of the system so that ordinary people can use them are expected to be the focus of soil research in Ethiopia. Since such approaches are data-intensive, continued generation and sharing of good quality in-situ and remote sensing data are anticipated.

13.4 Challenges and Opportunities for Advances in Soils

The challenges and opportunities for advances in soils of the country encompass soil-level and systemic challenges. Some of the major soil-level challenges for future advances in soils include the depletion of organic matter, erosion, acidity, salinity and/or sodicity, and pollution. The systemic challenges include lack of up-to-date information on soil resource data and exchange system; absence of soil technology registry; weak connection between research, extension, and academia; inefficient input value chain and lack of credit; limited human resources and infrastructure; and absence of proper land use planning (Sect. 9.5).

13.4.1 Challenges

Low organic matter contents of the soils, which is caused by insufficient use of organic inputs, excessive tillage, over grazing, deforestation, and complete removal of plant residues after harvest, is the major challenge for future advances in soils under different agroecologies. Organic residues are mainly used for feed, building materials, and fuel and, therefore, are not returned to the soil. The traditional application of livestock dung has been decreasing overtime due to the competing use of dung as an energy source. The removal of crop residues without sufficient replacement is a major reason for nutrient mining, causing nutrient deficiency and imbalance and low productivity of crops, particularly in erosion-prone regions (Tamene et al. 2017). There is strong competition in the use of crop residues as animal feed and cooking fuel that leaves very little for the soil (Zenebe 2007). Major drivers of this behavior include low availability of biomass and competing uses for this biomass (dung used as fuel and crop residues used as feed) (Zelleke et al. 2010). There is also strong competition for biomass, with about 63, 20, 10, and 7% of cereal straws being used for feed, fuel, construction and bedding purposes, respectively (Tamene et al. 2017).

The rate of soil losses due to erosion, 130 tons ha^{-1} for cultivated fields, is considered to be one of the highest in Africa (FAO 1986). The productive land has been exposed to degradation and menace both economic and survival of the people (Genene and Abby 2014). The soil loss due to erosion in the highlands is high, varying between 42 and 175.5 t/ha/yr (Adimassu et al. 2017). The rugged topography in different parts of the country contributes to soil loss by erosion from upslopes and sediment deposits in low-lying areas (Dinku et al. 2014; Mulugeta and Sheleme 2010) causes difficulties in combating soil losses in the affected areas. High population pressure, continuous and steep slope cultivation, low vegetation cover, deforestation and inadequate soil conservation practices cause annual soil loss of about 1.5 billion metric tons in the highlands (Girma 2001). The fragile volcanic soils in the rift valley create conditions ripe for excessive soil erosion, flooding and sediment deposits in low-lying areas also make some degradation problems.

Increasing areas of agricultural land are affected by soil acidity, the consequence of which resulted in declining the productivity of the soils. Vast areas of land in the western, southern, southwestern, northwestern, and even the central highlands of the country (which receive high rainfall) are affected by soil acidity (Mesfin 2007). Soil acidity affected about 41% of the cultivated land in the high rainfall areas (Agegnehu et al. 2021) whereby 13% is strongly acidic (pH < 4.5) and 28% is moderate to weakly acidic (pH 4.5–5.5) (Mesfin 2007). Although liming is recommended to reduce acidity, the level of current lime use is very low in the country. In GTP II, the Ethiopian government planned to rehabilitate about 226,000 ha of agricultural land by expanding the production, distribution, and promotion of lime by smallholder farmers (MoNR 2015). However, the lime distribution system is still inefficient resulting in price build-up from the production site to the farmers' fields, mainly due to price of transportation. Increasing the production of lime, improving infrastructure to reduce transportation cost, and awareness creation to enhance the adoption of lime use by farmers are required to tackle the problems of soil acidity.

The total land area of salt-affected soils is rapidly increasing (44 Mha), of which approximately 33 Mha is dominantly affected by salinity, 8 Mha by salinity-sodicity, and 3 Mha is dominantly affected by sodicity problems (Habtamu 2013). These are dominantly found in Rift Valley Zone, Wabi Shebelle River basin and various other lowlands and valley bottoms (Kefyalew and Kibebew 2016; Meron 2007). The main cause for the problem may be due to massive expansion of irrigated agriculture by saline water. Following poor practices of irrigation management, salinity has emerged as a major problem responsible for reduction of land productivity and natural resources degradation (Wondimu et al. 2022). For instance, it has been reported that of 4,000 ha of irrigated lands at Melka Sedi, where about 40, 16.9, and 0.02% are saline, saline-sodic and sodic soils, respectively. At present, most of the irrigated large state farms producing export crops are situated in these zones. Reclamation of these salt-affected soils requires huge investment posing a major problem in future issues of soils in the areas.

Increasing soil pollution caused by the presence of xenobiotic (human-made) chemicals or other alterations in the natural soil environment is an emerging problem in soils. This problem is increasingly of concern with increasing industrial activity, agricultural chemicals, and/or municipal waste disposals (EIP 2002). Looking at the current rate of urbanization, industrialization, and the extent of agro-chemical use, soil pollution will be a major concern in the future soil issues.

Comprehensive up-to-date data and knowledge base of the soil status across the nation are lacking. The soil resource data available at ATA are not widely available. The research institutes, soil testing laboratories, higher learning institutions, and various development partners have generated valuable soil data across. However, the weak connection between research and extension, particularly research-extension-academia is also a problem in exchange of the data. Additionally, the data are not centrally compiled, and hence not accessible among the various actors causing a significant challenge.

A soil technology registry for validation of the effectiveness of the technologies and their release has never been conducted. The absence of this system resulted in (i) lack of confidence in adopting the generated technologies, (ii) duplication of efforts that spend time and resource to understand technologies that have already been tested, and (iii) discouragement and reluctance by researchers.

Inefficiencies in the input value chain of inorganic and biofertilizers, and lime (e.g., weak distribution/handling system, capital provision, demand projection) pose major problems in advances of soils. Capital is the major resource that is constrained, and the underlying bottlenecks are the difficulty in accessing the capital, and the limited risk aversion mechanism for farmers. Credit and insurance services are critical in helping often cash-constrained farmers to purchase the required quantities of lime as well as complementary inputs (WLE 2017). The issue of lack of input credit for fertilizer for smallholders is a relatively recent problem that was caused by channeling input credit through 'regular' cooperatives.

The soil testing laboratories across the nation have human resources and infrastructure problems, and a lack of coherence in the management system also poses a significant problem on the laboratories. Acquiring modern equipment and retaining adequate skilled staff in the laboratories are challenges in future soil issues. Another problem is the absence of proper land use plan. The productive arable land is encroached by competing land uses, especially through urbanization and industrialization with the development of industrial focus. Changes in land use systems including severe deforestation, cultivation of steep slopes, excessive tillage, and pollution from urban and industrial wastes are also challenges in advance of soil issues.

13.4.2 Opportunities

The existence of conducive government policies, the soil data available at ATA, data generation through collaborative projects, the establishment of Coalition of the Willing to share available data, standardization of procedures for field data collection and soil testing, and the fair distribution of universities offering postgraduate soil science training and research institutions could be considered as opportunities for future advances in soil issues.

The government formulated policies and strategic frameworks to overcome bottle-necks, promote, and scale up sustainable land management; promote community-based participatory watershed management; protect the environment and the use of natural resources including soil.

- The Ethiopian Sustainable Land Management Investment Framework (ESIF) provides a holistic and integrated strategic planning framework to remove the barriers, and overcome the bottle-necks, to promoting and scaling up sustainable land management (SLM). The ESIF has been formulated with the goal of serving as a national level strategic planning framework that is to be used to guide the prioritization, planning, and implementation, by both the public and private sector, of current and future investments in SLM with the aim of addressing the interlinked problems of poverty, vulnerability, and land degradation at the rural community level (Daniel 2008).

- The Ministry of Agriculture and Rural Development developed a community-based participatory watershed guideline (MoARD 2005). The Guideline aims at harmonizing and consolidating planning procedures at the grass-roots level. The intent is to provide Development Agents (DAs) and rural communities with a workable and adaptable planning tool, as well as, to provide practical guidance on the correct selection of technologies under different conditions and their sequentially correct implementation.
- The formulated Environment Protection Policy includes, among others, promoting sound management and use of natural, human-made and cultural resources and the environment as a whole so as to meet the needs of the population. The Policy considers the adoption of 'conservation culture' in environmental matters and integration of natural resource and environmental management activities laterally across all sectors. Control of hazardous materials and pollution from industrial waste, atmospheric pollution and climate change, environmental economics and impact assessment and monitoring were also included.

The Agricultural Transformation Agency generated huge grid-based soil data from the whole country and presented also soil fertility maps of the different regions (EthioSIS 2016; ATA 2014–16). The digital soil fertility maps were developed using Digital Soil Mapping (DSM)-Soil Spatial Prediction Models (SSPM). The ATA/EthioSIS project finalized and released the digital soil fertility and fertilizer formulation maps/Atlas for almost all regions (ATA 2016). These data could be of fertile ground for future advances in soil issues provided that the data are made accessible.

Hundreds of postgraduate students and researchers at different universities generated huge soil data with the support of collaborative projects sponsored by international organizations including the Alliance for Green Revolution in Africa (AGRA), Swedish SIDA, NORAD, GAC/IDRC, BENEFIT CASCAPE and REALISE, CLIFOOD. Part of the generated data especially that of the CASCAPE project were used under Chap. 8, soil fertility under different agroecologies. Additionally, Elias (2016) authored a book on 'Soils of the Ethiopian highlands: Geomorphology and Properties', from the data generated by CASCAPE project indicating the potential of the collaborative projects in contributing to advances of future soil issues.

The establishment of 'Coalition of the Willing' by GIZ and CIAT paved the way in sharing available data at different research institutes, universities, international organizations, NGOs, and private workers. More than 20,000 soil profile data were collected from the work of different projects, institutions, and private workers and soil mapping started. This facilitates the challenge in data sharing and contributes to advances in soil issues. Manuals including 'Standardization of Field Data Collection and Soil Testing Procedures' were developed by the support of GIZ and CIAT that could enhance control of soil analytical data and lead to 'Sample Exchange Program'.

The universities offering graduate programs in Soil Science (Table 2.1) and Research Institutes are fairly distributed in the country. The soil science researchers at each university could work with the professional colleagues from the research institutes to generate soil-related data. Additionally, the graduate students could work under the supervision of soil scientists from both institutions. Such arrangements could help in covering the soil issues in the whole country within a very short period of time.

13.5 Conclusion

Population growth has direct effects on land and results in land fragmentation. Indirect effects also exist through the demand for food that puts pressure on land extensification and intensification, which affect soil quality. Agricultural depletion of soil nutrients through poor farming practices including continuous plowing, overgrazing, and poor irrigation management impacts soil health. It is therefore crucial to curb industrial farming and implement agricultural practices that enhance sustainable use of soil resources. Soils are also subjected to indirect impacts arising from human activity, such as acid deposition (for example, sulfur and nitrogen) and heavy metal pollution. Global efforts to reduce emissions of N and S to the atmosphere need to be extended to developing countries and continuous efforts should be in place to maintain or improve pH conditions for acidic agricultural soils so as to prevent further declines in productivity.

The precision agriculture requires enhanced description and understanding of the spatio-temporal soil variability using newly developed technologies. A remote sensing approach focusing on the spatial variability of the soils using spectroscopic techniques from visible to near-infrared light energy reflection is becoming a standard approach in soil science research. The advancement in data storage, processing, and sharing capacity as well as analytics and modeling competency are optimizing agricultural decision-making by predicting the various complexities that influence agricultural production processes. Pedometrics is also playing major roles in advancing the role of land evaluation, and the quantification of soil quality for land management, and sustainable use of land resources. Ethiopia can benefit from the opportunities through the adoption of global knowledge and skills and accessing the available open data sources in conjunction with the nationally generated data sets, on which advanced analytical approaches can be applied. Multivariate methods are available in the country for the analysis of high dimensional data such as those obtained from hyperspectral sensors in conjunction with data obtained directly from the field, for calibration and validation of the remote sensing data. A combined application of the interactive decision support tools within the framework of the Agronomic Response Unit and the socioeconomic circumstances including investment capacity of the farmers is believed to enhance resource use efficiency.

There exist multiple challenges for future advances in soil science. Nutritional deficiencies are posing a number of human and animal health concerns, while the importance of soils in alleviating these challenges is not well covered in the current courses or curricula of soil science. Future soil science education must respond to emerging societal challenges and sees soils beyond agriculture. The education must ensure that soil continues to be among the most important strategic resources that are central to solving societal challenges and make sure that the multiple functions of soils in an ecosystem are understood comprehensively. It has to impart a mix of basic and interdisciplinary knowledge and skills that enable graduates to view soil as a complex system that has multiple dimensions (social, economic, cultural, political, etc.) and attempt to address complex challenges through a holistic approach. Thus, there should be courses dealing with soil and society, economics of soil resources, soils and policy, soil politics, and legal aspects pertinent to soil resources. Additionally, the academia should team up with researchers in conducting researches, and sample exchange program should be implemented to enhance the quality of soil analytical data from different laboratories.

References

Abera D, Liben FM, Shimbir T, Balemi T, Erkossa T, Demiss M (2020) Guideline for agronomy and soil fertility data collection in Ethiopia: national standard. Ethiopian Institute of Agricultural Research (EIAR). Addis Ababa, 31 p

Abera W, Tamene L, Tesfaye K, Jiménez D, Dorado H, Erkossa T, Kihara J, Seid J, Amede T, Ramirez-Villegas J (2022) A data-mining approach for developing site-specific fertilizer response functions across the wheat-growing environments in Ethiopia. Exp Agric 2022:1–16. https://doi.org/10.1017/S0014479722000047

Adimassu Z, Langan S, Johnston R, Mekuria W, Amede T (2017) Impacts of soil and water conservation practices on crop yield, run-off, soil loss and nutrient loss in Ethiopia: review and synthesis. Environ Manag 59 (1):87–101. https://doi.org/10.1007/s00267-016-0776-1

Agegnehu G, Tilahun A, Teklu E, Chilot Y, Henry C, Tyler R, Nosworthy MG, Sheleme B, Gudeta WS (2021) Extent and management of acid soils for sustainable crop production system in tropical agrosystem: a review. Acta Agric Scand Sect B Soil Plant Sci 71. https://doi.org/10.1080/09064710.1954239

Ali A, Lulseged T, Teklu E (2020) Identifying, cataloguing, and mapping soil and agronomic data in Ethiopia. CIAT Publication no 506. Addis Ababa (Ethiopia): International Center for Tropical Agriculture (CIAT). 42 p

Amanullah I, Khalid S (2016) Integrated use of phosphorus, animal manures and biofertilizers improve maize productivity under semiarid condition. In: Larramendy ML, Soloneski S (eds) Organic fertilizers-from basic concepts to applied outcomes. https://doi.org/10.5772/61454

Amanullah I, Hidyatullah AF (2016) Influence of organic and inorganic nitrogen on grain yield and yield components of hybrid rice in northwestern Pakistan. Rice Sci 23(6):326–333

Amede T, Gashaw T, Legesse G, Tamene L, Mekonen K, Thorne P, Schultz S (2020). Landscape positions dictating crop fertilizer responses in wheat-based farming systems of east African Highlands. Renew Agric Food Syst 1–13. https://doi.org/10.1017/S1742170519000504

ATA, Agricultural Transformation Agency (2016) Soil fertility status and fertilizer recommendation atlas of amhara, oromiya and SNNP regional states, Addis Ababa

ATA, Agricultural Transformation Agency (2013) 5-year strategy for transformation of soil health and fertility in Ethiopia; vision, systemic bottlenecks, interventions, and implementation framework, Addis Ababa

Berthelin J, Babel U, Totain F (2006) History of soil biology. In: Warkentin BP (ed) Footprints in the soil. People and ideas in soil history. Elsevier, Amsterdam, pp 279–306

Bouma J (2014) Soil science contributions towards sustainable development goals and their implementation: linking soil functions with ecosystem services. J Soil Fertil Soil Sci 177:111–120

Brevik EC (2009) The teaching of soil science in geology, geography, environmental science, and agriculture programs. Soil Surv Horiz 50:120–123

Brevik EC (2013) Soils and human health: an overview. In: Brevik EC, Burgess LC (eds) Soils and human health. CRC Press, Taylor and Francis Group, Boca Raton, FL, USA, pp 29–50

Brevik EC, Hartemink AE (2010) Early soil knowledge and the birth and development of soil science. Catena 83:23–33

Chlingaryan A, Sukkarieh S, Whelan B (2018) Machine learning approaches for crop yield prediction and nitrogen status estimation in precision agriculture: a review. Comput Electron Agric 151:61–69

Churchman GJ (2010) The philosophical status of soil science. Geoderma 157:214–221

Daniel D (2008) Sustainable land management and Ethiopian strategic investment framework, nile BDC workshop, Addis Ababa, April 8, 2011

Darwish T, Atallah T, El Moujabber M, Khatib N (2005) Status of soil salinity in Lebanon under different cropping pattern and agro climatic zones. Agric Water Manage 78:152–164

Dinku D, Sheleme B, Nand R, Walley F, Teleab SG (2014) Effects of topography and land use on soil characteristics along a toposequence of ele subwatershed in southern Ethiopia. Catena 115 (2014):47–54

Dobrovolskii GV (2001) Soil science at the turn of the century: results and challenges. Eurasian Soil Sci 34:115–119

Dregne HE, Chou N-T (1992) Global desertification dimensions and costs. In: Degradation and restoration of arid lands. Texas Tech University, Lubbock

EIP, Environmental Impact Assessment (2002 Proclamation No 299/2002, Federal Negarit Gazette, 9th year no 11, Addis Ababa

Elias E (2016) Soils of the Ethiopian highlands: geomorphology and properties. CASCAPE Project, ALTERA, Wageningen University and Research Center (wgeningen UR). The Netherlands, 385 pp

Erkossa T, Leakemariam F, Abera W, Tamene L (2022) Evolution of soil fertility research and development in Ethiopia: From reconnaissance to data-mining approaches. Exp Agric vol 58(e4):1–12. https://doi.org/10.1017/S0014479721000235

EthioSIS, Ethiopian Soil Information System (2016) Soil fertility status and fertilizer recommendations Atlas of Ethiopia, Addis Ababa

Evtuhov C (2006) The roots of Dokuchaev's scientific contributions: cadastral soil mapping and agro-environmental issues. In: Warkentin BP (ed) Footprints in the soil. People and ideas in soil history. Elsevier, Amsterdam, pp 125–148

FAO, Food and Agriculture Organization (1986) The Ethiopian Highlands Reclamation Study (EHRS). Volume 1 main report. Food and Agriculture Organization, Rome

FAO, Food and Agriculture Organization (2015) Revised world soil charter, Food and Agriculture Organization, Rome, Italy

Feller C, Blanchart E, Yaalon DH (2006) Some major scientists (Palissy, Buffon, Thaer, Darwin and Muller) have described soil profiles and developed soil survey techniques before 1883. In: Warkentin BP (ed) Footprints in the soil. People and ideas in soil history. Elsevier, Amsterdam, pp 85–105

Field DJ, Koppi AJ, Jarrett L, McBratney A (2013) Engaging employers, graduates and students to inform the future curriculum needs of soil science. In: Proceedings of the Australian conference on science and mathematics education. Australian National University, September 19–21, pp 130–135

Field DJ, Yates D, Koppi AJ, McBratney AB, Jarrett L (2017) Framing a modern context of soil science learning and teaching. Geoderma 289:117–123

Genene T, Abiy GM (2014) Review on overall status of soil and water conservation system and its constraints in different agro ecology of southern Ethiopia. J Natl Sci Res 4(7):59–69

Girma T (2001) Land degradation: a challenge to Ethiopia. International Livestock Research Institute, Addis Ababa, pp 815–823

Gong Z, Zhang X, Chen J, Zhang G (2003) Origin and development of soil science in ancient China. Geoderma 115:3–13

Habtamu AA (2013) Impact of salinity on tolerance, vigor, and seedling relative water content of haricot bean (Phaseolus vulgaris L.) cultivars. J Plant Sci 1(3):22–27

Haregeweyn N, Nyssen A, Poesen J, Tsubo J, Tsegaye M, Meshesha D, Schüttm B, Adgo E, Tegegne F (2015) Soil erosion and conservation in Ethiopia. Prog Phys Geogr Earth Environ 39(6):750–774

Hartemink AE (2015) 90 years IUSS and global soil science. Soil Sci Plant Nutr 61(4):579–586. https://doi.org/10.1080/00380768.2015.1055699

Hasegawa S, Warkentin BP (2006) The changing understanding of physical properties of soils: water flow, soil architecture. In: Warkentin BP (ed) Footprints in the soil. People and ideas in soil history. Elsevier, Amsterdam, pp 339–365

Hendricks SB, Fry WH (1930) The results of X-ray and microscopical analyses. Soil Sci 29:457–478

Kaizzi KC, Mohammed MB, Nouri M (2017) Fertilizer use optimization: principles and approach. In: Wortmann CS, Sones K (eds) Fertilizer use optimization in sub-saharan Africa. CABI, Nairobi, pp 9–19

Kefyalew A, Kibebew K (2016) Evaluation of the effect of salt affected soil on selected hydraulic properties of soils in Meki Ogolcha area in east showa zone of oromia region, Ethiopia. J Bio Agric Healthc 6(21)

Koch A, McBratney AB, Adams M, Field DJ, Hill R, Lal R, Abbott L, Angers D, Baldock J et al (2013) Soil security: solving the global soil crisis. Glob Policy J 4:434–441

Krupenikov IP (1993) History OF SOIL SCIENCE. A.A Balkema, Rotterdam

McBratney A, Field DJ, Koch A (2014) The dimensions of soil security. Geoderma 213:203–213

McNeill JR, Winiwarter V (2004) Breaking the sod: humankind, history and soil. Science 304:1627–1629

Mesfin A (2007) Nature and management of acid soils in Ethiopia. Haramaya University, Haramaya, Ethiopia, 82 p

Meron N (2007) Characterization of salt affected soils in the central Rift Valley and assessing salt tolerance of different plants: a case study at the southwestern shore of Lake Ziway. MSc Thesis, Addis Ababa University, Ethiopia. 75p

Mnalku A, Demissie N, Assefa F (2020) Guideline for soil biology data collection in Ethiopia: national standard. Ethiopian Institute of Agricultural Research (EIAR). Addis Ababa, 29 p

MoARD, Ministry of Agriculture and Rural Development (2005) Agricultural sector policy and investment framework (PIF). In: Handbook of federal countries

MoANR, Ministry of Agriculture and Natural Resource (2015) Assessment of fertilizer consumption and use by crop in Ethiopia. IFRC, Addis Ababa

Mol G, Keesstra S (2012) Soil science in a changing world: editorial overview. Curr Opin Environ Sustain 4:473–477

Mulugeta D, Sheleme B (2010) Characterization and classification of soils along the toposequence of kindo koye watershed in southern Ethiopia. East Afr J Sci 4(2):65–77

Pachepsky Y, Rawls WJ (eds) (2005) Developments of pedotransfer functions in soil hydrology. developments in soil science, vol 30. Elsevier, Amsterdam

Pretty J, Bharucha ZP (2014) Sustainable intensification in agricultural systems. Ann Bot 114:1571–1596. https://doi.org/10.1093/aob/mcu205

Rossiter DG, Hewitt AE, Dominati EJ (2018) Pedometric valuation of the soil resource. In: Minasny B, McBratney AB, Stockmann U (eds) Pedometrics. Springer, Cham, pp 521–546

Sandor JA, WinklerPrins AMGA, Barrera-Bassols N, Zinck JA (2006) In: Warkentin BP (ed) Footprints in the soil. People and ideas in soil history. Elsevier, Amsterdam, pp 43–84

Tamene L, Amede T, Kihara J, Tibebe D, Schulz S (2017) A review of soil fertility management and crop response to fertilizer application in Ethiopia: towards development of site- and context-specifc fertilizer recommendation. CIAT Publication No 443. International Center for Tropical Agriculture (CIAT), Addis Ababa, Ethiopia, vol 443, p 86. http://hdl.handle.net/10568/82996

Tamene L, Abera W, Demissie B, Desta D, Woldearegay K, Mekonnen K (2022) J Soil Water Conserv March 77(2):144–157, https://doi.org/10.2489/jswc.2022.00002

Vohland M, Ludwig M, Thiele-Bruhn S, Ludwig B (2017) Quantification of soil properties with hyperspectral data: selecting spectral variables with different methods to improve accuracies and analyze prediction mechanisms. Remote Sens 9(11):1103. https://doi.org/10.3390/rs9111103

Webster R, Heuvelink GBM (2006) The Kalman filter for the pedologist's tool kit. Europ JSoil Sci 57:758–773

Wadoux A, Odeh IOA, Mcbratney AB (2022) Overview of pedometrics. In: Reference module in earth systems and environmental sciences. https://doi.org/10.1016/B978-0-12-822974-3.00001-X

WLE: Water, Land and Ecosystems (2017) CGIAR research program on water, land and ecosystems. Healthy soils for productive and resilient agricultural landscapes. (WLE Towards Sustainable Intensification: Insights and Solutions Brief 2) Colombo, Sri Lanka: International Water Management Institute (IWMI), 12 p. https://doi.org/10.5337/2017.211

Wogi L, Dechassa N, Haileselassie B, Mekuria F, Abebe A, Tamene L (2021) A guide to standardized methods of analysis for soil, water, plant, and fertilizer resources for data documentation and sharing in Ethiopia. CIAT Publication No 519. International Center for Tropical Agriculture (CIAT). Addis Ababa, Ethiopia, 41 p

Wondimu G, Daba A, Qureshi A (2020) Effects of salinity on producers' livelihoods and socio-economic conditions: the Case of Afar Region, Northeastern Ethiopia. J Sustain Agric Sci 46: 35–46

Yaalon DH (1997) History of soil science in context: international perspective. In: YaalonDH, Berkowicz S (eds) History of soil science-international perspectives: advances in geoecology, vol 29. Catena Verlag, Reiskirchen, pp 1–13

Yaalon DH, Berkowicz S (eds) (1997) History of soil science-international perspectives. Catena Verlag, Reiskirchen

Zenebe (2007) Household fuel consumption and resource use in rural-urban Ethiopia. http://www.efdinitiative.org/publications/household-fuel-consumption-and-resource-use-rural-urban-ethiopia

Zelleke G, Agegnehu G, Abera D, Rashid S (2010) Fertilizer and soil fertility potential in Ethiopia: constraints and opportunities for enhancing the system. https://doi.org/10.13140/RG.2.1.3632.5846

Appendix

See Appendix Tables A.1, A.2, A.3, A.4, A.5, A.6, A.7, A.8, A.9, A.10, A.11, A.12, A.13, A.14, A.15, A.16, A.17, A.18, A.19, A.20, A.21, A.22, A.23, A.24, A.25, A.26, A.27, A.28, A.29, A.30, A.31, A.32, A.33, A.34, A.35, A.36, A.37, A.38, A.39, A.40, A.41, A.42, A.43, A.44, A.45, A.46, A.47, A.48, A.49, A.50, A.51, A.52, A.53, A.54, A.55, A.56, A.57, A.58, A.59, A.60, A.61 and A.62.

Appendix Table A.1 Soil resource inventories by pioneer soil survey initiatives

Sn. no.	Soil resource inventories	Publication year	Scale	Area (km^2)
1	Geomorphology and Soils of Ethiopia	1984	1:1,000,000	11,230
2	Provisional Soil Association Map of Ethiopia	1984	1:2,000,000	11,230
3	Physiography and Soils of Haykoch and Butajira; Yerer and Kereyu Awrajas	1989	1:250,000	11,823
4	Soils of Lower Wabi Shebelle and Fafen River	1973	1:50,000	382
5	Soils of Tigray, North Mekelle	1976	1:50,000	3400
6	Soils of W&E Shores of Lake Abaya and Chamo	1977	1:100,000	8800
7	Soils of Kobo-Alamata Agri. Dev. Program	1976	1:50,000	800
8	Soils of Gambella Project Feasibility Report	1973	1:60,000	665
9	Soils of Southern Gamu-Gofa Project	1979	1:60,000	430
10	Soils of Weito Irrigation Project Feasibility Study	1982	1:50,000	60
11	Sils of Dabus Irrigation Project Feasibility Study	1982	1:50,000	239
12	Sils of Meki & Gelana Areas Land Eval. Survey	1982	1:50,000	329
13	Soils of Borkena Area	1985	1:50,000	3000
14	Soils of Bichen Area	1986	1:50,000	3655
15	Soils of Hosaina Area	1987	1:50,000	2296
16	Middle and Lower Awash Soil Survey	1975	1:250,000	196
17	Southern Rift Valley Soil Survey	1975	1:100,000	550
18	Tigrai Rural Development Study Soil Survey	1976	1:1,000,000	200
19	The Humera Report: Soil Survey	1973	1:50,000	100

Source Ali et al. (2020)

S. Beyene et al. (eds.), *The Soils of Ethiopia*, World Soils Book Series,
https://doi.org/10.1007/978-3-031-17012-6

Appendix Table A.2 Soil resource inventories of the major river basins of Ethiopia (Ali et al. 2020)

Sn. no.	Soil resource inventories	Publication year	Scale	Area (km^2)
1	Soils of Wabi-Shebelle River Basin	2004	1:250,000	202,220
2	Soils of Abay River Basin	1998	1:250,000	198,508
3	Soils of Genale Dawa	2001	1:250,000	171,000
4	Soils of Tekeze River Basin	1998	1:250,000	87,773
5	Soils of Omo-Gibe River Basin	1995	1:250,000	77,205
6	Soils of Baro-Akobo River Basin	1995	1:250,000	75,718
7	Soils of Central Rift Valley Lakes Basin	2008	1:250,000	52,000
8	Soils of Mereb River Basin	1998	1:250,000	23,455

Appendix Table A.3 Soil resource inventories of sub-basins and development corridors of Ethiopia (Ali et al. 2020)

Sr. no.	Soil resource inventories	Publication year	Scale	Area (km^2)
1	Soil Survey of GenaleDawa basin in Oromia National Regional State	2017	1:50,000	131,730
2	Soil Survey of OmoGebe basin in Oromia National Regional State	2017	1:50,000	20,181
3	Soil Survey of Baro basin in Oromia National Regional State	2017	1:50,000	24,957
4	Soil Survey of Awash basin in Oromia National Regional State	2018	1:50,000	25,986
5	Soil Survey of Abay basin in Oromia National Regional State	2018	1:50,000	64,206
6	Soil Survey of WabiShebele basin in Oromia National Regional State	2018	1:50,000	76,242
7	Dhihidhessa Sub-Basin Soil Survey/Dhihidhessa-Dabus Integrated Land Use Planning Study Project	2014	1:50,000	36,182
8	Dabus Sub-Basin Soil Survey/Dhihidhessa-Dabus Integrated Land Use Planning Study Project	2014	1:50,000	8837
9	Soil Survey of Oromia Special Zone/Oromia Special Zone Surrounding Finfine Land Use Plan Project	2014	1:50,000	4978
10	Soil Survey of Upper Wash Sub-Basin/Upper Awash Sub-Basin Integrated Land Use Planning Study Project	2014	1:50,000	19,373
11	Soil Survey of Sure_Lege Wata Sub-Basin/Borena Integrated Land Use Plan Project	2010	1:50,000	22,364
12	Soils of Rift Valley Lakes Sub-Basin/Borena Integrated Land Use Plan Project	2010	1:50,000	9767
13	Soil Resource Assessment of Lower Awash Sub-Basin	2012	1:50,000	23,000
14	Soils Resource Assessment of Terru Sub-Basin	2012	1:50,000	12,159
15	Soils of Dabus Irrigation Project	2013	1:10,000	208
16	Soils of Arjo Dhdesa	2007	1:50,000	368
17	Soils of Rib Irrigation Project	2008	1:10,000	209
18	Soils of Upper Belese Irrigation Project	2011	1:10,000	1,376
19	Soils of Megech (Robit) Irrigation Project	2010	1:25,000	96
20	Soils of Negeso Irrigation Project	2010	1:25,000	224
21	Soils of Amhara National Regional State	2004	1:250,000	157,638
22	Soils of Gambela National Regional State	2003	1:250,000	75,910
23	Soil Survey Final Report of Eastern Amhara Development corridor/East Amhara Development Corridor Land Use Planning and Environmental Study project	2011	1:50,000	18,773
24	Soil Resource Survey of North Western Amhara Development Corridor/Northwest Amhara Development Corridor Integrated Land Use Planning Study Project	2011	1:50,000	30,866
25	Soil Survey of Tekeze Sub-Basin/Tekeze Sub-Basin Land Use Planning and Environmental Study Project	2015	1:50,000	29,095
26	Soil Survey of Tana Sub-Basin/Tana Sub-Basin Land Use Planning and Environmental Study Project	2014	1:50,000	15,897
27	Wabishebele sub-basin Arsi zone soil survey and land suitability evaluation	2017	1:50,000	20,656
28	Soil Survey and Land Suitability Evaluation of Kesem-Bolahamo Sugar Development Project	2013	1:10,000	150
29	Soil Survey and Land Suitability Evaluation of Kuraz Sugar Development Project	2011–2015	1:10,000	2530
30	Soil Survey and Land Suitability Evaluation of Old Kuraz (Omo Ratte) Irrigation Project	2012	1:10,000	174

(continued)

Appendix Table A.3 (continued)

Sr. no.	Soil resource inventories	Publication year	Scale	Area (km^2)
31	Soil Survey and Land Suitability Evaluation of Lower Awash Flood Protection Multi-Purpose Dam Project	2014	1:10,000	120
32	Soil Survey and Land Suitability Evaluation of Lower Awash Multi-Purpose Project/*Alternative 2 command area/*	2016	1:10,000	150
33	Soil Survey and Land Suitability Evaluation of Rate Irrigation Development Project	2010	1:10,000	135
34	Soil Survey and Land Suitability Evaluation of Welenchiti Bofa Irrigation Design Expansion Project (Previous Study) Semi-Detail Soils Survey	2007	1:10,000	226
35	Soil Survey and Land Suitability Evaluation of Kessem-Kebena Dam & Irrigation Project	2005	1:10,000	218
36	Soil Survey and Land Suitability Evaluation of Welmel Irrigation Development Project	2009	1:10,000	124
37	Soil Survey and Land Suitability Evaluation of Welkayit Irrigation Project	2008	1:10,000	451
38	Soil Survey and Land Suitability Evaluation of Gumera Irrigation Project	2007	1:10,000	140
39	Soil Survey and Land Suitability Evaluation of Humera Irrigation Project	2006–2009/10	1:10,000	961
40	Soil Survey and Land Suitability Evaluation of Jema Irrigation Project	2009	1:10,000	116
41	Soil Survey and Land Suitability Evaluation of Raya Valley Pressurized Irrigation Project	2008	1:10,000	225
42	Soil Survey and Land Suitability Evaluation of Megech Irrigation Project	2009	1:10,000	175
43	Soil Survey and Land Suitability Evaluation of Kelafo/Wabishebele Multi-purpose (WS-18) Project	2010	1:10,000	412
44	Soils of Gilgel Abay Irrigation Project	2007–2009	1:10,000	163
45	Soil Survey and Land Suitability Evaluation of Gidabo Irrigation Development Project	2007	1:10,000	286
46	Soil Survey and Land Suitability Evaluation of Bilate Irrigation Development Project	2006/07	1:10,000	211
47	Soil Survey and Land Suitability Evaluation of Arjo Dedessa Irrigation Project	2007	1:10,000	267
48	Soil Survey and Land Suitability Evaluation of Ada'a Plains Ground Water Assessment Irrigation Project	2009	1:10,000	123
49	Soil Survey and Land Suitability Evaluation of Becho Plains Ground Water Assessment Irrigation Project	2009	1:10,000	112
50	Soil Survey and Land Suitability Evaluation of Tendaho Sugar Development Dam and Irrigation Project	2006	1:10,000	600
51	Soil Survey of Gilo Sub-Basin Land Use Planning Study Project	2012	1:50,000	6520
52	Soil Survey and Land Suitability of Erer-Dembel Sub-Basin	2011	1:50,000	29,136
53	Fentale Irrigation-based development project, feasibility study, soil survey and land evaluation report	2007	1:15,000	170
54	Soils of Sheneka irrigation project	1983	1:50,000	180
55	Soils of Weib Irrigation	2003	1:10,000	170
56	Shenen Dhungata sub-basin soil survey & Land Suitability Evaluation	2010	1:50,000	6704
57	Ramis sub-basin soil survey and land suitability evaluation	2010	1:50,000	12,388
58	Erer Mojo Gobale sub-basin soil survey and land suitability evaluation	2010	1:50,000	11,805
59	Dawa Dembel sub-basin soil survey and land suitability evaluation	2010	1:50,000	3557
60	Dakata sub-basin soil survey and land suitability evaluation	2010	1:50,000	1889
61	Fafen Dinti sub-basin soil survey and land suitability evaluation	2010	1:50,000	1712
62	Wabishebele sub-basin Bale zone soil survey and land suitability evaluation	2015	1:50,000	22,550

Appendix Table A.4 Digital soil property mapping missions in Ethiopia (Ali et al. 2020)

Sn. no.	Digital soil fertility maps	Publication year	Resolution (m)	Area (km^2)
1	Soil Fertility Status and Fertilizer Recommendation Atlas for Tigray Regional State	2014	250	Whole region
2	Soil Fertility Status and Fertilizer Recommendation Atlas for Amhara Regional State	2016	250	Whole region
3	Soil Fertility Status and Fertilizer Recommendation Atlas for Oromia Regional State	2016	250	Whole region
4	Soil Fertility Status and Fertilizer Recommendation Atlas for SNNP Regional State	2016	250	Whole region
5	Soil Fertility Status and Fertilizer Recommendation Atlas for DireDawa Administration	2017	250	Whole region
6	Soil Fertility Status and Fertilizer Recommendation Atlas for Harar Administration	2017	250	Whole region
7	Soil Fertility Status and Fertilizer Recommendation Atlas for Benshangul Regional State	2018	250	Whole region
8	Soil Fertility Status and Fertilizer Recommendation Atlas for Somali Regional State	2019	250	Whole region
9	Soil Fertility Status and Fertilizer Recommendation Atlas for Afar Regional State	2019	250	Whole region
10	Soil Fertility Status and Fertilizer Recommendation Atlas for Gambela Regional State	2018	250	Whole region

Appendix Table A.5 Soil-Geology relationships in Ethiopia

Parent material	Soil	References
Basalt rock	Vertisols, Leptosols, Phaeozems	Mesfin (1998)
Alluvial deposits	Vertisols	Mitiku (1987), Mesfin (1998)
Colluvial deposits and Quaternary volcanic materials	Cambisols, Regosols	FAO/UNDP (1984a, b)
Granites	Nitisols, Lixisols, Acrisols, Alisols	FAO/UNDP (1984a, b)
Pyroclastic and volcanic ash materials of the high mountains and rift valley zones	Andosols	Frei (1978), Bono and Sieler (1984), Weigel (1988), Belay (1995), Mohammed and Belay (2008), Venema and Paris (1987)
Salt flats, gypsum, calcium carbonate rich and sandy textured soils	Calcisols, Gypisols, Solonchaks and other arid/semiarid soils	FAO/UNDP (1984a, b)

Appendix Table A.6 Major soil types of Ethiopia

Soil type	Percent coverage	Soil type	Percent coverage	Soil type	Percent coverage
Leptosols	30	Alisols	3	Chernozems	0.12
Nitisols	12	Solonchaks	1.72	Gleysols	0.08
Vertisols	11	Acrisols	1.30	Solonetz	0.02
Cambisols	9	Regosols	1.13	Ferralsols	0.01
Calcisols	9	Andosols	0.62	Planosols	0.01
Luvisols	8	Arenosols	0.57	–	–
Gypsisols	8	Phaeozems	0.41	–	–
Fluvisols	4	Lixisols	0.18	–	–

Appendix Table A.7 Some characteristics of Eastern Ethiopian Leptosols (Mohammed et al. 2005)

Land use	Depth (cm)	pH (H_2O)	SOM (%)	N (%)	Total exchangeable base	Available P (Olsen, ppm)	CEC (Cmol c/kg soil)	Base saturation	Soil unit
Uncultivated	0–9	7.04	12.11	0.56	42	8	53	92	Lithic Leptosols
Uncultivated	0–10	6.70	5.59	0.23	40	8	49	83	Lithic Leptosols
Cultivated	0–8	6.29	2.07	0.12	36	4	55	67	Lithic Leptosols

Appendix Table A.8 Selected characteristics of Leptosols of central, northeastern, and northwestern highlands of Ethiopia

Soil group/unit, altitude, slope gradient (%)	Dystric Leptosols, 3020 masl, 25% slop		Umbric Leptosols, 3160 masl, 39% slope,		Haplic Leptosols–Gerado, 2413 masl, 33% slope		Mollic Leptosols Gerado), 2300 masl, 45% slope		Lithic Leptosols, 2400 masl, 50% slope
	Shimelis et al. (2007)				Asmamaw and Mohammed (2012)				
Horizons	Ap	C	Ah	R	Ah	R	Ah	R	Ah
Depth/thickness (cm)	0–17	>17	0–20	>20	0–25		0–20		0–9
Clay (%)	28	–	56	–	26		25		31
Silt (%)	40	–	36	–	35		40		55
Sand (%)	32	–	8	–	40		35		14
SOM (%)	5.80		19.80		5.01		4.96		12.10
Total N (%)	0.30		0.81		0.28		0.29		0.60
Avail P (ppm)	8.2		4.1	–	2		1		8
pH (H_2O)1:2.5	5.1		5.6		6.5		6.8		7.1
CEC (cmol$_c$ kg^{-1} soil)	30		24		52		52		52
Exch. Ca (cmol$_c$ kg^{-1} soil)	5.8		4		42		26		29
Exch. Mg (cmol$_c$ kg^{-1} soil)	1.4		3		12		5		17
Exch. K (cmol$_c$ kg^{-1} soil)	0.3		0.14		0.4		1.1		1.2
Exch. Na (cmol$_c$ kg^{-1} soil)	0.34		0.12		0.2		0.2		Trace
PBS	26		30		100		63		92

Appendix Table A.9 Some selected characteristics of Leptosolsof central, northeastern, and northwestern Ethiopian highlands

Soil group/unit, altitude, slope gradient (%)	Umbric Leptosols, 3160 masl, 39% (Shimelis et al. 2007)	Haplic Leptosols, 2413 masl, 33% slope	Mollic Leptosols, 2300 masl, 45% slope	Lithic Leptosols, 2400 masl, 50% slope	Eutric Leptosols (Engidawork 2002)	Lithic Leptosols, 2400 masl, 30% slope (Mohammed et al. 2005)	Eutric Leptosols, 2400 masl, 19% ay (Belay 1997)
		Asmamaw and Mohammed (2012)					
Horizons	Ah	Ah	Ah	Ah	Ap	Ah	Ah
Depth (cm)	0–20	0–25	0–20	0–9	0–15	0–24	0–15
Clay (%)	56	26	25	31	35	31	23
Silt (%)	36	35	40	55	36	55	30
Sand (%)	8	40	35	14	29	14	47
SOM (%)	19.8	5.0	5.0	12.1	3.0	12.0	4.0
Total N (%)	0.81	0.28	0.29	0.60	0.20	0.60	0.30
Avail P (ppm)	4.1	2	1	8	4	8	131
pH (H_2O)1:2.5	5.6	6.5	6.8	7.1	6.6	7.0	5.9
CEC (cmol$_c$ kg^{-1} soil)	24	52	52	52	55	52	157
Exch. Ca (cmol$_c$ kg^{-1} soil)	4	42	26	29	24	29	28

(continued)

Appendix Table A.9 (continued)

Soil group/unit, altitude, slope gradient (%)	Umbric Leptosols, 3160 masl, 39% (Shimelis et al. 2007)	Haplic Leptosols, 2413 masl, 33% slope	Mollic Leptosols, 2300 masl, 45% slope	Lithic Leptosols, 2400 masl, 50% slope	Eutric Leptosols (Engidawork 2002)	Lithic Leptosols, 2400 masl, 30% slope (Mohammed et al. 2005)	Eutric Leptosols, 2400 masl, 19% ay (Belay 1997)
		Asmamaw and Mohammed (2012)					
Horizons	Ah	Ah	Ah	Ah	Ap	Ah	Ah
Exch. Mg (cmol$_c$ kg^{-1} soil)	3	12	5	17	6	17	7
Exch. K (cmol$_c$ kg^{-1} soil)	0.14	0.4	1.1	1.2	0.6	1	2
Exch. Na (cmol$_c$ kg^{-1} soil)	0.12	0.20	0.20	Trace	0.60	Trace	1.0
PBS	30	100	63	92	56	92	88

Appendix Table A.10 Some selected characteristics of Nitisolsof south central highlands of Ethiopia

Soil group/unit, altitude, slope gradient (%)	Dystric Nitisols, 2960 masl, 10% slope (Shimelis et al. 2007)				Dystric Nitisols, 3040 masl, 37% slope (Shimelis et al. 2007)			
Horizons	Ap	Bt1	Bt2	Bt3	Ah	Bt1	Bt2	Bt3
Depth/thickness (cm)	0–20	20–60	60–135	135–200	0–25	25–70	70–110	110–200
Clay (%)	32	42	42	42	22	44	54	60
Silt (%)	50	42	40	32	42	36	36	28
Sand (%)	18	16	16	26	36	20	10	12
SOM (%)	3.8	2.9	3	1.7	7.5	1.4	1.1	0.5
Total N (%)	0.23	0.22	0.19	0.12	0.35	0.10	0.08	0.05
Avail P (ppm)	3.26	1.18	2.52	4.80	2.34	2.46	2.82	0.74
pH (H_2O)1:2.5	5.8	5.7	5.4	5.6	5.5	6.1	6.3	6.2
CEC (cmol$_c$ kg^{-1} soil)	34	24	26	23	31	51	32	23
Exch. Ca (cmol$_c$ kg^{-1} soil)	7	4	5	6	7	10	8	6
Exch. Mg (cmol$_c$ kg^{-1} soil)	2	1	1	2	2	3	7	3
Exch. K (cmol$_c$ kg^{-1} soil)	0.2	0.1	0.1	0.1	0.2	0.2	0.1	0.1
Exch. Na (cmol$_c$ kg^{-1} soil)	0.1	0.01	0.20	0.10	0.10	0.02	0.3	0.10
PBS	38	23	26	33	29	25	48	40

Appendix Table A.11 Some selected characteristics of Nitisols of northwestern highlands of Ethiopia

Soil group/unit, altitude, slope gradient (%)	Nitisols. 2960 masl, 10% slope (Getachew and Heluf 2007)				Nitisols. 2960 masl, 10% slope (Getachew and Heluf 2007)			
Horizons	Ap	Bt1	Bt2	Bt3	Ap	Bt1	Bt2	Bt3
Depth/thickness (cm)	0–25	25–50	50–100	100–200+	0–20	20–50	50–130	130–200+
Clay (%)	40	72	76	78	44	68	77	76
Silt (%)	57	24	20	19	54	30	20	22
Sand (%)	3	4	4	3	2	2	3	2
SOM (%)	4.5	2.6	1.7	1.0	4.8	3.3	2.2	1.9
Total N (%)	0.19	0.1	0.07	0.05	0.22	0.09	0.07	0.10
Avail P (ppm)	3	2	2	2	2	2	0.2	0.4
pH (H_2O)1:2.5	5.9	5.6	5.7	5.8	5.6	5.5	4.9	5.4
CEC (cmol$_c$ kg^{-1} soil)	34	30	28	23	33	27	20	26
Exch. Ca (cmol$_c$ kg^{-1} soil)	10	6	8	6	11	7	0.8	5
Exch. Mg (cmol$_c$ kg^{-1} soil)	4	2	3	3	3	2	1	3
Exch. K (cmol$_c$ kg^{-1} soil)	3	2	0.7	0.7	1.5	0.8	1.3	2.1
Exch. Na (cmol$_c$ kg^{-1} soil)	Trace	Trace	Trace	Trace	Trace	Trace	Trace	Trace
PBS	50	37	40	45	45	37	17	38

Appendix Table A.12 Some selected characteristics of Nitisols of southeastern highlands of Ethiopia

Soil group/unit, altitude, slope gradient (%)	Epidystric Nitisols, 2740 masl, 23% slope (Mohammed and Solomon 2010)					Eutric Nitisols, 2820 masl, 11% slope (Mohammed and Solomon 2010)			
Horizons	Ap	Bt1	Bt2	Bt3	Bt4	Ap	Bt1	Bt2	Bt3
Depth/thickness (cm)	0–18	18–66	66–90	90–135	135–200	0–20	20–76	76–142	142–200
Clay (%)	37	71	73	77	81	27	57	69	71
Silt (%)	34	12	20	20	14	44	32	20	18
Sand (%)	29	17	7	3	5	29	11	11	11
SOM (%)	4.8	4.2	3.5	2.9	2.2	5.3	4.5	2.6	1.7
Total N (%)	0.24	0.15	0.14	0.13	0.13	0.31	0.23	0.13	0.1
Avail P (ppm)	2	1	2	3	4	5	2	6	4
pH (H_2O)1:2.5	5.5	5.7	5.8	6.0	6.0	5.1	5.9	6.1	6.0
CEC ($cmol_c$ kg^{-1} soil)	27	24	24	24	21	28	28	24	27
Exch. Ca ($cmol_c$ kg^{-1} soil)	7	8	9	8	5	12	13	5	8
Exch. Mg ($cmol_c$ kg^{-1} soil)	1	2	2	3	2	0.8	2	3	3
Exch. K ($cmol_c$ kg^{-1} soil)	1.1	0.5	0.7	0.8	0.8	1.3	0.8	1.2	0.6
Exch. Na ($cmol_c$ kg^{-1} soil)	0.1	0.1	0.3	0.3	0.2	0.1	0.2	0.2	0.2
PBS	36	44	48	51	38	52	57	40	44

Appendix Table A.13 Selected characteristics of Vertisols of Chercher Mountain, Eastern Ethiopia (Mohammed et al. 2005)

Soil group/unit, altitude, slope gradient	Hypereustric Vertisols, 1810 masl, 1% slope				Chromic Vertisols, 1975 masl, 8% slope			Calcic Vertisols 1865, asl, 6% slope			
Horizons	Ap	A1	A2	AC	Ap	A1	Bw	Ap	A1	A2	Ac
Depth/thickness (cm)	0–15	15–40	40–150	150–200	0–10	10–130	130–200	0–15	15–45	45–90	90–200
Clay (%)	67	67	57	55	59	56	48	71	56	74	73
Silt (%)	30	30	40	42	32	38	45	25	42	24	25
Sand (%)	3	4	2	4	9	6	8	4	2	2	2
SOM (%)	2.53	1.03	1.03	0.10	0.64	2.59	0.38	2.74	1.41	2.47	0.95
Total N (%)	0.14	0.12	0.12	0.05	0.10	0.10	0.03	0.13	0.03	0.01	0.08
Avail P (ppm)	8	12	23	3	13	22	22	3	7	Trace	2
pH (H_2O)1:2.5	7.28	7.43	7.78	8.09	7.32	7.21	8.55	7.21	7.56	7.43	7.97
CEC ($cmol_c$ kg^{-1} soil)	59	58	59	54	43	48	49	57	58	57	58
Exch. Ca ($cmol_c$ kg^{-1} soil)	39	39	52	32	23	31	31	35	48	47	39
Exch. Mg ($cmol_c$ kg^{-1} soil)	5	7	14	10	17	16	8	6	11	10	16
Exch. K ($cmol_c$ kg^{-1} soil)	1.0	3.0	1.0	1.2	0.5	0.5	0.6	1.1	0.7	0.7	0.7
Exch. Na ($cmol_c$ kg^{-1} soil)	0.4	0.3	0.4	0.4	0.3	0.9	1.32	0.4	1.4	0.6	1.0
PBS	78	84	100+	80	95	100+	84	76	100	100	98

Appendix Table A.14 Selected characteristics of Vertisols of southeastern/Arsi highlands of Ethiopia

Soil group/unit, altitude, slope gradient (%)	Grumic-Mesotrophic Vertisols, 2885 masl, 2% slope (Mohammed and Solomn 2010)				Eutric Vertisols, 1% slope (Belay 1996)			Eutric Vertisols, 5% slope (Engidawork 2002)		
Horizons	Ap	A1	A2	A3	Ap	A2	A3	Ap	A2	A3
Depth/thickness (cm)	0–22	14/22–78	78–110	110–137	0–18	18–85	85–140	0–18	18–45	45–200
Clay (%)	37	57	77	77	46	70	66	51	67	69
Silt (%)	20	34	14	14	29	17	19	34	22	20
Sand (%)	43	9	9	9	25	13	15	15	11	11
SOM (%)	11	4	2.8	1.7	5	2	1.7	3	2	2
Total N (%)	0.40	0.13	0.1	0.1	0.3	0.1	0.1	0.2	1.3	1.0
Avail P (ppm)	1	2	1	0.2	5	2	2	6	4	1
pH (H_2O)1:2.5	5.2	5.3	6.1	6.3	8.2	8.6	8.5	6.1	6.4	6.6
CEC (cmol$_c$ kg^{-1} soil)	31	23	40	41	53	57	56	114	88	84
Exch. Ca (cmol$_c$ kg^{-1} soil)	8	9	23	27	43	40	32	29	35	39
Exch. Mg (cmol$_c$ kg^{-1} soil)	2	2	5	5	7	13	14	7	7	7
Exch. K (cmol$_c$ kg^{-1} soil)	0.4	0.3	0.7	0.7	0.8	0.4	0.3	0.5	0.5	0.5
Exch. Na (cmol$_c$ kg^{-1} soil)	0.1	0.3	0.5	0.5	1.2	3.2	9.3	0.4	0.6	1.0
PBS	84	40	51	54	99	100	100	57	69	77

Appendix Table A.15 Selected characteristics of Cambisols of central Shewa and northeastern highlands

Soil group/unit, altitude, slope gradient (%)	Dystric Gleyic Cambisols, 2870 masl, 1% slope (Shimelis et al. 2007)				Gleyic Cambisols 2230 masl, 1% slope (Asmamaw and Mohammed 2012)			Vertic Cambisols, 2230 masl, 4% slope (Asmamaw and Mohammed 2012)			
Horizons	Ah	Btg	Bwgl	Bg	Ah	Bt	Bwkg	Ah	Bt1	Bt2	Bw
Depth/thickness (cm)	0–25	25–85	85–130	130–200	0–10/20	10/20–95	95–200	0–28	28–57	57–95	95–140
Clay (%)	24	40	50	52	53	57	19	46	58	54	24
Silt (%)	40	20	24	28	28	22	52	30	26	30	26
Sand (%)	32	32	26	20	19	21	19	24	16	16	50
SOM (%)	4.7	5.5	1.5	1.1	9.0	2.5	0.6	11.4	4.4	1.8	0.7
Total N (%)	0.31	0.23	0.13	0.1	0.62	0.13	0.04	0.13	0.12	0.04	0.01
Avail P (ppm)	3	0.1	3	3	5	12	Trace	6	0.9	6	15
pH (H_2O)1:2.5	5.2	5.6	5.9	6.2	6.0	7.0	8.0	6.4	7.2	7.2	7.6
CEC (cmol$_c$ kg^{-1} soil)	28	37	21	36	43	48	43	53	57	52	30
Exch. Ca (cmol$_c$ kg^{-1} soil)	5	11	4	14	22	23	26	47	58	53	34
Exch. Mg (cmol$_c$ kg^{-1} soil)	2	5	2	7	9	9	8	13	14	14	9
Exch. K (cmol$_c$ kg^{-1} soil)	0.04	0.09	0.04	0.08	2	2	2	0.3	0.2	0.3	0.3
Exch. Na (cmol$_c$ kg^{-1} soil)	0.2	0.3	0.1	0.2	2	0.3	0.8	0.4	0.7	0.7	0.6
PBS	25	45	31	60	76	73	83	100	100	100	100

Appendix Table A.16 Selected characteristics of Cambisols of northeastern highlands and southern highlands, Ethiopia

Soil group/unit, altitude, slope (%)	Fluvic Cambisols			Vertic Cambisols				Dystric Fluvisols, 1970 masl, 0.5% slope						
	Asmamaw and Mohammed (2012)							Mulugeta and Sheleme (2010)						
Horizons	Ap	Btk	Bwk	Ah	Bt1	Bt2	Bw	A1	A2	A3	Ac1	Ac2	C1	C2
Depth/thickness (cm)	0–20	20–65	65–200	0–28	28–57	57–95	95–140	0–20	20–50	50–70	70–95	95–120	120–150	150–170
Clay (%)	33	41	35	46	56	54	24	45	43	41	43	35	29	31
Silt (%)	34	28	36	30	26	30	26	32	40	28	30	34	20	22
Sand (%)	33	31	29	24	18	16	50	23	17	31	27	31	51	47
SOM (%)	2.2	2.0	1.0	11.4	4.4	1.7	0.7	1.9	1.6	1.1	1.2	1.1	0.8	1.0
Total N (%)	0.20	0.10	0.09	0.73	0.12	0.04	0.01	0.19	0.19	0.17	0.14	0.17	0.11	0.13
Avail P (ppm)	40	69	37	6	1	6	16	2	2	1	2	2	4	4
pH (H_2O)1:2.5	6.7	6.8	7.1	6.4	7.2	7.2	7.6	6.2	5.6	6.2	6.0	5.9	6.3	6.1
CEC ($cmol_c$ kg^{-1} soil)	42	46	45	53	57	52	30	27	26	22	16	15	20	25
Exch. Ca ($cmol_c$ kg^{-1} soil)	24	22	24	47	58	53	33	7	4	7	2	2	1	8
Exch. Mg ($cmol_c$ kg^{-1} soil)	7	8	8	13	14	15	9	2	1	2	0.3	0.3	0.2	4
Exch. K ($cmol_c$ kg^{-1} soil)	1.5	1.3	2.2	0.3	0.2	0.3	0.3	0.1	0.1	0.1	1	1.4	1.9	0.1
Exch. Na ($cmol_c$ kg^{-1} soil)	Trace	2	0.3	0.4	0.7	0.7	0.6	0.24	0.2	0.27	0.98	0.98	0.96	0.38
PBS	75	70	76	100	100	100	100	35	21	41	23	28	22	47

Appendix Table A.17 Some selected characteristics of Cambisols of southeastern highlands of Ethiopia

Soil group/unit, altitude, slope gradient (%)	Hypereutric Cambisols, 2353 masl, 1.5% slope (Mohammed and Solomon 2012)				Endoglyic Cambisols, 2352 masl, 0.5% slope (Mohammed and Solomon 2012)				Mollic Cambisols, 2300 masl, 4% slope (Mohammed et al. 2005)			
Horizons	Ap	B1	Bw	Bw2	Ap	Bt1	Bt2	Bw	Ah	A1	Bw1	Bw2
Depth/thickness (cm)	0–20	20–70	70–100	100–170	0–25	25–50	50–80	80–115	0–25	25–40	40–70	70–160
Clay (%)	36	22	32	44	53	76	81	34	47	36	20	22
Silt (%)	54	61	57	49	34	18	14	59	41	40	40	65
Sand (%)	10	17	11	7	13	6	5	7	12	24	40	13
SOM (%)	3.4	1.6	1.6	0.5	4.5	3.2	1.7	0.8	9.0	3.0	1.0	0.9
Total N (%)	0.2	0.04	0.10	–	0.20	0.10	0.10	0.04	0.4	0.14	0.10	0.10
Avail P (ppm)	17	15	15	14	8	7	5	5	12	11	6	10
pH (H_2O)1:2.5	6.7	7.2	7.8	8.3	6.2	6.7	7.6	7.8	5.8	5.9	5.9	5.6
CEC ($cmol_c\ kg^{-1}$ soil)	58	58	58	58	50	49	54	53	52	51	49	48
Exch. Ca ($cmol_c\ kg^{-1}$ soil)	41	40	43	49	30	31	34	38	29	20	14	19
Exch. Mg ($cmol_c\ kg^{-1}$ soil)	7	11	13	10	9	10	15	16	3	7	18	18
Exch. K ($cmol_c\ kg^{-1}$ soil)	2.6	1.9	1.2	0.9	2.3	2.1	2.2	1.6	0.9	0.3	0.2	0.3
Exch. Na ($cmol_c\ kg^{-1}$ soil)	0.4	0.5	0.9	2	0.3	0.5	0.8	1.3	0.3	0.3	0.2	0.5
PBS	87	92	100	100+	84	88	96	100+	64	53	66	78

Appendix Table A.18 Some Cambisols of northeastern highlands

Soil group/unit, altitude, slope gradient (%)	Vertic Cambisols (Asmamaw and Mohammed 2012)						
Horizons	Ap	Btk	Bwk	Ah	Bt1	Bt2	Bw
Depth/thickness (cm)	0–20	20–65	65–200	0–28	28–57	57–95	95–140
Clay (%)	33	41	35	46	56	54	24
Silt (%)	34	28	36	30	26	30	26
Sand (%)	33	31	29	24	18	16	50
SOM (%)	2.2	2	1	11.4	4.4	1.7	0.7
Total N (%)	0.20	0.10	0.09	0.73	0.12	0.04	0.01
Avail P (ppm)	40	69	37	6	1	6	16
pH (H_2O)1:2.5	6.7	6.8	7.1	6.4	7.2	7.2	7.6
CEC ($cmol_c\ kg^{-1}$ soil)	42	46	45	53	57	52	30
Exch. Ca ($cmol_c\ kg^{-1}$ soil)	24	22	24	47	58	53	33
Exch. Mg ($cmol_c\ kg^{-1}$ soil)	7	8	8	13	14	15	9
Exch. K ($cmol_c\ kg^{-1}$ soil)	1.5	1.3	2.2	0.3	0.2	0.3	0.3
Exch. Na ($cmol_c\ kg^{-1}$ soil)	Trace	2	0.3	0.4	0.7	0.7	0.6
PBS	75	70	76	100	100	100	100

Appendix Table A.19 Some selected characteristics of Luvisols of central Shewa and northern highlands of Ethiopia

Soil group/unit, altitude, slope gradient (%)	Gleyic-Dystric Luvisols, 2910 masl, 6% slope (Shimelis et al. 2007)				Vertic Luvisols (Belay 1996)			Vertic Luvisols (Belay 1996)		
Horizons	Ap	Bt1	Bt2	Btg	Ap	Bt1	Bt2	Ap	Bt1	Bt2
Depth/thickness (cm)	0–20	20–70	70–130	130–200	0–16	16–41	41–79	0–12	12–66	66–91/99
Clay (%)	32	50	50	56	20	50	44	20	42	34
Silt (%)	46	30	30	26	21	19	19	27	19	21
Sand (%)	22	20	30	18	59	31	37	53	39	45
SOM (%)	6.02	1.3	0.8	0.4	1	1.3	1.3	1.3	1.4	0.9
Total N (%)	0.33	0.12	0.06	0.03	0.04	0.05	0.06	0.06	0.04	0.04
Avail P (ppm)	7	6	2	1	3	1	2	5	5	6
pH (H_2O)1:2.5	5.6	5.9	6.1	5.9	8.1	8.8	8.9	7.4	7.6	7.6
CEC (cmol$_c$ kg^{-1} soil)	84	43	51	53	17	32	34	19	34	34
Exch. Ca (cmol$_c$ kg^{-1} soil)	6	5	7	6	11	21	24	14	24	24
Exch. Mg (cmol$_c$ kg^{-1} soil)	2	3	2	5	2	4	5	3	5	5
Exch. K (cmol$_c$ kg^{-1} soil)	0.2	0.2	0.1	0.1	0.5	0.4	0.9	0.5	0.6	0.4
Exch. Na (cmol$_c$ kg^{-1} soil)	0.03	0.03	0.12	0.22	1.20	3.00	5.00	1.10	1.30	1.40
PBS	31	37	46	39	86	91	100	94	90	91

Appendix Table A.20 Some chemical characteristics of the Luvisols, northwestern Ethiopia

Horizon	Depth	pH H_2O	EC ds/m	Na	K	Ca	Mg	Sum	CEC cmol $_{(+)}$/ Kg^{-1}	Bas. Sa %	CEC/Clay	T.N %	O.C %	C/N	References
Profile YP-4				cmol (+)/Kg											Tibebu and Mohammed (2018)
Ap	**0–30**	5.8	0.185	0.09	0.47	15.66	4.35	20.57	26.08	78.86	2.8	0.08	1.11	13.88	
Bt1	30–65	5.9	0.174	0.11	0.28	11.36	3.60	15.35	21.70	70.74	3.73	0.05	0.49	10.36	
Bt2	65–130	6.1	0.123	0.09	0.24	10.71	3.32	14.36	21.90	65.61	3.7	0.02	0.31	14.06	
Bt3	130–200	5.9	0.042	0.11	0.41	16.77	4.35	21.63	22.48	96.23	3.6	0.02	0.23	13.03	

CEC = cation exchange capacity; TN = total nitrogen; EC = electric conductivity; OC = organic carbon; C: N = carbon: nitrogen

Appendix Table A.21 Particle size distribution, pH, organic matter, total nitrogen, and available phosphorus of Luvisols of Central Ethiopia

	Depth (cm)	sand	silt	Clay	Silt/clay	BD (gcm^{-3})	pH (H_2O) 1:2.5	pH (KCl) 1:2.5	OM (%)	OC (%)	Total N (%)	Available P (mg kg^{-1})
OR/BAK/BE/P1	0–20	41.5	15.7	42.9	Clay	1.13	6.36	5.43	5.52	3.2	0.31	60.5
	20–100	34.2	17.8	48.0	Clay	1.16	5.60	4.77	1.72	1.0	0.10	–
	100–150	31.1	18.8	50.1	Clay	1.03	4.73	3.91	1.07	0.6	0.06	–

Source Engdawork (2013)

Appendix Table A.22 Cation exchange capacity exchangeable basic cations, percentage base saturation Luvisols of central highlands

Depth	CEC (cmo$_{l(+)}$ kg^{-1})	Exchangeable (cmol$_c$ kg^{-1})					Sum of cations	BS %	Exchangeable Sodium % (ESP)	Available (%)S
		Na	K	Ca	Mg	Ca/Mg				
0–20	52.45	1.12	1	24.13	9.15	2.60	35.44	67.6	2.14	1.15
20–100	33.46	1.65	2.1	17.47	5.82	3.00	27.02	80.8	4.92	–
100–150	36.17	0.94	1.1	21.63	7.49	2.88	31.14	90.1	2.6	–

Source Engdawork (2013)

Appendix Table A.23 Some selected characteristics of Regosols of northeastern highlands

Soil group/unit, altitude, slope gradient (%)	Haplic Regosols, 2210 masl, 1% slope (Asmamaw and Mohammed 2012)		Eutric Regosols, 9% slope (Belay 1997)			Eutric Regosols, 9% (Belay 1997)			
Horizons	Ap	A2	Ap	CB	Bb	Ap	CB1	CB2	Bb
Depth/thickness (cm)	0–20	20–80	0–16	16–68	68–123	0–15	15–51	51–88	88–157
Clay (%)	31	23	20	14	17	29	36	34	46
Silt (%)	24	52	36	47	24	39	36	36	39
Sand (%)	45	25	44	36	20	32	28	30	15
SOM (%)	1.4	1.9	2.1	1.0	1.6	2.8	1.5	1.1	0.8
Total N (%)	0.10	0.10	0.17	0.07	0.07	0.12	0.11	0.08	0.08
Avail P (ppm)	12	2	28	14	16	33	12	23	28
pH (H_2O)1:2.5	7.5	7.7	6.3	6.7	6.7	6.5	6.5	7.1	6.9
CEC ($cmol_c$ kg^{-1} soil)	54	42	41	44	45	38	44	50	53
Exch. Ca ($cmol_c$ kg^{-1} soil)	31	24	30	35	36	27	31	37	36
Exch. Mg ($cmol_c$ kg^{-1} soil)	12	7	9	10	10	9	10	12	13
Exch. K ($cmol_c$ kg^{-1} soil)	1.5	1.3	0.7	0.2	0.8	0.6	0.3	0.3	0.3
Exch. Na ($cmol_c$ kg^{-1} soil)	0.6	0.9	0.8	0.6	0.7	0.6	0.8	0.9	0.8
PBS	85	78							

Appendix Table A.24 Some morphological and physical characteristics of Ethiopian Fluvisols, Middle Awash Valley

Profile name	Depth (cm)	Color Munsell value		Sand, %	Silt %	Clay %	Class	Bulk density (gm cm^{-3})	Total porosity (%)
		Moist	Dry						
WARC 105/106 Eutric Fluvisol	0–19	10YR 3/2	10YR 5/4	14	56	30	ZCL	1.4	47.17
	19–70	10YR 3/2	10YR 6/4	18.6	56.67	24.67	ZL	1.23	53.58
	70–95	10YR 3/4	10YR 4/4	11	56	33	ZCL	1.2	54.72
	95–115	10YR 3/2	10YR 5/3	14	51	35	ZCL	1.15	56.60
	130–168	10YR 3/2	10YR 3/4	10	28	61	C	1.2	54.72
	168–200	7.5YR 3/1	7.5YR 2/2	8	38	55	C	1.2	54.72
WARC 111/112 Eutric Fluvisol	0–17	10YR 4/2	10YR 4/4	16	54	31	ZCL	1.1	58.49
	17–65	10YR 4/2	10YR 5/4	18	53.3	29.67	ZCL	1.01	61.89
	65–83	10YR 3/2	10YR 4/4	12	50	39	ZCL	1.2	54.72
	83–135	10YR 3/2	10YR 5/4	18	60	23	ZL	1.0	62.26
	135–157	10YR 4/2	10YR 3/4	10	36	55	C	1.0	62.26
	157–200	10YR 3/2	10YR 4/3	20	46	36	ZCL	1.2	54.72
WARC 129/130 Salic Fluvisol	0–26	10YR 4/2	10YR 5/4	16	46	39	ZCL	1.2	54.72
	26–88	7.5YR 3/2	7.5YR 3/3	13.3	51.3	36.3	CL	1.2	54.72
	88–110	10YR 3/2	10YR 3/4	10	38	53	C	1.2	54.72
	110–135	7.5YR 3/2	7.5YR 3/4	8	38	54	C	1.3	50.94
	135–158	7.5YR 3/2	7.5YR 3/3	12	38	50	C	1.2	54.72
	158–200	7.5YR 3/2	7.5YR 3/3	10	38	25	C	1.2	54.72
WARC 213 Eutric Fluvisol	0–20	7.5YR 3/2	7.5YR 3/2	12	38	50	C	1.2	54.72
	20–60	10YR 3/2	10YR 3/4	12	34	54	C	1.35	49.06
	60–100	5YR 3/2	5YR 5/2	18	52	30	CL	1.0	62.26
	100–120	5YR 4/2	5YR 5/2	17	63	20	ZL	1.1	58.49
	120–155	5YR 3/3	5YR 5/2	10	68	22	ZL	1.1	58.49
	155–180	5YR 3/2	5YR 3/2	5	49	46	ZC	1.0	62.26

(continued)

Appendix Table A.24 (continued)

Profile name	Depth (cm)	Color Munsell value		Sand, %	Silt %	Clay %	Class	Bulk density (gm cm^{-3})	Total porosity (%)
		Moist	Dry						
WARC 229/230 Eutric Fluvisol	0–25	5YR 3/2	5YR 2/2	7	47	47	ZC	1.1	58.49
	25–46	7.5YR 3/2	7.5YR 2/3	7	51	51	ZC	1.3	50.94
	46–78	5YR 3/2	5YR 2/2	13	43	43	ZC	1.3	50.94
	78–95	7.5YR 4/2	7.5YR 4/3	9	45	45	ZC	1.3	50.94
	95–160	7.5YR 4/2	7.5YR 6/3	7	51	51	ZC	1.3	50.94
	160–195	7.5YR 4/2	7.5YR 2/2	9	51	51	ZC	1.2	54.72
MKSF F3/3/35 Eutic Fluvisol	0–20	10YR 3/1	10YR 3/1	8	39	53	C	1.4	47.17
	20–65	10YR 3/1	10YR 3/2	9	42	49	ZCL	1.2	56.60
	65–90	10YR 3/1	10YR 3/1	6	35	59	C	1.0	62.26
	90–123	10YR 3/2	10YR 4/2	14	45	41	ZC	1.1	58.49
	123–152	10YR 3/2	10YR 4/3	32	51	17	ZL	1.1	58.49
	152–180	10YR 3/2	10YR 4/3	8	53	39	ZCL	1.1	58.49
	180–225	10YR 3/2	10YR 4/2	8	47	45	ZC	1.1	58.49
MKSF F3/2/22 Salic Fluvisol	0–20	10YR 4/2	10YR 4/2	14	33	54	C	1.4	47.17
	20–33	10YR 4/3	10YR 5/2	6	53	41	ZC	1.4	47.17
	33–49	10YR 4/3	10YR 5/3	26	49	25	L	1.3	50.94
	49–68	10YR 4/3	10YR 5/3	20	51	29	CL	1.3	50.94
	68–113	10YR 4/3	10YR 5/3	52	29	19	ZL	1.2	54.72
	113–166	10YR 4/3	10YR 5/3	54	29	17	ZL	1.2	54.72
	166–200	10YR 4/2	10YR 5/2	84	5	11	LZ	1.2	54.72
MKSF F1/1/1 Eutric Fluvisol	0–16	10YR 3/2	10YR 4/2	8	43	49	ZC	1.3	50.94
	16–44	10YR 3/3	10YR 4/2	15	33	53	C	1.3	50.94
	44–65	10YR 4/2	10YR 5/2	6	47	47	ZC	1.2	54.72
	65–87	10YR 3/2	10YR 4/2	10	41	49	ZC	1.2	54.72
MKSF F1/28/49 Eutric Fluvisol	0–20	10YR 2/1	10YR 3/1	6	33	61	C	1.5	43.40
	20–45	10YR 2/1	10YR 3/1	6	31	63	C	1.5	43.40
	45–70	10YR 2/1	10YR 3/1	8	31	61	C	1.4	47.17
	70–105	10YR 2/1	10YR 3/1	8	27	65	C	1.4	47.17
	105–125	10YR 3/1	10YR 3/2	8	29	63	C	1.3	50.94
	125–151	10YR 3/1	10YR 3/2	4	39	58	C	1.2	54.72
	151–200	10YR 3/2	10YR 4/2	6	43	52	ZC	1.2	54.72
MKSF 2D/8 Eutric Fluvisol	0–19	10YR 3/2	10YR 4/2	22	37	42	C	1.4	47.17
	19–37	10YR 4/3	10YR 5/2	14	45	42	ZC	1.4	47.17
	37–60	10YR 3/2	10YR 4/2	12	53	36	ZCL	1.4	47.17
	60–90	10YR 4/3	10YR 5/3	22	53	26	ZL	1.3	50.94
	90–130	10YR 4/3	10YR 5/3	16	61	24	ZL	1.3	50.94
	130–200	10YR 4/3	10YR 5/3	32	31	38	CL	1.2	54.72

Appendix Table A.25 Some chemical characteristics of Fluvisols, Middle Awash Valley, central Ethiopian rift Valley (Ahmed et al. 2013)

Profile name	Depth (cm)	pH	ECe (dS m^{-1})	OM (%)	TN (%)	P (ppm)	Exchang, bases (meq/100g)				CEC	SAR	ESP (%)
							Na	K	Mg	Ca			
Eutic Fluvisol	20–65	8.4	0.55	3.65	0.12	29	5.5	3.15	4.7	69	65	2.2	12.0
	65–90	8.3	0.5	1.21	0.10	23	5.9	3.1	3.2	83.3	51	3.1	12.0
	90–123	8.3	0.5	1.09	0.09	22	7.4	2.5	4.1	86.4	44	4.0	17.0
	123–152	8.1	0.8	1.07	0.06	21	2.7	1.4	4.1	72.0	31	2.8	8.9
	152–180	8.4	0.7	1.30	0.05	21	5.3	1.0	4.1	97.2	44	4.0	12.0
	180–225	8.5	0.6	0.20	0.01	20	5.6	0.9	4.5	99.0	44	2.3	13.0
MKSF F3/2/22 Salic Fluvisol	0–20	8.1	26.0	4.95	0.18	23	13.0	3.2	3.5	73.5	34	0.2	38.0
	20–33	8	18.0	3.65	0.15	22	13.0	2.0	3.2	94.5	42	0.2	30.0
	33–49	7.9	14.0	3.56	0.07	17	11.0	1.2	3.1	80.1	36	0.3	30.0
	49–68	7.8	19.0	1.00	0.04	20	10.0	0.9	2.8	91.5	37	0.2	28.0
	68–113	7.9	14.0	1.09	0.02	17	7.1	1.0	3.2	38.3	28	0.2	25.0
	113–166	7.8	15.0	0.61	0.01	20	5.8	0.8	4.5	34.2	27	0.3	22.0
	166–200	7.9	3.8	0.31	0.01	18	4.0	0.6	1.4	29.3	18	0.2	22.0
MKSF F1/1/1 EutriFluvisol	0–16	8.7	1.0	6.21	0.30	38	5.2	4.2	2.7	54.9	53	0.2	9.9
	16–44	8.5	0.7	5.23	0.28	28	4.4	2.7	3.6	54.0	46	0.2	9.9
	44–65	8.4	1.1	5.12	0.22	24	4.8	2.1	1.8	54.0	50	0.3	9.6
	65–87	8.4	2.9	4.67	0.16	24	4.4	2.4	1.8	44.5	51	0.2	8.7
MKSF F1/28/49 Eutric Fluvisol	0–20	8.6	0.5	2.09	0.21	37	5.0	2.6	1.4	49.1	57	0.2	8.8
	20–45	8.5	0.5	2.02	0.11	36	4.4	2.3	3.2	48.6	56	0.2	7.9
	45–70	8.4	0.5	1.00	0.09	25	4.3	2	7.2	51.8	53	0.2	8.2
	70–105	8.4	0.6	0.40	0.09	13	4.6	1.9	8.1	51.8	48	0.2	9.7
	105–125	8.2	1.6	0.30	0.05	15	5.2	1.7	7.2	54.0	53	0.3	9.8
	125–151	8	4.6	0.20	0.03	22	5.6	1.6	7.7	51.8	51	0.5	11.0
	151–200	7.9	6.6	0.20	0.01	20	5.8	1.6	4.5	52.2	51	0.6	12.0
MKSF2D/8 Eutric Fluvisol	0–19	8.6	0.8	2.80	0.19	12	4.6	3.2	1.8	52.2	77	0.2	6.0
	19–37	8.6	0.6	1.90	0.18	10	4.2	2.7	2.7	50.9	57	0.1	7.3
	37–60	8.4	0.5	2.07	0.16	15	4.0	2.1	4.5	47.3	45	0.2	8.8
	60–90	8.3	0.5	1.64	0.11	16	3.3	1.3	2.3	47.7	49	0.2	6.8
	90–130	8.2	0.5	1.09	0.10	19	2.9	1.4	4.1	46.4	45	0.2	6.5
	130–200	8.2	0.4	1.10	0.03	20	3.5	1.3	2.3	45	41	0.1	8.6

ECe-electrical conductivity; OM-organic matter; TN-total nitrogen; P-available phosphorus; Na-sodium; Ca-calcium; K-potassium; Mg-magnesium; CEC-cation exchange capacity; SAR-sodium adsorption ratio; ESP-exchangable sodium percentage

Appendix Table A.26 Selected characteristics of Fluvisols of northeastern highlands and southern highlands

Soil group/unit, altitude, slope gradient (%)	Mollic Fluvisol, 2% slope			Mollic Fluvisol, 10% slope			Dystric Fluvisols, 1970 masl, 0.5% slope (Mulugeta and Sheleme 2010)						
	Engidawork (2002)												
Horizons	Ah1	1Bg	2Bg	Ah	1Bg	2Bg	A1	A2	A3	Ac1	Ac2	C1	C2
Depth/thickness (cm)	0–12	12–40	40–160	0–13	13–50	50–100	0–20	20–50	50–70	70–95	95–120	120–150	150–170
Clay (%)	23	41	29	27	19	31	45	43	41	43	35	29	31
Silt (%)	40	36	30	38	24	34	32	40	28	30	34	20	22
Sand (%)	37	23	41	35	57	35	23	17	31	27	31	51	47
SOM (%)	6.7	3.2	5.9	5.6	12.3	1.6	1.89	1.64	1.13	1.24	1.12	0.82	0.95
Total N (%)	0.5	0.5	0.6	0.5	1.2	0.5	0.19	0.19	0.17	0.14	0.17	0.11	0.13
Avail P (ppm)	8.8	5.4	17.6	14	12	11	2	2	1	2	2	4	4
pH (H_2O)1:2.5	5.1	6.2	5.0	5.1	6.2	5.0	6.2	5.6	6.2	6.0	5.9	6.3	6.1
CEC ($cmol_c$ kg^{-1} soil)	67	61	62	62	72	61	27	26	22	16	15	20	25
Exch. Ca ($cmol_c$ kg^{-1} soil)	26	24	21	31	32	28	7	4	7	2	2	1	8
Exch. Mg ($cmol_c$ kg^{-1} soil)	4	5	5	9	4	4	2	1	2	0.3	0.3	0.2	4
Exch. K ($cmol_c$ kg^{-1} soil)	0.2	0.3	0.3	0.5	0.2	0.3	0.1	0.1	0.1	1	1.4	1.9	0.11
Exch. Na ($cmol_c$ kg^{-1} soil)	0.7	0.5	0.6	1.2	0.6	0.5	0.24	0.2	0.2	0.27	0.98	0.98	0.96
PBS	46	52	43	67	52	52	35	21	21	41	23	28	22

Appendix Table A.27 Some selected characteristics of Phaeozems of eastern and northeastern highlands of Ethiopia

Soil group/unit, altitude, slope gradient (%)	Pachic Phaeozems, 2375 masl, 9%, Mohammed et al. 2005)		Luvic Phaeozems, 17% slope (Engidawork 2002)				Haplic Phaeozems, 26% slope (Engidawork 2002)				
Horizons	Ah	A1	Ap	A2	Bt1	Bt2	Ah	A2	Bw	Bb1	Bb2
Depth/thickness (cm)	0–25	25–95	0–15	15–67	67–100	100–145	0–18	18–57	57–90	90–100	1000–200
Clay (%)	30	48	33	35	41	41	47	45	41	45	45
Silt (%)	61	23	42	36	34	34	34	32	28	28	28
Sand (%)	9	29	25	29	25	25	19	23	31	27	27
SOM (%)	4.6	1.6	2.8	1.6	0.9	0.9	2	1.7	2.2	2.4	2.3
Total N (%)	0.2	0.1	0/11	0.09	0.06	0.06	0.14	0.11	0.14	0.15	0.16
Avail P (ppm)	16	11	17	5	4	9	20	25	19	15	5
pH (H_2O)1:2.5	5.4	5.7	6.1	6.4	6.6	6.8	6.5	6.3	6.6	7.5	7.7
CEC ($cmol_c$ kg^{-1} soil)	62	53	60	65	52	47	50	51	52	60	58
Exch. Ca ($cmol_c$ kg^{-1} soil)	45	30	33	38	31	42	32	36	35	35	37
Exch. Mg ($cmol_c$ kg^{-1} soil)	10	18	7	8	6	17	15	17	17	16	18
Exch. K ($cmol_c$ kg^{-1} soil)	0.6	0.6	0.5	0.3	0.3	0.3	0.7	0.6	0.6	0.7	0.7
Exch. Na ($cmol_c$ kg^{-1} soil)	0.8	0.7	0.4	0.4	0.4	0.3	0.3	0.4	0.7	0.5	0.9
PBS	91	92	68	73	73	100	96	100	100	88	97

Appendix Table A.28 Some selected physical and chemical characteristics of Calcisols of lowlands (Ahmed et al. 2013)

Location	Pedon	Horizon	Depth	Particle size distribution			Textural class[a]	pH (H_2O)	EC (dS/m)	OC (%)	TN (%)	C/N ratio	AvP[b] (mg/kg)	$CaCO_3$ (%)
				Sand	Silt	Clay								
T. Umbulo	1	Ap	0–26	48	34	18	L	8.08	0.06	4.1	0.41	10	12.98	1.2
		B	26–69	8	64	28	SiL	8.3	0.08	2.54	0.24	10.6	13.96	1.8
		C	69–116	62	26	12	SL	8.7	0.18	1.71	0.15	11.4	13.26	26
		Ck	116–156	68	20	12	SL	9.4	0.33	1.40	0.12	11.7	14.13	27
		2B	156–176	60	24	16	SL	9.5	0.39	1.29	0.10	12.9	12.59	18
		2Bn	176–209	36	40	24	L	9.7	0.47	1.63	0.13	12.5	14.23	7.6
	2	Ap	0–25	40	38	22	L	7.3	0.03	4.33	0.42	10.3	13.24	1.3
		B1	25–69	40	36	24	L	7.4	0.07	2.40	0.22	10.9	13.97	1.5
		B2	69–115	44	30	26	L	8.3	0.21	2.58	0.23	11.2	12.10	1.7
		Bt	115–131	34	34	32	CL	8.7	0.51	2.54	0.21	12.1	10.51	6.2
		Ck	131–157	50	40	10	L	9.6	0.62	1.25	0.11	11.4	15.65	17.0
		C	157–187	36	54	10	SiL	9.6	0.81	0.17	0.84	10.5	15.09	11

[a]CL = Clay loam; L = Loam; SiL = Silty loam; and SL = Sandy loam
[b]AvP= Available Phosphorus
Source Ahmed et al. (2013)

Appendix Table A.29 Selected physical and chemical characteristics of Calcisols of Central Rift Valley

Location	Pedon	Horizon	Depth particle size distribution				Textural class[a]	pH (H_2O)	EC (dS/m)	OC (%)	TN (%)	C/N ratio	AvP[b] (mg/kg)	$CaCO_3$ (%)
				Sand	Silt	Clay								
Kontela	1	Ak	0–67	25	56	19	SiL	7.94	0.58	3.51	0.45	7.80	13.29	30
		E	67–76	29	64	7	SiL	9.05	0.29	1.00	0.15	6.67	14.22	25
		Bkn	76–100	23	60	17	SiL	9.41	0.72	3.20	0.38	8.42	9.20	28
		C	100–107	29	62	9	Si	9.95	0.24	1.66	0.24	6.92	7.63	22
		2Bh	107–133	29	58	13	SiL	9.25	0.23	4.00	0.47	8.51	3.36	5.3
		2C	133–143	79	12	9	LS	9.80	0.69	1.53	0.21	7.29	1.82	7.4
	2	3C	143–169	25	66	9	SiL	9.12	0.84	3.70	0.41	9.02	9.84	8.1
		A	0–50	32	49	19	L	8.32	0.12	4.46	0.48	9.29	9.29	8.7
		B	50–80	25	56	19	SiL	8.50	0.20	1.96	0.28	7.00	14.02	26
		Bk	80–109	23	55	22	SiL	9.20	0.26	1.63	0.22	7.40	3.25	29
		Ck	109–140	78	12	10	LS	9.70	0.24	0.46	0.08	5.75	1.85	23
		2B	140–151	15	54	31	SiCL	9.55	0.24	0.19	0.02	9.50	3.58	5.1
		2C	151–187	47	42	11	L	9.54	0.36	0.16	0.02	8.00	0.67	5.8
Alage	1	Ap	0–20	66	12	22	SCL	6.90	0.15	1.88	0.21	8.95	12.62	4.6
		Bh	20–48	34	36	30	CL	7.65	0.13	2.43	0.31	7.84	9.94	6.9
		BC	48–87	42	38	20	L	8.22	0.11	1.17	0.17	6.88	9.81	7.4
		Ck	87–137	62	32	6	SL	8.46	0.14	0.71	0.11	6.45	10.06	28
		C_1	137–175	36	56	8	SiL	9.30	0.28	0.37	0.08	4.63	9.81	21
		C_2	175–197	44	48	8	L	9.50	0.41	0.28	0.05	5.60	10.86	26

[a]L = loam, C = clay, CL = clay loam; SiCL = silt clay loam; SCL = sand clay loam; SL = sandy loam; LS = loamy sand
[b]AvP = Available Phosphorus
Source Ahmed et al. (2013)

Appendix Table A.30 Selected characteristics of Calcisols of southeastern Ethiopia

Soil group/unit, altitude, slope gradient (%)	Gleyic Calcisols, 247 masl, 1.5% slope (Mohammed and Solomon 2012)			
Horizons	Ap	Bk1	Bk2	Bck
Depth/thickness (cm)	0–25	25–55/70	55/70–80	80–120/135
Clay (%)	19	19	27	25
Silt (%)	29	44	61	62
Sand (%)	52	37	12	13
SOM (%)	3.8	2.3	1	0.9
Total N (%)	0.3	0.1	0.1	0.1
Avail P (ppm)	15	11	7	10
pH (H_2O)1:2.5	7.9	8.1	8.1	8.3
CEC (cmol$_c$ kg^{-1} soil)	62	59	33	34
Exch. Ca (cmol$_c$ kg^{-1} soil)	44	49	29	29
Exch. Mg (cmol$_c$ kg^{-1} soil)	13	8	5	5
Exch. K (cmol$_c$ kg^{-1} soil)	2.4	0.6	0.3	0.4
Exch. Na (cmol$_c$ kg^{-1} soil)	0.2	0.3	0.4	1
PBS	96	100	100+	100+

Appendix Table A.31 Some characteristics of Alisols, western highlands of Ethiopia

Soil group/unit, altitude, slope gradient (%)	Gleyic Alisols (Asmamaw 2008)				Gleyic Alisols (Asmamaw 2008)			Gleyic Alisols (Asmamaw 2008)				
Horizons	H	Btg1	Btg2	Btg3	H	Bg1	Bg2	Ah	Btg1	Btg2	B3	B4
Depth/thickness (cm)	0–45	45–80	80–120	120–150	0–35	35–100	100–130	0–25	25–75	75–100	100–115	115–160
Clay (%)	19	51	35	33	19	21	29	33	47	41	67	23
Silt (%)	26	6	12	24	18	18	20	20	10	14	33	26
Sand (%)	55	43	53	43	63	63	51	47	43	45	22	47
SOM (%)	15.0	15.0	4.0	2.0	21.7	21.5	8.7	13.0	4.0	1.5	0.1	1.2
Total N (%)	0.74	0.43	0.17	0.07	1.10	0.90	0.30	0.60	0.18	0.06	0.12	0.06
Avail P (ppm)	9	3	1	4	7	4	17	16	16	7	5	20
pH (H_2O)1:2.5	4.2	4.4	4.2	4.8	4.3	4.5	4.3	4.3	4.4	4.1	–	4.1
CEC (cmol$_c$ kg^{-1} soil)	38	32	16	22	48	51	31	36	19	24	23	20
Exch. Ca (cmol$_c$ kg^{-1} soil)	6	5	4	5	8	6	5	4	1	4	5	5
Exch. Mg (cmol$_c$ kg^{-1} soil)	2	2	2	2	3	3	2	2	1	3	3	3
Exch. K (cmol$_c$ kg^{-1} soil)	0.6	0.2	0.1	0.2	0.4	0.2	0.2	0.4	0.2	0.4	0.4	0.3
Exch. Na (cmol$_c$ kg^{-1} soil)	0.5	0.4	0.4	0.4	0.5	0.4	0.5	0.4	0.5	0.4	0.5	0.5
PBS	25	24	41	34	25	19	23	18	16	28	39	43

Appendix Table A.32 Selected physical, bio, and micronutrients analytical data of Alisols, south central Shewa (Engidawork 2013)

Depth	Particle size classes (%)				Soil biology			Available Micronutrients (PPM)			
	Sa	C	Si	Si/C	SOM (%)	TN (%)	C/N	Cu	Zn	Fe	Mn
0–20	29	56	15	0.3	7	0.27	14	0.38	0.22	10.08	4.37
20–48	13	66	21	0.3	7	0.20	16	0.26	0.11	5.17	3.52
48–78/80	9	78	13	0.2	3	0.13	12	0.20	0.04	2.18	2.32
78/80–115	11	84	5	0.1	2	0.11	9	0.14	0.06	0.93	0.36
115+	9	83	8	0.1	2	0.10	9	0.11	0.82		0.07

Appendix Table A.33 Selected chemical analytical data of Alisols, south central Shewa (Engidawork 2013)

Depth	pH	Ava P	Exchangeable cations and CEC ($Cmol_c$ kg^{-1} soil)					CEC/Clay	PBS	Ca/Mg
			Ca	Mg	Na	K	CEC			
0–20	5.36	1.98	5.77	3.59	0.14	2.04	24.64	44	47	1.6
20–48	5.28	2.70	1.69	1.47	0.17	1.16	24.20	37	19	1.1
48–78/80	5.33	2.94	4.68	2.79	0.17	0.71	27.35	35	31	1.7
78/80–115	5.29	3.33	2.13	3.02	0.19	0.70	29.08	35	21	0.7
115+	5.46	2.22	3.26	3.02	0.19	0.67	21.53	26	33	1.1

Appendix Table A.34 Selected physical, biochemical data of South Western Ethiopian Alisols (Asmamaw 2008)

Depth	Particle size classes (%)				Soil biology			Available Micronutrients (PPM)			
	Sa	C	Si	Si/C	SOM (%)	TN (%)	C/N	Cu	Zn	Fe	Mn
0–20	10	90	0	0	9	0.21	24.58	0.31	0.59	9.93	22.28
20–38	12	66	22	0.3	8	0.19	25.41	0.34	0.58	9.74	17.95
38–65	16	62	22	0.4	4	0.18	11.86	0.32	0.51	8.44	17.06
65–105	6	84	10	0.1	2	0.12	11.19	0.15	0.11	1.99	3.72
105+	6	86	8	0.1	1	0.07	8.22	0.12	0.07	0.61	0.99
Depth	pH	Ava P	Exchangeable cations and CEC ($Cmol_c kg^{-1}$ soil)					CEC/ Clay	PBS	Ca/Mg	K/Mg
			Ca	Mg	Na	K	CEC				
Profile: ON/W/P1											
0–20	5.03	3.50	6.62	3.27	0.15	1.76	29.08	32	41	2	1.9
20–38	5.03	3.41	4.78	3.26	0.14	1.09	27.69	42	33	1.5	3.0
38–65	5.45	3.20	6.78	3.97	0.15	0.78	23.73	38	49	1.7	5.1
65–105	5.26	2.30	6.54	4.48	0.17	0.73	27.12	32	44	1.5	6.2
105+	5.70	2.40	2.78	4.13	0.20	0.75	22.68	26	35	0.7	5.5

Appendix Table A.35 Some selected characteristics of Alisols of southwestern Ethiopia (Asmamaw 2008)

Depth (cm)	Hor	Particle size (%)					Micronutrients (mg kg^{-1})			
		S	C	Si	Si/C	Class	Cu	Zn	Fe	Mn
0–19	Ap	21	62	17	0.3	C	0.71	1.68	14.31	40.2
19–45	Bt1	7	76	17	0.2	C	0.32	1.32	3.00	14.17
45–100	Bt2	44	46	10	0.2	C	0.50	0.12	1.89	9.32
100–140	Bt3	10	80	10	0.1	C	0.5	0.14	1.65	4.87
140–200	Bt4	10	80	10	0.1	C	0.34	0.12	2.30	6.86
Depth (cm)	Hor	pH (H_2O)	P (ppm)	Exchangeable cations and CEC ($cmol_c$ kg^{-1} soil)					CEC/C	BS (%)
				Ca	Mg	K	Na	CEC		
0–19	Ap	5.39	33.77	4.66	3.91	1.51	0.09	39.4	64	26
19–45	Bt1	5.62	3.97	5.41	5.67	0.68	0.07	38.35	50	31
45–100	Bt2	5.92	0.99	9.84	7.16	0.90	0.09	38.09	83	47
100–140	Bt3	6.04	2.84	8.64	6.98	1.09	0.09	51.92	65	32
140–200	Bt4	6.26	2.84	8.38	7.39	1.27	0.18	38.09	48	45
Depth (cm)	Hor	TN	OC	OM	C/N	Ca/CEC (%)		Mg/CEC	Ca/Mg	Mg/K
0–19	Ap	0.25	2.80	4.82	11	11.83		9.92	1.19	0.39
19–45	Bt1	0.15	0.91	1.57	6	14.11		14.78	0.95	0.12
45–100	Bt2	0.10	0.67	1.15	7	25.83		18	1.37	0.13
100–140	Bt3	0.09	0.47	0.82	5	16.64		13.4	1.24	0.16
140–200	Bt4	0.10	3.20	5.52	32	22		13.94	1.13	0.17

Appendix Table A.36 Surface physical and chemical properties of Cambic Arenosol, Hauzien (northeast Tigray) (Nyassen et al. 2019)

Depth	Texture			Class	Silt/clay	pH (H_2O)	OC (%)	TN (%)	Av P. (mg/100g)	$CaCO_3$	Reference
	Sand	Silt	Clay								
0–10	69	8	8	Loamy coarse sand	1	6.5	0.5	–	2.4	–	Munro (1975)
10–46	68	6	10	Coarse sandy loam y	0.6	6.7	0.4	–	0.4	–	
	Exchangeable bases (Cmol+/kg)					EC(dS/m, 1:5)	Ca/Mg	Micronutrients (mg kg^{-1})			
	Ca	Mg	Na	K	CEC soil			Zn	Mn	Cu	
0–10	2.9	0.6	–	0.2	4.6	0.1	4.3	1.6	73	0.7	
10–46	13.9	0.4	–	–	12.5	0.1	34.5	0.8	93	0.3	

Appendix Table A.37 Surface physical and chemical properties of depression Gleysol profile, northeastern highlands (South Wello)

Depth	Texture			Class	Silt/clay	pH (H_2O)	OC (%)	TN (%)	Av P. (mg/kg)	C/N	Reference
	Sand	Silt	Clay								
0–35	48	35	17	Silt clay	1.37	6.45	1.64	0.18	16.04	9	Engidawork (2002)
35–65	54	33	13	Silt clay	1.64	6.68	1.20	0.14	14.96	9	
	Exchangeable cations (Cmol+/kg)						PBS	ESP	Ca/Mg ratio		
	Ca	Mg	Na	K	CEC soil	CEC clay					
0–35	48.9	17.74	0.70	0.37	55.0	141	121	0.67	2.70		
35–65	53.9	18.74	0.70	0.35	54.6	153	134	0.64	0.64		

Appendix Table A.38 Surface physical and chemical properties of Gleysol profile, Eastern lowlands, Lower Wabishebelle, Basin

Depth	Texture			Class	Color	pH (H_2O)	OM (%)	TN (%)	$CaCO_3$ (%)	OC (%)	References
	Sand	Silt	Clay								
0–15	29	42	29	Clay loam	5 YR 3/1	7.9	11	0.57	15	6.63	FAO (1984a)
15–50	17	46	37	Silt loam	10 YR 3/2	8.0	1.5	0.1	17	0.9	
	Exchangeable cations (Cmol+/kg)					PBS	EC (dS/m)	Soluble salt (me/L, sat. ex)			
	Ca	Mg	Na	K	CEC soil			Na	Ca		
0–15	32	19	0.8	1.2	53.4	100	3.9	0.4	15.1		
15–50	30	20	1.3	0.5	51.0	100	3.2	0.7	10.3		

Appendix Table A.39 Oxide composition (%) of Feralsols

Pedon	Depth (m)	SiO_2	AL_2O_3	Fe_2O_3	SiO_2/R_2O_3
TP1	1.75	38.12	27.12	19.55	0.82 = 38.12/(27.12+19.55)
TP2	1.10	36.58	23.81	22.55	0.79

Appendix Table A.40 Mineralogical composition of Ferralsols, around Addis Ababa (Wossen 2009)

Pedon	Depth (m)	Mineral identification	Chemical formula	Identified minerals (%)
TP1	1.75	Kaolinite	$Al_4(OH)_8(Si_4O_{10})$	47.5
		Dickite	$Al_2Si_2O_5(OH)_4$	39.3
		Goethite	FeO (OH)	13.2
TP2	1.10	Kaolinite	$Al_4(OH)_8(Si_4O_{10})$	62.9
		Quartz	SiO_2	18.0
		Goethite	FeO (OH)	19.1

Appendix Table A.41 Selected morphological properties of the Ferralsol pedons, Gambella area (Teshome et al. 2016, symbols FAO 2006)

Pedon	Horizon	Depth (cm)	Boundary	Color		Texture (Feel method)	Structure[a]	Consistency[b]	Coarse fragment (%)
				Dry	Moist				
A-1	Ap	0–18	GS	10YR 3/6	10YR 3/3	CL	MO, VM, GR	SHA, FR, ST, SPL	–
	Bw1	18–25	GS	2.5YR 4/6	2.5YR 4/4	CL	MO, VM, AB	SHA, FR, ST, SPL	–
	Bw2	25–35	GS	2.5YR 4/8	2.5YR 4/6	C	MO, FM, AB	SHA, FR, ST, PL	–
	CB	35–110	–	2.5YR 4/6	2.5YR 4/6	C	MO, FM, AB	SHA, FR, ST, PL	47
A-2	A	0–18	GS	10YR 4/3	10YR 3/2	CL	MO, ME, GR	SHA, FR, ST, P	–
	Ad	18–27	GS	10YR 4/3	10YR 3/2	CL	MO, ME, AB	SHA, FR, ST, P	–
	Bt1	27–37	GS	2.5YR 4/4	2.5YR 4/3	C	MO, ME, AB	SHA, FR, ST, P	–
	Bt2	37–50	GS	2.5YR 4/6	2.5YR 4/5	C	MO, ME, AB	SHA, FR, ST, P	–
	BC	50–80	–	2.5YR 4/6	2.5YR 4/5	C	MO, ME, AB	SHA, FR, ST, P	21
A-3	Ap	0–22	CS	7.5YR 3/3	7.5YR 3/3	CL	MO, ME, GR	SHA, FR, ST, PL	–
	A	22–30	CS	7.5YR 3/3	7.5YR 3/2	C	MO, ME, AB	HA, FR, VST, VPL	–
	Bt1	33–45	GS	5YR 4/6	5YR 4/4	C	MO, ME, AB	HA, FR, VST, VPL	–
	Bt2	45–63	GS	5YR 4/4	2.5YR 4/3	C	MO, ME, AB	HA, FR, VST, VPL	–
	BC	63–85	–	5YR 3/6	5YR 4/4	C	MO, ME, AB	HA, FR, VST, VPL	19
A-4	Ap	0–10	CS	10YR 3/6	10YR 3/3	CL	WE, FI, GR	HA, FR, ST, PL	–
	A	10–19	CS	10YR 3/4	10YR 3/3	C	ST, FM, SB	VHA, FI, ST, PL	–
	Bt1	19–37	GS	5YR 4/4	7.5YR 3/3	C	ST, FM, AB	VHA, FI, ST, PL	–
	Bt2	37–75	GS	5YR 4/6	5YR 3/4	C	ST, FM, AB	VHA, FI, ST, PL	–
	BC	75–85	–	5YR 4/6	5YR 4/4	C	ST, FM, AB	HA, FR, ST, PL	23

[a] and [b] are as per FAO (2006) definition

Appendix Table A.42 Some profile site information and selected morphological and physical characteristics of Acriols of Bedele Area, western Ethiopian highlands (Abera and Kefyalew 2017, symbol according to FAO 2006)

Profile	Coordinates	Slope (%)	Slope position	Erosion evidence	Depth (cm)	Hor.	Color Munsell value		Struct.	Consistency (Dry, Moist, Wett)	Bound.
							Dry	Moist			
P1 (Forest land)	8° 27′ 3″ N and 36° 23′ 56″ E	5	Upper slope	Slight sheet erosion	0–26	A	7.5YR 3/3	7.5YR 2.5/3	mfc	Sh, fr, sssp	cs
					26–70	Bt1	5YR 4/4	7.5YR 2.5/3	mmsb	Sh, fr, sssp	cs
					70–120	Bt2	2.5YR 3/4	5YR 3/3	mmab	Mh, fr, msmp	gs
					120–180	Bt3	2.5YR 4/4	2.5YR 3/4	Wcab	Mh, fr, sp	–
P2 (Grazing land)	8° 27′ 4″ N and 36° 23′ 6″ E	2	Summit	Slight sheet erosion	0–23	A	7.5YR 3/4	5YR 3/4	smab	Mh, fr, sssp	cs
					23–52	Bt1	2.5YR 3/4	5YR 3/2	smab	Mh, fr, msmp	gs
					52–100	Bt2	10R 3/4	10R 3/2	wmsb	Mh, fr, msmp	ds
					100–190	Bt3	10R 3/6	10R 3/3	wfsb	mh, fr, sp	-
P3 (Cultivated field)	8° 23′ 3″ N and 36° 23′ 6″ E	5	Upper slope	Moderate sheet and rill erosion	0–25	Ap	5YR 4/4	5YR 3/3	sfsb	H, fr, sssp	cs
					25–65	AB	2.5YR 3/3	2.5YR 2.5/3	mmsb	H, fr, msmp	gs
					65–125	Bt1	2.5YR 3/4	2.5YR 2.5/3	sfsb	Sh, fr, msmp	ds
					125–190^{+}	Bt2	2.5YR 3/6	2.5YR ¾	wmsb	Sh, fr, msmp	–

Appendix Table A.43 Description of soil profile site characteristics in KindoKoye watershed, south central highlands (Wolayita area) (Mulugeta and Sheleme 2010)

Pedon	Slope (%)	Altitude (masl)	Surrounding landform	Physiographic position	Parent material
UEF	13	2010	Gentle slope	Upper slope	Basaltic
FEF	25	1984	Steep slope	Foot slope	Colluvium/Basaltic
UWF	10	2029	Gentle slope	Upper slope	Basaltic

UWF = Upper slope west-facing; UEF = Upper slope east-facing; FEF = Foot slope east-facing

Appendix Table A.44 Morphological features and physical properties of the soils of Kindo Koye watershed, south central highlands (Wolayita area) (Mulugeta and Sheleme 2010, symbol according to FAO 2006)

Profiles	Depth (cm)	Horizon	Color (moist)	Structure	Consistence (moist, wet)	Boundary	Particle size fraction (%)			Textural class	BD (g cm^{-3})
							Sand	Silt	Clay		
UEF	0–40	Ap1	7.5YR2.5/2	VW, FI, GR	VFI, SST-SPL	G-S	31	38	31	Clayloam	1.33
	40–64	Ap2	7.5YR2.5/2	MO, ME, SB	VFI, ST-PL	C-S	23	30	47	Clay	1.26
	64–90	BA	5YR3/3	ST, ME, AB	FI, VST-VPL	C-S	21	28	51	Clay	1.46
	90–136	Bt1	7.5YR2.5/2	ST, ME, AB	FI, VS-VP	G-S	11	12	77	Clay	1.42
	136+	Bt3	2.5YR2.5/4	ST, ME, AB	VFI, VS-VP	–	9	10	81	Clay	1.33
FEF	0–18	Ap1	7.5YR2/3	VW, ME, GR	FR, SST-SPL	G-S	32	30	38	Clayloam	1.34
	18–53	Ap2	5YR3/2	WE, ME, GR	FR, SST-SPL	G-S	26	34	40	Clay	1.21
	53–100	Bt1	2.5YR2.5/3	WE, ME, AB	FR, SST-SPL	G-S	22	20	58	Clay	1.49
	100–160	Bt2	2.5YR2.5/3	MO, ME, AB	FI, SST-SPL	D-S	18	20	62	Clay	1.42
	160+	Bt3	5YR3/4	ST, ME, AB	FI, SST-SPL	–	24	16	60	Clay	1.42
UWF	0–38	A1	7.5YR2.5/2	WE, FM, GR	VFR, ST-SS	G-D	31	34	35	Clayloam	1.25
	38–78	A2	7.5YR2.5/3	VW, FM, GR	VFR, ST-PL	G-S	31	36	33	Clayloam	1.36
	78–102	Bt1	5YR3/2	WE, FM, SB	VFI, VST-VPL	A-S	19	22	59	Clay	1.37
	102–171	Bt2	5YR3/3	FI, FM, AB	VFI, VST-VPL	G-S	19	22	59	Clay	1.26
	171+	Bt3	2.5YR2.5/4	FI, ME, AB	VFI, VST, VPL	–	15	20	65	Clay	1.43

Appendix Table A.45 Selected physical properties of Bedele area Acrisols (western highlands) (Abera and Kefyalew 2017)

Profiles	Depth (cm)	Particle size fraction (%)			Textural class	BD (g cm^{-3})	Total porosity (%)	Water content (%)		
		Sand	Silt	Clay				FC	PWP	AWC
P1 (Forest land)	0–26	53	34	13	SL	1.03	56.20	27.6	10.2	17.4
	26–70	25	34	41	C	1.12	52.70	25.3	10.9	14.4
	70–120	23	24	53	C	1.30	48.00	23.2	12.1	10.1
	120–180	25	16	59	C	–	–	–	–	–
P2 (Grazing land)	0–23	31	36	33	CL	1.23	51.20	25.7	10.4	15.3
	23–52	31	26	43	C	1.28	51.00	24.6	11.4	13.2
	52–100	22	21	57	C	1.34	45.30	23.5	13.7	9.8
	100–190	23	14	63	C	–	–	–	–	–
P3 (Cultivated field)	0–25	45	30	25	L	1.25	51.17	24.3	10.3	14.0
	25–65	33	32	35	CL	1.32	48.60	23.8	11.6	12.2
	65–125	19	34	47	C	1.34	45.50	23.7	11.9	11.8
	$125–190^{+}$	7	16	77	C	–	–	–	–	–

Appendix Table A.46 Bedele area Acrisols, western Ethiopian highlands (Abera and Kefyalew 2017)

Profiles	Depth	pH (1:2.5)		Exc. Acidity (cmol(+) kg^{-1})	Exc. Al (cmol (+) kg^{-1})	OC (%)	TN (%)	C:N	Available P (mg kg^{-1})-Olson	Exchang. bases cmolc kg^{-1}				CEC cmolc kg^{-1}		Ca: Mg	PBS (%)	Micronutrients (mg kg^{-1})			
		H_2O	KCl							Ca	Mg	K	Na	Soil	Clay			Fe	Mn	Zn	Cu
P1 (FL)	0–26	5.17	3.82	9.41	7.25	10.74	0.66	16.09	5.20	3.50	3.58	0.69	Trace	42.0	5.46	0.98	20.0	9.43	12.14	0.11	0.25
	26–70	5.23	3.85	9.33	7.14	2.24	0.24	10.04	2.00	0.73	3.45	0.32	Trace	22.5	9.23	0.21	21.7	1.33	10.34	trace	0.02
	70–120	5.24	3.91	8.11	6.78	1.59	0.17	9.32	1.80	0.66	2.39	0.30	Trace	15.4	8.16	0.28	15.4	1.05	8.58	0.03	0.03
	120–180	5.24	3.92	3.29	2.33	1.30	0.08	16.28	1.00	0.52	1.51	0.28	Trace	15.0	8.85	0.34	15.4	0.40	4.95	0.01	0.023
P2 (GL)	0–23	5.05	3.80	9.62	7.62	4.06	0.36	13.34	4.20	1.36	2.31	0.60	Trace	25.0	8.25	0.59	17.1	3.63	15.77	0.15	0.11
	23–52	5.09	3.85	8.58	7.08	2.00	0.17	11.88	1.19	1.04	2.25	0.37	Trace	20.0	8.60	0.46	18.3	1.49	6.38	0.01	0.06
	52–100	5.16	3.88	8.09	6.77	1.38	0.14	9.94	0.90	0.73	2.14	0.39	Trace	16.5	9.41	0.34	19.7	1.13	6.14	0.11	0.02
	100–190	5.17	3.92	4.35	3.02	0.49	0.07	7.38	1.02	0.73	0.83	0.43	Trace	16.2	10.21	0.88	12.3	0.42	2.67	0.13	0.03
P3 (CL)	0–25	4.92	3.75	10.86	9.62	3.09	0.34	9.00	3.20	1.20	0.97	0.58	Trace	21.0	5.25	1.24	13.0	4.83	17.42	0.05	0.15
	25–65	5.11	3.98	10.46	9.06	1.54	0.15	12.05	1.08	1.82	1.31	0.52	Trace	17.4	6.09	1.39	19.4	2.16	6.36	0.05	0.07
	65–125	5.23	4.01	8.35	6.94	0.93	0.08	11.64	0.72	2.34	0.26	0.45	Trace	15.8	7.43	9.00	19.8	1.81	5.50	Trace	0.04
	125–190$^+$	5.25	4.03	4.83	3.23	0.54	0.05	10.58	0.50	1.27	1.07	0.49	Trace	15.3	11.78	1.88	18.7	0.93	2.48	Trace	0.01

Appendix Table A.47 Cation exchange capacity (CEC), exchangeable bases, ESP, and pH of the soils at the Kindo Koye Watershed, south central highlands (Wolaita area) (Mulugeta and Sheleme 2010)

Profiles	Depth	pH (1:2.5)		OC (%)	TN (%)	C: N	Available P (mg kg^{-1})	Exchangeable bases (cmolc kg^{-1})				CEC cmolc kg^{-1}		PBS (%)	Micronutrients (mg kg^{-1})			
		H_2O	KCl					Ca	Mg	K	Na	Soil	ESP (%)		Fe	Mn	Zn	Cu
UEF	0–40	6.1	5.2	2.163	0.178	12	7.30	10.98	2.55	0.50	0.17	27.40	0.62	52	1.39	4.51	10.23	0.18
	40–64	5.8	5.3	1.344	0.140	10	1.40	9.18	1.91	0.50	0.21	27.60	0.76	45	0.59	0.53	3.94	0.11
	64–90	6.3	5.2	0.789	0.113	7	0.92	6.99	2.55	0.38	0.29	26.60	1.09	38	1.06	0.40	3.41	0.31
	90–136	6.0	4.8	0.481	0.070	7	0.92	5.49	3.54	0.41	0.33	27.60	1.19	35	0.59	0.95	0.79	0.18
	136+	6.1	4.8	0.585	0.090	7	1.66	5.84	4.31	0.41	0.37	28.80	1.28	37	0.37	0.44	0.48	0.15
FEF	0–18	5.8	4.2	1.743	0.217	8	2.30	1.36	0.69	2.33	0.73	19.50	3.74	26	1.03	8.78	0.70	0.20
	18–53	5.8	4.2	1.743	0.153	11	1.30	1.52	0.38	1.57	0.96	16.10	5.96	28	2.57	7.55	7.52	0.24
	53–100	5.8	4.3	0.905	0.137	7	0.34	1.01	0.41	2.75	0.97	18.00	5.39	29	0.66	0.81	1.76	0.15
	100–160	5.1	4.0	0.813	0.112	7	0.30	0.96	0.43	2.19	1.75	17.80	9.83	30	0.51	0.33	0.62	0.18
	160+	5.5	4.0	0.720	0.098	7	0.30	0.96	0.42	2.68	1.08	18.20	5.93	28	0.18	0.15	0.70	0.07
UWF	0–38	5.6	4.4	1.572	0.146	11	1.38	6.94	2.47	0.14	0.15	23.20	0.65	42	2.22	8.16	8.58	0.22
	38–78	5.9	4.4	1.407	0.115	12	1.20	7.24	1.88	0.21	0.20	24.40	0.82	37	2.09	3.21	8.07	0.24
	78–102	5.3	4.4	0.880	0.085	10	0.56	6.34	2.88	0.18	0.44	26.20	1.68	38	0.57	0.92	2.73	0.13
	102–171	5.2	4.1	0.657	0.092	7	0.56	3.64	1.23	0.17	0.52	25.00	2.08	22	0.44	0.37	2.24	0.13
	171+	5.6	3.8	0.491	0.070	7	0.28	6.29	1.56	0.17	0.40	24.40	1.64	35	0.33	0.68	1.28	0.18

Appendix Table A.48 Morphological properties of the Lixisols in Taba, in Huletegna Choroko in southern Ethiopia (Abay et al. 2015, symbol according to FAO 2006)

Depth (cm)	Horizon	Color (moist)	Structure grade/size/type	Consistence		Pores	Boundary
				Moist	Wet		
Pedon-1							
0–33	Ap	7.5YR3/1	–	LO	NST, SPL	M, C	G, W
33–73	AB	7.5YR4/2	–	FR	NST, SPL	F, F	C, S
73–94	ABg	GELEY16/10Y	M, M, G	FI	SST, VPL	V, F	C, S
94–157	Bti	GLEY14/N	ST, Platy	VRI	VST, VPL	N	–
Pedon-2							
0–47	Ap	GELEY2 4/10B	WE, M, G	VFR	SST, PL	M, C	D, W
47–94	E	10R 3/2	MO, FI, G	FI	SST, PL	M, C	C, W
94–170	Bt	10R 5/6	ST, VFI, G	VFI	ST, VPL	F, V	–

Appendix Table A.49 Physical and chemical properties of the Lixisol in Taba, in Huletegna Choroko in southern Ethiopia (Abay et al. 2015)

Depth (cm)	Sand	Silt	Clay	Tex cl	pH–H_2O, 1:2.5	EC (ms/cm) (1:2.5)	Org C (%)	Tot N (%)	C/N	Avail P mg (kg)	$CaCO_3$ (%)
Pedon-1											
0–33	30	38	32	CL	7.7	0.06	3.90	0.35	11.1	8.5	0.51
33–73	38	34	28	CL	7.0	0.03	3.10	0.26	11.9	6.8	0.54
73–94	32	40	28	CL	7.8	0.06	1.01	0.11	9.2	9.3	0.53
94–157	24	42	34	CL	7.6	0.11	1.03	0.15	6.9	9.9	0.64
Pedon-2											
0–47	28	40	32	CL	7.6	0.17	3.60	0.30	12.0	9.8	0.68
47–94	32	60	8	SiL	7.8	0.21	2.20	0.26	8.5	8.9	0.72
94–170	26	30	44	CL	7.9	0.16	1.80	0.14	12.9	10.2	0.81

Depth (cm)	Exchangeable bases (cmol (+) Kg^{-1})					Sum of basic cations	PBS (%)	Exchangeable Sodium %	Micronutrients (mg Kg^{-1})			
	CEC	Na	K	Ca	Mg				Zn	Mn	Cu	Fe
Pedon-1												
0–33	23.6	0.8	0.9	15.3	1.2	18.2	77.1	3.22	1.07	1.03	0.16	4.60
33–73	16.5	0.6	1.0	10.7	0.4	12.6	76.6	3.64	0.68	1.66	0.07	1.63
73–94	10.9	0.3	1.1	7.1	0.1	8.61	78.9	2.84	0.38	0.41	0.25	0.68
94–157	18.8	0.6	1.8	12.1	3.5	18.0	96.0	3.25	0.36	0.30	0.54	0.17
Pedon-2												
0–47	22.7	1.2	1.8	14.8	1.1	18.9	83.1	5.28	2.06	1.32	0.39	1.11
47–94	15.9	0.8	0.8	10.3	1.1	13.0	81.8	4.79	1.67	1.35	0.28	1.39
94–170	23.5	1.3	1.1	14.1	0.5	16.9	71.9	5.52	0.44	1.76	0.07	1.12

Appendix Table A.50 Selected physico-chemical properties of Planosols in Omo-Nada Woreda (Alemayehu 2019)

Horizon	Depth (cm)	% Sand	% Silt	% Clay	Silt/clay ratio	Tex. class	BD (Mgm^{-3})	pH (H_2O)	OC (%)	OM (%)	TN (%)	C/N	Exchangeable cations ($Cmol+ kg^{-1}$ soil)				CEC ($Cmol+ kg^{-1}$ soil)	PBS (%)
													Ca	Mg	Na	K		
Worabet Shuma																		
Apg	0–20	24	36	40	0.9	Clay	1.34	6.18	1.90	3.28	0.14	13.59	17.2	4.80	0.56	0.55	21.8	105.80
Eg	20–40	24	36	40	0.9	Clay	1.24	6.39	1.47	2.53	0.12	11.95	15.8	3.86	0.96	0.41	18.2	115.50
Bt1g	40–65	22	20	58	0.3	Clay	1.39	6.63	1.09	1.88	0.11	10.12	28.4	7.57	1.93	0.93	35.7	108.70
Bt2	65–100	40	14	46	0.3	Clay	1.48	7.22	1.03	1.78	0.11	9.81	34.9	9.20	2.33	1.17	43.1	109.30
Obiso Wacho																		
Apg	0–20	30	44	26	1.7	Loam	1.19	6.10	2.59	4.47	0.17	15.24	12.3	2.6	0.3	0.6	18.4	85.93
Eg	20–50	30	28	42	0.7	Clay	1.46	6.29	1.03	1.77	0.12	8.77	17.7	3.8	0.8	0.6	28.8	79.41
Btg	50–100	28	26	46	0.6	Clay	1.25	6.36	1.13	1.94	0.12	9.46	19.8	4.4	0.9	0.8	31.1	82.97
2Bssg	100–200	44	14	42	0.3	Clay	1.42	6.39	1.01	1.74	0.12	8.71	24.1	5.7	1.1	1.1	37.3	85.88
Fuge Achiraye																		
Ap	0–10	38	32	30	1.1	Clay Loam	1.19	5.99	4.36	7.52	0.28	15.46	13.8	3.4	0.2	0.6	22.0	81.87
Eg	10–25	28	36	36	1.0	Clay Loam	1.32	6.35	1.88	3.25	0.16	11.85	11.5	2.8	0.3	0.4	17.7	85.06
Bt1g	25–60	29	14	57	0.2	Clay	1.23	6.90	0.92	1.59	0.11	8.52	29.6	8.0	1.3	0.7	43.1	91.89
Bt2	60–150	31	10	59	0.2	Clay	1.27	7.50	0.68	1.17	0.09	7.19	42.1	9.6	1.7	0.9	53.4	101.59
Workat																		
Ap	0–5	18	37	45	0.8	Clay	1.28	6.00	2.34	4.03	0.20	11.70	9.9	2.9	0.2	0.5	19.0	70.92
EAg	5–25	16	39	45	0.9	Clay	1.20	6.00	1.63	2.81	0.16	10.17	8.3	2.5	0.2	0.4	18.2	63.08
Eg	25–65	16	41	43	1.0	Silty Clay	1.37	6.34	0.50	0.87	0.09	5.79	4.8	1.6	0.2	0.2	13.3	51.65
Bt1g	65–85	37	14	49	0.3	Clay	1.27	6.02	0.52	0.90	0.10	5.22	14.5	5.2	0.7	0.7	36.7	57.35
Bt2	85–150	31	14	55	0.3	Clay	1.12	5.95	0.50	0.86	0.09	5.55	15.4	5.4	0.7	0.7	40.5	54.53
Waktola																		
Ah	0–8	30	28	42	0.7	Clay	ND*	5.64	1.63	2.80	0.13	12.31	32.6	9.5	0.16	2.41	20.56	218.00
Eg	8–18	16	36	48	0.8	Clay	ND	5.40	1.20	2.06	0.10	11.95	21.5	5.8	0.54	2.33	28.46	106.00
Bg	18–60	36	0	64	0.0	Clay	ND	5.88	0.80	1.38	0.08	10.37	54.0	13.6	1.02	1.84	41.72	169.00
Bss1	60–95	26	14	60	0.2	Clay	ND	6.92	0.58	1.00	0.05	12.03	59.9	13.8	1.40	1.52	30.20	254.00
Bss2	95–150	28	14	58	0.2	Clay	ND	6.98	0.32	0.55	0.03	10.70	54.6	13.0	1.33	1.84	31.40	226.00

*Not determined

Appendix Table A.51 Selected nutrients of the studied Planosols of Gera Woreda, south central Ethiopian highlands (Alemayehu 2019)

Horizon	Depth (cm)	pH H_2O	Av.P (mg P_2O_5/kg soil)	Micro nutrients (mg/kg soil)			
				Cu	Zn	Fe	Mn
Worabet Shuma							
Apg	0–20	6.18	6.57	1.59	1.13	197.77	145.68
Eg	20–40	6.39	4.24	1.85	0.84	149.19	78.24
Bt1g	40–65	6.63	4.22	2.14	1.07	183.90	160.98
Bt2	65–100	7.22	7.10	2.23	0.95	188.92	168.96
Obiso Wacho							
Apg	0–20	6.10	10.27	1.72	8.21	229.94	226.04
Eg	20–50	6.29	6.82	1.78	7.12	253.80	274.43
Btg	50–100	6.36	5.77	1.96	4.85	289.97	305.85
2Bssg	100–200	6.39	8.59	1.61	3.77	361.89	437.52
Fuge Achiraye							
Ap	0–10	5.99	7.62	1.89	4.04	227.06	162.97
Eg	10–25	6.35	4.10	1.71	1.02	169.62	98.59
Bt1g	25–60	6.90	5.57	2.10	1.72	151.62	125.57
Bt2	60–150	7.50	5.71	2.36	1.59	129.04	132.19
Workat							
Ap	0–5	6.00	4.07	2.09	2.91	237.31	121.42
EAg	5–25	6.00	3.85	1.96	2.37	184.85	76.29
Eg	25–65	6.34	3.74	0.85	0.81	144.96	28.07
Bt1g	65–85	6.02	4.26	0.98	1.04	264.58	230.40
Bt2	85–150	5.95	4.26	0.77	0.96	299.75	309.53
Waktola							
Ah	0–8	5.64	8.50	0.62	0.57	335.60	226.40
Eg	8–18	5.40	5.40	0.59	0.40	221.20	197.10
Bg	18–60	5.88	4.33	0.51	0.34	139.90	88.60
Bss1	60–95	6.92	4.80	0.46	0.26	71.30	35.30
Bss2	95–150	6.98	3.01	0.40	0.29	53.30	22.10

Appendix Table A.52 Some selected characteristics of Simein Mountains Andosols, Ethiopia

Soil group/unit, altitude, slope gradient (%)	Umbric Andosols, 320nmasl, 4% slope (Mohammed and Belay 2008)								
Horizons	Ap	A2	C	Ah	Bw	C	Ah	Bw	R
Depth/thickness (cm)	0–23	23–71	>71	0–51	51–94	>94	0–39	39–70	>70
Clay (%)	25	27		17	17		13	17	
Silt (%)	56	54		48	30		56	48	
Sand (%)	19	19		35	53		31	35	
SOM (%)	18	14		12	0.83		24	9	
Total N (%)	0.8	0.59		0.61	0.05		0.78	0.66	
Avail P (ppm)	13	6		7	25		15	18	
pH (H_2O)1:2.5	5.1	5.7		5.6	6.1		5.2	5.5	
CEC (cmol$_c$ kg^{-1} soil)	52	59		35	50		49	55	
Exch. Ca (cmol$_c$ kg^{-1} soil)	20	21		9	30		15	18	
Exch. Mg (cmol$_c$ kg^{-1} soil)	2	3		1	7		3	4	
Exch. K (cmol$_c$ kg^{-1} soil)	1.2	0.8		1	0.7		0.5	0.6	
Exch. Na (cmol$_c$ kg^{-1} soil)	0.6	0.5		0.4	0.3		0.4	0.3	
PBS	46	42		33	76		37	42	

Appendix Table A.53 Some selected characteristics of Andosols of Ethiopia

Soil group/unit, altitude, slope gradient (%)	Luvic Andosols, 3440 masl, 1% slope (Mohammed and Belay 2008)									
Horizons	Ap	A2	Bt	Ah	A2	Bt	C	Ap	Bt	R
Depth/thickness (cm)	0–25	25–65	65–119	0–50	50–70	70–102/110	>110	0–22	22–41	>41
Clay (%)	19	17	45	13	25	49		27	27	
Silt (%)	60	52	26	56	38	34		58	38	
Sand (%)	21	31	29	31	37	17		35	35	
SOM (%)	16	11	1.9	16	9	3		3	3	
Total N (%)	0.8	0.6	0.1	0.8	0.3	0.1		0.1	0.1	
Avail P (ppm)	6	5	3	0	7	7		5	4	
pH (H_2O)1:2.5	5.2	5.4	5.0	5.6	6.0	5.6		6.1	6.3	
CEC ($cmol_c$ kg^{-1} soil)	36	37	63	45	45	43		57	61	
Exch. Ca ($cmol_c$ kg^{-1} soil)	10	12	26	16	14	18		40	37	
Exch. Mg ($cmol_c$ kg^{-1} soil)	1	2	8	3	3	5		11	10	
Exch. K ($cmol_c$ kg^{-1} soil)	1.1	0.8	0.6	2.4	2	0.9		2	2	
Exch. Na ($cmol_c$ kg^{-1} soil)	0.5	0.5	0.4	0.7	0.7	0.5		0.5	0.6	
PBS	36	40	56	47	45	56		93	81	

Appendix Table A.54 Some selected characteristics of Andosols of Ethiopian Rift Valley (Eylachew 2004)

Soil group/unit, altitude, slope gradient (%)	Haplic Andosols (2% slope)				Mollic Andosols (2% slope)					Haplic Andosols (2% slope)					
Horizons	Ap	Ab	Bw	C	Ap	AC1	AC2	C1	C2	Ap1	Ap2	BC	CB	C1	C2
Depth/thickness (cm)	0–15	15–60	60–102	120+	0–14	14–45	45–71	71–104	104–170	0–4	4–31	31–75	75–145	145–175	175–200
SOM (%)	1.6	1.3	1.0	0.6	6	6	9	5	1.2	6	4	3	3	1	1
Total N (%)	0.06	0.06	0.06	0.03	0.23	0.23	0.22	0.27	0.02	0.22	0.22	0.13	0.16	0.03	0.04
Avail P (ppm)	7	2	2	1	11	10	4	4	2	31	43	56	6	6	1
pH (H_2O)1:2.5	7.2	7.5	8.2	9.0	7.5	6.5	6.7	6.9	7.9	8.0	8.3	8.5	9.2	9.8	9.7
CEC ($cmol_c$ kg^{-1} soil)	22	26	34	27	12	14	13	13	4	22	22	22	12	8	5
Exch. Ca ($cmol_c$ kg^{-1} soil)	12	18	23	16	6	11	13	15	7	25	28	24	20		
Exch. Mg ($cmol_c$ kg^{-1} soil)	1	2	2	3	1	1	2	1	1	4	4	5	4		
Exch. K ($cmol_c$ kg^{-1} soil)	2.4	3.5	8	11	2	2	2	1	1	5	3	3	5		
Exch. Na ($cmol_c$ kg^{-1} soil)	0.6	1	4	6	0.2	1	1	3	2	2	2	2.4	7		
PBS	72	91	100+	100+	83	100+	100+	100+	100+	100+	100+	100+	100+		

Appendix Table A.55 Selected physical and chemical characteristics of Solonchaks of Ethiopia (Degife et al. 2019)

Pedon/LUT	Horizon	Depth (cm)	Sand (%)	Silt	Clay	Tex. class	BDgm cm^{-3}	pH (H_2O)	pH (KCl)	pH
Kulfo Watershed = Upper Chamo Basin (UCB) (6° 7′ N, 37° 29′ E)										
P-1	Ah	0–34	67.41	13.04	19.56	SL	1.49	7.54	6.87	0.67
	B	34–200	58.90	32.45	8.65	SL	1.26	7.48	6.56	0.92
P-2	Ah	0–90	10.14	63.56	26.30	SiL	1.52	6.77	6.04	0.73
	UU5	90–97	–	–	–	–	–	–	–	–
	Bw	97–200	10.98	55.64	33.38	SCL	1.34	8.48	7.57	0.91
P-3	Ah	0–80	37.57	55.97	6.46	SiL	1.35	7.10	6.46	0.64
	UU5	80–85	–	–	–	–	–	–	–	–
	B	85–200	33.58	58.92	7.50	SiL	1.42	6.79	6.07	0.72
Sille-Sego Watershed = Middle Chamo Basin (MCB) (5° 57′ N, 37° 19′ E)										
P-4	Ah	0–35	21.63	46.81	31.57	CL	1.32	7.52	6.64	0.88
	B	35–200	27.87	55.74	16.39	SiL	1.44	7.37	6.61	0.76
P-5	Ah	0–40	39.71	43.07	17.23	L	1.55	7.01	6.26	0.75
	B	40–200	18.36	62.89	18.76	SiL	1.65	7.08	6.24	0.84
P-6	Ah	0–50	20.74	59.98	19.28	SiL	1.52	7.70	6.87	0.83
	UU5	50–58	–	–	–	–	–	–	–	–
	B	58–200	8.77	69.50	21.72	SiL	1.59	7.56	6.94	0.62
Waseca-Doyso Watershed = Lower Chamo Basin (LCB) (5° 45′ N, 37° 26′ E)										
P-7	Ah	0–46	30.38	28.28	41.34	C	1.22	8.10	7.14	0.96
	Bw	46–140	28.85	27.36	43.78	C	1.33	8.05	7.14	0.91
P-8	Ah	0–30	10.93	39.10	49.97	C	1.45	7.33	6.45	0.88
	B	30–120	34.13	46.43	19.44	L	1.42	7.66	6.83	0.83
	C	120–200	81.15	14.66	4.19	SL	1.62	7.93	7.05	0.88
P-9	Ah	0–50	2.59	31.71	65.69	C	1.19	7.41	6.45	0.96
	UU5	50–54	–	–	–	–	–	–	–	–
	B	54–200	26.85	46.95	26.20	L	1.24	7.78	7.03	0.75
Arguba-Weseca Watershed = Toe Chamo Basin (TCB) (5° 44′ N, 37° 25′ E)										
P-10	A	0–30	34.08	46.47	19.45	L	1.16	7.90	7.14	0.76
	B	30–80	10.12	67.69	22.19	SiL	1.39	8.35	7.51	0.84
	C	80–200	47.45	45.98	6.57	SL	1.34	7.42	7.48	−0.06
P-11	Ah	0–46	8.80	47.77	43.43	SiC	1.26	6.74	5.91	0.83
	B	46–115	20.04	63.75	16.21	SiL	1.33	7.66	6.97	0.69
	C	115–200	81.15	14.66	4.19	SL	1.42	7.83	7.24	0.59
P-12	Ah	0–17	34.75	45.67	19.57	L	1.09	5.88	5.21	0.67
	B	17–70	24.71	49.11	26.19	L	1.39	7.48	6.64	0.84
	C	70–200	41.18	45.75	13.07	L	1.34	7.66	7.05	0.61

Notes L loam, *C* clay, *CL* clay loam, *SL* sandy loam, *SiL* silty clay, *BD* bulk density, *PD* particle density, *TP* total porosity

Appendix Table A.56 Ionic composition of the saturation extract and solid soils (Kidane et al. 2006)

Depth (cm)	Soil pH (1:1) soil: H_2O	ECe (dS^{-1} m)	Soluble cations (Cmolc kg^{-1})						
			Ca^2	Mg^2	Na+	K+	HCO_3-	Cl^-	SO_4^{2-}
0–20	7.20	16.68	103.79	3.29	54.37	14.36	1.20	119.4	55.69
20–40	7.18	17.88	95.21	4.28	65.46	10.26	0.90	126.20	55.69
40–64	7.20	18.23	63.62	4.11	73.68	10.77	1.10	133.30	59.98
64–88	7.23	18.57	41.48	4.93	130.75	1.28	1.10	133.30	59.98
88–97	7.45	18.30	28.14	2.30	84.38	0.64	0.80	113.00	51.41
97–140	7.38	19.79	33.83	4.44	202.27	1.41	0.90	139.80	85.68

Appendix Table A.57 Ionic composition of the soil solid surface

Depth (cm)	Soil pH KCl (1:2.5)	Exchangeable cations (Cmolc kg^{-1})				Exchangeable acidity (Cmolc kg^{-1})	CEC (Cmolc kg^{-1})	PBS	ESP
		Ca^2	Mg^2	Na^+	K^+				
0–20	6.90	43.89	5.76	2.23	3.23	0.15	55.26	99.73	4.03
20–40	6.90	43.79	5.66	2.07	2.08	0.15	53.75	99.72	3.85
40–64	6.80	47.37	5.95	3.19	0.90	0.20	57.61	99.65	5.53
64–88	7.00	36.59	6.64	3.63	0.41	0.15	47.52	99.68	7.64
88–97	7.00	27.16	22.68	3.05	0.26	0.10	53.25	99.81	5.73
97–140	7.00	32.28	41.10	2.96	0.46	0.05	77.85	99.93	3.80

Appendix Table A.58 Some selected chemical characteristics of solonchaks (Degife et al. 2019)

Pedon	Depth (cm)	SOM (%)	Av. N (mg/kg)	P_2O_5	K_2O
P-1	0–34	1.82	1164.24	30.01	166.58
	34–200	0.36	166.32	27.83	54.66
P-2	0–90	2.28	1175.02	161.53	228.54
	90–97	–	–	–	–
	97–200	0.82	508.20	160.79	164.36
P-3	0–80	1.62	988.68	48.29	257.87
	80–85	–	–	–	–
	85–200	1.22	659.12	48.47	183.09
P-4	0–35	4.32	1995.84	174.59	262.88
	35–200	0.72	335.72	58.00	154.99
P-5	0–40	2.82	1647.80	146.31	242.40
	40–200	0.54	335.72	40.75	194.39
P-6	0–50	1.14	329.56	153.16	110.88
	50–58	–	–	–	–
	58–200	0.62	166.32	153.13	78.08
P-7	0–46	2.48	1330.56	148.41	249.87
	46–140	2.40	671.44	145.74	485.98
P-8	0–30	2.38	1496.88	139.68	471.11
	30–120	0.56	329.56	33.51	90.25
	120–200	0.24	161.70	21.04	106.28
P-9	0–50	2.16	1207.36	156.35	383.29
	50–54	–	–	–	–
	54–200	0.54	332.64	154.41	88.50

(continued)

Appendix Table A.58 (continued)

Pedon	Depth (cm)	SOM (%)	Av. N (mg/kg)	P_2O_5	K_2O
P-10	0–30	0.84	329.56	52.07	252.71
	30–80	0.46	169.40	41.12	140.50
	80–200	0.34	167.86	41.12	288.96
P-11	0–46	1.88	1164.24	38.74	304.53
	46–115	0.62	329.56	32.25	195.98
	115–200	0.30	160.16	23.29	255.65
P-12	0–17	8.10	5155.92	266.62	242.06
	17–70	1.16	665.28	105.85	476.31
	70–200	0.42	166.32	19.82	78.08

Appendix Table A.59 Selected chemical characteristics of Chamo solonchaks, Upper rift valley of Ethiopia (Degife et al. 2019)

Pedon	Depth (cm)	OC (%)	TN	C/N	EC_e(dS/m)	EC2.5	Salt con. (%)	ESP	$CaCO_3$
Kulfo Watershed = Upper Chamo Basin (ULS)									
P-1	0–34	0.91	0.12	7.84	0.35	0.14	0.68	2.65	7.22
	34–200	0.18	0.02	10.60	0.23	0.09	3.32	3.82	7.00
P-2	0–90	1.14	0.12	9.67	0.83	0.33	5.40	2.50	5.56
	90–97	–	–	–	–	–	–	–	–
	97–200	0.41	0.05	8.15	1.70	0.68	12.36	4.48	7.01
P-3	0–80	0.81	0.10	8.22	0.25	0.10	1.60	1.60	2.89
	80–85	–	–	–	–	–	–	–	–
	85–200	0.61	0.07	9.25	0.20	0.08	1.15	1.85	7.15
P-4	0–35	2.16	0.20	10.81	0.83	0.33	2.10	1.86	5.01
	35–200	0.36	0.03	10.86	0.38	0.15	4.95	3.04	7.06
P-5	0–40	1.41	0.16	8.58	23.90	9.56	65.20	3.16	3.87
	40–200	0.27	0.03	8.19	13.78	5.51	150.40	2.86	5.06
P-6	0–50	0.57	0.03	17.15	18.05	7.22	61.50	2.38	7.26
	50–58	–	–	–	–	–	–	–	–
	58–200	0.31	0.02	18.46	15.33	6.13	147.68	2.37	6.00
P-7	0–46	1.24	0.13	9.36	0.55	0.22	1.84	2.08	6.11
	46–140	1.20	0.07	17.93	0.68	0.27	4.70	3.65	7.50
P-8	0–30	1.19	0.15	7.97	0.58	0.23	1.20	2.00	5.01
	30–120	0.28	0.03	8.52	0.35	0.14	1.80	2.79	7.80
	120–200	0.12	0.02	7.13	0.30	0.12	1.60	3.68	7.23
P-9	0–50	1.08	0.12	8.91	0.33	0.13	1.00	2.45	4.40
	50–54	–	–	–	–	–	–	–	–
	54–200	0.27	0.03	8.04	0.38	0.15	4.38	2.64	5.45
P-10	0–30	0.42	0.03	12.66	0.40	0.16	0.90	3.61	5.95
	30–80	0.23	0.02	13.61	0.63	0.25	2.00	2.69	7.12
	80–200	0.17	0.02	9.98	0.58	0.23	4.80	2.54	5.06
P-11	0–46	0.94	0.12	8.09	0.25	0.10	0.92	1.21	5.01
	46–115	0.31	0.03	9.36	0.30	0.12	1.38	2.32	5.18
	115–200	0.15	0.02	9.67	0.33	0.13	1.70	2.84	4.30
P-12	0–17	4.05	0.52	7.85	1.30	0.52	1.53	2.31	4.68
	17–70	0.58	0.07	8.69	5.55	2.22	20.14	4.14	8.87
	70–200	0.21	0.02	12.92	4.38	1.75	39.00	5.63	8.87

Appendix Table A.60 Chemical composition of typical saline-sodic soil profile sampled on uncultivated land, Melka Sedi-Amibara Plain of Middle Awash Valley (Kidane et al. 2006)

(A) Ionic composition of the saturation extract of soils

Depth (cm)	pH (1:1) soil: H_2O	ECe (dS^{-1} m)	Soluble cations (Cmolc kg^{-1})				Soluble ions (Cmolc kg^{-1})			
			Ca^2	Mg^2	Na+	K+	CO_3^{2-}	HCO_3-	Cl^-	SO_4^{2-}
0–22	7.20	16.24	2.10	4.63	62.64	4.49	Nil	0.90	19.50	68.50
22–50	7.60	17.65	1.85	3.37	84.38	1.28	Nil	1.40	26.40	70.52
50–63	7.90	17.79	1.55	3.29	94.21	1.15	Nil	1.20	27.60	72.06
63–103	7.90	18.24	1.85	6.91	241.91	1.92	Nil	1.00	29.00	183.10
103–160	7.60	20.73	2.00	8.06	304.48	2.05	Nil	1.30	44.00	228.52
160–180	7.70	18.90	1.55	0.74	118.31	1.15	Nil	2.00	64.00	102.82

(B) Ionic composition of the soil solid surface

Depth (cm)	Soil pH H_2O (1:2.5)	Exchangeable cations (Cmolc kg^{-1})				CEC (Cmolc kg^{-1})	pH KCl	PBS	ESP
		Ca^2	Mg^2	Na^+	K^+				
0–22	7.20	36.50	15.22	3.48	3.51	58.81	6.90		5.92
22–50	7.60	32.68	16.53	7.75	2.58	61.55	6.90		12.29
50–63	7.90	30.48	17.00	9.44	1.82	58.84	7.10		16.04
63–103	7.90	27.18	19.98	8.65	1.45	57.46	7.00		15.05
103–160	7.60	25.42	18.39	9.55	1.85	55.41	7.00		17.23
160–180	7.70	28.08	15.12	10.94	0.83	55.17	7.00		19.83

Appendix Table A.61 Sodic soils from Zeway flower farms 1 (Kidane et al. 2006)

(A) Ionic composition of the saturation extract of soils

Depth (cm)	Soil pH (1:1)	ECe (dS^{-1} m)	Soluble cations (Cmolc kg^{-1})				Soluble ions (Cmolc kg^{-1})				RSC (Cmolc kg^{-1})
			Ca^2	Mg^2	Na^+	K^+	CO_3^{2-}	HCO_3-	Cl^-	SO_4^{2-}	
0–28	7.5	2.09	9.72	3.03	10.12	3.31	2.0	5.65	10	7.01	–
28–70	7.9	0.68	2.72	0.65	5.08	0.62	4.0	2.00	4.5	–	2.63
70–96	8.1	0.58	2.29	3	3.57	0.55	2.6	3.23	3.3	0.36	0.54
96–126	8.0	0.56	1.58	0.79	3.33	0.56	1.7	1.18	3.0	0.18	0.51
126–166	8.0	0.52	1.31	0.76	2.88	0.64	–	3.22	2.5	–	1.15
166–200	8.7	0.58	2.10	0.26	5.26	0.59	5.0	1.50	2.5	0.10	4.14

(B) Ionic composition of the soil solid surface

Depth (cm)	Soil pH KCl (1:2.5)	Exchangeable cations (Cmolc kg^{-1})				CEC (Cmolc kg^{-1})	Delta pH	PBS	ESP	$CaCO_3$ (%)
		Ca^2	Mg^2	Na^+	K^+					
0–28	7.1	26.85	4.91	2.8	6.16	31	0.4	131	9.03	9.30
28–70	7.0	28.29	3.67	2.19	4.66	29.6	0.9	131	7.40	11.70
70–96	7.0	16.92	3.17	2.75	5.20	26	1.1	108	10.58	19.65
96–126	7.0	16.82	5.83	3.00	5.86	29.2	1.0	108	10.27	19.65
126–166	7.0	24.70	8.41	3.16	5.77	29.6	1.0	142	10.68	14.35
166–200	7.0	17.66	2.25	6.69	7.36	34.2	1.6	99	19.56	18.90

Appendix Table A.62 Sodic soils from Zeway flower farms 2 (Kidane et al. 2006)

(A) Ionic composition of the saturation extract of soils

Depth (cm)	Soil pH (1:1)	ECe (dS^{-1} m)	Soluble cations (Cmolc kg^{-1})				Soluble ions (Cmolc kg^{-1})				RSC (Cmolc kg^{-1})	SAR
			Ca^2	Mg^2	Na+	K+	CO_3^{2-}	HCO_3-	Cl^-	SO_4^{2-}		
0–28	8.0	0.49	4.50	0.45	2.37	0.74	2.39	4.11	2.37	0.16	1.55	1.51
28–84	9.6	0.80	1.71	0.31	11.80	0.20	5.00	5.43	2.51	0.26	8.41	11.74
84–144	9.5	0.99	3.82	1.47	10.05	0.53	5.60	6.40	2.51	0.73	6.71	6.18
144–200	9.7	1.46	0.89	0.85	14.15	0.53	9.57	2.50	3.64	0.55	10.33	15.17

(B) Ionic composition of the soil solid surface

Depth (cm)	Soil pH KCl (1:2.5)	Exchangeable cations (Cmolc kg^{-1})				CEC (Cmolc kg^{-1})	Delta pH	PBS	ESP	$CaCO_3$ (%)
		Ca^2	Mg^2	Na+	K+					
0–28	6.7	2.00	2.00	0.79	2.96	13.80	1.3	132	6.10	2.65
28–84	7.8	1.25	1.25	12.17	1.96	14.80	1.8	149	82.23	9.20
84–144	7.9	0.67	0.67	8.46	2.91	11.80	1.6	240	71.69	5.45
144–200	8.3	0.67	0.67	18.75	4.56	20.20	1.4	134	92.82	6.15

Index

S. Beyene et al. (eds.), *The Soils of Ethiopia*, World Soils Book Series,
https://doi.org/10.1007/978-3-031-17012-6

L

M

N

O

P

Q

R